Kazumi Nakamatsu, Gloria Phillips-Wren,
Lakhmi C. Jain, and Robert J. Howlett (Eds.)

New Advances in Intelligent Decision Technologies

Studies in Computational Intelligence, Volume 199

Editor-in-Chief
Prof. Janusz Kacprzyk
Systems Research Institute
Polish Academy of Sciences
ul. Newelska 6
01-447 Warsaw
Poland
E-mail: kacprzyk@ibspan.waw.pl

Further volumes of this series can be found on our homepage: springer.com

Vol. 176. Beniamino Murgante, Giuseppe Borruso and Alessandra Lapucci (Eds.)
Geocomputation and Urban Planning, 2009
ISBN 978-3-540-89929-7

Vol. 177. Dikai Liu, Lingfeng Wang and Kay Chen Tan (Eds.)
Design and Control of Intelligent Robotic Systems, 2009
ISBN 978-3-540-89932-7

Vol. 178. Swagatam Das, Ajith Abraham and Amit Konar
Metaheuristic Clustering, 2009
ISBN 978-3-540-92172-1

Vol. 179. Mircea Gh. Negoita and Sorin Hintea
Bio-Inspired Technologies for the Hardware of Adaptive Systems, 2009
ISBN 978-3-540-76994-1

Vol. 180. Wojciech Mitkowski and Janusz Kacprzyk (Eds.)
Modelling Dynamics in Processes and Systems, 2009
ISBN 978-3-540-92202-5

Vol. 181. Georgios Miaoulis and Dimitri Plemenos (Eds.)
Intelligent Scene Modelling Information Systems, 2009
ISBN 978-3-540-92901-7

Vol. 182. Andrzej Bargiela and Witold Pedrycz (Eds.)
Human-Centric Information Processing Through Granular Modelling, 2009
ISBN 978-3-540-92915-4

Vol. 183. Marco A.C. Pacheco and Marley M.B.R. Vellasco (Eds.)
Intelligent Systems in Oil Field Development under Uncertainty, 2009
ISBN 978-3-540-92999-4

Vol. 184. Ljupco Kocarev, Zbigniew Galias and Shiguo Lian (Eds.)
Intelligent Computing Based on Chaos, 2009
ISBN 978-3-540-95971-7

Vol. 185. Anthony Brabazon and Michael O'Neill (Eds.)
Natural Computing in Computational Finance, 2009
ISBN 978-3-540-95973-1

Vol. 186. Chi-Keong Goh and Kay Chen Tan
Evolutionary Multi-objective Optimization in Uncertain Environments, 2009
ISBN 978-3-540-95975-5

Vol. 187. Mitsuo Gen, David Green, Osamu Katai, Bob McKay, Akira Namatame, Ruhul A. Sarker and Byoung-Tak Zhang (Eds.)
Intelligent and Evolutionary Systems, 2009
ISBN 978-3-540-95977-9

Vol. 188. Agustín Gutiérrez and Santiago Marco (Eds.)
Biologically Inspired Signal Processing for Chemical Sensing, 2009
ISBN 978-3-642-00175-8

Vol. 189. Sally McClean, Peter Millard, Elia El-Darzi and Chris Nugent (Eds.)
Intelligent Patient Management, 2009
ISBN 978-3-642-00178-9

Vol. 190. K.R. Venugopal, K.G. Srinivasa and L.M. Patnaik
Soft Computing for Data Mining Applications, 2009
ISBN 978-3-642-00192-5

Vol. 191. Zong Woo Geem (Ed.)
Music-Inspired Harmony Search Algorithm, 2009
ISBN 978-3-642-00184-0

Vol. 192. Agus Budiyono, Bambang Riyanto and Endra Joelianto (Eds.)
Intelligent Unmanned Systems: Theory and Applications, 2009
ISBN 978-3-642-00263-2

Vol. 193. Raymond Chiong (Ed.)
Nature-Inspired Algorithms for Optimisation, 2009
ISBN 978-3-642-00266-3

Vol. 194. Ian Dempsey, Michael O'Neill and Anthony Brabazon (Eds.)
Foundations in Grammatical Evolution for Dynamic Environments, 2009
ISBN 978-3-642-00313-4

Vol. 195. Vivek Bannore and Leszek Swierkowski
Iterative-Interpolation Super-Resolution Image Reconstruction:
A Computationally Efficient Technique, 2009
ISBN 978-3-642-00384-4

Vol. 196. Valentina Emilia Balas, János Fodor and Annamária R. Várkonyi-Kóczy (Eds.)
Soft Computing Based Modeling in Intelligent Systems, 2009
ISBN 978-3-642-00447-6

Vol. 197. Mauro Birattari
Tuning Metaheuristics, 2009
ISBN 978-3-642-00482-7

Vol. 198. Efrén Mezura-Montes (Ed.)
Constraint-Handling in Evolutionary Optimization, 2009
ISBN 978-3-642-00618-0

Vol. 199. Kazumi Nakamatsu, Gloria Phillips-Wren, Lakhmi C. Jain, and Robert J. Howlett (Eds.)
New Advances in Intelligent Decision Technologies, 2009
ISBN 978-3-642-00908-2

Kazumi Nakamatsu, Gloria Phillips-Wren,
Lakhmi C. Jain, and Robert J. Howlett (Eds.)

New Advances in Intelligent Decision Technologies

Results of the First KES International Symposium IDT 2009

Springer

Prof. Dr. Kazumi Nakamatsu
School of Human Science and Environment
University of Hyogo
1-1-12, Shinzaike-honcho
Himeji, Hyogo 670-0092
Japan
E-mail: nakamatu@shse.u-hyogo.ac.jp

Prof. Dr. Gloria Phillips-Wren
Sellinger School of Business and Mgmt
Loyola College in Maryland
4501 N. Charles Street
Baltimore, MD 21210
USA
E-mail: gwren@loyola.edu

Prof. Dr. Lakhmi C. Jain
SCT-Building
University of South Australia
Adelaide City
The Mawson Lakes, SA 5095
Australia
E-mail: Lakhmi.Jain@unisa.edu.au

Prof. Dr. Robert J. Howlett
University Brighton
School of Engineering
Intelligent Signal Processing
Laboratories (ISP)
Moulsecoomb
Brighton BN2 4GJ
UK
E-mail: r.j.howlett@brighton.ac.uk

ISBN 978-3-642-10160-1 e-ISBN 978-3-642-00909-9

DOI 10.1007/978-3-642-00909-9

Studies in Computational Intelligence ISSN 1860949X

Typeset & Cover Design: Scientific Publishing Services Pvt. Ltd., Chennai, India.

Printed in acid-free paper

9 8 7 6 5 4 3 2 1

springer.com

Preface

IDT (Intelligent Decision Technologies) seeks an interchange of research on intelligent systems and intelligent technologies which enhance or improve decision making in industry, government and academia. The focus is interdisciplinary in nature, and includes research on all aspects of intelligent decision technologies, from fundamental development to the applied system.

It constitutes a great honor and pleasure for us to publish the works and new research results of scholars from the First KES International Symposium on Intelligent Decision Technologies (KES IDT'09), hosted and organized by University of Hyogo in conjunction with KES International (Himeji, Japan, April, 2009). The symposium was concerned with theory, design, development, implementation, testing and evaluation of intelligent decision systems. Its topics included intelligent agents, fuzzy logic, multi-agent systems, artificial neural networks, genetic algorithms, expert systems, intelligent decision making support systems, information retrieval systems, geographic information systems, and knowledge management systems. These technologies have the potential to support decision making in many areas of management, international business, finance, accounting, marketing, healthcare, military applications, production, networks, traffic management, crisis response, and human interfaces.

In addition to this preface, this book contains 62 chapters, each based on a paper selected from a large number submitted for consideration for the symposium, from various countries from all over the world. Each paper was peer reviewed by at least two independent referees. The best were finally accepted based on the recommendations of the reviewers, in some cases after required revisions had been undertaken by the authors.

The book is organized as follows. Chapters 1 – 15 are devoted to Engineering of IDTs for Knowledge Management Systems, ; Chapters 16 and 17 are devoted to Intelligent Data Processing Techniques for Decision Making, ; Chapters 18, 19, 20, 21, 22, 23, 24 and 60 are devoted to Decision Making in a Dynamic Environments, ; Chapters 25 and 26 are devoted to Decision and Health, Chapters 27, 28, 29, 30 and 31 are devoted to Foundations and Applications of Intelligent Systems, ; Chapters 32, 33, 34 and 35 are devoted to Non-Classical Logics for Intelligent Decision Technologies, ; Chapters 36, 37, 38, 39 and 40 are devoted to Knowledge - Based Interface Systems, ; Chapters 41, 42 and 43 are devoted to IDT Based Anomaly Detection, ; Chapters 44, 45, 46, 47, 48 and 49 are devoted to Knowledge-Based Software Engineering and Medical Decision Support Systems, ; Chapters 50, 51, 52, 53 and 54 are devoted to Rough Sets and Decision

Making, ; Chapters 55, 56, 57, 58, 59, 61 and 62 are devoted to Decision Making in a Changing Financial and Social Environment.

We wish to express our sincere gratitude to the plenary speakers, invited session chairs, delegates from all over the world, the authors of various chapters and reviewers for their marvelous contributions. For their sponsorship with the symposium, we express our great thanks to Himeji City and Himeji Convention Bureau. We would like to express our sincere thanks to Mr. Peter Cushion of KES International for his help with organizational issues. We would also like to express our special thanks to all of the Springer-Verlag editorial team members for their editorial support. Last, we would like to express our gratitude to the Local Organizing Committee and students at University of Hyogo for their assist.

We hope and believe that this volume will contribute to ideas for novel research and the advanced in the work of researchers, practitioners, professors and research students who are interested in knowledge-based and intelligent engineering systems.

Kazumi Nakamatsu
Gloria Phillips-Wren
Lakhmi C. Jain
Robert J. Howlett

KES IDT'09 Organization

Conference Chair

Professor Kazumi Nakamatsu	University of Hyogo, Japan

Conference General Chairs

Professor Gloria Phillips-Wren	Loyola College in Maryland, USA
Professor Lakhmi C. Jain	University of South Australia, Australia

Conference Executive Chair

Professor Robert J. Howlett	University of Brighton, KES International, UK

KES-IDT'09 International Program Committee Members

Abe, Jair M.(Brazil)
Abraham, Ajith (Norway)
Adam, Frederic (Ireland)
Akama, Seiki (Japan)
Al-Aubidy, Kasim (Jordan)
Badri, Lubna (Jordan)
Balas, Valentina (Romania)
Billington, David (Australia)
Brezillon, Patrick (France)
Bringay, Sandra (France)
Capretz, L.F.(Canada)
Cerrito, Patricia B.(USA)
Cervantes, Francisco (Mexico)
Chen, Heng-Shuen (Taiwan)
Chen, Yen-Wei (Japan)
Cheng, Jingde (Japan)
Cheng, Po-Hsun (Taiwan)
Chung, Cheng-Luen (Taiwan)
Corchado, Emilio (Spain)
Damper, Robert (UK)
Do Prado, Hércules A. (Brazil)
Doherty, Patrick (Sweden)
Druzdzel, Marek J. (USA)
Dubois, Daniel M. (Belgium)
Florez-Lopez, Raquel (Spain)
Forgionne,Guisseppi (USA)
Fulcher, John A (Australia)
Gabbar, Hossam A. (Canada)
Garrido, Leonardo (Mexico)
Gonzalez, Coty (USA)
Guo, Zhiling (USA)
Guzmán, Adolfo (Mexico)
Hall, David (USA)
Hassanien, Aboul E. (Egypt)
Hirokawa, Sachio (Japan)
Ichalkaranje, Nikhil (Australia)
Inuiguchi, Masahiro (Japan)
Ishii, Naohiro (Japan)
Ishikawa, Atushi (Japan)
Iwahori, Yuji (Japan)

Iwanaga, Saori (Japan)
Jain, Lakhmi C. (Australia)
Konar, Amit (India)
Kondo, Michiro (Japan)
Kountchev, Roumen (Bulgaria)
Kudo, Yasuo (Japan)
Laurent, Anne (France)
Lee, Zne-Jung (Taiwan)
Lim, C P (Australia)
Luh, Jer-Junn (Taiwan)
Maeda, Y. (Japan)
Marsden, James (USA)
McNelis, Paul D (USA)
Mesiar, Radko (Slovakia)
Mizuno, Tadahiro (Japan)
Mora, Manuel (Mexico)
Murai, Tetsuya (Japan)
Namatame, Akira (Japan)
Nishimura, Haruhiko (Japan)
Ohkura, Kazuhiro (Japan)
Oliveira, E. (Portugal)
Percannella, Gennaro (Italy)
Pereira, Luis M. (Portugal)
Phillips-Wren, Gloria (USA)
Poncelet, Pascal (France)
Power, Daniel (USA)
Roberts, Clive (UK)
Sakai, Hiroshi (Japan)
Saleem, Ashraf (Jordan)
Salo, Ahti (Finland)
Sansone, Carlo (Italy)
Sato, Aki-Hiro (Japan)
Sato-Ilic, Mika (Japan)
Silverman, Barry G (USA)
Smirnov, Alexander V.(Sweden)
Tanaka, Takushi (Japan)
Tanaka-Yamawaki, Mieko (Japan)
Teisseire, Maguelonne (France)
Teng, Wei-Chung (Taiwan)
Tolk, Andreas (USA)
Torra, Vicenc (Spain)
Tutunji, Tarek (Jordan)
Tweedale, Jeffrey (Australia)
Uchino, Eiji (Japan)
Virvou, Maria (Greece)
Wilson, Kevin (Greece)
Yamaguchi, Takahira (Japan)
Yamamoto, Shuichiro (Japan)
Yamawaki, Shigenobu (Japan)
Yuksel, Mehmet Emin (Turkey)
Zha, Xuan (USA)

Invited Session Chairs

Prof. Gloria Phillips-Wren
Prof. Lakhmi C. Jain
Prof. C.P. Lim
Dr. Jeffrey Tweedale
Dr. Sandra Bringay
Prof. Jair M. Abe
Prof. Seiki Akama
Prof. Yuji Iwahori
Prof. Naohiro Ishii
Prof. Wei-Chung Teng
Dr. Kevin Wilson
Prof. Po-Hsun Cheng
Dr. Heng-Shuen Chen
Dr. Jer-Junn Luh
Prof. Yasuo Kudo
Prof. Masahiro Inuiguchi
Prof. Tetsuya Murai
Prof. Mieko Tanaka-Yakawaki
Prof. Aki-Hiro Sato

Local Organizing Committee

Yutaka Yamamoto
Kazuo Ichikawa
Patrick T. Dougherty
Toshiaki Imai
Takehito Fukui
Youta Matsumoto

Sponsoring Institutions

University of Hyogo

University of Hyogo --- School of Human Science.and Environment

University of Hyogo --- Program of International Studies

Himeji City

Himeji Convention Bureau

KES International

Contents

Neural Network Inputs Selection for Breast Cancer Cells Classification

Harsa Amylia Mat Sakim, Nuryanti Mohd. Salleh, and Nor Hayati Othman

Abstract. Neural networks have been employed in many medical applications including breast cancer classification. Innovation in diagnostic features of tumours may play a central role in development of new treatment methods for earliest stage of breast cancer detection. Feature selection of neural network inputs is important for an accurate diagnosis application. Therefore, this study proposes elimination method for inputs feature selection of neural network to classify breast cancer cells. Morphological features were used as the inputs to several neural networks. The elimination method was employed on Hybrid Multilayer Perceptron (HMLP) network to investigate the diagnostic capability of features in combination and individually. Based on network performance resulted, the method was found practical for neural network inputs selection. Training the network with combination of dominant morphological features increased diagnosis capabilities and gave highest accuracy of 96%.

1 Introduction

Neural network methods which have become popular in medical applications have enabled diagnosis of breast cancer cells to be automated. Moreover, the ability of some of the methods has also been reported to be more accurate as compared to conventional method. Feature selection is crucial to achieve an accurate diagnosis. The diagnostic feature selection for neural network inputs is the process of finding a subset of dominant features from a larger set of overall acquired features. The selection is according to some predefined criteria, such as classification performance or class separability. During selection, redundant or irrelevant features will be removed from database. The proper method of selection may assist in

Harsa Amylia Mat Sakim and Nuryanti Mohd. Salleh
School of Electrical and Electronic Engineering,
Universiti Sains Malaysia, Engineering Campus Seri Ampangan,
14300 Nibong Tebal, Seberang Perai Selatan, Pulau Pinang, Malaysia
e-mail: harsaamylia@ieee.org, nuryantims@ieee.org

Nor Hayati Othman
School of Medical Science
Universiti Sains Malaysia, Health Campus, 16150 Kubang Kerian, Kelantan, Malaysia
e-mail: hayati@kb.usm.my

K. Nakamatsu et al. (Eds.): New Advan. in Intel. Decision Techno., SCI 199, pp. 1–11.
springerlink.com

increasing the speed of the learning algorithm, thus reduce running time of system. This has led to development of a variety of techniques for selecting an optimal subset of features from a larger set of possible features.

In this paper, the elimination method is proposed for the feature selection of neural network inputs in breast cancer cells classification. The selection is based on morphological features acquired from fine needle aspirated (FNA) cells of breast cancer. In this method, the impact of individual and combination of the feature is evaluated to determine the best subset of neural networks inputs.

2 Some Related Works

There has been a number of published works on feature selection methods. In recent discoveries, several algorithms have been applied for feature selection method to improve network classification accuracy. Redondo and Espinosa [10] have reviewed and compared feature selection procedures that were based on the analysis of trained multilayer feed-forward network.

Another research regarding the feature selection method using feed-forward neural networks was presented by Onnia et. al. [15]. In the paper, the network inputs selection are selected and added to the input layer based on the highest classification accuracy of the test data. The advantage of this feature selection algorithm is that it can be used together with other classifier. According to the experimental results, most of the classification of Monks problems with selected features gave the highest test set average accuracy of 100%. While almost over than 96% of average accuracy were achieved from the test data sets of real-world problems.

The feature selection method for artificial neural network inputs also can be considered by using genetic algorithms (GA). Researchers from Duke University, Durham [5] claimed that genetic algorithm is well-known and effective choice for feature selection in computer-aided diagnosis (CAD) applications. However, GA algorithms are computationally very demanding as the number of available features increases. Thus, in the research, information of theoretic approach for feature selection in CAD was investigated based on the mutual information (MI) concept. This MI concept focuses only on the linear relationships of variables to select the optimal subset of features. Based on the analysis results, the MI concept was shown as a promising feature selection criterion for nonlinear CAD models.

Previous discoveries have reported the successful implementation of feature selection methods for network inputs in medical applications. Tsymbal et. al. [2] had made comparison from four different search strategies for ensemble feature selection in medical diagnostics with a focus on the problems of the classification of acute abdominal pain. A research conducted by Hanczar et. al. [3] developed feature selection methods to classify lung cancer and leukemia data sets. The methods for cancer classification were applied based on combination of direct statistical gene selection method and ProGene reduction method. Both of the combination of feature selection methods gave potential results in classifying adenocarcinoma from gene expressions. Another method analyzing microarray data was using redundancy based feature selection which was reported by Yu and Liu [18].

Several approaches of feature selection methods have been employed in breast cancer diagnosis. In a research by Ghosh et. al. [12], a modular framework method was implemented for multi category feature classification in digital mammography. The canonical genetic algorithm based on modular feature selection approach was combined with Multilayer Perceptron (MLP) network and compared with general selection method. It was reported that the total number of feature selected by modular selection model is less than general selection model. Thus, the modular neural network runs almost 1.5 faster than general model. An overall test classification accuracy of 86.66% was achieved.

Abbass [1] introduced an evolutionary multi-objective approach for feature selection. The approach was based on pareto differential evolution algorithm which was named as memetic pareto artificial neural network (MPANN). Evolutionary algorithm was argued to be a better candidate and more efficient than genetic algorithms for evolving neural networks [17].

Recent studies have emphasized the usefulness of morphological features in automated breast cancer detection. Miscellaneous types of morphological features selection had been used for classification of breast cancer cells [6, 14, 16]. A research carried out by Tozaki et. al. [13] combined morphological features and kinetic information such as visual washout. The combination was very useful in differentiating between benign and malignant lesions of high-spatial-resolution MR imaging of focal breast masses.

This paper is an extension of preliminary study of features selection method for neural network inputs in breast cancer classification. The basic idea is to propose an elimination method as another practical technique in selecting the best feature subset for neural network inputs application.

3 Input Features

A clinical dataset of 804 cases has been provided by the Pathology Department, School of Medical Sciences, USM, Kubang Kerian, Kelantan, Malaysia. The data employed for this study was extracted from FNA smears of breast cell. The standard Papanicolou staining method as described by Bancroft and Stevens [4] was implemented to the smears before the cell images were viewed using Leica Qwin Image Analyser software (Leica, Cambridge, United Kingdom). The software was hosted by a computer which was attached to a digital microscope. The smears were generally viewed to identify its overall picture and its relative position was noted when a region of interest was identified. Magnification was fixed at x40 and constant light source was set before measurements were taken. Then, calculations were made on all smear slides. Only one experienced cytotechnologist was involved in capturing the images. This is to eliminate variations in visual subjectivity.

In this study, five morphological features were acquired for evaluation. The features chosen were those suggested by pathologists who are involved in breast cancer management. They are based on diagnostic features commonly looked for during diagnostic procedures. The features were Cellularity of Cells, Cells in Cluster, Cells in Discrete, Xlength and Ylength. Cellularity or cell density is a measure of distribution of breast cells on a smear. A smear with poor cell yield generally indicates normal breast gland tissue. Cellularity has been divided into three categories, ranked between 1 (poor density) up to 3 (highly dense). Cells visualized on a smear could be in clusters (grouped) or in discrete (individually placed, separated from other cells). Irregularity in distribution generally indicates abnormality. An estimated count of cells in discrete and cells in clusters are included as two separate inputs to the neural network. Xlength refers to the shortest while Ylength refers to the longest cells visualized. Both measurements were made in micrometers (μm).

Table 1 shows the features proposed as inputs for classification network. Corresponding input values were assigned accordingly. Suggested features were a mix between continuous and categorical values which were standardized according to respective categories.

4 Methodology

In this work, a preliminary investigation on several network types has been carried out to obtain the optimum network architecture for classifying breast cancer cells. The morphological features data were applied to six types of neural network and their performance compared. The trained networks were Multilayer Perceptron (MLP), Multilayer Perceptron Sigmoid (MLPSig), Hybrid Multilayer Perceptron (HMLP), Radial Basis Function (RBF), Hybrid Radial Basis Function (HRBF) [7] and Self-Organizing Maps (SOM).

Table 1 Proposed features

Input Marker	Categories	Neural Network Inputs
	High	3
Cellularity	Moderate	2
	Poor	1
	>100	3
Cells in Discrete	50-100	2
	<50	1
	>51	4
Cells in Cluster	31-50	3
	10-30	2
	<10	1
X Length	Mean of shortest length of cells in μm	
Y Length	Mean of longest length of cells in μm	
Output	Malignant	1
	Benign	0

Different network types have different optimum structure for solving a problem. To avoid complex network architecture and assure better run-time performance, each network was constructed based on standard network architecture. Therefore, the neural networks established consisted of three layers that were input, hidden and output layer. Only one hidden layer was employed for all network types. The output layer has only one node, which corresponds to classification of breast cancer cells. The network's output was graded to range from 0 to 1 and determined due to cut-off point of 0.5. A of high output (≥0.5) is considered to indicate malignancy, while a low output (<0.5) indicates benignity. The input layer has five nodes, which corresponds to the five morphological features under study.

Corresponding diagnostic results given by pathologist were assigned to 1 for malignant and 0 for benign. The actual output of neural network is compared to these values (0s and 1s) for comparative measurements. The number of training epochs and nodes in the hidden layer needs to be investigated to find the optimum network structure. Therefore, the trial and error method was carried out.

In general, each network type was initially trained with a fixed hidden node while the epoch was varied. Then, training and testing were repeated for increasing numbers of hidden nodes. Determining the optimum number of hidden nodes is crucial. In the literature, there is no specific rule on determining the number of nodes in the hidden layer. A network with a small number of hidden nodes will not have enough capability to represent the input-output mapping. While a network with a large number of hidden nodes may lead to overfitting where the network simply memorizes the training data [8, 9]. Nevertheless, in a work by Choong et al. [11] to predict recurrence of patients with node positive, a neural network with the least number of layers and nodes was defined as the optimum network.

Each of the 804 FNA samples was considered as individual cases. There were 538 (67%) benign and 266 (33%) malignant cases. For preliminary investigation of the suitable network type, the cases were randomly selected and grouped into two data sets. The first set was further partitioned into two subsets. The first subset was used for estimation of the network parameters (network training) while the second subset was used for evaluation of the performance of the model (network testing). In other words, the training and testing data sets were used to assess the performance of various candidate model structures. The second set is to validate (network validating) the resulting of the best network type for breast cells classification. When the network with best performing parameter values has been chosen, its generalization capability is assessed using the validating set. Division of the

Table 2 Division of data for preliminary investigation

	Total Cases = 804		
Data Set	**Training**	**Testing**	**Validating**
Benign	352	93	93
Malignant	152	57	57
Total	504	150	150

data sets is given in Table 2. In the training set, there are 152 (30%) malignant cases and 352 (70%) benign cases. In both the testing and validating sets, each has 57 (38%) malignant cases and 93 (62%) benign cases.

The best networks' architecture was determined based on results on testing set in terms of:

1. Classification accuracy, false positive, false negative, sensitivity and specificity percentage.
2. Mean Squared Error (MSE) nearest or equal to 0. For ease of comparison, the values of MSE were quoted in the unit of decibel (dB).

The optimum neural network architecture (and network type) identified is then validated and employed to investigate classification capabilities of features using the stepwise elimination method. Individual and various combinations of breast features were applied to the network to produce classification results. The process of finding subset of features using the stepwise elimination method is shown in Figure 1. During this process, all 804 data were employed.

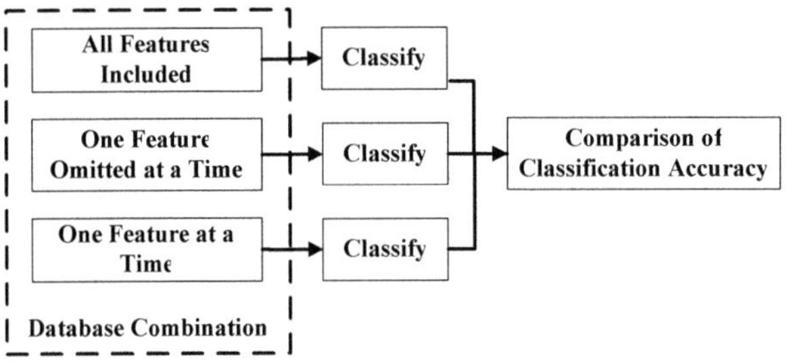

Fig. 1 Block diagram of stepwise elimination method

5 Experimental Results and Discussion

Results from preliminary investigation are tabulated in Table 3. The table included the best number of hidden node (HN) and epoch for each type of network investigated. It can be seen from the table that all the networks investigated were able to give more than 90% training and testing data accuracy except for SOM network, which achieved 84% for testing data accuracy. During training, HMLP network performed the best with classification accuracy of 96% and MSE of -30.81dB. During testing, the highest accuracy was 99% with least MSE of -36.66dB.

Based on this result, HMLP network was presented with the validating data set. HMLP network was able to give 96% accuracy, 91% sensitivity, 99% specificity, 1% false positive and 9% false negative. This shows that HMLP network is capable of classifying the breast cancer cells. Figure 2 illustrates the output of optimum structured HMLP network as compared to actual diagnosis of patients in the validating data set.

Table 3 Performance of all networks trained using morphological features

Network Type	No. of HN	No.of Epoch	Training MSE (-dB)	Training ACC (%)	Testing MSE (-dB)	Testing ACC (%)
MLP	10	11	27.43	94	33.35	98
MLPSig	8	6	29.26	95	34.21	98
HMLP	7	6	30.81	96	36.66	99
RBF	7	4	23.23	94	26.42	98
HRBF	13	2	27.84	94	32.58	98
SOM	10	50	14.09	92	12.05	84

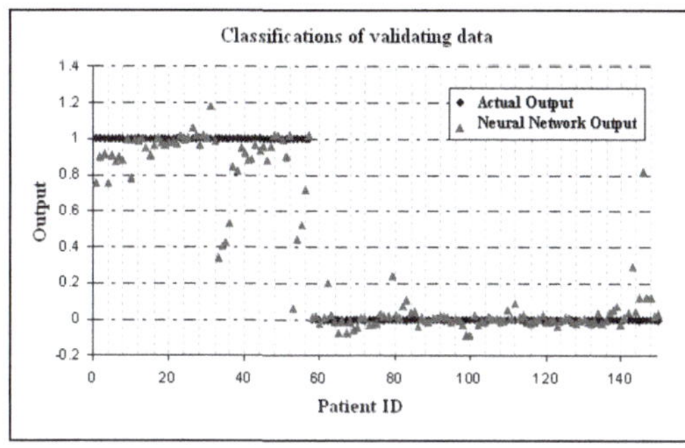

Fig. 2 Actual and neural network output of patients in the validating data set

The elimination method was employed on HMLP network with 7 hidden nodes and trained using 6 epochs. When the input features were omitted one at a time, the classification results are as shown in Table 4. FP%, FN%, SN%, SP% and ACC% are percentage of False Positive, False Negative, Sensitivity, Specificity and Accuracy respectively. From Table 4, some observations could be made. They were:

1. The network specificity and accuracy reduced the most when Ylength was omitted (using all features but Ylength). The increased in MSE and false positive percentage are too large. The network cannot identify benign cases correctly. Ylength is considered an important diagnostic feature.
2. Omitting Xlength affects the classification such that only benign cases were correctly identified. The network cannot detect any of malignant cases (100% false negative), thus reducing overall accuracy. The MSE was increased the most.
3. Omitting Cluster enabled the network to identify all malignant cases. However, there was a reduction in benign cases identified, thus reducing overall accuracy.
4. Omitting Discrete has positive effect on sensitivity but negative effect on specificity. The overall accuracy was not affected. Discrete can be considered comparatively, not to be a significant feature. However, MSE was increased. The feature Discrete may be useful for difficult cases.

Table 4 Results of classification when input features were omitted one at a time

Input Nodes	FP (%)	FN (%)	SN (%)	SP (%)	ACC (%)	Final MSE (-dB)
All features included	3	6	94	97	96	29.52
Cellularity omitted	3	18	82	97	92	23.9
Cluster omitted	26	0	100	74	83	18.31
Discrete omitted	4	3	97	96	96	28.13
Xlength omitted	0	100	0	100	67	3.94
Ylength omitted	93	3	97	7	37	9.42

Table 5 Results of classification when using combination of input features

Input Nodes	FP (%)	FN (%)	SN (%)	SP (%)	ACC (%)	Final MSE (-dB)
3 Input Nodes						
Cell,Clust&Discrete	0	100	0	100	67	7.80
Cell,Clust & XL	85	4	96	15	42	10.80
Cell,Clust & YL	0	100	0	100	67	4.95
Cell,Discrete & XL	96	3	97	4	35	8.58
Cell,Discrete & YL	0	100	0	100	67	6.73
Cell,XL&YL	34	0	100	66	77	16.18
Clust,Discrete &XL	47	3	97	53	67	14.43
Clust,Discrete &YL	26	100	0	74	50	-0.63
Clust,XL&YL	3	13	87	97	94	24.35
Discrete,XL&YL	22	2	98	78	85	19.08
2 Input Nodes						
Cell & Clust	0	100	0	100	67	8.15
Cell & Discrete	0	100	0	100	67	8.26
Cell & XL	92	4	96	8	37	9.00
Cell & YL	0	100	0	100	67	7.39
Clust & Discrete	0	100	0	100	67	6.58
Clust & XL	42	3	97	58	71	14.73
Clust & YL	26	100	0	74	49	-0.07
Discrete & XL	60	3	97	40	59	13.52
Discrete & YL	16	100	0	84	56	1.44
XL & YL	36	0	100	64	76	16.41

Cell=Cellularity, Clust=Cluster, XL=Xlength, YL=Ylength.

Various combinations of input markers were then presented to the HMLP network. The classification results on all the data are as shown in Table 5. Applying single input at a time to HMLP network gave classification results as shown in Table 6. Some observations made were:

1. Combination of the features Xlength and Ylength improved classification accuracy. The network achieved 100% sensitivity and no false negative presented. All malignant cases were correctly identified.
2. Combination of the features Cluster, Xlength and Ylength were able to correctly classify most of the patients. Compared to network with Xlength and Ylength only, more benign cases were correctly identified, although some malignant cases were missed.
3. Network with combination of Discrete, Xlength and Ylength was able to identify many of the malignant cases. However, many benign cases were missed, thus reducing overall accuracy.
4. Combination of Cellularity and Xlength was identified to be the lowest classification accuracy for two input nodes category. While combination of Cellularity, Discrete and Xlength was identified to be the lowest classification accuracy for three input nodes category. However, both of the combinations were more sensitive towards malignant cases than benign cases.
5. The feature Xlength on its own has positive effect on sensitivity and negative effect on specificity. Combination of this feature with others is important to detect malignancy.
6. Single input application for instance Cellularity, Cluster, Discrete or Ylength, was not sensitive towards malignant cases (only identified benign cases).

Table 6 Results of classification when using one feature at a time

Input Nodes	FP (%)	FN (%)	SN (%)	SP (%)	ACC (%)	Final MSE (-dB)
Cellularity only	0	100	0	100	67	9.05
Cluster only	0	100	0	100	67	6.13
Discrete only	0	100	0	100	67	7.02
Xlength only	53	3	97	47	63	13.87
Ylength only	0	100	0	100	67	2.17

6 Conclusion

Feature selection methods play a significant role for solving network classification problems. Proper neural network inputs selection could improve the performance of the diagnostic network. In this study, an elimination method was proposed and applied to the morphological features selection for breast cancer cells classification. In the method, various combinations of input features are presented to the neural network. Through this way, individual or combination of features for classification could be evaluated. Moreover, this feature selection method also enables to investigate the impact of features either individually or in combination with other features that are being considered.

The feature selection method began with classifying the breast cells from the total combination of acquired morphological features. Then, the feature was omitted or eliminated one by one from the previous combination until only a single feature was classified.

This study demonstrated that the elimination method is practical for neural network inputs selection.

7 Future Works

The stepwise elimination method may be useful in development of an expert system. The method may also be extended to investigation of optimum number of hidden nodes.

Acknowledgments. This research is funded by the Ministry of Science, Technology and Innovation (MOSTI) through the eScienceFund grant. The authors would like to thank the Pathology Department, School of Medical Science, USM, Kubang Kerian, Kelantan for providing the FNA smear samples for this study.

References

1. Abbass, H.A.: An Evolutionary Artificial Neural Networks Approach for Breast Cancer Diagnosis. Artif. Intell. Med. 25, 265–281 (2002)
2. Tsymbal, A., Cunningham, P., Pechenizkiy, M., Puuronen, S.: Search strategies for ensemble feature selection in medical diagnostics. In: Proceedings of the 16th IEEE Symposium on Computer-Based Medical Systems, pp. 124–129 (2003)
3. Hanczar, B., Courtine, M., Benis, A., Hannegar, C., Clement, K., Zucker, J.: Improving classification of microarray data using prototype-based feature selection. SIGKDD Explorations 5(2), 23–29 (2003)
4. Bancroft, J.D., Stevens, A.: Theory and Practice of Histological Techniques, 2nd edn., United Kingdom. Churchill Livingstone (1982)
5. Tourassi, G.G., Frederick, E.D., Markey, M.K., Floyd Jr., C.E.: Application of the mutual information criterion for feature selection in computer-aided diagnosis. Medical Physics 28(12), 2394–2402 (2001)
6. Sheshadri, H.S., Kandaswamy, A.: Detection of Breast Cancer Tumor based on Morphological Watershed Algorithm. ICGST-GVIP, International Journal on Graphic, Vision and Image Processing 5(5) (2005)
7. Mat Sakim, H.A., Mat Isa, N.A., Naguib, R.N.G., Sherbet, G.V.: Image Cytometry Data From Breast Lesions Analyzed using Hybrid Networks. In: Proceedings of the 27th Annual International Conference of the IEEE Engineering on Engineering in Medicine and Biology Society (EMBC 2005), Shanghai, China, paper 1710 (2005)
8. Huang, S.-C., Huang, Y.-F.: Bounds on the number of hidden neurons in Multilayer Perceptrons. IEEE Trans. On Neural Networks 2(1), 47–55 (1991)
9. Masters, T.: Practical Neural Network Recipes in C++. Academic Press, London (1993)
10. Redondo, M.F., Espinosa, C.H.: A comparison among feature selection methods based on trained networks. In: Proceedings of the 1999 IEEE Signal Processing Society Workshop, Neural Networks for Signal Processing IX, pp. 205–214 (1999)
11. Choong, P.L., DeSilva, C.J.S., Attikiouzel, Y.: Predicting local and distant metastasis for breast cancer patients using the Bayesian Neural Network. In: International Conference on Digital Signal Processing, vol. 1, pp. 83–88 (1997)

12. Ghosh, R., Ghosh, M., Yearwood, J.: A modular framework for multi category feature selection in digital mammography. In: Proceedings of European Symposium on Artificial Neural Networks (ESANN 2004), Bruges, Belgium, pp. 175–180 (2004)
13. Tozaki, M., Igarashi, T., Matsushima, S., Fukuda, K.: High-spatial-resolution MR Imaging of Focal Breast Masses: Interpretation Model Based on Kinetic and Morphological Parameters. Radiation Medicine 23(1), 43–50 (2005)
14. Wedegartner, U., Bick, U., Wortler, K., Rummeny, E., Bongartz, G.: Differentiation between benign and malignant findings on MR-mammography: usefulness of morphological criteria. Eur. Radiol. 11, 1645–1650 (2001)
15. Onnia, V., Tico, M., Saarinen, J.: Feature selection method using neural network. In: Proceedings of the IEEE International Conference on Image Processing, vol. 1, pp. 513–516 (2001)
16. Sheta, W., Eltonsy, N., Tourassi, G., Elmaghraby, A.: Automated Detection of Breast Cancer from Screening Mammograms using Genetic Programming. International Journal of Intelligent Computing and Information Sciences (IJICIS) 5(1) (2005)
17. Xin, Y., Yong, L.: A New Evolutionary System for Evolving Artificial Neural Networks. IEEE Transactions on Neural Networks 8(3), 694–713 (1997)
18. Yu, L., Liu, H.: Redundancy based feature selection for microarray data. In: Proceedings of the 10th ACM SIGKDD 2004 International Conference on Knowledge Discovery and Data Mining, Seattle, Washington, USA, pp. 737–742 (2004)

Cognitive Information Systems for Medical Pattern Analysis and Diagnosis Support Technologies

Lidia Ogiela, Ryszard Tadeusiewicz, and Marek R. Ogiela

Abstract. This publication presents a new concept of the intelligent analysis and interpretation of image-type data. The concept presented is based on a cognitive-semantic model of data learning and interpreting. The application of this model makes it possible to extract significant semantic information from the set of analysed data in order to understand it. It is also possible to execute the stage of reasoning based on the semantic contents of the analysed data. The proposed model for the cognitive analysis and interpretation of data is discussed using the example of a selected class of cognitive categorisation systems - UBIAS (Understanding Based Image Analysis Systems) analysing image-type data.

Keywords: cognitive informatics, cognitive image analysis, medical pattern interpretation, UBIAS systems.

1 Introduction

Information systems designed for the cognitive interpretation of data are based on psychological/cognitive subjects of which the cognitive process itself seems to the most important. This process is understood as a certain, rarely encountered ability of the IT system to acquire information from its surroundings and receive it from the outside. The ability itself to acquire and receive information is not enough to talk of intelligent cognitive information systems. This is because an important role

Lidia Ogiela
AGH University of Science and Technology
Faculty of Management,
Al. Mickiewicza 30, PL-30-059 Krakow, Poland
e-mail: logiela@agh.edu.pl

Ryszard Tadeusiewicz and Marek R. Ogiela
AGH University of Science and Technology
Institute of Automatics
Al. Mickiewicza 30, PL-30-059 Krakow, Poland
e-mail: {rtad,mogiela}@agh.edu.pl

K. Nakamatsu et al. (Eds.): New Advan. in Intel. Decision Techno., SCI 199, pp. 13–19.
springerlink.com

is played here by information processing, which is a stage in the effective control of the entire cognitive system operation. This is why cognitive information systems are designed not just to acquire information for its subsequent analysis, but their main purpose is the in-depth processing and interpretation of information acquired from the outside. What is also of great significance is acquiring information from inside the system - information, which is stored in the system but is frequently not directly disclosed. A cognitive data analysis and interpretation system can acquire such information using the processes of learning based on data contained in it. This type of data acquisition is much more complicated and requires an additional element in the system - the system learning stage.

A new class of systems which conduct meaning-based data analysis has been distinguished within the entire set of systems that analyse and interpret data. This type of systems belongs to the class of cognitive data interpretation and analysis systems. What is novel in the presented approach is using the approach of the purely human data analysis to system operation. Such a presentation of the problem is possible if we claim that the human brain is a system for processing, analysing and interpreting any type of data, and is also an exceptional reasoning system. This claim makes it possible to build information processing systems based on the human cognitive system.

The key to the presented approach is to show the similarities between the human reasoning system and an information system designed for executing the analysis and reasoning process. Information processing taking place in the human brain is described by all operations due to which the stimuli received by the individual are transformed, converted, reduced, amplified, stored (e.g. in memory) and recalled. The same stages of information (data) processing can be attributed to information systems, as every one of them can execute the above activities. Information systems which can execute these types of jobs based on activities taking place in the human brain are referred to as cognitive data analysis systems. If they can additionally conduct reasoning based on the semantic contents of the analysed data, they belong to the cognitive categorisation system group. Descriptions of this system type are presented in publications [8-13].

2 The Essence of In-Depth Data Analysis

Information systems conducting data analysis are currently described in great detail in various scientific publications. However, the processes of data analysis itself are so widespread that it is impossible to point out a significant difference between systems of this type. Consequently, it makes sense to approach the data analysis problem from a completely different angle: intelligent information systems should be built based on the cognitive informatics trend which is currently being developed [15, 16].

The essence of the announced scientific approach is to implement processes taking place in the human brain for data analysis and interpretation, as a result of which it will be possible to extend the entire analysis process to include the semantic reasoning stage.

Semantic data analysis is an extremely important type of analysis, as it is based on extracting the meaning contained in the analysed data set. The notion of meaning itself is understood as some cognitive information or data representation which can define the characteristics, function or the classification of the analysed data. The classification is very frequently the classification to a category which defines either some fragment of knowledge applied in the system or a type of knowledge referred to with certain notions or definitions.

What is very important in the reasoning process based on the aspects of semantic data analysis is to indicate the meaning/denotation bases for data analysis. This type of reasoning makes use of the knowledge possessed by the information system, derived from recognised sources (usually experts) and referring to the characteristics possessed and functions fulfilled (by the analysed data). We should remember that it would be very difficult to build cognitive data analysis systems based on meaning/connotation aspects. This type of meaning interpretation and reasoning would have to be based on subjective characteristics of the analysed data, and this would be quite complicated for information systems.

The process of in-depth data analysis itself is aimed at extracting information from the data set, which information describes that data in a significant way, has a major importance in the process of data characterisation, interpreting and analysis, and can also unambiguously indicate changes in the process of reasoning and projecting into the future.

3 Cognitive Reasoning for Medical Patterns

Computer understanding and cognitive analysis used in cognitive systems and intelligent information systems is very often based on the syntactic approach [0. In cognitive systems:

- it first uses a pre-processing operation usually composed of image coding using terminal symbols, shape approximation, as well as some kind of filtration or segmentation
- as a result of executing these stages it is possible to obtain a new image representation in the form of hierarchic semantic tree structures and subsequent production steps of this representation from the initial grammar symbol.

An intelligent cognitive system distinguishing image data at the stage of pre-processing must, in the majority of cases, perform image segmentation, identify primitive components and determine spatial as well as semantic relations between them. An appropriate classification is based on the recognition of whether a given representation of the actual image belongs to a class of images generated by languages defined by one of the possible number of grammars. Such grammars can be considered to belong to sequential, tree and graph grammars while recognition with their application is made in the course of a syntactic analysis performed by the system.

In cognitive systems intelligence understood is built on many levels (intelligence levels) determined by three parameters: the computational power and the memory capacity of the system, the automatic searching for data and the automatic selection of their processing routines when the system is used to find solutions to problems which are not completely known at the time the system is built, and the quality and quantity of information collected in the system.

A cognitive method of interpreting disease units and pathological lesions forms the main element of a correctly functioning IT system supporting medical image diagnostics. Further down, such an interpretation of changes occurring in foot bone pathologies will be presented.

4 Cognitive Interpretation Systems for Image Analysis

Cognitive data analysis systems are very often used for the meaning-based interpretation and analysis of image-type data. Medical images, due to their great variety, provide a very practical approach to the subject of cognitive system operation. However, as it is difficult to unambiguously indicate patterns which define selected lesions observed in various types of medical images, this type of data gives us very interesting experience with cognitive processes conducted by information systems.

Such difficulties and at the same time challenges mean that the constantly developed subject of cognitive data analysis yields new results and solutions. A class of cognitive categorisation systems, called UBIAS (*Understanding Based Image Analysis Systems*), has been developed for the cognitive analysis and interpretation of data which uses semantic data interpretation.

Cognitive data analysis systems have been used for analysing and interpreting medical data showing various types of pathologies and lesions of various organs. The most important of those were lesions in the central nervous system, wrist bone deformations, fractures of long bones as well as foot bone deformations.

The essence of this type of analysis is to extract medical content concerning the lesions present from the analysed image, which content is then subjected to analysis, interpreting, diagnostics, understanding and reasoning. Medical images very often contain various types of information noise, i.e. diverse interference that can impact the correct understanding and interpretation of the image. In addition, they contain a whole range of various medical information, which may not be significant in the process of the semantic analysis conducted. And so a medical image - an X-ray - may show not just the organ examined for the presence of a given lesion, but also a wider picture containing organs which are not subjected to semantic analysis. Obviously, from the point of view of the cognitive process itself, such an image is much more interesting than the classical presentation of the organ examined.

Semantic analysis conducted by cognitive systems is based on a linguistic apparatus very rich in meaning reasoning and meaning interpretation rules. This linguistic apparatus is usually contained in the appropriate grammar proposed for the analysis jobs (tree, sequential or graph) [14].

Cognitive categorisation systems for analysing images of the central nervous system and long bone fractures use a sequential grammar, while the grammar appropriate for analysing images showing wrist and foot bone deformations is a graph grammar. The right selection of grammatical methods depends on the type of organ analysed, its properties and characteristic features.

The main element of a correctly working information system to support medical image diagnostics is the development of a method for the cognitive analysis of disease entities and types of lesions which may appear in the case of a specific disease entity.

The cognitive analysis contained in a UBIAS system is aimed at proposing an automatic method for the correct interpretation of extremely complex medical images which are generated by imaging e.g. fragments of long bone fractures, wrist bone deformations, foot bone deformations or other lesions in other organs.

To illustrate the whole versatility of the method presented in this publication, below we show selected examples of UBIAS systems performing the automatic interpretation of image-type data and its meaning analysis. We have selected two medical images for this comparison.

The first represents UBIAS systems conducting cognitive analysis using sequential formalisms for images showing long bone fractures.

The second represents UBIAS systems conducting cognitive data interpretation using graph formalisms to analyse images showing foot bone deformations. Both results of the automatic analysis of data and reasoning based on it in UBIAS systems are shown in Fig. 1.

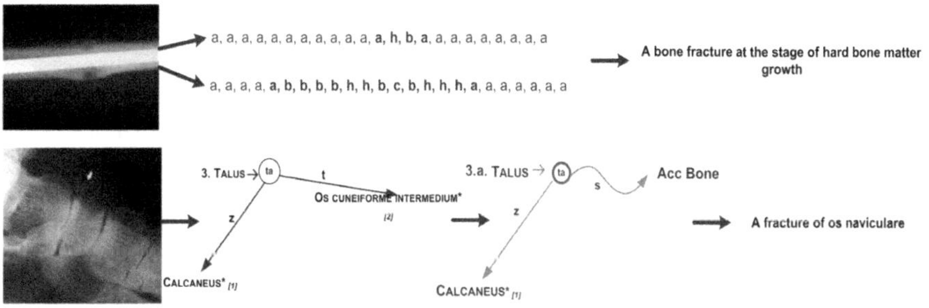

Fig. 1 Image-type data analysis using UBIAS cognitive systems for the semantic interpretation of data showing the bone fracture at the stage of hard bone matter formation in the case of a long bone fracture, and a fracture of *os naviculare* in the case of a foot bone deformation

The above figure shows the essence of the data analysis conducted by a UBIAS system. As mentioned above, a characteristic feature of the cognitive systems discussed is the correctly selected linguistic apparatus used to describe the data correctly. In our case, we have analysed two types from very different groups of medical images.

The first comprises long bone fractures. In this case, the appropriate sequential grammar was entered in the UBIAS system and this grammar supported the correct image analysis using defined terminal symbols and specific angle values expressed in degrees assigned to them. Figure 1 shows a sequence of individual terminals characteristic for the lower and upper edges of a long bone. These sequences, expressed in visible forms using a set of productions, are converted into semantic information, which for both of them is a statement that the analysed fracture type is a fracture at the stage of hard bone matter generation. Reasoning is conducted in a similar way for other types of fractures (transverse, longitudinal, spiral, bone during remodelling etc.).

The stage of image data analysis based on a linguistic description of data presented using a graph has a different form. This analysis type has been proposed for interpreting data in the case of foot bone deformation [8, 9]. The results of the analysis conducted for a selected foot bone image are also shown in Fig. 1, where a graph characterising the lesion being described is presented for a selected image.

It is worthwhile to emphasise the versatility of the presented method. Regardless of what formalism of the linguistic data description is selected, UBIAS systems perform the data identification, analysis and reasoning taking into account the semantic characteristics of this data. This is because all linguistic description formalisms can be used correctly and unambiguously to define and correctly identify the semantic contents of data.

5 Conclusion

The approach to the subject of cognitive data analysis systems presented here is illustrated with the semantic analysis of image-type data. The essence of the presented approach is to apply human data analysis processes and cognitive/ interpretation/ reasoning processes to the operation of systems.

Systems can be built based on human cognitive and decision-making processes only if the system will analyse and interpret data as well as conduct the reasoning and projecting stages using the semantic characteristics of data. The semantics of data supports their in-depth analysis and becomes the starting point for projecting changes that may take place in the future. Consequently, this type of system analysis makes it possible to at least partly eliminate errors that may occur in the future, which could not be identified in traditional data analysis systems.

So the characteristic feature which also distinguishes cognitive systems is the process of reasoning on the basis of analysed data and the process of projecting based on the data analysis conducted.

Acknowledgments. This work has been supported by the AGH University of Science and Technology under Grant No. 10.10.120.783

References

1. Albus, J.S., Meystel, A.M.: Engineering of Mind – An Introduction to the Science of Intelligent Systems. John Wiley & Sons Inc., Chichester (2001)
2. Branquinho, J. (ed.): The Foundations of Cognitive Science. Clarendon Press, Oxford (2001)

3. Burgener, F.A., Meyers, S.P., Tan, R.K., Zaunbauer, W.: Differential Diagnosis in Magnetic Resonance Imaging. Thieme (2002)
4. Davis, L.S. (ed.): Foundations of Image Understanding. Kluwer Academic Publishers, Dordrecht (2001)
5. Duda, R.O., Hart, P.E., Stork, D.G.: Pattern Classification, 2nd edn. John Wiley & Sons, Inc., Chichester (2001)
6. Khan, M.G.: Heart Disease Diagnosis and Therapy. Williams & Wilkins, Baltimore (1996)
7. Meyer-Baese, A.: Pattern Recognition in Medical Imaging. Elsevier, Amsterdam (2003)
8. Ogiela, L.: Cognitive Systems for Medical Pattern Understanding and Diagnosis. In: Lovrek, I., Howlett, R.J., Jain, L.C. (eds.) KES 2008, Part I. LNCS (LNAI), vol. 5177, pp. 394–400. Springer, Heidelberg (2008)
9. Ogiela, L.: Modelling of Cognitive Processes for Computer Image Interpretation. In: Al-Dabass, D., Nagar, A., Tawfik, H., Abraham, A., Zobel, R. (eds.) EMS 2008 European Modelling Symposium, Second UKSIM European Symposium on Computer Modeling and Simulation, Liverpool, United Kingdom, September 8-10, 2008, pp. 209–213 (2008)
10. Ogiela, L., Tadeusiewicz, R., Ogiela, M.: Cognitive Analysis In Diagnostic DSS-Type IT Systems. In: Rutkowski, L., Tadeusiewicz, R., Zadeh, L.A., Żurada, J.M. (eds.) ICAISC 2006. LNCS (LNAI), vol. 4029, pp. 962–971. Springer, Heidelberg (2006)
11. Ogiela, L., Tadeusiewicz, R., Ogiela, M.R.: Cognitive Approach to Visual Data Interpretation in Medical Information and Recognition Systems. In: Zheng, N., Jiang, X., Lan, X. (eds.) IWICPAS 2006. LNCS, vol. 4153, pp. 244–250. Springer, Heidelberg (2006)
12. Ogiela, L., Tadeusiewicz, R., Ogiela, M.R.: Cognitive Methods in Medical Image Analysis and Interpretation. In: The 4th International Workshop on Medical Image and Augmented Reality, MIAR 2008, August 1-2, 2008, The University of Tokyo, Tokyo, Japan (2008)
13. Ogiela, L., Tadeusiewicz, R., Ogiela, M.R.: Cognitive Categorizing in UBIAS Intelligent Medical Information Systems. In: Ogiela, L., Tadeusiewicz, R., Ogiela, M.R. (eds.) Advanced Computational Intelligence Paradigms in Healthcare 3. Studies in Computational Intelligence (SCI), vol. 107, pp. 75–94. Springer, Heidelberg (2008)
14. Tadeusiewicz, R., Ogiela, M.R.: Medical Image Understanding Technology. Springer, Heidelberg (2004)
15. Wang, Y.: On Cognitive Informatics. Brain and Mind: A Transdisciplinary Journal of Neuroscience and Neurophilosophy 4(2), 151–167 (2003)
16. Wang, Y.: The Theoretical Framework of Cognitive Informatics. International Journal of Cognitive Informatics and Natural Intelligence 1(1), 1–27 (2007)
17. Wilson, R.A., Keil, F.C.: The MIT Encyclopedia of the Cognitive Sciences. MIT Press, Cambridge (2001)
18. Zhong, N., Raś, Z.W., Tsumoto, S., Suzuki, E. (eds.): ISMIS 2003. LNCS, vol. 2871. Springer, Heidelberg (2003)

Development of an Intelligent Facial Expression Recognizer for Mobile Applications

Siu-Yeung Cho, Teik-Toe Teoh, and Yok-Yen Nguwi

Abstract. In the light of fast pace smart phone development, increasing numbers of applications are being developed to cater for portability. A real-time facial expression recognition application is develop that was tested in Windows Mobile environment. The underlying algorithm adopted in this work uses Boosting Naïve Bayesian (BNB) approach for recognition. We examine the structure of training data and the effect of attributes on the class probabilities through the use of Naïve Bayesian classifier (NBC). The experiments carried out show that we have achieved the important features of mobile application: speed and effi-ciency. This work is believed to be the first recorded initiative that de-ploys facial expression recognition into a mobile phone. It seeks to pro-vide a launching point for a sound and portable mobile application that is capable of recognizing different facial expressions.

1 Introduction

To date, facial expression recognition system becomes popular in many application domains. It can be applied in, for examples, when a facial expression recognition system is installed in a robot, it can communicate with human better. Robot/Human interaction is gaining focus nowadays. This contributes to the field of human computer interaction which will have great impact on the society following the speed advancement of robot. Innovation like household robot which can en-courage the owner when he is feeling down, comfort a child when he is crying, making a joke when the owner is bored and etc. The same concept can be further developed into an intelligent house that understands people's emotion through his face expression and entertain him. On the other hand, another possible application can be explored in an intelligent portable device for social communication. Eight out of ten people in Singapore own a mobile phone, be it a simple mobile phone or a high end 3G handheld device. Mobile phone becomes part of our lives that most people cannot live without. The development of mobile phone

Siu-Yeung Cho, Teik-Toe Teoh, and Yok-Yen Nguwi
School of Computer Engineering,
Nanyang Technological University
Singapore 639798
e-mail: `davidcho@pmail.ntu.edu.sg`

K. Nakamatsu et al. (Eds.): New Advan. in Intel. Decision Techno., SCI 199, pp. 21–29.
springerlink.com

hardware and soft-ware development becomes the hot pies in the commercial market. The commonly used smart phones are Windows Mobile Personal Digital Assistant (PDA), Black-Berry, Apply iPhone, Palm OS handhelds, Symbian OS handhelds and etc. Among the smart handheld devices, Windows Mobile platform is the most commonly used one.

The motivation behind this work is that we aim to develop a sound portable device that can help autistic children to understand the emotions of surrounding people. Emotion is a state of feeling involving thoughts, physiological changes, and an outward expression. There are five theories which attempt to understand the sequence of processes that we are experiencing when we are feeling a certain type of emotion. They are James-Lange theory, Cannon-Bard theory, Lazarus theory, Schachter-Singer theory, and Facial Feedback theory [1].

According to the facial feedback theory [1], emotion is the experience of changes in our facial muscles. In other words, when we smile, we experience pleasure or happiness. When we frown, we experience sadness. It is the changes in our facial muscles that direct our brains and provide the basis for our emotions. As there are many possibilities of muscle configurations in our face, there is seemingly unlimited number of emotions. Facial expression recognition is a challenging task. The challenges of such system are light variation, direction of subject face, the quality of image acquisition device, and occlusion problem. Recent research works have been done to contribute in this area, for examples, in [2-5].

This work attempts to use Naïve Bayes Classifier (NBC) to classify facial emotions. Bayesian classifier is the most popular classifier among the probabilistic classifiers used in the machine learning community and is often used for benchmarking. Naive Bayesian classifier is perhaps one of the simplest yet powerful techniques in constructing predictive models from labeled training sets. NBC can also provide valuable insights of the training data by exposing the relations between attribute values and classes. The NBC is robust and often outperform other more complex machine learning methods.

In this paper, we propose an approach to recognize emotion from facial images through the use of Boosting Naïve Bayesian (BNB) approach. The observation obtained through the classification process will be discussed. Four classes of facial emotion are investigated; they are namely neutral, joy, sad and surprise as shown in Fig 1. The focus of this work is to show how the BNB approach can be used to segregate the four classes of the facial emotion in a mobile application.

The paper is organized as follows: Section 2 provides an overview of our proposed system. Section 3 presents the experimental results and analysis. Finally, conclusion of this paper is drawn in Section 4.

Fig. 1 Four categories of facial expressions. (a) Neutral, (b) Joy, (c) Sad, and (d) Surprise

2 Our System

In this section, we outline the structure of facial expression recognizer. The proposed model is going to recognize four types of facial expression, namely, neutral, joy, sad and surprise. A probabilistic based approach using the Naïve Bayesian Boost is adopted to recognize these four types of facial expressions.

Figure 2 displays the block diagram of the entire system. The system is com-posed of three major blocks, the feature locator, the feature extractor, and the classifier. The feature locator finds crucial fiducial points for subsequent feature ex-traction processing. We adopted Gabor features and Gini extractions which will be discussed in the following sub-sections. Finally the meaningful features are classified into the corresponding class.

The fiducial points are located at the eyes, nose, and mouth. The fiducial points define the extended feature components. The component-based feature detector has two levels, namely, the micro SVM based independent component detector and the macro SVM based independent component detector. The micro level uses linear SVM based independent component detection. Each component classifier was trained on a set of extracted facial components (the 4 key fiducial components) and on a set of randomly selected non-face patterns. The macro level uses the maximum outputs of the component classifiers within rectangular search regions as inputs to a combination SVM classifier. The macro SVM performs the final detection of the face component regions.

The feature extractor adopted Gabor wavelet feature extraction incorporating with Gini-based feature selection. An image is first convoluted with Gabor wavelet filters so as to extract the facial features. We have chosen 7 orientations and 2 spatial frequencies; generating a total of 14 Gabor filtered images. The frequency selected is from 0.4 to 0.8, whereas the orientations are in the multiples of $\pi/7$ ranging from 0 to π. Gabor wavelet constructed 61 regions of 14 different features for each input data. The feature extraction process starts by marking the left eye, right eye and mouth positions. With the 854 (61x14) features obtained, we make

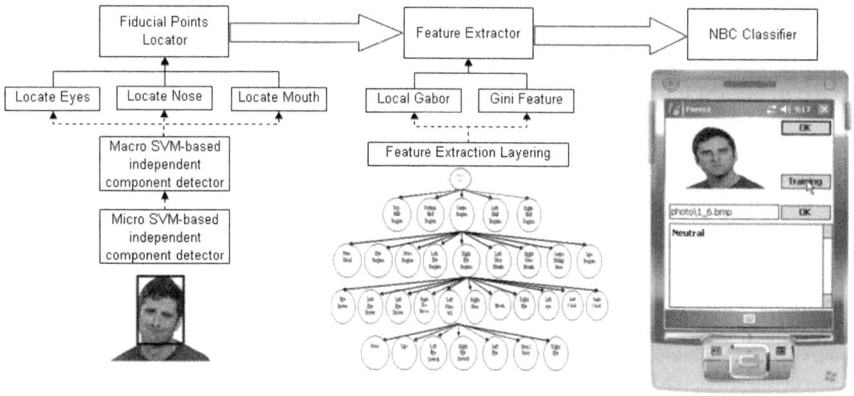

Fig. 2 The system block diagram

use of Gini to shrink down to ten features in which they are the best features to describe the face. The final stage is the probabilistic-based classification performed by the NBC classifier.

2.1 Gabor Feature Extraction

Gabor wavelet is a popular choice because of its capability to mimic mammals' visual cortex. The primary cortex of human brain interprets visual signals. It consists of neurons, which respond differently to different stimuli attributes. The receptive field of cortical cell consists of a central ON region surrounded by 2 OFF regions, each region elongated along a preferred orientation [6]. According to Jones and Palmer, these receptive fields can be reproduced fairly well using Daugman's Gabor function [7]. There is considerable evidence that the parameterized family of 2-D Gabor filters, proposed by Daugman in 1980, suitably models the profile of receptive cells in the primary visual cortex. Gabor filters models the properties of spatial localization, orientation selectivity, and spatial frequency selectivity and phase relationship of the receptive cells [8].

The Gabor wavelet function can be represented by:

$$g(x, y) = g_1(x, y)\exp(j2\pi Wx), \tag{2.1}$$

where

$$g_1(x, y) = \left(\frac{1}{2\pi\sigma_x\sigma_y}\right)\exp\left(-\frac{1}{2}\left(\frac{x^2}{\sigma_x^2}+\frac{y^2}{\sigma_y^2}\right)\right). \tag{2.2}$$

We consider that the receptive field (RF) of each cortical cell consists of a central ON region (a region excited by light) surrounded by two lateral OFF regions (excited by darkness) [18]. Spatial frequency W determines the width of the ON and OFF regions. σ_x^2 and σ_y^2 are spatial variances which establish the dimension of the RF in the preferred and non-preferred orientations.

2.2 Gini Index Feature Reduction and Selection

Gini Index selects features based on information theories [9, 10]. It measures the impurity for a group of labels. Gini Index for a given set s of points assigned to two classes C_1 and C_2 is given below:

$$GINI(s) = 1 - \sum_{j=1,2}\left[p\left(C_j \mid s\right)\right]^2. \tag{2.3}$$

With $p\left(C_j \mid s\right)$ corresponds to the frequency of class C_j at set *S*. The maximum $1-\frac{1}{nc}$ occurs when points are equally distributed among all classes, which implies less interesting information. On the other hand, the minimum 0.0 occurs when all points belong to one class which represents the most interesting information. We then sort the *n* features over different classes of samples in ascending order based on their best Gini index. Low Gini index corresponds to high ranking discriminative features.

2.3 Probabilistic-Based Classification

A Naive Bayesian (NB) classifier is a simple probabilistic classifier base on Bayesian' theorem with strong (Naive) independence assumptions [11, 12]. The NB classifiers often work much better in many complex real-world problems. The classifier requires only a small amount of training data to estimate the parameters necessary for classification. The independence assumption in Naïve Bayesian may lead to some unexpected results in the calculation of posteriori probability. The NB classifier has several attractive properties like the decoupling of the class conditional feature distributions which helps to alleviate problems stemming from the curse of dimensionality. The data arrives at the correct classification as long as the correct class is more probable than any other class. In overall, the classifier is robust enough to ignore serious deficiencies in the underlying Naive probability model.

Boosting Naïve Bayesian Algorithm:

GINI feature selections:

1. Assign each training sample with weight=1.
2. For ten iteration (ten features):
   - Sort features index S.
   - Split S.
   - Break if GINI criterion is satisfied.

BNB classification:

1. Apply simple Bayesian to weighted data set.
2. Compute error rate.
3. Iterate the training examples.
   - Multiply the weight by $\frac{e}{1-e}$.
   - Normalize the weight
4. Add $-\log\frac{e}{1-e}$ to weight of class predicted
5. Return class with highest sum

Fig. 3 The proposed Naïve Bayesian with boosting algorithm

The proposed algorithm adopted the Naïve Bayesian with boosting. Each iteration of boosting uses the Gini reduction and selection method and to remove redundant features. The main reason for using boosting NB approach is that the embedded feature selection technique makes the data more suitable for classification. The algorithm is summarized as shown in Fig. 3.

3 Experimental Results and Analysis

In this section, we will assess the ability of the system to recognize different facial expressions. We have adopted the Mind Reading DVD [13], a computer-based guide to emotions, developed by a team of psychologists led by Prof. Simon Baron-Cohen at the Autism Research Centre, University of Cambridge. The database contains images of approximately 100 subjects. Facial images are of size 320x240 pixels, 8-bit precision grayscale in PNG format. Subjects' age ranges from 18 to 30. Sixty-five percent were female, 15 percent were African-American, and three percent were Asian or Latino. Subjects were instructed by experimenter to perform a series of facial displays. Subjects began each display from a neutral face. Before performing each display, the experimenter described and modeled the desired display. The model recognizes four types of facial expression: neutral, joy, sadness and surprise. Twenty images were used for training, 5 images for each emotion.

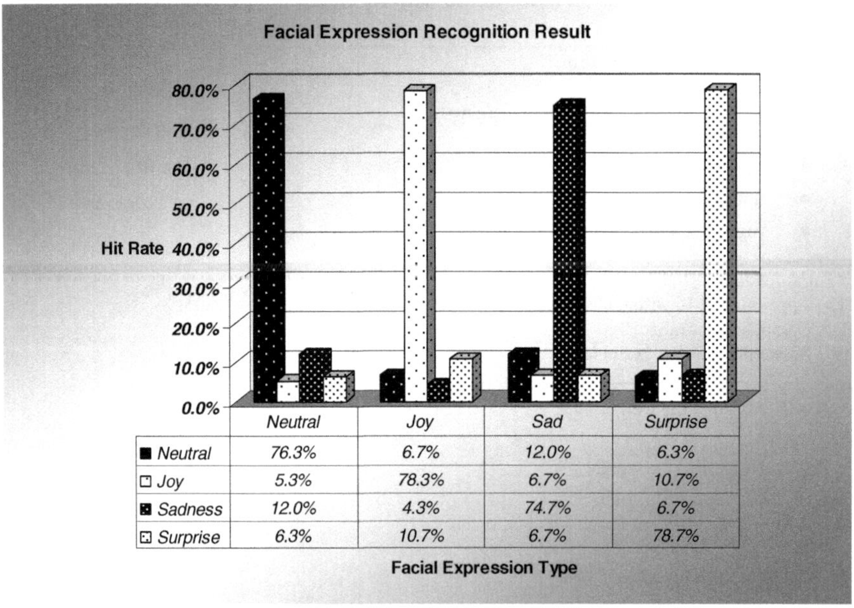

Fig. 4 Facial Expression Recognition Result of the System

The facial expression recognition result is shown in Fig. 4. The confusion matrix is included in the figure as well where the column of the matrix represents the instances in a predicted class, in which each row represents the instances in an actual class. The system correctly recognizes 76.3% of neutral, 78.3% of joy, 74.7% of sad and 78.7% of surprise expressions amongst 100 subjects in the database, although some facial expressions do get confused by the wrong class, however at an acceptable range of less than 12%.

In addition, comparisons with other approaches are necessary for us to investigate how the recognition performance of our approach can be benchmarked with others. Table 1 shows the recognition results for facial expression recognition using T-test, Euclidean, and K-nearest neighbour approaches. Only the average and maximum hit rates are included in the table. According to the results in the table, our approach achieves the most optimal result. The T-test assesses whether the means of different groups are statistically different from each other. K-nearest neighbour algorithm is a method for classifying objects based on closest training examples in the feature space. These approaches are generally used for benchmarking. Our approach that combines Gini and Boosting Naïve Bayesian achieves average of 75% and highest of 100% outperforms the rest.

The application is deployed in a Hewlett-Packard's IPAQ PDA phone with Windows Mobile operating system. Fig. 5 shows the screen capture of the system. Fig. 5(a) shows the main menu that provides selection for training and testing. The training can be done on the smart phone by loading different training images as shown in Fig. 4(b). The 4 classes of facial expressions are trained using the GUI as shown in Fig. 4(c). Lastly we can load the corresponding image for testing Fig. 4(d). The average recognition time is about 5 seconds.

Table 1 Result benchmarking with T-test, Euclidean and KNN approaches

Feature selection	Classifier	Average	Maximum
Gini	Naïve Bayesian	75%	100%
Gini	Euclidean	63%	83%
Gini	KNN	57%	99%
T-test	Euclidean	59%	69%
T-test	KNN	59%	89%
T-test	Naïve Bayesian	57%	100%
Euclidean	Euclidean	58%	69%
Euclidean	KNN	57%	100%
Euclidean	Naïve Bayesian	53%	99%

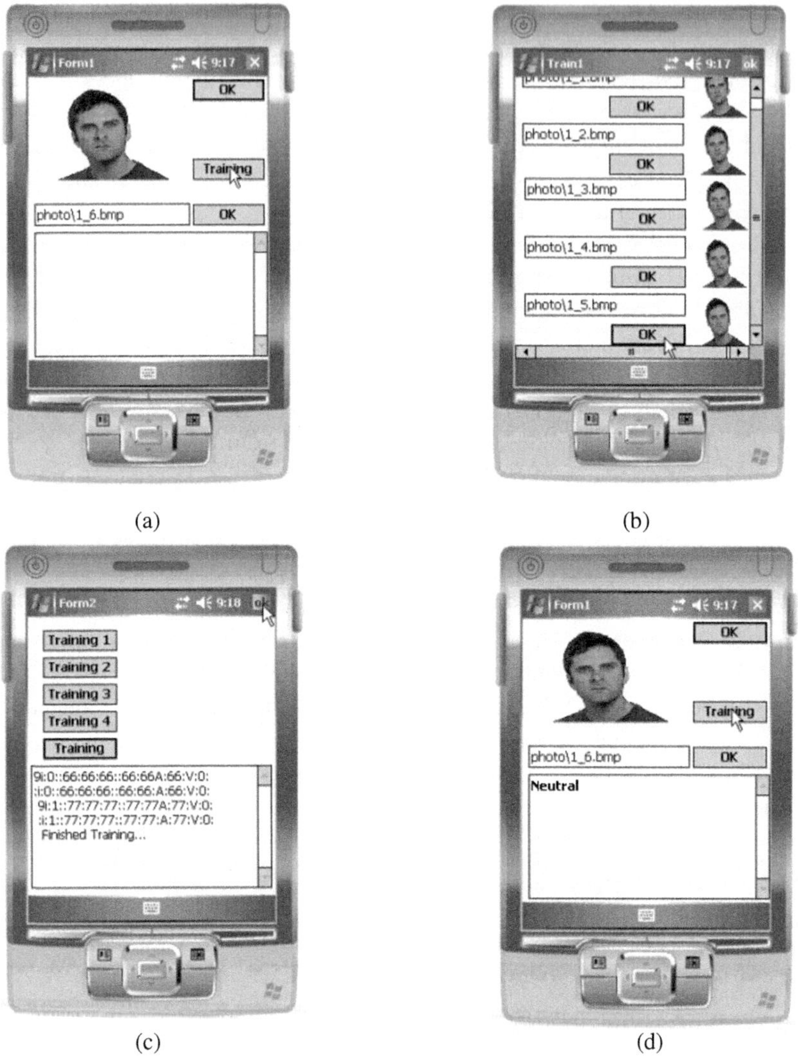

Fig. 5 Screenshot of facial expression recognizer in PDA. (a) Main menu. (b) Image loader (c) Training GUI (d) Test result

4 Conclusion

A real-time facial expression recognition system is presented in this paper. The system consists of fiduciary points localizer, features extractor, and the classifier. The fiduciary points localizer consists of both micro and macro SVM based independent component detectors. The feature extractor processes fiduciary points with Gabor wavelet convolution and Gini index sorting. The classifier segregates the facial expression based on Boosting Naïve Bayesian (BNB) algorithm.

The system is able to recognize four classes of facial expression: neutral, joy sad, and surprise. The experimental result shows that Gini indexing to the features plays a very important role in recognizing different facial emotion. The experimental result also suggests the combination of Gini and BNB achieves the most optimal result. The average hit rate is around 75%. The application is deployed in a Hewlett-Packard's IPAQ PDA phone. Various facial expressions can be trained in the smart phones and gets recognition within 5 seconds.

Currently, we are working on the improvement of the BNB classification in its structure and algorithm to obtain better accuracy and reduce the computation time for training. We are also looking into further enhance the system to obtain live images through the built-in camera in smart phones.

References

[1] Heffner, C.L.: Psychology 101, The virtual psychology classroom (2001), `http://allpsych.com/psychology101/emotion.html`

[2] Ji, Y.Z.Q.: Active and dynamic information fusion for facial expression understanding from image sequences. IEEE Transactions on Pattern Analysis and Machine Intelligence 27(5), 699–714 (2005)

[3] Wu, Y., Liu, H., Zha, H.: Modeling facial expression space for recognition. In: The IEEE/RSJ International Conference on Intelligent Robots and Systems (2005)

[4] Zeng, Z., Fu, Y., Roisman, G.I., Wen, Z., Hu, Y., Huang, T.S.: One-class classification for spontaneous facial expression analysis. In: The International Conference on Automatic Face and Gesture Recognition (2006)

[5] Xiang, T., Leung, M.K.H., Cho, S.Y.: Expression recognition using fuzzy spatio-temporal modeling. Pattern Recognition 41(1), 204–216 (2007)

[6] Jones, J.P., Palmer, L.A.: An evaluation of the Two-Dimensional Gabor Filter model of simple Receptive fields in cat striate cortex. J. Neurophysiol. 58(6), 1233–1258 (1987)

[7] Daugman, J.: Uncertainty relation for resolution in space, spatial frequency, and orientation optimized by two-dimensional visual cortical filters. J. Opt. Soc. Amer. 2(7), 1160–1169 (1985)

[8] La Cara, G.E., Ursino, M., Bettini, M.: Extraction of Salient Contours in Primary Visual Cortex: A Neural Network Model Based on Physiological Knowledge. In: Engineering in Medicine and Biology Society, 2003. Proceedings of the 25th Annual International Conference of the IEEE, September 17-21, 2003, vol. 3, pp. 2242–2245 (2003)

[9] Hall, M.A., Smith, L.A.: Feature Selection for Machine Learning: Comparing a Correlation-Based Filter Approach to the Wrapper. In: FLAIRS Conference, pp. 235–239 (1999)

[10] Zhou, X.J., Dillion, T.S.: A Heuristic - Statistical Feature Selection Criterion For Inductive Machine Learning In The Real World. In: Proceedings of the 1988 IEEE International Conference on Systems, Man, and Cybernetics, August 1988, vol. 1, pp. 548–552 (1988)

[11] Friedman, N., Geiger, D., Goldszmidt, M.: Bayesian network classifiers. Machine Learning 29, 131–163 (1997)

[12] Rish, I.: An empirical study of the naïve Bayes classifier. Technical Report RC 22230 (2001)

[13] Baron-Cohen, S., Golan, O., Wheelwright, S., Hill, J.J.: Mind Reading: The Interactive Guide to Emotions. Jessica Kingsley Publishers, London (2004)

INTSOM: Gray Image Compression Using an Intelligent Self-Organizing Map

Cheng-Fa Tsai and Jiun-Huang Ju

Abstract. The popularity of multimedia on the Internet has created a heavy load on bandwidth. Consequently, content compression and bandwidth reduction have become significant topics recently. An appropriate codebook design is an essential and valuable principle for Vector Quantization (VQ). This investigation presents a new image compression method called INTSOM, which relies on Hierarchical Self-Organizing Map (HSOM) and adopts LBG for speeding up. For a two-layer neural network, INTSOM first employs LBG to determine the number of first layer neurons, and uses an estimation function to determine the number of second layer neurons dynamically. A modified SOM is then performed to compress each sub-map. Experimental results indicate that INTSOM has better overall capability (time-cost and quality) than LBG, SOM and HSOM.

Keywords: image compression, vector quantization, LBG, SOM, HSOM.

1 Introduction

The development of information technology in the last few decades has led to increasing network services and user demands, especially for multimedia. However, this digital content may raise the need for storage capacity and network bandwidth. Therefore, data compression is an essential and significant process for those digital contents due to the smaller data size and transmission.

Image compression schemes can be categorized as lossy or lossless. Lossy compression approaches decrease redundancy and irrelevancy through a series of transform in frequency domain to reconstruct the image with the minimum

Cheng-Fa Tsai and Jiun-Huang Ju
Department of Management Information Systems,
National Pingtung University of Science and Technology,
91201 Pingtung, Taiwan
e-mail: {cftsai,m9656004}@mail.npust.edu.tw

K. Nakamatsu et al. (Eds.): New Advan. in Intel. Decision Techno., SCI 199, pp. 31–40.
springerlink.com

of errors. Redundancy reduction eliminates duplication from images, and irrelevancy reduction removes visually insignificant signals. The quality of lossy compression methods such as JPEG, JPEG2000, EZW [1] and SPIHT [2] has been studies. Lossless compression involves compressing an image such that it can be rebuilt to the original image identically. Lossless compression methods include PNG, JPEG-LS and LOCO-I [3].

Spatial domain is another model that differs from frequency domain intrinsically, and adopts Vector Quantization (VQ) commonly in a lossy compression. VQ is mainly applied to minimize the number of vectors in the codebook to approximate the original image. Restated, an image compression is made by preserving a codebook comprising several representative vectors and an index-book that records the index-values rather than raw vectors. Some VQ-based schemes, like LBG, SOM, HSOM and Tsai *et al.* [4], have been developed to design a codebook efficiently and appropriately.

LBG (proposed by **L**inde, **B**uzo and **G**ray) is fast and straightforward to implement, but has inconsistent coding, and is prone to becoming trapped in local optima. A neural network structure called Self-Organizing Map (SOM) applied to image compression can attain an excellent result, but it is very time-consuming. SOM trains the weights of several neurons, which it considers as codebook within specified training times. Since SOM gradually fine-tunes these neurons to obtain global optima, it is called a full-search algorithm. Hierarchical Self-Organizing Map (HSOM) has been proposed to lower the time complexity of SOM. Although HSOM divides the entire compression process into a few stages to simplify the large scale and accelerating compression, but generally yields slightly worse quality solutions than SOM.

This work presents a two-layer neuron network algorithm called INTSOM, based on LBG and HSOM. In the first layer, INTSOM adopts LBG to generate roughly the first layer neurons (the number of these neurons is the square root of codebook size) that form the representative vectors. In the second layer, INTSOM utilizes an estimation function to calculate the distortion degree of each representative vector to derive dynamically the number of second-layer neurons. Finally, a modified SOM is performed to compress each sub-map. Experimental results reveal that INTSOM has a higher quality and lower time-cost with a bigger codebook than other known compression algorithm.

The remainder of this paper is organized as follows. Section 2 describes VQ, LBG, SOM, HSOM and objective measures. Section 3 then presents the proposed algorithm, called INTSOM. Section 4 explains the experiments and analysis results. Conclusions are finally drawn in Section 5, along with recommendations for future research.

2 Preliminaries

This section describes in detail the principle of VQ and several VQ-based algorithms, namely LBG, SOM and HSOM, and presents the pros and cons of each of these algorithms. Finally, the measure functions of the image quality are introduced.

2.1 Vector Quantization

Vector Quantization (VQ), adopts few vectors (codewords) to approximate the original image, is commonly employed in a lossy compression to preserve an image with a good quality and few acceptable distortions [5]-[6].

In Fig. 1(a), if two neighboring pixels have similar RGB (a color model consists of **R**ed, **G**reen and **B**lue) distributions, then these pixels are classified into the same group, which has a representative vector $\boldsymbol{C}$ shown in Fig. 1(b). As listed in Fig. 1(c), the representative vectors of each group then become the codewords for the codebook.

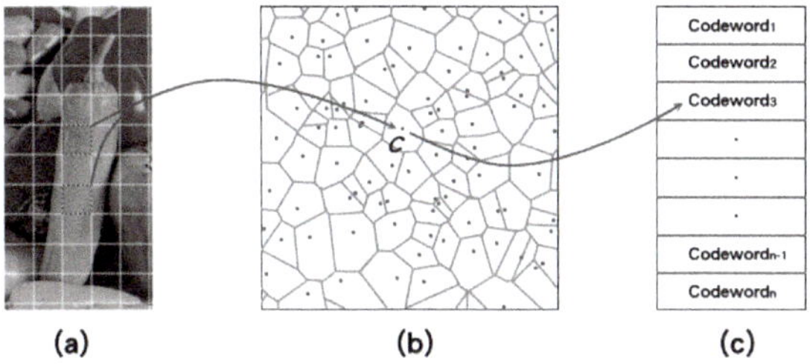

Fig. 1 The principle of Vector Quantization. (a) The blocked image (b) Codeword groups of image blocks (c) The codebook consists of n codewords

2.2 LBG

LBG, which was proposed by Linde, Buzo and Gray in 1980 [7], is the best-known VQ technique for image compression, and is conceptually similar to the K-means data clustering algorithm. LBG builds a codebook by continuously reassigning codewords until the average distortion is below a specified threshold (ϵ). The simple algorithm of LBG means that it can obtain the codebook promptly. However, the main limitations of LBG are that it may be apt to get trapped at local optima, or yield an unstable result because of selecting an initial codebook randomly.

2.3 Self-Organizing Map

Self-Organizing Map (SOM) is an unsupervised neural network algorithm developed by Kohonen in 1995 [8]-[9]. Applying SOM to image compression [10], it trains the weights of several neurons, which become the codebook.Consequently, SOM fine-tunes these neurons very slowly and precisely in order to approximate the global optima. Since SOM performs a full-search to obtain an optimal result, it is very time-consuming.

2.4 Hierarchical SOM

Although SOM has good image compression capability, its time cost is too high to design a codebook efficiently. Barbalho *et al.* first presented a hierarchy-based SOM (HSOM) scheme for image compression [11]. In contrast to SOM, HSOM employs a top-down structure to accelerate the convergence. Fig. 2 illustrates the principle of HSOM, which adopts a few neurons (in layer 1) to obtain the approximate position, and the remaindering neurons (in layer 2) are assigned to each group.

HSOM has a lower time complexity than SOM, but has two issues when applied to image compression. First, the number of neurons in each sub-group is fixed. Therefore, the number of neurons is insufficient to express all vector distributions appropriately. Second, each sub-group gains its neurons from its superior neuron, possibly causing those neurons to become trapped at local optima.

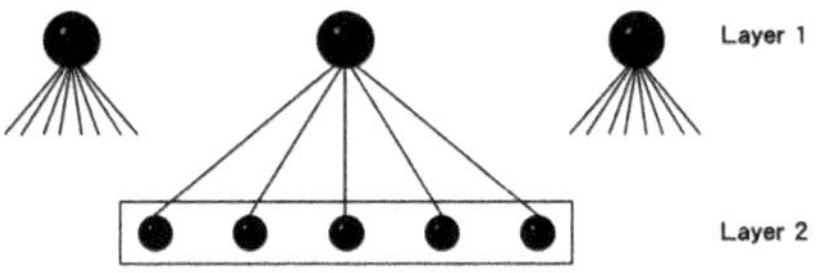

Fig. 2 A 2-layer neural network structure of HSOM

2.5 Measure Functions

The variation of a reconstructed image is easy for human to judge. However, objective measures are stipulated to measure the quality of a reconstructed image precisely. The Mean Square Error (MSE) and Peak Signal-to-Noise Ratio (PSNR) are typically used to measure quality efficiently, and are represented by the following equations:

$$MSE = \frac{1}{M \times N} \sum_{x=1}^{M} \sum_{y=1}^{N} [i(x,y) - \bar{i}(x,y)]^2, \tag{1}$$

where i and $\bar{i}$ denote the original and rebuilt pixel values of an image. It equals to

$$\sum_{g=1}^{Codewords} \frac{1}{GroupSize_g \times BlockSize} \sum_{n=1}^{GroupSize_g} \sum_{d=1}^{BlockSize} [i(x_{n_d}, y_{n_d}) - \bar{i}(x_{n_d}, y_{n_d})]^2 \tag{2}$$

$$PSNR = 10 \log_{10} \frac{255^2}{MSE} \tag{3}$$

The Bits per Pixel (BPP), refers to the number of bit representing each pixel. The BPP is calculated with Eqn. (4) below, where ***Codewords*** denotes the codebook size, and ***BlockSize*** indicates the dimension size of a codeword. For example, the BPPs of 16 dimensional images consisting of 1024 and 2048 codewords are 0.625 and 0.6875, respectively.

$$BPP = \frac{\log_2 Codewords}{BlockSize} \tag{4}$$

3 The Proposed INTSOM Algorithm

As we discussed earlier, although LBG is fast, it yields unstable clustering results because it randomly selects the initial codewords. Moreover, although SOM and HSOM yield excellent result, they are much slower than other algorithms. Accordingly, this work presents an image compression algorithm to combine the merits of different algorithms.

This section describes the principle of the proposed image compression algorithm, namely INTSOM. First, the INTSOM algorithm settles the layer-1 neurons by "root neuron distribution" where the neurons are determined using LBG. "sub-group scale estimation" is then adopted to derive the scale of each group in layer 2. A modified SOM then trains the neurons in the "neurons training" phase, to fine-tune the neurons of each sub-group. Finally, "neurons reassigning" is applied to enhance the quality of each sub-group. The specific implementation of INTSOM has four stages, which are described as follows:

(1) **Root neurons distribution:** In this stage, LBG is employed to train the image with all its vectors to obtain several representative vectors, which form a rough vector distribution in layer 1. The desired number of representative vectors in layer 1 of an image is the *square root* of the codebook size. Thus, an image composed of 256 codewords should have 16 vectors in layer 1.

(2) **Sub-group scale estimation:** After determining the layer 1 neurons, INTSOM computes the number of neurons of each sub-group based on a *Total Square Error (TSE)* measure as Eqn. (5):

$$TSE = \sum_{n=1}^{GroupSize} \sum_{d=1}^{BlockSize} [i(x_{n_d}, y_{n_d}) - \bar{i}(x_{n_d}, y_{n_d})]^2 \tag{5}$$

In Fig. 3, the MSE (noted in Eqn. (2)) of group O is given by $(L\times4)/4/16 = L/16$ and that of group P is $(L\times8)/8/16 = L/16$. Nevertheless, we believe the degree of distortion of these two groups should be different, since they cannot both be expressed by only one represented vector (the centered object in each

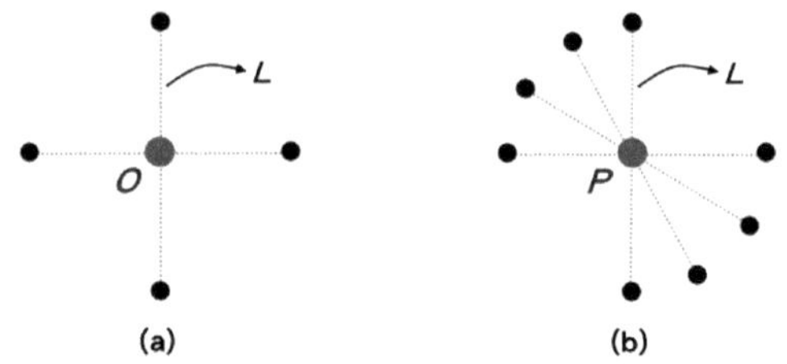

Fig. 3 Two 16-dimensional vector groups that are with different degree of distortion. (a) Vector group O consists of 4 vectors and the radius is L (b) Vector group P consists of 8 vectors and the radius is L

group). Hence, this stage adopts a TSE (observed in Eqn. (5)) function rather than the original MSE function. In Fig. 3, the TSE of group O is $(L \times 4) = 4L$, and that of group P is $(L \times 8) = 8L$. Consequently, group P has a larger TSE than group O, meaning that group P is assigned more neurons than group O in order to express vectors as appropriately as possible.

(3) **Neurons training:** The stopping criterion of a regular SOM is the fixed training times regardless of the steadiness. Therefore, SOM may spend an unnecessarily long time training its neurons to obtain an excellent quality. To accelerate the training process, INTSOM employs a dynamic stopping criterion like that of LBG. After training all neurons of a group, INTSOM calculates the current MSE value of that group, and compares it with the previous MSE value. The stopping criterion of INTSOM is that the variation of MSE value is below a specified and small threshold (ϵ). This threshold is obviously a trade-off between speed and quality. Although INTSOM is far faster than SOM, its solution quality may be lower. The equation of MSE variation is given as follows:

$$\varDelta D = |MSE(t) - MSE(t-1)|/MSE(t), \tag{6}$$

where $\boldsymbol{\varDelta D}$ represents the variation of MSE; $|\cdot|$ denotes the absolute value measure, and $\boldsymbol{t}$ indicates the iteration.

(4) **Neurons reassigning:** The purpose of neuron training is to approximate all objects by as few neurons as possible. Fig. 4 depicts a trained neural

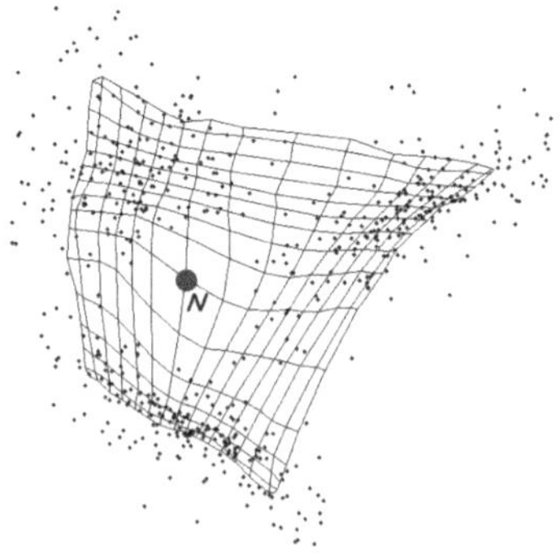

Fig. 4 A trained neural network map with an idle neuron N

network, where the intersections of lines denote the neurons, and they illustrate the approximation of the whole objects. Each neuron represents those objects that are closest to it. However, probably some idle neurons do not represent any objects, such as neuron N in Fig. 4.

To overcome this drawback, a reassigning approach is utilized to avoid idle neurons. For a trained neural network, INTSOM first calculates the numbers of idle neurons are idle. An LBG clustering operation is then performed on the raw objects to obtain K clusters, where K indicates the number of the idle neurons. Eventually, the cores of each cluster are substituted for those idle neurons. The quality of a reconstructed image is thus reinforced, because all neurons (codewords) are utilized effectively.

The pseudocode of INTSOM algorithm is given below.

```
input  : Data(X_i), ε, Codebook_size
output: Codebook(Clusters)
/*** Root neurons distribution ***/
rootClusters= LBG(Data, sqrt( Codebook_size ));
/*** Sub-group scale estimation ***/
for i ← 1 to rootClusters_size do
    estimate Sub_size[i] by Eqn. (5);
end
for i ← 1 to rootClusters_size do
    /*** Neurons training ***/
    while variation[i] > ε do
        SOM(rootClusters[i], Sub_size[i]);
        calculate variation[i] by Eqn. (6);
    end
    /*** Neurons reassigning ***/
    find out the idleNeurons in rootClusters[i];
    idleClusters= LBG(rootClusters[i], idleNeurons_size);
    for j ← 1 to idleNeurons_size do
        replace idleNeurons[j] by idleClusters[j]_core;
    end
end
```

Fig. 5 The pseudocode of INTSOM algorithm

4 Experimental Results

To verify the efficiency and quality of INTSOM, a series of 30 independent runs was undertaken for four well-known images, which are illustrated in Fig. 6. The algorithm was implemented in the Java programming language using a personal computer with Intel Pentium4 3.2GHz CPU and 1GB RAM. The images Lena, Airplane, Peppers and Boat were tested at a size of 512×512 pixels and grayscale with 8 bits per pixel (BPP), i.e. 256 levels. The block size was 4×4, namely 16 dimensions per vector. Significantly, HSOM cannot

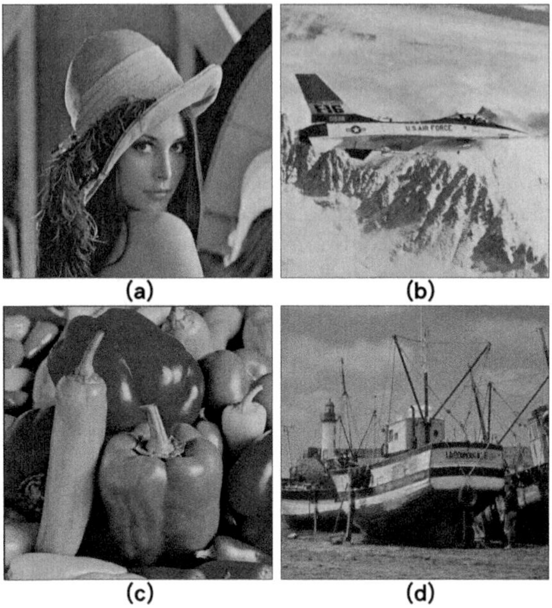

Fig. 6 The 2048-codebook and 16-dimensional gray images reconstructed by INTSOM. (a) Lena (b) Airplane (c) Peppers (d) Boat

Table 1 Comparison of the PSNR (in dB, higher is better) and time-cost (in second, lower is better) of reconstructed images for the proposed INTSOM and some existing approaches, and the bold types denote the best results

Image		Lena		Airplane		Peppers		Boat	
Codebook	Method	PSNR	Timecost	PSNR	Timecost	PSNR	Timecost	PSNR	Timecost
128	LBG	29.590	**9.91**	30.652	15.98	29.681	**10.08**	29.150	**15.00**
	SOM	**29.767**	194.06	**31.152**	199.55	**29.886**	197.16	**29.381**	196.71
	HSOM	N/A	N/A	N/A	N/A	N/A	N/A	N/A	N/A
	INTSOM	29.171	11.42	30.373	**11.83**	29.163	13.96	28.566	15.77
256	LBG	30.479	19.67	31.470	22.41	30.484	**17.52**	29.957	25.01
	SOM	**30.780**	390.50	32.094	396.06	**30.727**	392.84	**30.343**	393.66
	HSOM	30.642	67.58	**32.150**	68.44	30.613	68.95	30.175	67.88
	INTSOM	30.265	**15.25**	31.840	**15.92**	30.211	18.94	29.760	**19.13**
512	LBG	31.286	30.98	32.312	31.16	31.211	**25.79**	30.766	30.98
	SOM	31.603	776.81	33.002	789.05	31.456	793.88	**31.225**	792.11
	HSOM	N/A	N/A	N/A	N/A	N/A	N/A	N/A	N/A
	INTSOM	**31.608**	**20.12**	**33.104**	**18.57**	**31.502**	26.13	31.059	**26.34**
1024	LBG	32.129	45.15	33.291	40.44	31.975	36.82	31.985	36.96
	SOM	32.460	1575.06	33.961	1597.33	32.277	1555.05	32.164	1590.68
	HSOM	33.022	132.73	34.684	134.65	32.594	142.48	32.475	134.40
	INTSOM	**33.333**	**28.72**	**34.909**	**28.09**	**32.729**	**33.09**	**32.507**	**32.58**
2048	LBG	33.098	61.06	34.482	57.42	32.948	54.60	32.792	58.17
	SOM	33.467	3134.60	35.178	3133.12	33.295	3148.01	33.364	3131.83
	HSOM	N/A	N/A	N/A	N/A	N/A	N/A	N/A	N/A
	INTSOM	**35.440**	**39.96**	**37.157**	**40.87**	**34.158**	**43.25**	**34.549**	**44.52**

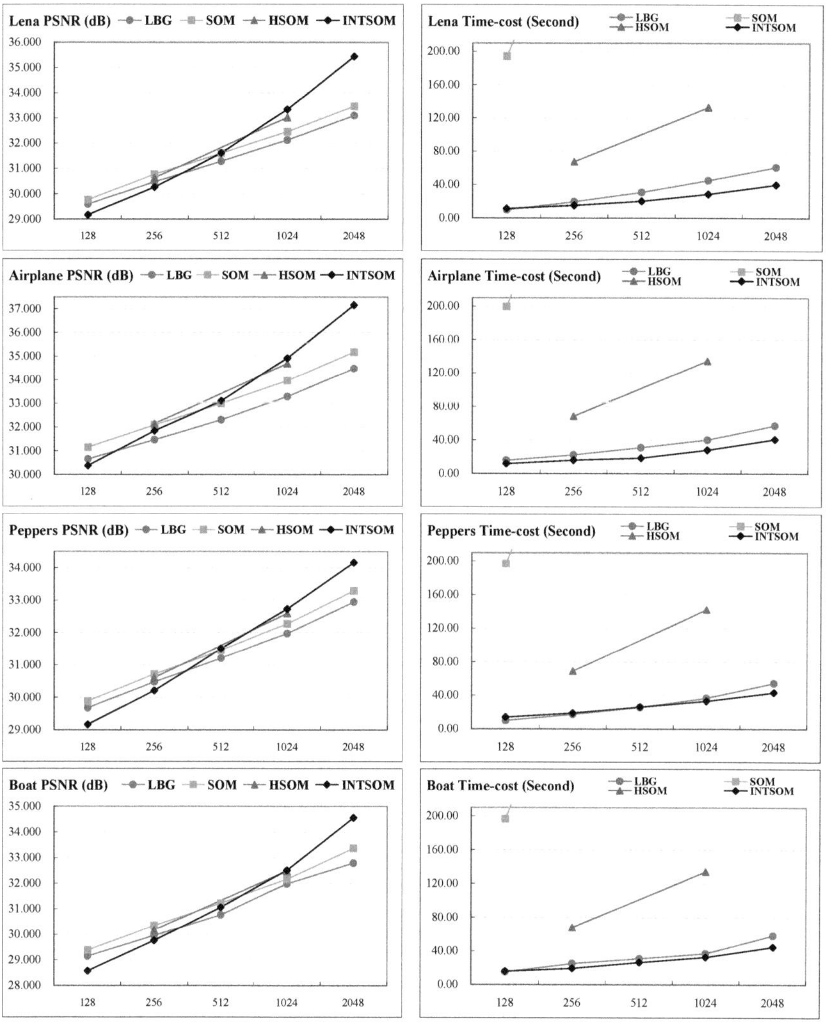

Fig. 7 An integral trend chart for the proposed INTSOM and some existing approaches

process an image with a codebook size that is not a square number, so its results were abandoned in some tests, given as N/A (Not Available). The experimental results are listed below.

As revealed in Table 1, INTSOM has the best compression capability for images with large codebook sizes. To help understand clearly the performance of all approaches, the trend charts of PSNR and time-cost for the proposed INTSOM and some existing approaches are illustrated as follows.

5 Conclusions

This investigation proposes a superior and efficient grayscale image compression algorithm. The proposed algorithm, named INTSOM, first adopts LBG to determine the layer 1 neurons, and then employs a modified SOM to train the dynamic sub-neurons. Finally, a reassigning step is performed to enhance the quality of the reconstructed image. Interestingly, although a smaller codebook size yields a higher compression ratio, it is not worth considering because the resulting quality is too low to preserve the image. Experimental results depict that the proposed INTSOM performs better and has a shorter processing time than LBG, SOM and HSOM, particularly with a large codebook size.

Acknowledgements. The author would like to thank the National Science Council of Republic of China, Taiwan for financially supporting this research under contract no. NSC 96-2221-E-020-027.

References

1. Shapiro, J.M., Center, D.S.R., Princeton, N.J.: Embedded Image Coding Using Zerotrees of Wavelet Coefficients. IEEE Trans. on Signal Processing 41(12), 3445–3462 (1993)
2. Said, A., Pearlman, W.A.: A New, Fast, and Efficient Image Codec Based on Set Partitioning in Hierarchical Trees. IEEE Trans. on Circuits and Systems for Video Technology 6(3), 243–250 (1996)
3. Weinberger, M., Seroussi, G., Sapiro, G.: The LOCO-I Lossless Image Compression Algorithm: Principles and Standardization Into JPEG-LS. IEEE Trans. on Image Processing 9, 1309–1324 (2000)
4. Tsai, C.F., Jhuang, C.A., Liu, C.W.: Gray Image Compression Using New Hierarchical Self-Organizing Map Technique. In: Proceedings of the 3rd International Conference on Innovative Computing Information and Control, pp. 544–549 (2008)
5. Gray, R.M.: Vector Quantization. IEEE ASSP Magazine 1(2), 4–29 (1984)
6. Sayood, K.: Introduction to Data Compression, 2nd edn. Morgan Kaufmann Publishers, San Francisco (2000)
7. Linde, Y., Buzo, A., Gray, R.: An Algorithm for Vector Quantizer Design. IEEE Trans. on Communications 28, 84–95 (1980)
8. Kohonen, T.: The Self-Organizing Map. Proceedings of the IEEE 78(9), 1464–1480 (1990)
9. Kohonen, T.: Self-Organizing Maps. Springer, Heidelberg (1995)
10. Madeiro, F., Vilar, R., Neto, B.: A Self-organizing Algorithm for Image Compression. IEEE Trans. on Neural Networks 28, 146–150 (1998)
11. Barbalho, M., Duarte, A., Neto, D., Costa, F., Netto, A.: Hierarchical SOM Applied to Image Compression. In: Proceedings of International Joint Conference on Neural Networks, vol. 1, pp. 442–447 (2001)

Learning a Selection Problem of Investment Projects and Capital Structure through Business Game

Yasuo Yamashita, Hiroshi Takahashi, and Takao Terano

Abstract. While the importance of financial education increases in recent years, the technique for deepening an understanding of finance theory is needed. In this research, we analyize learning method of the finance theory about the investment project selection and capital structure determination using the business game technique. As a result of analysis, the participant understood the investment project selection method and interesting phenomena – an understanding progresses about the method of determining the capital structure which raises capital stock value – were seen. These results show the effectiveness of the business game technique to study of finance theory.

1 Introduction

In recent years, as asset management market in Japan is increasing, Japanese asset management business is developing quickly. In addition to riskless assets, for example, cash, on asset management business, it is paid attention to invest to riskier financial products including an investment fund. In such current to the investment from the savings, in the asset management business industry, the need for personnel training is strongly recognized in order to heighten an asset management capability. The research [13] which focused on a financial investment theory centering on

Yasuo Yamashita
Department of Computational Intelligence and Systems Science,
Tokyo Institute of Technology, Japan
e-mail: yyasuo1@yahoo.co.jp

Hiroshi Takahashi
Graduate School of Business Administration, Keio University, Japan

Takao Terano
Department of Computational Intelligence and Systems Science,
Tokyo Institute of Technology, Japan

K. Nakamatsu et al. (Eds.): New Advan. in Intel. Decision Techno., SCI 199, pp. 41–50.
springerlink.com

risk return analysis is accomplished in the viewpoint of the personnel training in asset management business. But, there are not enough researches on learning of a finance theory. From a money manager's position, it is hard to notice the problem from a business administrator's view because money managers analyze corporate management from the view with security analysis, such as stocks and a corporate bond. Although especially known as knowledge by study by books and so on about the argument in finance theory, it is hard to say that an understanding is deepening enough to the application to actual decision-making[1]. It is required to deepen an understanding of main arguments in finance theory called the investment project selection and capital structure selection, which are the analysis objects of this research in order to raise management capabilities in severe investment competition.

On the other hand, it is hard to say that the research on finance with business game is done enough[4], and application of business game to finance research is expected. Business game is one field of gaming simulation[5], which is used to the research on management or marketing[12][11][10]. It is meaningful to do research on study of the finance using the business game technique for these reasons. In this research, it aims at showing the technique of learning the investment project selection and capital structure selection about finance theory using the framework of business game . Firstly, the model of the business game based on finance theory was built, and subsequently it experimented by using an actual human being as a player. In the following section, after explaining the model used for analysis, a result is shown in Section 3. Section 4 is a conclusion.

2 Method

2.1 System of Business Game

Environment required for development of the system in this research, and execution of an experiment is constituted by business model descriptive language (BMDL) and business model development system (BMDS)[9]. BMDL is the programming descriptive language of a short form. HTML files, CGI files, and so on can be created in BMDS[2] by describing BMDL for game managers (facilitator) and for game users (player). Fig. 1 shows development and execution environment of an experiment. Players input decision-making in each round through WWW browsers, and a facilitator also advances a game through a WWW browser.

2.2 Model of Business Game

Corporate management could be asked for various decision-making. In this research, the business game is built, which focused on the investment project selection

[1] About the argument in finance theory, it is referring to the reference[8][1].

[2] BMDL is changed and used in part so that the experiment of this research may be possible.

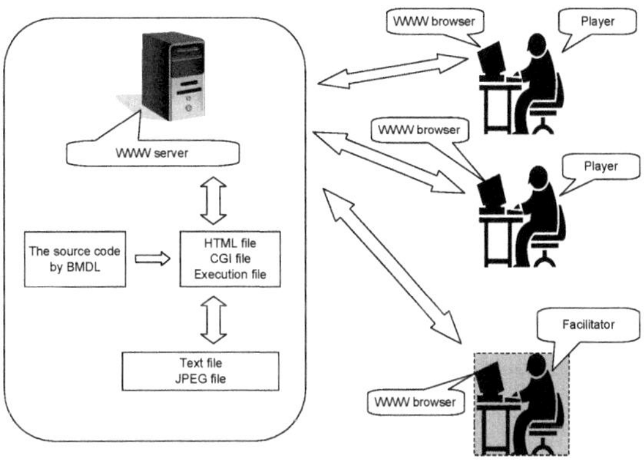

Fig. 1 Conceptual diagram of development and execution environment of experiment

and capital structure selection of the company which are the main problems of finance theory.

According to finance theory, it is known that the project selection method which makes stock value the maximum should just choose the project which makes the maximum the investment-amount weighted average expected return. The certain investment project exists in the investment project of a company. That investment project can usually be called main occupation and should be continued and invested. Therefore, in the model of this research, when choosing an investment project, there is a certain invested project[3], to be considered as a setup near the actual investment project selection situation.

In finance theory, there is Modigliani-Miller theory (MM theory) as most fundamental theory about the capital structure of a company[6][7]. According to MM theory, it is known that stock value will not be based on capital structure in a complete market. Here, a complete market is the market that does not have the asymmetric information of a tax, a bankruptcy risk between investors and managers. However, in an actual market, since there are a tax and a bankruptcy risk, stock value is influenced by capital structure. When there is a tax to the profits of a company, stock value can be raised by raising debt ratio. However, since there is bankruptcy cost in a debt financing, if not much many debts financing are performed, expense will increase, and on the contrary a result of stock-value reduction is brought[4].

According to finance theory, it is known that there is optimal capital structure which makes stock value the maximum. As decision-making, it carries out by specifying the amount of money of a debt and stocks. The kind of debt is assumed to be borrowing which does not specify but makes one round (one period) the due date, or a thing like a corporate bond. Moreover, in this experiment, there are asummed to be no transaction cost and no issuance cost for debts or stocks. Also in this experiment, there is the optimum capital structure which makes stock value the maximum in

[3] Although it is common to make decisions investment in the investment project in an actual company about the project from which an investment period differs, such analysis is a future subject.

[4] I would like to make analysis using the knowledge of behavioral finance[2][3] into a future subject about the case where there is asymmetric information of an investor and a manager.

raising the funds for an investment projects. In the business game of this experiment, closer to such optimal capital structure the player's one is, more the stock value can be raised, and made it possible to learn about the capital structure method which raises stock value. The business game in this experiment specifically advances as follows.

A player is provided with expected returns, risks and investment amounts and so on, of 11 investment projects. Those investment projects include an initial investment project as reference information at the beginning of every round. A player refers to such reference information and chooses investment projects which make stock value the maximum. Although the continuation investment project is considered as the always invested project, selection of the other investment project is arbitrary. It is also possible to choose all the investment projects or to choose no investment projects as an extreme example.

Each player decide whether an investment project is "invested" or "not invested". Next, a player determines the amount of financing according to an investment project amount as input of the capital structure "debt amount" and "stock amount." Extra-items are also created as input variables used inside the model of business game. The purpose is investigating the thinking of a players. A player selects the top 3 items from the 20 items when a player makes decisions from what has a high priority. Determination of the investment project of all the players and the input of capital structure will perform the update process which calculates a stock price, a debt ratio[5] and so on according to internal model.

Output informations are displayed on each player. Output informations are a stock price, a debt ratio, ranking and so on. Since ranking promotes competition between players, it is displayed on each player by making ranking based on the stock return of each round by a stock price, an accumulation stock return, ranking and so on into output information. Credit rating is estimated by the credit-spread calculated inside a model. Evaluation of the credit-spread is beforehand given to the model as exogenous information.

The player is given as a purpose raising the stock price of a company as much as possible. A stock price is calculated according to the input of a player based on finance theory inside a model. About selection of an investment project, there is the optimal solution which makes a stock price the maximum for every round. It is the structure where a stock price increases, so that selection of the investment project of a player is close to an optimal solution. In Fig. 2, the investment project 0 is a continuation investment project, and it is considered as the investment project which can be chosen from the investment project 1 to the investment project 10. About the 1-10 investment projects, it is assumed that a number is assigned to the high order of an expected return. It means that the number of the horizontal axis of Fig. 2 chooses from the investment project 1 to the investment project of the number in addition to the continuation investment project 0. A vertical axis is an amount-weighted expected return. In the example of Fig. 2, when the investment projects

[5] Debt ratio = Debt amount / Stock amount.

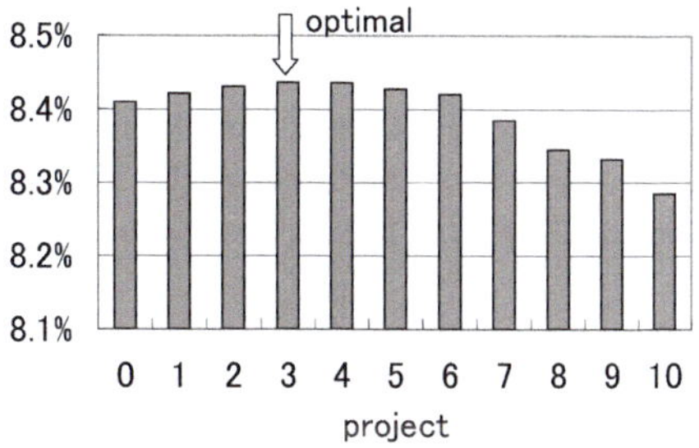

Fig. 2 Example of optimum investment project

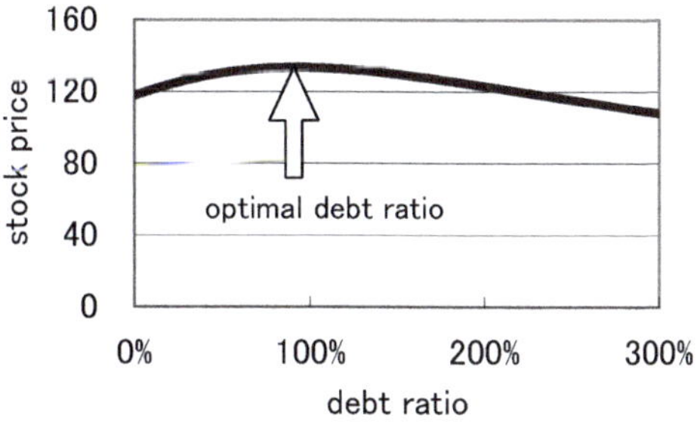

Fig. 3 Example of optimum debt ratio

1, 2, and 3 are chosen, a weighted average expected return serves as the maximum, and serves as an optimal solution.

Moreover, there is the optimal solution to which capital structure selection also makes a stock price the maximum according to the amount of an investment project. It is a setup with an increasing stock price, so that capital structure selection of a player is close to an optimal solution. In Fig. 3, at least 90% of the debt ratio from which the stock price is the maximum serves as an optimal solution.

All players made decisions under the same parameter conditions, and the parameter conditions during a round experimented by setup of being independent. The present decision-making is a setup which is not influenced by the past decision-making.

Business game of the above contents was carried out twice. Players are 4 institutional investor's affiliation members (all the members is financial analyst holder)[6]. The time for explanation of introduction and an experiment and decision-making of a round 1 is taken about 30 minutes. The players repeat decision-making of each round every about 10 minutes. One experiment took about around 2 hours[7]. A player was not told about the number of times of an end of a round, but it was decided to be an end in about 2 hours. Therefore, it is ended at eight rounds as a result.

[6] The experiment by 12 department-of-economics students is conducted twice as preparatory experiments. As for the decision-making item of this research, there were those who express the comment that it is more difficult than the experiment[13] only for securities investment theory among participants in order also to have to determine capital structure in addition to investment.

[7] This experiment is conducted as part of operating training, and the incentive to experiment participation of a player is maintained at the sufficiently high level.

3 Result

Explanation about an experimental result is given in this chapter. First, it explains that there was learning effect about investment project selection. Subsequently, it is explained that there was learning effect also about capital structure selection.

3.1 Investment Project Selection

This section explains that there was learning effect through this experiment about investment project selection.

Figs. 4 and 5 show transition of "the stock return by investment project selection" (it is called this "p return" hereafter) in experiment 1 and experiment 2. "The stock return by investment project selection" is the rate of change of the stock price evaluated in the market according to selection of the investment project which the player determined.

"Optim" in a figure is p return at the time of making investment project selection which makes a stock price the maximum. The investment project selection in this case is henceforth called an "optimum investment project". And "a" to "d" is p return of each player. P return of each player has deviated from p return of "optim" in Fig. 4. In Fig. 5, it has overlapped with p return of "optim" mostly. Although

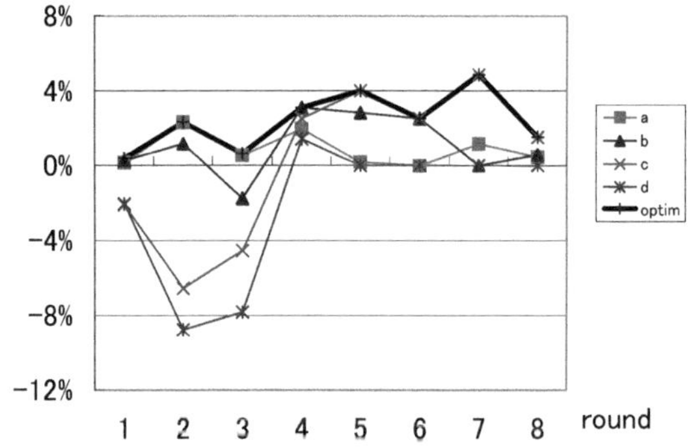

Fig. 4 p return of experiment 1

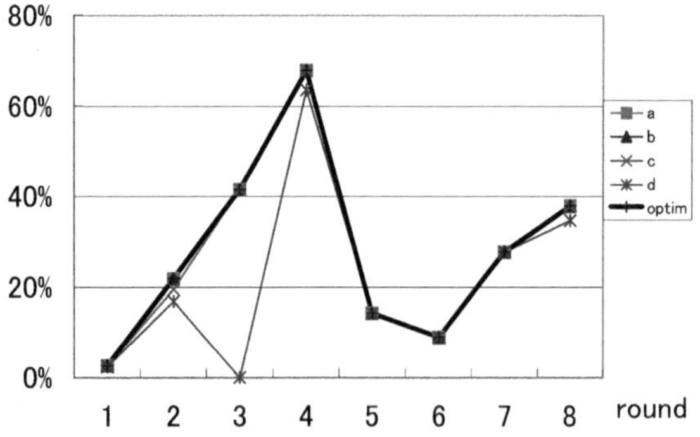

Fig. 5 p return of experiment 2

Table 1 Average of p return rate of deviation

Experiment	a	b	c	d	median	Difference of a median
1	1.51	1.29	2.11	4.47	1.81	–
2	0.00	0.00	0.24	4.81	0.12	1.69*

Unit:%.
*It is significant at a 5% level. median test (one side).

investment project selection to which a player raises a stock price was not made of experiment 1 as for this, it shows that it was learned in the experiment 2. This is an interesting result.

Table 1 is the verification that the players learn the investment project selection method effectively through the experiment 1 and the experiment 2. A significant difference is observed in the median of all the players of the rate of deviation[8] of p return by the level 5% in experiment 1 and experiment 2. And it shows that the experience of the investment project selection through business game has an effect by this experiment.

About the selection method of an investment project, if the player is an institutional investor's employment section affiliation member, he or she should usually know the selection method with books and so on as basic knowledge. However, as checked in the experiment 1 of this research, when it was shown as a problem of "reality", there was the phenomenon in which the right investment project could not immediately be chosen even if the player was knowledgeable. It is surmised by experiencing quasi-reality through the technique of business game that the player's understanding is deepening about the selection method of an investment project so that it may be suggested in experiment 1 and experiment 2. There is meaning which just uses the business game technique in such a case.

From the questionnaire after an experiment, we has also actually obtained the reply that an understanding deepened about the selection method of the investment project from all the players. It seems to be effective technique by this research to learn investment project selection through business game.

3.2 Capital Structure Selection

This section explains that there is the learning effect through this experiment about capital structure selection.

Figs. 6 and 7 show transition of the debt ratio of the player of experiment 1 and experiment 2, respectively. "Optim" in a figure is a debt ratio (the optimum debt ratio) which makes a stock price the maximum. And "a" to "d" is a debt ratio of each player. As compared with Fig. 6, it seems that deviation of the debt ratio of each player from the optimum debt ratio has decreased in Fig. 7.

[8] "Rate of deviation of p return" =("p return of optim" - "p return of player")/(1+ "p return of optim") defines the rate of deviation of p return.

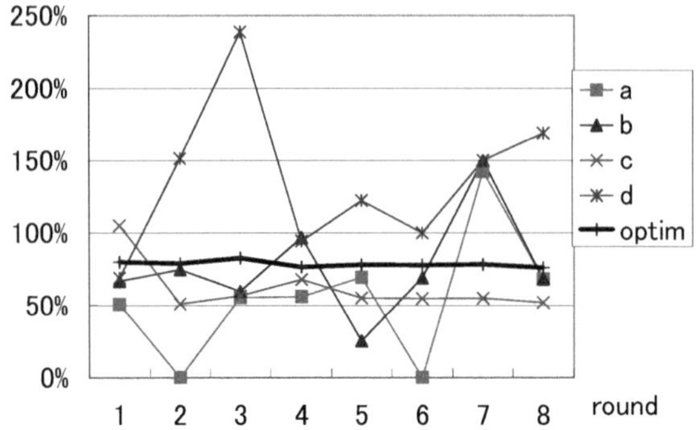

Fig. 6 Debt ratio of experiment 1

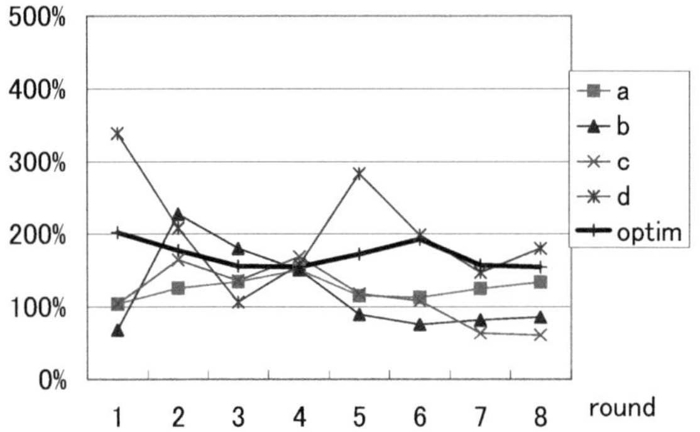

Fig. 7 Debt ratio of experiment 2

Table 2 Tracking error of debt ratio rate of deviation

Experiment	a	b	c	d	median	Difference of a median
1	61.3	43.2	29.6	94.9	52.2	–
2	29.2	44.3	39.9	35.9	37.9	14.3*

Unit: %.
*It is significant at a 5% level. median test (one side).

Table 2 is the verification that the players learn the capital structure method effectively through the experiment 1 and the experiment 2. The significant difference is observed in the median of all the players of the tracking error of a debt ratio rate of deviation[9] by the level 5% in the experiment 1 and the experiment 2. And it shows that the experience of the capital structure selection through business game has an effect by this experiment.

Although it is known that there is the optimum debt ratio to which a player makes a stock price the maximum about a capital structure selection method from before experiment implementation, the concrete determination method cannot have been

[9] "Rate of deviation of debt ratio" = ("debt ratio of player" - "optimum debt ratio") / "optimum debt ratio" defines the rate of deviation of a debt ratio.

learned. Inside the model of business game, stock price evaluation is performed based on finance theory. Therefore, if a player gets to know details and there is sufficient time calculable by self, it will be thought that the optimum debt ratio was able to be drawn. However, the time given to the player for every round in the experiment is about 10 minutes. Hence, it is thought that the players use the different methods of calculating the optimum debt ratio.

Actually, there are some players who answer that they think the past data of other teams as important as items when players make decisions. It is thought that the basis of restrictions that it must make decisions for a short time, and the player learned the decision-making method that the optimum debt ratio was approached while applying a trial-and-error method for restrictive information. Also in an actual market, it is common that the quality and quantity of information are restricted, and referring to indexes , such as a debt ratio of the other company within a same sector, is often used in business. In this experiment, the interesting result that the repeatability of a market is realizable, is obtained with the point that the players use heuristics (which is used in such an actual market).

4 Conclusion

It is shown that the use of the business game technique is effective in study of the finance theory about investment project selection and capital structure selection. About investment project selection, the player's understanding is deepened by experiencing business game, and it is shown that the experience of the investment project selection through business game is effective. About capital structure selection, it is shown that the experience of the capital structure selection through business game is effective.

References

1. Brealey, R., Myers, S., Allen, F.: Principles of Corporate Finance. McGraw-Hill, New York (2006)
2. Constantinides, G.M., Harris, M., Stulz, R.M. (eds.): Handbook of the Economics of Finance: Financial Markets and Asset Pricing. North-Holland, Amsterdam (2003)
3. Eckbo, B.E. (ed.): Handbook Of Corporate Finance: Empirical Corporate Finance. North-Holland, Amsterdam (2007)
4. Faria, A.J.: Business simulation games: Current usage levels—an update. Simulation & Gaming 29, 295–308 (1998)
5. Greenblat, C.S.: Designing games and simulations. Sage Publications, Inc., Thousand Oaks (1988)
6. Modigliani, F., Miller, M.: The cost of capital, corporation finance and the theory of investment. American Economic Review 48, 655–669 (1958)
7. Modigliani, F., Miller, M.: Corporate income taxes and the cost of capital: A correction. American Economic Review 53, 433–443 (1963)
8. Ross, S., Wester, R., Jaffe, J.: Corporate finance. McGraw-Hill, New York (2005)

9. Terano, T., Suzuki, H., Kuno, Y., Fujimori, H., Shirai, H., Nishio, H., Ogura, N., Takahashi, M.: Understanding your business through home-maid simulatior development. Developments in Business Simulation and Experimential Learning (26), 65–71 (1999)
10. Tompson, G.H., Dass, P.: Improving students' self-efficacy in strategic management: the relative impact of cases and simulations. Simulation & Gaming 31(1), 22–41 (2000)
11. Walters, B.A., Coalter, T.M., Rasheed, A.M.A.: Simulation games in business policy courses: is there value for students? Journal of Education for Business 72(3), 170–174 (1997)
12. Wolfe, J.: The effectiveness of business games in strategic management course work. Simulation & Gaming 28(4), 360–376 (1997)
13. Yamashita, Y., Takahashi, H., Terano, T.: The development of the financial learning tool through business game. In: Lovrek, I., Howlett, R.J., Jain, L.C. (eds.) KES 2008, Part II. LNCS, vol. 5178, pp. 986–993. Springer, Heidelberg (2008)

Evolution Prospection

Luís Moniz Pereira and Han The Anh

Abstract. This work concerns the problem of modelling evolving prospective agent systems. Inasmuch a prospective agent [1] looks ahead a number of steps into the future, it is confronted with the problem of having several different possible courses of evolution, and therefore needs to be able to prefer amongst them to decide the best to follow as seen from its present state. First it needs a priori preferences for the generation of likely courses of evolution. Subsequently, this being one main contribution of this paper, based on the historical information as well as on a mixture of quantitative and qualitative a posteriori evaluation of its possible evolutions, we equip our agent with so-called evolution-level preferences mechanism, involving three distinct types of commitment. In addition, one other main contribution, to enable such a prospective agent to evolve, we provide a way for modelling its evolving knowledge base, including environment and course of evolution triggering of all active goals (desires), context-sensitive preferences and integrity constraints. We exhibit several examples to illustrate the proposed concepts.

1 Introduction

Prospective agent systems [1] address the issue of how to allow evolving agents to be able to look ahead, prospectively, into their hypothetical futures, in order to determine the best courses of evolution from their own present, and thence to prefer amongst those futures. In such systems, *a priori* and *a posteriori* preferences,

Luís Moniz Pereira
Centro de Inteligência Artificial (CENTRIA), Universidade Nova de Lisboa,
2829-516 Caparica, Portugal
e-mail: `lmp@di.fct.unl.pt`

Han The Anh
Centro de Inteligência Artificial (CENTRIA), Universidade Nova de Lisboa,
2829-516 Caparica, Portugal
e-mail: `h.anh@fct.unl.pt`

K. Nakamatsu et al. (Eds.): New Advan. in Intel. Decision Techno., SCI 199, pp. 51–63.
springerlink.com

embedded in the knowledge representation theory, are used for preferring amongst hypothetical futures, or scenarios. The *a priori* ones are employed to produce the most interesting or relevant conjectures about possible future states, while the *a posteriori* ones allow the agent to actually make a choice based on the imagined consequences in each scenario. ACORDA [1] is a prospective logic system that implements these features. It does so by generating scenarios, on the basis only of those preferred abductions able to satisfy agents' goals, and further selecting scenarios on the basis of the immediate side-effects such abductions have within them.

However, the above proposed preferences have only local influence, i.e. for example, immediate *a posteriori* preferences are only used to evaluate the one-state-far consequences of a single choice. They are not appropriate when evolving prospective agents want to look ahead a number of steps into the future to determine which decision to make from any state of their evolution. Such agents need to be able to evaluate further consequences of their decisions, i.e. the consequences of the hypothetical choices abduced to satisfy their goals. Based on the historical information as well as quantitative and qualitative a posteriori evaluation of its possible evolutions, we equip an agent with so-called evolution-level preferences mechanism.

For evolving agents, their knowledge base evolves to adapt to the outside changing environment. At each state, agents have a set of goals and desires to satisfy. They also have to be able to update themselves with new information such as new events, new rules or even change their preferences. To enable a prospective agent to evolve, we provide a way for modelling its evolving knowledge base, including the environment and course of evolution triggering of all active goals (desires), of context-sensitive preferences and of integrity constraints. To further achieve this, immediate *a posteriori* preferences are insufficient.

After deciding on which action to take, agents evolve by committing to that action. Different decision commitments can affect the simulation of the future in different ways. There are actions that, if committed to, their consequences are nevermore defeated and thus permanently affect the prospective future. There are also actions that do not have any inescapable influence on the future, i.e. committing to them does not permanently change the knowledge base, like the previously described "hard" commitments – they are "ongoing". They may be taken into account when, in some following future state, the agents need to consider some evolution-level preferences trace. Other action commitments are "temporary", i.e. merely momentary.

In addition, we specifically consider so-called inevitable actions that belong to every possible evolution. By hard committing to them as soon as possible, the agent can activate preferences that rule out alternative evolutions that are ipso facto made less relevant.

The rest of the paper is organized as follows. Section 2 discusses prospective logic programs, describing the constructs involved in their design and implementation. Section 3 describes evolving prospective agents, including single-step and multiple-step look-ahead, and exhibits several examples for illustration. The paper ends with conclusions and directions for the future.

2 Prospective Logic Programming

Prospective logic programming enables an evolving program to look ahead prospectively into its possible future states, which may include rule updates, and to prefer among them to satisfy goals [1]. This paradigm is particularly beneficial to the agents community, since it can be used to predict an agent's future by employing the methodologies from abductive logic programming [2, 4] in order to synthesize, prefer and maintain abductive hypotheses. We next describe constructs involved in our design and implementation of prospective logic agents and their preferred and partly committed but still open evolution, on top of Abdual [3] - a XSB-Prolog implemented system which allows computing abductive solutions for a given query.

2.1 *Language*

Let $\mathscr{L}$ be a first order language. A domain literal in $\mathscr{L}$ is a domain atom A or its default negation *not* A. The latter is used to express that the atom is false by default (Closed World Assumption). A domain rule in $\mathscr{L}$ is a rule of the form:

$$A \leftarrow L_1, \ldots, L_t \qquad (t \geq 0)$$

where A is a domain atom and $L_1, \ldots, L_t$ are domain literals. An integrity constraint in $\mathscr{L}$ is a rule with an empty head. A (logic) program P over $\mathscr{L}$ is a set of domain rules and integrity constraints, standing for all their ground instances.

2.2 *Preferring Abducibles*

Every program P is associated with a set of abducibles $\mathscr{A} \subseteq \mathscr{L}$. These, and their default negations, can be seen as hypotheses that provide hypothetical solutions or possible explanations to given queries. Abducibles can figure only in the body of program rules.

An abducible A can be assumed only if it is a considered one, i.e. if it is expected in the given situation, and, moreover, there is no expectation to the contrary [6].

$$consider(A) \leftarrow expect(A), not\ expect\_not(A), A$$

The rules about expectations are domain-specific knowledge contained in the theory of the program, and effectively constrain the hypotheses available in a situation. Handling preferences over abductive logic programs has several advantages, and allows for easier and more concise translation into normal logic programs (NLP) than those prescribed by more general and complex rule preference frameworks. The advantages of so proceeding stem largely from avoiding combinatory explosions of abductive solutions, by filtering irrelevant as well as less preferred abducibles [5].

To express preference criteria among abducibles, we envisage an extended language $\mathscr{L}^\star$. A preference atom in $\mathscr{L}^\star$ is of the form $a \triangleleft b$, where a and b are abducibles. It means that if b is assumed (i.e. abduced), then $a \triangleleft b$ forces a to be assumed too (b can only be abduced if a is as well). A preference rule in $\mathcal{L}^\star$ is of the form: $a \triangleleft b \leftarrow L_1, ..., L_t \ (t \geq 0)$, where $L_1, ..., L_t$ are domain literals over $\mathcal{L}^\star$. *A priori* preferences are used to produce the most interesting or relevant conjectures about possible future states. They are taken into account when generating possible scenarios (abductive solutions), which will subsequently be preferred amongst each other a posteriori.

2.3 A Posteriori Preferences

Having computed possible scenarios, represented by abductive solutions, more favorable scenarios can be preferred a posteriori. Typically, *a posteriori* preferences are performed by evaluating consequences of abducibles in abductive solutions. An *a posteriori* preference has the form:

$$A_i \ll A_j \leftarrow holds\_given(L_i, Ai),\ holds\_given(L_j, A_j)$$

where A_i, A_j are abductive solutions and L_i, L_j are domain literals. This means that A_i is preferred to A_j a posteriori if L_i and L_j are true as the side-effects of abductive solutions A_i and A_j, respectively, without any further abduction. Optionally, in the body of the preference rule there can be any Prolog predicate used to quantitatively compare the consequences of the two abductive solutions.

2.4 Active Goals and Context Sensitive Integrity Constraints

In each cycle of its evolution the agent has a set of active goals or desires. We introduce the $on\_observe/1$ predicate, which we consider as representing active goals or desires that, once triggered by the observations figuring in its rule bodies, cause the agent to attempt their satisfaction by launching the queries standing for them. The rule for an active goal AG is of the form:

$$on\_observe(AG) \leftarrow L_1, ..., L_t \ (t \geq 0)$$

where $L_1,...,L_t$ are domain literals. During evolution, an active goal may be triggered by some events, previous commitments or some history-related information. We differentiate events that have temporary influence, i.e. affect only the current cycle and thus are entered into its knowledge base as facts and removed when the influence is finished, from ones that have permanent influence, i.e. affect every cycle issuing from the current one and thus are entered to the knowledge base as facts and stay there forever. Respectively, we provide two predicates, *event/1* and *asserts/1*.

When starting a cycle, the agent collects its active goals by finding all the $on\_observe(AG)$ that hold under the initial theory without performing any abduction, then finds abductive solutions for their conjunction.

Context sensitive integrity constraints

When finding abductive solutions, all integrity constraints in the knowledge base must be satisfied. However, when considering an evolving agent, there is a vital need to be able to code integrity constraints dependent on time points and external changing environment. A context sensitive integrity constraint with the name *icName* and a non-empty context is coded by using an active goal as follows:

$$on\_observe(not\ icName) \leftarrow L_1, ..., L_n\ (t \geq 0)$$

$$icName \leftarrow icBody$$

where $L_1,...,L_t$ are domain literals which represent the triggering context of the integrity constraint. Whenever the context is true, the active goal *not icName* must be satisfied, which implies that the integrity constraint $\leftarrow icBody$ must be satisfied. When the context is empty ($t = 0$) the integrity constraint becomes a usual one which always must be satisfied.

2.5 *Levels of Commitment*

Each prospective cycle is completed by registering any surviving abductive solutions (represented by their abducibles) into the knowledge base and moving to the next cycle of evolution. Committing to each alternative abductive solution will create a new branch of the so-called evolution tree. The history of the evolution is kept by setting a time stamp for the abducibles that the agent commits to in each cycle.

As a program is evolving, the commitment can affect the future in different ways. Based on their influence, we classify commitments in three categories. Firstly, there are abducibles, representing actions or other options that, after committed to in a state, will not be subsequently defeated, i.e. a commitment to reverse the committed to actions is not allowed. This kind of commitment inscribes a permanent consequence on the future and therefore plays the role of a fact in the knowledge base for all future evolution states issuing from that state. Commitments of this sort are called *hard*. In addition, there are commitments that, when committed to in a state, can nevertheless be defeated by committing to their opposite abducibles at some future state, but will keep on affecting the future (by inertia) up until then. Commitments of this kind are called *ongoing*. Lastly, the weakest kind of commitments are those immediately withdrawn in the following state and so have direct influence only on the transition from the current state. They can have indirect influence when in some future state the history of the evolution needs to be taken into account. We call this kind *temporary*.

3 Evolving Prospective Agents

Informally, an evolution of a prospective agent is a sequence of time stamped sets of commitments at each cycle of the evolution. The agent self-commits to abducibles, which are used to code available and preferred decision choices on all manner of options. Depending on the capabilities and need, at each time point in the evolution the agent acts just to satisfy the active goals and integrity constraints at hand, or needs to look ahead a number of steps into the future in order to satisfy its long-term and context triggered goals and constraints in a prospective way, taking into account its possible futures and evolution-sensitive reachable decision choices.

3.1 Single-Step Prospective Agent

Each cycle ends with the commitment of the agent to an abductive solution. Alternative commitments can be explored by searching the space of evolutions.

Example 1. Suppose agent John is going to buy an air ticket for traveling. He has two choices, either buying a saver or a flexible ticket. He knows the flexible one is expensive, but, if he has money, he does not wish a saver ticket because, if he bought it, he would not be able to change or return it in any circumstance. The saver ticket is one that, when committed to, the reverse action of returning is not allowed (a hard commitment thus). However, if John does not have much money, he is not expected to buy something expensive. Later, waiting for the flight, John finds out that his mother is ill. He wants to stay at home to take care of her, thus needing to cancel the ticket. This scenario can be coded as in Figure 1.

Line 1 is the declaration of program abducibles, and of which of these are ongoing and hard commitments. The abducibles in the *abds/1* predicate not declared as ongoing or hard are by default temporary. Line 2 says there is unconditional expectation for each abducible declared.

When John wants to travel, specified by entering *event(travel)*, i.e. the fact *travel* is temporarily added, which, in turn, triggers the only active goal *ticket*. *empty_pocket* is false and *have_money* is true, hence there is expectation to the contrary of *saver_ticket* but not of *flexible_ticket* (lines 5-6). Thus, there is only one abductive solution: [*flexible_ticket*]. The cycle ends by committing to this abductive solution. Since *flexible_ticket* is an ongoing commitment, it will be added in every abductive solution of the following cycles until John knows that his mother is ill, entering *event*(*mother_ill*). The only active goal *stay_home* now needs to be satisfied.

In addition, the event mother being ill triggers *saver_ticket_ic* and *cancel_ticket_ic*, context-sensitive integrity constraints in line 10. There is no expectation to the contrary of *cancel_ticket* and *lose_money*, and the ongoing commitment *flexible_ticket* is defeated, there being now three minimal abductive solutions: [*cancel_ticket*, *not flexible_ticket*], [*lose_money*, *not flexible_ticket*], [*lose_money*, *not cancel_ticket*].

In the next stage, *a posteriori* preferences are taken into account. Considering the only *a posteriori* preference in line 11, the two abductive solutions that include *lose_money* are ruled out since they lead to the consequence *lose_money*, which is less preferred than the one that leads to *cancel_ticket*. In short, agent John bought a

```
1. abds([saver_ticket/0, flexible_ticket/0,
         cancel_ticket/0, lose_money/0]).
   ongoing_commitment([flexible_ticket]).
   hard_commitment([saver_ticket]).
2. expect(saver_ticket).    expect(flexible_ticket).
   expect(cancel_ticket).   expect(lose_money).
3. on_observe(ticket) <- travel.
   ticket <- saver_ticket.        ticket <- flexible_ticket.
4. expensive(flexible_ticket).
5. expect_not(saver_ticket) <- have_money.
6. expect_not(X) <- empty_pocket, expensive(X).
7. empty_pocket  <- buy_new_car.
   have_money    <- not empty_pocket.
8. on_observe(stay_home) <- mother_ill.
9. stay_home <- cancel_ticket.   stay_home <- lose_money.
10.change_ticket <- mother_ill.
   on_observe(not saver_ticket_ic)  <- change_ticket.
   on_observe(not cancel_ticket_ic) <- change_ticket.
   saver_ticket_ic  <- saver_ticket, cancel_ticket.
   cancel_ticket_ic <- cancel_ticket, ticket.
11.Ai << Aj <- holds_given(cancel_ticket, Ai),
               holds_given(lose_money, Aj).
```

Fig. 1 Ticket example

flexible ticket to travel, but later he can cancel the ticket to stay at home to take care of his mother because the flexible ticket is a defeasible ongoing commitment.

Next consider the same initial situation but suppose John just bought a new car, by entering *asserts(buy_new_car)*. *empty_pocket* becomes true and *have_money* becomes false. Hence there is expectation to the contrary for *flexible_ticket* (line 7) and no expectation to the contrary for *saver_ticket* (line 6). Therefore, the only abductive solution is [*saver_ticket*]. Since *saver_ticket* is a hard commitment, it is not defeated and later on, during evolution, it will always be added to every abductive solution. Even when the mother is ill, *saver_ticket_ic* will prevent having *cancel_ticket* (line 10). Thus, the only abductive solution is the one including *lose_money*.

In short, John made a hard commitment by buying a saver ticket, and later on, when his mother is ill, he must relinquish the ticket and lose money to stay at home.

Inevitable Actions

There may be abducibles that belong to every initial abductive solution (before considering *a posteriori* preferences). These abducibles are called *inevitable* and will be committed to whatever the final abductive solution is. Realizing that actually committing to some abducible changes the knowledge base, and may trigger preferences that subsequently might help to rule out some irrelevant abductive solutions (or even to provide the final decision for the current active goals), our agent is equipped with the ability to detect the inevitable abducibles, committing to them. Doing the inevitable first can lead to further inevitables.

Example 2. Suppose agent John wants to take some money. He can go to one of three banks: *a*, *b* or *c*. All the banks are at the same distance from his place. In addition, John needs to find a book for his project work. The only choice for him is to go to the library. At first, John cannot decide which bank to go to. After a moment, he realizes that in any case he must go to the library, so does it first. Arrived there he notices that bank *c* is now the nearest compared to the others. So he then decides to go to *c*. This scenario can be coded with the program in Figure 2.

```
1. abds([lib/0, a/0, b/0, c/0]).
2. expect(lib). expect(a). expect(b). expect(c).
3. on_observe(take_money).
   take_money <- a, not b, not c.
   take_money <- b, not a, not c.
   take_money <- c, not b, not a.
4. on_observe(find_book).     find_book <- lib.
5. Ai << Aj <- dif_distance, hold(dist(Di), Ai),
                hold(dist(Dj), Aj), Di < Dj.
6. dist(10) <- prolog(current_position(lib)), a.
   dist(5)  <- prolog(current_position(lib)), b.
   dist(0)  <- prolog(current_position(lib)), c.
   beginProlog.
7. dif_distance :- current_position(lib).
8. go_to(lib) :- commit_to(lib).
   current_position(C) :- go_to(C).
   endProlog.
```

Fig. 2 Inevitable action example

There are two active goals *take_money* and *find_book*, and hence, three strict abductive solutions (i.e. consider only positive abducibles) that satisfy them: $[a, lib]$, $[b, lib]$, $[c, lib]$. Since the abducible *lib* belongs to all abductive solutions, it is an inevitable one. Thus, the actual commitment to *lib*, i.e. the action of going to the library, is performed. This changes John's current position (line 8). John's new position is at different distances from the banks (line 7) which triggers the *a posteriori* preference in line 5. This preference rules out the abductive solutions including *a* and *b* since they lead to the consequences of having further distances in comparison with the one including *c* (line 6). In short, from this example we can see that actually committing to some inevitable action may help to reach a decision for a problem that could not determinedly and readily be solved without doing that first, for there were three equal options competing.

3.2 *Multiple-Step Prospective Agent*

While looking ahead a number of steps into the future, the agent is confronted with the problem of having several different possible courses of evolution. It needs to be able to prefer amongst them to determine the best courses from its present state (and any state in general). The (local) preferences, such as the *a priori* and *a posteriori* ones presented above, are no longer appropriate enough, since they can be used to evaluate only one-step-far consequences of a commitment. The agent should be able to also declaratively specify preference amongst evolutions through their available historical information as well as by quantitatively or qualitatively evaluating the consequences or side-effects of each evolution's choices.. We equip our agent with two kinds of evolution-level preferences: *evolution result a posteriori preference* and *evolution history preference*.

3.2.1 Evolution Result a Posteriori Preference

A posteriori preference is generalized to prefer between two evolutions. An *evolution result a posteriori preference* is performed by evaluating consequences of following some evolutions. The agent must use the imagination (look-ahead capability) and present knowledge to evaluate the consequences of evolving according to a particular course of evolution. An *evolution result a posteriori preference* rule has the form:

$$Ei \lll Ej \leftarrow holds\_in\_evol(L_i, Ei), holds\_in\_evol(L_j, Ej)$$

where E_i, E_j are evolutions and L_i, L_j are domain literals. This preference implies that E_i is preferred to E_j if L_i and L_j are true as side-effects of evolving according to E_i or E_j, respectively. Optionally, in the body of the preference rule there can be recourse to any Prolog predicate, used to quantitatively compare the consequences of the two evolutions for decision making.

Example 3. During war time agent David, a good general, needs to decide to save one city, a or b, from an attack. He does not have enough military resources to save both. If a city is saved, citizens of the city are saved. Normally, a bad general, who just sees the situations at hand would prefer to save the city with more population, but a good general would look ahead a number of steps into the future to choose the best strategy for the war as a larger whole. Having already scheduled for the next day that it will be a good opportunity to make a counter-attack on one of the two cities of the enemy, either a small or a big city, the prior action of first saving a city should take this foreseen future into account. In addition, it is always expected a successful attack on a small city, but the (harder) successful attack on the big city would lead to a much better probability of making further wins in the war. It is expected to successfully attack the big city only if the person who knows the secret

```
1. abds([save/1, big_city/0, small_city/0]).
   on_going_commitment([save(_)]).
2. expect(save(_)).
3. on_observe(save_place) <- be_attacked.
   save_place <- save(a).     save_place <- save(b).
4. on_observe(not save_atmost_one_ic) <- lack_of_resources.
   save_atmost_one_ic <- save(a), save(b).
5. save_men(P) <- save(City), population(City, P).
   alive(X) <- person(X), live_in(X, City), save(City).
6. population(a, 1000).  population(b, 2000).
   person(john). live_in(john, a).  knows(john, secret_inf).
7. Ai << Aj <- holds_given(save_men(Ni), Ai),
               holds_given(save_men(Nj),Aj), Ni > Nj.
8. on_observe(attack) <- good_opportunity.
   attack <- big_city.          attack <- small_city.
9. expect(small_city).
   expect(big_city) <- alive(Person),knows(Person, secret_inf).
10.pr(win, 0.9) <- big_city.     pr(win, 0.01) <- small_city.
11.Ei <<< Ej <- holds_in_evol(pr(win,Pi), Ei),
                holds_in_evol(pr(win,Pj), Ej), Pi > Pj.
```

Fig. 3 Saving a city example

information about the enemy (John) is alive in the city to be saved beforehand. The described scenario is coded with the program in Figure 3.

Line 1 is the declaration of abducibles. Save a city is an ongoing commitment since it has direct influence on the next state, but is defeasible. The context sensitive integrity constraint in line 4 implies that at most one city can be saved since the lack of resources is a foreseen event. Thus, there are two abductive solutions: $[save(a), not\ save(b)]$ and $[save(b), not\ save(a)]$.

If the general is a bad one, i.e. is a single-step prospective agent, the *a posteriori* preference in line 7 would be immediately taken into account and rule out the abductive solution including $save(a)$, since it leads to the saving of 1000 people, which is less preferred than the one including $save(b)$ which leads to the saving of 2000 people (lines 5-6). Then, on the next day, he can attack the small city, but leads to the consequence that the further wining of the whole conflict is very small.

Fortunately David is a good general, capable of prospectively looking ahead, at least two steps in the future. David sees three possible evolutions:

$E_1 = [[save(a), not\ save(b)],\ [big\_city, save(a)]]$

$E_2 = [[save(a), not\ save(b)],\ [small\_city, save(a)]]$

$E_3 = [[save(b), not\ save(a)],\ [small\_city, save(b)]$

In the next stage, the *evolution result a posteriori preference* in line 11 is taken into account, ruling out E_2 and E_3 since both lead to the consequence of a smaller probability to win the whole conflict when compared to E_1.

In short, the agent with better capability of looking ahead will provide a more rational decision for the long term goals.

3.2.2 Evolution History Preference

This kind of preference takes into account information from the history of evolutions. The information can be quantitative, such as having in the evolution a maximal or minimal number of some type of commitment, or having the number of commitments greater, equal or smaller than some threshold. It also can be qualitative, such as time order of commitments along an evolution. Such preferences can be used *a priori* upon the process of finding possible evolutions. However, if all preferences (of every kind) coded in the program have been applied but there is still more than one possible evolution, an interaction mode with the user is turned on to ask for user's additional preferences. Similarly, if no solution can satisfy the preferences, the user may be queried about which might be relaxed, or which relaxation option to consider. Now the *evolution history preferences* are used *a posteriori*, given by the user in a list, so as to choose the most cherished evolutions. An evolution history preference can exhibit one of these forms, where C is an abducible:

1. *max(C)/min(C)/greater(C,N)*: find the evolutions having number of commitments to C *maximal/minimal/greater* than N.
2. *smaller(C,N)/times(C,N)*: find the evolutions having number of commitments to C *smaller than/equal* to N.
3. *prec(C1,C2)/next(C1,C2)*: find the evolutions with commitment C1 *preceding/ next to* C2 in time.

Example 4. Agent John must finish a project. He has to schedule his everyday actions so that he can finish it on time. Everyday he either works or relaxes. He relaxes by going to the beach, to a movie or watching football. Being a football fan, whenever there is a football match on TV, John relaxes by watching it. The described scenario is coded in Figure 4.

In line 5 we can see how an *evolution history preference* is used *a priori* in the predicate $working\_days/2$. There are two reserved predicates $plan\_pref/1$ and $plan\_ending/1$ that allow for asserting *a priori* evolution history preferences and the necessary number of look ahead steps. At the beginning, the agent tentatively runs the active goals to collect all *a priori evolution preferences* and decide how many steps are needed to look ahead. In this case, the agent will look ahead five steps taking into account the *a priori evolution history preference times(work,2)*. There are six possible evolutions: $E_1 = [[beach],[football],[football],[work],[work]]$,
$E_2 = [[movie],[football],[football],[work],[work]]$,
$E_3 = [[work],[football],[football],[beach],[work]]$,
$E_4 = [[work],[football],[football],[movie],[work]]$,
$E_5 = [[work],[football],[football],[work],[beach]]$,
$E_6 = [[work],[football],[football],[work],[movie]]$
Since there are several possible evolutions, the interaction mode is turned on for John to give a list of evolution history preferences. Suppose, he prefers the evolutions with maximal number of goings to the beach, entering the list $[max(beach)]$. Three possible evolutions E_1, E_3 and E_5 remain. John is asked again for preferences. Suppose he likes going to the beach after watching football, thereby entering $[next(football,beach)]$. Then the only possible evolution is now E_3.

```
1. abds([beach/0, movie/0, work/0, football/0]).
2. expect(beach).  expect(movie).   expect(work).
3. on_observe(everyday_act).
   everyday_act <- work.     everyday_act <- relax.
   relax <- beach.  relax <- movie.  relax <- football.
4. expect(football)  <- prolog(have_football).
   expect_not(beach) <- prolog(have_football).
   expect_not(work)  <- prolog(have_football).
   expect_not(movie) <- prolog(have_football).
5. on_observe(on_time).
   on_time <- deadline(Deadline), project_work(Days),
              prolog(working_days(Deadline, Days)).
   deadline(5).   project_work(2).
6. beginProlog.
   :- import member/2 from basics.
   have_football :- current_state(S), member(S, [1,2]).
   working_days(Deadline, Days) :-
         assert(plan_pref(times(work, Days))),
         assert(plan_ending(Deadline)).
   endProlog.
```

Fig. 4 Football example

4 Conclusions and Future Work

We have shown how to model evolving prospective logic program agent systems, including single-step and multiple-step ones. Besides declaratively specifying local preferences such as *a priori* and *a posteriori* ones, in order to let a prospective agent look ahead a number steps into the future and prefer amongst their hypothetical evolutions, we provide a new kind of preference, at evolution level, that can evaluate long-term consequences of a choice as well as analyze different kinds of information about the evolution history, which is kept by annotating such information with time stamps for each evolution cycle. In addition, active goals triggered by external events and context sensitive integrity constraints provide flexible ways for modelling the changing knowledge base of an evolving prospective agent. We exhibited several examples to illustrate all proffered concepts. By means of them, we have, to some degree, managed to show multiple-step prospective agents are more intelligent than the single-step ones, in the sense that they are able to give more reasonable decisions for long-term goals. In addition, the decision making process at each cycle during an evolution of our agent was, in many cases, enhanced by committing to so-called inevitable abducibles.

There are currently several possible future directions to explore. First of all, in each cycle the agent has to satisfy a set of active goals and sometimes he cannot satisfy them all. There are goals more important than others and it is vital to satisfy them while keeping the others optional. The agent can be made more focussed by setting a scale of priorities for the active goals so it can focus on the most important ones. We can make a scale by using preferences over the *on_observe*/2 predicates that are used for modelling active goals. Similarly, since there are integrity constraints that must be satisfied and there are also ones that are less important, we can prefer amongst integrity constraints by making them all context-sensitive and then prefer amongst the *on_observe*/2 predicates used for modelling them.

When looking ahead, the prospective agent has to search the evolution tree for the branches that satisfy his goals and preferences. From this perspective we can improve our system with heuristic search algorithms such as best-first search, i.e. the most promising nodes will be explored first. We also can improve the performance of the system by using multi-threading which is very efficient in XSB, from version 3.0 [7]. Independent threads can evolve on their own and they can communicate with each other to decide whether some thread should be canceled or kept evolving, based on the search algorithm used.

On a more general note, it appears the practical use and implementation of abduction in knowledge representation and reasoning, by means of declarative languages and systems, has reached a point of maturity, and of opportunity for development, worthy the calling of attention of a wider community of potential practitioners.

References

1. Pereira, L.M., Lopes, G.: Prospective Logic Agents. In: Neves, J., Santos, M.F., Machado, J.M. (eds.) EPIA 2007. LNCS (LNAI), vol. 4874, pp. 73–86. Springer, Heidelberg (2007)
2. Kakas, A., Kowalski, R., Toni, F.: The role of abduction in logic programming. Handbook of Logic in Artificial Intelligence and Logic Programming 5, 235–324 (1998)
3. Alferes, J.J., Pereira, L.M., Swift, T.: Abduction in Well-Founded Semantics and Generalized Stable Models via Tabled Dual Programs. Theory and Practice of Logic Programming 4(4), 383–428 (2004)
4. Kowalski, R.: The logical way to be artificially intelligent. In: Toni, F., Torroni, P. (eds.) CLIMA 2005. LNCS (LNAI), vol. 3900, pp. 1–22. Springer, Heidelberg (2006)
5. Pereira, L.M., Lopes, G., Dell'Acqua, P.: On Preferring and Inspecting Abductive Models. In: Gill, A., Swift, T. (eds.) PADL 2009. LNCS. Springer, Heidelberg (2009)
6. Dell'Acqua, P., Pereira, L.M.: Preferential theory revision. Journal of Applied Logic 5(4), 586–601 (2007)
7. XSB-PROLOG system freely, `http://xsb.sourceforge.net`

A Proposal of Context-Aware Service Composition Method Based on Analytic Hierarchy Process

Yusuke Koumoto, Hidetoshi Nonaka, and Takuto Yanagida

Abstract. We propose a new service composition method with the analytic hierarchy process and discuss its availability. The concept of context-aware services has been attracting attention as an approach to improving the usability of computer-mediated services. In ubiquitous computing environments, there are several means to provide services for users, and thus, to select an appropriate mean among them is a challenge. Our method for context-aware service composition determines service behaviors by context data. Through the implementation and examination of the method, we have found that the method can output reasonable results.

1 Introduction

The concept of *context-aware* services has been attracting attention as an approach to improving the usability of computer-mediated services [1]. Especially in ubiquitous computing environments [2], the context-aware services are expected to become more effective and beneficial. The term, context-aware or context awareness is used for meaning that specific services are provided for users' desired results according to the users' contexts. For taking the usability of services into account, we have to consider not only what services are provided, but also how they are provided.

In ubiquitous computing environments, there are several ways to provide services for users, and thus, to select an appropriate mean among them is a challenge. When a call is coming in a user's mobile phone, for example, the phone can notify the user of the call with several means, such as its vibrator, speaker, or display device in general. For getting the notification with an appropriate mean, the user needs to configure which device is to be used in advance. However, in ubiquitous computing

Yusuke Koumoto, Hidetoshi Nonaka, and Takuto Yanagida
Graduate School of Information Science and Technology, Hokkaido University, Kita 14, Nishi 9, Sapporo, 060-0814, Japan
e-mail: {koumoto,nonaka,akty}@main.ist.hokudai.ac.jp

K. Nakamatsu et al. (Eds.): New Advan. in Intel. Decision Techno., SCI 199, pp. 65–71.
springerlink.com

environment, it is difficult to configure all such means (devices) beforehand, because there might be too many services and means there. Thus, we have to consider how to combine previously created service components (service contents and means) automatically depending on user contexts. To achieve this automation, it is necessary to select most suitable combination of service components in accordance with service contents and user contexts.

As one of previous studies for the *service composition* for use in ubiquitous computing environments, Yamato et al. proposed a system named service composition engine [3, 4, 5]. The system discovers suitable service components from network based on a user context and binds them dynamically in accordance with a semantic-level service scenario. In this study, different evaluation functions are assigned to each evaluation items such as proximity, price , and set evaluation value, and the total of the functions is addressed as the overall score of the current context. As an evaluation standard for service composition, the system calculates the overall score according to service elements or components and user contexts. However, it is difficult both to evaluate qualitative factor such as taste, feeling, and sensibility; and to investigate the weight of each evaluation function for the overall score.

In this paper, we propose a new service composition method with the analytic hierarchy process (AHP) [6] and discuss its availability. With introducing the AHP, we can deal with qualitative factor and treat wide variety of context entities. Moreover, we can facilitate finding and fixing flaw of evaluation function. In our method, service-device compositions are considered as hierarchical structures of the AHP, and their overall score are derived with these hierarchies. We claim that our method might enable to develop efficient context-aware services.

2 Context-Aware Service

Context-aware services dynamically recombine their components depending on the surrounding contexts of the services and users. The context used here is some information that characterizes the situation of an entity, such as a person, an object, or a place that relevant to the interaction between a user and a service. For offering context-aware services, a system needs to

1. acquire some sort of data from circumstances with its sensors,
2. abstract the data and interpret its meanings for comprehending the contexts, and
3. determine service compositions and behaviors based on the contexts.

Our method is used at the second and third step, for comprehension of contexts and composite service components.

The term, service composition means that composing some combinable services, contents, and devices developed in advance, according to dynamically-changed contexts. This enables to utilize dynamically-composed and suitable service provided in ubiquitous computing environments.

3 Service Composition Method

Our method for context-aware service composition determines service behaviors based on context data. The method derives the most suitable mean to provide service from context data of services, devices, surroundings, etc. After that, it outputs derived service composition policy as XML descriptions.

3.1 Analytic Hierarchy Process

The analytic hierarchy process (AHP) is a structured technique for helping people to deal with complex decisions relative to a wide variety of situation and various criteria. When evaluating alternative solutions, it is not necessarily the case that they satisfy all criteria. The AHP decompose their decision problem into a hierarchy of more easily comprehended sub-problems, each of which can be analyzed independently. An AHP hierarchy consists of an overall *goal*, *alternatives* for reaching the goal, and *criteria* that relate the alternatives to the goal. Each alternatives and criteria is compared by means of pairwise comparison method, and the obtained pairwise comparison matrix is converted to the importance degree of the alternatives and criteria. Next, the importance degree of each element of the hierarchy is calculated from the pairwise comparison matrix using the eigenvalue calculation. When A is the obtained pairwise matrix and W is the importance degree of each element, the eigenvalue expression is given by

$$AW = \lambda_{max} W, \tag{1}$$

where λ_{max} is maximum eigenvalue. Therefore, W is an eigenvector corresponding to the maximum eigenvalue of A. This enables to calculate the unknown value of W. The importance degree of the goal X is derived from applying this calculation for each element and multiplying mutually. The X, defined by

$$X = [W_1^T W_2^T \dots W_n^T]W, \tag{2}$$

is computed by multiplying by importance degrees of alternatives $[W_1^T W_2^T \dots W_n^T]$, importance degrees of criteria W. In this way, the alternative priorities are calculated by the method.

3.2 Method Procedure

We accounted for the method procedure with the example of a context-aware service scenario which is to notify a user of receiving a phone call. We determined this example's goal, alternatives and criteria (Table 1).

In accordance with the procedure of the AHP, the user creates a hierarchical structure of the criteria and the alternatives. Firstly, the user derives the importance

Table 1 The goal, criteria, and alternatives for the service composition example

Goal	To notify a user of receiving a phone call
Criteria	User-device distance (far or near) Awareness of incoming (high or low) User's taste
Alternatives	Speaker Display Vibrator

degrees of criteria by pairwise comparison method. In this case, the user determines which one of the criteria should be emphasized for notifying the call. Next, the user derives the importance degrees of alternatives by pairwise comparison method as well. For example, the user compares alternatives, which one of the alternatives (devices) should be emphasized from the viewpoint of fitting it to the user's taste. In this example, the user derives the importance degrees as shown in Fig. 1. In addition, with introducing dynamically-changed context data, context factors can be added to the AHP. The dynamically changed context data is converted to the importance degree of criteria, and introduce into the AHP expression. The expression is given by

$$X = [W_1^T W_2^T \dots W_n^T]CW, \tag{3}$$

where the importance degree of the goal X is computed by multiplying by importance degrees of alternatives $[W_1^T W_2^T \dots W_n^T]$, importance degrees of criteria W and dynamically-changed context data C.

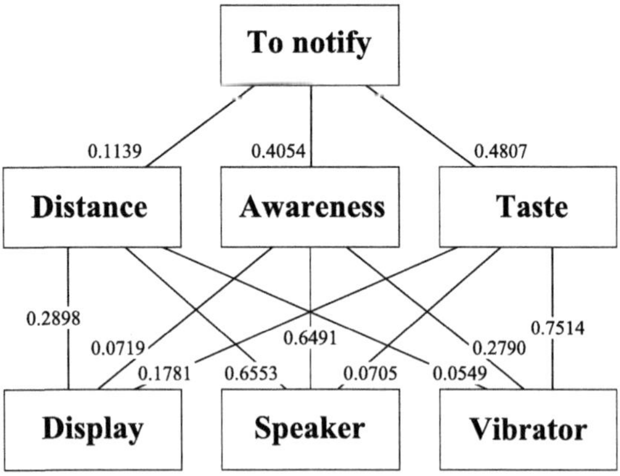

Fig. 1 AHP hierarchy for the sample scenario, with some associated priorities. The numbers show importance degrees of respective pairs of elements. $W^T_{Distance}$ = (0.2898, 0.6553, 0.0549), $W^T_{Awareness}$ = (0.0719, 0.6491, 0.2790), W^T_{Taste} = (0.1781, 0.0705, 0.7514) and W^T = (0.1139, 0.4054, 0.4807), which are obtained by respective pairwise matrices $A_{Distance}$, $A_{Awareness}$, A_{Taste} and A

4 Experiments and Results

We implemented a prototype system of our service composition method, and had it calculate a representative service example of notifying a user of receiving a phone call we mentioned above. We created context data as an input, and examined that our method can evaluate each devices.

A user determines the importance degrees of alternatives by pairwise comparisons in accordance with the service. The system receives a context data as an input, which is created by various sensors in real time. The importance degrees of criteria and final evaluation value for an overall goal vary according to the real time data.

We evaluate the proposed method by examining the result for a sample scenario and some sample context data. Four sorts of sample data expressed by matrix are created according to whether user-device distance is far or not, and user awareness level is high or low (Fig. 2). Each diagonal value represents the weight corresponding to the criteria in the context. In reality, these values are dynamically-changed in real time. In this instance, four representative values are prepared and calculated. The device which has the maximum score is the most suitable mean for providing the service in the context (Fig. 3). Since distance and awareness vary according to surrounding contexts, the importance degrees also change. The result shows that when the user-device distance is near and the user can awake easily, the method put much weight in the taste, while when the user-device distance is far and the user can hardly awake, the method put much weight in distance and awareness to do reliably. This relation fits in the way of thinking that a user selects the favorite device when either notifying device is located near or surroundings are not loud, because which device can notify user of receiving a phone call.

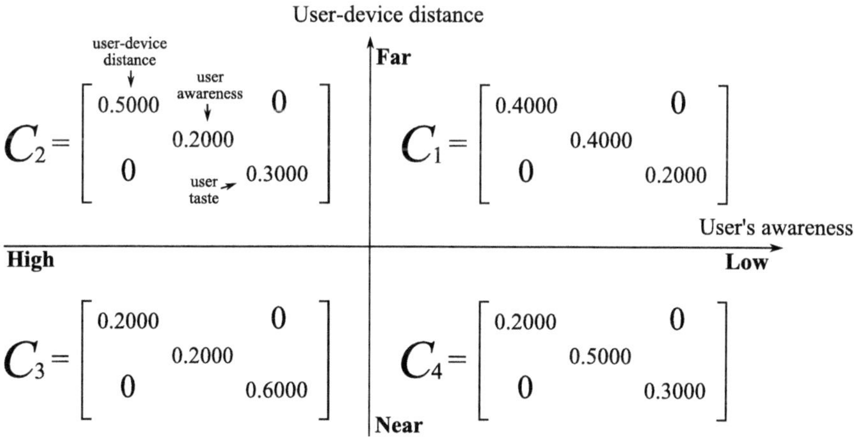

Fig. 2 Each criterion's weight calculated by the sample data under each context, and expressed by matrix $C_1,\ldots,C_4$. Diagonal values represent the weights corresponding to the criteria, such as distance, awareness, taste in the context

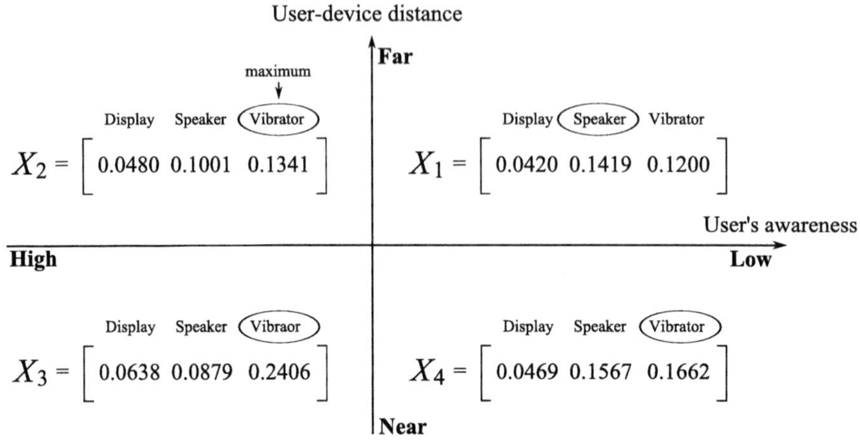

Fig. 3 The Results of the method. $X_1,\ldots,X_4$ are calculated from $C_1,\ldots,C_4$. Elements of vector $X_1,\ldots,X_4$ represent the importance of degrees of alternatives. The device which has the maximum score is the most suitable mean

5 Conclusion and Future Work

In this paper, we have presented the method for context-aware service composition based on AHP for use in ubiquitous computing environments. We have performed the preliminary experiment and examined the availability of the method using sample data, and have confirmed that the method can output reasonable results for each sample context. The advantage of our method is that it can deal with qualitative factors in consideration of the weight through the pairwise comparison method, and can interpret each factor's weight and modify it by reconsidering the evaluation process. Further work is to apply this method to real-world situations, such as home network, wireless mesh network, and, intelligent building, to compare it to other methods, and to evaluate the effectiveness quantitatively.

References

1. Baldauf, M., Dustdar, S., Rosenberg, F.: A survey on context-aware systems. International Journal of Ad Hoc and Ubiquitous Computing 2(4), 263–277 (2007)
2. Weiser, M.: The Computer for the 21st Century, pp. 94–104. Scientific American (1991)
3. Yamato, Y., Nakatsuji, M., Sunaga, H.: Ubiquitous Service Composition Technology for Ubiquitous Network Environments. Transactions of Information Processing Society of Japan 48(2), 562–577 (2007) (in Japanese)
4. Yamato, Y., Tanaka, Y., Sunaga, H.: Context-aware Ubiquitous Service Composition Technology. In: The IFIP International Conference on Research and Practical Issues of Enterprise Information Systems (CONFENIS 2006), pp. 51–61 (2006)

5. Takemoto, M., Oh-ishi, T., Iwata, T., Yamato, Y., Tanaka, Y., Shinno, K., Tokumoto, S., Shimamoto, N.: A Service-Composition and Service-Emergence Framework for Ubiquitous-Computing Environments. In: Proceedings of the 2004 International Symposium on Apprications and the Internet Workshops (SAINTW 2004), pp. 313–318 (2004)
6. Saaty, T.L.: The Analytic Hierarchy Process. McGraw-Hill, New York (1980)

Flexible Widget Layout Formulated as Fuzzy Constraint Satisfaction Problem

Takuto Yanagida and Hidetoshi Nonaka

Abstract. We show an improvement of our previous work, a formulation of the flexible widget layout (FWL) problem as a fuzzy constraint satisfaction problem (FCSP) and a method for solving it. The automation of widget layout is one of the most important challenges for the generation of graphical user interfaces (GUIs). In the field of model-based user interface design, widget layout is more complicated because a layout system needs to select widgets. FWL is the automatic GUI generation requiring (1) deciding which widgets are used and (2) completing the layout immediately. We formulate the desirability of selection as fuzzy constraints; thus, we can utilize existing techniques of FCSP without extending its framework. We divide the layout process into three phases, and realize the automatic layout in feasible time.

1 Introduction

The automation of widget layout is one of the most important challenges [5] in the context of dynamic generation of graphical user interfaces (GUIs). By *widget layout* we mean the process of determining the positions and the sizes of widgets on a dialog box; it is also used for mentioning the result of the process. The layout has a significant impact on the usability of applications and services using GUIs, and it determines the ease of tasks which can be accomplished with them.

In the field of model-based user interface (UI) design [2, 16], widget layout is more complicated because a layout system needs *to select widgets*. Many systems in the field automatically generate GUIs based on *logical descriptions* through layout processes. The logical descriptions specify common *UI functions* independently of

Takuto Yanagida and Hidetoshi Nonaka
Graduate School of Information Science and Technology, Hokkaido University, Sapporo 060–0814, Japan
e-mail: {takty,nonaka}@main.ist.hokudai.ac.jp

K. Nakamatsu et al. (Eds.): New Advan. in Intel. Decision Techno., SCI 199, pp. 73–83.
springerlink.com

specific devices and platforms, instead of specifying widgets to be used. Although that is useful for realizing the diversity of UIs [17, 19], the systems must select widgets according to the UI functions before they place widgets. In addition, these widgets are sometimes not uniquely determined.

There are many related studies. As an application of the facility layout problem [12] for UI, the system of UI components layout was proposed [10], but it handles the layout of a UI model, not widgets. Many studies for the LSI layout problem have been reported, but they do not consider the speed of layouts because it is unnecessary for their purpose [4]. Researches in the field of model-based UI design [7–9] handle mainly how to realize the use of various devices and platforms rather than how to select widgets and to perform layouts. Many GUI toolkits [6, 14, 15] offer layout managers, which decide the positions and sizes of widgets at run time, but they do not handle the selection of suitable widgets.

Automatic GUI generation from logical descriptions requires both (1) deciding which widgets are used and (2) completing the layout immediately especially when a system does this at run time. We call the layout satisfying the conditions, *the flexible widget layout* (FWL), and we call the problem of determining a layout which fulfills conditions of a dialog box and widgets, *the FWL problem*. By solving this problem, a system can select small widgets with less usability for small screens, or large widgets with enough usability for large screens.

In this paper, we show an improvement of our previous work [18, 20], in which we proposed a formulation of the FWL problem as a *fuzzy constraint satisfaction problem* (FCSP) [11] and a method for solving the FWL problem. In our formulation, *fuzzy constraints* express subjective constraint conditions involved with usability, sensitivity, etc. such as *desirable* widgets and *conspicuous* layout. Moreover, our method can realize the layout in a practical time by dividing the layout process into three phases. In this paper, these basic ideas are the same as the previous ones, but the formulation is improved so that it coincides more strictly with FCSP.

In the following sections, we define and formulate the FWL problem, show a layout system, discuss some consideration for it, and conclude our work.

2 Flexible Widget Layout Problem

The flexible widget layout problem is a solution search problem for finding better combinations of widgets. Each widget is selected from *a widget candidate set* including some widgets, which represent the same UI function, but have different *size* and *desirability*. The sizes of widgets are constraint conditions which decide possibility of the layout, and their desirabilities are constraint conditions which decide its suitability. In this paper, UI functions are modeled as *selection act model*, and described in the *abstract interaction description language* (AIDL) [17, 19].

The complexity of the FWL is caused by the tradeoff between widget usability and the ease of layout. For example, we can use a list box or a drop down list box for selecting one item from an item list. For usability, since users can view many items at once, the former is better, but it needs a larger area and may not be placed in a small dialog box (or a small screen).

As a UI model, we adopt the selection act model, where UI functions are represented as *selection acts*, and which are grouped to make a tree graph. A UI element set in the model is expressed as $S = S_S \cup S_G \cup S_D$, where S_S, S_G, and S_D are respectively selection, group, and description elements. Selection element $s_i \in S_S$ is represented here as a 5-tuple: $s_i = \langle L_i, e_i, t_i, o_i, r_i \rangle$, where L_i is the list of choices, and $|L_i|$ is the number of choices. $e_i \in \{\texttt{single}, \texttt{multiple}\}$ is the selection size, $t_i \in [1, 10]$ is the importance, and $o_i \in \{\texttt{true}, \texttt{false}\}$ is the flag meaning whether its choices are opposite when they have two choices. r_i is a flag which is true if the type of choice list is `Numeric` and constrained by `Range`, and otherwise, it is false. Relevant UI elements including group elements are grouped by group elements, and compose a tree graph of UI functions. This tree corresponds to a dialog box to be generated, and its root element is a group. In addition, each description element can be attached to the selections and groups for their explanations.

The UI elements of the model are represented as widgets. All widgets used here $W = W_N \cup W_C$ are rectangular and have their own presentations; also they are divided into two categories: normal widgets W_N for representing selection elements and container widgets W_C for group elements and labeling. As the normal widgets, since they are adopted by many toolkits, we use the eight widgets for representing selection elements (Table1). Here, a check box and check boxes are distinguished because they are used for different functions. In addition, for representing description elements, we use *caption label* (CL) and *abbreviation label* (AL) as the normal widgets. On the other side, as the container widgets, we use *vertical array*, *horizontal array*, and *tab pages* (Fig. 1) for representing group elements; and *left labeling* and *top labeling* (Fig. 2) for representing the positioning of description elements.

We defined the desirability $\alpha \in [0, 1]$ corresponding to the normal widgets as follows: α_{CB} for a check box, α_{RBS} for radio buttons, α_{DLB} for a drop-down list box, α_{CBS} for check boxes, α_{LB} for a list box, α_B for a button, α_{SP} for a spinner, and α_{SL} for slider. The desirability of a list box is defined as mutable according to the rate of its visible items, and it has the range $\alpha_{LBmin} \leq \alpha_{LB} \leq \alpha_{LBmax}$. The desirability of container widgets are defined as α_{VA}, α_{HA}, and α_{TP} for vertical array, horizontal

Table 1 Selection widget candidates

Candidate $w \in W_i$	Selection size e_i	Item size $\lvert L_i \rvert$	Is opposite o_i	Is ranged r_i
Check box	`single`	$\lvert L_i \rvert = 2$	`true`	-
Radio buttons	`single`	$0 < \lvert L_i \rvert < 6$	-	-
List box	-	$6 \leq \lvert L_i \rvert$	-	-
Drop-down list box	`single`	$0 < \lvert L_i \rvert$	-	-
Spinner	`single`	$0 < \lvert L_i \rvert < 10$	-	`true`
Slider	`single`	$10 \leq \lvert L_i \rvert$	-	`true`
Check boxes	`multiple`	$0 < \lvert L_i \rvert$	-	-
Button	-	$\lvert L_i \rvert = 0$	-	-

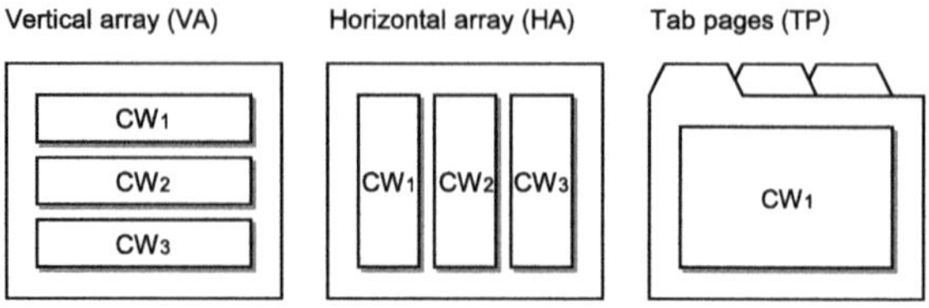

Fig. 1 Three container widgets for positioning child widgets of a group element (CW stands for child widget)

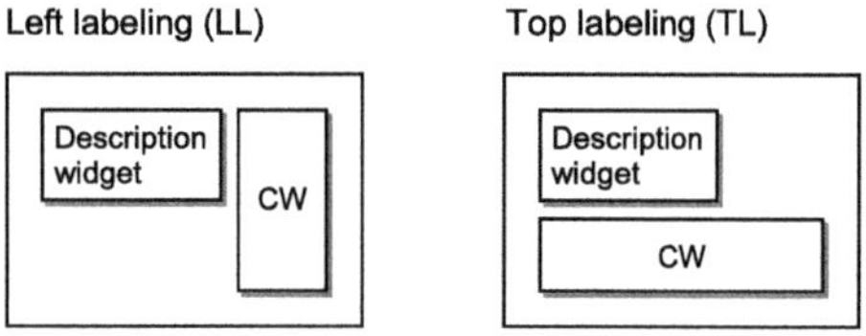

Fig. 2 Two container widgets for positioning a widget and its description (CW stands for child widget)

array, and tab pages; α_{LL} and α_{TL} for left labeling and top labeling. Referring to [1, 3], in terms of the usability of the widgets, we defined the order of the desirability.

Selection elements and description elements are mapped to the corresponding sets of normal widget candidates $W_i \subset W_{\mathrm{N}}$. Selection element s_i is expressed with widget $w \in W_i$ chosen from its corresponding widget candidate set (Table 1). We created the widget candidates table by also referring to [1, 3], but it is possible to apply users' preferences there. Based on the table, for example, widget candidates W_i corresponding to s_i, where $e_i = \texttt{single}$ and $|L_i| = 5$, are determined as a set of a drop down list box and radio buttons. In the layout process, it is decided which one of the candidates are used. Each widget candidate $w \in W_i$ has uniquely a minimum size $ms_w = \langle ms.width_w, ms.height_w \rangle$, which is defined by corresponding UI element.

Group elements and positioning of description elements are mapped to a set of container widget candidates $W_i \subset W_{\mathrm{C}}$, and they are expressed with widget $w \in W_i$. A group element is represented as a set containing a vertical array, a horizontal array, and (a container of) tab pages. A positioning of a description is represented as a set containing a left labeling and a top labeling. Each container widget candidate $w \in W_i$ also has a unique minimum size $ms_w = \langle ms.width_w, ms.height_w \rangle$. The minimum sizes of container widgets are defined by the minimum sizes of its child elements.

Based on the minimum sizes of widgets, a system evaluates whether it is possible to execute the layout defined by the selections from widget candidates, where the possibility means that the child elements of a container can be placed in its rectangle. We mention the condition expression for that possibility in the next section.

3 Formulation

A FCSP consists of a finite set of variables $X = \{x_i\}_{i=1}^{m}$, a finite set of domains of values $D = \{D_i\}_{i=1}^{m}$ associated with the each variable, and a finite set of constraints $C = \{c_k\}_{k=1}^{r}$. Constraint c_k denotes a fuzzy relation μR_k on a subset $S_k (S_k \subset X)$ of X. S_k is called the *scope* of R_k. We introduce the framework of FCSP to the FWL problem, and represent the desirability α of widgets with satisfaction degrees of fuzzy constraints. After the formulation mentioned below, the FWP is completely represented as a FCSP. It can be solved with general-purpose FCSP algorithms.

In the formulation, variable $x_i \in X$ corresponds to widget candidate set W_i and the value assigned in it expresses a selected candidate from the set. Variables X_{N} and X_{C} express the set of variables for the normal widget candidates and container widget candidates respectively. For example, UI element $s_i \in S$ corresponds to widget candidate set $W_i \subset W$, and next, it corresponds to variable $x_i \in X$. Note that widget candidate sets of other UI elements, groups, descriptions, and the positioning of the descriptions, are also expressed with the variables.

The values of domains are tuples, which are calculated form the bottom to the top of the tree structure of the variables. The domain of normal widget candidates variable $x_i \in X_{\mathrm{N}}$ is a set of the tuples, which consists of normal widget w and its minimum size $ms_w = \langle ms.width_w, ms.height_w \rangle$ as follows:

$$D_i \ (\in D_{\mathrm{N}}) = \left\{ \langle w, ms_w \rangle \mid w \in W_i \subset W_{\mathrm{N}} \right\}. \tag{1}$$

The minimum width of a normal widget is the sum of the width of the widest one of its items, and the width of the control parts such as a vertical scroll bar, a radio button, and a check box. The minimum height is defined by the type of widget, its item size, and the item height *item_h* (Table 2).

The domain of container widget candidates variable $x_i \in X_{\mathrm{C}}$ is a set of the tuples, which consists of container widget w, a combination of values of child widget candidates M, and its minimum size $ms_{w,M} = \langle ms.width_{w,M}, ms.height_{w,M} \rangle$ as follows:

$$\begin{aligned} D_i \ (\in D_{\mathrm{C}}) = \big\{ \langle w, M, ms_{w,M} \rangle \mid \ & w \in W_i \subset W_{\mathrm{C}}, \\ & M \in D_{\mathrm{child}(i,1)} \times \cdots \times D_{\mathrm{child}(i,\mathrm{cn}_i)}, \mathrm{checksize}(W_i, ms_{w,M}) \big\}, \end{aligned} \tag{2}$$

where $\mathrm{child}(i,j)$ is the function for obtaining the index of jth child of W_i, cn_i is the number of children of W_i, and $\mathrm{check\_size}(W_i, ms)$ is the function, which checks that the combination of its parameters is available or not with estimated minimum size (ems) (Function 1). This function prunes the domains for container variables when

Table 2 Minimum widget heights

Widget	Minimum height (without edges)
Check box	*item_h*
Drop-down list box	
Spinner	
List box	$\min(\lvert L \rvert, 4)\, item\_h$
Radio buttons	$\lvert L \rvert\, item\_h$
Check boxes	
Slider	*slider_h*
Button	*button_h*

Function 1. checksize$(W_{i,j}, ms)$

if W_i is root **then**
  if $ms.width \le given\_width$ **and** $ms.height \le given\_height$ **then**
    return true
  else
    return false
  end if
else
  $ms' \leftarrow \mathrm{ems}'(W_i, ms)$
  return checksize(W_i, ms')
end if.

constructing the problems. In the function, *given_width* and *given_height* are the size of the client area of the dialog box, and function ems is defined as follows:

$$W_i \subset W_N,\ \mathrm{ems}'(W_i, ms_{w\in W_{i,j}}) = \min_{w\in W_i}\left(ms''_{w,\{\mathrm{ems}(W_{i,1}),\ldots,ms_w,\ldots,\mathrm{ems}(W_{i,\mathrm{cn}(i)})\}}\right), \quad (3)$$

$$W_i \subset W_N,\ \mathrm{ems}(W_i) = \min_{w\in W_i}\left(ms'_{w,\{\mathrm{ems}(W_{i,1}),\ldots,\mathrm{ems}(W_{i,\mathrm{cn}_i})\}}\right), \quad (4)$$

$$W_i \subset W_N,\ \mathrm{ems}(W_i) = \min_{w\in W_i}(ms_w) = \left\langle \min_{w\in W_i}(ms.width_w), \min_{w\in W_i}(ms.height_w)\right\rangle. \quad (5)$$

The container widgets have different sizes of child widgets; therefore, the sizes of tuples of their domains are also different. The minimum size of vertical array (VA), horizontal array (HA), and tab pages (TP) is calculated based on the minimum sizes of its child widgets ($ms_{W_{i,j}} = \langle ms.width_{W_{i,j}}, ms.height_{W_{i,j}}\rangle$) as follows (where gaps among child widgets and tabs space are omitted):

$$ms_{\mathrm{VA}\in W_i} = \left\langle \max_j(ms.width_{W_{i,j}}),\ \textstyle\sum_j ms.height_{W_{i,j}}\right\rangle, \quad (6)$$

$$ms_{\mathrm{HA}\in W_i} = \left\langle \textstyle\sum_j ms.width_{W_{i,j}},\ \max_j(ms.height_{W_{i,j}})\right\rangle, \quad (7)$$

$$ms_{\mathrm{TP}\in W_i} = \left\langle \max_j(ms.width_{W_{i,j}}),\ \max_j(ms.height_{W_{i,j}})\right\rangle. \quad (8)$$

The minimum sizes of a left labeling (LL) and a top labeling (TL) are calculated based on the size of their description widget ($ms_{W_{i,\mathrm{D}}}$) and the minimum sizes of their one child widget ($ms_{W_{i,\mathrm{C}}}$) as follows (where gaps are omitted):

$$\mathrm{ms}_{\mathrm{LL}\in W_i} = \left\langle ms.width_{W_{i,\mathrm{D}}} + ms.width_{W_{i,\mathrm{C}}},\ \max(ms.height_{W_{i,\mathrm{D}}}, ms.height_{W_{i,\mathrm{C}}})\right\rangle, \quad (9)$$

$$\mathrm{ms}_{\mathrm{TL}\in W_i} = \left\langle \max(ms.width_{W_{i,\mathrm{D}}}, ms.width_{W_{i,\mathrm{C}}}),\ ms.height_{W_{i,\mathrm{D}}} + ms.height_{W_{i,\mathrm{C}}}\right\rangle. \quad (10)$$

When the estimated minimum sizes of container widgets are calculated, the above equations are also used, but, instead of actual widget sizes, the estimated minimum sizes of child widgets are used there.

In this formulation, crisp and fuzzy constraints are used accordingly, where crisp constraints can be handled as particular cases of fuzzy constraints. Each variable is connected by a unary constraint for expressing the desirability, and arbitrary two variables are connected by a binary constraint for expressing a parental relationship. Unary constraint $c_k \in C_D$ denote the desirability of the value of its scope x_{k_1} as their satisfaction degrees. If the scope of c_k is $S_k = \{x_{k_1}\}$ and the value of x_{k_1} is $v\ (\in D_{k_1}) = \langle w, \ldots \rangle$, $w \in W_{k_1}$, the satisfaction degree of c_k is calculated as follows:

$$c_k(v)\ (\in C_\mathrm{D}) = des(w), \tag{11}$$

where des is the projection from the widget candidates to their desirability α. A binary constraint $c_k \in C_\mathrm{P}$ denotes whether the assignments of the variables of its scope correspond with each other. As mentioned above, each value is a tuple of a combination of widget and its minimum size, and thus, this constraint accords this combination with the actual combination of its child widgets. If the scope of c_k is $S_k = \{x_{k_1}, x_{k_2}\}$, the value of x_{k_1} is $v_p\ (\in D_{k_1}) = \langle w, M, ms_w \rangle$, and the value of x_{k_2} is $v_c \in D_{k_2}$, the satisfaction degree of $c_k(v_p, v_c)$ is calculated as follows:

$$c_k(v_p, v_c)\ (\in C_\mathrm{P}) = \begin{cases} 1 & \text{if } v_c = M\left[\mathrm{childindex}(x_{k_1}, x_{k_2})\right] \\ 0 & \text{otherwise} \end{cases}. \tag{12}$$

where $\mathrm{childindex}(x_1, x_2)$ is the projection from pairs of variables to the index of the widget candidates (corresponding to $x_2 \in X$) as a child of the parent widget candidates (corresponding to $x_1 \in X_\mathrm{C}$).

4 Layout Method

We developed a layout system, which consists of three phases: (1) creating a FCSP from an AIDL document, (2) solving the problem, and (3) performing an actual layout based on the result. The system receives an AIDL document as input and outputs a dialog box with Swing. The desirability α_* of the widget candidates are given empirically assuming typical applications.

In the first phase, a FCSP is generated from a given AIDL document—a tree structure having one-to-one correspondence to the model of the document. Group, selection, and description elements in a document are seen as each corresponding typed variable in a FCSP, and their parental relationships are seen as composition constraints. All variables have unary constraints for expressing the desirability of the current assignments of the variables. The minimum sizes of widgets, which are the values of the domains of the variables, are determined by the parameters of the corresponding elements. As mentioned above, infeasible elements of the domains are pruned in parallel with the construction of the FCSP.

In the second phase, for a better assignment of the variables, the system iterates solving the FCSP generated in the previous phase using the *forward checking algorithm*, which is extended for improving the worst constraint satisfaction degree. The system prunes the domains according to a *worst satisfaction degree*. The process of solving the FCSP is as the following 4 steps: (Step 1) Before the iteration, the system prepares satisfaction degree set A by collecting possible satisfaction degrees

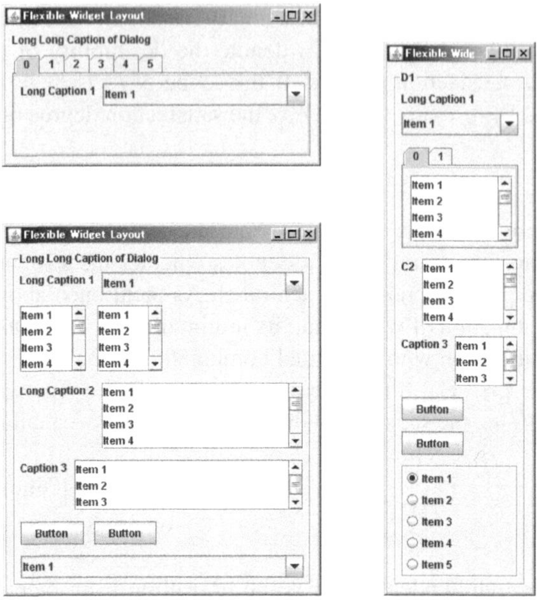

Fig. 3 Results of flexible widget layout from the same AIDL document (UI model) with different dialog sizes

from all unary constraints. Since all unary constraints express the desirability α of each widget candidates, A is a discrete set. (Step 2) The system chooses and removes the maximum one $a \in A$ as the worst satisfaction degree, and prunes values in the domains worse than a based on the unary constraints. At this step, the minimum heights of list boxes in the domains are reset as follows:

$$ms.height_{\mathrm{LB}\in W_i} = \min(|L_i|, 4)\, item\_h \left(1 - \frac{\alpha_{LBmax} - a}{\alpha_{LBmax} - \alpha_{LBmin}}\right)^{\frac{1}{t_i}}. \tag{13}$$

(Step 3) The system solves the FCSP with the algorithm. (Step 4) If the system can find an assignment of the variables, it moves to the next phase; otherwise, it moves back to the step 2. If A is empty, the system stops in failure.

In the last phase, based on the assignments of the variables, the system decides the positions and the sizes of the selected widgets, and then it places them. In the FWL, the variables express the selections of widget candidates; therefore, the solution of a FCSP is not an actual layout yet. The system generates widgets adopted in the solution, and then it sets their concrete positions (pixels) and sizes (pixels) based on the selected container widgets. In our implementation, the resize of the dialog box is reflected in the re-layout of the widgets (Fig. 3).

For solving the problem rapidly, the system prunes the domains based on giving worst constraint satisfaction degrees before applying the algorithm. The forward checking algorithm searches systematically through the search space of the possible combinations of the assignments until it finds a solution. It is guaranteed to find a solution if one exists but has the disadvantage that it requires large cost of time;

hence, pruning the domains and reducing the problem scale is effective. We mention the effect of pruning in the next section.

5 Discussion

In the previous formulation of the FWL problem [18, 20], its domains are not statically decided, but dynamically change when searching solutions. Therefore, binary constraints checking these dynamic domains are in fact n-ary constraints under the influence of more than two variables. To improve this problem, the easiest way is to introduce n-ary constraints, but, this is difficult because to implement algorithms accepting n-ary constraints is difficult. In the current formulation, this problem is resolved using the technique of binarization of n-ary constraints, as well as pruning in the problem construction phase.

We need to evaluate our system in detail, but for now, we have confirmed that it can perform the layout of the example (Fig. 3) fast enough for dynamic generation of GUI; the system can complete the layout example within 650 milliseconds using Java 6 on a PC (AMD Turion 64 CPU 2.0 GHz, 768 MB main memory, and Windows XP Professional edition). We also executed the system with the same layout without pruning and found that it takes more than 50000 milliseconds. That shows that pruning is effective in the system.

6 Conclusion and Future Work

In this paper, we have shown an improved formulation of the flexible widget layout problem (FWL). The formulation proposed here coincides more strictly with the FCSP framework, and thus, the possibility of extending this work with other techniques for FCSP is increased. The widget selection before doing layout is general; it is not specific for model-based GUI generations, because GUI designers also need to select widgets when they do layouts by hand. The constraints in the problem are used for only parent-child compositions and desirability of widgets, however, this limitation is posed just in the current implementation, but not in our approach itself.

For advancing this work, it is necessary to add other layout rules, to evaluate the relation between problem scales and solving times, and to consider other FCSP algorithms. For example of layout rules, constraints between sibling widgets can be added for keeping the same types and states among them. For evaluating the availability of our method, we are considering extracting GUI dialog boxes from existing applications and reconstructing them with our method. In the current implementation, we use the forward checking algorithm, but our method is not limited to it. Formulated as FCSP, the layout problem allows us to use suboptimal results; thus, we can also consider adopting local search or hybrid algorithms such as SRS [13].

References

1. Apple Inc.: Apple human interface guidelines (2008), `http://developer.apple.com/documentation/UserExperience/Conceptual/AppleHIGuidelines/OSXHIGuidelines.pdf`
2. Eisenstein, J., Vanderdonckt, J., Puerta, A.: Applying model-based techniques to the development of UIs for mobile computers. In: Proceedings of the 2001 International Conference on Intelligent User Interfaces (IUI 2001), pp. 69–76. ACM Press, Santa Fe (2001)
3. Fowler, S.L.: GUI Design Handbook. Mcgraw-Hill Companies, Inc., New York (1997)
4. Kitazawa, H.: Overview of the lsi layout cad algorithms and their applications to image processing. In: Technical Report of the Institute of Electronics, Information and Communication Engineers VLD 2006–38, vol. 106, pp. 25–30 (2006) (in Japanese)
5. Lok, S., Feiner, S.: A survey of automated layout techniques for information presentations. In: Proceedings of the 1st International Symposium on Smart Graphics, pp. 61–68. ACM, Hawthorne (2001)
6. Microsoft Corporation: Windows forms, `http://msdn2.microsoft.com/en-us/netframework/aa497342.aspx`
7. Nichols, J., Myers, B.A., Higgins, M., Hughes, J., Harris, T.K., Rosenfeld, R., Pignol, M.: Generating remote control interfaces for complex appliances. In: Proceedings of the 15th Annual ACM Symposium on User Interface Software and Technology (UIST 2002), pp. 161–170. ACM, Paris (2002)
8. Nylander, S., Bylund, M., Waern, A.: The ubiquitous interactor–device independent access to mobile services. In: Proceedings of the 4th International Conference on Computer-Aided Design of User Interfaces (CADUI 2004), pp. 274–287. Kluwer, Funchal (2004)
9. Olsen, D.R., Jefferies, S., Nielsen, S.T., Moyes, W., Fredrickson, P.: Cross-modal interaction using XWeb. In: Proceedings of the 13th Annual ACM Symposium on User Interface Software and Technology (UIST 2000), pp. 191–200. ACM, San Diego (2000)
10. Peer, S.K., Sharma, D.K.: Human-computer interaction design with multi-goal facilities layout model. Computer and Mathematics with Applications 56, 2164–2174 (2008)
11. Ruttkay, Z.: Fuzzy constraint satisfaction. In: Proceedings of the 3rd IEEE Conference on Fuzzy Systems, pp. 1263–1268. IEEE, Orlando (1994)
12. Singh, S.P., Sharma, R.R.K.: A review of different approaches to the facility layout problems. The International Journal of Advanced Manufacturing Technology 30, 425–433 (2006)
13. Sudo, Y., Kurihara, M.: Spread-repair-shrink: A hybrid algorithm for solving fuzzy constraint satisfaction problems. In: Proceedings of the 2006 IEEE International Conference on Fuzzy Systems (FUZZ IEEE 2006), pp. 2127–2133. IEEE, Vancouver (2006)
14. Sun Microsystems, Inc.: JDK 6 swing (java foundation classes) (2005), `http://java.sun.com/javase/6/docs/technotes/guides/swing/index.html`
15. Trolltech ASA: Qt, `http://trolltech.com/products/qt/`
16. Vanderdonckt, J.M., Bodart, F.: Encapsulating knowledge for intelligent automatic interaction objects selection. In: Proceedings of CHI 1993, pp. 424–429. ACM, Amsterdam (1993)
17. Yanagida, T., Nonaka, H.: Architecture for migratory adaptive user interfaces. In: Proceedings of the IEEE 8th International Conference on Computer and Information Technology (CIT 2008), pp. 450–455. IEEE, Sydney (2008)

18. Yanagida, T., Nonaka, H.: Flexible widget layout with fuzzy constraint satisfaction. In: Proceedings of the 2008 IEEE Conference on Soft Computing in Industrial Applications (SMCia 2008), pp. 387–392. IEEE, Muroran (2008)
19. Yanagida, T., Nonaka, H.: Interaction description with service-specific meanings. In: Proceedings of the 5th International Conference on Cybernetics and Information Technologies, Systems and Applications (CITSA 2008), pp. 185–188. IIIS, Orlando (2008)
20. Yanagida, T., Yasuhiro, S., Nonaka, H.: Flexible widget layout based on fuzzy constraint satisfaction. Journal of Japan Society for Fuzzy Theory and Intelligent Informatics 20(6) (2008) (to be published in Japanese)

Modelling Probabilistic Causation in Decision Making

Luís Moniz Pereira and Carroline Kencana Ramli

Abstract. Humans know how to reason based on cause and effect, but cause and effect is not enough to draw conclusions due to the problem of imperfect information and uncertainty. To resolve these problems, humans reason combining causal models with probabilistic information. The theory that attempts to model both causality and probability is called probabilistic causation, better known as Causal Bayes Nets.

In this work we henceforth adopt a logic programming framework and methodology to model our functional description of Causal Bayes Nets, building on its many strengths and advantages to derive a consistent definition of its semantics. ACORDA is a declarative prospective logic programming which simulates human reasoning in multiple steps into the future. ACORDA itself is not equipped to deal with probabilistic theory. On the other hand, P-log is a declarative logic programming language that can be used to reason with probabilistic models. Integrated with P-log, ACORDA becomes ready to deal with uncertain problems that we face on a daily basis. We show how the integration between ACORDA and P-log has been accomplished, and we present cases of daily life examples that ACORDA can help people to reason about.

1 Introduction

Humans reason basically based on cause and effect. David Hume described causes as objects regularly followed by their effects [1]. Hume attempted to analyse

Luís Moniz Pereira
CENTRIA, Departamento de Informática, Faculdade de Ciências e Tecnologia, Universidade Nova de Lisboa, 2829-516 Caparica, Portugal
e-mail: lmp@di.fct.unl.pt

Carroline Kencana Ramli
CENTRIA, Departamento de Informática, Faculdade de Ciências e Tecnologia, Universidade Nova de Lisboa, 2829-516 Caparica, Portugal
e-mail: carroline.kencana@googlemail.com

K. Nakamatsu et al. (Eds.): New Advan. in Intel. Decision Techno., SCI 199, pp. 85–95.
springerlink.com

causation in terms of invariable patterns of succession that are referred to as "regularity theories" of causation. The difficulty with regularity theories is that most causes are not invariably followed by their effects. For example, it is widely acceptable that smoking is a cause of lung cancer, but not all smokers have lung cancer. By contrast, the central idea behind probabilistic theories of causation is that causes raise the probability of their effects; an effect may still occur in the absence of a cause or fail to occur in its presence. The probabilistic theorem of causation helps in defining a pattern in problems with imperfect regularities [2]. This approach is called Causal Bayes Nets.

Translation of human reasoning using causal models and Bayes Nets into a computational framework is possible using logic programming. The main argument is that humans reason using logic. There is an obvious human capacity for understanding logic reasoning, one that might even be said to have developed throughout our evolution. Logic itself can be implemented on top of a symbol processing system like a computer. Thus, in this work we henceforth adopt a declarative logic programming framework and methodology to model our functional description of causal models and Bayes Nets, building on its many strengths and advantages to derive both a consistent definition of its semantics and a working implementation with which to conduct relevant experiments.

The paper is organized as follows: the next Section provides the background of prospective logic programming and probabilistic logic programming. Section 3 provides the explanation of our implementation and the paper continues with an example of how our implementation can be used in Sect. 4. The paper finishes with conclusions and directions for future work in Sect. 5.

2 Prospective and Probabilistic Logic Programming

2.1 Prospective Logic Programming

Prospective logic programming is an instance of an architecture for causal models, which implies a notion of simulation of causes and effects in order to solve the choice problem for alternative futures. This entails that the program is capable of conjuring up hypothetical *what-if* scenaria and formulating abductive explanations for both external and internal observations. Since we have multiple possible scenaria to choose from, we need some form of preference specification, which can be either a priori or a posteriori. A priori preferences are embedded in the program's own knowledge representation theory and can be used to produce the most relevant hypothetical abductions for a given state and observations, in order to conjecture possible future states. A posteriori preferences represent choice mechanisms, which enable the program to commit to one of the hypothetical scenaria engendered by the relevant abductive theories. These mechanisms may trigger additional simulations, by means of the functional connectivity, in order to posit which new information to acquire, so more informed choices can be enacted, in particular by restricting and committing to some of the abductive explanations along the way.

Definition 1 (Language). Let $\mathcal{L}$ be a *first order language*. A domain literal in $\mathcal{L}$ is a domain atom A or its default negation *not* A, the latter expressing that the atom is false by default. A domain rule in $\mathcal{L}$ is a rule of the form:

$$A \leftarrow L_1, \ldots, L_t \ \ (t \geq 0)$$

where A is a domain atom and $L_1, \ldots, L_t$ are domain literals.

Definition 2 (Integrity Constraint). An *integrity constraint* in $\mathcal{L}$ has a form:

$$\perp \leftarrow L_1, \ldots, L_t \ \ (t > 0)$$

where $\perp$ is a domain atom denoting falsity, and $L_1, \ldots, L_t$ are domain literals.

A *(logic) program* P over $\mathcal{L}$ is a set of domain rules and integrity constraints, standing for all their ground instances.

Each program P is associated with a *set of abducibles* $\mathcal{A}_P \subseteq \mathcal{L}$. Abducibles can be seen as hypotheses that provide hypothetical solutions or possible explanations of given queries. An abducible a can be assumed in the program only if it is a considered one, i.e. if it is expected in the given situation, and moreover there is no expectation to the contrary [3]. The atom $consider(a)$ will be true if and only if the abducible a is considered. We can define it by the logic programming rule:

$$consider(A) \leftarrow expect(A), not\ expect\_not(A)$$

where A stands for a logic programming variable. The rules about expectations are domain-specific knowledge contained in the theory of the program, and effectively constrain available the hypotheses .

To express preference criteria amongst abducibles, we introduce the language $\mathcal{L}^*$. Let $\mathcal{L}^*$ be a language consisting of logic program and relevance rules.

Definition 3 (Relevance Rule). Let a and b are abducibles. A *relevance atom* $a \lhd b$ means abducible a is more relevant or preferred than abducible b, i.e. one cannot have b without also having a. A *relevance rule* is one of the form:

$$a \lhd b \leftarrow L_1, \ldots, L_t \ \ (t \geq 0)$$

where $a \lhd b$ is a relevance atom and every $L_i (1 \leq i \leq t)$ is a domain literal or a relevance literal.

Example 1 (Tea 1). Consider a situation where an agent Claire drinks either tea or coffee (but not both). She does not expect to have coffee if she has high blood pressure. Also suppose that agent Claire prefers coffee over tea when she is sleepy. This situation can be represented by a program Q with the set of abducibles $\mathcal{A}_Q = \{tea,\ coffee\}$:

```
falsum ← not drink.
drink ← consider(tea).
drink ← consider(coffee).
```

```
4 constrain(1,[tea,coffee],1).
5 expect(tea). expect(coffee).
6 expect_not(coffee) ← blood_pressure_high.
7 coffee ◁ tea ← sleepy.
```

The query is triggered by an integrity constraint coded with `falsum/0`. ACORDA has to fulfill the integrity constraint by trying to find all the explanations to the atom *drink*. We codified the preference using the `</2` built–in ACORDA predicate. The exclusivity of the two abducibles is coded with `constrain/3` on line 4. So there are two abductive solusions: $A_1 = \{coffee\}$, $A_2 = \{tea\}$. We should choose our solution from those two explanations based on our preferences. If we assert *sleepy* is true, the only explanation left is A_1 because of the preference rule – in order to have the abducible tea we should have the abducible *coffee*. On the other hand, if we assert *blood_pressure_high* is true, the only remaining solution is A_2 because the abducible *coffee* becomes not expected.

Having the notion of expectation allows one to express the preconditions for an expectation or otherwise about an abducible a, and expresses which possible expectations are confirmed (or assumed) in a given situation. If the preconditions do not hold, then abducible a cannot be considered, and therefore a will never be assumed. By means of `expect_not/1` one can express situations where one does not expect something. In this case, when blood pressure is high, coffee will not be considered or assumed because of the contrary expectation arising as well (and therefore tea will be assumed).

2.2 Probabilistic Logic Programming

Probabilistic logic programming (P-log) was introduced for the first time by Chitta Baral et.al [4]. P-log is a declarative language that combines logical and probabilistic reasoning, and uses Answer Set Programming (ASP) as its logical foundation and Causal Bayes Nets [5] as its probabilistic foundation. P-log can also represent a mechanism for updating or reassessing probability [6].

The declaration part of a P-log program Π contains sorts and attributes. A sort c is a set of terms. It can be defined by listing all its elements: $c = \{x_1, x_2, \ldots, x_n\}$. Given 2 integers $L \leq U$ we can also use 2 shortcut notations: $c = \{L..U\}$ for the sort $c = \{L, L+1, \ldots, U\}$ and $c = \{h(L..U)\}$ for the sort $c = \{h(L), h(L+1), \ldots, h(U)\}$. We are also able to define a sort by arbitrarily mixing the previous constructions, e.g. $c = \{x_1, .., x_n, L..U, h(M..N)\}$. In addition, it is allowed to declare union as well as intersection of sorts. A union sort is represented by $c = union(c_1, ..., c_n)$ while an intersection sort by $c = intersection(c_1, ..., c_n)$, where $c_i, 1 \leq i \leq n$ are declared sorts.

Definition 4 (Sorted Signature). The *sorted signature* Σ of a program Π contains a set of constant symbols and term-building function symbols, which are used to form terms in the usual way. Additionally, the signature contains a collection of special function symbols called attributes.

Definition 5 (Attribute). Let $c_0, c_1, \ldots, c_n$ be sorts. An *attribute* a with the domain $c_1 \times ... \times c_n$ and the range c_0 is represented as follows:

$$a : c_1 \times ... \times c_n \ \rightarrow \ c_0$$

If attribute a has no domain parameter, we simply write $a : c_0$. The range of attribute a is denoted by $range(a)$. Attribute terms are expressions of the form $a(\bar{t})$, where a is an attribute and $\bar{t}$ is a vector of terms of the sorts required by a.

The *regular part* of a P-log program Π consists of a collection of rules, facts and integrity constraints formed using literals of Σ.

Definition 6 (Random Selection Rule). A *random selection rule* has a form:

$$random(RandomName, a(\bar{t}),\ DynamicRange) \ :- \ Condition$$

This means that the attribute instance $a(\bar{t})$ is random if the conditions in $Condition$ are satisfied. The $DynamicRange$ allows to restrict the default range for random attributes. The $RandomName$ is a syntactic mechanism used to link random attributes to the corresponding probabilities. The constant $full$ is used in $DynamicRange$ to signal that the dynamic domain is equal to $range(a)$.

Definition 7 (Probabilistic Information). Information about probabilities of random attribute instances $a(\bar{t})$ taking particular value y is given by probability atoms (or simply pa-atoms) which have the following form:

$$pa(RandomName, a(\bar{t}, y), d\_(A, B)) :- \ Condition$$

It means if $Condition$ were to be true, and the value of $a(\bar{t})$ were selected by a rule named $RandomName$, then $a(\bar{t}) = y$ with probability $\frac{A}{B}$.

Definition 8 (Observations and Actions). *Observations* and *actions* are, respectively, statements of the forms $obs(l)$ and $do(l)$, where l is a literal. Observations are used to record the outcomes of random events, i.e. random attributes and attributes dependent on them.

Example 2 (Tea 2). (Continuation of Example 1) Suppose if we know that the availability of tea for agent Claire is around 60%.

```
beginPr.
    beverage = {tea, coffee}.
    available : beverage.
    random(rd, available, full).
    pa(rd, available(tea), d_(60, 100)).
endPr.
```

The probabilistic part is coded in between `beginPr/0` and `endPr/0`. There are two kinds of beverage: tea and coffee and both are randomly available. We get the information the availability of tea is 60%, coded by $pa(rd, available(tea), d\_(60, 100))$.

3 Implementation

ACORDA [7, 8][1] is a system that implements prospective logic programming. ACORDA is the main component of our system with P-log used as probabilistic support in the background. Computation in each component is done independently but they can cooperate in providing the information needed. Both ACORDA and P-log rely on the well-known Stable Model semantics to provide meaning to programs (see their references for details).

The system architecture follows this separation in a very explicit way. There is indeed a necessary separation between producer and consumer of information. ACORDA and P-log can both act as a producer or consumer of information. The interfaces between the various components of the integration of ACORDA with P-log are made explicit in Fig. 1[2]. Each agent is equipped with a Knowledge Base and a Bayes Net as its initial theory. The first time around, ACORDA sends Bayes Net information to P-log and later P-log translates all the information sent by ACORDA and keeps it for future computation. The problem of prospection is then that of finding abductive extensions to this initial theory which are both relevant (under the agent's current goals) and preferred (w.r.t. the preference rules in its initial theory). The first step is to select the goals that the agent will possibly attend to during the prospective cycle. Integrity constraints are also considered to ensure the agent always performs transitions into valid evolution states.

Once the set of active goals for the current state is known, the next step is to find out which are the relevant abductive hypotheses. At this step, P-log uses the abductive hypothesis for computing probabilistic information to help ACORDA do a priori preferences. Forward reasoning can then be applied to abducibles in those scenaria to obtain relevant side-effect consequences, which can then be used to enact a posteriori preferences. These preferences can be enforced by employing utility theory that can be combined with probabilistic theory in P-log. In case additional information is needed to enact preferences, the agent may consult external oracles. This greatly benefits agents in giving them the ability to probe the outside environment, thus providing better informed choices, including the making of experiments. Each oracle mechanism may have certain conditions specifying whether it is available for questioning. Whenever the agent acquires additional information, it is possible that ensuing side-effects affect its original search, e.g. some already considered abducibles may now be disconfirmed and some new abducibles are triggered. To account for all possible side-effects, a second round of prospection takes place.

If everything goes well and only a single model emerges from computation of the abductive stable models, the ACORDA cycle terminates and the resulting abducibles are added to the next state of the knowledge base. In most cases, however, we cannot guarantee the emergence of a single model, since the active preferences may not be sufficient to defeat enough abducibles. In these situations, the ACORDA system has

[1] The integration of ACORDA with P-log can be accessed at `http://sites.google.com/site/acordaplog/Home`

[2] Please note the dashed lines represent communication between ACORDA and P-log.

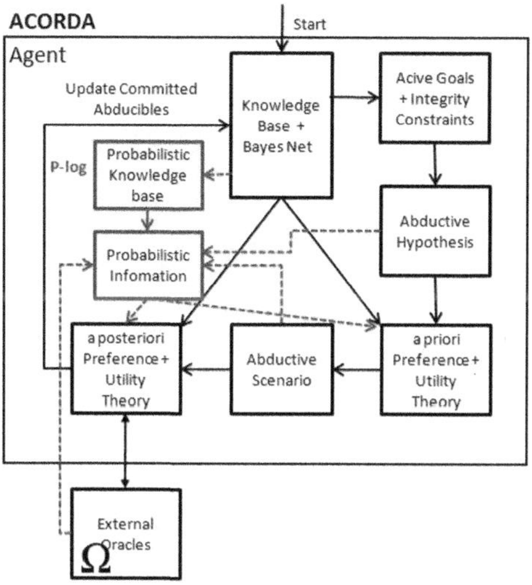

Fig. 1 Integration Architecture

to resort on additional information for making further choices. If no model emerges, relaxation of constraints and preferences may be in order.

Indeed, a given abducible can be defeated in any one of two cases: either by satisfaction of an `expect_not/1` rule for that abducible, or by satisfaction of a preference rule that prefers another abducible instead. However, the current knowledge state may be insufficient to satisfy any of these cases for all abducibles except one, or else a single model would have already been abduced. It is then necessary that the system obtains the answers it needs from somewhere else, namely from making experiments on the environment or from querying an outside entity.

ACORDA consequently activates its a posteriori choice mechanisms by attempting to satisfy additional selecting preferences. There are two steps: first ACORDA computes the utility value for each abducible. ACORDA selects amongst them based on the preference function defined. This step is coded below in the meta predicate `select/2`.

```
select(M, NewM) :- select1(M, M1), select2(M1, NewM).
select1(M, M1)  :- addUtilityValue(M, M1).
select2(M1, NewM) :-%use preference function to select the model.
```

Example 3 (Tea 3). (Continuation of Example 2) Consider now we introduce the utility rate for coffee and tea for agent Claire based on her preference. What do we suggest for agent Claire's beverage when the utility probability value is taken into account?

```
beginProlog.
   select(M, Mnew) :-
      utilityRate(tea, 0.8), utilityRate(coffee, 0.7),
      select1(M, M2),
      select2(M2, 0, [], Mnew).
% Add utility value to each model
   select1([], []).
```

```
21     select1([X|Xs], [Y|Ys]) :-
22        addUtilityValue(X, Y), select1(Xs, Ys).
23  % A posteriori preference models
24     select2([], _, M, M).
25     select2([M|Ms], Acc, OldM, NewM) :-
26        member(utilityModel(U), M),
27        U > Acc -> select2(Ms,U,M,NewM); select2(Ms,Acc,OldM,NewM).
28     addUtilityValue([X], [utilityModel(UModel)|[X]]) :-
29        holds utilityRate(X, R),
30        pr(available(X), P), UModel is R * P.
31     addUtilityValue(_, []).
32  endProlog.
```

First, ACORDA launches oracles to acquire information about Claire's condition – whether she is sleepy and whether she has high blood pressure. If there is no contrary expectation for any of the beverages, i.e. agent Claire is not sleepy and also does not have high blood pressure, we will have two different abductive solutions: $M_1 = \{coffee\}$, $M_2 = \{tea\}$. Next, ACORDA performs a posteriori selection. Our selecting preference amongst abducibles is codified on lines 14-32. First we initialize the utility rate for both beverages. In the `addUtilityValue/2` predicate, we define our utility function. In this example, we define the utility value for each beverage as its probability of availability times its utility rate. After the computation we get the result: $M_1 = \{utilityModel(0.2800),\ coffee\}$, $M_2 = \{utilityModel(0.4800),\ tea\}$. Predicate `select2/2` will select the highest utility model and the final result is $M_2 = \{utilityModel(0.4800),\ tea\}$ which means that based on the a posteriori preference using our utility function, agent Claire is encouraged to have tea.

4 Example: Risk Analysis

The economics of risk[9] has been a fascinating area of inquiry for at least two reasons. First, there is hardly any situation where economic decisions are made with perfect certainty. The sources of uncertainty are multiple and pervasive. They include price risk, income risk, weather risk, health risk, etc. As a result, both private and public decisions under risk are of considerable interest. This is true in positive analysis (where we want to understand human behaviour), as well as in normative analysis (where we want to make recommendations about particular management or policy decisions). Second, over the last few decades, significant progress has been made in understanding human behaviour under uncertainty. As a result, we have now a somewhat refined framework to analyse decision-making under risk.

In a sense, the economics of risk is a difficult subject; it involves understanding human decisions in the absence of perfect information. In addition, we do not understand well how the human brain processes information. As a result, proposing an analytical framework to represent what we do not know seems to be an impossible task. In spite of these difficulties, much progress has been made. First, probability theory is the cornerstone of risk assessment. This allows us to measure risk in a fashion that can be communicated amongst decision makers or researchers. Second, risk preferences are better understood. This provides useful insights into the economic rationality of decision-making under uncertainty. Third, over the last decades, good insights have been developed about the value of information. This helps us to better understand the role of information and risk in private as well as public decision-making.

We define risk as representing any situation where some events are not known with certainty. This means that one cannot influence the prospects for the risk. It can also relate to events that are relatively rare. The list of risky events is thus extremely long. First, this creates a significant challenge to measure risky events. Indeed, how can we measure what we do not know for sure? Second, given that the number of risky events is very large, is it realistic to think that risk can be measured? We will present a simple example about decision–making in a restaurant where risk is taken into account.

Example 4 (Mamamia). The owner of the Restaurant "Mamamia" wants to offer a new menu. Before the launch of the new menu, he performed some research. Based on it, 75% of teenagers prefer the new menu and 80% of adults prefer the original menu. Around 40% of his costumers are teenagers. If he offers the new menu, each new menu will return 5 Euros. The basic cost that he should spend for 100 new menus is 200 Euros. What is your suggestion for the owner of the Restaurant "Mamamia"? The owner's utility function is approximately represented by $U(X) = 2 * X - 0.01X^2$, $(X \leq 100)$, X being his income.

```
expect(decide(launch_new_menu)). expect(decide(not_launch)).
constrain(1,[decide(launch_new_menu), decide(not_launch)], 1).
decision ← consider(decide(X)).
decide(launch_new_menu) ◁ decide(not_launch) ← not_take_risk,
    prolog(pr(menu(new), PN)), prolog(pr(menu(original), PO)),
    prolog(PN > PO).
decide(not_launch) ◁ decide(launch_new_menu) ← not_take_risk,
    prolog(pr(menu(original), PO)), prolog(pr(menu(new), PN)),
    prolog(PO > PN).
not_take_risk ← not take_risk.
falsum ← not decision.
beginPr.
  age = {teenager, adult}.   offer = {original, new}.
  customer : age.
  random(rc, customer, full).
  pa(rc, customer(teenager), d_(60,100)).
  menu : offer.
  random(ro, menu, full).
  pa(ro, menu(new), d_(75,100)) :- customer(teenager).
  pa(ro, menu(original), d_(80,100)) :- customer(adult).
endPr.
beginProlog.
  :- import member/2, length/2 from basics.
  select(M, Mnew) :- select1(M, Mnew), select2(_,0,[],_).
  select1([],[]).
  select1([X|Xs], [Y|Ys]) :-
     addUtilityValue(X,Y), select1(Xs,Ys).
```

```
    select2([],_,M,M).
    select2([M|Ms], Acc,OldM,NewM):-
       member(utilityModel(U), M),
       U > Acc → select2(Ms,U,M,NewM); select2(Ms,Acc,OldM,NewM).
    addUtilityValue([decide(X)],
       [utilityModel(EU), expectedProfit(Profit),decide(X)]) :-
          expectedProfit(X, Profit), expectedUtility(X, EU).
    expectedProfit(Action, P) :-
       return(Action, R), cost(Action, C), P is R - C.
    expectedUtility(Action, EU) :-
       pr(menu(new), PrN), expectedProfit(Action, Pf),
       EU is (2*Pf*PrN - 0.01*Pf*Pf*PrN).
    return(launch_new_menu, 5). return(not_launch, 0).
    cost(launch_new_menu, 2).   cost(not_launch, 0).
endProlog.
```

In Example 4, the expected return value depends on the probability of the number of new menu offers. Given probability information about the age of customers and their behaviour in choosing a menu offer, we compute the expected probability of new menu choices. In this case, the probabilty of new menu choices is 0.42 an the probability of original menu choices is 0.58. The probability of original menu choices is higher than the probability of new menu choices. Furthermore, if we compute the expected return more comprehensively, we will get the result:
$M_1 = \{utilityModel(3.1323),\ expectedProfit(3),\ decide(launch\_new\_menu)\}$,
$M_2 = \{utilityModel(0.0000),\ expectedProfit(0),\ decide(not\_launch)\}$
Based on the computation, it is better if he launches a new menu even though it is risky.

5 Conclusion and Future Work

Humans reason using cause and effect models with the combination of probabilistic models such as Bayes Nets. Unfortunately, the theory of Bayes Nets does not provide tools for generating a scenario that is needed for generating several possible worlds. Those are important in order to simulate human reasoning not only about the present but also about the future. On the other hand, ACORDA is a prospective logic programming language that is able to prospect possible future worlds based on abduction, preference and expectation specifications. But ACORDA itself cannot handle probabilistic information. To deal with this problem, we integrated ACORDA with P-log, a declarative logic programming language based on probability theory and Causal Bayes Nets. Now, using our new system, it is easy to create models using both causal models and Bayes Nets. The resulting declarative system can benefit from the capabilities of the original ACORDA for generating scenaria, and is equipped with probabilistic theory as a medium to handle uncertain events. Using probability theory and utility functions, the new ACORDA is now more powerful in managing quantitative as well as qualitative a priori and a posteriori preferences.

Often we face situations when new information is available that can be added to our model in order to perform our reasoning better. Our new ACORDA also makes it possible to perform a simulation and afterwards add more information directly to the agent. It can also perform several steps of prospection in order to predict the future better. For illustrative understanding, we presented taking a risk example in which the system can help people to reason and rationally decide.

Applying utility functions to decision-making under uncertainty requires having good information about the measurement of the probability disribution and the risk preferences of the decision-maker. It is possible to conduct risk analysis without precise information about risk preferences using stochastic dominance. The latter sees the elimination of "inferior choices" without strong a priori information about risk preferences. For future work, we can extend P-log to handle stochastic processes.

References

1. Hume, D.: An Enquiry Concerning Human Understanding: A Critical Edition. Oxford Philosophical Texts (1748)
2. Mackie, J.: The Cement of the Universe. Clarendon Press, Oxford (1974)
3. Dell'Acqua, P., Pereira, L.M.: Preferential theory revision (extended version). J. Applied Logic (2007)
4. Baral, C., Gelfond, M., Rushton, N.: Probabilistic reasoning with answer sets. In: Lifschitz, V., Niemelä, I. (eds.) LPNMR 2004. LNCS, vol. 2923, pp. 21–33. Springer, Heidelberg (2004)
5. Pearl, J.: Causality: Models, Reasoning, and Inference. Cambridge University Press, Cambridge (2000)
6. Gelfond, M., Rushton, N., Zhu, W.: Combining logical and probabilistic reasoning. In: AAAI Spring Symposium (2006)
7. Pereira, L.M., Lopes, G.: Prospective logic agents. In: Neves, J., Santos, M.F., Machado, J.M. (eds.) EPIA 2007. LNCS, vol. 4874, pp. 73–86. Springer, Heidelberg (2007)
8. Pereira, L.M., Lopes, G.: Prospective logic agents. In: International Journal of Reasoning-based Intelligent Systems (IJRIS) (to appear, 2009)
9. Chavas, J.-P.: Risk Analysis in Theory and Practice. Academic Press, London (2004)

A Library Marketing System for Decision Making

Toshiro Minami

Abstract. The major aim of library marketing system is to help the library and its patrons with providing useful information and various kinds of knowledge, which are extracted from the data that are collected by the library system. In this paper we lay heavy stress on the use of such information and knowledge for assisting decision making of the library and of its patrons. Furthermore our main concern is to investigate how much we can utilize the usage data of materials obtained in the intelligent bookshelves (IBSs), i.e. the bookshelves equipped with RFID antennas and their reader/writer controllers (R/Ws). In this paper we propose some analysis methods of these usage data, alone and combining with other library data, and demonstrate their potential importance for library marketing.

Keywords: Library Marketing, Data Analysis, Data Mining, RFID (Radio Frequency Identification), Intelligent Bookshelf.

1 Introduction

The most important mission of libraries is to provide services to their users, or patrons, so that they are very helpful and meet the patrons' needs in our knowledge society. Patron orientedness is well expressed in "The five laws of library science" advocated by the famous Indian library scientist S. R. Ranganathan [13], which was published way back in 1960s. They are: (1) Books are for use, (2) Every reader his book, (3) Every book his reader, (4) Save the time of the reader, (5) The library is a growing organism. Our surrounding society has changed vigorously in these several decades. However, these phrases are still true as we rephrase them by replacing "book" with "information," "material," or "service." They are still the central dogma for libraries and will keep being the same in the

Toshiro Minami
Kyushu Institute of Information Sciences, Faculty of Management and Information Sciences, 6-3-1 Saifu, Dazaifu, Fukuoka 818-0117 Japan

Kyushu University Library, Research and Development Division, 6-10-1 Hakozaki, Higashi, Fukuoka 812-8581 Japan
e-mail: minami@kiis.ac.jp, minami@lib.kyushu-u.ac.jp
http://www.kiis.ac.jp/~minami/ (in Japanese)

K. Nakamatsu et al. (Eds.): New Advan. in Intel. Decision Techno., SCI 199, pp. 97–106.
springerlink.com

future. In order to fulfill the patrons' needs, libraries have been trying to change themselves and to provide better services; based on their intuition.

In a decade or two we have experienced a rapid advance in technology; especially in information and communication technology (ICT). Internet should be the most prominent ICT technology that gives big influence to our society and our everyday life. Marketing methods by collecting and analysing data from the Internet services, mostly with Web-based mechanisms, are very popular now.

Now is a high time for libraries to introduce such methodology to their services. By collecting and analysing the data created in the library services, libraries can get useful tips for them. Such technology and methodologies would change the ways the libraries have been taking so that many decisions made in library jobs are done based not on the subjective intuition but on the objective data. Not only for libraries but also for their patrons such methodologies give much benefits by getting new services from their libraries.

In this paper we describe the concept of library marketing and library marketing system, and we also demonstrate its usefulness by taking some examples supporting functions for decision making by the libraries and their patrons.

The rest of this paper is organized as follows: First of all in Section 2, we explain the concept of library marketing. We note that it has a self-growing nature so that it will solve the cold start, or the free-riding, problem. Then we show a model of library marketing system with RFID (Radio Frequency Identification) technology [5]. The library materials can be classified into two types; physical media and net media. The former ones are printed books, magazines, casette tapes, CDs, DVDs, and so on, which are the major materials used in library services so far, and will be so in the near future. The latter one is the materials used in network services such as e-journals, e-books, and so on. The log data, or usage data, are collected as the libraries provide services. RFID is useful for collecting log data of physical media. For net media, the service log data are collected automatically. We show an experimental library marketing system with RFID and explain how to extract the data that are used in data analysis, or data mining, which is the most important and essential part for library marketing system. In Section 3, we would like to demonstrate the importance and usefulness of the library marketing system by showing examples of information or knowledge that are extracted by analysing the usage data. Lastly in Section 4, we conclude our discussions and present some of our future plans.

2 Library Marketing System

This section starts with describing the concept of library marketing [7, 8]. We emphasize the utilization of the usage data of library materials. Intelligent bookshelf, or IBS in short, is a key tool for library marketing for physical media. RFID is an AIDC (Automatic Identification and Data Capture) technology [1] so that the data collection is performed automatically and therefore saves the librarians' time.

Then we describe a model of library marketing system, which we use in an experiment. The original log data is converted to the event data, which is more suitable for analysis. We define some terms such as a session, session (duration) time, which are extracted from the event data.

2.1 Library Marketing and Library Marketing System

Figure 1 illustrates the concept of the library marketing. It is the mutual depending structure between databases and services. In the right part of the figure are library services provided by the library.

OPAC (Online Public Access Catalog) is a search engine for library materials. Reference service is a consulting service to the patrons who want to know how to find appropriate material, or need some advices. These days many libraries provide this service online, which is often called "Ask a Librarian" service. "My Library" is a relatively new service that provides patrons with personalized information such as which books they are borrowing and have been borrowed, a list of books the patrons might like to read, and so on.

These services depend on the databases that consist of catalog data, circulation data, patrons' profile data, etc. These data are collected at a circulation counter, at a self checkout machine, and in other ways. The log data, which are collected for recording how the system provides services to patrons, are stored in the service log database and used for various services.

The collected data are supposed to be analyzed, or data mined. The results obtained by analysis should be reflected to the services in order to improve the quality of patron services. If the quality of services gets higher, then more patrons will use the library and the system. If many patrons use them, then the system gets more service log data. If the system is able to get more data, then they can be used for better services. Then the customer satisfaction (or rather patron satisfaction) will rise. Such a natural self-growing mechanism is very important for library marketing with databases and services as has been pointed out.

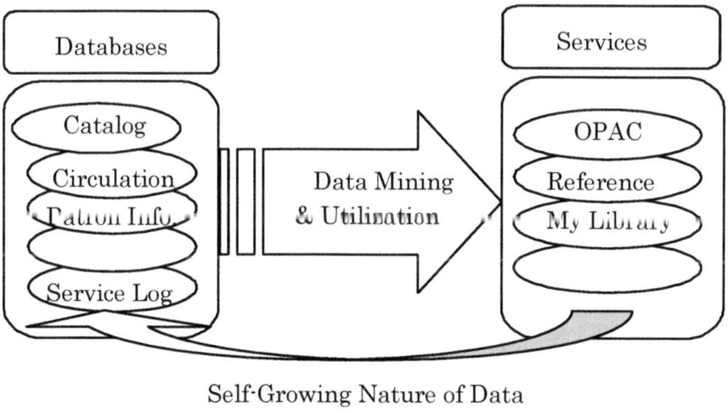

Fig. 1 Library Marketing with Databases and Services

2.2 Library Marketing System with RFID

For the services using the net media, it is easy to collect not only provinding materials for the patrons but also to collect log data. Everything can be done

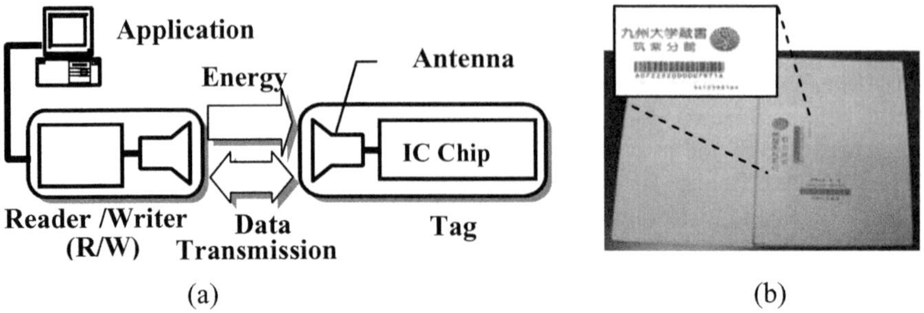

Fig. 2 (a) Principle of RFID (b) RFID Tag Attachment on a Book

automatically. For physical media such as printed books, magazines, audio visual materials like CDs and DVDs, it is a difficult task to do the job and collect the usage data, bacause these jobs are supposed to be done by the library staff.

AIDC technology [1] is a great help in such jobs. So far the barcode system has been used in libraries. RFID is getting to be more and more popularly used by libraries in these years because it is faster and easier to be used for doing library jobs [6, 11, 15]. Furthermore there are more advantages in RFID such as the circulation processing becomes faster and easier, the patrons are happy to use self-checkout machines and do the processing themselves, inventory of library materials becomes faster. Considering these benefits it is worth putting the costs on equipments and RFID tags.

The principles of RFID tag system is illustrated in Figure 2(a). The RFID tag system consists of two major components; tags and reader/writers (R/Ws). A tag is able to communicate with an R/W when they are located sufficiently close to each other. Figure 2(b) shows a tag put on a book.

From the library marketing standpoint intelligent bookshelf (IBS) (Figure 3) [3] is a very useful tool for collecting data. An IBS is a bookshelf which is equipped with RFID antennas so that the book IDs shelved in it are detected and recorded

Fig. 3 Intelligent Bookshelf: A Tool for Library Marketing

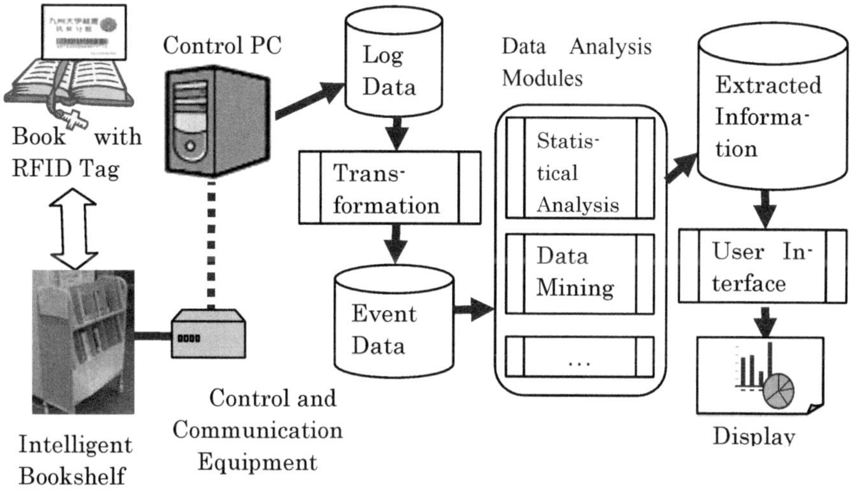

Fig. 4 An Experimental Library Marketing System with RFID

automatically. The antennas and their controller, i.e. R/W in all, transmit the data to the controller PC and the data will be analyzed either real time or later on.

Figure 4 illustrates a model for library marketing system with RFID, which is developed for a feasibility study of IBS systems [9, 10, 16]. Original log data obtained from the IBS are transmitted to the controller PC and are saved as log data. The log data consist of the fields of data type, timestamp (yymmdd and hhmmss), shelf ID, the number of detected tag IDs, followed with the list of tag IDs. A tag ID looks like "E00401000314E148" and "E004010003150040."

The log data are redundant because the status will not change for a long time and thus the detected tag IDs are the same ones in a long line of data. So it is more convenient to transform this log data format into the format that is easy to deal with. We call the transformed data the event data. The transformation program converts the original log data into the event data.

The event data can be represented as a list of events. They consist of the timestamp data, shelf ID, i.e. tier ID and shelf ID, one column of tag ID, and the type of event, or the change of status, which is either IN or OUT. The status IN means that the book is put on the denoted shelf and OUT means it is taken out from the shelf.

Data analysis modules use the event data as input and extract information and knowledge. They analyse the data in statistical and/or data mining methods.

Lastly the extracted information is presented to the librarians and users according to the situation.

2.3 Data in Library Marketing System

The original log data from an R/W are the list of a line in the following format:

O = (Time, ShelfID, Number of Tags, List of TagIDs) where *Time* is a timestamp, *ShelfID* consists of tier ID and the shelf ID in the tier.

From these data we get event data set, which is a list of event data, where *i*-th event data collected in this IBS are represented in the following format:

d_i = *(InOut, ShelfID, TagID, Time)* where *InOut* specifies if the book is stored in a bookshelf or removed from a bookshelf, *ShelfID* and *TagID* specify which shelf the book is stored to or removed from and which book is the one, respectively. The *Time* is the timestamp which specifies date and time of this status change.

By specifying a book, or its corresponding tag, with its ID, we can define a session of usage of the book. A session starts when the book is taken out from a bookshelf and ends when it is returned to a bookshelf, which may be the same one or a different one.

Let TID be a book ID. The event data of the book is defined:

Event(TID) = {(InOut, ShelfID, Time) | d_i=(InOut,ShelfID, TID, Time) for some InOut, ShelfID, and Time where d_i is an event data }

We can define a session of the book with TID as follows:

Session(TID) = {(OutTime, InTime, OutShelf, InShelf) | there exist e1=(InOut1, OutShelf, OutTime) and e2=(InOut2, InShelf, InTime) in Event(ID) such that InOut1="OUT", InOut2="IN", OutTime<InTime, and there is no data e3=(InOut3, ShelfID3, Time3) in Event(TID) that satisfies OutTime<Time3<InTime }

The duration time of a session is defined as follows:

Duration((OutTime, InTime, OutShelf, InShelf)) = InTime – OutTime

If *OutShelf* = *InShelf*, it means that a patron has put the book back in the same shelf and if not, he or she has put it in a different shelf, probably erroneously.

Suppose we set a time period *p=(StartTime, EndTime)*, where the start and end times may be in a same day or in different days, weeks, or months. We would define that a session *s=(OutTime, InTime, OutShelf, InShelf)* of a book is in this time period as follows: *s is in p iff StartTime<=OutTime<InTime<=EndTime* where <= means "less than or equals to."

In this definition we choose to say that the session occurs in the specified period only when it starts and ends in the period. Note that we can define this concept in some different ways.

3 Library Marketing System as Decision Support System

As was pointed out, the most important application of library marketing system is to help the librarians and their patrons with decision making on collection of books, shelving policy, shelf arrangement, planning learning process, and others. In this section we investigate several possible applications that should be very useful for the library, for librarians, and for library patrons.

3.1 Decisions of Learning Materials for Patrons

First of all we investigate how to help patrons with deciding which material to read, or study.

Statistical Method

Patrons will be happy if they are able to know what materials are popularly used and checked-out. Some of the typical data in this direction, for example, are the rankings of the books which are used in the library, which are borrowed from the library, which are popular among the patrons in general, and so on.

From the original usage data obtained from the IBSs, we can extract various statistical data on each of the book shelved on the IBSs, such as frequency of sessions, average session time, calculated time of sessions, etc.

Furthermore, because the library books have their catalog data, each book has its classification data; in DDC (Dewey Decimal Classification) or in its modified way in most libraries. By summing up the frequencies of the books belonging to a classification number, we can get the frequency data of each classification area. Not only the frequencies but other number data can also be processed in the same way.

In such a way, we can obtain many kinds of number data that indicate how the books are popularly used in the library. By applying this method to the circulation data, where one session starts when a book is borrowed and ends when it is returned to the library, we can obtain similar data about the borrowed materials.

In addition to them we can combine these data and other data from other libraries and from the Internet, we can provide the library patrons with additional information. One possible candidate for another data might be the one obtained from the blogs, the articles provided by the net people. For example we can obtain the popularity data of a book by using a blog search engine and get the number of the articles that mention the book as their subjects.

Even though the statistical methods are simple in one point of view, they are very valuable tools in order to extract useful and valuable data and information to the library patrons. We have to keep investigating further.

Collaborative Filtering

Collaborative filtering (CF), or social filtering, is now well known as the amazon.com's "Customers Who Bought This Item Also Bought"-type recommendation function [2]. This function has applied to other fields as well [12]. In library marketing we can use it in the format "who borrowed this book also borrowed," "who belong to your same department also borrowed," and furthermore "who come in this time zone of the day prefer to use," etc.

3.2 *Decisions on Book Selection for Libraries*

The collaborative filtering method can be applied to the selection of library materials as well. In this section we rather take another method that might be more appropriate in order to demonstrate the usefulness of the usage data obtained from the IBSs. It is a method for evaluating the appropriateness of the books for purchasing by combining the usage data from IBSs and the circulation data from the counters or from the self-checkout machines.

Table 1 Comparing the Frequencies of Usage data and Borrow Data

Usage / Borrow	High	Middle	Low
High	HH		HL
Middle			
Low	LH		LL

In Table 1, suppose the books ranked in High are those in highest 20% of usage frequency and circulation frequency in a specified time interval, respectively and those ranked in Low are lowest 20%.

Then the books labeled HH are popular ones in terms of both usage and borrow. They are the first candidates to purchase. On the other hand the library would not purchase the books with LL if they have not other reasons to purchase.

The ones labeled with LH, i.e. low usage and high borrow, might be the books that patrons have already decided to borrow when they visit the library, or those they have found them already as they used the OPAC (Online Public Access Catalog) system and have decided to borrow. These are good candidates for purchase.

The ones labeled with HL, i.e. high usage and low borrow, are difficult to decide immediately. The books that are prohibited to borrow may be set in this category; or HN, high usage and no borrow. Other books that are substantially used as reference materials may be in this category.

Also some books in this category might be something that look like interesting but actually they are not so much interesting nor useful as the patrons read some part of the book. If we check the average session time of the books in this category we may know better about such books. So we need to have more research on this.

3.3 Decisions of Book/Shelf Arrangement for Libraries

This is another interesting application of IBS. The system analyzes the usage data and extracts the correlation of two books in terms of usage time and others. Firstly we define the event time of an OUT event data *(OUT, ShelfID, Time)* by the timestamp *Time.*

Suppose TID1 and TID2 are different book IDs. We define the succeeding event time of TID2 of TID1 by:

SuccEventTime(TID1, TID2) = Time2 – Time1 where there exist event data e1=(OUT, ShelfID1, Time1) in Event(TID1) and e1=(OUT, ShelfID2, Time2) in Event(TID2) so that there does not exist any data e3=(OUT, ShelfID2, Time3) in Event(TID2) such that Time1 < Time3 < Time2.

Let us define the average time of *SuccEventTime(TID1, TID2)* in a specified time interval by the average succeeding event time from TID1 to TID2. Then we define the correlation index of two books TID1 and TID2 by adding the average succeeding event times from TID1 to TID2 and the time TID2 to TID1. The correlation index is a measure for closeness of two books in terms of usage pattern.

Using the correlation indexes we can make a special corner of books that are related closely in this sense so that the patrons can access them in one place.

4 Concluding Remarks

In this paper we tried to convince the importance of the concept of marketing for libraries because library marketing methodologies have essential importance for the future libraries. The libraries should have good reputation and they succeed in surviving in the competitive society by utilizing them.

The methodology of utilizing the data is the same one with OLAP (Online Analytical Processing). However we put more emphasis on the point that the information and knowledge are useful not only for the libraries but also for their patrons in library marketing. Our approach is different to that of OLAP in this point.

We have presented a model of library marketing system with intelligent bookshelves, which utilizes the advantage of RFID technology. With IBS we can automatically collect the usage data of physical materials in a library. The only usage data libraries can get so far are circulation data. So the librarians do not have data for materials that are not allowed to borrow, e.g. reference tools. Now it is possible to obtain those data with IBSs. Also we presented several examples of data that might be useful for librarians and patrons. Such services could not be realized without library marketing tools. So far almost all libraries with RFID tag system use HF (13.56MHz). Chiyoda Library in Tokyo has decided to introduce an UHF (950MHz) tag system [4, 14] because it has better potential performance.

Even with such high potential of the library marketing system, we are in the first stage towards the goal. Some of our future plan includes (1) developing a library marketing system for experimenting more sophisticated functions, (2) developing a lot more analysis methods so that they can be applied to a wide variety of libraries, and (3) evaluate the usefulness of the concept of library marketing.

Our eventual goal is to apply such library marketing system to many libraries so that the system contributes to the people as they learn and do research works.

Last of all the author would like to express his gratitude to his daughter, Mariko Minami for her valuable comments for the draft of this paper.

References

1. AIM Global, `http://www.aimglobal.org/`
2. Amazon.com, `http://www.amazon.com/`
3. Central Engineering Co. Ltd. (in Japanese), `http://www.central-eng.co.jp/`
4. Chiyoda Library (in Japanese), `http://www.library.chiyoda.tokyo.jp/`
5. Finkenzeller, K.: RFID Handbook, 2nd edn. John Wiley & Sons, Chichester (2003)
6. Kyushu University Library, `http://www.lib.kyushu-u.ac.jp/`

7. Minami, T.: RFID Tag based Library Marketing for Improving Patron Services. In: Hoffmann, A., Kang, B.-h., Richards, D., Tsumoto, S. (eds.) PKAW 2006. LNCS (LNAI), vol. 4303, pp. 51–63. Springer, Heidelberg (2006)
8. Minami, T.: On-the-site Library Marketing for Patron Oriented Services. Bulletin of Kyushu Institute of Information Sciences 8(1), 15–33 (2006) (in Japanese)
9. Minami, T.: A Perspective to the Library in Network-Oriented Society – Ubiquitous Library Services through PDAs. Bulletin of Kyushu Institute of Information Sciences 10(1), 1–17 (2008) (in Japanese)
10. Minami, T.: A Design for Library Marketing System and its Possible Applications. In: Proc. Pacific Rim Knolwdge Acquisition Workshop 2008 (PKAW 2008), pp. 187–200 (2008)
11. National Library Board Singapore, `http://www.nlb.gov.sg/`
12. Oda, M., Minami, T.: From Information Search towards Knowledge and Skill Acquisition with SASS. In: Proc. 6th Pacific Rim Knowledge Acquisition Workshop (PKAW 2000), pp. 245–260 (2000)
13. Ranganathan, S.R.: The Five Laws of Library Science. Bombay Asia Publishing House (1963)
14. SOFEL Co. Ltd., `http://www.sofel.co.jp/en/index.html`
15. University of Nevada, Las Vegas Libraries, `http://www.library.unlv.edu/`
16. Zhang, L., Minami, T.: Library Marketing that Boosts Education with RFID. In: Proc. 6th International Conference on Information Technology Based Higher Education and Training (ITHET 2007) (2007)

Analysis of Four Wheeled Flexible Joint Robotic Arms with Application on Optimal Motion Design

M.H. Korayem, H.N. Rahimi, and A. Nikoobin

Abstract. Designing optimal motion is critical in several applications for mobile robot from payload transport between two given states in a prescribed time such that a cost functional is minimized. This paper deals with the problem of path design of wheeled non-holonomic robots with flexible joints, based on Pontryagin's minimum principle. The simplified case study of a Four Wheeled, two-link manipulator with joint elasticity is considered to study the method in generalized model. Nonlinear state and control constraints are treated without any simplifications or transforming them into sequences of systems with linear equations. By these means, the modeling of the complete optimal control problem and the accompanying boundary value problem is automated to a great extent. Performance of method is illustrated through the computer simulation.

Keywords: Flexible Joint Robotic Arm, Non-holonomic Constraints, Optimal Motion Design, Pontryagin's Minimum Principle.

1 Introduction

Mobile robotic arms consist of a mobile platform equipped with mechanical manipulators. If assumed that the mobile base does not slide then a non-holonomic constraint is imposed on the system hence the system is considered as a non-holonomic platform. Besides, in many industrial applications such as high-speed assembly and heavy load carrying, the joint flexibility exists in most manipulators in the drive transmission systems (transmission belts, gears, shafts, etc.), that usually neglected to analysis of such flexible joint systems [1].

Several research works have been carried out for mobile robot arms. A comprehensive literature survey on nonholonomic systems can be found in [2].

M.H. Korayem, H.N. Rahimi, and A. Nikoobin
Robotic Research Lab, Mechanical Engineering Department, Iran University of Science and Technology
e-mail: hkorayem@iust.ac.ir, hamedrahimi.n@gmail.com, anikoobin@iust.ac.ir

K. Nakamatsu et al. (Eds.): New Advan. in Intel. Decision Techno., SCI 199, pp. 107–116.
springerlink.com

However, little work has been reported on a comprehensive model that Encompass mobile manipulator with flexible joint. In [3] a computational technique for obtaining the maximum load-carrying capacity of robotic manipulators with joint elasticity is described while different base positions are considered, so finding the optimal path is not considered in it. Another algorithm to the maximum load determination via linearizing the dynamic equation and constraints in [4] is studied on the basis of Iterative Linear Programming (ILP) method for flexible mobile manipulators. But, because of difficulties of ILP method the flexibilities of joints are neglected either in the dynamic equation or simulation procedure.

Nowadays the advantages of optimal control theory are well established and a host of issues related to this technique have been studying specially in the field of optimal motion planning of robots [5]. Accordingly indirect solution of open-loop optimal control method was proposed to trajectory optimization of flexible link mobile manipulator in point-to-point motion [6]. In mentioned work, despite ILP based studies, boundary conditions are satisfied exactly, in addition the complete form of the obtained nonlinear equation is used. However, the joints are assumed rigid in addition to kinematics and dynamics of wheels are not considered in it.

In this paper with applying the optimal control approach path generating of four wheeled mobile manipulators is done by modeling the elasticity at each joint as a linear torsinal spring. The remaining of the paper is organized as follows. The development of kinematics and dynamics of wheeled mobile robotic arms for four wheels-two links robot is deal with, considering the joint flexibility in each joints in section 3. Subsequently, the optimal control problem that with implementing of Pontryagin's Minimum Principle supports the execution of the optimization solution of model is expressed as a brief review in section 4. Simulation is done and results are discussed in Section 5. Finally, the concluding remarks by highlighting the main advantages of the presented method are expressed in the last section.

2 Kinematic and Dynamic Analysis of System

The mobile manipulator consisting of differentially driven vehicle with flexible revolute joints robot arms is expressed in this section. With decide on choosing X_0Y_0 as the inertial co-ordinate frame (CF) and selecting proper subsequent frames then writing the transition equation between them, the kinematic equations are obtained.

The nonholonomic constraints states that the robot can only move in the direction normal to the axis of the driving wheels, in addition, the acceleration constraint should be considered in order to avoid slippage during the robot navigation. For a mobile robot with four differential wheels move on a planar surface these constraints can be derived as:

$$\begin{aligned} \dot{x}\cos(\alpha+\theta+\beta)+\dot{y}\sin(\alpha+\theta+\beta)+l\dot{\theta}\sin\beta+d(\dot{\theta}+\dot{\beta})=0 \\ -\dot{x}\sin(\alpha+\theta+\beta)+\dot{y}\cos(\alpha+\theta+\beta)+l\dot{\theta}\cos\beta+r\dot{\phi}=0 \end{aligned} \tag{1}$$

where θ is the robot's heading angle, l is the width of the platform, r is the radius of the wheels and ϕ is angular displacement of wheels.

Dynamic equation of rigid mobile manipulator is obtained in compact form as:

$$M(q)\ddot{q} + H(q,\dot{q}) + G(q) = U, \tag{2}$$

where M is the inertia matrix, H is the vector of Coriolis and centrifugal forces, G describes the gravity effects and U is the generalized force inserted into the actuator. Their details are omitted.

To model a flexible joint manipulator (FJM) the link positions are let to be in the state vector as is the case with rigid manipulators. Actuator positions must be also considered because in contradiction to rigid robots these are related to the link position through the dynamics of the flexible element. By defining the link number of a flexible joint manipulator is m, position of the ith link is shown with $\theta_{2i-1} : i = 1,2,...,m$ and the position of the ith actuator with $\theta_{2i} : i = 1,2,...,m$, it is usual in the FJM literature to arrange these angles in a vector as follows:

$$Q = [\theta_1, \theta_3, ...\theta_{2m-1} \mid \theta_2, \theta_4, ...\theta_{2m}]^T = [q_1^T, q_2^T]^T \tag{3}$$

So by adding the joint flexibility with considering the elastic mechanical coupling between the ith joint and link is modeled as a linear torsional spring with constant stiffness coefficient k_i, the set of equation of motion comprising mobile base with both link and joint flexibility can be rearranged into the following form:

$$\begin{aligned} M(q_1)\ddot{q}_1 + H(q_1,\dot{q}_1) + G(q_1) + K(q_1 - q_2) = 0 \\ J\ddot{q}_2 + K(q_2 - q_1) = U \end{aligned} \tag{4}$$

where K=diag[$k_1, k_2, ..., k_m$] is a diagonal stiffness matrix which models the joint elasticity, J=diag[$J_1, J_2, ..., J_m$] is the diagonal matrix representing motor inertia

3 Defining of the Optimal Control Problem

This section identifies the basic content of the optimization problem in order to be deal with in path optimization procedure. In summary the optimal control problem which that uses for optimization solution has basic statements as:

- Finding a state function X and actuating inputs, i.e. the control variables U during the overall time t_f where the state equation describing the dynamic evolution of the multibody system over this time interval be specified as:

$$\dot{X}(t) = f(X(t), U(t)) \tag{5}$$

- Minimizing a performance criterion

$$J(u) = \int_{t_0}^{t_f} L(X(t), U(t))dt \tag{6}$$

This can combine, for instance, energy consumption, actuating torques, traveling time or bounding the velocity magnitude or maximum payload.

By defining $\overline{U}$ as a set of admissible control torque over the time interval the imposed bound of torque for each motor can be expressed as:

$$\overline{U} = \{U^- \le U \le U^+\} \tag{7}$$

The optimization problem is completed by the boundary conditions

$$X(t_0) = X_0 \quad , X(t_f) = X_f \tag{8}$$

which represent the characteristics of each joint at initial and final time.

By implementing Pontryagin's minimum principle for solving optimization problems the necessary conditions for optimality are obtained as stated on the basis of variational calculus. Defining the Hamiltonian function as:

$$H^*(X,U,Y,m_p,t) = Y^T f(X,U,t) + L(X,U,m_p,t) \tag{9}$$

in addition to costate time vector-function $Y(t)$ that verifying the costate vector-equation (or adjoint system)

$$\dot{Y}^T = -\partial H^* / \partial X \tag{10}$$

and the minimality condition for the Hamiltonian as:

$$\begin{cases} \partial H^* / \partial U = 0 \ , \\ \dot{X} = \partial H^* / \partial Y \end{cases} \tag{11}$$

leads to transform the problem of optimal control into a non-linear multi-point boundary value problem, that it can be solved by numerical techniques.

4 Simulations

4.1 Background, Deriving the Equations

A mobile manipulator as depicted in Fig. 4.1 is considered to simulate the model. It consists of a symmetric wheeled mobile platform and a two-link manipulator mounted on top of the platform.

The platform moves by driving four independent wheels. The manipulator is constructed as a two-link planar arm with attached at the flexible joints. A concentrated payload of mass mp is connected to the second link.

This nonholonomic robot arm for motion in the plane has nine degrees of freedom that are arose from movement of platform- 3 DOFs-, with arms -2 DOFs- and wheels -4 DOFs-. Hence the question which naturally arises is: how to guarantee soft and well-organized movement encounter of seven redundant DOFs in system? A common answer is: by prescribing five kinematical constraints the redundancy resolution of system is ensured as well as previously specification of base trajectory during the motion. These constraints are consisted of four conditions

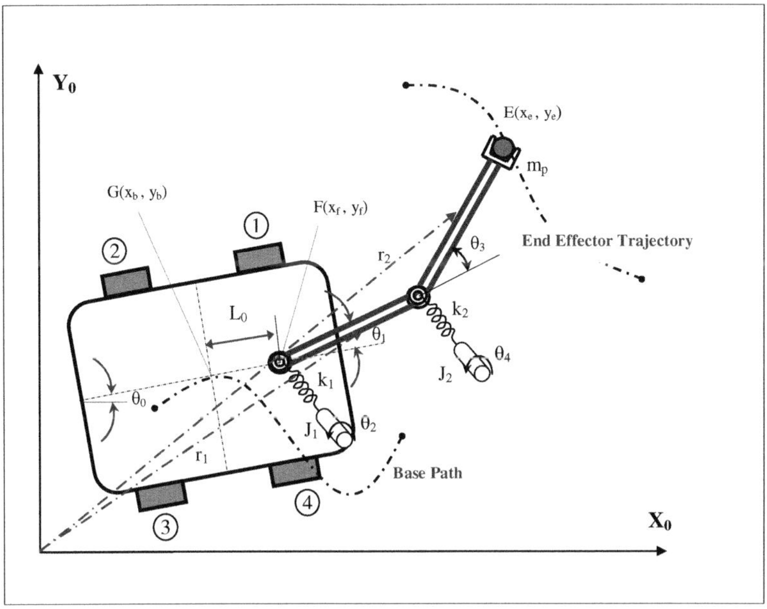

Fig. 4.1 Four-wheeled mobile manipulator with two flexible joint

associated to the avoidance of slippage of wheels during the robot navigation and last is related to the rolling without slipping condition Eq. (1).

The physical parameters of arms are summarized in the Table 4.1.

Table 4.1 Simulation parameters

Parameter	Value	Unit
Length of Links	$L_1 = L_2 = 0.7$	m
Mass of Links	$m_1 = m_2 = 6$	Kg
Spring constant	$k_1 = k_2 = 2000$	N/m
Moment of inertia	$J_1 = J_2 = 2$	Kg. m^2
Max. no Load Speed of Actuators	$w_{s1} = w_{s2} = 3$	rad/s
Actuator Stall Torque	$\tau_{s1} = \tau_{s2} = 70$	N.m

Also each wheel has 0.3 kg mass, 0.3 m length, 0.3 m radius and the mass of base is 2 kg.

By decomposing the system dynamic Eq. (4) into redundant and non-redundant parts and considering components associated with non-redundant ones the resultant equations in state space formed as:

$$\dot{x}_{2i-1} = x_{2i}\ ,\ \dot{x}_{2i} = f_2(i)\ \ ;\ i = 1, \cdots, 4 \tag{12}$$

where $X = [X_1 \quad X_2 \quad X_3 \quad X_4]^T$ is a state vector as:

$$X_1 = Q^T = [x_1 \quad x_3 \quad x_5 \quad x_7]^T, \\ X_2 = \dot{Q}^T = [x_2 \quad x_4 \quad x_6 \quad x_8]^T. \tag{13}$$

The boundary condition can be expressed as follows:

$$x_1(0) = x_5(0) = 1.5\, rad\,,\ x_3(0) = x_7(0) = 2\, rad\,; \\ x_1(f) = x_5(f) = -1\, rad\,, x_3(0) = x_7(f) = 1\, rad\,; \\ x_{2i}(0) = x_{2i}(f) = 0, \quad i = 1...4 \tag{14}$$

By controlling all active joints so as to achieve the best dynamic coordination of joint motions, while minimizing the actuating inputs together bounding the velocities can ensure soft and efficient functioning while improving the manipulator working performances. For this reason the objective function is formed as:

$$L = \frac{1}{2}\left(r_1 u_1^2 + r_2 u_2^2 + \sum_{i=1}^{8} w_i x_i^2 \right) \tag{15}$$

Then, by considering the costate vector as $Y = [y_1 \quad y_2 \quad \ldots \quad y_8]$, the Hamiltonian function can be expressed from as:

$$H = \frac{1}{2}\left(r_1 u_1^2 + r_2 u_2^2 + \sum_{i=1}^{8} w_i x_i^2 \right) + \sum_{i=1}^{8} y_i \dot{x}_i\,, \tag{16}$$

where $\dot{x}_i, i = 1,\ldots,8$ can be substituted from Eq. (12).

In order to derive the equations associated with optimality conditions, penalty matrices can be selected as follows:

$$W_1 = diag(w_1, w_3, w_5, w_7) \\ W_2 = diag(w_2, w_4, w_6, w_8) \\ R = diag(r_1, r_2) \tag{17}$$

Using Eq. (10), differentiating the Hamiltonian function with respect to the states, result in costate equations as follows:

$$\dot{y}_i = -\frac{\partial H}{\partial x_i}\,, \ i = 1,\cdots,8 \tag{18}$$

Control functions are computing by differentiating the Hamiltonian function with respect to control and setting the derivative equal to zero. After using the extreme bound of control for each motor, by substituting the obtained control equations into (12) and (18), obtained 16 nonlinear ordinary differential equations that with

16 boundary conditions given in (14), constructs a two point boundary value problem that can be solved using the BVP4C command in MATLAB®.

4.2 Path Optimization

These simulations are carried out at three cases. In the first case, path planning is performed for generating the minimum vibration trajectory; second case is deal with the designing of minimum effort path, the effect of joint stiffness in performance characteristics of robot is investigated in last simulation.

In this simulations the payload is considered to be 3 kg and it must be carried from an initial to final point during the overall time t_f = 1.5s, while the mobile base is initially at point ($x_0 = 0.6\text{m}$, $y_0 = 0.8\text{m}$, $\theta_0 = 0$) and moves to final position ($x_f = 1.4\text{m}$, $y_f = 1.2\text{m}$, θ_0 (end) = 0.44 rad).

4.3 Minimum Effort Trajectory

In this case finding the optimal path with minimum effort is considered, therefore, because of the matter that increasing W_2 decreases the proportion of weighting matrix R and the result of this is increasing the control values, penalty matrices can be considered to be R=diag (1), W_1=[1] and W_2=[0]. End effector trajectories in the XY plane are shown in Fig. 4.2. Figs. 4.3 and 4.4 show the angular position and velocity of joints with respect to time respectively. The computed torque is plotted in Fig. 4.5. Angular position and velocity of wheels are shown in figure 4.6. Wheels 1 with 2 also 3 with 4 have the same figures because of symmetry in system.

4.4 Minimum Speed Trajectory

In this case study, the problem of bounding the velocity of joints in minimum magnitude is considered and the obtained results are compared with minimum effort ones. Simulations are performed for selecting weighting factors as: R=diag (1), W_1 = [1] and W_2 = [1000]. The results of these simulations are illustrated in same figures of case 1. as shown in figures: a) by increasing the W_2 the angular position change to approach approximately to a straight line, b) as it can be seen, for the minimum speed trajectory, the oscillation amplitudes in velocity curves has been reduced considerably, but the magnitude of motor torques has been increased. It means that for achieving a smoother path, more effort must be applied, c) it is clearly observed that the minimum speed path is smoother than the minimum effort path, d) As shown in figures, in this method, via changing the penalty matrices values, various optimal trajectories with different specifications can be obtained, Consequently, this method enables the designer to compromise between different objectives by considering the proper penalty matrices.

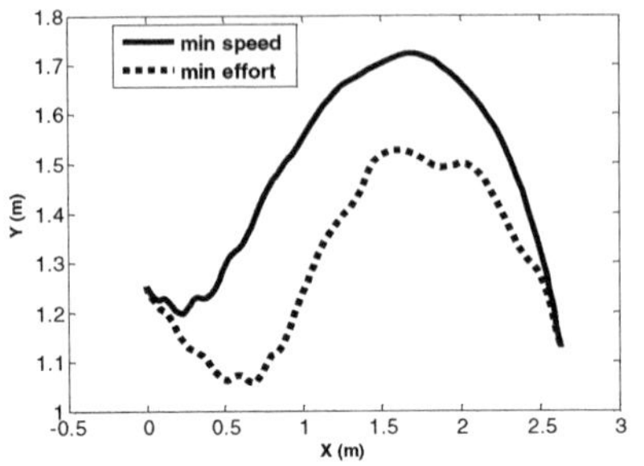

Fig. 4.2 End effector trajectory in XY plane

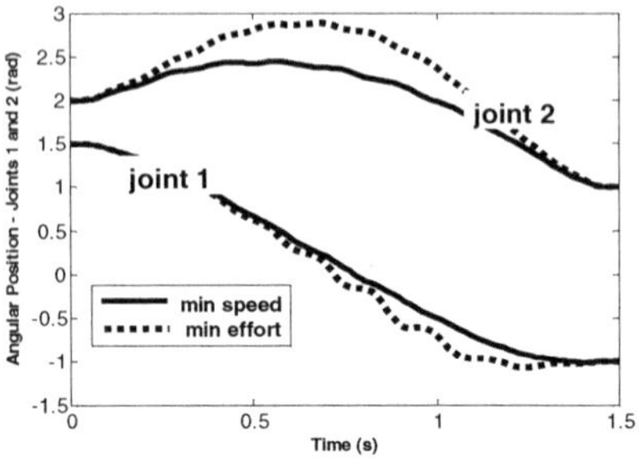

Fig. 4.3 Angular positions of joints

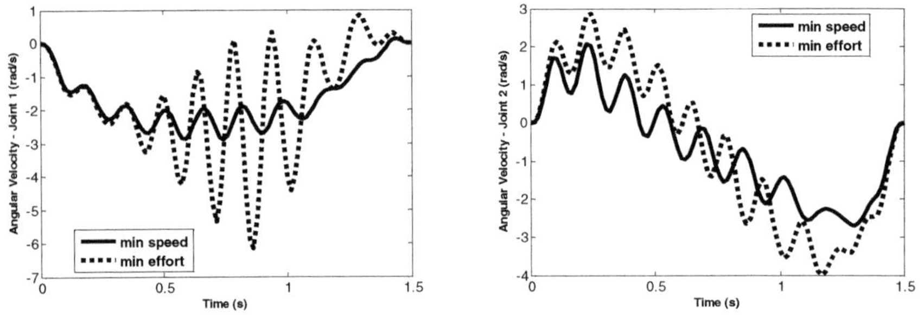

Fig. 4.4 Angular velocities of joints

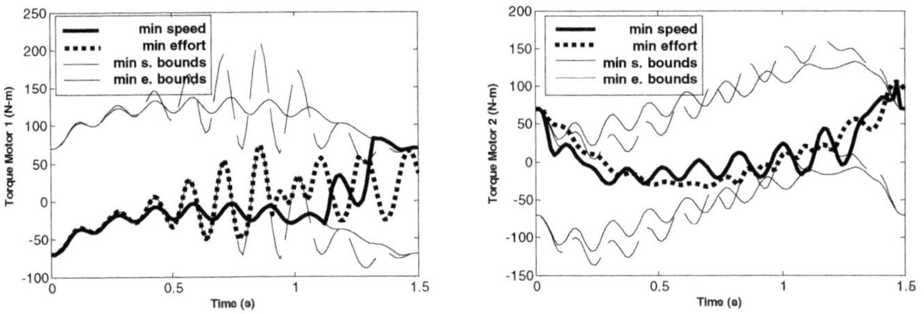

Fig. 4.5 Torque of motor 1

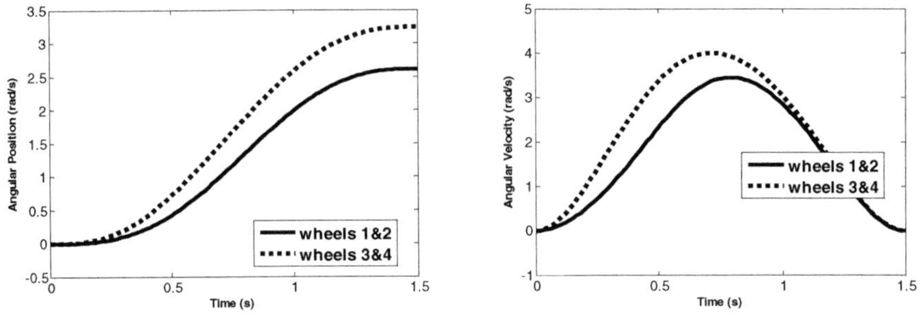

Fig. 4.6 Angular position and velocity of wheels

4.5 Different Joint Stiffness Trajectory

In this section, the effect of joint stiffness in performance characteristics of robot is investigated. Penalty matrices are considered to be :W_1=W_2=[1] and R=diag(1). The values of K using in simulation are given in Table 4.2.

Table 4.2 The values of K used in simulation

case	1	2	3
K	Diag(500)	Diag(1000)	Diag(5000)

Fig. 4.7 presents angular velocities of the results. It is observed from figure that increasing the joint stiffness caused the reducing oscillatory behavior of system. In addition, it can be seen growing the elasticity in joints enlarges bounds of velocity. The angular velocities of links and motors in case 2 are given in Fig. 4.8. It shows that both the link angular velocities have deviations from their respective motor angular velocities. Thus, it is clear that joint flexibility significantly affects the link vibrations.

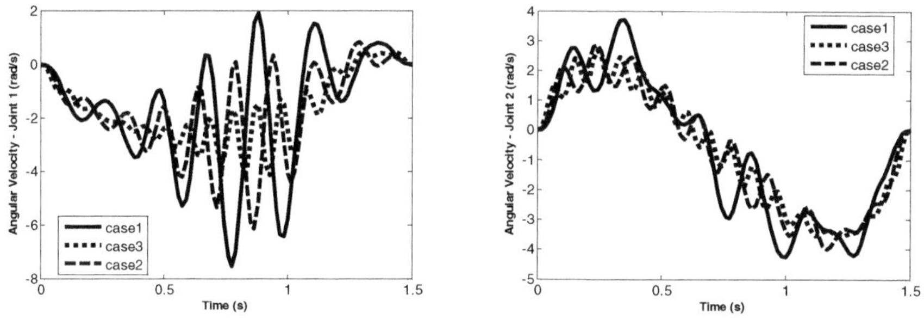

Fig. 4.7 Angular velocities of joints

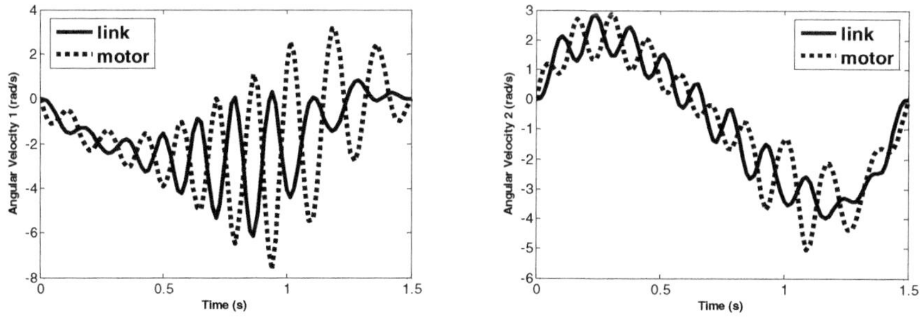

Fig. 4.8 Angular velocities of links and motors

5 Conclusions

In this paper, formulation of designing the path for wheeled mobile flexible joints manipulator in point-to-point motion is presented based on Pontryagin's minimum principle. Therefore, an efficient solution on the basis of TPBVP solution is proposed to optimize the path in order to achieve the predefined objective. The simulations on path planning of the minimum vibration trajectory, designing of minimum effort path and investigate on effect of joint stiffness in performance characteristics of robot are performed. Results show by defining the proper objective function and changing the penalty matrices can be achieve the desired requirements. The obtained results illustrate the power and efficiency of the method to overcome the high nonlinearity nature of the optimization problem which with other methods, it may be very difficult or impossible. The optimal trajectory and corresponding input control obtained using this method can be used as a reference signal and feed forward command in control structure of such manipulators.

References

1. Desai, J.P., Kumar, V.: Motion planning of nonholonomic cooperating mobile manipulators. Journal of Robotic Systems 16(10), 557–579 (1999)
2. Bloch, A.M.: Nonholonomic mechanics and control. Springer, Heidelberg (2003)
3. Korayem, M.H., Ghariblu, H., Basu, A.: Dynamic load-carrying capacity of mobile-base flexible joint manipulators. Int. J. Adv. Manuf. Technol. 25, 62–70 (2005)
4. Korayem, M.H., Ghariblu, H.: Analysis of wheeled mobile flexible manipulator dynamic motions with maximum load carrying capacities. Robot. auton. syst. 48(2-3), 63–76 (2004)
5. Bessonnet, G., Chessé, S.: Optimal dynamics of actuated kinematic chains, Part 2: Problem statements and computational aspects. European Journal of Mechanics A/Solids 24, 472–490 (2005)
6. Korayem, M.H., Rahimi Nohooji, H.: Trajectory optimization of flexible mobile manipulators using open-loop optimal control method. In: Xiong, C., Liu, H., Huang, Y., Xiong, Y. (eds.) ICIRA 2008, Part I. LNCS (LNAI), vol. 5314, pp. 54–63. Springer, Heidelberg (2008)

Finite Element Method and Optimal Control Theory for Path Planning of Elastic Manipulators

M.H. Korayem, M. Haghpanahi, H.N. Rahimi, and A. Nikoobin

Abstract. Planning of robot trajectory is a very complex task that plays a crucial role in design and application of robots in task space. This paper is concerned with path planning of flexible robot arms for a given two-end-point task in point-to-point motion, based on indirect solution of optimal control problem. We employ the finite element method to modeling and deriving the dynamic equations of robot manipulator with flexible link, so in the presence of all nonlinear terms in dynamic equations open loop optimal control approach is a good candidate for generating the path that optimizes the end effector trajectory. Then the Hamiltonian function is formed and the necessary conditions for optimality are derived from the Pontryagin's minimum principle. The obtained equations establish a two point boundary value problem which is solved by numerical techniques. Finally, simulations for a two-link planar manipulator with flexible links are carried out to investigate the efficiency of the presented method. The results illustrate the power and efficiency of the method to overcome the high nonlinearity nature of the problem.

Keywords: Flexible Manipulator, Finite Element, Optimal Trajectory, Optimal Control.

1 Introduction

Flexible robot arms have some advantages over rigid ones, such as their capability to assure faster motions and a higher ratio of payload to arm weight. However, due to the flexible nature of the system, their dynamic equations are highly non-linear and complex.

M.H. Korayem, H.N. Rahimi, and A. Nikoobin
Robotic Research Lab, Mechanical Engineering Department, Iran University of Science and Technology
e-mail: hkorayem@iust.ac.ir, hamedrahimi.n@gmail.com, anikoobin@iust.ac.ir

M. Haghpanahi
Biomechanics Lab, Mechanical Department, Iran University of Science and Technology
e-mail: mhaghpanahi@yahoo.com

K. Nakamatsu et al. (Eds.): New Advan. in Intel. Decision Techno., SCI 199, pp. 117–126.
springerlink.com

Wang et al. have solved the optimal control problem with direct method using the B-Spline functions in order to determine the maximum payload of a rigid manipulator [1]. The assumed mode expansion method is used by Sasiadek and Green [2, 3] to derive the dynamic equation of fixed base flexible manipulator. In [4, 5] a formulation based on Iterative Linear Programming (ILP) is presented to determine the Maximum Allowable Dynamic Load (MADL) of flexible manipulators. Indeed, the linearizing procedure and its convergence to the proper answer is a challenging issue, especially when nonlinear terms are large and fluctuating, e.g. in problems with consideration of flexibility in links or having high speed motion. As a result, in none of the previous mentioned work which is based on the ILP method, the link flexibility has not been considered either in the dynamic equation or simulation procedure.

None of these published works have used Finite Element Method (FEM) to model and analysis for their systems. One of the main advantages of FEM over the most of other approximate solution methods to modeling the flexible links is the fact that in FEM the connection are supposed to be clamp-free with minimum two mode shape per link. Another significant advantage of FEM, especially over analytical solution techniques is the ease with which nonlinear conditions can be handled. The finite element method has been used to solve very complex structural engineering problems during the past years. The maximum payload of flexible mobile manipulator is determined along the given trajectory by using the finite element approach in [6], so finding the optimal path is not considered in it.

Optimal control can be used in both open loop and close-loop strategies. However, because of the off-line nature of the open loop optimal control in spite of the close-loop ones, many difficulties such as system nonlinearities and all types of constraints may be catered for and implemented easily, so it generally used in analyzing nonlinear systems such as trajectory optimization of different types of robots [7, 8]. It solved by direct and indirect approaches. But, since direct method leads to the approximate solution and this approach is time consuming and quite ineffective due to the large number of parameters involved [9], indirect methods is a good candidate for the cases where the system has a large number of degree of freedom or optimization of the various objectives is targeted [10].

Open-loop optimal control method is proposed as an approach for trajectory optimization of flexible link mobile manipulator for a given two-end-point task in point-to-point motion [11]. But in mentioned paper combined Euler–Lagrange formulation and assumed modes method is used for driving the equation of motions with considering the simply support mode shape and one mode per link. So beside the advantages of this paper over than ILP based ones it can not expressed realistically the behavior of links besides it connection to the motors.

In this paper, for path planning of Elastic manipulators, an indirect solution of the optimal control problem is employed. Dynamic equations are derived using the FEM. Hamiltonian function is formed, and necessary conditions for optimality are obtained from the Pontryagin's minimum principle. These equations establish a Two Point Boundary Value Problem (TPBVP) solved by MATLAB. In comparison with other method the open-loop optimal control method does not require linearizing the equations, differentiating with respect to joint parameters and using of a fixed-order polynomial as the solution form. Finally, a two-link elastic manipulator is simulated to demonstrate the capability of the method.

2 Modeling of Robot Arms with Multiple Flexible Links

The finite element method is used to derive dynamic equations of flexible manipulators. The overall approach involves treating each link of the manipulator as an assemblage of n elements of length L_i. For each of these elements the kinetic energy T_{ij} and potential energy V_{ij}, (where i and j are referred to the number of links and the number of elements respectively) are computed in terms of a selected system of n generalized variables $q = (q_1, q_2, \ldots, q_n)$ and their rate of change $\dot{q}$. These energies are then combined to obtain the total kinetic energy, T, and potential energy, V, for the entire system. Finally, using Lagrange equations the equations can be written in compact form as:

$$M(q)\ddot{q} + C(q,\dot{q}) + G(q) = U, \tag{1}$$

By defining the state vector as:

$$X = [X_1 \quad X_2]^T = [q \quad \dot{q}]^T, \tag{2}$$

Eq. (1) can be rewritten in state space form as:

$$\dot{X} = [\dot{X}_1 \quad \dot{X}_2]^T = [X_2 \quad N(X) + D(X)U], \tag{3}$$

where $N = -M^{-1}(C(X_1, X_2) + G(X_1))$ and $D = M^{-1}$. Then optimal control problem is imposed to determine the position and velocity variable $X_1(t)$ and $X_2(t)$ and the joint torque $U(t)$ which optimize a well-defined performance measure when the model is given in Eq. (3)

3 Formulation of the Optimal Control Problem

The basic idea to improve the formulation is to find the optimal path for a specified payload. For the sake of this, the following objective function is considered

$$\underset{U(t)}{Minimize}\, J_0 = \int_{t_0}^{t_f} L(X,U)dt, \tag{4}$$

where

$$L(X,U) = \frac{1}{2}\left\|X_2\right\|_W^2 + \frac{1}{2}\left\|U\right\|_R^2. \tag{5}$$

Integrand *L(.)* is a smooth, differentiable function in the arguments, $\| X \|_K^2 = X^T KX$ is the generalized squared norm, *W* is symmetric, positive semi-definite ($k\times k$) weighting matrix and *R* is symmetric, positive definite ($k\times k$) matrix. The objective function specified by Eqs. (4) and (5) is minimized over the entire duration of the motion. The designer can decide on the relative importance among the angular position, angular velocity and control effort by the numerical choice of *W* and *R* which can also be used to convert the dimensions of the terms to consistent units. According to the Pontryagin's minimum principle, the following conditions must be satisfied,

$$\dot{X} = \partial H/\partial \psi \tag{6}$$

$$\dot{\psi} = -\partial H/\partial X \tag{7}$$

$$0 = \partial H/\partial U \tag{8}$$

where by defining the nonzero costate vector $\psi = \left[\psi_1^T \quad \psi_2^T\right]^T$, the Hamiltonian function can be obtained as:

$$\mathrm{H}(\mathrm{X},\mathrm{U},\psi) = 0.5(\left\|X_2\right\|_W^2 + \left\|U\right\|_R^2) + \psi_1^T X_2 + \psi_2^T\left[N(X) + D(X)U\right]. \tag{9}$$

So, according to Eq. (7), the optimality conditions can be obtained by differentiating the Hamiltonian function with respect to states, costates and control as follows:

$$\left[\dot{X}_1 \quad \dot{X}_2\right]^T = \left[X_2 \quad N(X) + D(X)U\right]^T \tag{10}$$

$$\left[\dot{\psi}_1 \quad \dot{\psi}_2\right]^T = -\left[\partial H/\partial X_1 \quad \partial H/\partial X_2\right]^T \tag{11}$$

$$RU + D^T \psi_2 = 0 \tag{12}$$

The control values are limited with upper and lower bounds, so using Eq. (12) the optimal control are given by:

$$U = \begin{cases} U^+ & -R^{-1}D^T\psi_2 > U^+ \\ -R^{-1}D^T\psi_2 & U^- < -R^{-1}D^T\psi_2 < U^+ \\ U^- & -R^{-1}D^T\psi_2 < U^- \end{cases}. \tag{13}$$

The actuators which are used for medium and small size manipulators are the permanent magnet D.C. motor. The torque speed characteristic of such D.C. motors may be represented by the following linear equation:

$$U^{+} = K_1 - K_2 X_2 \ , U^{-} = -K_1 - K_2 X_2 \, . \tag{14}$$

Where $K_1 = [\tau_{s1} \ \tau_{s2} \cdots \tau_{sn}]^T$, $K_2 = dig[\tau_{s1}/\omega_{m1} \cdots \tau_{sn}/\omega_{mn}]$, $\dot{\theta} = [\dot{\theta}_1 \ \ \dot{\theta}_2 \ \ \ldots \ \ \dot{\theta}_n]^T$, τ_s is the stall torque and ω_m is the maximum no-load speed of the motor. The boundary values will be expressed as below:

$$\begin{aligned} &X_1(0) = X_{10}, \ X_2(0) = X_{20} \ ; \\ &X_1(t_f) = X_{1f}, \ X_2(t_f) = X_{2f} \end{aligned} \tag{15}$$

In this formulation, for a specified payload value, $4m$ differential equations given in Eq. (11) are used to determine the $4m$ state and costate variables. The set of differential Eq.(11), the control law Eq. (13), and the boundary conditions construct a standard form of TPBVP, which is solvable with available commands in different software such as MATLAB or MATEMATHICA.

4 Simulation for a Flexible Planar Manipulator

In this Section, simulations are carried out for a two-link planar flexible manipulator as shows in figure 4.1. This manipulator must carry a concentrated payload with mass of 1 kg during the overall time t_f = 1 second.

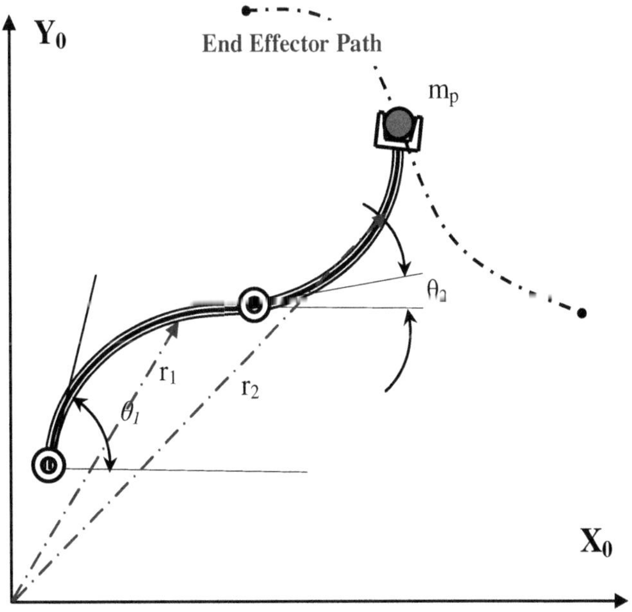

Fig. 4.1 Two-link manipulator with flexible links

The necessary parameters of the flexible manipulator are summarized in the Table 4.1.

Table 4.1 Simulation parameters

Parameter	Value	Unit
Length of Links	$L_1 = L_2 = 1$	m
Mass of Links	$m_1 = m_2 = 5$	Kg
Moment of Area of Links	$I_1 = I_2 = 5e\text{-}9$	m^4
Module of Elasticity of Links	$E_1 = E_2 = 2e10$	$Kg.m^2$
Max. no Load Speed of Actuators	$w_{s1} = w_{s2} = 3.5$	rad/s
Actuator Stall Torque	$\tau_{s1} = \tau_{s2} = 30$	N.m

By defining the state vectors as follows:

$$X_1 = Q^T = [x_1 \quad x_3 \quad x_5 \quad x_7 \quad x_9 \quad x_{11}]^T,$$
$$X_2 = \dot{Q}^T = [x_2 \quad x_4 \quad x_6 \quad x_8 \quad x_{10} \quad x_{12}]^T. \tag{16}$$

The sate space form of Eq. (16) can be written as:

$$\dot{x}_{2i-1} = x_{2i} \ , \ \dot{x}_{2i} = F_2(i) \quad ; i = 1 \ldots 6 , \tag{17}$$

where $F_2(i)$ can be obtained from Eq. (3). And the boundary condition can be expressed as:

$$\begin{aligned} &x_1(0) = \pi/2 \ \text{rad} , \ x_3(0) = 2\times\pi/3 \ \text{rad} \\ &x_1(t_f) = \pi/6 \ \text{rad}, \quad x_3(t_f) = \pi/3 \ \text{rad} \\ &x_{2i}(0) = x_{2i}(t_f) = 0, \quad i = 1 \ldots 6 \\ &x_5(0) = x_5(t_f) = x_7(0) = x_7(t_f) = 0 \end{aligned} \tag{18}$$

In order to derive the equations associated with optimality conditions, penalty matrices can be selected as:

$$\begin{aligned} &W = diag(w_1, w_2, w_3, w_4, w_5, w_6); \\ &R = diag(r_1, r_2). \end{aligned} \tag{19}$$

So the objective function is obtained by substituting Eq. (19) Into Eq. (5) as below

$$L = \frac{1}{2}\left(r_1 u_1^2 + r_2 u_2^2 + \sum_{i=1}^{6} w_i x_{2i}^2 \right). \tag{20}$$

Then, by considering the costate vector as $\varphi = [\varphi_1 \quad \varphi_2 \quad \ldots \quad \varphi_{12}]$, the Hamiltonian function can be expressed from as:

$$H = \frac{1}{2}\left(r_1\tau_1^2 + r_2\tau_2^2 + \sum_{i=1}^{6} w_i x_{2i}^2 \right) + \sum_{i=1}^{12} \varphi_i \dot{x}_i \,, \tag{21}$$

where $\dot{x}_i, i = 1,\ldots,12$ can be substituted from Eq. (17). Using Eq. (11) differentiating the Hamiltonian function with respect to the states, result in costate equations as follows:

$$\dot{\varphi}_i = -\frac{\partial H}{\partial x_i} \,,\; i = 1,\cdots,12 \tag{22}$$

The control function in the admissible interval can be computed using Eq. (11), by differentiating the Hamiltonian function with respect to the torques (τ_1, τ_2) and setting the derivative equal to zero.

Then, by applying motors torque limitation, the optimal control becomes:

$$PU_i = \begin{cases} U_i^+ & U_i > U^+ \\ U_i & otherwise \;\; ; i = 1,2 \,. \\ U_i^- & U_i < U^- \end{cases} \tag{23}$$

After that, from Eq. (14) the extrimal bound of control for each motor becomes:

$$\begin{aligned} U_1^+ &= k_{11} - k_{12}x_2 \quad ; \quad U_1^- = -k_{11} - k_{12}x_2 \\ U_2^+ &= k_{21} - k_{22}x_4 \quad ; \quad U_2^- = -k_{21} - k_{22}x_4 \end{aligned} \tag{24}$$

Consequently, substituting computed control equations (23) and (24) into Eqs. (17) and (22), obtain 16 nonlinear ordinary differential equations that with 16 boundary conditions given in Eq. (18), constructs a two point boundary value problem. A number of methods exist for solving these problems including shooting, collocation, and finite difference methods. In this study, BVP4C command in MATLAB® which is based on the collocation method is used to solve the obtained problem. The details of the numerical technique used in MATLAB to solve the TPBVP are given in [12]. This problem can be solved using the BVP4C command in MATLAB®.

In this simulation, the payload is considered to be 1 kg and the purpose is to find the optimal path between initial and final point of payload in such a way that the minimum amount of control value can be applied and the angular velocity values of motors be bounded in 2.

By considering the penalty matrices as *W*= (2, 2, 0, 0, 0, 0) and *R*=diag (0.1), the first path with appropriate amount of control value is determined, but the angular velocities are greater than 2 rad/s. Therefore for decreasing the velocities, *W* must be increased. A range of values of *W*= (*w*, *w*, 0, 0, 0, 0) used in simulation are given in Table 4.2.

Table 4.2 The values of w used in simulation

Case	1	2	3
w	2	50	500

Figs. 4.2 and 4.3 show the angular velocities of the first and second joints. It can be found that by increasing the w, maximum values of angular velocity reduce from -8.6 rad/s to -1.6 rad/s.

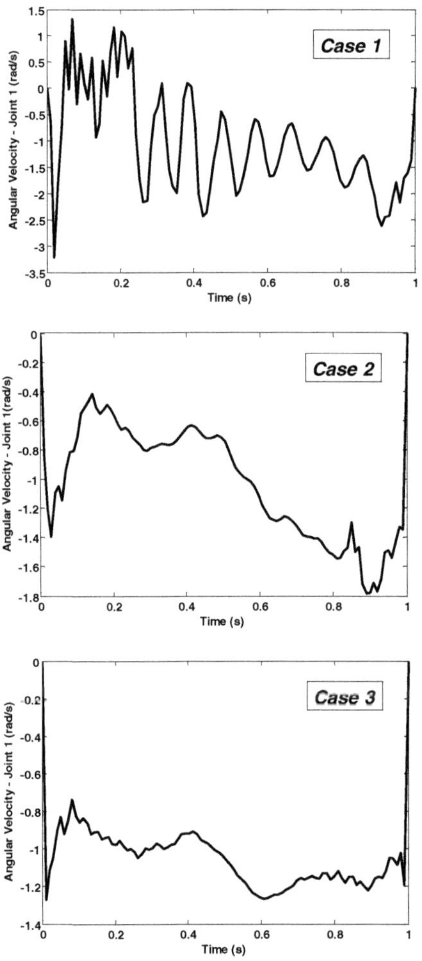

Fig. 4.2 Angular velocity-joint 1

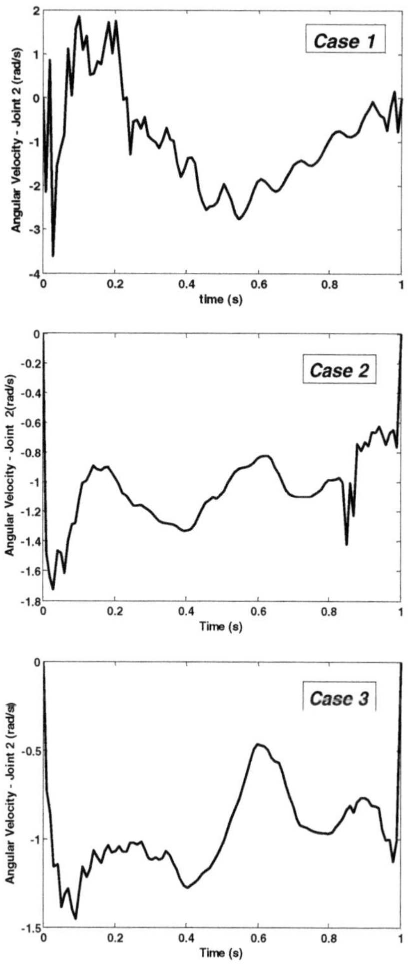

Fig. 4.3 Angular velocity-joint 2

Computed torqueses of motors are plotted in Figs. 4.4 and 4.5. As it can be seen, increasing the *w* causes to raise the torques. This result is predictable, because increasing the *w*, decreases the proportion of R and the result of this is increasing the control values.

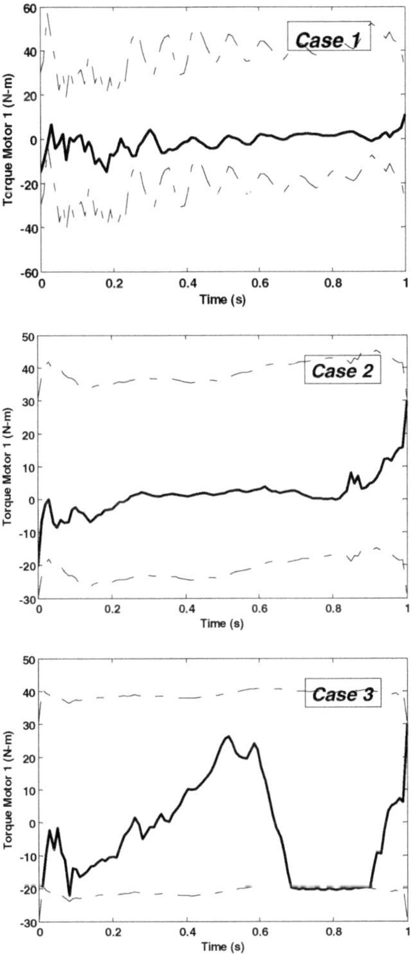

Fig. 4.4 Torque of motor 1

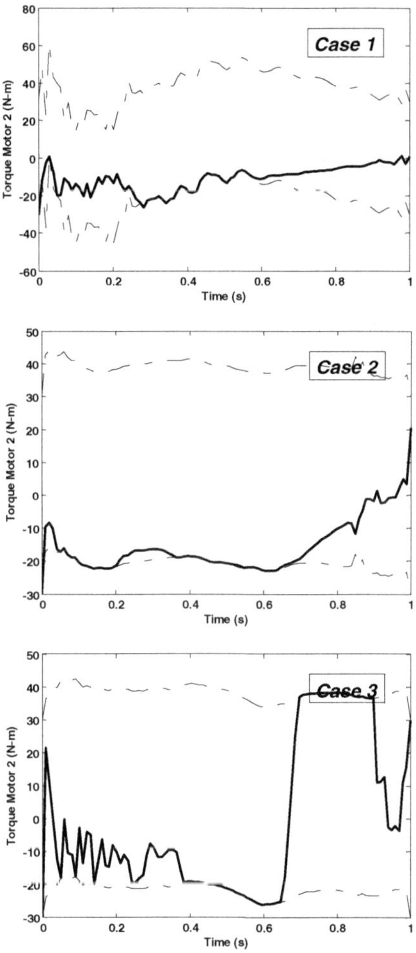

Fig. 4.5 Torque of motor 2

As show in figures (4.2- 4.5), there is not the solution that satisfies all the desired objectives simultaneously, e.g. the optimal path with minimum effort has maximum velocity and the optimal path with minimum velocity has maximum effort. Consequently, in this method, designer compromises between different objectives by considering the proper penalty matrices.

5 Conclusions

In this paper, formulation for the path planning of flexible robot arms in point-to-point motion, based on the open-loop optimal control approach is presented. Dynamic equations are derived using finite element method and an efficient solution on the basis of TPBVP is proposed. In comparison with other method the open-loop optimal control method does not require linearizing the equations, differentiating with respect to joint parameters and using of a fixed-order polynomial as the solution form. Moreover via changing the penalty matrices values, various optimal trajectories with different specifications can be obtained which able the designer to select a suitable path through a set of obtained paths. By a simulation study the application and validity of the algorithm is investigated. Results illustrate the power and capability of the method to overcome the high nonlinearity nature of the optimization problem in spite of using complete form of obtained nonlinear equations.

References

1. Wang Chia-Yu, E., Timoszyk, W.K., Bobrow, J.E.: Payload maximization for open chained manipulator: Finding motions for a puma 762 robot. IEEE Transactions on Robotics and Automation 17(2) (2001)
2. Green, A., Sasiadek, J.Z.: Dynamics and trajectory tracking control of a two-link robot manipulator. Journal of Vibration and Control 10(10), 1415–1440 (2004)
3. Green, A., Sasiadek, J.Z.: Robot manipulator control for rigid and assumed mode flexible dynamics models. In: AIAA Guidance, Navigation, and Control Conference and Exhibit (2003)
4. Korayem, M.H., Ghariblu, H.: Analysis of wheeled mobile flexible manipulator dynamic motions with maximum load carrying capacities. Robot auton. syst. 48(2-3), 63–76 (2004)
5. Gariblu, H., Korayem, M.H.: Trajectory optimization of flexible mobile manipulators. Robotica 24(3), 333–335 (2006)
6. Korayem, M.H., Heidari, A., Nikoobin, A.: Maximum Allowable Load of Flexible Mobile Manipulators Using Finite Element Approach. Journal of AMT 36(5-6), 606–617 (2008)
7. Wilson, D.G., Robinett, R.D., Eisler, G.R.: Discrete dynamic programming for optimized path planning of flexible robots. In: International Conference on intelligent Robot and Systems, pp. 2918–2923 (2004)
8. Korayem, M.H., Nikoobin, A.: A Maximum payload for flexible joint manipulators in point-to-point task using optimal control approach. International Journal of AMT 38(9-10), 1045–1060 (2008)
9. Hull, D.G.: Conversion of optimal control problems into parameter optimization problems. J. Guid. Control. Dynam. 20(1), 57–60 (1997)
10. Kirk, D.E.: Optimal control theory, an introduction. Prentice-Hall Inc., Upper Saddle River (1970)
11. Korayem, M.H., Rahimi Nohooji, H.: Trajectory optimization of flexible mobile manipulators using open-loop optimal control method. In: Xiong, C., Liu, H., Huang, Y., Xiong, Y. (eds.) ICIRA 2008, Part I. LNCS (LNAI), vol. 5314, pp. 54–63. Springer, Heidelberg (2008)
12. Shampine, L.F., Reichelt, M.W., Kierzenka, J.: Solving boundary value problems for ordinary differential equations in MATLAB with bvp4c tutorial, `http://www.mathworks.com/bvp`

Belief-Based Stability in Non-transferable Utility Coalition Formation with Uncertainty

Chi-Kong Chan and Ho-fung Leung

Abstract. Coalition stability is an important concept in coalition formation. One common assumption in many stability criteria in non-transferable utility games is that the preference of each agent is publicly known so that a coalition is said to be stable if there is no objections by any sub-group of agents according to the publicly known preferences. However, in many software agent applications, this assumption is not true. Instead, agents are modeled as individuals with private belief and decisions are made according to those beliefs instead of common knowledge. There are two types of uncertainty here. First, uncertainty in beliefs regarding the environment means that agents are also uncertain about their preference. Second, an agent's actions can be influenced by his belief regarding other agents' preferences. Such uncertainties have impacts on the coalition's stability which is not reflected in the current stability criteria. In this paper, we extend the classic stability concept of the core by proposing new belief based stability criteria under uncertainty, and illustrate how the new concept can be used to analyze the stability of a new type of belief-based coalition formation game.

Keywords: Coalition Formation, Uncertainty, Private Beliefs. NTU Games.

1 Introduction

There are some commonly agreed upon principles of multi-agent systems. For instance, intelligent agents are *autonomous*, meaning that they can be viewed as independent entities that are able to achieve goals with only minimal supervision; computations and knowledge in multi-agent systems are typically *distributed* in nature and there are no central-controlling agents, and that the knowledge and beliefs of the agents are private and not accessible by other agents; agents are *not omniscient* and their belief depend on their past perception, which is both limited

Chi-Kong Chan and Ho-fung Leung
Department of Computer Science and Engineering
The Chinese University of Hong Kong
Hong Kong
e-mail: {chanck,lhf}@cse.cuhk.edu.hk

K. Nakamatsu et al. (Eds.): New Advan. in Intel. Decision Techno., SCI 199, pp. 127–138.
springerlink.com

and fallible; finally, intelligent agents have *social ability* and they are able to *form coalitions* in order to achieve tasks that cannot be done alone.

One important class of agent coalition formation problems are the semi-competitive systems, where the individual agents cooperate only if it is beneficial for them to do so. In these systems, the agents are assumed to be rational and self-interested. This means that each agent will act to maximize its own benefit only, based on its perceptions, beliefs and goals. An implication of this is that an agent can pursue any action to achieve its goals, even if that is achieved at a cost to other fellow agents. Such agents will not care about the overall system efficiency.

The problem of semi-competitive agent cooperation is studied in both computer science and economics. In computer science, protocols and mechanisms are created so that interaction and cooperation between agents can be made possible [3, 4]. However, having proposed the mechanisms, people need a way to measure the quality of the produced solutions of these mechanisms. To do this, many researchers rely on the stability concepts from game theory [2], such that a mechanism is considered to be good if it can produce many stable solutions. Several stability concepts exists, with the strictest and most common one being the core[1, 5].

Game theoretic concepts such as the core can make sure that any solutions produced by the mechanism satisfying those criteria are indeed stable even if the agents are self-interested. However, there is a drawback when they are applied to multi-agent coalition formation problems: most of these stability concepts have a *common knowledge* assumption, meaning that various characteristics of the game, including each individual agent's preference, are assumed to be known, with certainty, to both the agent himself and all other agents. However, this assumption is in conflict with the principle that agents' beliefs are both private and not omniscient. For example, agents in the commercial world may be reluctant to disclose their preferences to other agents and there may be uncertainties in their beliefs.

Such private and fallible beliefs result in two types of uncertainty. First, an agent's preference depends on the agent's beliefs, but in the real world, beliefs are sometimes uncertain, which means an agent is often uncertain about his preference as well. As an example, suppose that the preference of a company is to form a partnership with another company if they believe the economy will be good in the following year. However, they are uncertain about the economy as they are receiving conflicting forecasts from different experts. Then in this case, their preference regarding the partnership is uncertain also. We label this *internal uncertainty.*

Second, an agent's action depends also on his belief of the other agent's preference. Continuing with the above example, suppose that we have two such companies, with both of them believing that the economy will be good and hence prefer to form a coalition. But even in such a case, one more condition is required before the coalition will take place, Otherwise, even if this is the preferred option for both agents, as making such a proposal would be inconsistent with his belief regarding the other agents. We label this *external uncertainty.*

Thus, we see that such uncertainties have effects on the stability of a coalition formation game, but they are not modeled in the traditional game stability concepts such as the core and there is actually a discrepancy in assumptions when we apply these traditional game stability concepts to multi-agents coalition formation problems. Thus, we see that such uncertainties have effects on the stability of a

State s_1: 3G phones are well received.
a's preference: $\{a,b\} \succ_a \{a,c\} \succ_a \{a\} \succ_a \{a,b,c\}$
b's preference: $\{b\} \succ_b \{a,b\} \succ_b \{b,c\} \succ_b \{a,b,c\}$
c's preference: $\{c\} \succ_c \{a,c\} \succ_c \{b,c\} \succ_c \{a,b,c\}$

State s_2: 3G phones are badly received.
a's preference: $\{a,b,c\} \succ_a \{a,c\} \succ_a \{a,b\} \succ_a \{a\}$
b's preference: $\{a,b,c\} \succ_b \{b\} \succ_b \{b,c\} \succ_b \{a,b\}$
c's preference: $\{a,b,c\} \succ_c \{c\} \succ_c \{a,c\} \succ_c \{b,c\}$

Fig. 1 Agent Preferences in example 1

a's belief of b's preference: $\{b\} \succ \{a,b,c\}$; $\{a,b\} \succ \{b\}$
a's belief of c's preference: $\{c\} \succ \{a,b,c\}$; $\{b,c\} \succ \{a,c\}$
b's belief of a's preference: $\{a\} \succ \{a,b,c\}$; $\{a,b\} \succ \{a\}$
b's belief of c's preference: $\{c\} \succ \{a,b,c\}$
c's belief of a's preference: $\{a\} \succ \{a,b,c\}$; $\{a,b\} \succ \{a,c\}$
c's belief of b's preference: $\{b\} \succ \{a,b,c\}$

Fig. 2 Agents Mutual Beliefs in example 1

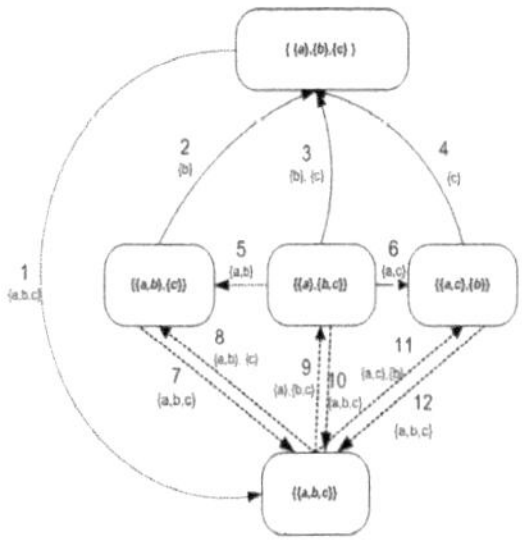

Fig. 3 Definite objections and potential objection

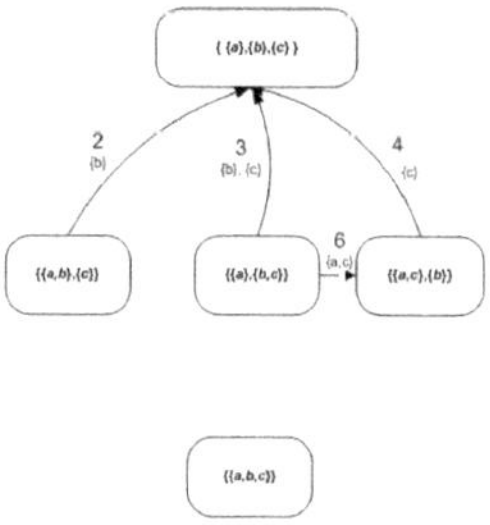

Fig. 4 Definite objections

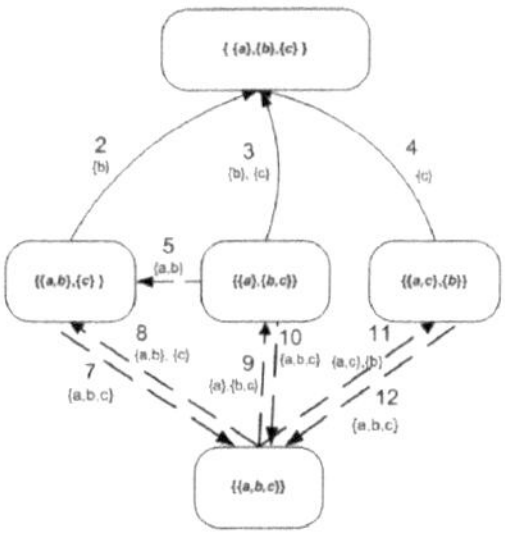

Fig. 5 Raisable objections

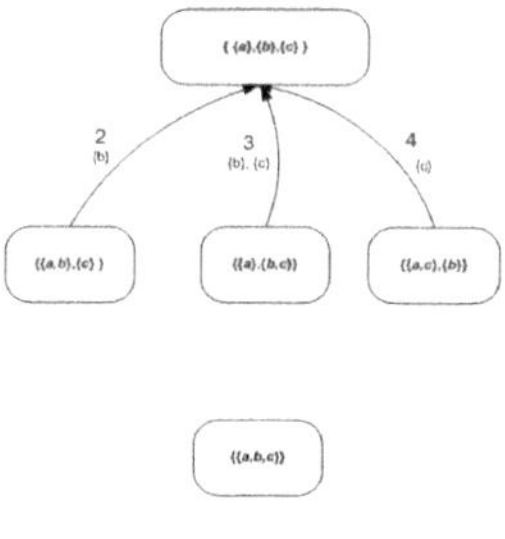

Fig. 6 Raisable definite objections

coalition formation game, but they are not modeled in the traditional game stability concepts such as the core and there is actually a discrepancy in assumptions when we apply these traditional game stability concepts to multi-agents coalition formation problems. Instead, what we need for multi-agent applications is a stability model that is based on private beliefs instead of common knowledge, and that agent omniscience is not assumed. We will develop such models in this paper. Our contribution in this work is as follows. We extend the classic stability concept of the non-transferable utility core by proposing a new stability criterion that are based on private beliefs instead of common knowledge and we develop a new stability concept where possible uncertainty in agents beliefs are accounted for. To do this, we propose a new type of game called *belief based non-transferable utility games with uncertainty*. Because of space limitation, we assume the reader is already familiar with the basic concepts in non-transferable utility games [1].

2 Motivating Example

Example 1. Three agents a, b, and c, representing three companies, are considering to enter the 3G phone market in a country, but they are not sure whether the new technology will be well received by the public. For simplicity, we assume there are only two possibilities: that the phones are either well-received or not, which are represented by environmental states s_1 and s_2 respectively. The agents have different preferences depending on which state they believe the world is in. The preferences of the agents, as described in figure 1, are totally ordered relations on the set of coalitions of which the agents are members. For instance, if agent b believes the phones will be well-received, then its first choice is to enter the market on his own (represented by the singleton coalition $\{b\}$), and his second choice is teaming up with a (represented by $\{a, b\}$), followed by teaming up with c (represented by $\{b, c\}$), and so on. But if he expects them to be badly-received, his first choice would be to form a coalition of all three agents for cost-saving purposes.

We assume that all three agents are unsure whether the phones will be well-received or not, which means that they may or may not be certain about their preferences. For example, although agent a certainly knows that he prefers the coalition $\{a, b\}$ to the coalition $\{a\}$, as his preference order for these two coalitions is the same regardless of the states, he is indecisive regarding the two coalitions $\{a, b\}$ and $\{a, c\}$, since that depends on something that he is unsure.

Apart from their preferences, each agent also maintains partial ordered relations representing his belief regarding other agents' preferences (figure 2), which are called the agents' external beliefs. For example, the second last entry in figure 2 says that agent c believes agent a certainly prefers the coalition $\{a\}$ to the three agents coalition $\{a, b, c\}$, and the coalition $\{a, b\}$ to $\{a, c\}$. However, each agent makes no further assumptions regarding the others' preferences that are not part of his external beliefs. For example, agent c has no opinion on whether agent a prefers $\{a, b, c\}$ to $\{a, c\}$ or not. External beliefs are not required to be correct.

We now discuss different types of objections that can arise out of this coalition formation game. Refer to figure 3, where all five possible coalition structures are shown as nodes, and possible objections indicated by labels on the edges. The edges are numbered for ease of description. First, consider edge 2. We see that the singleton coalition $\{b\}$ is certainly an objection to $\{\{a, b\}, \{c\}\}$ since agent b certainly prefers the coalition $\{b\}$ to coalition $\{a, b\}$ no matters what states it is in, thus resulting in the coalition structure $\{\{a\}, \{b\}, \{c\}\}$. The same can be said about edges 3 and 4, since agent c will certainly break away by similar argument, and so are edge 1 and 6. We label these *definite objections*, which are denoted by solid lines in figure 3.

However, edge 5, which represents the case that agent b breaks its partnership with agent c and teams up with agent a instead, is more problematic. According to the preference of agent b, he will make such a switch only if he believes the world is in state s_1, but this is something he is uncertain. Therefore, the validity of this objection depends on how agent b handles his internal uncertainty. The same can be said about edges 7 to 12. We label such cases *potential objections*, meaning that it is possible for these objections to arise, but conditional on how the internal uncertainties are resolved by at least one agent. In figure 3, potential objections are denoted by dashed lines.

So far we have not considered the agents' external beliefs. In some cases, the external beliefs will not affect the validity of the objections. Consider edge 2 again. Although agent a (incorrectly) believes that agent b prefers to stay in coalition $\{a, b\}$ instead of deviating and from coalition $\{b\}$, agent b will not be affected as he will raise an objection to the coalition $\{a, b\}$ regardless of agent a's (incorrect) beliefs. On the other hand, edge 1 is problematic if the external beliefs are taken into accounts. Although all three agents prefer the coalition $\{a, b, c\}$ to their original corresponding coalition in $\{\{a\}, \{b\}, \{c\}\}$ in one of the states (thus making it a potential objection), no agent can realize this because each of them happens to believe (incorrectly) that one or more of this potential partners will not agree to switch. Thus in this case, no agent is likely to raise any objections even though the objection is a valid one, because raising such an objection will be in conflict with their beliefs. Similar analysis can be applied to edge 6. We call this type of objections "non-raisable objections", and the rest "raisable objections". The raisable objections for this game are shown in figure 5. Again, there are two types of raisable objections depending on whether the objecting agents' preference are certain or not, and they are labeled as raisable definite objections if the objecting agents' preferences are certain, and raisable potential objections otherwise.

The stability of this game can be analyzed as follows. First, in cases where the agents do not require the objections to be consistent with the external beliefs, no coalition structure is stable if both potential and definite objections are allowed (this can be seen by the fact that each node in figure 3 has at least one out-going solid or dashed edge). However, there are two stable outcomes, namely $\{\{a, b, c\}\}$ and $\{\{a\}, \{b\}, \{c\}\}$ if only definite objections are considered (figure 4).

Second, in cases where the agents do not raise any objections that are inconsistent with their external beliefs (figure 5), we see that there is one stable outcome, namely $\{\{a\}, \{b\}, \{c\}\}$, if both raisable potential objection and raisable definite objections are allowed, and again there are two stable outcomes, namely $\{\{a\}, \{b\}, \{c\}\}$ and $\{\{a, b, c\}\}$ if we limit ourselves to raisable definite objections only (figure 6). This is a point worth mentioning as we now see that some "stable" outcomes are more stable than others: on one hand, we have outcomes such as $\{\{a\}, \{b\}, \{c\}\}$, which is definitely stable, and on the other hand, we have the definitely not stable coalition structure (CS) such as $\{\{a\}, \{b, c\}\}$, and somewhere in between, we have the "perhaps stable" outcomes such as $\{\{a\}, \{b\}, \{c\}\}$.

Thus, we see that the traditional stability concepts, which classify each CS as either stable (*e.g.*, in the core) or not, are insufficient in describing games such as this one. The reason is that we are facing a new type of games, where the agents' decisions are based on private and uncertain beliefs.

In the coming sections, this idea will be illustrated formally, and their significance will be explained. To do this, we will first define a new type of games called BNTUU games, and then we will define new stability concepts to capture the idea illustrated by the above example.

3 BNTUU Games

In this section we define a new type of coalition formation game called *belief based non-transferable utility games with uncertainty* (BNTUU games) which is an extension of the traditional NTU games. A BNTUU game is defined as a tuple $g = (N, S, (\succ_i^s), P, B, s^*)$, as follows. $N = \{1,...,n\}$ is a set of agents. Any subset $C \subseteq N$ is called a coalition. The goal of a BNTUU game is to partition the set of agents into a coalition structure (CS) of exhaustive and non-overlapping coalitions $CS = \{C_1,\ldots,C_k\}$. Given a coalition structure CS, we use $coal_i(CS)$ to denote the coalition in CS where the agent i is a member. There is a set S of environmental states and one of which is the prevailing state s^*. We assume that the outcome of a coalition is decided only by the coalition itself (*i.e.*, who its members are) and which state is the current the prevailing state s^*. For each state $s \in S$, each agent i has a strict total ordered *true preference* relation $\succ_i^s$ on the set of coalitions of which it is a member of, so that for any two coalitions C_1 and C_2, $i \in C_1 \cap C_2$, we have $C_1 \succ_i^s C_2$ if agent i prefers C_1 to C_2 when the current state is s.

It is assumed that the prevailing state is not observable by the agents. Instead, each agent i is associated with a set of states $S_i \subseteq S$, meaning that agent i has reasons to believe the world is in one of the states in S_i. Each S_i is called the agent i's environmental beliefs. The collection of all environmental beliefs of the agent is called an environmental belief profile $P = \{S_1,...,S_n\}$. In general, we do not require s^* to be in S_i. In example 1, we have $S_i = \{s_1, s_2\}$ for each agent $i \in \{a,b,c\}$.

To capture the uncertainty in an agent's preference, we define an agent's *certain preferences* and *uncertain preferences* as follows. For any two coalitions C_1 and C_2, we write $C_1 \rhd_i C_2$ if agent i certainly prefers C_1 to C_2 without uncertainty, and write $C_1 \lhd\rhd_i C_2$ if the agent may prefer C_1 to C_2, but this is uncertain and may be conditioned on other factors such as the agent's decision making strategy amidst his internal uncertainty, as follows:

Definition 1 (Agent's Certain Preference). Given an agent i and two coalitions C_1 and C_2, $i \in C_1 \cap C_2$, we say agent i certainly prefers C_1 to C_2, written $C_1 \rhd_i C_2$, if *i)* there exists $s \in S$ such that $s \in S_i$ and $C_1 \succ_i^s C_2$ holds, and *ii)* there does not exist any $s_2 \in S$, $s_2 \neq s$, such that both $s_2 \in S_i$ and $C_2 \succ_i^{s_2} C_1$ hold. ■

Definition 2 (Agent's Uncertain Preferences). Given an agent i and two coalitions C_1 and C_2, $i \in C_1 \cap C_2$, we say agent i is uncertain about his preferences regarding C_1 and C_2, written $C_1 \lhd\rhd_i C_2$, if there exists $S' \subseteq S$ and $s_1, s_2 \in S'$, such that $S_i = S'$ and both $C_1 \succ_i^{s_1} C_2$ and $C_2 \succ_i^{s_2} C_1$ holds. ■

The agents' external beliefs regarding other agents' preferences are represented by a relation bel_i, so that for two agents i and j, we write $bel_i(C_1 \rhd_j C_2)$ if agent i believes that agent j certainly prefers coalition C_1 to coalition C_2:

Definition 3 (Agent's External Beliefs). Given two agents i and j, and two coalitions C_1 and C_2, $j \in C_1 \cap C_2$, we write $bel_i(C_1 \rhd_j C_2)$ if agent i believes agent j certainly prefers C_1 to C_2 and we write $bel_i(C_1 \lhd\rhd_j C_2)$ if agent i believes agent j is uncertain regarding C_1 and C_2. ■

The set of external beliefs of all agents is represented by an external belief profile $B = \{bel_1, bel_2, ..., bel_n\}$.

We assume the following consistency requirements for the agents' external beliefs. First, each of an agent's certain preferences is represented by a corresponding "external" belief for itself. That is, $C_1 \rhd_i C_2 \Leftrightarrow bel_i(C_1 \rhd_i C_2)$. Second, conflicting external beliefs are not allowed. That is, for any two agents i and j, and two coalitions C_1 and C_2, $bel_i(C_1 \rhd_j C_2)$ implies that both $bel_i(C_2 \rhd_j C_1)$ and $bel_i(C_1 \lhd\rhd_j C_2)$ does not hold.

We are now ready to define our stability concepts. First, the core of a BNTUU game is defined just as its counterpart in NTU game:

Definition 4 (Core of BNTUU game). A coalition structure $CS=\{C_1,...C_k\}$ is in the core of a BNTUU game if there does not exist a coalition C', $C' \notin CS$, such that for all agents $i \in C'$, we have $C' \succ_i^{s^*} coal_i(CS)$ where s^* is the prevailing state. ■

The problem here is that the prevailing state is not observable to the agents, which limits the applicability of core in BNTUU games. Moreover, the agents' beliefs, which will affect the decisions of the agents, are not considered by the core, so that a CS that is in the core may not be stable in practice, and vice versa.

So instead, we need to define new belief-based stability concepts as follows. First, regarding the more simple case where internal uncertainties are considered but not external uncertainties, let us label the objections that involve uncertain preference as potential objections, and objections that involve only certain preference as definite objections. The idea is that, instead of using the traditional approach where every CS is either in the core or not, we now classify a CS as follows: We say a CS is in the *weak core* if it has no definite objections. We say it is in the *strong core* if it has no definite or potential objections. More formally:

Definition 5 (w-core: weak core of BNTUU game). A coalition structure $CS=\{C_1,...C_k\}$ is in the weak core (w-core) if there does not exist another coalition $C' \notin CS$, such that for all agents $i \in C'$, we have $C' \rhd_i coal_i(CS)$. ■

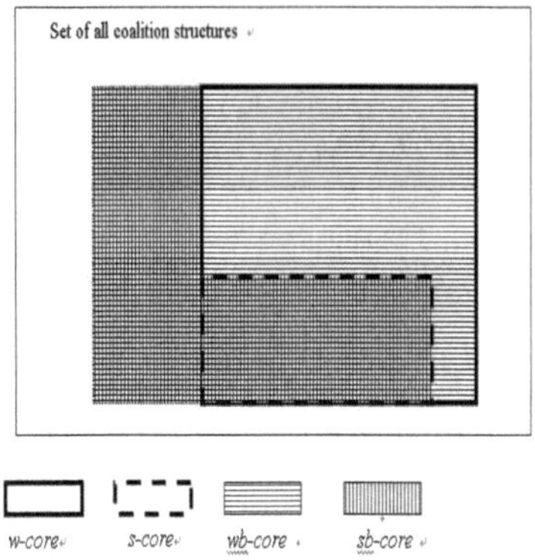

Fig. 7 The Belief Based Cores

Definition 6 (s-core: strong core of BNTUU game). A coalition structure $CS=\{C_1,\ldots C_k\}$ is in the strong core (s-core) if there does not exist another coalition $C' \notin CS$, such that for all agents $i \in C'$, we have $C' \rhd_i coal_i(CS)$ or $C' \lhd\rhd_i coal_i(CS)$. ■

Example 2. Continuing with example one, the strong core (s-core) is empty, and the coalition structures $\{\{a, b, c\}\}$ and $\{\{a\}, \{b\}, \{c\}\}$ are in the weak core (w-core). ■

We represent the *w-core* and *s-core* of a BNTUU game *g* as *w-core(g)* and *s-core(g)* respectively

Next, we are ready to define the external belief based equivalence of the above concepts, namely, the *external-belief based weak core* (wb-core) and the *external-belief based strong core* (sb-core). However, before we present the definitions, we first discuss the rationale behind them. The intuitive idea is that, for any objection to be considered as valid, in addition to the above conditions given in definition 5 and 6, we also require the objection to be "consistent" with the belief of at least one member of the deviating coalition. Otherwise, if the objection is inconsistent with every member's beliefs, then no agent will realize there is an objection and it will never be raised. Now, the remaining problem is that how to define an objection as being "consistent" with the agents' beliefs. Before we continue, we have the following intuitive assumption.

Assumption 1. In games where the external beliefs of each agent is empty (*i.e.*, each agent has no belief on the other agents' preferences), the external belief based weak core (see below) should reduce to the weak core (w-core) and the external belief based strong core (which we are going to define below) should reduce to the strong core (s-core) ■

This assumption is straightforward enough. If the external belief of an agent is empty, then in such a case, we expect the agent to behave just as when external beliefs are not considered in the coalition formation process.

Based on this assumption, we define the wb-core and sb-core as the set of solutions with *no conflicting raisable definite objections*, meaning that we require that for at least one member of the deviating coalition, the objection does not conflict with his external beliefs. That is, let *CS* be a coalition structure and let $C' \notin CS$ be a coalition (an objection), we call C' a raisable objection if there exists agent *j* in C' such that for all agent $k \in C'$, $bel_j(coal_k(CS) \rhd_k C')$ does *not* hold. We define the following external-belief based core, in accordance to option 3, as follows.

Definition 7 (wb-core: external-belief based weak core of BNTUU Game). A coalition structure $CS = \{C_1,\ldots,C_k\}$ is in the external-belief based weak core (wb-core) if there does not exist another coalition $C' \notin CS$ that satisfy the following two conditions: *i)* for all agents $i \in C'$, we have $C' \rhd_i coal_i(CS)$, and *ii)* there exists an agent $j \in C'$ such that, for all agents $k \in C'$, we do *not* have $bel_j(coal_k(CS) \rhd_k C')$. ■

Definition 8 (sb-core: external-belief based strong core of BNTUU Game). A coalition structure $CS=\{C_1,\ldots,C_k\}$ is in the external-belief based strong core (sb-core) if there does not exist another coalition $C' \notin CS$ that satisfy the following two conditions: *i)* for all agents $i \in C'$, we have $C' \rhd_i coal_i(CS)$ or $C' \lhd\rhd_i coal_i(CS)$, and *ii)* there exists an agent $j \in C'$ such that, for all agents $k \in C'$, we do *not* have $bel_j(coal_k(CS) \rhd_k C')$. ■

The wb-core requires every objection to fulfill two conditions. First, as in w-core, every agent in the deviating coalition must prefer the new coalition to the coalition of which he is currently a member. Second, the objection must not be in conflict with the external-beliefs of all members of the coalition simultaneously. That is, at least one agent in the coalition can see that the objection is a viable option and has a chance to succeed, and this agent will therefore raise the objection, which will then be accepted by all other members because of the first condition. The idea behind the sb-core is similar, except that potential objections are also allowed in the first condition. (The second condition, on the other hand, needs not be extended to cover potential objections for reasons already discussed above).

The four proposed concepts (s-core, w-core, sb-core, wb-core) are labeled collectively as the belief based cores.

4 Properties

Theorem 1. The wb-core is a superset of the sb-core. Similarly, the w-core is a superset of the s-core. The sb-core is a superset of the s-core whereas the wb-core is a superset of the w-core. (Refer to figure 7 for the complete picture).

The reason that, in general, the w-core of an BNTUU differs from the s-core, and the wb-core differs from the sb-core, is that the environmental beliefs of the agents are often imprecise, meaning that there are often more than one states that the agents consider to be the possible current state, *i.e.*, $|S_i| > 1$ for some $i \in N$. To understand the effects of such imprecision, we have the followings.

Definition 10 (Precision relation of environmental beliefs). Given two environmental beliefs $S_1, S_2 \subseteq S$, we say S_1 is *more precise* than S_2 if $|S_1| < |S_2|$.

Definition 11 (Precision relation of environmental belief profiles). Given two environmental beliefs profiles $P = \{ S_1, S_2, ..., S_n)$, $P' = \{ S_1', S_2', ..., S_n')$, we say P is *more precise* than P' if there exists $i \in N$ such that S_i is more precise than S_i', and either $S_j \equiv S_j'$ or S_j is more precise than S_j' for all $j \in N - \{i\}$.

Theorem 2. Given two games $g = (N, S, (\succ_i), P, B, s^1)$ and $g' = (N, S, (\succ_i), P', B, s^1)$, the following holds if P is *more precise* than P':

$$\text{w-core}(g) \subseteq \text{w-core}(g'), \qquad \text{wb-core}(g) \subseteq \text{wb-core}(g'),$$
$$\text{s-core}(g) \supseteq \text{s-core}(g'), \qquad \text{sb-core}(g) \supseteq \text{eb-core}(g')$$

Definition 12 (Accuracy relation of external beliefs). Given two external belief relations bel_1 and bel_2, we say bel_1 is *more accurate* than bel_2 if for all $j \in N$, $x, y \subseteq N$ we have the followings:

i) If $bel_1(x \rhd_j y)$ holds but $bel_2(x \rhd_j y)$ does not hold, then $x \rhd_j y$ holds.
ii) If $bel_2(x \rhd_j y)$ holds but $bel_1(x \rhd_j y)$ does not hold, then $x \rhd_j y$ does not holds.
iii) If $bel_1(x \lhd\rhd_j y)$ holds but $bel_2(x \lhd\rhd_j y)$ does not hold, then $x \rhd_j y$ holds.
iv) If $bel_2(x \lhd\rhd_j y)$ holds but $bel_1(x \lhd\rhd_j y)$ does not hold, then $x \lhd\rhd_j y$ does not holds.

Definition 13 (Accuracy relation of external beliefs profiles). Given two external beliefs profiles $B = \{ bel_1, bel_2, ..., bel_n)$, $B' = \{ bel_1', bel_2', ..., bel_n')$, we say B is *more accurate* than bel' if there exists $i \in N$ such that bel_i is more precise than bel_i', and either $bel_j \equiv bel_j'$ or bel_j is more accurate than bel_j' for all $j \in N - \{i\}$.

Theorem 3. Given two games $g = (N, S, (\succ_i), P, B, s^1)$ and $g' = (N, S, (\succ_i), P, B', s^1)$, if B is *more accurate* than B', then $\text{wb-core}(g) \subseteq \text{wb-core}(g')$ and $\text{sb-core}(g) \subseteq \text{sb-core}(g')$.

5 Related Works

A Bayesian-core concept is proposed in [6] and [7] where the agents are assumed to belong to various types which are unknown to other agents. The agents are required to estimate the value of potential coalitions by maintaining a Bayesian belief system regarding the possible types of their potential partners. Our work differs from theirs in that we do not pre-assume any probabilistic models in particular, and that our model assumes the more general problem of non-transferable utilities games, instead of transferable ones.

A solution concept for coalition game with stochastic payoff is presented in [8]. In this approach, the coalitional payoffs are assumed to be stochastic variables, and agents preferences over those stochastic variables are used to determine the stability. Thus, their work is on stochastic games, whereas our focus is on a more general class of non-transferable utility games that are not necessarily probabilistic in nature.

6 Conclusion

Classical solution concepts in cooperative game theory rely on a common information assumption. Multi-agents in a semi-competitive applications, on the other hands, are typically modeled self-interested entities that are far from being omniscient: they act according to private beliefs and they have to rely on whatever evidences they can perceive during their decision making. The basic assumptions of these two disciplines do not match, despite that game theoretical solution concepts have been used to evaluate the qualities of multi-agent coalition formation mechanisms.

In this paper, we propose a new type of game which we label belief based non-transferable utility games with uncertainty, and provide new solution concepts for describing the stability of coalitions these games. First, in order to handle internal uncertainties in the agents belief, instead of classifying each coalitions as simply "stable" (*i.e.*, in the core) or not, as done in the traditional approaches, we divide the obtainable consequences of a coalition into three stability classes: those that are *certainly* stable (s-core), those that *may* be stable (w-core), depending on the agents' conflict resolution strategy, and those that are certainly not stable. We also see that how the s-core, and the w-core are affected by the precision of the agents' beliefs regarding the environment: the more precise are the beliefs, the smaller are the weak-cores and the larger are the strong cores. Later, we expanded the idea to include the effect of the agents' mutual beliefs regarding other agents, and the result is the concepts of the sb-core and the wb-core. By doing so, we are able to provide useful stability concepts for this new type of game which otherwise cannot be analyzed properly using the traditional approaches. We believe our model provide a useful tool in evaluating coalition formation algorithms for agent based cooperative games.

References

1. Osborne, M.J., Rubinstein, A.: A Course in Game Theory. MIT Press, Cambridge (1994)
2. Gerber, A.: Coalition Formation in General NTU Games. Review of Economic Design 5, 149–175 (2000)

3. Sandholm, T.: Distributed rational decision making. In: Weiss, G. (ed.) Multiagent Systems: A Modern Approach to Distributed Artificial Intelligence. MIT Press, Cambridge (1999)
4. Ketchpel, S.: Forming coalitions in the face of uncertain rewards. In: Proceedings of National Conference on Artificial Intelligence (AAAI 1994), pp. 414–419 (1994)
5. Gillies, D.B.: Solutions to general non-zero-sum games. In: Tucker, A.W., Luce, R.D. (eds.) Contributions to the Theory of Games, vol. IV. Princeton University Press, Princeton (1959)
6. Chalkiadakis, G., Boutilier, C.: Bayesian reinforcement learning for coalition formation under uncertainty. In: Proceedings of the Third International Joint Conference on Autonomous Agents and Multiagent Systems, pp. 1090–1097 (2004)
7. Chalkiadakis, G., Boutilier, C.: Coalition Formation under Uncertainty: Bargaining Equilibria and the Bayesian Core Stability Concept. In: Proceedings of the 2007 International Joint Conference on Autonomous Agents and Multiagent Systems, pp. 1090–1097 (2007)
8. Suijs, J., Borm, P., De Waegenaere, A., Tijs, S.: Cooperative Games with Stochastic Payoffs. European Journal of Operational Research 133 (1999)

Side-Effect Inspection for Decision Making

Luís Moniz Pereira and Alexandre Miguel Pinto

Abstract. In order to decide on the course of action to take, one may need to check for side-effects of the possible available preferred actions. In the context of abduction in Logic Programs, abducible literals may represent actions and assumptions in the declarative rules used to represent our knowledge about the world. Besides finding out which alternative sets of actions achieve the desired goals, it may be of interest to identify which of those abductive solutions would also render *true* side-effect literals relevant for the decision making process at hand, and which would render those side-effects *false*. After collecting all the alternative abductive solutions for achieving the goals it is possible to identify which particular actions influence inspected side-effect literals' truth-value.

To achieve this, we present the concept of Inspection Point in Abductive Logic Programs, and show how, by means of examples, one can employ it to investigate side-effects of interest (the *inspection points*) in order to help evaluate and decide among abductive solutions. We show how this type of reasoning requires a new mechanism, not provided by others already available. We furthermore show how to implement this new mechanism it on top of an already existing abduction solving system — ABDUAL — in a way that can be adopted by other systems too.

Keywords: Abduction, Side-Effects, Decision Making, Posteriori Preferences.

1 Introduction

In this paper we present a new decision-making-aid reasoning mechanism for abductive logic programs — the inspection points (IPs). The IPs permit the declarative

Luís Moniz Pereira and Alexandre Miguel Pinto
Centro de Inteligência Artificial (CENTRIA)
Universidade Nova de Lisboa
2829-516 Caparica, Portugal
e-mail: {lmp,amp}@di.fct.unl.pt

K. Nakamatsu et al. (Eds.): New Advan. in Intel. Decision Techno., SCI 199, pp. 139–149.
springerlink.com

specification and implementation of more efficient abductive logic programming based decision-making agents. The efficiency comes from that using IPs allows the agent to selectively specify the relevant side-effect consequences of abductions instead of computing all possible abductions and all their consequences, only to subsequently having to filter out irrelevant abductions and to ignore irrelevant side-effects.

Typically, in a logic programming setting, consequences of abductions are computed by some forward-chaining mechanism. The problem with such an approach is the waste in time and computing resources that comes from computing *all* the consequences of the adopted abductions, and not just the consequences relevant for the task at hand. A kind of "selective forward chaining" is what would be desired. The IPs we present and implement efficiently exactly enact such forward chaining. This is accomplished by first selecting, in a top-down fashion, the rules necessary for forward propagation, from choices made, to the side-effects whose inspection is desired. Posterior preferences among alternative solutions can then take the observed side-effects into account.

We begin by presenting the motivation, and some background notation and definitions follow. The general problem of reasoning with logic programs is addressed in section 2; in particular, we take a look at the nature of backward and forward chaining and their relationship to query answering in an abductive framework. In section 3 we very briefly describe our implementation of the IPs.

Further elaboration on possibilities of use of IPs is sketched, and conclusions and future work close the paper.

1.1 Motivation

When faced with some situation where several alternative courses of actions are available a rational agent must decide and choose which action to take. *A priori* preferences can be applied before choosing in order to reduce the number of considerable possible actions curtailing the explosion of irrelevant combinations of choices, but still several (possibly exclusive) may remain available. To make the best possible informed decision, and commit to a course of action, the agent must be enabled to foresee the consequences of its actions and then prefer on the basis of those consequences (with *a posteriori* preferences). Choosing which set of consequences is most preferred corresponds to an implicit choice on restricting which course of action to take. But only the consequences relevant to the *a posteriori* preferences should be calculated: there are virtually infinitely many consequences of a given action, most of which are completely irrelevant to the preference-based decision making. Other consequences may be just predictions about the present state of the world, and observing whether they are verified can eliminate hypothetical scenarios where certain decisions would appear to make sense. Not all consequences are experimentally observable though, hence IPs may serve to focus on the ones that are, and thus guide the experimentation required to decide among competing hypothesis, as in medical diagnosis say. That is, IPs can be put to the service of sifting through competing explanations, prior to any acting but albeit in preparing and configuring the context for it. In science, such decisive consequences are often know as "crucial" side-effects, because hopefully they can guarantee excluding untoward possibilities.

Computationally too, there are many advantages as well to preferring a posteriori, i.e. to enact preferences on the computed models, after the consequences of opting for one or another abducible are known, by means of inspection points that examine specific side-effects of abduction. The advantages of so proceeding stem largely from avoiding combinatory explosions of abductive solutions, by filtering both irrelevant as well as less preferred abducibles.

We code the agent's knowledge in the form of a Normal Logic Program (NLP), and its possible actions or hypotheses as abducibles. When trying to find an existential answer to a query where some desired goal is achieved, the answers will include the abductions produced at top-down query-answering time. Instead of computing whole models to find out the side-effect consequences of the abductions, we introduce — and use — the new mechanism of *inspection point* whereby we declaratively specify which side-effect consequences of the possible answers are of interest. In the usual abductive reasoning setting, normal top-down query answering resorts to performing abduction to construct hypothetical answers. However, when using inspection points we just want to check if some literal is a consequence of the abductions made when finding an answer to the query. Thus, further abduction is disabled when inspection point checking.

We show how this type of reasoning requires a new mechanism, not provided by others already available. We furthermore show how to implement this new mechanism on top of an already existing abduction solving system — ABDUAL [3] — in a way that can be adopted by other systems too.

Example 1. **Relevant and irrelevant side-effects.** Consider this logic program where $drink\_water$ and $drink\_beer$ are abducibles.

$\leftarrow thirsty, not\ drink.$ % This is an Integrity Constraint
$wet\_glass \leftarrow use\_glass.$ $use\_glass \leftarrow drink.$
$drink \leftarrow drink\_water.$ $drink \leftarrow drink\_beer.$
$thirsty.$ $drunk \leftarrow drink\_beer.$
$unsafe\_drive \leftarrow inspect(drunk).$

Suppose we want to satisfy the Integrity Constraint, and also to check if we get drunk or not. However, we do not care about the glass becoming wet — that being completely irrelevant to our current concern. In this case, full forward-chaining or computation of whole models is a waste of time, because we are interested only in a subset of the program's literals. What we need is a selective ersatz forward chaining mechanism, an inspection tool which permits to check the truth value of given literals as a consequence of the abductions made to satisfy a given query plus any Integrity Constraints.

Moreover, in this example, if we may simply want to know the side-effects of the possible actions in order to decide (to drive or not to drive) **after** we know which side-effects are true. In such case, we do not want to the IC $\leftarrow not\ unsafe\_drive$ because that would always impose $not\ drink\_beer$. We want to allow all possible solutions for the single IC $\leftarrow thirsty, not\ drink$ and then check the side-effects of each abductive solution.

1.2 Background Notation and Definitions

Definition 1. Logic Rule. A (Normal) Logic Rule has the general form
$A \leftarrow B_1, \ldots, B_n, not\ C_1, \ldots, not\ C_m$
where A and the B_i and C_j are atoms.

A is the head of the rule, and $B_1, \ldots, B_n, not\ C_1, \ldots, not\ C_m$ its body. 'not ' denotes default negation. When the body of a rule is empty, we say its head is a fact and write the rule simply as A.

Definition 2. Logic Program. A (Normal) Logic Program (LP for short) P is a (possibly infinite) set of Logic Rules, standing for all its ground instances.

Definition 3. Integrity Constraint. An Integrity Constraint (IC) is a logic rule, expressing a denial, whose head is the reserved atom '*false*'.

A simpler way of writing an IC is by omitting the head of its rule. An example is the first rule of the program in example 1, meaning that '$thirsty$' cannot be *true* whenever '$drink$' is *false*, and vice-versa.

In the next sections, we focus on abductive logic programs, i.e., those with abducibles. Abducibles are literals that are not defined by any rules and correspond to hypotheses that one can independently assume or not — apart from eventual Integrity Constraints. Abducibles or their default negations can appear in bodies of rules just like any other literal. They are specified along with the LP.

2 Reasoning with Logic Programs

Recall that when finding an abductive solution for a query, one may want to check whether some other literals become *true* or *false* strictly within the abductive solution found (i.e., whether they are consequences, or side-effects, of such abductions), but without performing additional abductions, and without having to produce a complete model to do so. This type of reasoning requires a new mechanism. To achieve it, we introduce the concept of inspection point, and show how one can employ it to investigate side-effects of interest. Procedurally, the checking of an inspection point corresponds to performing a top-down query-proof for the inspected literal, but with the specific proviso of disabling new abductions during that proof. The proof for the inspected literal will succeed only if the abducibles needed for it were already, or will be adopted, in the present ongoing solution search for the top query. Consequently, this check is performed after a solution for the query has been found. At inspection-point-top-down-proof-mode, whenever an abducible is encountered, instead of adopting it, we simply adopt the intention to *a posteriori* check if the abducible is part of the answer to the query (unless of course the negation of the abducible has already been adopted by then, allowing for immediate failure at that search node.) That is, one (meta-)abduces the checking of some abducible A, and the check consists in confirming that A is part of the abductive solution by matching it with the object of the check. In our method, the side-effects of interest are explicitly indicated by the user by wrapping the corresponding goals subject to inspection mode within a reserved construct $inspect/1$.

2.1 Backward and Forward Chaining

Query-answering is intrinsically backward-chaining as it is a top-down dependency-graph oriented proof-procedure. And all the more so of typically by need abductive query-answering. Finding the side-effects of a set of assumptions is conceptually envisaged as forward-chaining as it consists of progressively deriving conclusions from the assumptions until the truth value of the chosen side-effect literals is determined. The problem with full-fledged forward-chaining is that too many (often irrelevant) conclusions are derived. Since efficiency is always a concern, wasting time and resources on deriving conclusions only to be discarded afterwards, is a flagrant setback. Even worse, in combinatorial problems, there may be many alternative solutions whose differences repose just on irrelevant conclusions. So, the unnecessary computation of irrelevant conclusions in full forward-chaining may be multiplied, leading to immense waste. A more rational solution, when one is focused on some specific conclusions of a set of premises, is afforded by selective top-down ersatz forward-chaining. In such a setting, ideally, the user would be allowed to specify the conclusions she is focused on, and only those would be computed in a backward-chaining fashion, checking whether they are consequences of desired abductions, but without abducing — hence their ersatz character. Combining backward-chaining with forward-chaining (and in particular with (ersatz) selective forward-chaining) allows for a greater precision in specifying what we wish to know, and altogether improve efficient use of computational resources.

Significantly, if abduction is enabled, the computation of side-effects should take place without further abduction, passively (though not destructively) just "consuming" abducibles that are "produced" elsewhere by abduction for the top query.

In the sequel, we shall show how such a selective forward chaining from a set of hypotheses can actually be prepared by backward chaining from the focused on conclusions — the inspection points — by virtue of a controlled form of abduction.

2.2 The Use of Stable Models

When we need to know the 2-valued truth value of all the literals in the program for the problem we are modeling and solving, the only solution is to produce complete models. In such a case, tools like *SModels* [13] or *DVL* [5] are adequate because they can indeed compute whole and all models for the (finitely grounded) program. Typically, each abducible is then coded as a pair of rules, that form an even loop over default negation between the abducible and a representation of its negation. One may discuss other alternative semantics (2-valued and 3-valued) that can also be used in this situation, and compare them with Stable Models (SM) semantics [10]. In an abductive reasoning situation, however, computing the whole model entails pronouncement about each of the abducibles, whether or not they are relevant to the problem at hand, and subsequently filtering the irrelevant ones. When we simply want to find an answer to a query, we either compute a whole model and check if it entails the query (the way SM semantics does), or, if the underlying semantics we are using enjoys the *relevancy* property — which SM semantics do not — we can simply use a top-down proof-procedure (*à la* Prolog). In this second case, the user does not pay the price of computing a whole model, nor the price of abducing all possible abducibles or their negations, since the only abducibles considered will be those needed for answering the query.

2.3 Abduction

Abduction ([1, 2, 3, 6, 7, 8, 9, 11, 12]) can naturally be used in a top-down query-oriented proof-procedure to find an (abductive) answer to a query, where abducibles are leafs in the call-dependency graph. The Well-Founded Semantics (WFS), which enjoys relevancy, allows for abductive query answering. We used it in the implementation briefly described in section 3. Though WFS is 3-valued, the abduction mechanism it employs can be, and in our case is, 2-valued.

2.4 Inspection Points

From a practical perspective, under the abductive logic programming setting, a typical query with side-effect checking via IPs could look like $main\_query, inspect(checked\_side\_effect)$. If there is an answer to such query then we know that the abductions made to satisfy the $main\_query$ are sufficient to prove also the $checked\_side\_effect$.

2.4.1 Meta-abduction

Intuitively, meta-abduction, in this setting, consists in abducing the intention of *a posteriori* checking for the abduction of some abducible, i.e. the intention of verifying that the abducible is indeed adopted. In practice, when we want to meta-abduce some abducible 'x', we abduce a literal '$abduced(x)$', which represents the intention that 'x' is eventually abduced along the process of finding an answer. The check is performed after a complete abductive answer to the top query is found. Operationally, 'x' has been or will be abduced as part of the ongoing solution to the top goal. Meta-abduction can be implemented by any abduction capable system.

When using a system that allows only for the top-down dependency-graph-oriented abductive query-solving we must *simulate* this *selective* bottom-up *forward chaining* by means of a top-down query where making actual (non-meta) abductions is disallowed. In this setting, *inspection points* are the literals whose truth value we are interested in.

Example 2. **Inspection Points.** Consider this NLP, where '$tear\_gas$', '$fire$', and '$water\_cannon$' are the only abducibles.

$\leftarrow police, riot, not\ contain.$	% this is an Integrity Constraint
$contain \leftarrow tear\_gas.$	$contain \leftarrow water\_cannon.$
$smoke \leftarrow fire.$	$smoke \leftarrow inspect(tear\_gas).$
$police.$	$riot.$

Notice the two rules for '$smoke$'. The first states that one explanation for smoke is fire, when assuming the hypothesis '$fire$'. The second states '$tear\_gas$' is also a possible explanation for smoke. However, the presence of tear gas is a much more unlikely situation than the presence of fire; after all, tear gas is only used by police to contain riots and that is truly an exceptional situation. Fires are much more common and spontaneous than riots. For this reason, '$fire$' is a much more plausible explanation for '$smoke$' and, therefore, in order to let the explanation for '$smoke$' be '$tear\_gas$', there must be a plausible reason — imposed by some other likely phenomenon. This is represented by $inspect(tear\_gas)$ instead of simply '$tear\_gas$'. The '$inspect$' construct disallows regular abduction — only meta-abduction — to be performed whilst trying to solve '$tear\_gas$'. I.e., if we take tear gas as an abductive solution for smoke, this rule imposes that the step where we abduce '$tear\_gas$' is performed elsewhere, not under the derivation tree for '$smoke$'. Thus, '$tear\_gas$' is an *inspection point.*

The Integrity Constraint, because there is '$police$' and a '$riot$', forces '$contain$' to be *true*, and hence, '$tear\_gas$' or '$water\_cannon$' or both, must be abduced. '$smoke$' is only explained if, at the end of the day, '$tear\_gas$' is abduced to enact containment.

Abductive solutions should be plausible. '$smoke$' is plausibly explained by '$tear\_gas$' if there is a reason, a best explanation, that makes the presence of tear gas plausible; in this case the riot and the police. Plausibility is an important concept in science which lends credibility to hypotheses. Assigning plausibility measures to situations is an issue orthogonal to the problem.

Example 3. **Nuclear Power Plant Decision Problem.** This example was extracted from [16] and adapted to our current issues. In this example the abducibles do not represent actions. In a nuclear power plant there is decision problem: cleaning staff can clean the power plant on cleaning days, but only if there is no sound alarm. The alarm sounds when the temperature in the main reactor rises above a certain threshold, or if the alarm itself is faulty. When the alarm sounds everybody must evacuate the power plant immediately! Abducible literals are $cleaning\_day$, $temperature\_rise$ and $faulty\_alarm$.

$$\begin{array}{lll} dust & \leftarrow & cleaning\_day, inspect(not\ sound\_alarm) \\ sound\_alarm & \leftarrow & temperature\_rise \\ sound\_alarm & \leftarrow & faulty\_alarm \\ evacuate & \leftarrow & sound\_alarm \\ & \leftarrow & not\ cleaning\_day \end{array}$$

Satisfying the unique IC imposes *cleaning_day true* and gives us three minimal abductive solutions: $S_1 = \{dust, cleaning\_day\}$,
$S_2 = \{cleaning\_day, sound\_alarm, temperature\_rise, evacuate\}$, and $S_3 = \{cleaning\_day, sound\_alarm, faulty\_alarm, evacuate\}$. If we pose the query $? - not\ dust$ we want to know what could justify to the cleaners dusting not to occur given that it is a cleaning day (this last is enforced by the IC). However, we do not want to abduce the rise in temperature of the reactor nor to abduce the alarm to be faulty in order to prove *not dust*. Any of these justifying two abductions must result as a side-effect of the need to explain something else, for instance the observation of the sounding of the alarm, expressed by the IC $\leftarrow not\ sound\_alarm$, which would then abduce one or both of those two abducibles as plausible explanations. The $inspect/1$ in the body of the rule for *dust* prevents any abduction below *sound_alarm* to be made just to make *not dust* true. One other possibility would be for the observations coded by ICs $\leftarrow not\ temperature\_rise$ or $\leftarrow not\ faulty\_alarm$ to be present in order for *not dust* to be true as a side-effect. A similar argument can be made about evacuating: one thing is to explain why evacuation takes place, another altogether is to justify it as necessary side-effect of root explanations for the alarm to go off. These two pragmatic uses correspond to different queries: $? - evacuate$ and $? - inspect(evacuate)$, respectively.

A declarative semantics of Inspection Points is presented and detailed in `http://centria.fct.unl.pt/ lmp/publications/online-papers/ IP08.pdf` but, due to lack of space, we omit it here.

3 Implementation

We based our practical work on a formally defined, implemented, tried and true abduction system: Abdual [3]. Meta-abduction is implemented adroitly by means of a new reserved abducible predicate which engages the abduction mechanism to try and discharge any meta-abductions by means of the corresponding abducible. The approach taken can easily be adopted by other abductive systems, as we had the occasion to check [4].

Abdual is composed of two modules: the preprocessor which transforms the original program by adding its dual rules, plus specific abduction-enabling rules; and a meta-interpreter allowing for top-down abductive query solving. When solving a query, abducibles are dealt with by means of extra rules the preprocessor added to that effect. These rules just add the name of the abducible to an ongoing list of current abductions, unless the negation of the abducible was added before to the list of false abducibles in order to ensure abduction consistency.

3.1 Abdual with Inspection Points

Inspection Points in Abdual function mainly by means of controlling the general abduction step, which involves very few changes, both in the pre-processor and the meta-interpreter. Whenever an '$inspect(X)$' literal is found in the body of a rule, where 'X' is a goal, a meta-abduction-specific counter — the '$inspect\_counter$' — is increased by one, in order to keep track of the allowed character, active or passive, of performed abductions. The top-down evaluation of the query for 'X' then proceeds normally. Actual abductions are only allowed if the counter is set to zero, otherwise only meta-abductions are allowed. After finding an abductive solution for the query 'X' the counter is decreased by one. Backtracking over counter assignations is duly accounted for.

Of course, this way of implementing the Inspection Points (with just one $inspect\_counter$) presupposes the abductive query answering process is carried out "depth-first", guaranteeing the order of the literals in the bodies of rules actually corresponds to the order they are processed.

The present implementation of Abdual with Inspection Points is available on request and is detailed in `http://centria.fct.unl.pt/~lmp/publications/online-papers/IP08.pdf`

In case the "depth-first" discipline is not followed, either because goal delaying is taking place, or multi-threading, or co-routining, or any other form of parallelism is being exploited, then each queried literal will need to carry its own list of ancestors with their individual $inspect\_counters$. This is necessary so as to have a means, in each literal, to know which and how many $inspects$ there are between the root node and the literal currently being processed, and which $inspect\_counter$ to update; otherwise there would be no way to determine if abductions or else meta-abductions should be performed.

An alternative implementation would rely instead not on ABDUAL but on XSB-prolog's XASP, to let Smodels produce partial models where abducibles and meta-abducibles are coded as even loops over default negation, as in ACORDA [14].

3.2 Comparing to Other Systems

We briefly compared our abduction system with inspection points to HyProlog [4].The HyProlog system supports abduction and additionally a system of assumptions, which differs from abducibles in that they are explicitly produced and explicitly applied. Creating an assumption in parallel with mentioning an abducible makes it possible, as Veronica Dahl showed us, to check the state of abducibles and thus provide a HyProlog implementation of inspection points.

In general, there is not such a big difference between the operational semantics of HyProlog and the Inspection Points implementation we present; however, there is a major functionality difference: in HyProlog we can only require consumption directly on abducibles, and with Inspection Points we can inspect any literal, not just abducibles. Moreover, the part of HyProlog concerned with abduction has a standard first order semantics, whereas assumptions+expectations rely on a resource-oriented semantics *à la* linear logic.

4 Conclusions and Future Work

In the context of abductive logic programs, we have presented a new mechanism of inspecting literals, which corresponds to a selective forward chaining, that can be used to check for side-effects. The implementation of side-effects inspection, relying on the meta-abduction principle, which is at the heart of this contribution, permits consequence assessment which can be used to identify basic causality relationships usable for decision making. We have implemented the inspection mechanism within the Abdual [3] meta-interpreter and checked that it can easily be ported to other systems [4]. The semantics underlying Abdual (and, therefore, the current implementation of inspection points) is WFS with abduction. However, in case we need a total 2-valued semantics we need recourse to Layered Models [15] if we want to retain advantage of relevancy for top-down querying. Hence, our future work directions include extending the LM semantics with abduction, and, of course, meta-abduction. An efficient implementation of this semantics is also under way.

Acknowledgements. We thank Robert A. Kowalski and Verónica Dahl and Henning Christiansen for their insightful discussions and references to related works [4, 16].

References

1. Alferes, J., Leite, J., Pereira, L., Quaresma, P.: Planning as abductive updating. In: Kitchin, D. (ed.) Procs. of AISB 2000 (2000)
2. Alferes, J.J., Pereira, L.M., Swift, T.: Well-founded abduction via tabled dual programs. In: Intl. Conf. on Logic Programming, pp. 426–440 (1999)
3. Alferes, J.J., Pereira, L.M., Swift, T.: Abduction in well-founded semantics and generalized stable models via tabled dual programs. TPLP 4(4), 383–428 (2004)
4. Christiansen, H., Dahl, V.: Hyprolog: A new logic programming language with assumptions and abduction. In: Gabbrielli, M., Gupta, G. (eds.) ICLP 2005. LNCS, vol. 3668, pp. 159–173. Springer, Heidelberg (2005)
5. Citrigno, S., Eiter, T., Faber, W., Gottlob, G., Koch, C., Leone, N., Mateis, C., Pfeifer, G., Scarcello, F.: The dlv system: Model generator and advanced frontends (system description). In: Workshop Logische Programmierung (1997)
6. Console, L., Theseider Dupre, D., Torasso, P.: On the relationship between abduction and deduction. J. of Logic and Computation 1(5), 661–690 (1991)
7. Denecker, M., De Schreye, D.: Sldnfa: An abductive procedure for normal abductive programs. In: Apt (ed.) Procs. of the Joint Intl. Conf. and Symposium on Logic Programming, Washington, USA, pp. 686–700. The MIT Press, Cambridge (1992)
8. Dung, P.M.: Negations as hypotheses: An abductive foundation for logic programming. In: ICLP, pp. 3–17. MIT Press, Cambridge (1991)
9. Eiter, T., Gottlob, G., Leone, N.: Abduction from logic programs: semantics and complexity. Theoretical Computer Science 189(1–2), 129–177 (1997)

10. Gelfond, M., Lifschitz, V.: The stable model semantics for logic programming. In: ICLP/SLP, pp. 1070–1080. MIT Press, Cambridge (1988)
11. Inoue, K., Sakama, C.: A fixpoint characterization of abductive logic programs. Journal of Logic Programming 27(2), 107–136 (1996)
12. Kakas, A.C., Riguzzi, F.: Learning with abduction. In: Džeroski, S., Lavrač, N. (eds.) ILP 1997. LNCS, vol. 1297, pp. 181–188. Springer, Heidelberg (1997)
13. Niemelä, I., Simons, P.: Smodels - an implementation of the stable model and well-founded semantics for normal logic programs. In: Fuhrbach, U., Dix, J., Nerode, A. (eds.) LPNMR 1997. LNCS, vol. 1265, pp. 420–429. Springer, Heidelberg (1997)
14. Pereira, L.M., Lopes, G.: Prospective logic agents. In: Neves, J., Santos, M.F., Machado, J.M. (eds.) EPIA 2007. LNCS (LNAI), vol. 4874, pp. 73–86. Springer, Heidelberg (2007)
15. Pereira, L.M., Pinto, A.M.: Layered models top-down querying of normal logic programs. In: Proceedings of the Practical Aspects of Declarative Languages. LNCS. Springer, Heidelberg (to appear, 2009)
16. Sadri, F., Toni, F.: Abduction with negation as failure for active and reactive rules. In: Lamma, E., Mello, P. (eds.) AI*IA 1999. LNCS, vol. 1792, pp. 49–60. Springer, Heidelberg (2000)

Knowledge Mining with a Higher-Order Logic Approach

Kittisak Kerdprasop and Nittaya Kerdprasop

Abstract. Knowledge mining is the process of deriving new and useful knowledge from vast volumes of data and patterns previously discovered and stored as background knowledge. We propose a knowledge-mining system as a repertoire of tools for discovering strong and useful patterns. A pattern is strong if it represents frequently occurring relationships. Usefulness is achieved through constraints guided by users. To be able to derive strong and useful patterns from underlying data and background knowledge we consider employing the concept of higher-order logic as a major approach of our implementation. Higher-order logic can greatly reduce the burden of programmers as it is a very high level programming scheme suitable for the development of knowledge-intensive tasks. We have shown in this paper frequent pattern mining implemented with higher-order logic. The implementation is applied to mine breast cancer data. Our design of a logic-based knowledge-mining system is intended to support higher-order and constraint mining which is the next step of our research direction.

1 Introduction

Knowledge is a valuable asset to most organizations as a substantial source to support better decisions and thus to enhance organizational competency. Researchers and practitioners in the area of knowledge management view knowledge in a broad sense as a state of mind, an object, a process, an access to information, or a capability [2, 11]. The term *knowledge assets* [17, 19] is used to refer to any organizational intangible assets related to knowledge such as know-how, expertise, intellectual property. In clinical companies and computerized healthcare applications knowledge assets include order sets, drug-drug interaction rules, guidelines for practitioners, and clinical protocols [10].

Knowledge assets can be stored in data repositories either in implicit or explicit form. Explicit knowledge can be managed through the existing tools available in the current database technology. Implicit knowledge, on the contrary, is harder to

Kittisak Kerdprasop and Nittaya Kerdprasop
Data Engineering and Knowledge Discovery Research Unit,
School of Computer Engineering, Suranaree University of Technology,
Nakhon Ratchasima 30000, Thailand

K. Nakamatsu et al. (Eds.): New Advan. in Intel. Decision Techno., SCI 199, pp. 151–159.
springerlink.com

achieve and retrieve. Specific tools and suitable environments are needed to extract such knowledge.

Implicit knowledge acquisition can be achieved through the availability of the knowledge-mining system. *Knowledge mining* is the discovery of hidden knowledge stored possibly in various forms and places in large data repositories. In health and medical domains, knowledge has been discovered in different forms such as association rules, classification, clustering, trend or temporal pattern analysis [20]. The discovered knowledge facilitates expert decision support, diagnosis and prediction.

In this paper we present the design of a complete knowledge-mining system to support a high-level decision not only in medical domains but also in any domain that requires a knowledge-based decision support. A rapid prototyping of the proposed system is also provided to highlight the fact that higher-order logic is an appropriate approach to the implementation of a complex knowledge-mining system. The intuitive idea of our design and implementation is that for such a complicated knowledge-based system program coding should be done declaratively at a high level to alleviate the burden of programmers. The declarative style of programming also eases the future extension of our system to cover the concepts of higher-order mining [16] and constraint programming [6] that should naturally applied to the task of knowledge mining.

The rest of this paper is organized as follows. Section 2 reviews related works. Section 3 is the architecture of SUT-Miner, the proposed knowledge-mining system. Section 4 shows the implementation of association mining [1] using higher-order logic programming scheme with some running examples. Section 5 concludes the paper and discusses our future research directions.

2 Related Works

In recent years we have witnessed increasing number of applications devising database technology and machine learning techniques to mine knowledge from biomedicine, clinical and health data. Roddick et al [15] discussed the two categories of mining techniques applied over medical data: explanatory and exploratory. Explanatory mining refers to techniques that are used for the purpose of confirmation or making decisions. Exploratory mining is data investigation normally done at an early stage of data analysis in which an exact mining objective has not yet been set.

Explanatory mining in medical data has been extensively studied in the past decade employing various learning techniques. Bojarczuk et al [3] applied genetic programming method with constrained syntax to discover classification rules from medical data sets. Thongkam et al [21] studied breast cancer survivability using AdaBoost algorithms. Ghazavi and Liao [7] proposed the idea of fuzzy modeling on selected features medical data. Huang et al [9] introduced a system to apply mining techniques to discover rules from health examination data. Then they employed a case-based reasoning to support the chronic disease diagnosis and treatments. The recent work of Zhuang et al [25] also combined mining with case-based reasoning, but applied a different mining method. They performed data clustering based on self-organizing maps in order to facilitate decision support on solving new cases of pathology test ordering problem. Biomedical discovery support systems are recently proposed by a number of researchers [4, 5, 8, 23, 24]. Some works [14, 18] extended medical databases to the level of data warehouses.

Exploratory, as oppose to explanatory, is rarely applied to medical domains. Among the rare cases, Nguyen and Kawasaki [13] introduced knowledge visualization in the study of hepatitis patients. Palaniappan and Ling [14] applied the functionality of OLAP tools to improve visualization.

It can be seen from the literature that most medical knowledge discovery systems have applied only some mining techniques such as classification rules mining, association mining, data clustering to discover hidden patterns and knowledge. We, on the contrary, design a knowledge-mining system aiming at providing a suite of tools to facilitate users and medical practitioners on discovering different kinds of knowledge from their data and background knowledge repositories.

3 SUT-Miner: A Knowledge-Mining system

Knowledge mining is a complex and possibly iterative process that involves many different steps. The main input to the process is data from heterogeneous sources, and the final output is the useful information desired by the users. For the medical domains, we design the system to be composed of two main phases as shown in figure 1. Knowledge induction phase is the back-end of the system responsible for acquiring and discovering new and useful knowledge. Usefulness is to be validated at the final step by human experts. Discovered knowledge is stored in the knowledge base to be applied to solve new cases in knowledge inferring phase which is the front-end of the proposed system.

SUT-Miner in a knowledge induction phase is comprised of three main modules: pre-DM, DM, post-DM. The term data mining (DM) means automatic learning of patterns or models from specific data. *Pattern* is an expression describing a subset of the data, *e.g.* $f(x) = 3x^2 + 3$ is a pattern induced from a given dataset $\{(0,3), (1,6), (2,15), (3,30)\}$, whereas the term *model* refers to a representation of the source generating the data, *e.g.* $f(x) = ax^2 + b$. In his paper we refer to both patterns and models as new knowledge discovered from data sources.

The pre-DM module performs data preparation tasks such as to locate and access relevant data sets, transform the data format, clean the data if there exists noise and missing values, reduce the data to a reasonable and sufficient size with only relevant attributes. The DM module performs mining tasks including classification, prediction, clustering, and association. We adopt the ontology concept at this step to guide the mining methodology selection. A simple form of mining method ontology is shown in figure 2. The post-DM module is composed of two main components: knowledge evaluator and knowledge integrator. These components perform major functionalities aiming at a feasible knowledge deployment which is important for the applications in medical diagnosis and predicting. Knowledge evaluator involves evaluation, based on corresponding measurement metrics, of the mining results. Knowledge integrator examines the induced patterns to remove redundant knowledge. Ontology has also been applied at this step to provide essential semantics regarding the domain problems.

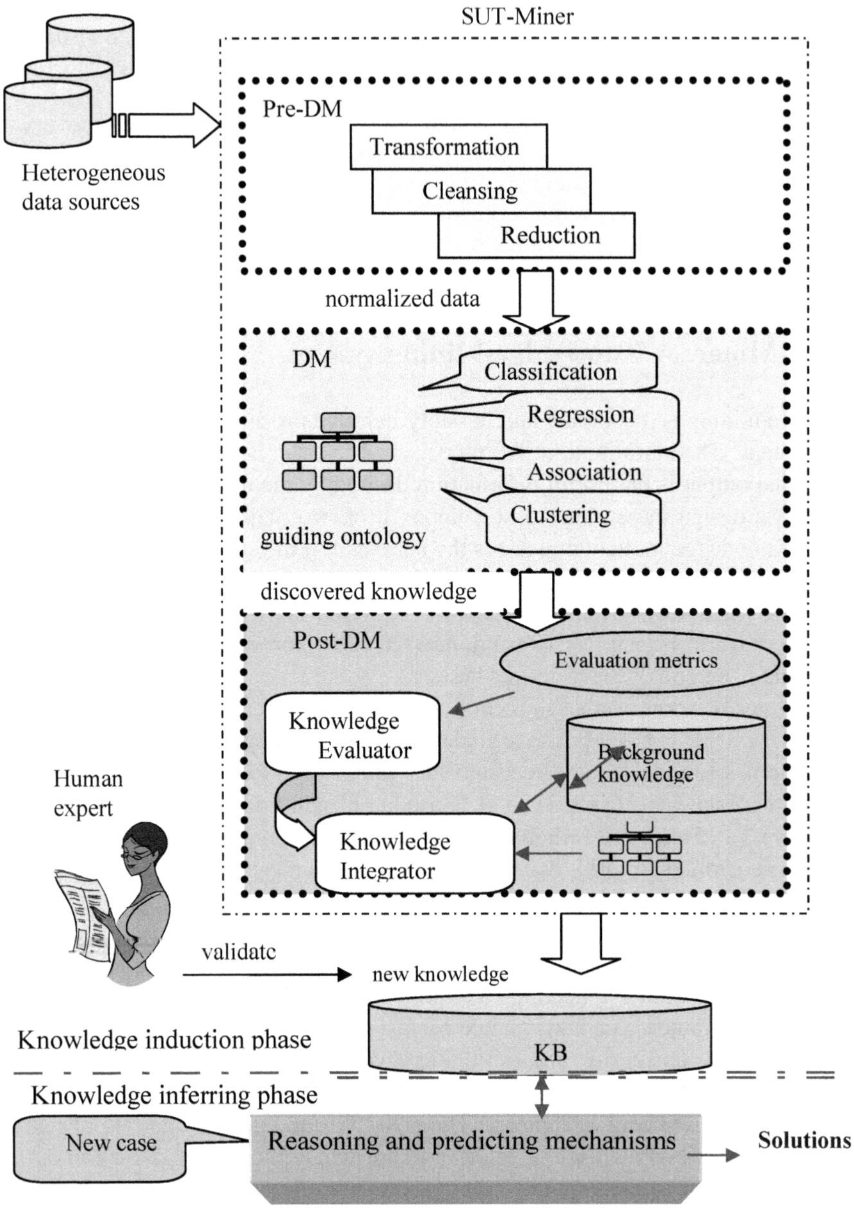

Fig. 1 Architecture of a knowledge-mining system

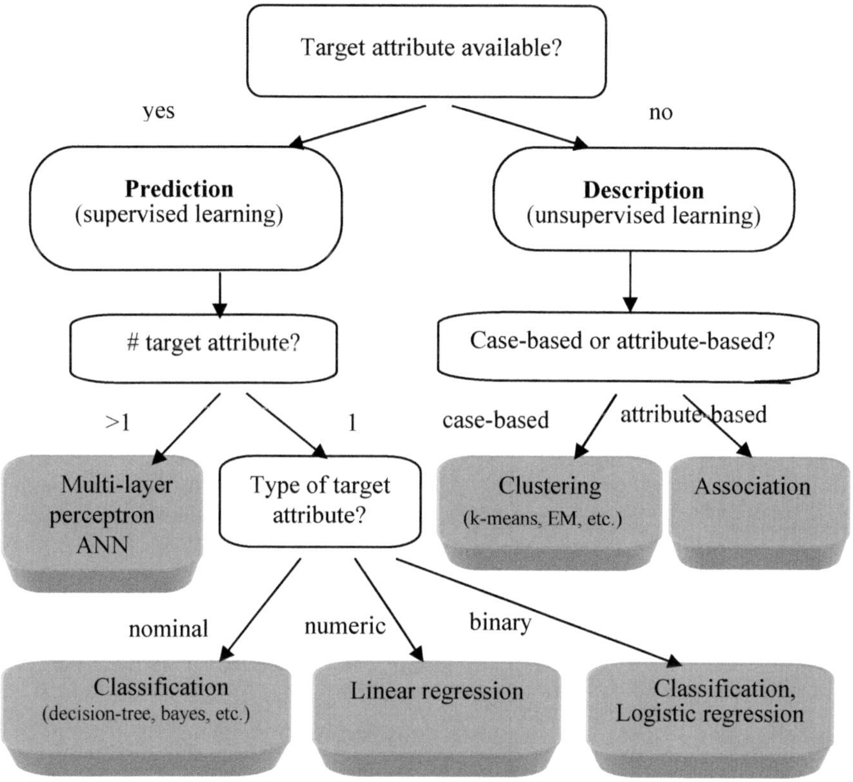

Fig. 2 Ontology for guiding mining method selection at the DM step

4 Implementation and Running Example

The SUT-Miner system has been implemented based on the concept of higher-order logic which is an extension of first-order logic to be more expressive and more powerful. First-order logic has been extensively used in intelligent systems [22] as an inference mechanism to deduce new facts. A classic example is that given a general rule $\forall x$ *Man(x) => Mortal(x)* and a known fact *Man(Socretes)*, we can deduce new fact that *Mortal(Socretes)*.

Despite its successful application to many computational problems such as natural language processing, first-order logic poses a restriction on the type of variables appearing in quantifications to exclude predicates. Higher-order logic [12], on the other hand, allows variables to quantify over predicates. With such relaxation, higher-order logic facilitates the implementation of a knowledge-intensive system that takes other knowledge as its input in a closed form. An exemplar in figure 3 demonstrates a higher-order logic-based implementation of frequent pattern mining problem. The coding is based on the syntax of SWI prolog (www.swi-prolog.org).

```
frequent_pattern_mining:-
   min_support(V),                      % set minimum support
   makeC1(C),                           % create candidate 1-itemset
   makeL(C,L),                          % compute large itemset
   apriori_loop(L,1).                   % recursively run apriori

apriori_loop(L,N):- length(L) is 1,!.  % base case of recursion
apriori_loop(L,N):-                     % inductive step
   N1 is N+1,
   makeC(N1,L,C),    makeL(C,Res),
   apriori_loop(Res,N1).

makeC1(Ans):-
   input(D),                      % input data as a list, e.g. [[a], [a,b]]
   allComb(1,ItemList,Ans2),            % make combination of itemset
   maplist(countSS(D),Ans2,Ans).        % scan database
                                  % countSS is predicate passing as argument
                                  % to a higher-order predicate maplist
makeC(N,ItemSet,Ans):-
   input(D),
   allComb(2,ItemSet,Ans1),
   maplist(flatten,Ans1,Ans2),
   maplist(list_to_ord_set,Ans2,Ans3),
   list_to_set(Ans3,Ans4),
   include(len(N),Ans4,Ans5),       % include is also a higher-order predicate
   maplist(countSS(D),Ans5,Ans). % scan database find: List+N

makeL(C,Res):-                    %for all L creation
   include(filter,C,Ans),         % call higher predicates include
   maplist(head,Ans,Res).         %       and maplist

filter(_+N):- input(D),length(D,I),min_support(V), N>=(V/100)*I.
head(H+_,H).                      % filter and head are for pattern matching

% an arbitrary subset of the set containing given number of elements.
comb(0,_,[]).
comb(N,[X|T],[X|Comb]):-N>0,N1 is N-1,comb(N1,T,Comb).
comb(N,[_|T],Comb):-N>0,comb(N,T,Comb).

allComb(N,I,Ans):-
   setof(L,comb(N,I,L),Ans).      % setof is a second-order predicate

countSubset(A,[],0).
countSubset(A,[B|X],N):-not(subset(A,B)),countSubset(A,X,N).
countSubset(A,[B|X],N):-subset(A,B),countSubset(A,X,N1),N is N1+1.

countSS(SL,S,S+N):-countSubset(S,SL,N).
len(N,X):-length(X,N).
```

Fig. 3 Frequent-pattern mining implemented with a higher-order logic approach

```
[tumor-size=20-24, inv-nodes=0-2] =>[irradiate=no]                            conf:(1)
[age=50-59, inv-nodes=0-2, node-caps=no] =>[irradiate=no]                     conf:(1)
[age=50-59, menopause=ge40, node-caps=no] =>[irradiate=no]                    conf:(1)
[menopause=ge40, inv-nodes=0-2, deg-malig=1]=>[irradiate=no]                  conf:(1)
...
[node-caps=yes, breast=left, recurrence=yes]=>[irradiate=yes]                 conf:(0.63)
[breast-quad=left_low, recurrence=yes]=>[irradiate=yes]                       conf:(0.5)
[menopause=premeno, node-caps=yes]=>[irradiate=yes]                           conf:(0.5)
[node-caps=yes, breast=left]=>[irradiate=yes]                                 conf:(0.5)
...
[tumor-size=10-14, node-caps=no]=>[recurrence=no]                             conf:(1)
[tumor-size=10-14, irradiate=no]=>[recurrence=no]                             conf:(1)
[tumor-size=10-14, inv-nodes=0-2]=>[recurrence=no]                            conf:(1)
...
[node-caps=yes, breast=left, irradiate=yes]=>[recurrence=yes]                 conf:(0.91)
[menopause=premeno, node-caps=yes, irradiate=no]=>[recurrence=yes] conf:(0.83)
[menopause=premeno, node-caps=yes, breast=left]=>[recurrence=yes] conf:(0.83)
[menopause=premeno, node-caps=yes, deg-malig=3]=>[recurrence=yes] conf:(0.79)
[breast=left, breast-quad=left_low, irradiate=yes]=>[recurrence=yes]   conf:(0.77)
```

Fig. 4 Some part of a breast-cancer frequent pattern mining result

Frequent pattern mining is the discovery of relationships or correlations between items in a database. Let $I = \{i_1, i_2, i_3, ..., i_m\}$ be a set of m items and $DB = \{C_1, C_2, C_3, ..., C_n\}$ be a database of n cases or observations and each case contains items in I. A *pattern* is a set of items that occur in a case. The number of items in a pattern is called the length of the pattern. To search for all valid patterns of length 1 up to m in large database is computational expensive. For a set I of m different items, the search space for all distinct patterns can be as huge as 2^m-1. To reduce the size of the search space, the *support* measurement has been introduced [1]. The function *support*(P) of a pattern P is defined as a number of cases in DB containing P. Thus, *support*(P) = $|\{T \mid T \in \text{DB},\ P \subseteq T\}|$. A pattern P is called *frequent pattern* if the support value of P is not less than a predefined minimum support threshold *minS*. It is the *minS* constraints that help reducing the computational complexity of frequent pattern generation. The *minS* metric has an anti-monotone property and is applied as a basis for reducing search space of mining frequent patterns in algorithm Apriori [1].

We have tested our implementation with the breast cancer dataset taken from the UCI repository (http://www.ics.uci.edu/~mlearn/MLRepository.html). This medical data set is a collection of 191 observations on follow-up patients examining on a recurrence of breast cancer. Each patient's record contains ten attributes: age, menopause, tumor-size, inv-nodes, node-caps, deg-malignant, breast (left, right), breast-quad, irradiate (yes, no), recurrence (yes, no). We run frequent-pattern mining with minimum support 0.001 and show some of the discovered association rules in figure 4. Each rule is attached with the confidence value as a metric to evaluate accuracy of the induced rule. For the rule $A \Rightarrow B$, confidence is computed from proportion of *support(A&B)* to *support(A)*. The confidence value 1.0 thus implies a 100% accurate association rule.

The mining results shown in figure 4 are in the form of implication. Rule interpretation is straightforward. Taking the last rule in the figure as an example, it can be interpreted as "a patient who has cancer at her left breast in left lower quadrant position and had been treated with radiation implies that she will have a 77-percent chance of cancer recurrence." During the process of searching for frequent patterns, the support values of frequently occurred patterns can also be out put to quantify supporting evidence of the induced association rules.

5 Conclusions and Discussion

We have proposed the design of SUT-Miner, a knowledge-mining system. The system is intended to support knowledge acquisition in medical data and other domains that require new knowledge to support better decisions. The proposed system is an environment for knowledge discovery composing of tools and methods suitable for various kinds of mining such as data classification, regression, clustering, association mining. The intelligence of the system is supported by ontology technology to provide semantics for mining technique selection as well as for knowledge integration at a post-data mining step.

The implementation of the proposed system has been done based on the concept of higher-order logic which is an extension of the well-known first-order logic. With the expressive power of higher-order logic, program coding of the designed system is very concise as demonstrated in the paper. Program conciseness directly contributes to program verification and validation which are important issues in software engineering. The closed form of higher-order logic also supports constraint mining and higher-order mining, i.e. mining from previously discovered knowledge. Our future research is to extend the design of SUT-Miner to facilitate constraint and higher-order mining. Implementing a mechanism for knowledge inferring is also one of our research plans.

Acknowledgments. This work has been fully supported by research fund from Suranaree University of Technology (SUT) granted to the Data Engineering and Knowledge Discovery research unit. This research is also partly supported by grants from the National Research Council of Thailand (NRCT) and the Thailand Research Fund (TRF) under grant number RMU 5080026.

References

[1] Agrawal, R., Srikant, R.: Fast algorithm for mining association rules. In: Proc. VLDB, pp. 487–499 (1994)
[2] Alavi, M., Leidner, D.E.: Review: Knowledge management and knowledge management systems: Conceptual foundations and research issues. MIS Quarterly 25(1), 107–136 (2001)
[3] Bojarczuk, C.C., Lopes, H.S., Freitas, A.A., et al.: A constrained-syntax genetic programming system for discovering classification rules: Application to medical data sets. Artificial Intelligence in Medicine 30, 27–48 (2004)

[4] Bratsas, C., Koutkias, V., Kaimakamis, E., et al.: KnowBaSIGS-M: An ontology-based system for semantic management of medical problems and computerised algorithmic solutions. Computer Methods and Programs in Biomedicine 83, 39–51 (2007)
[5] Correia, R., Kon, F., Kon, R.: Borboleta: A mobile telehealth system for primary homecare. In: Proc. ACM Symposium on Applied Computing, pp. 1343–1347 (2008)
[6] De Raedt, L., Guns, T., Nijssen, S.: Constraint programming for itemset mining. In: Proc. KDD, pp. 204–212 (2008)
[7] Ghazavi, S., Liao, T.W.: Medical data mining by fuzzy modeling with selected features. Artificial Intelligence in Medicine 43(3), 195–206 (2008)
[8] Hristovski, D., Peterlin, B., Mitchell, J.A., et al.: Using literature-based discovery to identify disease candidate genes. Int. J. Medical Informatics 74, 289–298 (2005)
[9] Huang, M.J., Chen, M.Y., Lee, S.C.: Integrating data mining with case-based reasoning for chronic diseases prognosis and diagnosis. Expert Systems with Applications 32, 856–867 (2007)
[10] Hulse, N.C., Fiol, G.D., Bradshaw, R.L., et al.: Towards an on-demand peer feedback system for a clinical knowledge base: A case study with order sets. J. Biomedical Informatics 41, 152–164 (2008)
[11] Kakabadse, N.K., Kouzmin, A., Kakabadse, A.: From tacit knowledge to knowledge management: Leveraging invisible assets. Knowledge and Process Management 8(3), 137–154 (2001)
[12] Nadathur, G., Miller, D.: Higher-order Horn clauses. J. ACM 37, 777–814 (1990)
[13] Nguyen, D., Ho, T., Kawasaki, S.: Knowledge visualization in hepatitis study. In: Proc. Asia-Pacific Symposium on Information Visualization, pp. 59–62 (2006)
[14] Palaniappan, S., Ling, C.S.: Clinical decision support using OLAP with data mining. Int. J. Computer Science and Network Security 8(9), 290–296 (2008)
[15] Roddick, J.F., Fule, P., Graco, W.J.: Exploratory medical knowledge discovery: experiences and issues. ACM SIGKDD Explorations Newsletter 5(1), 94–99 (2003)
[16] Roddick, J.F., Spiliopoulou, M., Lister, D., et al.: Higher order mining. ACM SIGKDD Explorations Newsletter 10(1), 5–17 (2008)
[17] Ruppel, C.P., Harrington, S.J.: Sharing knowledge through intranets: A study of organizational culture and intranet implementation. IEEE Transactions on Professional Communication 44(1), 37–51 (2001)
[18] Sahama, T.R., Croll, P.R.: A data warehouse architecture for clinical data warehousing. In: Proc. 12th Australasian Symposium on ACSW Frontiers, pp. 227–232 (2007)
[19] Satyadas, A., Harigopal, U., Cassaigne, N.P.: Knowledge management tutorial: An editorial overview. IEEE Transactions on Systems, Man and Cybernetics, Part C 31(4), 429–437 (2001)
[20] Shillabeer, A., Roddick, J.F.: Establishing a lineage for medical knowledge discovery. In: Proc. 6th Australasian Conf. on Data Mining and Analytics, pp. 29–37 (2007)
[21] Thongkam, J., Xu, G., Zhang, Y., et al.: Breast cancer survivability via AdaBoost algorithms. In: Proc. 2nd Australasian Workshop on Health Data and Knowledge Management, pp. 55–64 (2008)
[22] Truemper, K.: Design of logic-based intelligent systems. John Wiley & Sons, New Jersey (2004)
[23] Uramoto, N., Matsuzawa, H., Nagano, T., et al.: A text-mining system for knowledge discovery from biomedical documents. IBM Systems J. 43(3), 516–533 (2004)
[24] Zhou, X., Liu, B., Wu, Z.: Text mining for clinical Chinese herbal medical knowledge discovery. In: Discovery Science 8th Int. Conf., pp. 396–398 (2005)
[25] Zhuang, Z.Y., Churilov, L., Burstein, F.: Combining data mining and case-based reasoning for intelligent decision support for pathology ordering by general practitioners. European J. Operational Research 195(3), 662–675 (2009) doi: 10.1016/j.ejor.2007.11.003

A New Pruning Technique for the Fuzzy ARTMAP Neural Network and Its Application to Medical Decision Support

Shahrul N.Y., Lakhmi Jain, and C.P. Lim

Abstract. This paper describes a neural network-based classification tool that can be deployed for data-based decision support tasks. In particular, the Fuzzy ARTMAP (FAM) network is investigated, and a new pruning technique is proposed. The pruning technique is implemented successively to eliminate those rarely activated nodes in the category layer of FAM. Three data sets with different characteristics are used to analyze its effectiveness. In addition, a benchmark medical problem is used to evaluate its applicability as a decision support tool for medical diagnosis. From the experiment, the pruning technique is able to improve classification performances, as compared with those of to the original FAM network, as well as other machine learning methods. More importantly, the pruning technique yields more stable performances with fewer nodes, and results in a more parsimonious FAM network for undertaking data classification and decision support tasks.

Keywords: Fuzzy ARTMAP, pruning, classification, decision support, medical diagnosis.

1 Introduction

Advances in computer technologies have opened up the possibility of everyone in the use of computerized decision support systems (DSSs) in various activities. Traditionally, DSSs are mainly used for supporting decision making in management problems. Currently, they are widely used as a strategic tool by many organizations in different domains.

A DSS normally consists of several building components that are integrated together. The model or solver component [7] is one of the important modules of a

Shahrul N.Y., Lakhmi Jain, and C.P. Lim*
Knowledge-Based Intelligent Engineering Systems (KES) Centre,
School of Electrical and Information Engineering, University of South Australia, Australia
e-mail: Yaasn001@postgrads.unisa.edu.au,
Lakhmi.Jain@unisa.edu.au, cp.lim@unisa.edu.au
* School of Electrical and Electronic Engineering, University of Science Malaysia, Malaysia

K. Nakamatsu et al. (Eds.): New Advan. in Intel. Decision Techno., SCI 199, pp. 161–169.
springerlink.com

DSS. In this component, the decision is generated by the DSS based on the knowledge and information gathered a problem domain. There are various techniques that can be used to develop this component. This paper focuses on the use of neural network to design and develop a data-based classification and decision making model for DSSs. There are many types of neural network that have been used in many applications especially for recognition and classification purposes. One of them is the Fuzzy ARTMAP (FAM) [5] network. FAM has several salient properties as a learning system as compared with other types of neural networks. One of the main advantages of FAM is the ability to construct its network structure automatically with both online and offline learning capabilities. FAM has been known to produce a good classification performance with simpler training steps in many applications. For example, Wee *et al.* [12] used FAM to classify rice grain images. They demonstrated that the performance of FAM, as compared with that of the Multi-Layer Perceptron (MLP) network, with back propagation learning, can yield a high classification rate.

Besides, FAM has been identified as an incremental learning model. This is because FAM has the ability to learn how to learn incoming patterns on a one-by-one basis as the patterns become available for learning [8]. This feature makes FAM suitable for performing data classification tasks in on-line learning environments. During on-line learning, new input patterns can be learned by the FAM network without re-training using old and new data patterns [13]. In addition, FAM is a transparent learning model since it is easy to explain why an input pattern x produces a particular output y [2]. Here, transparency refers to the ability of the designer to explain the reason why the network responds to a specific input pattern in a particular way. Despite the advantages, there are issues that must be considered during the development and implementation of FAM as a pattern classification tool. One of these is that FAM is a growing network, and its keep adding nodes into its structure incrementally. As such, the network complexity increases with time, as more and more nodes are created. Several researchers have proposed methods to remedy this problem. One approach is by allowing some training error for FAM, e.g. Micro ARTMAP (μAM) [9] and Bayesian ARTMAP (BAM) [11]. This paper introduces a new pruning technique to reduce the number of nodes in FAM, in an attempt to reduce the FAM network complexity.

The organizations of this paper are as follows: In Section 2, we discuss the dynamics of FAM and the problem of excess nodes in FAM. Section 3 explains the data sets used, cross validation, and setting of FAM parameters, as well as the experimental results and discussion. Section 4 is about an analysis of the effectiveness of FAM in medical diagnosis application. Section 5 gives the conclusions and suggestions for further work.

2 Fuzzy ARTMAP (FAM)

FAM was introduced by Carpenter, G. *et al.* in 1992 [5]. FAM is a supervised version of *ART* that consists of two blocks of unsupervised *Fuzzy ART* and a Map Field. The map field links *ARTa* and *ARTb* together. Normally, *ARTa* receives the stream of input patterns and *ARTb* receives the stream of the target outputs associated with the input patterns, and FAM is able to process both the analogue and binary input patterns.

Fig. 1 shows the ART architecture. It consists of two layers of nodes, i.e. recognition/category and input layers. The input layer comprises nodes corresponding to the number of input dimensions with complement coding [5]. The recognition layer is where all the nodes or categories are created in an incremental manner. These nodes are essential because they represent prototypes of the input patterns, and are used to recall a predicted output during the test phase. In the F_0 layer, the original input vector a goes through a normalization technique called complement coding [5]. Weights W_{ji}^{a} represent the weight vector between the F_2 (recognition/category) layer and the F_1 layer, with j refers to the number of nodes or categories in F_2, and i is the number of nodes in F_1, while $_a$ refers to the vigilance parameter of *ARTa*.

Fig. 1 also illustrates the architecture of the map field, i.e., the connection between F_2 of *ARTa* and F_2 of *ARTb*. Weights W_{jk}^{ab} characterize the weight vectors that connect F_2 of *ARTa* with the map field, with k defines the number of categories in the map field. The number for categories in the map field and F_2 of *ARTb* are same. The output vector of the map field is X^{ab}. During the training phase, an input vector and its desired output vector are presented to *ARTa* and *ARTb*, respectively. *ARTa* and *ARTb* modules classify the input and desired output vector into categories. Then, the input to the map field module uses a map field vigilance parameter to determine whether the *ARTa* winning category is linked to correct target category in *ARTb*, and corrections are made accordingly.

The first step involved in the learning phase is the computation of the category choice in the Fuzzy ART module. The Category Choice Function (CCF), T_j, for each input I and $F2$ node j is defined using Equation (1), where the fuzzy *AND* operator ^ is defined in Equation (2) and the norm |.| in Equation (3).

$$T_j = \frac{|I \wedge w_j|}{\alpha + |w_j|} \tag{1}$$

$$(I \wedge w)_i = \min(I_i, w_i) \tag{2}$$

$$|I| \equiv \sum_{i=1}^{M} |I_i| \tag{3}$$

The next step is to find the maximum value of T_j, a competitive process, using equation (4). Only one node in $F2$ with the highest value of T_j is selected. Resonance is said to occur if the vigilance test (Equation (5) and (6)) is satisfied. When the Jth category is chosen, $y_j = 1$ and $y_j = 0$ for $j \neq J$. If V_j does not meet the vigilance test, choice function T_j is set to zero for the duration of the input presentation, and a new index j is chosen using Equation (4). All these operations occur simultaneously in *ARTa* and *ARTb*.

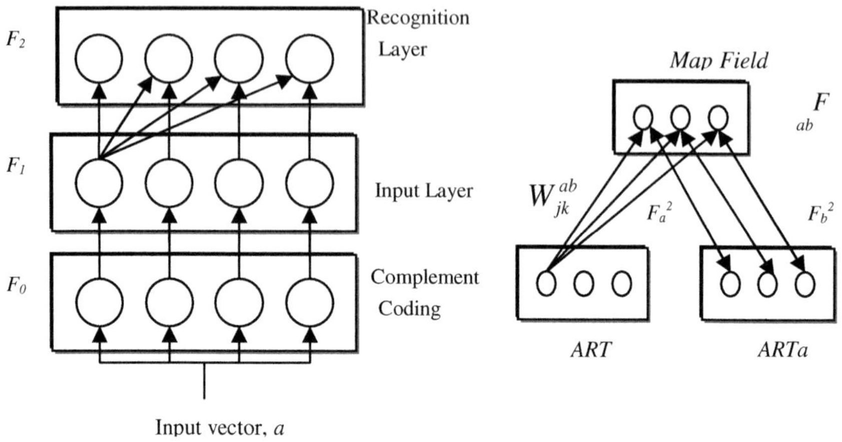

Fig. 1 Fuzzy ART architecture (right) and Map Field architecture (left)

$$T_j = \max\{T_j : j = 1.....N\} \tag{4}$$

$$V_j = \left|I \wedge w_j\right| / \left|I\right| \tag{5}$$

$$V_j \geq \rho \tag{6}$$

The next stage is to find the value of X^{ab}. If the Jth of F^a_2 node is active and K^{th} of F^b_2 is active, the value of X^{ab} can be found using Equation (7). When both *ARTa* and *ARTb* are active, and $X^{ab} \neq 0$, then the weight vector between F^a_2 layer and map field F^{ab} is set according to Equation (8). If $X^{ab} = 0$, then the vigilance parameter of *ARTa* is increased such that the competitive process starts again, and a new category that satisfies Eq. (5) is found. If there is no winner, a new node is created and the value of input vectors is assigned as its weight. Learning ensues by using Equation (9) whereby the weight vector w_j of *ARTa* is updated to encode the input pattern.

$$X^{ab} = y^b \wedge w_j^{ab} \tag{7}$$

$$w_j^{ab} = \begin{cases} 1 & j = J; k = K \\ 0 & otherwise \end{cases} \tag{8}$$

$$w_j^{(new)} = \beta\left(1 \wedge w_j^{(old)}\right) + (1 - \beta) w_j^{(old)} \tag{9}$$

In the test phase, Fuzzy *ARTa* receives an input vector. The category choice and category match computations are the same as in the training phase. Therefore, the output of map field X^{ab} related to the Jth category of *ARTa* can be defined using Equation (10). A link traced from the map filed to *ARTb* leads to the predicted target output.

$$X^{ab} = w_j^{ab} \tag{10}$$

2.1 The Proposed Pruning Technique

In the pruning stage, every node j in the recognition layer is assigned with a parameter, Ω_j. The pruning process only occurs during the training phase of FAM. Each time when a new learning epoch begins, the value of Ω_j is set to zero. During the learning epoch, Ω_j changes to one when the associated node learns an input pattern. This value remains as one irrespective of the number of times the node becomes the winning node whereby its weight is updated. At the end of each learning epoch, any node with Ω_j equal to zero is pruned. These steps are repeated for the next learning epoch until the learning phase is completed.

3 Experiments with Benchmark Problems

To analyze the effectiveness of the new pruning technique, three artificial data sets are used. They are the Gaussian 2 Dimensional, Concentric, and Clouds data sets from the ELENA project databases [10]. Fig. 2 and Fig. 3 are graphical illustrations of Gaussian 2-D, Concentric, and Clouds, respectively. The Gaussian 2-D data set demonstrates a densely overlapped data distribution. The Concentric data set has nested classes without overlapping. The Clouds data set shows intersection of the class distribution, and has a high degree of nonlinearity in the class boundaries. The details of the data sets are summarized in Table 1.

Table 1 Characteristics of ELENA dataset

	Instance/Size	Attributes	Classes
Gaussian 2D	1000	2	2
Concentric	1000	2	2
Clouds	1000	2	2

In this work, 10-folds cross validation is used where each data set is divided into 10 mutually exclusive subsets, designated as G_1, G_2, G_3,....G_{10}, of equal size. Then, FAM is trained and tested 10 times. The Percentage of Correct

Classification (*PCC*) is defined as the number of correct classification divided by the number of data available in the data set, as in Eq. (11). Note that $(x,y)_t=1$ for correct prediction, otherwise $(x,y)_t=0$, and n refers to the number of data sample tested.

$$PCC_k = 100\frac{1}{G}\sum_{k=1}^{10} NCC_k \tag{11}$$

$$NCC_k = \sum_{t=1}^{n} \sigma(x,y)_t \tag{12}$$

The Standard Deviation Value (STD) is also calculated to reflect the distribution of FAM results. Other FAM parameters are set as in Table 2.

Table 2 FAM Parameter setting

Parameter	Value
Baseline Vigilance	0.5
Vigilance of ARTb	1
Learning,	1
Choice,	0.0001

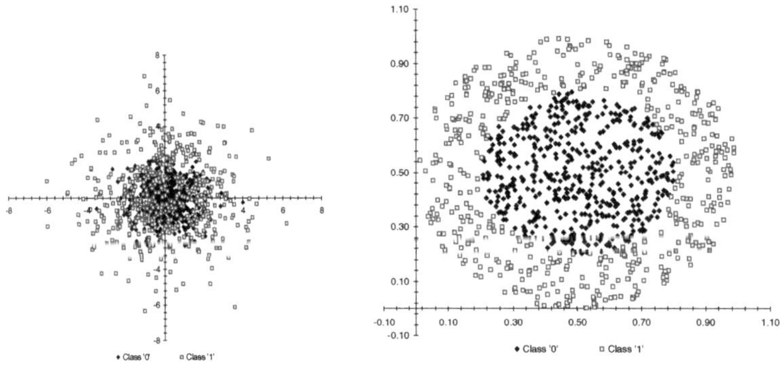

Fig. 2 The Gaussian 2D data (left) and Concentric data (right)

3.1 Results and Discussion

The FAM classification results, both the original and pruned versions, are shown in Table 3. The pruning technique is able to improve accuracy rate by 5.1% to 6.3%, as compared with those of original FAM. For example, the classification

Fig. 3 The illustration of Clouds data

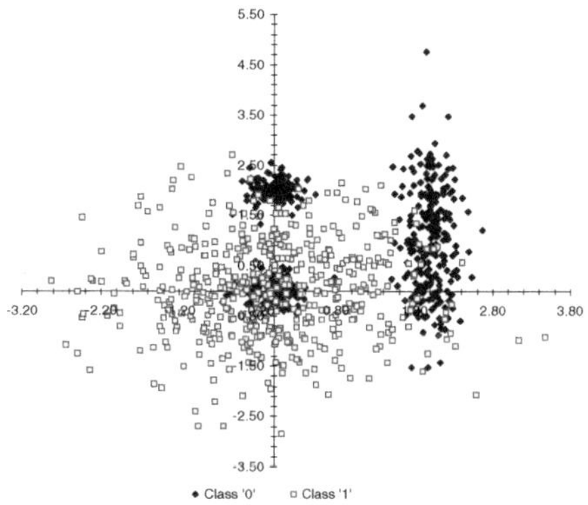

rate of Concentric increased by 6.3% to 97.6%. More importantly, the pruning technique is able to reduce 34 to 136 nodes, as compared with those of original FAM. This shows the effectiveness of the proposed pruning technique in producing a compact FAM network structure with improved classification performance.

Table 3 Classification results

		Nodes		Classification Result, *PCC*[b]	
Dataset	FAM	Average	*STD*[a]	Average	*STD*
GUSSIAN 2-D	Original	933.9	3.21	58.60	0.050
	Pruned	899.4	1.27	63.70	0.042
CONCENTRIC	Original	967.9	79.39	91.30	0.043
	Pruned	831.8	42.67	97.60	0.011
CLOUDS	Original	960.8	11.47	80.40	0.054
	Pruned	894.1	4.12	86.20	0.025

[a] Standard Deviation [b] Percentage Correct Classification

The lower STD values also suggest the stability of the performance of the pruned FAM network. In other words, the pruning technique makes the original FAM less sensitive to the order of data presentation. This will make FAM more suitable in on-line learning environment [4]. Note that the results of Gaussian 2-D are low. This is caused by the densely overlapped region in data distribution. As illustrated in Fig. 2, there is no clear delineation between the boundaries of the two Gaussian 2-D classes. Nevertheless, the pruning technique is able to increase the classification performance with fewer numbers of nodes.

4 Application to Medical Diagnosis

In this work, the Wisconsin Breast Cancer (WBC) [3] data set is used to evaluate the applicability of FAM with the proposed pruning technique as a medical decision support tool. The WBC data set contained 699 records of virtually assessed nuclear features of fine needle aspirates from patients, with 458 benign and 241 malignant cases of breast cancer. The same training and test procedures as in Section 3 were adopted to assess the FAM performance. Table 4 shows the classification results of FAM (both original and pruned versions). It can be clearly seen that while original FAM produced the lowest classification rate of 91.88%, the pruned technique is able to improve its performance to 96.76%, which is the highest as compared with those of other machine learning methods (Table 5). The results demonstrate the effectiveness of the pruning technique in producing a parsimonious FAM network with improved classification performance.

Table 4 The classification of WBC data set

Method	Classification Result (%)
C4.5 [8]	94.74
Optimized-LVQ [8]	96.70
Supervised Fuzzy Clustering [1]	95.57
Perceptron Decision Tree (FAT) [9]	96.45
Fuzzy ARTMAP (Original)	91.88
Fuzzy ARTMAP (Pruned)	96.76

5 Summary

In this paper, we have introduced a new pruning technique to reduce the number of category nodes in the FAM network. The results obtained indicate that the pruning technique can improve the classification performance result of FAM with a more compact network structure. The results also are more stable (in terms of STD) as compared with those from original FAM. Applicability of the proposed approach as a decision support tool to medical diagnosis is also demonstrated using the WBC problem. The results, again, positively demonstrate that the pruned FAM network, as compared with other machine learning methods, is able to produce high performance with a less complex network structure. The proposed pruning technique can be incorporated with other FAM-based networks, e.g. Micro ARTMAP and Bayesian ARTMAP. In addition, other pruning strategies can be implemented and compared with the proposed technique. Besides, more benchmark and real data sets can be used to further ascertain the performance and stability of the pruned network. All these constitute the direction of further work of this research.

References

1. Abonyi, J., Szeifert, F.: Supervised fuzzy clustering for the identification of fuzzy classifiers. Pattern Recognition Letter 24, 2195–2207
2. Anagnostopoulos, G.C., et al.: Reducing generalization error and category proliferation in ellipsoid ARTMAP via tunable misclassification error tolerance: boosted ellipsoid ARTMAP. In: International Joint Conference on Neural Networks, vol. 3, pp. 2650–2655 (2002)
3. Blake, C.L., Merz, C.J.: UCI Repository of Machine Learning Databases, University of California at Irvine (cited March 15, 2008), `http://www.ics.uci.edu/~mlearn/MLRepository.html`
4. Bouchachia, A., Gabrys, B., Sahel, Z.: Overview of Some Incremental Learning Algorithms. In: IEEE International Fuzzy Systems Conference, 2007, pp. 1–6 (2007)
5. Carpenter, G.A., Grossberg, S., Markuzon, N., Reynolds, J.H., Rosen, D.B.: Fuzzy ARTMAP: A neural network architecture for incremental supervised learning of analog multidimensional maps. IEEE Transactions on Neural Networks 3, 698–713
6. Carpenter, G.A., Grossberg, S., Markuzon, N., Reynolds, J.H., Rosen, D.B.: Fuzzy ARTMAP: An adaptive resonance architecture for incremental learning of analog maps. In: International Joint Conference on Neural Networks, vol. 3, pp. 309–314 (1992)
7. Dong, C.S.J., Loo, G.S.L.: Flexible web-based decision support system generator (FWDSSG) utilising software agents. In: 12th International Workshop on Database and Expert Systems Applications, pp. 892–897 (2001)
8. Le, Q., Anagnostopoulos, G.C., Georgiopoulos, M., Ports, K.: An experimental comparison of semi-supervised ARTMAP architectures, GCS and GNG classifiers. In: IEEE International Joint Conference on Neural Networks, 2005, vol. 5, pp. 3121–3126 (2005)
9. Sanchez, E.G., Dimitriadis, Y.A., Cano-Izquierdo, J.M., Lopez-Coronado, J.: μARTMAP: use of mutual information for category reduction in Fuzzy ARTMAP. IEEE Transactions on Neural Networks 13, 58–69
10. Verleysen, M., Bodt, E.D., Wertz, V.: UCL Neural Network Group, Université catholique de L-ouvain (Cited March 15, 2008), `http://www.dice.ucl.ac.be/neural-nets/Research/Projects/ELENA/elena.htm`
11. Vigdor, B., Lerner, B.: The Bayesian ARTMAP. IEEE Transactions on Neural Networks 18, 1628–1644
12. Wee, C.Y., Paramesran, R., Takeda, F., Tsuzuki, T., Kadota, H., Shimanouchi, S.: Classification of rice grains using Fuzzy Artmap neural network. In: Asia-Pacific Conference on Circuits and Systems, vol. 2, pp. 223–226
13. Zhong, M., Rosander, B., Georgiopoulos, M., Anagno-stopoulos, G.C., Mollaghasemi, M., Richie, S.: Experiments with Safe ARTMAP and Comparisons to Other ART Networks. In: International Joint Conference on Neural Networks, pp. 720–727 (2006)

Image Color Space Transform with Enhanced KLT

Roumen Kountchev and Roumiana Kountcheva

Abstract. The use of the Karhunen-Loève Transform (KLT) for the processing of the image primary color components gives as a result their decorrelation, which ensures the enhancement of such operations as: compression, color-based segmentation, etc. The basic problem is the high computational complexity of the KLT. In this paper is offered a simplified algorithm for the calculation of the KL color transform matrix. The presented approach is based on non-recursive approach for the color covariance matrix eigenvectors detection. The new algorithm surpasses the existing similar algorithms in its lower computational complexity, which is a prerequisite for fast color segmentation or for adaptive coding of color images aimed at real time applications.

Keywords: Karhunen-Loève transform, color transforms, color space models, color covariance matrix, eigenvalues and eigenvectors, angles of Euler rotation, color image compression.

1 Introduction

The methods for color image space transform recently attract significant interest because they influence the efficiency of the processing aimed at the information redundancy reduction and color features extraction. Many attempts have been made to model the color perception by researchers working in various fields: psychology, computer vision, image processing and retrieval, computer graphics, etc. The color spaces used in image processing are derived from visual system models, adopted from technical domains or developed especially for image processing. The commonly used color space for image representation is based on RGB, XYZ, YCbCr, YUV, YCoCg, Lab, HSV/HSL, etc. color models [1-5]. All these color transform techniques do not depend on the image content. Unlike them, the Karhunen-Loeve Transform (KLT) [6] highly depends on the image content. The

Roumen Kountchev
Technical University – Sofia, Department of Radio Communications and Video Technologies, Boul. Kl. Ohridsky 8, Sofia 1000, Bulgaria
e-mail: rkountch@tu-sofia.bg

Roumiana Kountcheva
T&K Engineering Co. Mladost 3, POB12, Sofia 1712, Bulgaria
e-mail: kountcheva_r@yahoo.com

K. Nakamatsu et al. (Eds.): New Advan. in Intel. Decision Techno., SCI 199, pp. 171–182.
springerlink.com

KLT is utilized as a tool to eliminate the image redundancy, because the transformation components are highly uncorrelated. The KLT has found many applications in traditional fields such as communications [7-9] and computer vision. In computer vision, it is used for a variety of tasks such as image segmentation, face and object recognition, motion estimation, object tracking, etc. [10-13]. The KLT is however used infrequently as it is dependent on the image statistics, i.e. when the statistics change the KLT matrix should be changed as well. Because of this signal dependence, general fast algorithm is not developed. Additional disadvantage is its higher computational complexity in comparison with the deterministic image color space transforms. Some solutions of the problem have already been proposed. The basic methods used for the matrix eigenvectors calculation are given in [14]. One or the relatively simple methods is the iterative approach proposed by Jacobi. In accordance with it, the KLT matrix is presented as a product of rotation matrices, which consecutively rotate the image vector around the corresponding coordinate axis in the vector space. While this technique is quite simple, for big matrices it can take a large number of calculations. A more efficient approach for larger, symmetric matrices divides the problem into two stages. The Householder algorithm can first be applied to reduce a symmetric matrix into a tridiagonal form in a finite number of steps. Once the matrix is in this simpler form, an iterative method such as QR factorization (the matrix is represented by the product of upper triangular and orthogonal matrices) can be used to generate the eigenvalues and eigenvectors. The advantage of this approach is that the factorization of the simplified tridiagonal matrix requires fewness iterations than the Jacobi method. Significant interest attracted the iterative methods for principal components extraction with neural networks which do not require the calculation of a covariance matrix [15-17]. These techniques update the estimate of the eigenvectors for each input training vector. While these algorithms have some advantages over covariance-based methods, there are still some concerns over stability and convergence [18]. In particular, analytical solution for the components of the covariance matrix eigenvectors exists for the case when the color vectors depict first order stationary Markov process [6].

The goal of this work is to present a simplified method for the calculation of the KLT matrix for the RGB image color space. The paper is arranged as follows: section 2 introduces the principle of the direct calculation of the KLT matrix; in section 3 is evaluated the computational complexity of the new method and are given some results obtained by means of the method modeling and the comparison with the determined transform YCbCr; section 4 is the Conclusion.

2 Direct Calculation of the KLT Matrix

The color of every pixel in the digital image is represented in the color space by the vector $\vec{C} = [C_1, C_2, C_3]^t$, whose coordinates (C_1, C_2, C_3) correspond to the chosen primary colors (for example, RGB, etc.). The general algorithm for efficient coding of digital color images based on preliminary color space transform comprises the following basic operations:

1. Direct (linear or nonlinear) transform of the vector $\vec{C}$ into a new color space aimed to obtain maximum uncorrelated components. When linear transforms are used, the corresponding transformed color vector $\vec{L}=[L_1,L_2,L_3]^t$ is defined by the relation $\vec{L}=[T]\vec{C}$, where [*T*] is the color transform matrix of size 3×3;

2. Coding of the transformed vector $\vec{L}$ components in correspondence with the relations $L_1^q=\psi_1(L_1)$, $L_2^q=\psi_2(L_2)$, $L_3^q=\psi_3(L_3)$ were $\psi_1(*),\psi_2(*)$ and $\psi_3(*)$ are functions, which correspond to the selected coding method for every component (aimed at the reduction of its psycho-visual and/or statistical redundancy). As a result is obtained the coded vector $\vec{L}_q=\psi(\vec{L})=\left[L_1^q,L_2^q,L_3^q\right]^t$, whose components are transferred to the decoder;

3. Decoding of the vector components in correspondence with the relations $L_1'=\psi_1^{-1}(L_1^q)$, $L_2'=\psi_2^{-1}(L_2^q)$, $L_3'=\psi_3^{-1}(L_3^q)$, where $\psi_1^{-1}(*),\psi_2^{-1}(*)$ and $\psi_3^{-1}(*)$ are the inverse transform functions. As a result is obtained the vector $\vec{L}'$;

4. Inverse transform of the decoded vector $\vec{L}'=[L_1',L_2',L_3']^t$. When orthogonal transforms are used, the restored color vector $\vec{C}'=[C_1',C_2',C_3']^t$ is defined by the relation $\vec{C}'=[T]^{-1}\vec{L}'$, where $[T]^{-1}$ is the inverse transform matrix of size 3×3.

The choice of the color space transform depends on the primary colors and the restored image quality, the compression ratio and the computational complexity required. In order to obtain high efficiency of the color image coding the components of the transformed color vector $\vec{L}$ for every pixel should be uncorrelated. This requirement is satisfied by the KLT model only.

In correspondence with this model the RGB color vector for the sth pixel $\vec{C}_s=[R_s,G_s,B_s]^t$ is transformed into the vector $\vec{L}_s=[L_{1s},L_{2s},L_{3s}]^t$ using the matrix [Φ] of the linear orthogonal KLT. The elements of the matrix Φ_{ij} are defined by the statistical characteristics of the image pixels colors and could be calculated in the way described below. The covariance color matrix $[K_C]$ of size 3×3, is calculated first:

$$[K_C]=\frac{1}{S}\sum_{s=1}^{S}(\vec{C}_s\vec{C}_s^t)-\vec{m}_c\vec{m}_c^t=\begin{bmatrix}k_{11} & k_{12} & k_{13}\\ k_{21} & k_{22} & k_{23}\\ k_{31} & k_{32} & k_{33}\end{bmatrix}, \tag{1}$$

where $\vec{m}_c=[\overline{R},\overline{G},\overline{B}]^t$ is the mean color vector. Let $\overline{x}=E(x_s)=\frac{1}{S}\sum_{s=1}^{S}x_s$ is the operator for the calculation of the mean value of x_s, for $s=1, 2, \ldots, S$. Then the elements of the vector $\vec{m}_c$ and of the matrix $[K_C]$ could be represented as follows:

$$\overline{R} = E(R_s), \quad \overline{G} = E(R_s), \quad \overline{B} = E(B_s), \tag{2}$$

$$k_{11}= k_1= E(R_s^2)-(\overline{R})^2,\ k_{22}= k_2= E(G_s^2)-(\overline{G})^2,\ k_{33}= k_3= E(B_s^2)\text{-}(\overline{B})^2, \tag{3}$$

$$k_{12}= k_{21}= k_4= E(R_sG_s)-(\overline{R})(\overline{G}),\quad k_{23}= k_{32}= k_6= E(G_sB_s)-(\overline{G})(\overline{B}), \tag{4}$$

$$k_{13}= k_{31}= k_5= E(R_sG_s)-(\overline{R})(\overline{G}). \tag{5}$$

Here $S = M{\times}N$ is the number of the pixels in the image with components $R_s\ G_s\ B_s$, whose mean values are correspondingly $\overline{R},\ \overline{G},\ \overline{B}$.

In this work below is presented one new approach for the calculation of the eigenvalues and eigenvectors of the covariance matrix $[K_C]$. The eigenvalues $\lambda_1,\lambda_2,\lambda_3$ of the matrix $[K_C]$ are the solution of its characteristic equation:

$$det|\, k_{ij}-\lambda\delta_{ij}\, |=\lambda^3 + a\lambda^2 + b\lambda + c = 0, \tag{6}$$

where:

$$\delta_{ij}=\begin{cases} 1, & i=j, \\ 0, & i\neq j. \end{cases} \quad \begin{aligned} & a=-(k_1+k_2+k_3),\ b=k_1k_2+k_1k_3+k_2k_3-(k_4^2+k_5^2+k_6^2), \\ & c=k_1k_6^2+k_2k_5^2+k_3k_4^2-(k_1k_2k_3+2k_4k_5k_6), \end{aligned} \tag{7}$$

The matrix $[K_C]$ is symmetrical and its eigenvalues are always real numbers. They can be defined using the Cardano relations [19] for the "casus irreducibilis" (or the so-called "trigonometric solution"):

$$\lambda_1=2\sqrt{\frac{|p|}{3}}\cos\left(\frac{\varphi}{3}\right)-\frac{a}{3};\ \lambda_2=-2\sqrt{\frac{|p|}{3}}\cos\left(\frac{\varphi+\pi}{3}\right)-\frac{a}{3};\ \lambda_3=-2\sqrt{\frac{|p|}{3}}\cos\left(\frac{\varphi-\pi}{3}\right)- \tag{8}$$
$$\text{for } \lambda_1 \geq \lambda_2 \geq \lambda_3 \geq 0,$$

$$\varphi=\arccos\left[-q/2\Big/\sqrt{(|\,p\,|/3)^3}\,\right],\ q=2(a/3)^3-(ab)/3+c,\ p=-(a^2/3)+b\,\cdot \tag{9}$$

The eigenvectors $\vec{\Phi}_1,\vec{\Phi}_2,\vec{\Phi}_3$ of the covariance matrix $[K_C]$ are the solution of the set of equations:

$$[K_C]\,\vec{\Phi}_m=\lambda_m\vec{\Phi}_m \ \text{ and } \left|\vec{\Phi}_m\right|^2=\sum_{i=1}^{3}\Phi_{mi}^2=1 \quad \text{for } m = 1, 2, 3. \tag{10}$$

The last equation derives from the conditions for orthogonality and normalization of the obtained eigenvectors:

$$\vec{\Phi}_s^t \vec{\Phi}_k = \sum_{i=1}^{3} \Phi_{is}\Phi_{ik} = \begin{cases} 1 & \text{for } s = k ; \\ 0 & \text{for } s \neq k . \end{cases} \quad \text{for } s,\ k = 1,2,3. \tag{11}$$

Solving Eq. (10) are defined the components of the mth eigenvector $\vec{\Phi}_m = [\Phi_{m1}, \Phi_{m2}, \Phi_{m3}]^t$, which corresponds to the eigenvalue λ_m (Eq. 8). Then:

$$\Phi_{m1} = A_m / P_m;\ \Phi_{m2} = B_m / P_m;\ \Phi_{m3} = D_m / P_m \quad \text{for } m = 1,2,3 \tag{12}$$

$$A_m = (k_3 - \lambda_m)[k_5(k_2 - \lambda_m) - k_4 k_6],\ \mathrm{B_m} = (k_3 - \lambda_m)[k_6(k_1 - \lambda_m) - k_4 k_5], \tag{13}$$

$$D_m = k_6[2k_4k_5 - k_6(k_1 - \lambda_m)] - k_5^2(k_2 - \lambda_m),\ P_m = \sqrt{A_m^2 + B_m^2 + D_m^2} \neq 0. \tag{14}$$

The matrix $[\Phi]$ whose rows comprise the components Φ_{ms} of the eigenvectors $\vec{\Phi}_m$ is:

$$[\Phi] = \begin{bmatrix} \vec{\Phi}_1^t \\ \vec{\Phi}_2^t \\ \vec{\Phi}_3^t \end{bmatrix} = \begin{bmatrix} \Phi_{11} & \Phi_{12} & \Phi_{13} \\ \Phi_{21} & \Phi_{22} & \Phi_{23} \\ \Phi_{31} & \Phi_{32} & \Phi_{33} \end{bmatrix} \text{ for } m = 1,2,3 \tag{15}$$

The color vector $\vec{C}_s = [R_s, G_s, B_s]^t$ is then transformed into the vector $\vec{L}_s = [L_{1s}, L_{2s}, L_{3s}]^t$ using the direct KLT:

$$\begin{bmatrix} L_{1s} \\ L_{2s} \\ L_{3s} \end{bmatrix} = \begin{bmatrix} \Phi_{11} & \Phi_{12} & \Phi_{13} \\ \Phi_{21} & \Phi_{22} & \Phi_{23} \\ \Phi_{31} & \Phi_{32} & \Phi_{33} \end{bmatrix} \left(\begin{bmatrix} R_s \\ G_s \\ B_s \end{bmatrix} - \begin{bmatrix} \overline{R} \\ \overline{G} \\ \overline{B} \end{bmatrix} \right) \text{ for } s = 1, 2, \ldots, S. \tag{16}$$

The components of the vector $\vec{L}_s = [L_{1s}, L_{2s}, L_{3s}]^t$ could be coded using various methods (decimation and interpolation, filtration, orthogonal transforms, quantization, etc.) in correspondence with relations $L_{1s}^q = \psi_1(L_{1s})$, $L_{2s}^q = \psi_2(L_{2s})$, $L_{3s}^q = \psi_3(L_{3s})$ and then is obtained the coded vector $\vec{L}_s^q = \psi(\vec{L}_s) = [\psi_1(L_{1s}), \psi_2(L_{2s}), \psi_3(L_{3s})]^t$.

For the restoration of the vector $\vec{L}_s^q$ components, are applied inverse functions in correspondence to relations $\hat{L}_{1s} = \psi_1^{-1}(L_{1s}^q)$, $\hat{L}_{2s} = \psi_2^{-1}(L_{2s}^q)$, $\hat{L}_{3s} = \psi_3^{-1}(L_{3s}^q)$ and is obtained the decoded vector $\vec{\hat{L}}_s = [\hat{L}_{1s}, \hat{L}_{2s}, \hat{L}_{3s}]^t$. Using the inverse KLT the vector $\vec{\hat{L}}_s$ is transformed into the restored color vector $\vec{\hat{C}}_s = [\hat{R}_s, \hat{G}_s, \hat{B}_s]^t$:

$$\begin{bmatrix} \hat{R}_s \\ \hat{G}_s \\ \hat{B}_s \end{bmatrix} = \begin{bmatrix} \Phi_{11} & \Phi_{21} & \Phi_{31} \\ \Phi_{12} & \Phi_{22} & \Phi_{32} \\ \Phi_{13} & \Phi_{23} & \Phi_{33} \end{bmatrix} \begin{bmatrix} \hat{L}_{1s} \\ \hat{L}_{2s} \\ \hat{L}_{3s} \end{bmatrix} + \begin{bmatrix} \overline{R} \\ \overline{G} \\ \overline{B} \end{bmatrix} \text{ for } s = 1, 2, \ldots, S, \tag{17}$$

where

$$[\Phi]^{-1} = [\Phi]^t = [\vec{\Phi}_1, \vec{\Phi}_2, \vec{\Phi}_3] = \begin{bmatrix} \Phi_{11} & \Phi_{21} & \Phi_{31} \\ \Phi_{12} & \Phi_{22} & \Phi_{32} \\ \Phi_{13} & \Phi_{23} & \Phi_{33} \end{bmatrix} \tag{18}$$

is the matrix of the inverse KLT.

Unlike the deterministic transforms, the restoration of the primary color vectors $\vec{\hat{C}}_s = [\hat{R}_s, \hat{G}_s, \hat{B}_s]^t$ with the inverse KLT (Eq. 17) needs not only the transformed color vectors $\vec{L}_s = [L_{1s}, L_{2s}, L_{3s}]^t$, but the elements Φ_{ij} of the matrix $[\Phi]$ as well. The number of these elements is reduced when the ability of the matrix $[\Phi]$ to represent the transform of the initial color space in correspondence with Eq. 17 as three rotations around each coordinate axis *R*, *G* and *B*, is used. The three angles of the Euler rotation (α, β, γ) define the position of the transform coordinate axes (L_1, L_2, L_3) in respect to the original color space. Using this quality of the matrix $[\Phi]$ it is represented by the product of the following 3 rotation matrices for the axes *(R, G, B)* [19]:

$$[\Phi] = \begin{bmatrix} \Phi_{11} & \Phi_{12} & \Phi_{13} \\ \Phi_{21} & \Phi_{22} & \Phi_{23} \\ \Phi_{31} & \Phi_{32} & \Phi_{33} \end{bmatrix} = [\Phi_3(\alpha)][\Phi_2(\beta)][\Phi_3(\gamma)] = [\Phi_{32}(\alpha, \beta, \gamma)] \tag{19}$$

where

$$[\Phi_3(\alpha)] = \begin{bmatrix} \cos\alpha & -\sin\alpha & 0 \\ \sin\alpha & \cos\alpha & 0 \\ 0 & 0 & 1 \end{bmatrix}; \quad [\Phi_2(\beta)] = \begin{bmatrix} \cos\beta & 0 & -\sin\beta \\ 0 & 1 & 0 \\ \sin\beta & 0 & \cos\beta \end{bmatrix}; \tag{20}$$

$$[\Phi_3(\gamma)]=\begin{bmatrix}\cos\gamma & -\sin\gamma & 0\\ \sin\gamma & \cos\gamma & 0\\ 0 & 0 & 1\end{bmatrix}$$

$$\begin{aligned}&\Phi_{11}=\cos\alpha\cos\beta\cos\gamma-\sin\alpha\sin\gamma;\ \Phi_{12}=-(\cos\alpha\cos\beta\sin\gamma+\sin\alpha\cos\gamma);\\ &\Phi_{21}=\sin\alpha\cos\beta\cos\gamma+\cos\alpha\sin\gamma;\ \Phi_{22}=-\sin\alpha\cos\beta\sin\gamma+\cos\alpha\cos\gamma;\\ &\Phi_{13}=-\cos\alpha\sin\beta;\ \Phi_{23}=-\sin\alpha\sin\beta;\ \Phi_{31}=\sin\beta\cos\gamma;\ \Phi_{32}=-\sin\beta\sin\gamma;\\ &\Phi_{33}=\cos\beta.\end{aligned} \tag{21}$$

The determinant of the KLT matrix should be checked carefully, since the Euler theorem applies only to matrices with a determinant of 1, but the determinant of the KLT matrix may be -1 as well. In this case, one of the rows should be inverted to produce the matrix of determinant 1 prior to decomposition. The matrix of the inverse KLT in this case is defined by the relation:

$$[\Phi]^{-1}=[\Phi_3(-\gamma)][\Phi_2(-\beta)][\Phi_3(-\alpha)] \tag{22}$$

Hence, in order to define $[\Phi]^{-1}$ for the decoding are needed the Euler angles α, β and γ only. These angles are calculated using Eqs. (19 - 21), when the elements of the matrix $[\Phi]$ are known:

$$\alpha=-\arcsin\left(\frac{\Phi_{23}}{\sqrt{1-\Phi_{33}^2}}\right);\quad \beta=\arccos(\Phi_{33});\quad \gamma=\arccos\left(\frac{\Phi_{31}}{\sqrt{1-\Phi_{33}^2}}\right). \tag{23}$$

The elements of the matrix $[\Phi]^{-1}$ are then restored using Eqs. (21-23) and knowing the angles α, β and γ. As a result to the decoder are transferred the values of angles α, β, γ only, instead of the 9 elements of the matrix $[\Phi]^{-1}$, i.e. the number of needed coefficients is 3 times smaller.

3 Estimation of the Enhanced KL Color Transform

On the basis of Eqs. (1-15) used for the direct calculation of the eigenvalues and eigenvectors of the color covariance matrix and the Euler angles (Eq. 23), was developed the algorithm for enhanced KL image color space transform. It offers acceleration of the transform calculations because of the reduced number of mathematical operations in comparison with the numerical methods used for the general approach with covariance matrix of arbitrary size [6, 19].The computational complexity of the new algorithm was compared with that of the QR algorithm for calculation of the eigenvalues $\lambda_1, \lambda_2, \lambda_3$. The number of the operation needed for their calculation for a square matrix of size $n \times n$ when QR decomposition of the matrix in the step k is used [20] is given in Table 1.

Table 1 Computational complexity of the QR decomposition in the k-th step

Operation	Number of operations
multiplications	$2(n-k+1)^2$
additions	$(n-k+1)^2+(n-k+1)(n-k)+2$
divisions	1
square roots	1

Summing these numbers over the $(n-1)$ steps for equal weight of the operations in Table 1 is obtained:

$$SS(n)=\sum_{k=1}^{n-1}[3(n-k+1)^2+(n-k+1)(n-k)+4]=(n-1)(\frac{4}{3}n^2+\frac{17}{6}n+7). \tag{24}$$

Then, for $n=3$ from Eq. (24) follows that $SS(3) = 55$.

For the case with the direct calculation of $\lambda_1, \lambda_2, \lambda_3$ using Eqs. (7-9) is obtained:

$$SS(3) = 28\ multiplications + 19\ additions = 47. \tag{25}$$

Hence, the computational complexity for the direct calculation of the matrix $[K_C]$ eigenvalues in respect of the QR decomposition is 14.5 % lower. Further reduction is obtained as a result of the direct calculation of the $[K_C]$ eigenvectors in accordance with Eqs. (10-14).

The efficiency of the new method regarding the bit-rate and the restored image quality was evaluated for RGB color images (24 bpp) of various sizes. For the evaluation was used DCT-based JPEG-like compression algorithm. Some of the results obtained are given in Table 2. It is obvious that for same (or very close) quality the bit-rate obtained with the KL transform is lower (in average more than 10%).

For the evaluation of the influence of the new approach on the color components' energy distribution was used a database of 4 image groups. Some of the test images are shown in Fig.1: deserts and beaches (a, b); forest (c, d); interiors (e, f) and city (g, h). The goal of the experiments was to prove the efficiency of the proposed algorithm: the energy concentration mainly in one component, the decorrelation of the three components and the MSE minimization.

The algorithm performance was also compared with that of YCbCr and HSV color transforms. For this the relative energy for each color component of the transformed image was calculated. The energies concentrated in each of the three

Table 2 Comparative results for the bit-rate/PSNR relation for the direct KLT approach versus YCbCr (Rec. 601.2) for a group of RGB color test images (24 bpp)

No.	Image (pixels)	KLT		YCbCr	
		Bit-rate bpp	PSNR dB	Bit-rate bpp	PSNR dB
1	Amber (88x128)	1,36	29,06	1,42	29,05
2	Chris (88x128)	1,46	28,99	1,53	28,98
3	Canyon (600x480)	1,63	26,37	1,69	26,37
4	Lena (512x512)	1,11	30,90	1,48	30,85
5	Myanmar (1024x768)	1,45	26,10	1,50	26,09
6	Phong (800x608)	0,14	37,36	0,18	37,68

transformed components were represented as percentage of their sum. The results obtained for the energy distribution within individual image groups were very close. In each group was observed that the principal eigenvector of each image is close to the principal eigenvector of the group. So instead of calculating a KLT for each image of each group we could calculate a single transform for all images in each group. The energy distribution for one of the tested image groups (forest) is given in Table 3.

The color transforms YCbCr and HSV were performed for same database of color images and the relative power for each of the three components of the color transforms was calculated. In Fig. 2a is shown the relation of the relative energies of the first and the second component and Fig. 2b represents the relation of the relative energies of the first and the third component. The graphics show that the relationship between the energies of the first and the second component for the color transforms YCbCr and HSV is close but this is not the same for the KL color transform. As opposed to the other color transforms the energies of the three bands are strongly uncorrelated: most energy is concentrated in the first component; the energies in the second and third component are much smaller. From the data given in Table 2 for the tested image groups is assumed that the relation between the components in the K-L color space is roughly 4:2:1, this means that the first component can be left untouched, the second is twice as small as the original and the third is four times as small as the original.

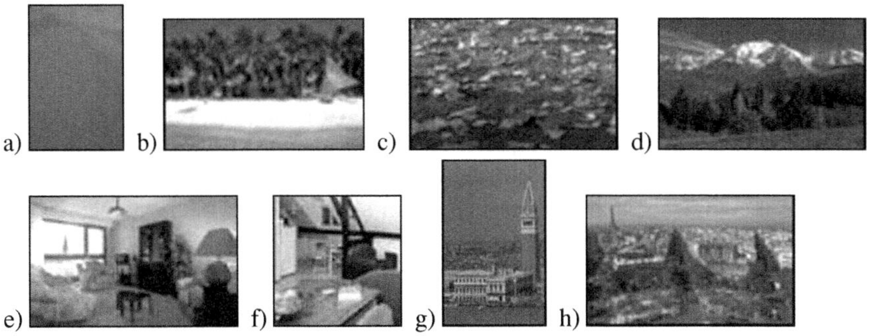

Fig. 1 Some of the test images used for the experiments

Table 3 Energy distribution of the components L_1, L_2, L_3 for each image group

Group	Energy of L_1 [%]	Energy of L_2 [%]	Energy of L_3 [%]
Desert/Beach	64.30	18.40	17.30
Forest	76.99	21.63	1.38
Indoor	48.66	35.51	15.82
City	58.53	23.66	17.81

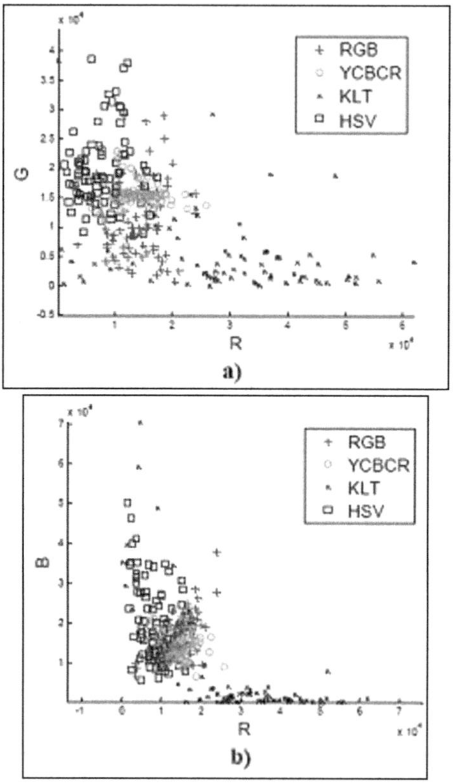

Fig. 2 Distribution of the color components' energies

4 Conclusion

The presented algorithm for direct calculation of the KLT matrix for the RGB image color space does not require the use of numerical methods for the calculation of the eigenvalues and eigenvectors of the color covariance matrix. Further enhancement of the KLT algorithm performance can be achieved if the principles of distributed arithmetic are used. More experiments could be performed concerning the use of the integer numbers arithmetic for the presented approach.

The experimental results obtained confirmed the direct KLT efficiency: it offers lower computational complexity and bit rate for same quality. Another advantage is that the algorithm is universal in respect of the initial color space.

One more advantage is that the algorithm permits the corresponding transform matrices for some large image classes to be calculated in advance. Then, on the basis of the color histogram analysis the processed images could be easily classified and the most suitable matrix to be used.

The main application areas of the direct KL color transform, presented above, are:

- fast color segmentation for automatic target recognition and image classification (image data mining);
- adaptive and more efficient coding of color images.

Acknowledgement. This paper was supported by the National Fund for Scientific Research of the Bulgarian Ministry of Education and Science (Contr. VU-I 305/2007)

References

1. Fairchild, M.: Color and Image Appearance Models. John Wiley & Sons, Chichester (2005)
2. Pratt, W.: Digital Image Processing. Wiley Interscience, New York (2007)
3. Gonzalez, R., Wood, R.: Digital Image Processing. Prentice-Hall, Englewood Cliffs (2001)
4. Malvar, H., Sullivan, G.: Transform, Scaling and Color Space Impact of Professional Extensions, ISO/IEC JTC1/SC29/WG11 and ITU-T SG16 Q.6 Document JVT-H031, Geneva (2003)
5. Hoffmann, G.: CIELab Color Space, Wikipedia, the free encyclopedia.mht
6. Dony, R.: The Transform and Data Compression Handbook. In: Rao, K., Yip, P. (eds.) Karhunen-Loève Transform. CRC Press LLC, Boca Raton (2001)
7. Carevic, D., Caelli, T.: Region based coding of color images using K-L transform. Graphical Models and Image Processing 59(1), 27–38 (1997)
8. Hao, P., Shi, Q.: Reversible Integer KLT for progressive-to-lossless compression of multiple component images. In: ICIP, vol. 1, pp. 633–636 (2003)
9. Chen, Y., Hao, P., Dang, A.: Optimal transform in perceptually uniform color space and its application in image coding. In: Campilho, A.C., Kamel, M.S. (eds.) ICIAR 2004. LNCS, vol. 3211, pp. 269–276. Springer, Heidelberg (2004)
10. Devaux, J., Gouton, P., Truchetet, F.: Karhunen-Loeve transform applied to region-based segmentation of color aerial images. Optical Engineering 40(7), 1302–1308 (2001)
11. Ionita, M., Corcoran, P.: Benefits of Using Decorrelated Color Information for Face Segmentation/Tracking. Advances in Optical Technologies (2008) article ID 583687
12. Kouassi, R., Devaux, J., Gouton, P., Paindavoine, M.: Application of the Karhunen-Loeve transform for natural color images analysis. In: Proceedings of the 31st Asilomar Conference on Signals, Systems & Computers (ACSSC 1997), USA, vol. 2, pp. 1740–1744 (1997)
13. Devaux, J., Gouton, P., Truchetet, F.: Aerial color image segmentation by Karhunen-Loeve transform. In: Proceedings of the 15th International Conference on Pattern Recognition (ICPR 2000), Spain, vol. 1, pp. 309–312 (2000)
14. Jolliffe, I.: Principal Component Analysis, 2nd edn. Springer, NY (2002)

15. Dony, R., Haykin, S.: Optimally adaptive transform coding. IEEE Trans. Image Processing 4(10), 1358–1370 (1995)
16. Gorban, A., Kegl, B., Wunsch, D., Zinovyev, A. (eds.): Principal Manifolds for Data Visualisation and Dimension Reduction. LNCSE, vol. 58. Springer, Berlin (2007)
17. Diamantaras, K., Kung, S.: Principal Component Neural Networks: Theory and Applications. John Wiley & Sons, New York (1996)
18. Solo, V., Kong, X.: Performance analysis of adaptive eigen analysis algorithms. IEEE Trans. Signal Processing 46(3), 636–645 (1998)
19. Korn, G., Korn, T.: Mathematical Handbook for Scientists and Engineers. Mc Graw-Hill Book Company, NY (2000)
20. QR decomposition - Wikipedia, the free encyclopedia.mht

Interoperable Intelligent Agents in a Dynamic Environment

Mohammad Khazab, Jeffrey Tweedale, and Lakhmi Jain

Abstract. Current research conducted at the Knowledge-Based Intelligent Information and Engineering Systems (KES) Centre aims to improve/develop the communication aspects of an agent-oriented architecture that enables agents to automatically adapt their functionality at runtime based on message flows. Rigid design-time constraints can be replaced by a flexible plug-and-play componentized capability. Intelligent Agents (IAs) must possess interoperability and capability to share knowledge and context in order to achieve their goal(s). A concept demonstrator is being developed, using a number of dynamic distributed environments, to show how interoperable Multi-Agent Systems (MASs) can improve data flow in a distributed environment. The agents in this MAS are equipped with a number of sensors that provide data from the environment, which is fused to produce knowledge. The fused information is fed into an inference engine which contains the Subject Mater Expert (SME) knowledge-based required to make decision(s) and/or change some course of action.

1 Introduction

The area of agent technology is growing rapidly and being used in many areas such as Information Technology (IT), e-Commerce, and other industries [1]. MASs consist of IAs that can interact, communication, and cooperate with each other to achieve their goal(s). Using communication techniques, IA can exchange information and knowledge to all components within the environment. In a MAS, IAs can

Mohammad Khazab, Jeffrey Tweedale, and Lakhmi Jain
School of Electrical and Information Engineering,
Knowledge Based Intelligent Engineering Systems Centre,
University of South Australia, Mawson Lakes, SA 5095, Australia
e-mail: `mohammad.khazab@postgrads.unisa.edu.au`, `jeffrey.tweedale`, `lakhmi.jain@unisa.edu.au`

K. Nakamatsu et al. (Eds.): New Advan. in Intel. Decision Techno., SCI 199, pp. 183–191.
springerlink.com

break down a task into subtasks and allocate each subtask to an IA that has the required capabilities. The IAs can also plan a series of tasks and cooperate as a team to solve a problem. The key to a successful interaction and cooperation between IAs is a good communication and knowledge sharing.

This research aims to develop an interoperable Multi-Agent System (MAS) and agent teaming architecture that enables their entities to automatically communicate and adapt their functionality at runtime. A concept demonstrator is developed using Java to simulate this scenario and investigate the interaction, communication, and knowledge-sharing activities among agents within MASs.

The next section covers a literature review on IA and MAS. Section three describes the concept demonstrator and explains the decision making, dynamic environment, and communications within the system. The last section discusses the future of this research.

2 Agent Architectures

Agents can be software or hardware entities that can perceive their environment and perform operations in order to achieve a goal on behalf of their users [2]. Since agents are not generally intelligent, they need to possess a capability to interoperate. For that reason, the area of agent technology has expanded to include the concept of intelligence, learning and the means to communicate.

IAs are described as autonomous computational entities that can respond to the changes in their environment, plan a series of tasks accordingly, and interact with each other to achieve their goals [1]. They generally possess characteristics that include: autonomy, social ability, reactivity, and pro-activity [3]. Autonomy is the ability for an agent to control its actions and internal states and operate on its own without the direct intervention of human or other machines. Social ability of an IA allows the interaction with other entities via predefined communication languages that enable them to cooperate in performing a common task [1]. Reactivity refers to an IA's ability to react to the changes of the environment. IA can also plan their activities to achieve their goals and this is referred to as pro-activity.

IAs can have the ability to learn from interactions and the changes in the environment by experience in order to improve their performance over time [1]. The reasoning ability enables IAs to make decisions dynamically based on the environment situation. Mobile IAs can distribute across a network to perform some of their task on the client machine and return the result to the host. This feature increases the processing speed of the system and reduces network traffic. When an agent includes a combination of these characteristics it is called a hybrid agent.

With intelligence and interaction, IAs act as a bridge between humans and machines. The development of social intelligent agent is not quite achieved due to the lack of characteristics like learning and teaming. It is needed to include a human in the supervisory loop and develop human-centric MAS because the dynamic environment has many external and changing constraints [4].

2.1 Multi-Agent Systems

A MAS is a group of IAs that interact with each other and the environment to achieve their goal(s) [5]. IAs in a MAS can be competing with each other to achieve their own goal(s) or cooperating to achieve a common goal. In the latter case, IAs need to interact, share their knowledge and intelligence, and use each other's capabilities to solve problems [6].

Cooperation of IAs enables the establishment of a voluntarily relationship with each other to adopt each other's goals or form a team [4]. IAs collaborate with each other when they have positive motivations and recognize the need for cooperative actions [7]. The need for cooperation is recognized by an IA when it realizes that its goal cannot be achieved without another IA's resources. The IA would search for other IAs that can assist in accomplishing its goal and form a group with those that have the required capabilities. IAs would then collaborate and negotiate with each other within the team to plan and take actions to solve the problem.

The technology of MAS is becoming more popular and is used to tackle difficult tasks and problems. In MAS, the inter-operability of IAs is an important issue in establishing multi-agent cooperation and teamwork. A diverse range of knowledge sharing techniques and communication protocols have been developed to describe the environment within computer systems and help interchanging knowledge between agents in order to improve their interoperability. Some of the Agent Communication Languages (ACLs) include: Knowledge Query Manipulation Language (KQML)[1], Foundation of Intelligent Physical Agents (FIPA) ACL[2], and Simple Object Access Protocol (SOAP)[3]. IAs require knowledge which is communicated using a predetermined taxonomy to interact successfully.

The dynamic environment of a MAS changes in time and the IAs influence it and change its state. In such environments, the interaction among IAs and using each other's knowledge is very important and useful to learn more about the environment in a compressed period [6]. An example of this is an environment where agents are required to find specific targets or take a predefined route in a region. When a group of agents are used, the task can be decomposed to produce search plans for different regions independently and shared via a facilitator agent. By sharing knowledge, the region can be explored more efficiently. Replacing a single complex unit with a swarm of units reduces the design complexity of each unit and makes the system more economical. The system become more scalable and the overall failure rate decreases because the destruction of single units would not stop the whole system functioning.

---

[1] KQML is a high-level, message-oriented communication language developed by Knowledge Sharing Effort (KSE) that is based on Speech Act theory [8].

[2] FIPA promotes agent-based technology and produce specifications on ACL [8]. FIPA ACL's syntax is similar to KQML and its specification is expressed using the Beliefs, Desires, Intentions (BDI) model.

[3] SOAP is a simple and flexible inter-application cross-platform communication protocol that uses Extensible Markup Language (XML) to package and transmit messages between applications[9].

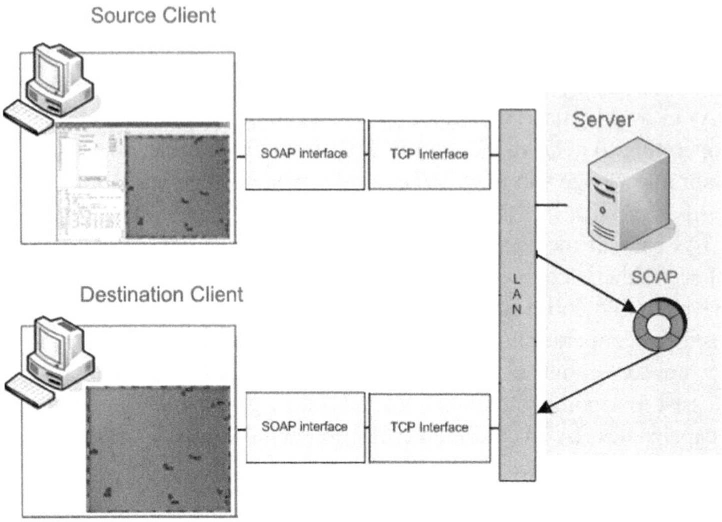

Fig. 1 Developing Distributed System using SOAP as communication protocol

3 Concept Demonstrator

As part of this research, a Concept Demonstrator is being developed simulating the real world to investigate the interactions between agents and dynamically provide capabilities. The system consists of three main parts, as shown in Figure 1, 1) a server, 2) source client(s), and 3) destination client(s).

The source client portrays a MAS with agents that can hold different capabilities such as navigation in a simulation environment, collision avoidance, and finding optimum routes to specific targets. Agents can share knowledge and team up based on their knowledge and expertise in order to conduct a series of tasks. The capabilities of agents can be extended by appending other agents' capabilities until the goal is achieved through cooperation.

The server is the centre of system that translates and transmits data between the destination and source client via SOAP messages. The server can also run the source client on the same machine.

The destination client runs a simulation environment where its entities are guided by the MAS in the source client to implement a distributed system over network. The entities in the source client's simulation environment publish their geographical location to the destination client's entities, which move in sympathy with the source client when connected via the server's translation component. Every set amount of time the destination client requests the source client via the server for new position coordinates to navigate its entities based on agent's movement in the MAS.

During a recent collaborative research, the system was also tested with a destination client simulating a virtual sea with fish schools [10]. This simulation implements the Boid model [11] to add steering behaviour to each individual fish based on the movement of its nearby fish swarm. One feature of the simulation program allows navigating a fish swarm to a target point and getting them to follow a path by moving the target. The coordination for this target point can be obtained from the source client via the server's translation component.

3.1 Dynamic Environment

The dynamic environment consists of an overall root system (Figure 2), which holds a MAS, Knowledge Based System (KBS), physics engine, navigation tactics, and mathematics models to enable planing and replanning capabilities. In the simulation environment the physics engine adds the effects of laws of physics such as mass, velocity, and the forces applied to the entities, such as gravity and friction, in order to approximate movements as it happens in real life.

The MAS comprises of different layers that form a hierarchical structure. In the bottom layers of the hierarchy, entities exist. An entity itself is a subsystem within the system because it has in-built capabilities to retain its autonomous ability. It is also equipped with a number of sensors that provide data from the environment. The entity controller facilitates navigation and calculates the intercept angles and distance from other entities or objects. The entities with mutual goal(s) can form a team. In a higher layer, team leader facilitates coordination of team members and controls them to conduct a task. The team leaders communicate with the supervisor also referred to as facilitator agent.

3.2 Dynamic Communication

Communication and cooperation is important for teamwork. In this simulation environment, when agents reach certain distance apart they can start communicating and sharing their knowledge. The facilitator agent is used to facilitate the communications and interactions among other agents. The facilitator agent offers various useful communication and network services such as [12] [13]: registering and identifying agents, and passing or translating information between agents. When agents are created at run-time they inform the facilitator agent of their name, address, and capabilities or services they provide. The facilitator agent then registers the created agent in its database based on this information, which then can be used to identify the agent. The facilitator agent can also forward information between agents and act as a translator between agents with different semantic and ontology in the knowledge content of their messages. As a result, a better team work is achieved by the MAS.

KQML messages are used to transmit information among agents. Data was structured into XML documents and transmitted between server-client applications via SOAP messages (Figure 1). This information include information transmitted between agents, data obtained from the simulation environment, and the requests and replies communicated between the server and client applications. Transmitting this data at once can exceed the communication channel overhead and cease the system from its other operations and networking. Two styles of communication are

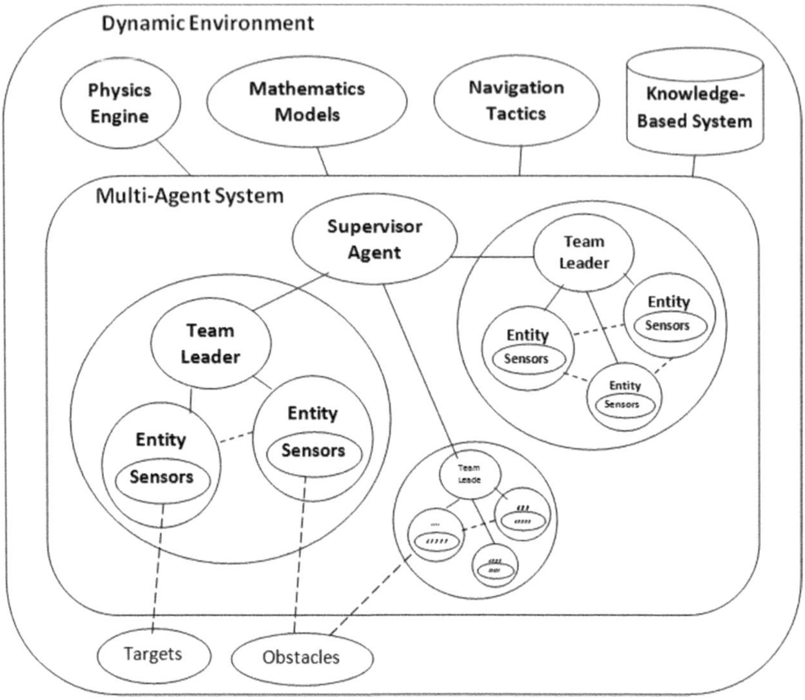

Fig. 2 Hierarchical Structure of the MAS in the Dynamic Environment

considered in simulating this dynamic environment to handle the data load. One style is used when an event requires immediate attention, e.g. an agent finds a target. The other style deals with less frequent communication which provides agents with the slower update about the environment, e.g. an agent is moving from point A to B. Parsing the information into XML documents enabled client and server applications to extract and process only the related information that was required.

Dividing the tasks into subtasks also helps with the issue of data management. For example, few agents are used to listen for network connection and communication requests from client-side applications. The other agents are decomposed into subgroups, each conducting a different task such as finding an optimum route to the target, and sending its coordinates to clients. Supervisor agent also has an important role in coordinating these teams by annunciating keeping a record of their capabilities and function status.

3.3 Decision Making

KBS, or in other words Expert System, is a rule-based processing system that consists of a collection of if-then rules and an inference engine for processing the data based on those rules [14]. All the logics and decision process required for agent's capabilities and actions is decision support. The decision support might be used to find an optimum path to the target(s).

The concept demonstrator is used to show how interoperable MAS can improve data flow in a distributed environment. The entities in MAS team gather data from the environment. The data is fused by the entities and team leader agents to produce knowledge. The obtained knowledge is fed into an inference engine which possesses a database of if-then rules. The inference engine processes rules and data to produce new rules, or make decision and change some course of action.

Agent teaming architecture divides the tasks into subtasks and allocates them to agents that have the capabilities to solve the problem. Agents in a MAS can be cooperating or competing with each other to achieve their goal(s). Multi-agent Planning specifies the actions for agents to perform [15]. Multi-body planning enables agents to find the possible joint plans to achieve the goal. The cooperation of agents requires them to agree on the same goal, which can be achieved by communication.

A BDI framework in a MAS architecture that operates in a constrained environment is an effective model for this research. Beliefs are the understandings of the agents from the environment, desires are the goals of the agent to achieve, and intentions are its plans or actions for achieving the goal [16]. The BDI model originated from philosophical theory of practical reasoning [17]. Practical reasoning involves two aspects of deliberation and reasoning (planning) that refer to decisions of what to be achieved and how to achieve it respectively. Based on the concept of BDI, agents are assumed to have beliefs and understanding about other agents and the state of environment. Agents also have plans for taking the required actions to achieve a particular goal and change the state of environment.

In the scenario of this research, agents are equipped with navigation skills to be able to patrol the virtual simulation environment and find the path to particular targets in a certain time while avoiding obstacles. Agents are capable of detecting objects in certain distance from them and can also realize the type of those objects. The decision making feature enables agent to use their rule-base to react accordingly based on the objects they reached to. For example, if the object they hit is an obstacle like a wall they would using their collision avoidance rule and turn away, or if the object is the target they are searching for they store its position coordinates and send the information to the team leader agent. The decision support system is being extended to enable agents remember the actions they take that lead to achieving their goal, e.g. the direction they take to reach the target. This information can be used by the agent or other agents to achieve their goal(s) quicker the next time they encounter the same situation.

4 Conclusion and Future Research

We managed to maintain the dynamic component of the process by being able to run and link two separate applications which were written on different programming languages, and were running on different operating systems, and platform, with different character sets. The server-side application was developed in Java language running on Windows XP SP2 with English based character set, and the client-side application was developed in C language running on Windows Vista with Japanese based character set.

We are attempting to extend the functionality of system by enabling the creation and allocation of new tasks to agents at run-time. Future decision support work might be to enable agents replan their tasks based on the situations occurring at run-time and perform joint plans. We are looking into refining the communication between agents, also establishing a communication link between the two applications via Internet. We are also looking at the dynamic nature of being able to plug one or more distributed simulations together and then dynamically addressing environment such that they might rescale in size allowing entities to travel across the real state available to all simulations.

Acknowledgements. I would wish to appreciate the contribution made by Hisashi Kawabayashi in this project. I would also like to give my ardent appreciation to all the members of the KES Centre at the University of South Australia for their inspirational, continued and unconditional assistance in all matters concerning this study.

References

1. Jennings, N., Wooldridge, M.: Software agents. IEE Review, The Institution of Engineering and Technology (IET) 42(1), 17–20 (1996)
2. Chira, O., Chira, C., Roche, T., Tormey, D., Brennan, A.: An agent-based approach to knowledge management in distributed design. Journal of Intelligent Manufacturing 17(6), 737–750 (2006)
3. Wooldridge, M., Jennings, N.: Intelligent agents: theory and practice. Knowledge Engineering Review 10(2), 115–152 (1995)
4. Tweedale, J., Ichalkaranje, N., Sioutis, C., Jarvis, B., Consoli, A., Phillips-Wren, G.: Innovations in multi-agent systems. Journal of Network and Computer Applications 30(3), 1089–1115 (2006)
5. Panait, L., Luke, S.: Cooperative multi-agent learning: the state of the art. Autonomous Agents and Multi-Agent Systems 11(3), 387–434 (2005)
6. Dudek, G., Jenkin, M., Milios, E., Wilkes, D.: Taxonomy for swarm robots, Yokohama, Jpn, pp. 441–447 (1993)
7. Wooldridge, M.: Verifying that agents implement a communication language, Orlando, FL, USA, pp. 52–57 (1999)
8. Labrou, Y., Finin, T., Peng, Y.: Agent communication languages: The current landscape. IEEE Intelligent Systems and Their Applications 14(2), 45–52 (1999)
9. Seely, S., Sharkey, K.: SOAP: Cross Platform Web Services Development Using XML. Prentice Hall, Upper Saddle River (2001)
10. Kawabayashi, H., Chen, Y.-W.: Interactive system of artificial fish school based on the extended boid model, Harbin, China, pp. 721–724 (2008)
11. Reynolds, C.W.: Steering behaviors for autonomous characters. Sony Computer Entertainment America, Foster City (1999)
12. Patil, R., Fikes, R., Patel-Schneider, P., McKay, D.P., Finin, T., Gruber, T., Neches, R.: The darpa knowledge sharing effort: Progress report. In: Nebel, B. (ed.) Proceedings of the Third International Conference on Principles Of Knowledge Representation And Reasoning. Morgan Kaufmann Publishers Inc, San Fransisco (1992)

13. Finin, T., Fritzson, R., McKay, D., McEntire, R.: Kqml as an agent communication language. In: CIKM 1994. Proceedings of the Third International Conference on Information and Knowledge Management, pp. 456–463. ACM, Gaithersburg (1994)
14. Bigus, J.P., Bigus, J.: Constructing Intelligent Agents Using Java, 2nd edn. Wiley, New York (2001)
15. Russell, S.J., Norvig, P.: Artificial intelligence: A Modern Approach. Prentice Hall, Pearson Education, Inc. (2005)
16. Wooldridge, M., Muller, J., Tambe, M.: Agent theories, architectures, and languages: a bibliography, Montreal, Que., Canada, pp. 408–431 (1996)
17. Bratman, M.E.: Intention, Plans, and Practical Reason. Harvard University Press, US (1999)

Multilingual Agents in a Dynamic Environment

Jeffrey Tweedale and Lakhmi Jain

Abstract. Experiments conducted by the Knowledge-Based Intelligent Information and Engineering Systems (KES) Centre use Java to gain its many advantages, especially in a distributed and dynamically scalable environment. Interoperability within and across ubiquitous computing operations has evolved to a level where *plug 'n' play* protocols that invoke common interfaces, provide the flexibility required for effective multi-lingual communications. One example includes: dynamic agent functionality within simulations that automatically adapt to incoming data and/or languages via scripts or messaging to achieve data management and inference. This has been shown using demonstrations at the Centre herein. Many aspects of the model involve web centric transactions, which involve data mining or the use of other types of Intelligent Decision Support System (IDSS). Section One of this paper provides an introduction, Section Two introduces the basic concepts of Decision Support System (DSS), Section Three discusses Intelligent Decision Support System (IDSS) enhancements, Section Four explains how agents use a multi-lingual dynamic environment, while Section Five highlights conclusions and future research direction.

Keywords: Multi-Agent System, Intelligent Decision Support System, Universal Plug and Play, XML Object Model.

1 Introduction

As technology advances, humans are increasingly introducing delays and/or errors through the lack of response within prescribed system limits. Natural language

Jeffrey Tweedale and Lakhmi Jain
School of Electrical and Information Engineering,
Knowledge Based Intelligent Engineering Systems Centre,
University of South Australia, Mawson Lakes, SA 5095, Australia
e-mail: `Jeffrey.Tweedale@unisa.edu.au`, `Lakhmi.Jain@unisa.edu.au`

K. Nakamatsu et al. (Eds.): New Advan. in Intel. Decision Techno., SCI 199, pp. 193–202.
springerlink.com

interfaces are slowly gaining acceptance; however many enhancements are still required before these become reliable enough for widespread use. Researchers also need to maintain the support of operators, therefore human involvement must be integrated to avoid system conflicts and isolate the operators from dirty, dull or dangerous tasks. Conversely, automation assists humans in becoming more productive, increases safety and reduces the cost of manufacturing or equipment operation. This research aims to develop an agent learning and teaming architecture suitable for controlling distributed systems or systems-of-systems as plug-able components in a Service-Oriented Architecture (SOA) using Web Services via common interfaces.

In 1980 Tim Berners-Lee enabled seamless connectivity using the World Wide Web (WWW)[1][1], while Marc Andreesen and Eric Bina coded Mosaic in (1992)[3][2] using HyperText Transfer Protocol (HTTP) to facilitate the presentation of requested information [4]. A series of follow-on developments resulted in a logic based protocol that could be cataloged into libraries and distributed in the format derived for Web-Services Description Language (WSDL) using an Extensible Markup Language (XML) protocol over HTTP. "This rich service provides a powerful mechanism of assembling information resources in context that require the agile construction of virtual organizations [5]". Many applications now use Web-Service style components, technology and type safe languages (like SOAP), to preserve the semantics, ontology and intent of the message transaction. AMAZON, YAHOO and GOOGLE all use electronic kiosk style applications to transform requests from semi-intelligent information retrieval, into information rich responses, that is tailored using some form of server-side Decision Support System (DSS) technology.

2 Decision Support System

DSS use "Systems" that value add to the data being collected (using a knowledge-base to create an inference), prior to promulgating a response. A simple thermostat uses a temperature sensor to extract environmental conditions (precise data), while a comparator[3] sets the threshold used to switch between heating or cooling modes. When not switching, the DSS simply waits for the temperature to exceed a limit set by the operator (rules). More complex environmental control

[1] Tim Berners-Lee is acknowledged as the founder of the public version of the WWW, however the original architects where a panel of 12 members board called the Internet Architecture Board (IAB), employed by Defense Advanced Research Projects Agency (DARPA) and headed by Vince Surf which first convened in 1974. This board created standards, using instruments called Request For Comments (RFCs). The first topics raised included: Domain Name Servers (DNSs), Network File Systems (NFSs), email, Universal Resource Locators (URLs) and a host of subsequent protocols [2].

[2] Purchased by Microsoft in 1994 and released as its Internet Explorer.

[3] This is the logic center containing actuators and reference limits which are used to decide to heat or cool.

systems may be required, where that system controls a building, a complete facility or even a complex variety of environments that are dispersed geographically. Regardless of the technology used to generate the decision, the outputs relate to a predefined set of judgments based upon readings changing in the environment being controlled.

A DSS uses Computation Intelligence to collect, fuse and analyze data in order to derive a decision to enhance a human's ability in a given environment[4]. An expansion of research into a variety of DSSs created greater confidence in the decision being generated by computers [6]. The context could be used to determine the problem solving technique used. Examples include: communication-driven, data-driven, document-driven, knowledge-driven and model-driven architectures.

The growing volume of data had an overall effect on the efficiency of these systems. A series of tools where developed to measure the validity and results of many DSS. Factors such as; availability, accuracy, user response and operator ability, were raised as constraints to be considered before specifying courses of action [7]. The history of these systems has bee well documented. Power (2007) [8] for instance gives an excellent historical description of DSSs, which discusses many of the fundamental issues, however misses the vital human aspects. Keen reports on a study relating to "the theoretical aspects of organizational decision making". A preliminary study was done at Carnegie in the 1950's, while a more concise study from MIT was completed in the early 1960s [9].

3 Intelligent Decision Support System (IDSS)

Intelligent agents are described as computational systems that have properties such as autonomy, social ability, reactivity, and pro-activity [10]. Agent technology has expanded to include the concept of IDSS as shown in Figure 1. Intelligence can be described as a computational process that enables intelligence and interaction. Agents are increasingly required to bridge the gap between humans and machines and can be integrated to form social structures, that include teaming and learning. To support a human in the supervisory loop, a dynamic interface has to constructed to overcome the external constraints [11]. According to Tweedale et al. (2006) [12], new research and technologies are beginning to focus on developing human-centric multi-agent system in order to achieve mutual goals [13]. To accomplish an automated task effectively computers, hardware, software and even firmware agents are normally employed [14]. They must react to changes in the environment (reactivity) and plan activities to solve their goals (pro-activity).

[4] McCarthy developed LISP, enabling Bachman to create the first database management system which was converted into an Intelligent Decision Support (IDS) application called SAGE, while Feighambaum labeled his expert system DENDRAL.

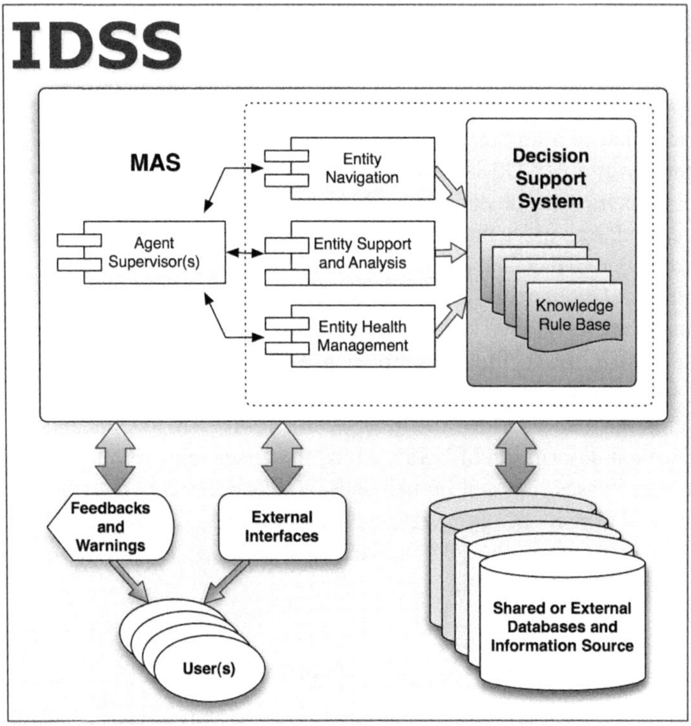

Fig. 1 The Principle of MAS within IDSS

4 Demonstration Examples

4.1 Teaming

Teams can be represented in many ways, however, experience from previous experiments indicates that Intelligent Agents using MAS architecture in a Beliefs, Desires, Intentions (BDI) framework that operates in a constrained environment is an effective model for this research. Agents have been defined in variety of ways as their functionalities have grown in different fields [12].

The technology of MAS is becoming more popular because agents can be ignored, informed, or forced to intelligently cooperate with one another, especially in the field of e-commerce where large-scale data mining is employed [15]. According to Dudek et al. (1993) [16], using MAS has advantages over a single agent. For example, replacing one complex robot with a group of simple robots that effortlessly explore an unknown area. This reduces the design complexity of each robot, is more economical, robust and scalable. The overall rate of failure would also decrease because if one or more robots fail, the whole system would still be able to function. The shared knowledge among agents should contain the knowledge of meaningful words and phrases the contextual information about particular issues or items of interest to one or more agent [17].

Adaptability refers to the MASs capacity to inter-operate and reorganize to form a new subgroup or dynamically adopt the functionality required to complete other tasks. For example, if the environment is small, it would limit the movement of agents within a specified hierarchy. Dynamic filters or components can be instantiated to achieve specific connectivity or customized configuration(s).

4.2 Component Messaging

Specialized techniques have been developed to enable the communication of agent based software components and the exchange of information between specific applications. Agents can communicate by sending messages directly to one and other using a prescribed format with rigid semantics. Theses messages are accepted by all the agents within the MAS and actioned as required [18]. Standards are required in order to establish compatible communication among any agent that gains access to that distributed MAS environment. Several common standards include: Agent Communication Languages (ACL), Knowledge Query Manipulation Language (KQML), Foundation of Intelligent Physical Agents (FIPA) and Simple Object Access Protocol (SOAP). Communication includes the delivery of complex attitudes or behaviours [17] through a specified architecture [12]. Therefore there is a requirement to identify the: syntax, semantics and pragmatism used, especially the symbols. Agents are able to communicate with other agents, using a specified precedence and hierarchy [16], however the need for immediate and broadcast messaging is becoming commonplace.

4.3 Decision Support

In some MAS, agents would be able to send direct messages to any individual agent by knowing their address and Identification [16]. Other topologies include agents only communicating with agents that are linked to them, for example, through a graph network or a hierarchy structure. In a hierarchical structure, there might be some controller agents who command and give tasks to other agents lower the hierarchy where those agents are only able to communicate with the controller agent. There are advantages and disadvantages to this type of structure. One advantage is that it reduces and simplifies the interactions between the agents because agents only interact with certain agents. However, one disadvantage is that communication among the agents in this type of structure is sensitive to the failure of one or few of the agents as this could cut the linkage of some group of agents from the rest of the swarm.

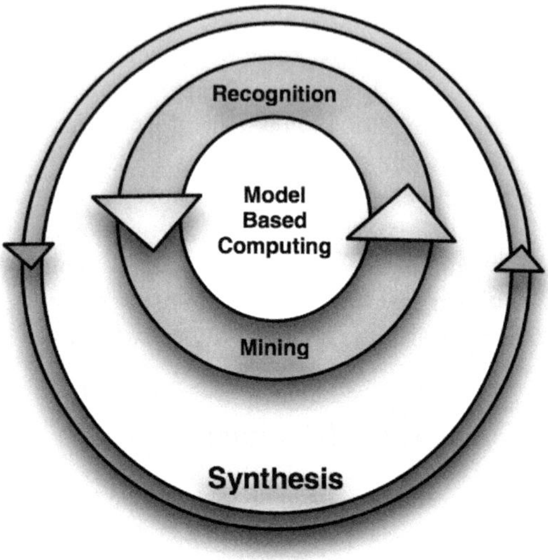

Fig. 2 The Concept of Micro-Simulated Decision Support [19]

4.4 Micro-simulated Decision Support

People rely on automated systems to make decisions, especially when time lines or expectations are compressed. There level of confidence and expertise in the system being used will also effect their response to any inference generated. Humans prefer simplicity and they attempt to streamline most processes to avoid stress, This may result in them skipping the "cognitive effort required to gather and process information [20]", and rely on intuition [21]. This method of coping often leads to mistakes via mode confusion, especially during intense periods of mission critical operation. *Automation bias* best explains the misuse of automated decision aids, although it cannot account for the lack of use in cases were it may have assisted [22].

Better training techniques and alternate models are being researched to solve some of these issues. The concept of using simulation to replicate the outcomes of a virtual reality is not new. The practice of using simulation to rehearse a number of possible outcomes, is becoming achievable, although real-time speeds are elusive. Through a series of sequencing, appending or combining the results of previous simulation, the system is able to recommend more precise courses action which would normally enhance the solution being sought. Doing this in parallel, at speeds that are faster than real-time (micro-simulations), data can be extrapolated to backfill missing variables or segments of a specific situation in order to gain an advantage in a variety of situations performed. To achieve this milestone, a combination of carefully chosen operator estimates and/or beliefs could be used as stimulants during specified scenario cycles. Each cycle must conform to a collection of rules, repeatable system states and consistent algorithms to stimulate the reality being attempted.

Thus using a combination of data analysis, mining and synthesis techniques in micro-simulated scenarios, researchers can employ an iterative approach to problem solving, using a variety of technologies or scenario instantiations that operate in parallel across a distributed environment. This promotes the enhanced concept of IDSS, which was initially described as Micro-Synthesized Simulated or Micro-Simulated Decision Support [19], however at its latest research day Intel combined a number of processes adopted above, they now call Model-Based Computing (MBC) as shown in Figure 2. The technique involves: the analysis of a volume of data to create a feasible model, use data to test instances of a parameter against a model and create an instance of a potential model. By increasing the number of processing elements within a system, analysts will be able to increase the number of parallel variables or constraints used to solve the problem [23].

5 Distributed Architecture

It is important for agents to gather knowledge about its environment to enable it to make intelligent decisions and actions [24]. Agents can be equipped with sensors to collect information. Communication between agents can be conducted directly between two agents or through a facilitator or interpreter. The exchange of information in a particular domain of knowledge that requires each agent to have shared knowledge of concepts in a particular domain is known as ontology. Each domain of knowledge would have its own ontology, although this can become flexible using more modern protocols (like SOAP)[5]. Figure 3 displays the physical architecture invoked in each computer/component to achieve distributed messaging based on this framework.

SOA models are constructed using loosely coupled components based on Web Services (Beans, Components or Enterprise Applications). Java classes are deployed as Web Services to consume targeted objects. Monson-Haefel (2003) postulated that Web Services, like distributed computing, is: "simply the hardest problem in computer science [28]". SOA provides an abstracted service at any granularity, which is great for hiding an entire system or sub-system. It can also become hierarchically dynamic using web service. Figure 3 displays the physical architecture invoked in each computer to achieve distributed messaging based on the SOAP framework. SOAP should have been a simple answer to interoperability, however the lack of proper standards initially made the job of distributing objects more frustrating than necessary. XML, SOAP, WSDL and Web Services retain the key topics in making successful distributed applications, although seamless integration of JAX-WS and JAXB is required to achieve the low-level functions and interfaces.

The SOA capability was included in Java 5.0 (Mar 2006) and was updated in Java 6.0 (Dec 2006). These Application Program Interfaces (APIs) reduced the expertise required by programmers to use Java Web Services in their SOA applications

[5] SOAP is an inter-application cross-platform communication mechanism. It defines a simple and flexible communication format that is based on XML documents passed across HTTP [25]. XML is used to represent data as an object instead of symbols and is passed in a standardized text-based format [26]. Using the SOAP protocol, information is packaged in SOAP messages that consists an envelope and body, with an optional header elements [27].

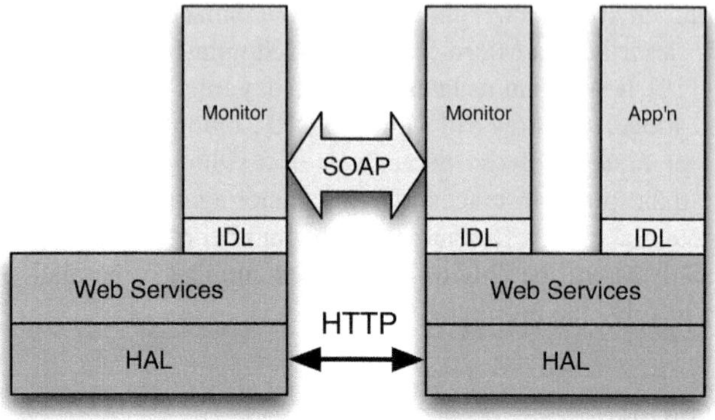

Fig. 3 Application Based on SOA Structure

(Knowledge about *Generics* and *Annotations* would aid productivity). Although developers will need to their update your knowledge on SOA, cross compilation of WSDL to Java and deployment before launching into creating a distributed application. Remote procedure code was includes in Java 4.0 (2002) with Web Services, Binding, Meta-data and Run-time behavior undergoing significant revision to trigger the *tipping point* of acceptance and wider spread implication.

6 Conclusion

Automation provides a path to increased productivity of many new machines and applications. Humans have traditionally been a bottleneck that impedes that goal, however technology is being used to erode that barrier. Agents, components, teams, DSS, IDSS, MBC and modern distributed computing techniques are being used to solve many of the issues encountered by researchers. Web services and SOAP are elements of that future and have evolved to a point were its evolution is nearing maturity. SOA makes it easier to product functional designs with limited functionality. Future testing and development is required to integrate these concepts with inter-operable computing technologies before world class applications mature commercially. The research conducted so far by KES has developed a blackboard design upon which segregated functions can be integrated into an application of aimed at achieving this goal.

References

1. Berners-Lee, T., Masinter, L., McCahill, M.: Uniform Resource Locators, RFC 1738. Internet Engineering Task Force, Internet Society, Geneva, Switzerland (1994)
2. Zittrain, J.: Interview: J. zittrain and d. clark, issues of domain names. Research Seminar on the Internet and Society. Harvard Law School, Cambridge (1997)

3. Salus, P.H.: Casting the Net: From ARPANET to Internet and Beyond.. Addison-Wesley Longman Publishing Co., Inc., Boston (1995); foreword By-Vinton G. Cerf
4. Mitchell, H.: Internet browser technology. Inventor of the Week, Massachusetts Institute of Technology, Cambridge, MA (Janurary 2001)
5. Spooner, J.G.: Intel breaks 3ghz speed barrier. ZDNet Australia, San Francisco, California, p. 1 (2002)
6. Yen, J., Fan, X., Sun, S., Hanratty, T., Dumer, J.: Agents with shared mental models for enhancing team decision makings. Decis. Support Syst. 41(3), 634–653 (2006)
7. Dhar, V., Stein, R.: Intelligent decision support methods: the science of knowledge work. Prentice-Hall, Inc., Upper Saddle River (1997)
8. Power, D.J.: A Brief History of Decision Support Systems. In: Quorum Books Division. Greenwood Publishing (March 2007) no. 156720497X
9. Keen, P.G.W., Morton, M.S.S.: Decision Support Systems: An Organizational Perspective. Addison-Wesley, Reading (1978)
10. Wooldridge, M., Jennings, N.: Intelligent agents: theory and practice. Knowledge Engineering Review 10(2), 115–152 (1995)
11. Finn, A., Kabacinski, K., Drake, S., Mason, K.: Design challenges for an autonomous cooperative of UAVs. In: Information Decision and Control, IDC 2007, Adelaide, Australia, February 2007, vol. 4477 (2007)
12. Tweedale, J., Ichalkaranje, N., Sioutis, C., Jarvis, B., Consoli, A., Phillips-Wren, G.: Innovations in multi-agent systems. Journal of Network and Computer Applications 30(3), 1089–1115 (2006)
13. Parsons, S., Pettersson, O., Saffiotti, A., Wooldridge, M.: Artificial Intelligence Today: Robots with the best of intentions. In: Veloso, M.M., Wooldridge, M.J. (eds.) Artificial Intelligence Today. LNCS, vol. 1600, pp. 329–338. Springer, Heidelberg (1999)
14. Nwana, H.S.: Knowledge engineering review, pp. 1–40 (1996)
15. Wooldridge, M.: Verifiable semantics for agent communication languages. In: Proceedings. International Conference on Multi Agent Systems, pp. 349–356. IEEE Computer Society, France (1998)
16. Dudek, G., Jenkin, M., Milios, E., Wilkes, D.: Taxonomy for swarm robots. In: International Conference on Intelligent Robots and Systems 1993, IROS 1993. Proceedings of the 1993 IEEE/RSJ, Yokohama, Jpn, July 26-30, 1993, vol. 1, pp. 441–447. IEEE, Piscataway (1993)
17. Labrou, Y., Finin, T., Peng, Y.: Agent communication languages: The current landscape. IEEE Intelligent Systems and Their Applications 14(2), 45–52 (1999)
18. Vaucher, J., Ncho, A.: Jade tutorial and primer (April 2003-2004)
19. Siotus, C., Tweedale, J.: Challenges and opportunities for a complex and networked world: Mission systems and simulators. In: Bhalla, J. (ed.) SimTecT 2006 Simulation Conference, SIAA, Melbourne, Australia, 29 May - 1 June 2006, pp. 1–6 (2006)
20. Mosier, K.L., Skitka, L.J.: Human decision makers and automated decision aids: made for each other. In: Cohen, M.S., Parasuraman, R., Freeman, J.T. (eds.) Automation and Human Performance: Theory and Applications. Lawrence Erlbaum Associates, Mahwah (1996)
21. Tversky, A., Kahneman, D.: The framing of decisions and the psychology of choice. Science 211, 453–458 (1981)
22. Dzindolet, M.T., Beck, H.P., Pierce, L.G.: Encouraging human operators to appropriately rely on automated decision aids. In: Research, O.H., Directory, E. (eds.) The 6th International Command and Control Research Program (ICCRTS), DTIC, Annapolis (2002)
23. Koehl, S.: What would you do with 80 cores? Research@Intel, Santa Clara, CA, USA, pp. 1–6, Tech. Rep. (2007)

24. Bigus, J.P., Bigus, J.: Constructing Intelligent Agents Using Java: Professional Developer's Guide, 2nd edn. Wiley, Chichester (2001)
25. Seely, S., Sharkey, K.: SOAP: Cross Platform Web Services Development Using XML. Prentice Hall PTR, Upper Saddle River (2001)
26. Benz, B., Durant, J., Durant, J.: XML Programming Bible. Wiley Publishing, Inc., New York (2003)
27. Graham, S., Davis, D., Simeonov, S., Daniels, G., Brittenham, P., Nakamura, Y., Fremantle, P., Koenig, D., Zentner, C.: Building Web services with Java: making sense of XML, SOAP, WSDL, and UDDI. Developer's Library (2002)
28. Hansen, M.: SOA Blog. Service Centric, San Fransisco, USA (2006)

Defining a Decision-Support Framework in AC^{3}M

Angela Consoli, Jeff Tweedale, and Lakhmi Jain

Abstract. The prime objective of intelligent agents in Multi-Agent Systems (MAS) is to act. An effective action results from a solid decision-making process. Decision-Support Systems (DSS) are used in MASs to assist in the development of a course of action for an individual or system goal. To ensure decision-making processes between agents remain objective and coherent, coordination model and cooperative problem-solving methodologies need to be implemented. Presently, coordination models have been developed as data-driven, process or control-driven, or hybrid models. Cooperative problem-solving methodologies have been designed to solely focus on allowing agents to share their knowledge which assists in achieving an individual goal or a course of actions. Although coordination and cooperation has been successfully implemented as separate frameworks within intelligent MASs, there is a significant limitation: cognitive modeling within each framework is limited or non-existent. This is a major obstacle within dynamic or unknown environments, as these cognitive environments heavily depend on precise information being made available to make well-informed and instantaneous decisions. This paper shows how the relationship between Belief-Desire-Intention (BDI) and Observe-Orient-Decide-Act (OODA) architectures, coordination and cooperation can promote decision-making processes in MASs. The linking of the decision-making process with coordination and cooperation can ameliorate their lack of cognitive processes. This enhancement is demonstrated by the decision support framework within the Agent Coordination and Cooperation Cognitive Model, or AC^{3}M.

1 Introduction

The main objective of an entity within an organisational-type system is to complete an assigned task. In order to engage in an activity, an entity must act as a result of

Angela Consoli, Jeff Tweedale, and Lakhmi Jain
School of Electrical and Information Engineering, Knowledge Based Intelligence Engineering Systems Centre
University of South Australia, Mawson Lakes, SA 5095, Australia
e-mail: angela.consoli@unisa.edu.au

K. Nakamatsu et al. (Eds.): New Advan. in Intel. Decision Techno., SCI 199, pp. 203–212.
springerlink.com

some decision [3], [6], [8]. Saaty (1986) discusses the structure of a decision and its components, which include: i) details about the goal, ii) actors that are involved, iii) the objective and finally iv) the contributing influences of the decision. To ensure an objective decision, individually or as a group, procedures need to be exercised for an effective outcome [18]. Shim *et al.* (2002) support the argument that decisions made from organisational-type systems need to be self-regulating and self-contained with a task-oriented approach. To guarantee the influences of a group decision are beneficial and the objectives are met, the 'procedures' that are required are cooperation and coordination [19], [21]. Organisational-type systems work in a cooperative manner to assure that the individual entities work in a coherent, connected and common production process. This cooperative manner promotes the task-oriented approach for a group decision. To regulate the intercommunication between entities within these systems, coordination is required [19].

The paper is structured with section 2 discussing current developments in decision support systems, coordination and cooperation models. Section 3 shows how BDI and OODA is incorporated into cooperation and coordination and how it can affect decision-making processes. Section 4 discusses the decision-making framework within AC^3M at an entity and system level and illustrates how the decision framework can assist in implementing cognitive processes within coordination and cooperation models. This paper concludes with a discussion on future development on decision-making in AC^3M.

2 Background to Decision-Making, Coordination and Cooperation Models

A course of action is seen as the chosen outcome of a decision-making process. This outcome is commonly known as a decision. Hollnagel (2007) relates decision-making as an interface which links the evaluation of a situation and the execution of actions. To instantiate the decision-making process within this interface, a cyclical set of events, intentions and actions must be provided from the information gathered [11].

The decision-making process becomes complex in dynamic and unknown environments [6]. An agent can alter their perception on the information being presented, which may simplify the reality of the decision needed to complete their goal. In addition is the stability of the decision from the time taken to observe and analyse the information gathered. Hollnagel (2007) and Schwenk (1984) agree with these complexities and highlights the need for particular mechanisms to be implemented simultaneously with a decision-making system. These mechanisms are coordination and cooperation [11], [20]. Noble (2002) articulates that coordination and cooperation promotes an entity's perspective and ability to reason about their environment, encourages entities to construct other possible courses of action and shows the consequences on applying alternative decisions for a goal [14].

2.1 Structure of Decision-Support Systems

Gorry and Morton (1971) define a DSS as a computer program that is capable of solving a problem that possesses a semi or unstructured component. Furthermore, they state that a DSS should assist in the efficiency and effectiveness of a decision [9]. Early DSSs were structured around the concepts of management activities and decisional types [2], [21], [22]. Anthony (1965) elaborates on the former by suggesting that these activities involve strategic planning, management and operational control. In terms of decisional types, Simon (1960) categorises decisions into programmed, or routine and well-defined decisions and non-programmed, or novel and ill-structured decisions. Simon (1960) describes that the foundation of decision-making processes includes intelligence and the ability for alternatives to be designed. These factors will assist entities to chose and implement the best alternative. In addition, these foundational concepts show DSSs possessing human-centric characteristics [21].

By the virtue of being human-centric, DSSs encapsulate cognitive processes and classifications. Zachary (1986) agrees and defines six classifications. They are i) process, ii) choice, iii) information-controlled, iv) representational, v) analysis and vi) reasoning and judgment. For the purpose of this paper, the classification of analysis and reasoning only will be considered. Zachary (1986) describes this taxonomy as computational techniques that support expert reasoning procedures, and contains two main techniques: numerical and symbolic representation. The advantage of these techniques are their generic analytical methods, which can be applied to more than a single problem [26]. Symbolic representation technique is suited to Artificial Intelligence (AI) since it uses inferences that simulate human reasoning processes. This technique consists of three inferences: goal, data and process. Goal inferences are where decisions are processed in terms of goals or a set of conditions and requires knowledge from the decision-maker on how to transform the problem from the initial state to the goal state. Data-driven inferences make decisions based on the current set of data available and process-driven inferences are concerned with the reasoning from a causal or temporal relationship [26].

2.2 Structure of Coordination Models

Coordination in AI is viewed as the process which builds programs that fuse active components together. To exhibit this behaviour, coordination models are designed to provide the paradigm where separate activities are bound together [15], [16]. Papadopoulos and Arbab (1998) describe coordination models as possessing three components: i) the entities, ii) medium being used to coordinate and finally iii) the semantic framework. In addition is the function of time, as it is needed since the nature and often the rationale of decisions made by entities are often needed for further decisions [14]. Coordination models are classified as either data or control-driven [16]. Data-driven models use values and data that is being received and sent from the architecture and the configuration of the coordinated components when computing the state of coordination. This is seen as where a coordinator is responsible for examining and using data from the environment and also being coordinated at the same time. This type of coordination model is heavily dependent on rules and regulations and thus, does not endorse the implementation of cognitive modeling [6], [7]. This results in a lack of decision-making in a multi-agent system.

To overcome limitations of data-driven models, control, or process-driven coordination models give coordinated frameworks the means of observing changes in processes and broadcast these changes to the system [16]. Even though this quality favours cognitive modeling, the control aspect entrenched is in opposition to an agent's autonomy and reactivity. These models are also seen to be heavily dependent of rules and regulations and with the data manipulation being completely segregated, an agent may not be able to react as the data is not readily available [7], [13]. This means knowledge between the environment and the agent is ceased, rendering coordination, and any decision-making, invalid.

Hybrid coordination models have been developed to surmount the limitations of data-driven and control-driven architectures. These models allow the exchange of data structures via a shared dataspace that is able to detect communication events and allowing agents to react. Hybrid models are primarily concerned with how effectively the data can be manipulated and communicated between coordinables as well as provide rules and regulations on their autonomy, and reactivity. A major limitation of hybrid models is the lack of human-centric coordinables, which are now essential in open and dynamic systems [7], [16].

2.3 Structure of Cooperative Models

Unlike coordination models, there are no explicit categorisation of cooperative models. Instead a cooperative model is described as the cooperative problem-solving methodology of a multi-agent system [25]. For the purposes of this paper, cooperative problem solving methodology will be known as cooperative modeling.

Jones and Jacobs (2000) outline the functional purpose of cooperative modeling. These models are required to ensure the information between agents is augmented, integrated and deliberate and allows goal sharing and coordinated activities within a system. By combining these functions and attributes of a cooperation model, van Bruggen *et al.* (2003) show the three main frameworks of a cooperative model consists of orientation (or the perception of the problem), solution (or the transformation of a problem state into a goal state by knowledge) and evaluation, where the solution is evaluated based on agents' activities [23]. These frameworks possess further attributes defined by Wooldridge and Jennings (1999) and Jones and Jacobs (2000). Jones and Jacobs (2000) believe the method which information is shared between agents is through three mechanisms: i) Request-Based where agents communicate their goals with each other, ii) Inference-Based where agents infer their goals from their behaviours and iii) Structurally-Based where the structure of the MAS endorses and sustains cooperative behaviours between agents [12]. These mechanisms promotes the human-centric attributes that agents need to possess in a MAS, which include autonomy, reactivity, proactivity and social ability [12], [25]. By including human-centric factors into a cooperative model shows significant problem-solving and decisional-making capabilities which helps in sharing an agent's cognitive workload. This is important in dynamic environments since the beliefs are constantly changing at any given time [12].

3 BDI and OODA in Coordination and Cooperation

Intelligent agents have matured from possessing basic human-centric characteristics to cognitive agents that incorporate their beliefs, desires, intentions and goal-oriented behaviours [6]. In order to successfully utilise these behaviours, the management of their interdependencies within and between the cognitive agents is required so they can successfully decide on a course of action to satisfy their coordinated goal [5]. Furthermore, this management of interdependencies must facilitate the voluntary relationship between agents, where one agent may require the resources of another to complete their goal [5].

Consoli *et al.* (2008) not only show this link, but state that the beliefs, desires and intentions as well as cognitive processes of perception and orientation are linked to decision making. Therefore, not only is there a link between coordination and cooperation, but there exists a link between coordination, cooperation and the decision-making processes.

3.1 BDI in Coordination and Cooperation

In terms of coordination and cooperation, a BDI system consists of three main events. They include i) the observation and updating of an agents internal state, ii) generation of new plans for execution and iii) selecting a plan for execution [1]. Ancona *et al.* (2003) state that BDI systems and agents do not allow for the exchanging of plans between agents, however, do have the ability for cooperation. Their concept of agent cooperation is the ability to help agents achieve their desires. A Cooperation-BDI engine takes into account the desires as well as the active and suspended intentions of the other agents [1]. This is consistent with the description provide by Consoli *et al.* (2006), where they state that once the desire and intentions are equal, cooperation can be realised [5].

BDI in coordination focuses on the beliefs of an agent and the environment. Wobcke *et al.* (2005) agree and show that a coordinator must maintain the knowledge of the belief state and identify an agents intentions [24]. Chalmers and Gray (2001) not only agree, but shows the act of coordination involves informing agents about changes in the environment (Beliefs), notifying a task's status and completions to help scheduling amongst agents (Desires), and delegating tasks to agents based on their composition (Intentions) [4]. Furthermore, situation awareness is required as it promotes accurate and up-to-date information so an agent has the ability to make a well-informed decision prior to engaging in a course of action.

3.2 OODA in Coordination and Cooperation

Traditionally, OODA was not intended for coordination and cooperation [10]. OODA has now matured to be cognitive, dynamic and integrates coordination, cooperation and decision-making processes. Rousseau and Breton (2004) have adjusted the OODA loop from being static to dynamic by including decision-making [17]. Within their model exists three main modules. Firstly is data gathering or Observe, where information from the environment of the system is gathered from the entities within the team. Situation Understanding or the Orientation module, is where mental models are created from the information gathered. The final module, Action Selection encompasses a Decision-like component, which evaluates and lists potential courses of action and their associated risks from the mental models. It is at this point where coordination and cooperation are needed. Consider a course of action. Within this list, estimation of gaining a goal and the managing of options needs to be explicitly defined for an entity [17]. If an entity cannot complete their course of action, another resource may be needed to complete the task. The former is coordination and the latter is cooperation. Once complete, the final process, Action Implementation can be executed. With the assistance of decision-making processes, not only is the action executed, but the link between coordination and cooperation is further strengthened.

4 Decision-Making Processes in AC^{3}M

The architecture of AC^{3}M uses the theory of hybrid coordination models, cooperative problem-solving methodology and the cognitive architectures of BDI and OODA to support the link between agent coordination and cooperation. AC^{3}M consists of four main frameworks: i) Knowledge, ii) Stimulus/Perceptor, iii) Situational Assessment and iv) Decisional Support. The Knowledge Framework is where the information gathered from the environment is stored and analysed. Stimulus/Perceptor framework is where observations from the Knowledge framework are processed and where agents will design their beliefs. Situational Assessment framework is where the orientation from the information from Stimulus/Perceptor framework begins. Coordination is also initiated within this framework.

The Decisional Support Framework is where coordination and cooperation is executed. In this framework, information gathered from the Knowledge and Stimulus/Perceptor frameworks, and the cognitive and decision-making processes discussed previously are used to fire a Coordinative Event to initiate Coordinative Cooperation. This will ensure informative decisions about the orientation, decision and execution of actions are produced to successfully complete cooperation. Furthermore, the lead agent uses this framework to coordinate the agents based on new perceptions, that may have altered their goals and beliefsets.

4.1 Decisional Support Framework in AC^{3}M

AC^{3}M makes possible the decision-making process at a system and entity level. In both cases, AC^{3}M begins with perception analysis from new environmental events. This perception analysis consists of evaluating and assessing the events for new information they may have altered for any agent or system goals. This is the first step of each of the decision-making processes and is consistent with the decisional cycle defined by Hollnagel (2007). AC^{3}M extends on this decisional cycle to include the renewal of the beliefs of each agent within the perception analysis. It is important to note that the 'intention' component for AC^{3}M is different for the system and the agent. The 'intention' of the lead agent is supervisory and used solely to monitor and coordinate within AC^{3}M.

4.2 System and Entity-Level Decision-Making in AC^{3}M

As stated previously, the decision-making process in this framework is separated into two categories. At the system level, upon receiving information gathered from Evaluate/Assess Situation stage, or Observation, all agents will re-evaluate their goals and beliefs based on new events and perceptions of the environment. AC^{3}M allows agents to process these observations and thus, the beliefset, desires and intentions of each agent will be updated. In addition, the acts of coordination, cooperation and the decision-making and eventual execution of an action are derived the system level. However, not all of these acts are completed by all agents.

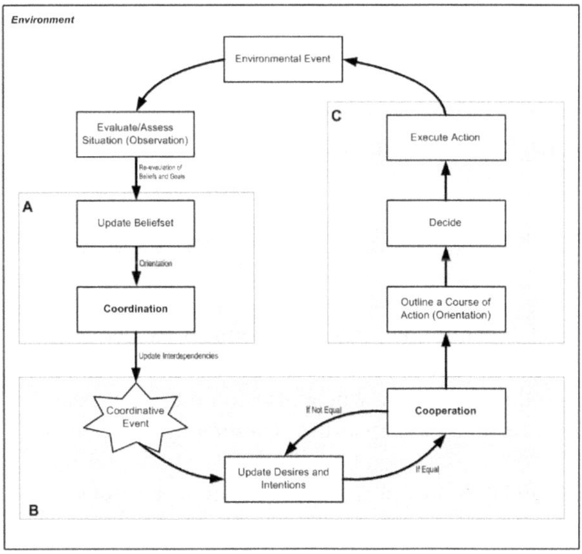

Fig. 1 System and Entity-Level Decision-Making Processes in AC^{3}M

Fig. 1A and Fig. 1B shows the difference between the decision-making processes of the subordinate and lead agents. In Fig. 1A, once the re-evaluation of an agent's beliefs and goals are complete, the lead agent will also complete their re-evaluation, but in terms of effectively managing the interdependencies of each agent and the system as a whole. With the sole intention to coordinate, the lead agent will use the information to the update an agent's beliefset. This is orientation since the lead agent takes the perspectives of each agent and creates mental models from the information that they had gathered. The decision process is completed in AC^{3}M by updating their resource allocation or constraints. Once this has occurred, the lead agent can then coordinate an agent to a new goal or amend their previous goal to ensure it is up to date with the new environmental events. This decision is actioned in Decisional Support Framework by a Coordinative Event. This event links and initiates the decision-process within and between subordinate agents and is seen as the entity level decision-making processes in AC^{3}M.

Upon receiving a Coordinative Event, an agent's decision-making process may be altered, hence modifying how an agent may complete its goal. This modification can result in an agent requiring the assistance of another with a certain expertise. In Fig. 1B, the desires and intentions of each agent is updated upon receiving their Coordinative Event. This is the decisional process AC^{3}M uses to assist in deriving the best possible decision from the event. AC^{3}M requires both the desires and intentions to initiate not just cooperation, but Coordinative Cooperation between agents. This differs from the Hollengal (2007) decision cycle, where he shows only intentions are required for decisions to derive a course of action. This is not the case in AC^{3}M. In order to cooperate, the desires and intentions must be equal for each agent. If the desires and intentions are not equal, they must be re-evaluated until this condition is satisfied. Once equal, Coordinative Cooperation is said to have been completed.

Coordinative Cooperation is viewed as the entity-level execution of an entity-based action resulting from the decision-making process between agent. This is seen in Fig. 1C and as the result of Coordinative Cooperation, a system decision can then be made to outline a course of action for the execution and eventual completion and success of a system goal.

5 Future Work

The very nature of an unknown and dynamic environment can hinder the decision-making processes for a group of entities. This impedance can start as early as the information gathering stage, which then makes the whole decision-making process ineffective. An inadequate decision made by a human-user within a dynamic environment could prove to be fatal.

The importance of BDI, OODA and their link in agent coordination and cooperation, as well as their importance in decision-making within dynamic environments, has been shown using AC^{3}M. This model illustrates this new generation of cognitive MASs in dynamic environments by successfully demonstrating situational awareness and assessment, decision-making and coordination and cooperation for dynamic environments. Further development of the decision-making framework of AC^{3}M includes using concepts from blackbox architectures and a 'system of systems' approach to complex decision-making. This will promote superior cognitive problem-solving techniques for dynamic environments and will also enhance the BDI and OODA architectures and their link with coordination, cooperation and decision-making.

Acknowledgements. I wish to extend my gratitude to BAE Systems Australia for their support. I would like to give my ardent appreciation to the KES Centre at the University of South Australia and Stephen Robinson for their continued assistance.

References

1. Ancona, D., Mascardi, V.: Coo-BDI: Extending the BDI model with cooperativity. In: Leite, J., Omicini, A., Sterling, L., Torroni, P. (eds.) DALT 2003. LNCS, vol. 2990, pp. 109–134. Springer, Heidelberg (2004)
2. Anthony, R.: Planning and control systems: A framework for analysis. Harvard Graduate School of Business Administration, Tech. Rep. (1965)
3. Castelfranchi, C.: Modelling social actions for AI agents. Artificial Intelligence 103, 157–182 (1998)
4. Chalmers, S., Gray, P.: BDI agents and constraint logic. AISB Journal 1(1) (2001)
5. Consoli, A., Tweedale, J., Jain, L.: The link between agent coordination and cooperation. In: Shi, Z., Shimohara, K., Feng, D. (eds.) ICIIP 2006 - 3rd International Conference on Intelligent Information Processing, vol. 228, pp. 11–19. Springer, Adelaide (2006)
6. Consoli, A., Tweedale, J., Jain, L.: Aligning cognitive models using AC^{3}M. In: ICETET 2008 - First International Conference on Emerging Trends in Engineering and Technology, pp. 882–886. IEEE Computer Society, G. H. Raisoni College of Engineering (2008)
7. Decker, K.: Coordinating human and computer agents. In: Conen, W. (ed.) ASIAN-WS 1996. LNCS, vol. 1364, pp. 77–97. Springer, Heidelberg (1998)
8. Fox, M.: An organizational view of distributed systems. IEEE Transactions on Systems, Man and Cybernetics 11(1), 70–80 (1981)
9. Gorry, G., Morton, M.: A framework for management information systems. Sloan Management Review 13(1), 50–70 (1971)
10. Grant, T.: Uinifying planning and control using an ooda-based architecture. In: SAICSIT 2005 - Annual Research Conference of the South African Institute of Computer Scientists and Information Technologists. ACM, White River (2005)
11. Hollnagel, E.: Decisions about 'what' and decisions and 'how'. In: Cook, M., Noyes, J., Masakowski, Y. (eds.) Decision-Making in Complex Systems. Ashgate Publishing Company (2007)
12. Jones, P., Jacobs, J.: Cooperative problem solving in human-machine systems: Theory, models, and intelligent associate systems. IEEE Transactions on Systems, Man, and Cybernetic 30(4), 397–407 (2000)
13. Klein, M.: Coordination science: Challenges and directions. In: Conen, W., Neumann, G. (eds.) ASIAN-WS 1996. LNCS, vol. 1364, pp. 161–176. Springer, Heidelberg (1998)
14. Noble, D.: A cognitive description of collaboration and coordination to help teams identify and fix problems. In: 7th International Command and Control Research Technology Symposium. Canadian Department of National Defense, Quebec (2002)

15. Ossowski, S., Menezes, R.: On coordination and its significance to distributed and multi-agent systems. In: Concurrency and Computation: Practice and Experience, vol. 18, pp. 359–370 (2006)
16. Papadopoulos, G., Arbab, F.: Coordination models and languages. Centrum voor Wiskunde en Inforfmatica, Tech. Rep. SEN-R9834 (December 31, 1998)
17. Rousseau, R., Breton, R.: The M-OODA: A model incorporating control functions and teamwork in the OODA loop. In: CCRTS 2004 - 2004 Command and Control Research Technology Symposium, Coronado Bay Resort, San Diego, California, United States of America (2004)
18. Saaty, T.: How to make a decision: The analytic hierarchy process. European Journal of Operational Research 48, 9–26 (1990)
19. Salas, E., Guthrie, J., Burke, S.: Why training team decision making is not as easy as you think: Guiding principles and needs. In: Cook, M., Noyes, J., Masakowski, Y. (eds.) Decision-Making in Complex Systems. Ashgate Publishing Company (2007)
20. Schwenk, C.: Cognitive simplification processes in strategic decision-making. Strategic Management Journal 5, 111–128 (1984)
21. Shim, J., Warketntin, M., Courtney, J., Power, D., Sharda, R., Carlsson, C.: Past, present and future decision support technology. Decision Support Systems 33, 111–126 (2002)
22. Simon, H.: The New Science of Management Decisions. Harper Brothers, New York (1960)
23. van Bruggen, J., Boshuizen, H., Kirschner, P.: A cognitive framework for cooperative problem solving with argument visualization. In: Kirschner, P., Buckingham Shum, S., Carr, C.S. (eds.) Visualizing argumentation: software tools for collaborative and educational sense-making, pp. 25–47 (2003)
24. Wobcke, W., Ho, V., Nguyen, A., Krzywicki, A.: A BDI agent architecture for dialogue modelling and coordination in a smart personal assistant. In: IAT 2005 - 2005 IEEE/WIC/ACM International Conference on Intelligent Agent Technology. IEEE Computer Society, France (2005)
25. Wooldridge, M., Jennings, N.R.: The cooperative problem-solving process. Journal of Logic and Computation 9(4), 563–592 (1999)
26. Zachary, W.: A cognitively based functional taxonomy of decision support techniques. Human-Computer Interaction 2, 25–63 (1986)

An Intelligent Decision Support System Using Expert Systems in a MAS

Kamal Haider, Jeffrey Tweedale, and Lakhmi Jain

Abstract. Safety onboard airborne platforms rests heavily on the way they are fixed. This fact includes repairs and testing, to reduce its down time. Maintenance practices using these components are achieved using generic and specific test equipment within the existing Maintenance Management System (MMS). This research paper reports the work performed to improve reliability and maintainability of Avionics Systems using an Intelligent Decision Support System (IDSS). In order to understand the shortcomings of the existing system, the prevalent practices and methodologies are researched. The paper reports the significant improvements made by integrating autonomous information sources as knowledge into an IDSS. Improvements are made by automating the existing data collection to create an expert system using intelligent agents. Data Mining techniques and intelligent agents are employed to create an expert system. Using feedback, the IDSS generates forecasts, alerts and warnings prior to system availability being compromised. If the data was stored electronically, Data Mining techniques and intelligent agents could be employed to create an expert system. Using feedback, an IDSS should generate forecasts or warnings prior to system availability being compromised. A Knowledge Base of all aspects of the logistics cycle is created as the system ages, to help make informed decisions about the platform, the Unit Under Test (UUT) or even the environment that supports it.

Keywords: Intelligent Decision Support System (IDSS), Mean Time Between Failure (MTBF), Maintenance Management System (MMS), Maintenance Management Information Systems (MMIS), Integrated Logistics Support (ILS), Unit Under Test (UUT).

Kamal Haider, Jeffrey Tweedale, and Lakhmi Jain
School of Electrical and Information Engineering,
Knowledge-based Intelligent Engineering Systems Centre,
University of South Australia, Mawson Lakes, SA 5095, Australia
e-mail: {Kamal.Haider,Jeffrey.Tweedale,Lakhmi.Jain}@unisa.edu.au

K. Nakamatsu et al. (Eds.): New Advan. in Intel. Decision Techno., SCI 199, pp. 213–222.
springerlink.com

1 Introduction

The tempo of Defence Operations have grown and increased over the past two decades. This increases the demand for operational reliability and increased expectations from maintainers or Defence contractors. As a result, to deliver the ex-pected availability targets, Integrated Logistics Support (ILS) needs to become more sophisticated. Decision Support System (DSS) are required to provide adaptive automated responses for provisioning and maintenance of an increasing number of Defence Platforms. Many are now being supported by an increasing number of longterm maintenance contracts from within the private sector. Traditional methods of repair are already being modified to include automated testing although legacy platforms which still rely heavily on manual maintenance techniques. Regardless of the testing regime being employed in defence maintenance, a large volume of paper based maintenance data has already been and is still being generated. Despite the availability of such enormous amounts of data, there is very limited access of what could be treated as valuable and potential information. In this paper we present the development and implementation of a multi-agent expert system; referred to as an Intelligent Decision Support System (IDSS). This paper reports the progress made in the design and development of the agent-based IDSS. The developed IDSS is capable of providing cognitive feedback to support reliability predictions and informed decision making to proactively minimise the issues of Logistics down time, obsolescence and any associated risks. The project is undertaken in collaboration of Industry (Raytheon Australia), Department of Defence (Defence Science and Technology Organisation) and Academia (University of South Australia). We discuss a brief history of maintenance methodology and present concepts perceived within the ILS system followed by a brief discussion on some of the challenges with the existing system. After summarising our previous related research, we explain the newly developed system, highlighting some of its salient features and capabilities. Finally we discuss our future strategy and direction of this research.

2 Maintenance Methodologies

2.1 Standard Maintenance Environment

In the last few years, electronics, and in particular, avionics have experienced exponential advances in technology. In today's modern aircraft, avionics equipment consists of the major part of the platform. With recent advancements in electronics technology an increased number of platforms are fitted with and rely on sophisticated avionics equipment. Documented advancements include modular form factor and circuit design based on solid state electronic, VLSI and tools using FPGA derived components. Similarly the concept of Modular test equipment like Automated Test Equipment (ATE) has emerged using extensive Commercial-off-the-shelf (COTS) equipment which has introduced turnkey solutions that are cost effective. Some of the older airframes like the Hercules (C-130) and Orion (P3) have undergone significant avionics upgrades to enhance the aircraft capabilities. Although component count has increased significantly, the reliability has also increased due to improved component tolerances, sophistication and manufacturing techniques. The trend is to install more advanced avionics systems, although the logistics systems remain unchanged. The survey [1] conducted by Chris Guy of Defence Science and Technology Organisation in 1989 clearly shows that the number of aircraft Combat Systems have tripled during the period from 1940 to 1990. This is to say that the number of Unit Under Test (UUT) in a mission system grows with complexity hence statistically, failures increase and the complexity of testing also increases as new technology is introduced. Host of military and commercial aviation standards are put in place to protect the standardized and robust framework to ensure avionics equipment reliability and in turn the platform reliability. These standards include environment, electrical, mechanical and support standards. Some of the standards are Testability Handbook for Systems and Equipment [2], Avionics Equipment Maintainability (IEC 60300),Functional safety of electrical / electronic / programmable electronic safety-related systems-Overview of Techniques and Measures [3] and IEEE Standard for Signal and Test Definition [4].

2.2 *Maintenance and Integrated Logistics Support*

Effective and efficient logistics management system plays a vital role in ensuring availability and maintainability of avionics equipment. Recent maintenance methodologies have introduced concepts which aim at the supportability of the product throughout its anticipated and planned designed life [5]. This gives an incentive for the system designers to develop more maintainable systems with inherently reduced logistics downtime. The ILS paradigm [6] forms an integral part of system readiness, maintainability and supportability. As a result, to deliver the expected availability targets, ILS needs to become more sophisticated. The ILS involves the key concepts of Maintenance management and planning to support the avionics assemblies and sub-assemblies, Supply support, Technical Data, Design interface, Database management systems and Support equipment. DSS [7, 8] are required to provide adaptive automated response for provisioning and maintenance of defence platforms. Within the Maintenance Management System (MMS), preventive and corrective types of maintenances are performed on repairable avionics items [6, 9]. This maintenance is either performed through manual testing techniques or through ATE which is already an established maintenance practice around the world. Regardless to the maintenance technique being used, the maintenance data needs to be fed back into a central repository to enable MMS to aid future decisions. Legacy avionics platforms make a large proportion of the inventory employing manual test equipment. The test results produced are generated manually by technicians or operators. Often due to time or other resource constraints, these results are not generated and contain insufficient or unreliable information. The importance of avionics platforms in today's operational environment highlights this issue which becomes critical in sustaining military equipment. Sustainment considers the issues of maintainability and availability of logistic support for the life of the product, otherwise called Through Life Support (TLS). Efficient and effective maintenance become critical when considering TLS [10].

3 Problem Statement

In the traditional maintenance management approach, there is only a manual paper trail between the UUT, test results and the LMS unless collected manually through log books or workflow systems. Regardless of whether maintenance is performed manually or automatically, the test results have to be collated manually to the Logistics Management System (LMS) or MMS. This activity relies heavily on the availability of time and resources. Currently most maintenance data is post processed and subject to very limited analysis. Some of the other issues with this traditional approach are:

- extensive human-machine interaction;
- non-availability of maintenance data and defect trend data; and
- non-availability of reliability information.

Most legacy platforms constitute a large proportion of repairable avionics inventory in operations and equipment downtime. Many faults are caused by obsolescence and attract no support available from the Original Equipment Manufacturer (OEM). Although some of the existing Maintenance Management Information Systems (MMIS) retain certain information in one form or another, more often than not, this information is of no real value as it does not offer any value to the overall decision making process. The maintenance data in most of the existing MMIS is in the raw form and is rarely transformed into information or knowledgebase. In absence of any such knowledgebase, the analysis becomes evermore tedious and practically impossible. In the current maintenance management approach, there is a manual paper trail of results between the Unit Under Test (UUT) and the LMS. These results are collected manually through log books or workflow systems. These test results are then manually collated and fed to the LMS or MMS. This activity relies heavily on the availability of time and resources and is subject to very limited analysis. Other issues with this traditional approach include access to results, accuracy and integration.

4 Related Research

The ILS approach, in modern logistic operations, has tried to address the issue of wasting existing opportunities of collecting the available corporate knowledge. As discussed in the above paragraphs, the ILS environment provides a single platform to store the related maintenance and logistics information. Modern ILS tools lack the capability of using existing data to provide the managers with an insight into likely future trends. They fail to embed critical information including

Interface Data, Technical Drawings, Failure Reporting, Analysis and Corrective Action System (FRACAS) [2], Failure Mode Effect Criticality Analysis (FMECA) and Mean Time Between Failure (MTBF) [5]. Therefore, for the system to work, an automated method of collating and uploading these missing threads is vital. Until then, all affected parties will be unable to make informed decisions. Most of the maintenance data in existing systems is in the raw form and is rarely transformed into information or knowledge. In the absence of a knowledge, the analysis becomes 'man-draulic' which is unfeasible. Efficient and consistent decision making in today's complex environment is not an easy task without the decision support tools. Tools, such as the IDSS, have supported military decision-making in a complex environment characterised by high uncertainty and rapidly changing conditions. The architectures have been developed for various military applications to demonstrate possible advances of such systems. A number of these applications and associated research has been reported in our earlier papers [11, 12, 13]. Our research is aimed at enhancing the ILS paradigm and its utilisation for aircraft avionics by developing an IDSS to improve availability and reliability predictions for aircraft avionics and potentially other systems. The model developed in this research might help in the future development and utilisation of decision support systems to predict likelihood of obsolescence in avionics and other aircraft systems and components.

5 System Development

The IDSS being developed and reported in this paper is a Multi-Agent System (MAS) [14, 15, 16]. It consists of multiple intelligent software agents that are loosely coupled to solve problems that are beyond the individual capacities or knowledge of each individual agent. Each agent in the IDSS possesses diverse and unique capabilities required for their unique task. In the developed system, the agents support a modular, extensible approach to design a complex information system. Some of the useful behaviours demonstrated by these individual agents include Inter-agent communication through sharing knowledge, intentions and beliefs, Inter-agent collaboration, coordination and cooperation.

6 IDSS Architecture

Within the developed IDSS, the MAS consists of four agents namely MMS, LMS, EMS and Inference Agent. Three agents (MMS, LMS and EMS) act as information collection agents which continually observe the environment for a potential change and report any such change back to the Inference Agent. These collection agents also process information and convert it into knowledge (as a set of beliefs about their perceived environment). The Inference Agent accesses the expert knowledge in terms of production rules and continually interprets the knowledge created and updated by the collection agents. The three collection agents interact with their environment which, for this demonstration, consists of information stored on various diverse independent information storage platforms and information servers. For the purpose of this demonstration we have used three information servers, which include: My SQL, Oracle and ODBC servers [17]. The IDSS also contains a user interface to enable user-requested analysis. The outputs of the IDSS are the generated reports, warnings and other relevant feedback about the current state of the system.

7 Implementation

While developing the IDSS, platform independence and reduced development time were achieved using COTS software. To fulfil these two considerations, two software packages were selected to implement the IDSS. Java [18] was chosen as a development language for following reasons:

- Portability;
- Platform Independence; and
- Extensive development support.

To implement the software agents framework, we have used JACK Intelligent Agents [19] for the following reasons:

- Extension of Java;
- Graphical representation of agents; and
- Specifically written to implement software intelligent agents.

JACK is an event driven software development platform where software agents are developed to implement proactive and reactive strategies. In the developed IDSS, each agent is developed to continually monitor the assigned environment. Each agent consists of one or more plans to execute certain logical operation based on particular events or based on its own beliefs. The explanation for these agents is provided in following paragraphs.

7.1 Intelligent Agents

The developed IDSS constitutes of four intelligent agents developed in JACK software development environment. These agents are namely MMS, LMS, EMS and INF Agent. These agents are programmed to continually monitor their respective environment. The respective environment for each agent is unique and may vary significantly from one agent to the other in terms of type and extent of data it contains. Some of the information held in the environment includes maintenance history of avionics components, repair venues, avionics components repaired and replaced, servicing dates, maintenance technician details, inventory, batch numbers, receipt and issue, turn-around times, suppliers, lead times, Interface Control Documents (ICD) and OEM provided information like MTBF and Mean Time To Repair (MTTR) etc. Being programmed to monitor the environment continually, the information and facts collected by the agents are kept up to date in form of its belief-sets which represent agents perceived state about its environment. Figure1 represents the schematic of all four agents. The pictorial view of these agents highlights that each agent contains certain *Capabilities, Plans and Events*. Every change in the agents perceived environment is handled by the agent as an *Event*. For every event, the agent has a *Plan* to handle that particular event. The Plan specifies the particular actions that need to be taken in case of certain *Events*. A collection of *Plans* constitutes a *Capability*. For example MMS agent has the *Maintenance* capability to monitor the maintenance environment and to observe any changes in the environment. With in this Capability, the agent has *Readbeliefdata* Plan and *test Plan*. Every time a test is conducted, the test tables are updated reflecting a change in the environment. These changes may include new entries in any table and addition or removal of data from any of the observed tables. This change triggers an Event called *Addbelief*. This event is handled by *test Plan* resulting in Agents belief-set update. These belief-sets, within each agent, represent the knowledge.

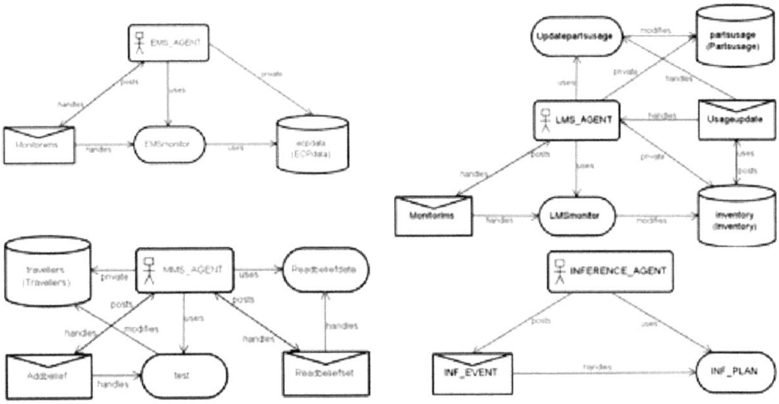

Fig. 1 Intelligent Agents Schematics

The MMS, LMS and EMS agents monitor the environment, constantly updating their belief-sets. As soon as a change in the environments is observed, a trigger is generated by the agent. This trigger contains the up-to-date knowledge about the environment. The information is passed to the Inference agent for analysis. Inference agent contains a set of predefined production rules which represent the expert knowledge. Inference agent performs a GO-NOGO test on the available information by comparing it with the stored production rules. Based on the success or failure of particular tests, it then generates alerts, warnings and recommendations.

8 Preliminary Findings

For the purpose of analysing the efficiency of the developed IDSS, we have used three separate sets of test databases, stored on the MySQL server. Each database constitutes of three separate elements being an MMS, LMS and an Engineering

Management Systems (EMS). Each of these elements includes multiple tables containing vital information about the system. In following paragraphs we discuss the different scenarios tested with our developed system.

8.1 Interpreted, Alerts, Warnings and Recommendations

In this scenario, we used the test database with parts maintenance and usage history. The data in the maintenance management system was analysed for the maintenance history of all avionics parts within a specified period. For the purpose of trials, one specific avionics part Front End Receiver (FER) was selected for its simpler built structure and relatively higher usage. The analysis of maintenance history data, clearly indicated an increase in failures and inturn the repair effort for FER. This increase however, apparently, could not be attributed to a particular reason by looking at the MMS data alone. We then analysed the same maintenance history data using the newly developed IDSS. The MMS agent successfully established the fact of higher failure rate of these avionics components and triggered an event for the inference agent to analyse and interpret. Inference agent in turn triggered two events for LMS and EMS agents respectively to retrieve relevant information about the specific avionics component being FER. LMS and EMS agents, which are already observing their respective environment, have accumulated knowledge about the LMS and EMS in a structured format. The already collated knowledge is returned to Inference agent for interpretation. In this particular scenario, the EMS agent does not return any information as it does not have any knowledge about the specific part at this stage. Using the available knowledge provided by LMS agent, and using the interpretation rules, Inference agent is able to relate the available knowledge and thus generates an interpreted Alert for the maintainers. The interpreted output alert generated by the IDSS, successfully correlated and interpreted the available information and assisted us in understanding the system performance and possible shortfalls in time. These warnings could be utilised to initiate investigations which may improve overall system performance. In another scenario, we used the test database to analyse the usage of Break-Down Spares (BDS) and possibly highlight any trends in the usage. The BDS are generally referred to as the sub-components in an avionics part. An example could be a resistor or a transistor with in an avionics component. The LMS agent within the developed IDSS continually monitors the logistics environment and updates its knowledgebase to reflect the current usage of all sub-components. The Inference agent consistently monitors the progressive usage history for each individual sub-component and compares this to previous usage history. To test the effectiveness of the developed IDSS to report any abnormalities in parts consumptions, we increased the consumption of a particular transistor in the test database. As anticipated, the Inference agent was able to instantly recognise this increase. Further analysis was instantiated by the Inference agent at this point. Additional information regarding the transistor is retrieved by the Inference agent from the already available and updated knowledgebase with in LMS and MMS agents. Some of the pertinent information retrieved from the knoweledgebase includes target levels, lead times and if the part is discontinued or not. Using the available knowledge and comparing it with the usage rates, the Inference agent was able to create an interpreted early warning of a possible shortage of transistors. This early warning could be employed to place an automated order for the transistors. Various other warnings and recommendations are successfully being generated by the developed IDSS. These alerts, warnings and recommendations assist maintainers and managers to enable them in making informed decisions before the system availability is compromised.

9 Future Research

The initial tests have shown that by applying a systems perspective to available information, we can greatly improve our ability to understand and interpret the available information and make informed decisions [20]. At present, we are integrating the developed system with a real test database EAGLE [21] to test it with a fully populated set of targeted maintenance data. The developed IDSS design will be further tested to create additional rules and to enhance its decision making capabilities. These rules will be packaged as capabilities and will become part of the IDSS. Further testing will be performed to improve the quality and range of reports and warnings. Ultimately, the developed IDSS may be packaged and integrated within real MMS and LMS to enhance the ILS operations. We anticipate that the developed IDSS using MAS will significantly improve the information sharing and introduce automated efficiencies into the current ILS paradigm.

References

1. Guy, C.: Avionics: An overview of history and trends. In: Proceedings, UK (1989)
2. DoD, Mil-Hdbk-2165: Testability Handbook For Systems and Equipment (1995)
3. IEC, International Standards IEC 61508-7, Functional safety of electrical/electronic/programmable electronic safety-related systems: Overview of techniques and measures (2003)
4. Cornish, M., Brown, M.: Implementing IEEE 1641 - A demonstration of portability, IEEE CNF. Autotestcon, pp. 144–152 (September 2005)
5. Rausand, M., Hoyland, A.: System Reliability Theory: Models, Statistical Methods, and Applications, 2nd edn. John Wiley and Sons, Chichester (2004)
6. Dhillon, B.: Engineering Maintainability: How to design for reliability and easy maintenance. Gulf Professional Publishing (1999)
7. Feigenbaum, E.A., Buchanan, B.G., Lederberg, J.: On generality and problem solving: a case study using dendral program. In: Machine Intelligence, vol. 6 B, pp. 165–190. Meltzer and D. Michie, Edinburgh University Press, Edinburgh, Scottland (1971)
8. Russel, S., Norvig, P.: Artificial Intelligence: A Modern Approach, 2nd edn. Prentice-Hall, Eaglewood Cliffs (2003)
9. Langford, J.: Logistics: Principles and Applications. McGraw-Hill Professional, New York (1998) (special reprint edn.)
10. Eisner, H.: Essentials of Project and Systems Engineering Management, 2nd edn. John Wiley and Sons, Chichester (2002)

11. Haider, K., Tweedale, J., Urlings, P., Jain, L.: Intelligent decision support system in defence maintenance methodologies. In: 2nd IEEE International Conference on Emerging Technologies, pp. 560–567. IEEE, Los Alamitos (2006)
12. Haider, K., Tweedale, J., Urlings, P., Jain, L.: Intelligent decision support feedback using multi-agent system in defence maintenance environment 1(3/4), 311–324 (2007)
13. Haider, K., Tweedale, J., Urlings, P., Jain, L.: Enhancing avionics systems reliability and maintainability through development of idss - a systems approach. In: SysCon, I. (ed.) IEEE International Systems Conference, pp. 280–285. IEEE, Montreal (2008)
14. Wooldridge, M., Jennings, N.: Agent theories, architectures, and languages: A survey. In: Wooldridge, M.J., Jennings, N.R. (eds.) ECAI 1994 and ATAL 1994. LNCS, vol. 890, p. 403. Springer, Heidelberg (1995)
15. Wooldridge, M., Jennings, N.: Intelligent agents: Theory and practice. The Knowledge Engineering Review 10(2) (1995)
16. Ferber, J.: Multi-Agent Systems. Addison-Wesley, New York (1999)
17. Harnandez, M.: Database Design for mere Mortals: A Hands-on Guide to Relational database Systems, 2nd edn. Addison-Wesley, New York (2003)
18. Savitch, W.: Absolute Java, 2nd edn. Addison-Wesley, New York (2005)
19. A. O. S. Inc, JACK Intelligent Agents Manual, 2nd edn., Agent Oriented Software, PO Box 639, Carlton South, Victoria, Australia (June 2005)
20. Tweedale, J., Ichalkaranje, N., Sioutis, C., Urlings, P., Jain, L.: Future directions: Building a decision making framework using agent teams, ch. 14, pp. 381–402. Springer, Heidelberg (2008)
21. R. C. USA, MMIS EAGLE software, RAytheon Company (2008), `http://www.raytheoneagle.com`

Decentralized Real-Time Control Algorithms for an AGV System

Alexander Klaas, Mark Aufenanger, Nando Ruengener, and Wilhelm Dangelmaier

Abstract. Automated guided vehicles (AGVs) are increasingly being used to transport goods or people. Navigation is a core issue in such a system. The AGVs use real-time control algorithms to reach their assigned destination autonomously. For reasons like scalability and flexibility, it is beneficial that the shuttles compute the necessary calculations decentrally. In this paper, we present such decentralized algorithms for conflict-free routing in a specific AGV system. Based on existing algorithms for deadlock handling in theory and routing in computer networks, we implemented three different sets of algorithms of varying sophistication in a logistics simulator. Evaluation reveals their functionality and relative performance.

1 Introduction

Automated Guided Vehicles (AGVs) have been used for cost efficient transportation of goods or people. One example for an AGV system is the research concept "RailCab", which is being developed at the University of Paderborn [7]. In this scenario, small autonomous vehicles (so called shuttles) operate using the existing rail infrastructure. The shuttles run on demand and work on individual tasks. They must be able to navigate themselves to an arbitrary target within the network. It is beneficial to let each shuttle compute the route to the destination itself decentrally. This

Alexander Klaas
Heinz Nixdorf Institute
Business Computing, esp. CIM
University of Paderborn
Fuerstenallee 11, 33102 Paderborn, Germany
e-mail: a-klaas@upb.de

Mark Aufenanger, Nando Ruengener, and Wilhelm Dangelmaier
Heinz Nixdorf Institute
e-mail: Mark.Aufenanger@hni.upb.de, Nando.Ruengener@hni.upb.de, Wilhelm.Dangelmaier@hni.upb.de

K. Nakamatsu et al. (Eds.): New Advan. in Intel. Decision Techno., SCI 199, pp. 223–233.
springerlink.com

would enable the shuttles to spontaneously react to unforeseen changes. Opposed to a centralized algorithm, the computation costs of such a decentralized approach, by nature, scale well with the size of the rail network. Conflicts between shuttles also need to be handled. This paper presents algorithms for online control. We have developed three sets of algorithms as a first solution of solving the specific problem of autonomous routing and deadlock handling. The algorithms are experimentally evaluated in the logistics simulation software d^3Fact.

2 Problem Definition

We abstract the rail network as a planar undirected graphVertices of degree 3 represent switches. The tracks are split into distinct zones and each zone corresponds to a vertex of degree 2. This ensures that shuttles maintain a certain safety distance, as each shuttle occupies a vertex but vertices cannot be occupied by more than one shuttle. Edges in the graph intuitively represent that tracks and switches are connected. The shuttles need to reach a specific vertex (their destination) reliably and autonomously.

3 Deadlocks

In any system with parallel processes and shared resources, research has shown that there are four necessary conditions for deadlocks to be possible [2]:

1. *mutual exclusion*: One process occupies a resource exclusively.
2. *no preemption*: A process must release resources itself.
3. *hold and wait*: A process occupies a resource and waits when it has requested another.
4. *circular wait*: Two or more processes wait for each other to release a resource.

These conditions are fulfilled by our given AGV system. For example, when two or more shuttles try to occupy the same vertex, the request is denied for one of them (*mutual exclusion*).

There are traditionally three different approaches to deal with deadlocks [8]: Avoidance, Prevention and Detection with Resolution. When using avoidance, the state of the system is monitored and no situations are allowed that could potentially lead to a deadlock. This is not practical in a decentralized system as global information is required. Deadlock prevention, for which the system's general conditions are altered so that deadlocks cannot occur, can also hardly be realized here due to physical constraints. Deadlock detection and resolution however is a feasible approach for decentralized systems. We present an algorithm that guarantees that all involved shuttles detect when a deadlock occurs and decentrally resolve the situation. To improve performance, a second algorithm was developed that heuristically predicts deadlocks. It can be used on top of the first algorithm.

3.1 Deadlock Detection and Resolution

Shuttles must be able to detect a deadlock situation locally. In a decentralized system, that is not a trivial problem. We assume communication between shuttles on neighboring vertices is possible. A simple form of *edge chasing* is employed [9]. The idea is that shuttles send messages along the edges of the abstract *dependency graph*. Once a message returns to the original shuttle, it can infer that there is a cycle in the dependency graph, which is equivalent to a deadlock. For our AGV system,

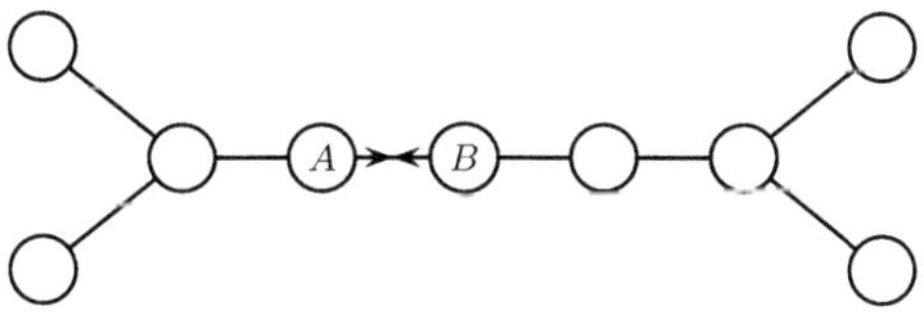

Fig. 1 Case 1: Two shuttles ***A*** und ***B*** face each other heads on

we need to consider only the cases that cycles exist between two neighboring shuttles that request each other's vertex. Whenever a shuttle's request to lock a vertex is denied, the shuttle will send a message to the occupying shuttle. Each shuttle maintains a list of shuttles it received such a message from. If an occupying shuttle is in that list, there must be a deadlock.

To resolve the situation, shuttles must consider whether the deadlock occurred at a switch or at a section of straight tracks. We classify the possible situations into three cases. In case 1, two shuttles face each other heads on (figure 1). The shuttles can detect the distance to the two adjacent switches and the shuttle closest to a switch voluntarily turns around and runs towards that switch. We have reduced case 1 to the next case: a shuttle ***A*** on a switch tries to leave it, but another shuttle ***B*** is occupying the vertex in that direction (see figure 2). The other two neighboring vertices are not occupied and are used for sidestepping. Again, some communication is required. Shuttle ***A*** needs to receive the direction ***B*** will take so it can sidestep using the other remaining vertex. Assuming there are other shuttles waiting behind ***B***, this sidestep

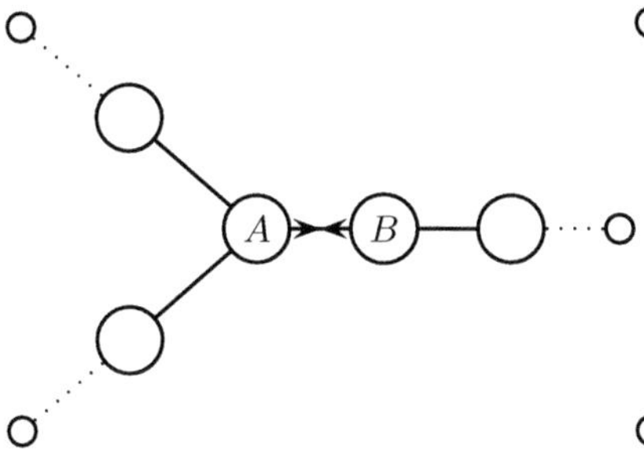

Fig. 2 Case 2: Shuttle ***A*** occupies a switch, while ***B*** tries to reserve it. ***A*** however seeks to leave it in ***B***'s direction

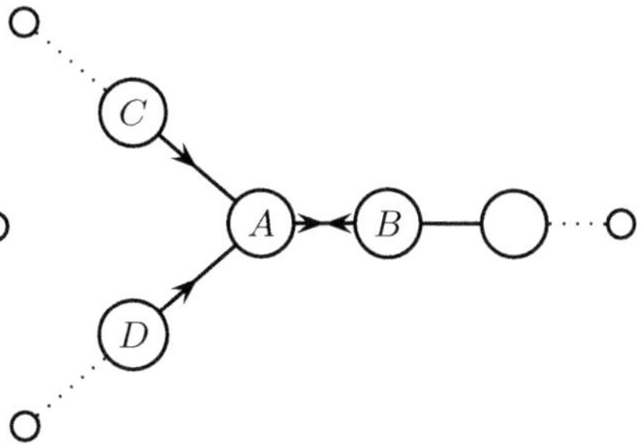

Fig. 3 Case 3: Shuttle ***A*** occupies a switch und and seeks to leave it in ***B***'s direction while ***B***, ***C*** und ***D*** try to reserve the switch

process can be repeated until ***A*** is able to leave the switch. By induction, this proves that the deadlock can be resolved.

Even when there are always new shuttles coming from the direction ***A*** tries to leave the switch in, we could still solve the situation. We would introduce a mechanism that over time prioritizes ***A*** in the occurring case 1 deadlocks.

Case 3 is the worst case: contrary to the previous situation, all neighboring vertices are occupied and there is no room for shuttle ***A*** to step aside (figure 3). We introduce another type of message: ***A*** can send a backoff command to the occupying shuttles ***C*** and ***D***. They may need to forward this message, but eventually they will leave their vertices and this reduces case 3 to case 2. When many shuttles are involved in the situation, this method queues them up along the tracks.

As a precondition, we assume that between any two switches there are enough zones for all shuttles in the system so that no two queues can run into each other.

In summary, the shuttles can locally detect a deadlock and quickly decide their next action to resolve it. The algorithm has constant complexity regarding shuttles or vertices in the system as only neighboring, i.e. three vertices, need to be considered. Obviously, resolving a deadlock is costly in terms of travel duration. In total and from a global perspective, it takes the n shuttles that are involved in the deadlock $\mathcal{O}(n^2)$ steps (moving from one vertex to another) to resolve it.

While we have concluded that we cannot avoid all deadlocks, fewer occurrences should improve our travel times and therefore, we developed a method to reduce their probability.

3.2 Deadlock Prediction

This algorithm is run locally at each vertex (switch) within the network. The idea is that using information from within a local zone around the vertex, we predict that a deadlock will occur with a certain probability. When a shuttle arrives at a vertex, the algorithm will determine a risk factor of a deadlock for each of the three possible paths the shuttle can take.

We consider the shuttles that are within the local zone. Specfically, whenever a shuttle within the local zone is moving on an arc towards the vertex, it will contribute $\frac{1}{3^n}$ to the risk factor, where n is the number of vertices between the shuttle and the central vertex. Naively, that is the probability that the shuttle will actually arrive at the central vertex as at each of the vertices in between, the chance is $\frac{1}{3}$ it will continue travelling towards it. If a shuttle is moving away from the vertex, we add $\frac{1}{3^{n+1}}$.

Additionally, when a shuttle is moving towards the vertex, the switch will communicate with it to determine what direction it will take at the next vertex. It requests its destination and other parameters and transmits them to the next vertex, prompting the routing result. According to how the shuttle will prospectively move, its contribution to the risk factor is altered.

Consider the situation illustrated in figure 4. We assume that ***B*** will continue moving towards switch W1 while ***C*** will turn around. There will be no switches between ***A*** and ***B*** so ***B*** adds $\frac{1}{3^{1-1}} = 1$ to the risk factor. C adds $\frac{1}{3^{2+1}}$ as both switches W2 and

W3 lie in between ***A*** and ***C*** and additionally, ***C*** will move away from switch W1. The result is that for ***A***, moving towards switch W2 carries a deadlock risk of $\frac{28}{27}$.

The resulting factor is a quite rough approximation and not an expected value in the stochastic sense. However, using only local information it is not possible to calculate the probability of a deadlock exactly. Shuttles from outside the local zone could cause a deadlock within it and unforeseen events like failures could happen. Thus, the risk factor indicates the likeliness of a deadlock well enough for our application.

How to use the approximation depends on the employed routing algorithm. We will reexamine the question later.

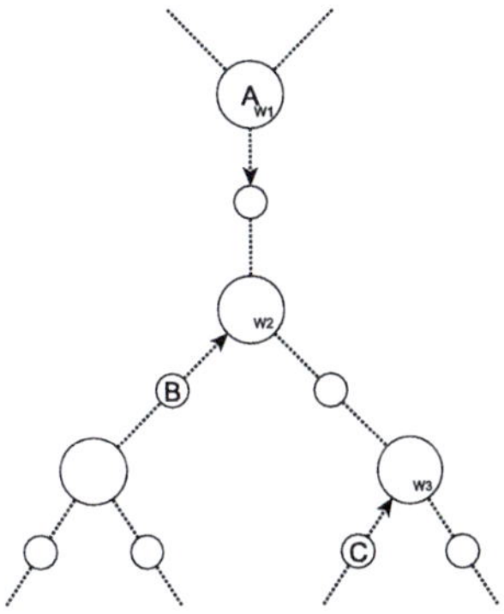

Fig. 4 Example for the deadlock prediction algorithm with a local zone radius of 3. Shuttle ***A*** calculates the risk factor for moving towards W2. It must consider the two shuttles ***B*** ($n = 1$) and ***C*** ($n = 2$) that are within the local zone

4 Routing Algorithms

In the well known routing problem, we seek the optimal path between two vertices in a graph. When the graph is changing however, it is beneficial to compute the path dynamically in contrast to a one time static computation. In computer networks, routing is a fundamental issue and therefore many solutions have been developed and refined over time. Packet based networks, in which data streams are split up into atomic pieces, use hop-to-hop routing. It is a form of dynamic route computation as routers forward the packet only to an adjacent router; the whole route is not precomputed [10].

There are striking similarities between routing in networks and routing in transport systems such as ours. Switches and routers, shuttles and data packets naturally match. Differences lie in error handling (shuttles cannot be dropped and retransmitted like data packets) and time scales (shuttles move much slower than data packets). Vertices of degree 2 (safety zones between switches) are not relevant to the problem.

We found two routing algorithms to be especially suited for our application: Geographic routing and the Enhanced Interior Gateway Routing Protocol.

4.1 Geographic Routing

Geographic routing was initially developed for wireless adhoc networks [1]. Each switch, which acts as a router, must know its geographic coordinates and the shuttles must carry the coordinates of their targets.

The algorithm uses a simple greedy strategy, which is to strictly reduce the distance of the shuttle to the target in each step, as much as possible. The switches can do this locally using the coordinates of their three neighbors.

The strategy may fail however, when all of the three adjacent vertices have a greater or equal distance to the destination. This is a sort of local minimum and a typical problem for greedy strategies. To overcome the minimum, we use a method called face routing. Both strategies combined have been proven to guarantee delivery [4], a crucial requirement for our transport system.

Whenever the greedy strategy fails, the shuttle saves the coordinates of the vertex where the failure occurred. Face routing moves the shuttles along the edges of so called faces within the graph, until the shuttle arrives at a switch which is closer to the destination than the saved vertex. The local minimum is then overcome.

Face routing moves the shuttles along a face using the right hand rule. The name is derived from an exploring strategy, which is to walk through a maze with the right hand touching the wall. It can be implemented locally by calculating the angles between neighboring vertices. The next hop is the first vertex in clockwise direction, relative to the vertex the shuttle came from. This way, the shuttle will travel to each vertex in the face.

We need to consider the case that none of the vertices in the face are closer to the destination than the one where face routing began. The algorithm plots a line from that vertex to the destination. Should the edge between the current vertex and the next hop intersect the plotted line, we route the shuttle along the edges of the adjacent face. In that case, the shuttle should travel through the face in opposite direction and follow the left-hand rule. Intuitively, the routing follows the line towards the destination. These calculations can also be done using local coordinates.

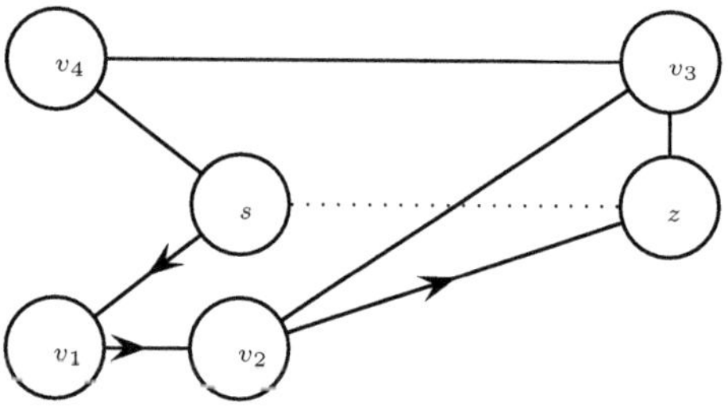

Fig. 5 The greedy strategy fails at vertex s, face routing routes the shuttles along the edges of the face towards z using the right hand rule

Figure 5 illustrates face routing. At vertex s, the greedy strategy fails to route the shuttle closer to the destination z. Face routing routes the shuttle along the edge of the face s,v_1 - v_4. Note the plotted line from s to z. It is intersected by the edge $\{v_2,v_3\}$. In this case, the algorithm switches to the adjacent face z,v_2,v_3 and the shuttle has already reached its destination z.

To realize face routing, our shuttles merely need to maintain two coordinates: Those of the switch where the local minimum was first met and the coordinates of their destination. Additionally, they must know which direction (left hand or right hand rule) they need to be routed in and transmit this data to the current vertex.

The advantage of geographic routing is that the calculations are very simple and each vertex only needs to know the coordinates of itself and its neighbors. In our application, we also have the benefit that the switches do not move. The run time and storage requirements of the algorithm are independent of the size of the network or number of shuttles; it scales very well. Communication between switches is not required.

The main drawback is that the greedy strategy will not always find the optimal path and recovering from a local minimum adds to the travel time unnecessarily. Additionally, it is sensible to dynamic effects like link failure, which it cannot react to.

4.2 *Enhanced Interior Gateway Routing Protocol (EIGRP)*

EIGRP is a routing algorithm actually used in computer network devices [6]. Routers store information about the network and exchange these by communicating with each other. EIGRP is based off the distance vector routing protocol, which employs the bellman-ford algorithm for calculating shortest paths [6]. Basically, each router maintains a routing table and for each other router in the network, the cost of the shortest path and the next hop on that path is stored. Routers are able to detect changes (i.e., edge weights have been modified) and inform their neighbors via *update* messages, which they will also forward. Note that we now view our rail network as a weighted graph, where the edge weights represent geographic distance for now.

A lot of attention has to be paid to the communications protocol for updating graph information however. EIGRP uses the diffusing update algorithm (DUAL) as it is robust against issues like the *count to infinity* problem [5]. Every time a router detects a change, it will broadcast a *query* message for each affected destination in the routing table. It will wait until it receives all *reply* messages with new shortest paths. Before it receives the replies, all other update messages concerning affected destinations are ignored, which is the main reason why DUAL avoids inconsistencies.

EIGRP is well suited for our application as it is a genuine decentralized algorithm and proven to find the shortest path, unlike geographic routing. It can quickly adapt to changes in the rail network, which propagate rapidly. When a shuttle arrives at a switch and requests the next hop, a quick look up in the routing table is sufficient. The protocol creates many messages however. Whenever one edge weight changes, we have shown that a worst case maximum of $6 \times |V|^2$ messages are exchanged in total between the $|V|$ switches. Additionally, the algorithm requires $\mathcal{O}(|V|)$ space at each switch, as each vertex in the network has an entry in each routing table. As hardware requirements should be kept to a minimum, we developed a method to reduce the space demand.

4.2.1 Subnetting

The idea here is to reduce the number of rows in the routing table by aggregating a group of vertices to one entry in form of a subnet. This is a common practice in computer networks [6] and again, we adapt it to our application. Our approach is based on the geographical location of the vertices. We define a subnet as the vertices that lie in a specific rectangular area. Every switch then maintains one entry for each vertex in its subnet and one for each of the other subnets.

Optimally, we can reduce the space requirements from $\mathcal{O}(|V|)$ to $\mathcal{O}(\sqrt{|V|})$ if each subnet includes $\sqrt{|V|}$ vertices. Furthermore, as communication generally needs to happen only within the subnets, less messages are being sent. The impact on the travel times is expected to be minimal. Two vertices within the same subnet, but not close to a switch lie in the same general direction, so the next hops on the shortest paths to those two vertices are very likely to be the same.

5 Agent Design

Based on the developed algorithms, we specify three different sets of agents. It was necessary to include the result of the deadlock prediction algorithm in the routing calculations.

Our first agent system is the most basic type, as it simply realizes geographic routing with the deadlock detection and resolution algorithm.

The second agent system is utility based. At a switch, geographic routing ordinarily computes the distance of the three neighboring vertices to the destination and compares them. We define a utility function: $utility : V \rightarrow \mathbb{R}$. The value for each vertex is defined as the negative sum of distance to destination and result from the deadlock prediction algorithm. The routing algorithm then chooses the vertex with the highest utility. If routing to the vertex closest to the destination means a high risk of deadlock occurrence, the algorithm will select a different vertex. The deadlock prediction value must be weighted, how much needs to be determined experimentally. When face routing needs to be employed, we cannot consider deadlock prediction to ensure delivery.

The third agent system is the most hardware intensive and uses EIGRP. Instead of using geographic distance for the edge weights, we add the result of a simplified deadlock prediction algorithm. We simply count the number of shuttles on the tracks forming an edge. As the edge weights propagate, this approach potentially means that the shuttles avoid crowded tracks. Additionally, when tracks are closed, due to e.g. construction work, we can set the edge weight to infinity and shuttles automatically choose alternative routes. As edge weights constantly change, many messages are being sent. These messages travel a lot faster than the shuttles however.

Again, we must experimentally determine the right weight for the deadlock prediction result. We developed a simple genetic algorithm (not detailed here) to find the optimal weight automatically.

6 Simulation

6.1 Implementation

We implemented the three types of agent systems in the logistics simulation software d^3Fact. It provided us with an event discrete simulation kernel [3]. By altering the speed of the simulation, we could gather data about the long-term behavior of the systems. We tested the algorithms on two rail networks that were created using a separately developed custom editor. The first network contains 40 vertices that were placed in relation to 40 German cities. The second network is much bigger and consists of 944 switches that are aligned in a randomized pattern.

6.2 Results

We evaluated the systems by placing a variable amount of shuttles onto the tracks and continuously assigning them random destinations. After a certain amount of time (typically after each shuttle finished 1500 tasks), we stop such a simulation run and compare key metrics such as average travel time and deadlock occurrences.

Figure 6 shows the relative performance of the three agent systems. Using EIGRP instead of geographic routing reduces the average travel time by 6.3% in this scenario. The main reason for the performance increase is that the geographic routing algorithm spent 5.8% of the travel time using face routing to overcome local minima. The simplest agent, which does not use deadlock prediction, lags behind EIGRP 10.8%, which demonstrates that deadlock prediction is effective.

Fig. 6 Comparison of the three agent systems with 10 shuttles in the smaller network

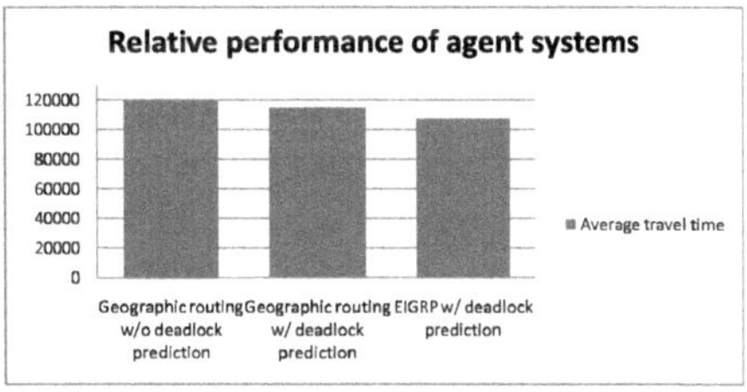

These values were obtained using the optimal weight for the deadlock prediction result. Figure 7 shows that the higher we factor in the risk of a deadlock, the fewer deadlocks actually occur - average travel times however do not decrease in the same manner. Apparently, at some point shuttles spend too much time avoiding each other where it would be quicker to let the deadlocks happen and then resolve them.

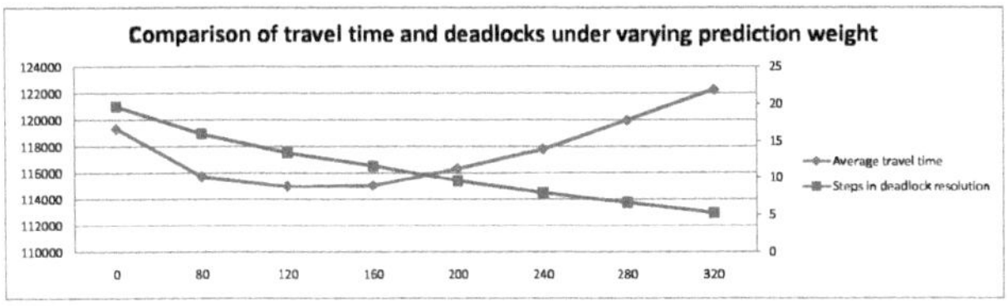

Fig. 7 The weight for the deadlock prediction algorithm is plotted on the horizontal axis. The blue line shows the average travel time (left scale), the right line shows how many deadlocks occurred (right scale)

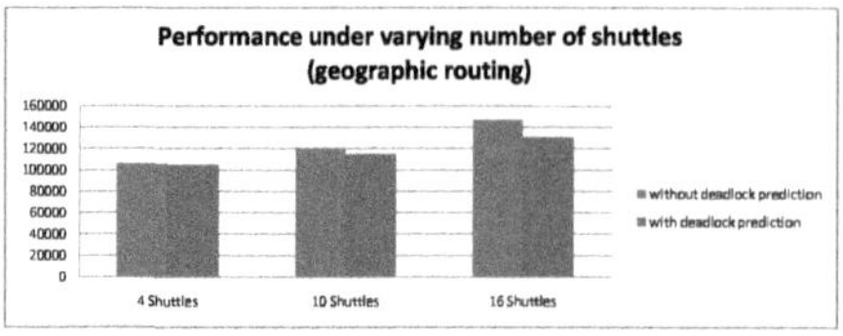

Fig. 8 The amount of shuttles in the system is plotted on the horizontal axis. The bars represent the average travel time, with and without deadlock prediction

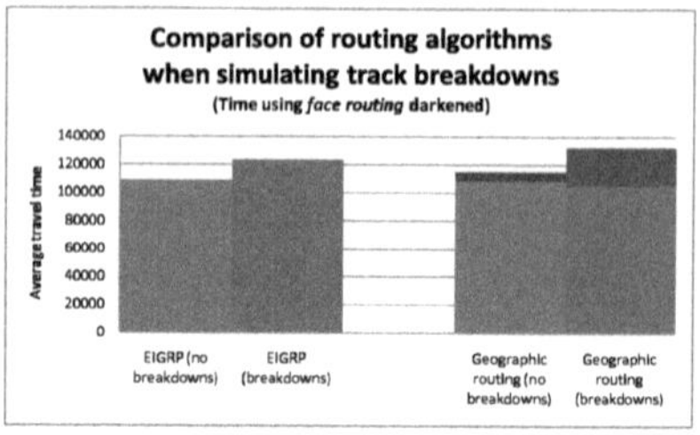

Fig. 9 The average travel times of both routing algorithms with and without track breakdowns. The darkened parts of the bars illustrate how long face routing was employed

Figure 8 illustrates the impact of deadlocks on the travel times with a varying amount of shuttles in the system. When there are 16 shuttles, deadlock prediction yields an 11.6% reduction in travel time - in fact, with the optimal weight, half of the deadlocks do not occur anymore.

Testing on the larger rail network delivered basically the same results. We varied the size of the local zone for the deadlock prediction algorithm and found that increasing the size was beneficial. The performance stagnates when considering shuttles more than three vertices away however.

We tested how the system deals with changes in the network. We randomly removed arcs from the graph for a certain amount of time to simulate unforeseen events like accidents on a track. Figure 9 shows the result of the simulation. As EIGRP propagates the changes through the network instantly, the travel times do not increase as much as with geographic routing. Analysis shows that face routing is used a lot more often than without arc removal.

7 Conclusion

We have demonstrated a functional decentralized system for routing AGVs in a rail network without typical constraints like unidirectional tracks. Our approaches to improve travel times using an advanced routing algorithm and the deadlock prediction method were successful. Remarkably, we were able to adapt two algorithms designed for computer networks. Results show that the algorithms solve the specific problem, assumingly not as well as centralized approaches however.

Our system allows little control over travel times; currently there is no way to prioritize specific shuttles. Interestingly, similar problems have been solved for computer networks with quality of service algorithms, which may be suited for AGV routing as well.

References

1. Bose, P., Morin, P., Stojmenović, I., Urrutia, J.: Routing with guaranteed delivery in ad hoc wireless networks. In: DIALM 1999: Proceedings of the 3rd international workshop on Discrete algorithms and methods for mobile computing and communications, pp. 48–55. ACM Press, New York (1999), http://doi.acm.org/10.1145/313239.313282
2. Coffman, E.G., Elphick, M., Shoshani, A.: System deadlocks. ACM Comput. Surv. 3(2), 67–78 (1971), http://doi.acm.org/10.1145/356586.356588
3. Dangelmaier, W., Laroque, C.: d3fact insight - Immersive Ablaufsimulation von richtungsoffenen und wahlweise zeitorientierten Materialflussmodellen. Industrie Management (2), 73–76 (2007)
4. Frey, H., Stojmenovic, I.: On delivery guarantees of face and combined greedy-face routing in ad hoc and sensor networks. In: MobiCom 2006, pp. 390–401. ACM, New York (2006)
5. Garcia-Lunes-Aceves, J.J.: Loop-free routing using diffusing computations. IEEE/ACM Trans. Netw. 1(1), 130–141 (1993), http://dx.doi.org/10.1109/90.222913
6. Medhi, D., Ramasamy, K.: Network Routing. In: Algorithms, Protocols, and Architectures, pp. 60–82. Morgan Kaufmann, San Francisco (2007)
7. Rustemeier, C., Nachtigal, V.: Railcab a new modular rail system. In: 4th international scientific conference Trans-Mech-Art-Chem Moscow, Russia (2006)
8. Sánchez, C.: Deadlock avoidance for distributed real-time and embedded systems. Ph.D. thesis, Stanford University (2007)
9. Sanders, B.A., Heuberger, P.A.: Distributed deadlock detection and resolution with probes. In: Proceedings of the 3rd International Workshop on Distributed Algorithms, pp. 207–218. Springer, Heidelberg (1989)
10. Scholz-Reiter, B., Freitag, M., Rekersbrink, H., Wenning, B.L., Gorldt, C., Echelmeyer, W.: Auf dem Weg zur Selbststeuerung in der Logistik - Grundlagenforschung und Praxisprojekte. In: Intelligente Logistikprozesse - Konzepte, Lösungen, Erfahrungen, Magdeburg. Magdeburger Logistiktagung, vol. 11, pp. 166–180. Logisch-Verlag, Magdeburg (2005)

Adaptive Decision Support for Dynamic Environments

Gloria Phillips-Wren

Abstract. Decision support systems are designed to assist a user by delivering targeted support for a specific decision problem. The application domain is an important consideration in developing the system, and the decision problem defines needed data, models and processing. Dynamic environments offer unique challenges to the system designer since the situation can change rapidly, requiring modification to the supporting technology. Modern intelligent techniques make it possible to envision a new type of adaptive decision support system (ADSS) that can perceive context and associate relevant models and data with the decision problem. In this paper we propose a framework for ADSS that utilizes intelligent methods to contextualize the decision problem and provide appropriate support to the user.

Keywords: adaptive system, decision support, decision making, artificial intelligence, intelligent agents adaptive system, decision support, decision making, artificial intelligence, intelligent agents.

1 Introduction

Dynamic environments offer unique challenges to the system designer for decision support since the situation can change rapidly. The user may need to make a decision considering variables that were not important in a previous timeframe or may need to consider new ways to view the decision choices. Current systems are designed to deliver targeted support for a specific decision problem. The application domain is an important consideration in developing the system, and the decision problem defines data, models and processing needs. Although intelligent methods can assist with such tasks as updating data in real-time, the structure of the decision paradigm in terms of models and data types is defined ahead of time. Modern intelligent techniques, however, make it possible to envision a new type of adaptive decision support system (ADSS) that can perceive context and associate relevant models and data with the decision problem.

Gloria Phillips-Wren
The Sellinger School of Business and Management
Loyola College in Maryland, Baltimore, MD 21210 USA
e-mail: `gwren@loyola.edu`

K. Nakamatsu et al. (Eds.): New Advan. in Intel. Decision Techno., SCI 199, pp. 235–243.
springerlink.com

In addition to dynamic environments, Internet-based, distributed systems make more data and information available to businesses and decision makers. The result is an increase in complexity. External variables as well as internal variables need to be considered, and dynamism associated with the environment may create an environment requiring time-pressured decisions. Information can be inaccurate or obtained from multiple sources that are inconsistent with each other, resulting in uncertainty and information overload for the user. Data can be difficult to access or require analysis for interpretation. ADSS can assist the decision maker by filtering and focusing the decision problem and the necessary data. Computational techniques such as intelligent agents and neural networks can enable ADSS to become powerful aids to decision makers by mimicking human behavior in some limited yet meaningful way (Pedrycz et al., 2008). These systems attempt to assist the decision maker overcome cognitive limitations in making a decision while possibly creating and learning useful knowledge for the future (Burnstein, 2008). The human decision maker can be helped by improving both the outcome from, and process of, decision making (Phillips-Wren, Hahn and Forgionne, 2004).

In this paper we propose a framework for an ADSS that utilizes intelligent methods to contextualize the decision problem and provide appropriate support to the user. The paper is organized in the following manner. We describe data types needed for decision making in section 2, and the role of context in section 3. Adaptive systems are discussed in section 4 with our proposed framework. The final section gives our conclusions and implications for the practitioner.

2 Data Types

ADSS utilize analysis to predict outcomes and optimize solutions for decision problems. To develop the analysis, it is convenient to structure information into a hierarchy usually ascribed to Ackoff (1989), and sometimes called the knowledge hierarchy, with each level having additional characteristics over previous levels. The levels are not well-defined, although general characteristics can be given.

At the lowest level is data, a collection of facts with no meaning in and of themselves. For example, a company may store data about their customers and sales in a database. Data can be inaccurate, invalid, incomplete, and not useful in analyzing a particular problem. The next level is information, obtained by converting or relating data and giving it some meaning. For example, we can derive average sales per month of a product. Information can be inaccurate if it is based on faulty data. Knowledge comes from interpreting and understanding a body of information. For example, by combing information about company sales and finances, we know that the company may have future financial problems. Knowledge often requires judgment and expertise, and its reliability and validity may be questioned. At the top of the hierarchy is wisdom, a synthesis of knowledge from various sources. Wisdom is often formed by expert judgment, and validity is informed by the results. For example, based on the data, information and knowledge we have from a variety of sources, we conclude that the company should consider a merger. Whether this conclusion is in fact wise will only be known based on future company health.

The data, information, knowledge, wisdom paradigm forms the foundation for the field of knowledge management (KM). KM is a general set of strategies and processes that companies use to capture, store, share, disseminate and use knowledge internally and/or externally. Knowledge is a primary asset of any company since it encompasses its intellectual property in a manner that can be used for competitive advantage (Hedlund, 1994; Hsu, 2008). As companies have become global and diverse, individuals within a company increasingly have differentiated knowledge, and profitable companies will need to have effective knowledge management strategies aided by information technology (Bresman, Birkinshaw and Nobel, 1999). Technologies such as discussion boards, peer to peer networks, blogs, and wikis have broadened the reach of KM activities to the individual level. While sharing knowledge in a one-to-one encounter is straightforward, capturing and disseminating to a larger audience is much more difficult. New technologies such as intelligent systems can help reach this goal.

Table 1 Knowledge hierarchy

Element	Description	Example
Data	Collection of facts	Number of units sold
Information	Converted or related data that have meaning in context	Average sales per month.
Knowledge	Interpreted information based on expertise	Company may have future financial problem.
Wisdom	Knowledge synthesized and applied to make a decision	Company should consider a merger.

Knowledge is an elusive concept that generally encompasses both explicit know-how and tacit know-what (Hedlund, 199; Bresman, Birkinshaw and Nobel, 1999). Explicit knowledge can be articulated and codified as, for example, a set of rules or guidelines, and it can be captured in KM systems. Tacit knowledge is more difficult to quantify. It is based more on experience, culture and social norms. Some authors have proposed that operations and management activities of a company are primarily task-based and knowledge-intensive, and that, in these cases, tacit knowledge can be codified if it is extracted from previous tasks (Liua and Wu, 2008). The implication is that under some circumstances tacit knowledge as well as explicit knowledge can be stored and used for decision making.

3 Intelligent KM and DSS

What is intelligence in the context of KM and DSS? Characteristics of intelligent behaviors are learning, making sense out of ambiguous situations, proactiveness, using reasoning, applying knowledge, dealing with complexity, and recognizing the relative importance of different aspects in a situation (Turban and Aronson, 1998).

Intelligent behaviors can be delivered in DSS (Phillips-Wren, Ichalkaranje and Jain, 2007) and KM systems using artificial intelligence methods, as shown by recent examples from the literature (see Table 2). DSS are based on Simon's (1977) decision-making paradigm of intelligence, design and choice. During intelligence the user collects information and focuses the decision problem with the relevant data and models. The design phase consists of determining the decision variables and criteria. During choice the user explores alternatives and makes a selection. Several of the examples demonstrate that knowledge-based DSS are possible that combine KM and DSS capabilities into an integrated system (Wen, Chen and Pao, 2007; Xie, 2006; Ho et al., 2006). However, decision support systems are still designed to assist users for a specific decision problem.

Table 2 Examples of Intelligent KM and DSS Systems

Author(s)	Year	Description
Liu and Wu	2008	Modified relevance feedback technique, integrated with task-relevance assessment method, to derive the task profile for the task-at-hand.
Wen, Chen and Pao	2008	KM decision support system using multi-agent technology for managing an electronic business.
Castiello, Castellano and Fanelli	2008	Meta-learning framework founded on the integration of connectionist paradigms and fuzzy KM, accumulating learning experience in cross-task contexts.
Apostolou et al.	2008	Exchange system couples case-based reasoning with ontologies to assist match-making between knowledge offers and knowledge demands in an interorganizational context.
Loia et al.	2007	Experimental agent based framework skilled to help the user both in managing achieved information and in personalizing web searching activity; proactive support to the searching on the web by suggesting pages, which are selected according to the user's behavior shown in his navigation activity
Yang and Reidsema	2007	Knowledge-based intelligent design system using search algorithm based on the computation of a similarity index.
Xie	2006	Knowledge-based DSS to support rapid one-of-a-kind product design.
Ho et al.	2006	Process quality control during production using ANN and fuzzy logic to measure, predict, control and act.

4 Context and Adaptive Decision Support

Decision makers need relevant data, information and knowledge to make wise decisions. The appropriate mix of information and knowledge depends on the characteristics of the decision making context (Zack, 2007), and context particularly affects value-based decisions (Hosack, 2007). The current approach of developing DSS for a specific decision problem and tailoring the information available to the system requires the developer to determine *à priori* the information needs and perform the appropriate filtering and focusing of the material. On the other hand, KM systems typically manage large amounts of information that may or may not be related. Users of KM systems are faced with the task of digging though massive amount of data to determine the information needed for decision making, and the use of context knowledge may provide a solution to user's information overload (Bobillo, Delgado, Gomez-Romero, 2008).

In dynamic environments, systems need the capability to adapt to changes. Michalewicz et al. (2007) give four characteristics of complex problems as: (1) the number of possible solutions is so large that all possibilities cannot be examined; (2) the environment changes with time; (3) the problem is heavily constrained; and, (4) there are many possible conflicting objectives. Adaptive business intelligence then uses prediction and optimization to recommend near-optimal solutions followed by a learning adaption module to improve future recommendations. The context of the decision problem is known by the system designer.

Intelligent techniques make it possible to envision a system that can place the decision into context. Some progress toward this goal has been made for applied problems. In dynamic disaster situations such as a pandemic outbreak, people need to interact to address the problem. Fuzzy logic can be used to determine the awareness level and subsequently specify the communication requirement for each human role involved in the interaction (Ray and Chattopadhyay, 2009). Multi-agent planning and coordination techniques assist in resource management in command and control systems in which several combat ships need to defend against incoming threats (Beaumont and Chaib-draa, 2007).

A system is needed that can integrate DSS and KM by perceiving the context of the decision problem for a general class of decision problems and translate from the intelligence phase to the design phase of the decision. Figure 1 shows the proposed architecture of such a system. It combines the KM process of transforming data into information into knowledge into wisdom with adapting to a specific decision problem defined by the user, and based on general adaptive business systems proposed by Michalewicz et al. (2007). The adaption module is implemented with intelligent techniques. It serves as an interface between the decision maker and the requisite model. The model defines the data and processing needs, providing information. To convert the information into knowledge, the adaptation module compares the proposed solution to information in the KM system by mining previous experience. Knowledge is passed to the adaption module to assess and modify it for the specific decision problem. The system provides a proposed decision to the human user, who may accept, reject or modify the parameters. The user is central to the system and has the final decision on accepting or not accepting the recommended course of action and implementing it, as well as interacting with the system to provide user-specific domain knowledge or preferences.

Intelligent techniques can be used in the adaption module to interact with the user to define the decision problem, choose appropriate models, consider previous experience, and assess the validity of the proposed solution in terms of the original problem. Agents and multi-agent systems are mature enough to make them ideal for use in ADSS (Padgham and Winikoff, 2004). For example, a Belief-Desire-Intention (BDI) agent system was used to link a diverse set of AI techniques for an intelligent personal assistant for time and task management that interacts with the user, is responsive to user inputs, and adapts to user working style and preferences (Myers et al., 2007). In this mixed-initiative planning system (Cox and Zhang, 2007), the user and the system are loosely coupled, so that neither needs full awareness of what the other is doing.

In addition to agents, other computational techniques can assist the decision maker process in ADSS. Once models and data produce information, the adaption module can be used to identify empirical experience relevant to the decision problem. Empirical data can be mined in the data mining module to suggest solutions and to produce knowledge. The adaption module then compares suggested solutions from all sources, determines which one is best, and delivers a decision to the user, who may accept the decision or modify the problem and initiate another interaction with the ADSS.

In the data mining module, a variety of techniques are possible. For example, Artificial neural networks (ANN) use a framework inspired by human brain functioning rooted in learning (Pedrycz et al., 2008). One of the characteristics of ANN is that they can find and represent nonlinear relationships discovered in data or information naturally since they do not pre-suppose a particular structure. ANN can be trained using a variety of methods that generally can be classified as either supervised learning or unsupervised learning (Michalewicz et al., 2007). Supervised learning gives the ANN both input and output values which are used to correct errors. Unsupervised learning is applied when there are data without output values, and correction is based on clustering and data analysis. In this way, ANN can learn from empirical data.

Genetic algorithms and other evolutionary computing start with a population of possible solutions and evolve new solutions based on a measure of quality (Chien, Tseng and Chen, 2008). Fuzzy logic provides a way to represent imprecise and uncertain knowledge, and it can be combined with other methods such as crisp values to improve accuracy (Petrovic, Ying and Burnham, 2006). Spatial methods such as self organized feature maps are able to represent dimensionality (Ceglowski and Churilov, 2008). Case-based reasoning can be used to process experience in the presence of incomplete knowledge (Jahnke, Chwolka and Simons, 2005). Methods can be combined by applying first one, and then another, or combing the results from different methods (Phillips-Wren, Sharkey and Dy, 2008). In dynamic situations, data can be modified rapidly so that the results reflect the most recent data.

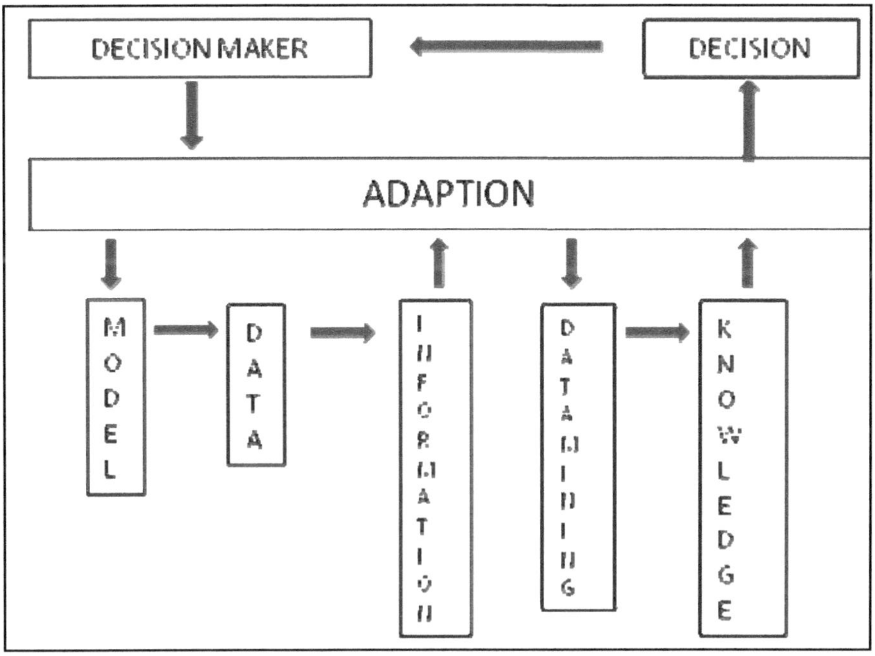

Fig. 1 Architecture of an adaptive decision support system (ADSS)

The ADSS would either select the data analysis technique or assist the user with a selection. The resulting knowledge would consist of the results from different algorithms. In the adaption module potential solutions to the decision problem can be updated to reflect the most recent data in dynamic environments and assessed to determine the best solution. Various methods have been developed for assessment such as ROC curves (Phillips-Wren, Sharkey and Dy, 2008). Historical data can be used as a benchmark for validation.

5 Future Trends

This paper has presented a conceptual ADSS. Future research needs to focus on instantiating the architecture and applying it to specific complex decision situations. In particular, measurement criteria for good decisions needs to be researched so that decisions are well-framed by context, and so that decisions are responsive to changes in the decision environment.

We expect partnerships to continue to develop between humans and computers as complementors of each other; humans are strong in areas such as communication and learning, and computers bring strength in areas such as memory and computational reasoning (Pohl, 2008). Computers are parallel processors that are fast, accurate and unemotional. By contrast, human decision makers can be slow, inaccurate, emotional, and single-focused, especially in stressful situations that can occur in dynamic situations. In intelligent agent research, some researchers use the term human-centric to describe teaming between humans and computers (Tweedale et al., 2007). Adaptive decision support systems that personalize for different users, recognize context, identify and update data, utilize various analysis techniques, perceive user intention, and assist the user in identifying a desirable solution to a decision problem are on the horizon.

Acknowledgments. The author would like to thank several anonymous reviewers for their helpful comments and suggestions.

References

Ackoff, R.: From Data to Wisdom. Journal of Applied Systems Analysis 16, 3–9 (1989)

Apostolou, D., Mentzas, G., Klein, B., Abecker, A., Maass, W.: Interorganizational knowledge exchanges. IEEE Intelligent Systems 23(4), 65–74 (2008)

Beaumont, P., Chaib-draa, B.: Multiagent coordination techniques for complex environments: the case of a fleet of combat ships. IEEE Transactions on Systems, Man and Cybernetics, Part C (Applications and Reviews) 37(3), 373–385 (2007)

Bobillo, F., Delgado, M., Gomez-Romero, J.: Representation of context-dependant knowledge in ontologies: a model and an application. Expert Systems with Applications 35(4), 1899–1908 (2008)

Bresman, H., Birkinshaw, J., Nobel, R.: Knowledge transfer in international acquisitions. Journal of International Business Studies 30(3), 439–462 (1999)

Castiello, C., Castellano, G., Fanelli, A.M.: Mindful: a framework for meta-inductive neuro-fuzzy learning. Information Sciences 178(16), 3253–3274 (2008)

Ceglowski, A., Churilov, L.: Using self organizing feature maps to unravel process complexity in a hospital emergency department: A decision support perspective. In: Phillips-Wren, G., Ichalkaranje, N., Jain, L. (eds.) Intelligent Decision Making: An AI-Based Approach, pp. 365–385. Springer, Berlin (2008)

Chien, C.-F., Tseng, F.-P., Chen, C.-H.: An evolutionary approach to rehabilitation patient scheduling: a case study. European Journal of Operational Research 189(3), 1234–1253 (2008)

Cox, M., Zhang, C.: Mixed-initiative goal manipulation. AI Magazine 28(2), 62–74 (Summer 2007)

Hedlund, G.: A model of knowledge management and the N-form corporation. Strategic Management Journal 15, 73–91 (1994)

Ho, G., Lau, H., Lee, C., Ip, A., Pun, K.: An intelligent production workflow mining system for continual quality enhancement. International Journal of Advanced Manufacturing Technology 28(7-8), 792–809 (2006)

Hosack, B.: The effect of system feedback and decision context on value-based decision-making behavior. Decision Support Systems 43(4), 1605–1614 (2007)

Hsu, I.-C.: Knowledge sharing practices as a facilitating factor for improving organizational performance through human capital: a preliminary test. Expert Systems with Applications 35(3), 1316–1326 (2008)

Jahnke, H., Chwolka, A., Simons, D.: Coordinating Service-Sensitive Demand and Capacity by Adaptive Decision Making: An Application of Case-Based Decision Theory. Decision Sciences 36(1), 1–32 (2005)

Liu, D.-R., Wu, I.-C.: Collaborative relevance assessment for task-based knowledge support. Decision Support Systems 44(2), 524–543 (2008)

Loia, V., Pedrycz, W., Senatore, S., Sessa, M.I.: Interactive knowledge management for agent-assisted Web navigation. International Journal of Intelligent Systems 22(10), 1101–1122 (2007)
Michalewicz, Z., Schmidt, M., Michalewicz, M., Chiriac, C.: Adaptive Business Intelligence. Springer, Berlin (2007)
Myers, K., Berry, P., Blythe, J., Conley, K., Gervasio, M., McGuinness, D., Morley, D., Pfeffer, A., Pollack, M., Tambe, M.: An intelligent personal assistant for task and time management. AI Magazine 28(2), 47–62 (Summer 2007)
Padgham, L., Winikoff, M.: Developing Intelligent Agent Systems. John Wiley & Sons Ltd., West Sussex (2004)
Pedrycz, W., Ichalkaranje, N., Phillips-Wren, G., Jain, L.: Introduction to computational intelligence for decision making. In: Phillips-Wren, G., Ichalkaranje, N., Jain, L. (eds.) Intelligent Decision Making: An AI-Based Approach, pp. 79–96. Springer, Berlin (2008)
Petrovic, D., Ying, X., Burnham, K.: Fuzzy decision support system for demand forecasting with a learning mechanism. Fuzzy Sets and Systems 157(12), 1713–1725 (2006)
Phillips-Wren, G., Hahn, E., Forgionne, G.: A multiple criteria framework for the evaluation of decision support systems. Omega 32(4), 323–332 (2004)
Phillips-Wren, G., Ichalkaranje, N., Jain, L. (eds.): Intelligent Decision Making: An AI-Based Approach. Springer, Berlin (2007)
Phillips-Wren, G., Sharkey, P., Dy, S.: Mining lung cancer patient data to assess healthcare resource utilization. Expert Systems with Applications 35(4), 1611–1619 (2008)
Pohl, J.: Cognitive Elements of Human Decision Making. In: Phillips-Wren, G., Ichalkaranje, N., Jain, L. (eds.) Intelligent Decision Making: An AI-Based Approach, pp. 41–76. Springer, Berlin (2008)
Ray, P., Chattopadhyay, S.: Fuzzy Awareness Model for Disaster Situations. Intelligent Decision Technologies (forthcoming) (2009)
Simon, H.: Administrative Behavior, 4th edn. The Free Press, New York (1997) (Original publication date 1945)
Turban, E., Aronson, J.: Decision Support Systems and Intelligent Systems. A. Simon and Schuster Company, Upper Saddle River (1998)
Tweedale, J., Sioutis, C., Phillips-Wren, G., Ichalkaranje, N., Urlings, P., Jain, L.: Future directions: Building a decision making framework using agent teams. In: Phillips-Wren, G., Ichalkaranje, N., Jain, L. (eds.) Intelligent Decision Making: An AI-Based Approach, pp. 387–408. Springer, Berlin (2008)
Wang, C.-Y., Yang, H.-Y., Chou, S.-c.T.: Using peer-to-peer technology for knowledge sharing in communities of practices. Decision Support Systems 45(3), 528–540 (2008)
Wen, W., Chen, Y.H., Pao, H.H.: A mobile knowledge management decision support system for automatically conducting an electronic business. Knowledge-Based Systems 21(7), 540–550 (2008)
Xie, S.Q.: A decision support system for rapid one-of-a-kind product development. International Journal of Advanced Manufacturing Technology 28(7-8), 643–652 (2006)
Yang, Q., Reidsema, C.: Design information handling in a knowledge-based intelligent design system. Cybernetics and Systems 38(5-6), 549–573 (2007)
Zack, M.H.: The role of decision support systems in an indeterminate world. Decision Support Systems 43(4), 1664–1674 (2007)

Evaluation of the Continuous Wavelet Transform for Feature Extraction of Metal Detector Signals in Automated Target Detection

Minh Dao-Johnson Tran and Canicious Abeynayake

Abstract. Landmines pose a significant problem in many countries around the world. Although technological systems such as metal detectors have been employed to combat these threats, many of these still require significant human interaction especially in the area of target and clutter discrimination. The aim of this research is to develop an automated decision making system for landmine detection. The initial stages of the research involves comparing various techniques for feature extraction to determine which methods provide the best representation for metal detector data to achieve improved target discrimination from background noise. This paper will focus on evaluating a technique utilizing the Continuous Wavelet Transform with false alarm rate and probability of detection used as performance measures.

Keywords: Continuous Wavelet Transform, feature extraction, metal detector, landmines.

1 Introduction

Landmines pose a significant problem in many countries around the world. Many of these threats have no means of self-neutralization or self-destruction and hence can remain active long after the conflicts have ceased. As a result an estimated 26,000

Minh Dao-Johnson Tran
Knowledge Based Intelligent Engineering Systems Centre,
School of Electrical and Information Engineering,
University of South Australia, Mawson Lakes, SA 5095, Australia
e-mail: Minh.D.Tran@postgrads.unisa.edu.au

Canicious Abeynayake
Threat Mitigation Group, Weapon Systems Division,
Defence Science and Technology Organisation (DSTO),
Edinburgh, Australia
e-mail: Canicious.Abeynayake@dsto.defence.gov.au

K. Nakamatsu et al. (Eds.): New Advan. in Intel. Decision Techno., SCI 199, pp. 245–253.
springerlink.com

people are maimed and killed each year, many of whom are civilians [1]. They also deny entire regions of basic services by impeding infrastructure repairs, disrupting humanitarian aid shipments, and prohibiting civilians from possible land for agriculture and housing. Many mine-plagued countries with fragile economies cannot support the cost of either mine clearance or victim rehabilitation.

Although systems such as metal detectors have been employed to combat these threats, many of these still require significant human interaction especially in the area of target and clutter discrimination. The aim of this research is to develop an automated Decision Making System (DMS) for landmine detection.

A DMS is usually composed of many individual processes such as 'Region Of Interest' (ROI) detection, feature extraction, classification and decision fusion. Figure 1 depicts a simplistic form of a generic DMS. ROI detection involves processing the input signal data and isolating suspicious areas that may indicate the presence of a target. Data from these ROI are fed into the feature extraction stage that extracts information that can be used to identify any target objects within that region. Finally the classification stage processes the extracted features to determine the class of objects within an ROI. This determination is essentially a method to distinguish the ROI that correspond to either targets (i.e. landmines) or clutter (natural and artificial).

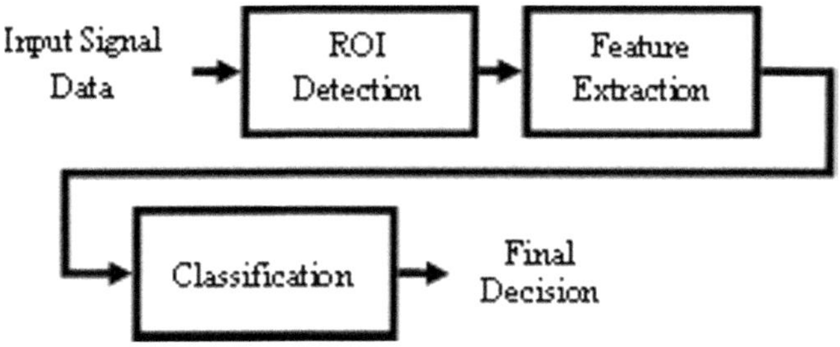

Fig. 1 Simplistic Decision Making System

This paper focuses on the feature extraction stage. However, the purpose of this paper is not to present a fully functioning feature extractor, but to evaluate the potential of the feature extraction technique used. In other words, the aim is to determine if the chosen technique produces features suitable for target detection, and its potential effectiveness in performing this task (i.e. proof of concept).

The input data used for the experiments was obtained from a metal detector (MD) which produces a time series of sampled data. Due to the behavioral nature of the data, inspiration was drawn from electrocardiogram (ECG) signal representation for possible feature extraction methods. A literature review revealed wavelet transforms (WT) are a successful method of signal representation of time series data such as ECG waveforms [2, 3, 4, 5, 6]. Other researchers have incorporated the discrete WT (DWT) in their feature extractors [2, 4, 5, 6], but this current research utilizes the continuous WT (CWT). This approach was chosen to enable the use of the wavelet coefficient power spectrum. The motivation behind this is explained in Section 2.

The organization of this paper is as follows: Section 2 provides brief background information of the WT and the motivation behind its use in this research. Section 3 describes the data set used, the implementation of key modules, and the simulation procedure adopted. Section 4 discusses the results obtained from the simulations. Finally, Section 5 summaries the paper and poses suggestions for further research.

2 Wavelet Transform

The origins of wavelet analysis dates back to the mid-eighties and was originally driven by application needs to analyze seismic signals more sensitively than with Fourier techniques [7]. The WT can be thought of as an extension of the classic Fourier transform except operating on a multi-resolution basis [2]. The multi-resolution feature of the WT enables a signal to be decomposed into a number of resolutions or 'scales' via dilation and translation of a specified analyzing or 'mother' wavelet. Each resolution represents a particular coarseness of the signal. A feature of the WT is its ability to preserve spatial information after the transformation [7], which is essential for locating landmine threats. Equation 1 represents the WT of a signal $f(x)$ where $\Psi(x)$ is the dilation of a wavelet function by the scale factor s. Due to the nature of the WT equation, the WT can also be seen mathematically as the convolution of the wavelet function with the signal $f(x)$ [4, 8].

$$W_s f(x) = f(x) * \psi_s(x) = \frac{1}{s}\int_{-\infty}^{+\infty} f(t)\psi\left(\frac{x-t}{s}\right)dt \tag{1}$$

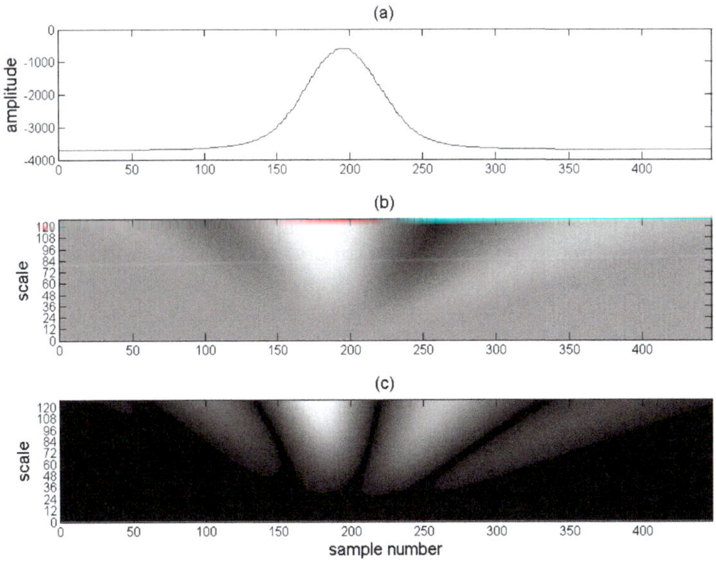

Fig. 2 (a)Original Target Signal (b)Target Signal Wavelet Coefficients (c)Power Spectrum of Target Signal Wavelet Coefficients

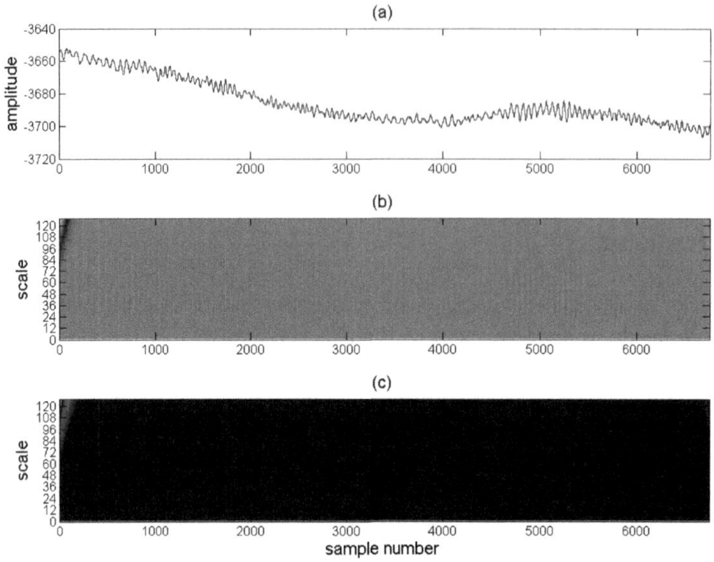

Fig. 3 (a)Original Background Noise Signal (b)Background Noise Wavelet Coefficients (c)Power Spectrum of Background Noise Signal Wavelet Coefficients

The core motivation behind the use of the CWT in this research arose from the analysis of the power spectrum of the wavelet coefficients decomposed across multiple scales. It was found that at certain scales the areas associated with the target signals were distinctively emphasized in the power spectrum while areas associated with background noise were significantly attenuated. This attenuation lead to the idea that emphasized areas in the power spectrum could contain data that uniquely distinguished the target signals from the background clutter. Hence it was decided that this motivation would be utilized in the construction of the feature extractor. Figures 2 and 3 depict the original signal against its generated wavelet coefficients and power spectrum for a target and background noise signal respectively. Note that the lack of symmetry in the power spectrum for the target signal is due to the original signal not being perfectly symmetrical about its maximum.

3 Simulation Procedure

Prior to performing any simulations, the raw MD data was analyzed visually to locate areas of known target and background noise occurrence for each signal channel (refer to section 3.1). Once located, these areas were manually extracted from each raw data file to form the input data sets of target and background noise regions. The target data set was then randomly divided into two groups. One group was used as the training set and the other, together with the background noise set, used as the testing set.

One reason for only training with target related data was the lack of artificial clutter signals in the available data. Another reason was related to the fact that clutter encompasses a wide variety of objects. Thus the aim was to achieve a DMS trained to detect most types of targets, while assuming everything else not conforming to the system's target criteria as non-targets (i.e. clutter and background noise).

The following subsections describe in more detail the input data set used and the implementation of the feature extractor and target detector.

3.1 *Metal Detector Data*

The MD used was a Minelab Generation 1 (G1) array. The array is composed of 16 coils arranged in a linear fashion (i.e. straight line) with each coil operating three channels labeled '1', '2', and 'ground'. The array was arranged in the cross-track direction to the test lane and used to scan at relatively constant speed along an area of laterite soil containing a number of flush and depth buried metallic targets. These targets were buried in the down-track direction. Figure 4 gives a pictorial description of the operating environment.

Fig. 4 Operating Environment of MD Array

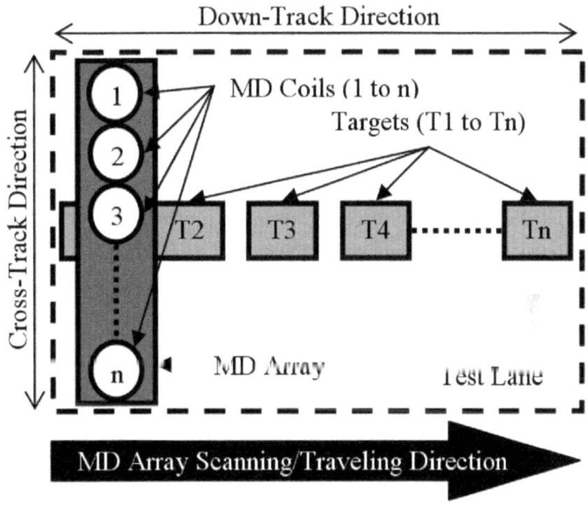

Using the described operating configuration a number of trials were executed with the MD array traveling at two different speeds. As a result two sets of data for the same test lane were acquired. Since all the targets were located in the centre of the scan lane, only the data obtained from the central four coils (i.e. coils '7 through to 10') were used in the simulations.

3.2 *Feature Extractor*

The feature extractor based on the CWT was implemented in MATLAB using the Wavelet toolbox. Selecting a wavelet function that closely matches the signal to be processed is of utmost importance in wavelet applications [4]. For this particular extractor the inbuilt Daubechies wavelet (see [8]) of order '4' was chosen due to its resemblance to the target signals. This wavelet was used to decompose the ROIs from scales '1 through to 128'. By analyzing the generated wavelet coefficients, the maximum coefficient across all scales was located and a narrow down-track zone surrounding the peak value was isolated. The means of the coefficients for each scale within this area were then calculated and used as the extracted features. This resulted in each ROI being represented by a vector composed of 128 elements.

3.3 Target Detector

Ideally the features extracted uniquely represent each ROI. Hence in order to evaluate the feature extractor, a basic target detector was implemented. The detector is essentially the initial phase of the classification stage where the DMS decides whether the ROI contains a possible target.

The target detector was based on a simple thresholding method. This technique was chosen for its simplicity and to maintain a level of conformity in the comparative analysis between this and possible future feature extraction methods. The target detector was implemented in MATLAB.

The detector is composed of a training phase and a detection phase. During the training phase the detector is presented with the feature vectors generated from the training set (i.e. training vectors). It then determines the minimum value obtained by individual features across all training vectors and stores this information in memory for future use. This information will be referred to as the threshold vector (see equations 2 and 3 where equation 2 represents the training set with each row corresponding to one training vector and equation 3 represents the threshold vector with each value corresponding to the minimum value for each feature).

$$\begin{bmatrix} f_{11} \cdots f_{1n} \\ \vdots \quad \ddots \quad \vdots \\ f_{k1} \cdots f_{kn} \end{bmatrix} \tag{2}$$

$$\Downarrow$$

$$\left[min\,(f_{11} : f_{1n}) \ldots min\,(f_{k1} : f_{kn}) \right] \tag{3}$$

During the detection phase, the detector is individually presented with the feature vectors generated from the testing set (i.e. testing vectors). It then performs a threshold operation with each feature value against its corresponding value within the threshold vector. If the feature value is greater than or equal to its corresponding threshold value, a '1' is generated or a '0' otherwise. The pseudo code below gives an algorithmic description of the threshold procedure.

```
let threshold_vector = [tv_1 , ... , tv_n]
let feature_vector = [fv_1 , ... , fv_n]
for(i = (1 : n))
    if(fv_i >= tv_i)
        return 1
    else
        return 0
    end
end
```

Once the initial threshold operations are completed, the total number of '1's produced are summated. This value then undergoes another threshold operation with a different predefined threshold to determine whether a target is present. This predefined threshold value will be referred to as the final threshold. The final threshold represents the number of features that must pass their initial individual thresholds before a target is declared. Hence the maximum that this value can attain is equivalent to the total number of features generated (i.e. 128).

4 Results and Discussion

To analyze the potential of the CWT method as a feature extraction method, the probabilities of detection (PD) and false alarm rate (FAR) were calculated and used

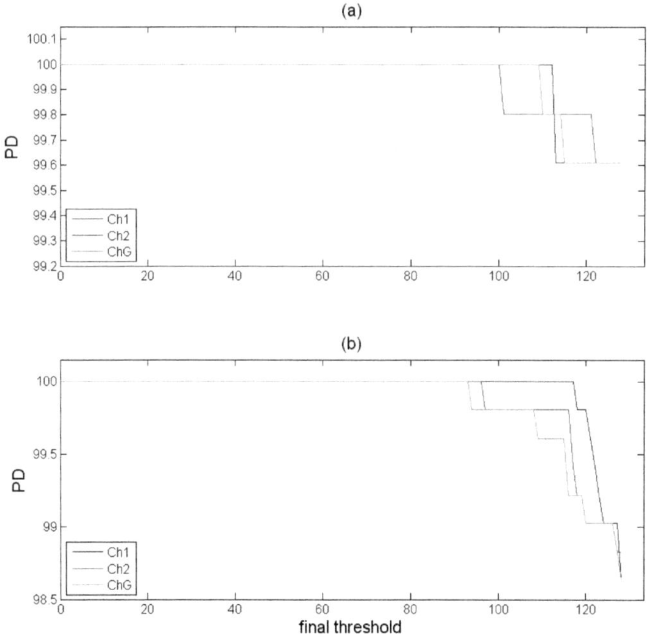

Fig. 5 PD against Final Threshold for (a)First Data Set and (b)Second Data Set

as the performance measures. These were calculated separately and averaged across each data set for the target detector operating at final threshold values between '0 and 128'.

Figure 5 displays the PDs achieved at each final threshold for the data sets acquired from the MD operating at two different speeds. The graphs clearly show that by utilizing the extracted features generated by the CWT method, the target detector was able to achieve PDs of approximately 100% for almost all final thresholds used for both data sets. Although there is a slight decline after the final threshold of 90, these are still relatively high and above 98%.

Figure 6 displays the corresponding FARs achieved at each final threshold for both data sets. The graphs clearly show that the target detector is able to achieve FARs of approximately '1' false alarm per correct detection (FA/CD) for the majority of the final thresholds and for both data sets. As the final threshold increases the FARs also begin to decrease below '1' FA/CD, and all channels appear to attain a minimum within the range of 0.6 and 0.8 FA/CD. One case in particular is 'channel 1' for the first data set which was able to achieve a minimum FAR below 0.2 FA/CD.

The results confirm that the CWT technique has the potential to produce features suitable for the detection of Minelab G1 MD metallic target signals in laterite soils. Although the false alarms were not completely removed even at the highest final threshold, it is likely that this can be significantly improved by incorporating additional features to the current feature set. Possible additional features that may contribute to minimizing the false alarms include the morphological properties of either the wavelet coefficient power spectrum, or the original data signal, or a combination of properties from both signals.

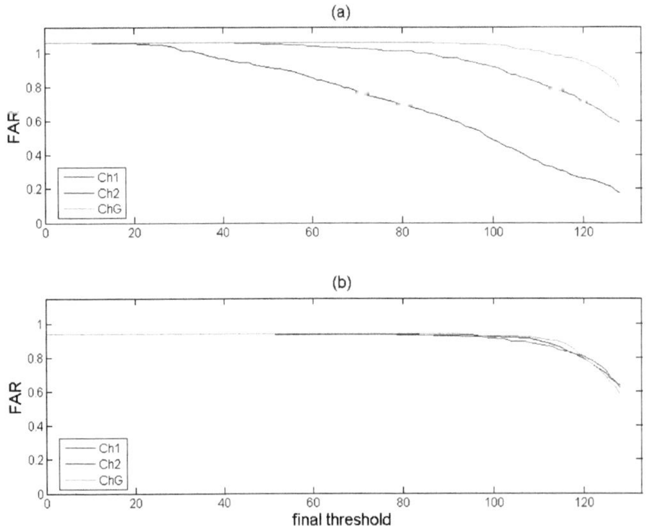

Fig. 6 FAR against Final Threshold for (a)First Data Set and (b)Second Data Set

5 Summary and Future Research

In this paper the evaluation of a feature exaction technique based on CWT has been presented. The purpose of the assessment was to determine whether this method produced suitable features for target detection, and its potential effectiveness in executing this task. The results obtained verify that the CWT based technique produces features that are suitable for target detection. These findings apply to signal data obtained using the Minelab G1 MD of metallic targets in laterite soil environments.

The CWT based feature extractor may possibly be further improved by incorporating the morphological properties of the wavelet coefficient power spectrum and/or the original data signal as additional features.

The logical direction of future work in this research would be to test the technique with data sets containing targets with both artificial and natural clutter. The results could then be evaluated against the earlier incarnations of the system and/or systems utilizing other feature extraction techniques for comparative analysis.

Acknowledgements. Gratitude goes to the reviewers, and Prof. Lakhmi Jain and Prof. C.P. Lim from The University of South Australia, for their invaluable advice in the production of this paper.

References

1. Office of International Security & Peacekeeping Operations, Hidden killers: The global landmine crisis, United States Department of State. Bureau of Political-Military Affairs, Tech. Rep. 10225, record Number: 77 (December 1994)
2. Ubeyli, E.D.: Analysis of ecg signals by diverse and composite features. Journal of Electrical & Electronics Engineering 7(2), 393–402 (2007)
3. Zhu, J., Zhang, X., Wang, Z., Wang, X.: Preprocessing and analysis of the ecg signals. In: Seventh International Symposium on Instrumentation and Control Technology: Sensors and Instruments, Computer Simulation, and Artificial Intelligence, vol. 7127(1), pp. 71272H–1–5. SPIE (2008)
4. Mahmoodabadi, S.Z., Ahmadian, A., Abolhasani, M.D., Eslami, M., Bidgoli, J.H.: Ecg feature extraction based on multiresolution wavelet transform. In: 27th Annual International Conference of the Engineering in Medicine and Biology Society, IEEE-EMBS 2005, pp. 3902–3905 (2005)
5. Dokur, Z., Olmez, T., Yazgan, E.: Comparison of discrete wavelet and fourier transforms for ecg beat classification. Electronics Letters 35(18), 1502–1504 (1999)
6. Zhao, Q., Zhang, L.: Ecg feature extraction and classification using wavelet transform and support vector machines. In: International Conference on Neural Networks and Brain, 2005. ICNN&B 2005, vol. 2, pp. 1089–1092 (2005)
7. Stark, H.-G.: Wavelets and signal processing: an application-based introduction. Springer, Berlin (2005)
8. Torrence, C., Compo, G.P.: A practical guide to wavelet analysis. Bulletin of the American Meteorological Society 79(1), 61 (1998)

Hypertension Detection Using a Case-Based Reasoning Approach

Kuang-Hung Hsu, Chaochang Chiu, Nan-Hsing Chiu, Po-Chi Lee, Wen-Ko Chiu, Thu-Hua Liu, Yi-Chou Juang, Chorng-Jer Hwang, and Chi-I Hsu

Abstract. The exploration of three-dimensional (3D) anthropometry scanning data along with other existing subject medical profiles using data mining techniques becomes an important research issue for medical decision support. This research attempts to construct a classification approach based on the hybrid of the case-based reasoning (CBR) and genetic algorithms (GAs) approach for hypertension detection using anthropometric body surface scanning data. The experiment showed that our proposed approach is able to improve the effectiveness of case matching of hypertension disease.

1 Introduction

Hypertension may lead to some major causes of death such as cardiovascular diseases and is deemed as factors for Syndrome X that has been investigated for years in epidemiologic studies [3],[4]. Earlier identification of this disease is gaining concerns for issues in clinical research. As the recent development of a new 3D scanning technology that has many advantages over the old system of anthropometric measurements [5], the Department of Health Management of Chang Gung Medical Center (CGMC) at Taiwan is able to collect 3D anthropometric body surface scanning data [7]. The exploration of these 3D data along with other existing subject medical profiles using data mining techniques becomes an

Kuang-Hung Hsu
Dept. of Health Care Management, Chang Gung University, Taiwan, Republic of China

Chaochang Chiu*, Po-Chi Lee, Yi-Chou Juang, and Chorng-Jer Hwang
Dept. of Information Management, Yuan Ze University, Taiwan, Republic of China
e-mail: imchiu@saturn.yzu.edu.tw

Nan-Hsing Chiu
Dept. of Information Management, Ching Yun University, Taiwan, Republic of China

Wen-Ko Chiu and Thu-Hua Liu
Dep. of Industrial Design, Chang Gung University, Taiwan, Republic of China

Chi-I Hsu
Dept. of Information Management, Kainan University, Taiwan, Republic of China

* Corresponding author.

K. Nakamatsu et al. (Eds.): New Advan. in Intel. Decision Techno., SCI 199, pp. 255–263.
springerlink.com

important research issue for medical decision support [8]. CBR is one of the most common approaches of knowledge discovery. GA is a searching technique based on the mechanism of natural evolution of species. It has been increasingly aware for its searching ability for many optimization problems [1],[6]. In the present paper, GA is adapted in CBR learning processes to derive suitable feature weights of similarity measures. Two differently weighted CBR approaches are applied to derive the feature weights for similarity measure in CBR classification approaches for predicting hypertension disease.

2 Literature Review

Classification by CBR is a technique that has been proposed for a long time as a valid alternative to expert judgment. The uses of CBR classification approaches have been evaluated and confirmed in many studies [11],[12]. A distance measure in the CBR approach is the degree of dissimilarity between two cases in terms of their features. Generally, the performance of the similarity measure and the weighting of features are keys to this reasoning process [13]. The Euclidean distance, as one type of distance measure, assumes that all features are equally relevant. Appropriately setting the feature weights in the similarity measure can improve the estimate performance [13]. Essentially, more important features should be assigned with larger weights than less important features. Other, the determination of feature weights depended largely on the subjective opinions of experts. To effectively determine the feature weights of similarity measure is not easy because it is difficult to give the appropriate values.

GA is a fast search algorithm that uses an adaptive approach for solving problems [14]. Recently, several studies showed that one could set feature weights in CBR classification approaches using GA. In Liao et al. [15], the GA selected important features and subsequently determined their weights. Chiu et al. [16] employed GA to enhance dynamic weight and the design of non-similarity functions. Beddoe and Petrovic [10] presented a feature selection and weighting approach that a GA is developed for capturing nurse rostering decisions.

3 The Weighted CBR

3.1 Equally Weighted CBR

Traditionally, equally weighted CBR (EWCBR) approach applies an equal weight to distance measure for each feature in classification approaches. Equally weighted Euclidean distance is adapted to measure the dissimilarity between two features in the EWCBR approach. EWCBR measures the straight-line distance between two features using the values of features normalized between 0 and 1. The sum of the squares of the distance for each feature is the square of distance between two cases. Consequently, the closest analogue to a case x is the case with the minimum Euclidian distance. Eq. (1) shows the overall distance between two cases of the EWCBR approach for continuous data type, where n is the number of features. P_x is case x to be estimated and P_y is case y in the historical case base. P_{xi} is the value of feature i in case x and P_{yi} is the value of feature i in case y. The distance increases as similarity decreases; the distance metric measures the dissimilarity between two cases in terms of their features. For category data type, the distance between two features P_{xi} and P_{yi} is 0, if feature values of two features are the same; otherwise, it is 1.

$$D(P_x, P_y) = \sqrt{\sum_{i=1}^{n} W_i (P_{xi} - P_{yi})^2} \quad (1)$$

Where W is the feature weight and $W_i = W_j$ $(i, j = 1 \text{ to } n)$

3.2 Differently Weighted CBR

Differently weighted CBR are adapted when features are weighted to reflect the relative importance of each feature. GA is adapted to search for the suitable feature weights for distance measures in CBR approaches. The weight of distance measure for each feature is encoded in genes. Each chromosome is composed of a set of genes, and collections of chromosomes make up a population. GA can be thought of as an evolutionary process, where a population of weights evolves over a sequence of generations. During each generation, the fitness of each weight is calculated, and weights are selected for reproduction on the basis of their fitness. The probability of survival of a weight is proportional to its fitness values. Then the reproduced weights undergo recombination, which consists of crossover and mutation. These iterative processes optimize the weights of distance measures for each feature. The fitness function is shown in Eq. (2). The weights of distance measures of each feature are treated as inputs, and the fitness function is defined as minimum of the total number of misclassification (*MIC*). For the jth examples, the misclassification result is 1 if the actual category for dependent variable is equal to the predicted category; otherwise, 0. Among m examples, the fitness function is to minimize the total misclassification results.

$$Fitness\ \ Fuction = Min \sum_{j=1}^{m} MIC_j \quad \ldots\ldots \quad (2)$$

$$MIC_j = \begin{cases} 1, & if \text{ the Acual Category is Equal to the Predicted Category} \\ 0, & if \text{ the Acual Category is Unequal to the Predicted Category} \end{cases}$$

According to Eq. (1), a weighted equation of the UWCBR approach is shown in Eq. (3). The value of variable W_i depends on the searching results of GA after several learning cycles. Consequently, each feature has a different weight obtained with respective distance measure.

$$D(P_x, P_y) = \sqrt{\sum_{i=1}^{n} W_i \times (P_{xi} - P_{yi})^2} \quad (3)$$

Where W_i is a variable weight of feature i.

NWCBR utilizes a separate nonlinear equation as the weight of distance measure for each feature. A nonlinear equation of NWCBR approach is shown in Eq. (4), where coefficient a_i, exponent b_i and constant c_i are for the feature i. The values of a_i, b_i and c_i are confirmed using GA after several learning cycles. The computed outcome of the nonlinear equation, W_i, is the feature weight of distance measure for feature i. Therefore, each feature has a different nonlinear equation as its feature weight obtained from the corresponding distance measure that varies depending on the feature value P_{xi} of the case x to be classified.

$$D(P_x, P_y) = \sqrt{\sum_{i=1}^{n} W_i \times (P_{xi} - P_{yi})^2} \tag{4}$$

Where $W_i = a_i \times P_{xi}^{\ bi} + c_i$

4 Dataset Description

This study has collected 1152 subjects from Department of Health Examination of CGMC. BMI [weight (kg)/height (m)^2] was used as an indicator of obesity. Hypertension was defined as a systolic blood pressure (SBP) of 140 mmHg or greater and/or a diastolic blood pressure (DBP) of 90 mmHg or greater. The Whole Body 3D Laser Scanner scans a cylindrical volume 1.9 meters high and 1.0 meter in diameter in CGMC. Those 3D factors included in this research are body mass index (BMI), left arm volume (LAV), trunk surface area (TSA), weight (W), waist circumference (WC), waist-hip ratio (WHR), and waist width (WW). In Table 1 each data record contains 5 categorical attributes and 11 continuous attributes for the input part. The output part is one categorical attribute whose value is either normal or abnormal. 489 subjects have hypertension (abnormal); 663 are without hypertension (normal).

5 The Experiments and Results

For the comparing purpose, artificial neural networks (ANN) [18], C5.0 [20] and classification tree and regression trees (CART) [19] were applied to all the data sets. The evaluation of EWCBR, UWCBR, NWCBR, ANN, C5.0 and CART approaches were based on a 3-fold cross validation. We aggregated the accuracy across all testing datasets as test result and the accuracy across all training datasets as training results. Table 2 summarized the GA parameter settings in this study. The GA parameters chosen for UWCBR and NWCBR approaches were 10*V organisms in the population. V is the number of variables (problem size) to be explored for GA. The entire learning process stopped after 1000*V trials or when the best result did not change in the last 1000*V trials.

Table 1 The Features for Hypertension Detection Model

Feature Type	Description	Abbreviation	Data Type
Three Dimension Anthropometry	Body Mass Index	BMI	Continuous
	Left Arm Volume	LAV	
	Trunk Surface Area	TSA	
	Weight	W	
	Waist Circumference	WC	
	Waist-Hip Ratio	WHR	
	Waist Width	WW	
Risk Factors	Number of Cigarettes/per day	SMOK	
	Number of Cups in Drinking Wine/per day	WINE	
	Dietary Pattern (Yes/No)	Diet	Category
Family History of Disease	Family Hypertension History (Yes/No)	FHHT	
	Family Paralysis History (Yes/No)	FHSK	
Biochemical Tests	Urine Turbidity (Level 1,2, or 3)	UT	
	Total Cholesterol	TC	Continuous
Demographic Data	Age	Age	
	Sex (Male/Female)	Sex	Category

Table 2 The GA Parameter

GA Parameter			Value
Population Size			10*V
Basic Generic Operators	Selection Method		Roulette Wheel
	Crossover	Rate	0.6
		Method	Uniform
	Mutation Rate		0.06

Table 3 shows the accuracy rate at both the training and testing stages of these six hypertension classification approaches based on the 3-fold cross validation. Among these three CBR approaches (EWCBR, UWCBR and NWCBR), applying GA to explore the suitable weight for CBR (i.e. UWCBR and NWCBR) produced better accuracy rate than the CBR with equal weights (i.e. EWCBR) at both training and test stages. These two differently weighted CBR approaches also

Table 3 Accuracy Rates of Different Approaches

Stage	Dataset	Approach					
		C5.0	ANN	CART	EWCBR	UWCBR	NWCBR
Training (%)	Dataset-1	96.1	83.6	89.8	79.8	93.6	95.6
	Dataset-2	97.4	87.8	91.0	79.2	93.9	93.4
	Dataset-3	96.2	88.9	88.9	79.2	91.1	92.4
	Average	96.6	86.8	89.9	79.4	92.9	93.8
Test (%)	Dataset-1	88.3	81.5	83.6	77.3	94.8	93.5
	Dataset-2	85.2	84.1	83.3	76.3	87.0	94.5
	Dataset-3	84.6	87.0	85.9	79.4	94.0	90.4
	Average	86.0	84.2	84.3	77.7	91.9	92.8

outperformed the C5.0, ANN and CART approaches at test stage for each fold (i.e. Dataset-1, Dataset-2 and Dataset-3). For these two differently weighted CBR approaches, the accuracy rate revealed a slight difference for each fold at both training and test stages. However, applying GA for exploring feature weights enhanced the classification performance of the equally weighted CBR approach.

The average accuracy rate (i.e. the mean of Dataset-1, Dataset-2 and Dataset-3) at the training and testing stages for each approach are shown in Figure 1. Applying different weighted mechanisms to the CBR, i.e., unequal weight and nonlinear weight, presents a slight difference for hypertension classification at both training and test stages. It also reveals a trivial divergence between the two differently weighted CBR approaches (UWCBR and NWCBR). However, applying GA to investigating the feature weights of CBR enhanced the classification performance as compared to the CBR with equal weight, i.e., EWCBR. These differently weighted CBR approaches are also superior to the CART, ANN and C5.0 at the test stage. In general, applying GA to search feature weights for CBR enhanced the classification performance as compared to CBR with equal weight at both the training and test stages for hypertension classification with the data sets used in this research. It is also better than those widely used hypertension classification approaches, i.e., ANN, CART and C5.0.

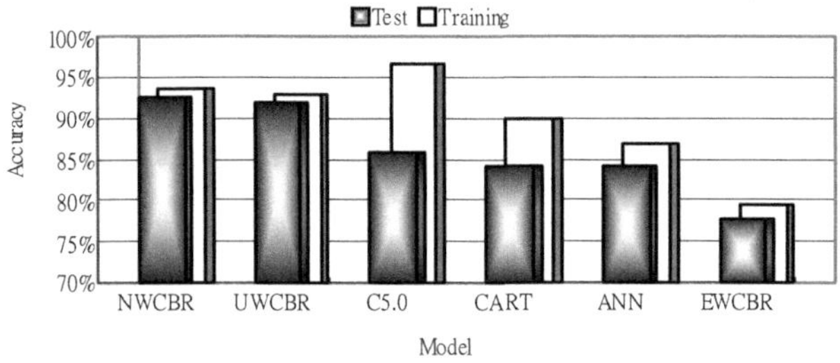

Fig. 1 Classification Results from the Mean of Different Datasets

Table 4 shows the improvement of differently weighted CBR approaches between the training and the test stages. For the improvement from the equally weighted CBR to the unequally weighted CBR, the UWCBR approach produces the highest improvement by 18.3% at training stage and 22.6% at test stage. The UWCBR approach demonstrates the mean improvement by 17% at training stage and 18.4% at test stage. For the improvement from the equally weighted CBR to the nonlinearly weighted CBR, the NWCBR approach displays the highest improvement by 19.7% at training stage and 23.9% at test stage. The NWCBR approach shows the mean improvement by 18.1% at training stage and 19.5% at test stage. Each differently weighted CBR approach provides improvement based on the equally weighted CBR approach at both training and testing stage. For the

Table 4 The Improvement of the Weighted CBR Approaches

Stage	Dataset	Equal vs. Unequal			Equal vs. Nonlinear			Unequal vs. Nonlinear		
		EWCBR	UWCBR	IMP*	EWCBR	NWCBR	IMP*	UWCBR	NWCBR	IMP*
Training (%)	Dataset-1	79.8	93.6	17.3	79.8	95.6	19.7	93.6	95.6	2.1
	Dataset-2	79.2	93.9	18.6	79.2	93.4	17.9	93.9	93.4	-0.6
	Dataset-3	79.2	91.1	15.1	79.2	92.4	16.7	91.1	92.4	1.4
	Average	79.4	92.9	17.0	79.4	93.8	18.1	92.9	93.8	1.0
Test (%)	Dataset-1	77.3	94.8	22.6	77.3	93.5	20.9	94.8	93.5	-1.4
	Dataset-2	76.3	87.0	14.0	76.3	94.5	23.9	87.0	94.5	8.7
	Dataset-3	79.4	94.0	18.3	79.4	90.4	13.8	94.0	90.4	-3.9
	Average	77.7	91.9	18.4	77.7	92.8	19.5	91.9	92.8	0.9

*: IMP denotes Improvement

improvement from the unequally weighted CBR to the nonlinearly weighted CBR, the NWCBR approach reveals the mean improvement by 1% at training stage and 0.9% at test stage. The NWCBR presents a slight improvement from the UWCBR approach for the hypertension classification with the data sets used in this research.

6 Conclusions

Predicting a subject's hypertension disease from the 3D anthropometry data and other related profile information is a novel way for medical decision support. The present paper investigates the impacts of differently weighted CBR similarity measures for deciding the appropriate feature weights on the accuracy improvement in the classification of hypertension disease approaches. Experimental results were encouraging when GA was applied on CBR to improve the accuracy rate by the data sets used in this research. Differently weighted CBR approaches have generated better accuracy rate of hypertension disease in comparison with the equally weighted approach and widely used classification approaches. The above-mentioned results demonstrate that applying GA to CBR is a feasible approach that can provide objective weights for features rather than the subjective weights assigned by experts. It also reveals that the CBR approach with nonlinear weight presents slightly better accuracy rate over the results from the unequally weighted CBR. Classification results from the integration GA with CBR are also promising and innovative in field of biomedical sciences. CBR classification approaches are intolerant of noise and of irrelevant features. We are encouraged by the results of the present study and are interested in investigating whether further improvements can be made by another design of the CBR classification approaches. In addition to the hypertension disease classification, the proposed approaches in the present paper could also be further employed in different disease for medical decision support in the future.

References

1. Ombuki, B., Ross, B.J., Hanshar, F.: Multi-objective genetic algorithms for vehicle routing problem with time windows. Applied Intelligence 24(1), 17–30 (2006)
2. Mykkanen, L., Haffner, S.M., Ronnemaa, T., Bennemaa, T., Bergman, R.N., Laakso, M.: Low insulin sensitivity is associated with clustering of cardiovascular disease risk factors. Am. J. Epidemiol. 146, 315–321 (1997)
3. Chen, W., Bao, W., Begum, S., Elkasabany, A., Srinivasan, S.R., Berenson, G.S.: Age-related patterns of the clustering of cardiovascular risk variables of syndrome X from childhood to young adulthood in population made up of black and white subjects: The Bogalusa Heart Study. Diabetes 49, 1042–1048 (2000)
4. Jeppesen, J., Hein, H.O., Suadicani, P., Gyntelberg, F.: High triglycerides and low HDL cholesterol and blood pressure and risk of ischemic heart disease. Hypertension 36, 226–232 (2000)
5. Coombes, A.M., Moss, J.P., Linney, A.D., Richards, R., James, D.R.: A mathematical method for the comparison of 3D changes in the facial surface. Eur. J. Orthodontic. 13, 95–110 (1991)
6. Kaya, M., Alhajj, R.: Utilizing genetic algorithms to optimize membership functions for fuzzy weighted association rules mining. Applied Intelligence 24(1), 7–15 (2006)
7. Lin, J.D., Chiou, W.K., Weng, H.F., Tsai, Y.H., Liu, T.H.: Comparison of three-dimensional anthropometric body surface scanning to waist–hip ratio and body mass index in correlation with metabolic risk factors. Journal of Clinical Epidemiology 55, 757–766 (2002)
8. Jones, P.R.M., Rioux, M.: Three-dimensional surface anthropometry: applications to the human body. Opt. Laser. Eng. 28, 89–117 (1997)
9. Beddoe, G.R., Petrovic, S.: Selecting and weighting features using a genetic algorithm in a case-based reasoning approach to personnel rostering. European Journal of Operational Research 175, 649–671 (2006)
10. Subcommittee, G.: World health organization–international society of hypertension guidelines for the management of hypertension. J. Hypertens 17, 151–183 (1999)
11. Bichindaritz, I.: Case-based reasoning in the health sciences. Artificial Intelligence In Medicine 36(2), 121–125 (2006)
12. Yager, R.R.: Soft aggregation methods in case based reasoning. Applied Intelligence 21(3), 277–288 (2004)
13. Auer, M., Trendowicz, A., Graser, B., Haunschmid, E., Biffl, S.: Optimal project feature weights in analogy-based cost estimation: improvement and limitations. IEEE Transactions on Software Engineering 32(2), 83–92 (2006)
14. Holland, J.: Adaptation in natural and artificial systems. University of Michigan Press, Ann Arbor (1975)
15. Liao, T.W., Zhang, C.R., Mount, C.R.: A cased-based reasoning system for identifying failure mechanisms. Engineering Applications of Artificial Intelligence 13(2), 199–213 (2000)
16. Chiu, C., Chiu, N.H., Hsu, C.I.: Intelligent aircraft maintenance support system using genetic algorithms and case-based reasoning. The International Journal of Advanced Manufacturing Technology 24(5-6), 440–446 (2004)
17. Chiu, C., Hsu, K.H., Hsu, P.L., Hsu, C.I., Lee, P.C., Chiou, W.K., Liu, T.H., Chuang, Y.C., Hwang, C.J.: Mining three-dimensional anthropometric body surface scanning data for hypertension detection. IEEE Transactions on Information Technology in Biomedicine 11(3), 264–273 (2007)

18. Wu, J.D., Ye, S.H.: Driver identification based on voice signal using continuous wavelet transform and artificial neural network techniques. Expert Systems with Applications 36, 1061–1069 (2009)
19. Rothwell, J.J., Futter, M.N., Dise, N.B.: A classification and regression tree model of controls on dissolved inorganic nitrogen leaching from European forests. Environmental Pollution 156, 544–552 (2008)
20. Ture, M., Tokatli, F., Kurt, I.: Using Kaplan–Meier analysis together with decision tree methods (C&RT, CHAID, QUEST, C4.5 and ID3) in determining recurrence-free survival of breast cancer patients. Expert Systems with Applications 36, 2017–2026 (2009)

Improving Medical Database Consistency with Induced Trigger Rules

Nittaya Kerdprasop, Sirikanjana Pilabutr, and Kittisak Kerdprasop

Abstract. The concept of triggers has been around for more than two decades. Despite their diverse potential usages, trigger rules are difficult to define correctly and have to be carefully hand-coded by database programmers. We suggest an automatic way of trigger rule creation by the advanced technology of data mining. We propose a framework of trigger rule induction as well as a method for trigger conflict resolution. On trigger firing the problem may arise if several trigger rules are eligible for execution. We propose a conflict resolution scheme that incorporates derived knowledge as a major part of the trigger rule prioritization. By means of trigger scheduling, deterministic behavior of the trigger processing can be guaranteed. We demonstrate the utilization of our proposed method on enhancing medical database consistency.

1 Introduction

A database is a collection of objects such as patient records together with a set of integrity constraints on these objects. Integrity constraints are predicates defined by the database designer as a requirement for database to be true on any database state. The values of objects in the database at any given time determine the state of the database. The state changes if there is a modification in the value of a database object. A database state is *consistent* if the values of the objects satisfy the specified integrity constraints. Database consistency is an important property to guarantee its reliability on any application.

To prevent the database from being inconsistent, data objects have to be accessed and modified only through the transactions. A transaction is a set of operations such as INSERT, DELETE, UPDATE that causes the database to change from one consistent state to another. On transaction processing, integrity constraints pertaining to the transaction are evaluated. If the constraints evaluate to false, called constraint violation, then the transaction that causes this event is undone. Integrity constraints are, however, capable of ensuring simple events such as domain integrity, referential integrity. To impose complex enforcement such as

Nittaya Kerdprasop, Sirikanjana Pilabutr, and Kittisak Kerdprasop
Data Engineering and Knowledge Discovery (DEKD) Research Unit,
School of Computer Engineering, Suranaree University of Technology,
Nakhon Ratchasima 30000, Thailand

K. Nakamatsu et al. (Eds.): New Advan. in Intel. Decision Techno., SCI 199, pp. 265–274.
springerlink.com

business rules or complicated update constraints across applications, trigger rules are deployed as a powerful and expressive tool to enforce integrity checking and thus enabling the filtering of state changes that violate database consistency.

Triggers, also known as event-condition-action (ECA) rules [26], are one major concept of active databases which extend traditional database systems with the mechanism to respond automatically to some specific events. The events may take place either inside or outside the database system. Upon the occurrence of the specified event, the rule condition is evaluated. If the condition is satisfied, then some actions are performed.

Although triggers are regarded as an important database feature on consistency monitoring, their deployment is still limited. This is due to the fact that creating complex trigger rules is not an easy task [9, 17, 19]. Tools and environments to aid users and database programmers are certainly needed. It is thus our aim to provide a method to automatically generating trigger rules from current database contents by means of data mining techniques. The induced trigger rules can be viewed as supplementary constraints to help increasing database consistency.

This paper is organized as follows. After the introduction section, we review some background on triggers and their related issues. Section 3 is the proposed framework of inducing trigger rules from database contents and its application to medical database. We also provide an algorithm to solve trigger conflict problem together with detailed explanation of the algorithm in section 4. Section 5 discusses other work related to ours. Section 6 concludes the paper.

2 Triggers and Related Issues

In SQL standard [11, 14], triggers are expressed by means of event-condition-action rules, as presented in figure 1. Each trigger is identified by a name. It is possible to specify whether a trigger must be executed BEFORE or AFTER its triggering event. SQL triggers allow only the INSERT, DELETE, and UPDATE as triggering event, and limit to a single event be monitored per single trigger rule.

```
<trigger definition> ::= CREATE TRIGGER <trigger name>
                         { BEFORE | AFTER } <trigger event> ON <table>
                         [ REFERENCING <tran_table or var_list> ]
                         <triggered action>
<trigger event>    ::= INSERT | DELETE | UPDATE [ OF < column list> ]
<triggered action> ::= [ FOR EACH { ROW | STATEMENT } ]
                       [ WHEN ( <condition> ) ] <triggered SQL statement>
```

Fig. 1 Definition of SQL triggers

The WHEN clause specifies an additional condition to be checked once the trigger rule is fired and before the action is executed. Conditions are predicates over the database state. If the WHEN clause is missing, the condition is supposed to be true and the trigger action is executed as soon as the trigger event occurs. The action is executed when the rule is triggered and its condition is true. Actions are stored procedures and may include SQL statements, control constructs, and calls to user-defined functions. The following example shows a trigger rule to impose a constraint on the database that the age of any person may never decrease.

Example 1: Trigger rule to guarantee no decrease on age value.

```
CREATE TRIGGER age-no-decrease BEFORE UPDATE OF Patient
FOR EACH ROW
   WHEN (new.Age < old.Age) begin log the event; signal error condition; end
```

The potential applications of triggers are significant [9, 19, 23]: signal integrity constraint violation and force rollbacks of the violating transactions, maintain consistency across system catalogs or other metadata, notify users in the form of messages, implement business rules or workflow management, and many more.

The behavior of triggers is defined as the "execution model." It specifies how trigger rules are evaluated and treated at runtime. Figure 2 illustrates the steps in processing triggers [22]. The signaling phase detects and signals the occurrence of an event. The event activates the corresponding trigger rules in the triggering phase, and the condition parts of the triggered rules are evaluated in the evaluation phase. The trigger conflict problem occurs when the conditions of more than one trigger rules are evaluated to be true. The scheduling phase indicates the order to process conflict triggers. The execution phase processes the scheduled trigger rules. On processing rule's action, the change in a database state may trigger another or even the same set of rules.

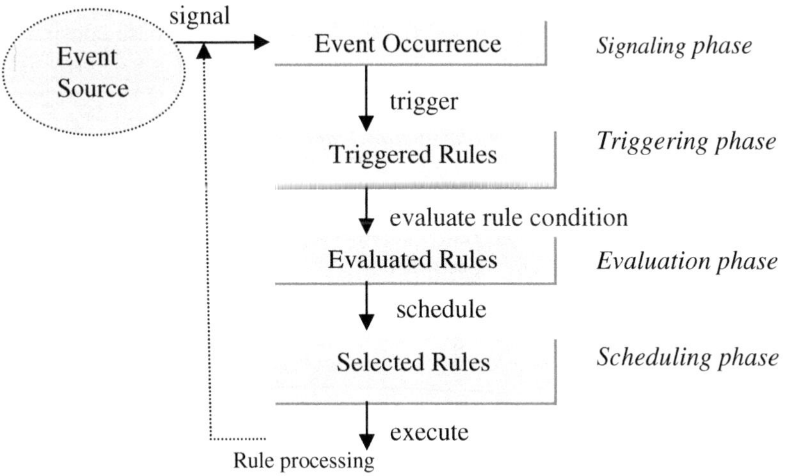

Fig. 2 Trigger rule execution steps

After the evaluation phase, more than one rule may be eligible for execution. This problem is known as *trigger rule conflict* [4, 22]. To solve the problem, the database management system must provide a *conflict resolution policy* to select a trigger rule for execution. The common conflict resolution policy adopted by most systems is assigning rule priority [21, 26]. The rule prioritization is either assigning a high priority to the rule that is most recently fired, or setting priority on the specificity of the rule. The two approaches are dynamic, thus less practical in the system with large and complex trigger rule set. When deterministic behavior is highly desirable, the scheme to associate rules with priority statically is more appropriate. Static priority mechanism determines order of trigger rules either by the system (e.g., based on rule creation time) or by the user (e.g., explicitly associate each rule with a numeric value). We propose a conflict resolution mechanism (in section 4) to incorporate derived knowledge (i.e., the knowledge obtained from the database content) into the rule prioritization scheme.

3 The Trigger Induction Framework and Its Application

We design the framework to add active behavior to the medical database through the induced trigger rules and rule processing module as shown in figure 3. There are three major components in our model: mining, trigger generation and conflict resolution components. Mining component induces knowledge in form of rules, association and classification, from the database contents. The data repository contains both base data and trigger rules. Trigger generation component is responsible for converting induced classification/association rules into trigger format then stores generated triggers in the repository. In case of trigger rule application and rule conflict occurs, conflict resolution component will handle the situation. We demonstrate the application of our proposed framework towards database consistency enforcement through example 2.

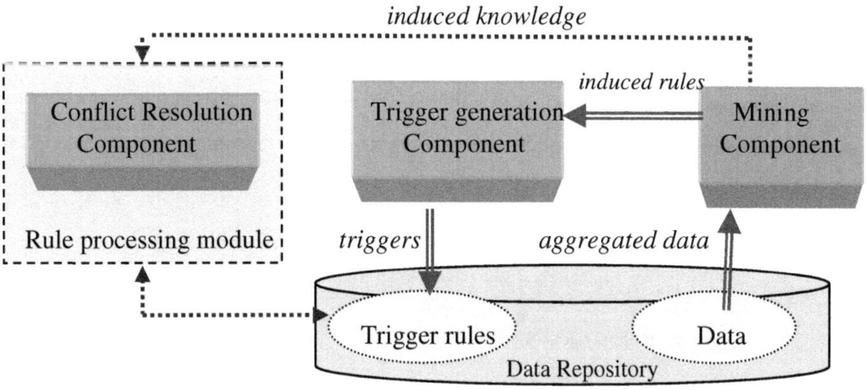

Fig. 3 A framework of active medical database containing trigger rule induction and trigger conflict resolution components

Example 2: Trigger induction on Diabetes database.

For a given medical database containing base tables related to patient personal information and treatment records, the aggregated information collected from related base tables with selected features suitable for mining component is shown schematically as follows.

Diabetes (Patient_ID, Name, Sex, Age, Temperature, Blood_Pressure_Upper, Blood_Pressure_Low, Diabetes_family, Weight, Height, BMI, Blood_Sugar, Diabetes, Insulin_level)

These data are input to the mining component to induce the following classification rules:

1. If Diabetes_family=yes and BMI>24.9 Then Diabetes=yes
2. If Diabetes_family=no and blood_sugar>128 Then Diabetes=yes

The trigger generation component then converts these rules into SQL triggers.

```
CREATE TRIGGER rule_1 ON diabetes FOR UPDATE, INSERT
    AS IF (SELECT COUNT(*) FROM diabetes
WHERE (diabetes_family = 'yes') and (BMI > 24.9) and (diabetes <> 'yes') ) > 0
BEGIN ROLLBACK TRAN; RAISERROR ('diagnose error'); END

CREATE TRIGGER rule_2 ON diabetes FOR UPDATE, INSERT
    AS IF (SELECT COUNT(*) FROM diabetes
WHERE (diabetes_family = 'no') and (blood_sugar > 128) and (diabetes <> 'yes')) > 0
BEGIN ROLLBACK TRAN; RAISERROR ('diagnose error'); END
```

Suppose there is an attempt to insert the following information into the database.

```
INSERT INTO Diabetics (Patient_ID, Name, Sex, Age, Temperature,
        Blood_Pressure_Upper, Blood_Pressure_Low, Diabetes_family, Weight,
        Height, BMI, Blood_Sugar, Diabetes, Insulin_level)
Values(P021, Amitta, Female, 33, 37.6, 130, 80, no, 72, 1.62, 27.4, 130, no, 0)
```

This new information violates Trigger rule_2 since *Diabetes_family* = *'no'* and *Blood_Sugar* = *130*, but the diagnosed *Diabetes* = *'no'*. Therefore, this transaction has to be undone and the database remains in a consistent state.

The proposed framework is semi-automatic in that the trigger induction process has to be invoked by the database administrator. After database contents have been created and modified and numerous new contents might have been inserted into the database, the administrator may consider activating the trigger induction process. Upon activation the previous induced trigger rules are removed as they may not be relevant to the current database state. The mining component has been set to induce only rules with 100% accuracy rate since precision is critical criteria in medical domain. Induced rules are prioritized in descending order of their support (or coverage) values. The *top-k* rules will be sent to trigger generation component; the *k* value is adjustable by the database administrator. At the moment we do not consider the issue of rule conflict since we assume that database integrity constraints are powerful mechanism sufficient to prevent conflicting cases inserted into database contents.

4 A Method for Trigger Conflict Resolution

After triggers have been created, the problem of trigger conflict might occur during trigger processing phase. In this section, we define an algorithm (in figure 4) to handle trigger conflict by reorganizing the trigger rules into different layers, or strata. Then, associate each stratum with a numeric priority. The major mechanism leading to priority assigning is the induced knowledge regarding the database state modification.

Input: an unordered stratum set S, a database R and its metadata
Output: an ordered stratum set O with assigned priority on each stratum
Steps:

1. ***Active_Tuple_Set$_i$* ← *Activate(S$_i$, E)***
   /* Activate every stratum S_i , $S_i \in$ S, such that the occurrence of an event *E* can invoke its trigger rule(s), and record all affected tuples in the corresponding *Active_Tuple_Set*. */
2. ***K$_i$* ← *Induce(Active_Tuple_Set$_i$)***
   /* The knowledge induction method (such as rule learning, decision-tree induction) is applied to induce knowledge from the content of each *Active_Tuple_Set*. The induced knowledge is stored in K_i. */
3. ***q$_i$ = Degree_of_Constraint (K$_i$, metadata)***
   /* Calculate the value *q*, or the *Degree_of_Constraint*, of each set of induced knowledge *K* comparing to the integrity constraints given as a metadata. */
4. ***sort(i, q)***
   /* Apply any sorting algorithm on the *q*-value associated with each stratum S_i. */
5. ***return an ordered stratum set O = { S$_i$ | S has been sorted by its index i }***

Fig. 4 Trigger conflict resolution algorithm

We have applied the concept of stratum [3] to guarantee that trigger rule execution eventually terminates. A *stratum* is an ordered set of trigger rules that locally converge. A stratum locally converges if after any transaction invoking trigger rules, rule processing terminates in a final state in which the set of triggered rules is empty. *Stratum set* is an unordered set of strata in which each stratum is independent from other strata so that the trigger rule execution in one stratum does not affect rules in other strata. The example of non-terminate trigger execution due to the cycle in action-triggering events is shown in figure 5. Breaking the cycle into different layers, depicted in figure 6, and the local convergence within each stratum are two sufficient conditions for termination on trigger execution.

Fig. 5 The cycle among trigger rules

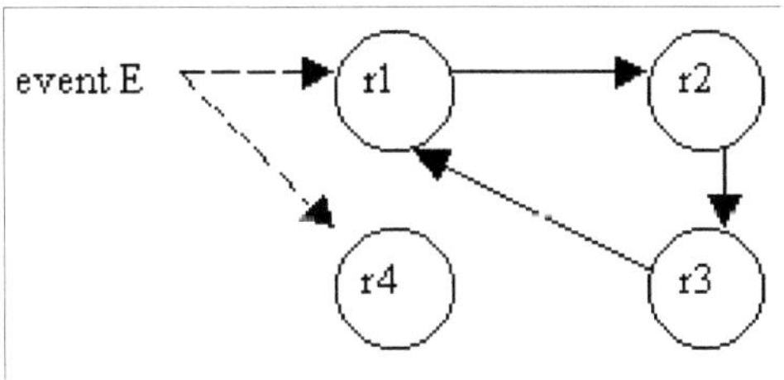

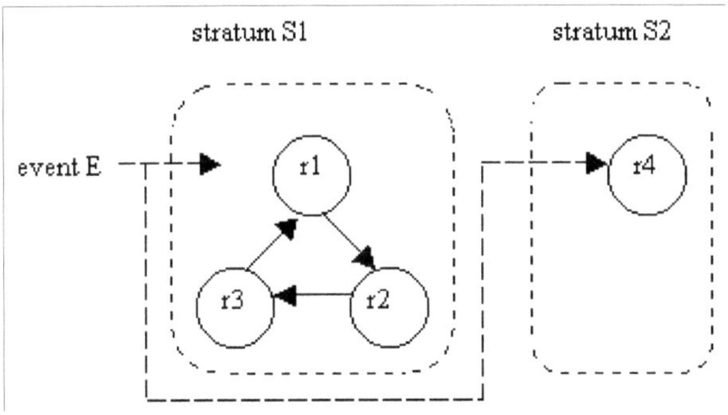

Fig. 6 Organizing cyclic rules into strata

Step 3 of the proposed algorithm is to set priorities among strata when several strata are activated by the occurrence of an event. We propose the trigger conflict resolution algorithm to solve the problem of multiple stratum activation on an event E. The key concept is to apply, in step 2, knowledge induced from the database content and the metadata (i.e., the set of integrity constraints) to guide the priority assigning scheme. The function to compute priority for each stratum is defined as:

$$\mathit{Degree\_of\_Constraint}(\mathrm{K}, \mathrm{IC}) = \frac{\#\mathit{matched\_rules}}{\#\mathit{total\_rules}} + \alpha$$

where # *matched_rule* is the number of induced knowledge, represented in the format of rule, completely matching with the integrity constraint rule (i.e., NOT (k) AND c yields the contradiction when $k \in$ K and $c \in$ IC),
total_rules is the total number of induced knowledge represented as rules,

α is the average accuracy of the induced knowledge rules normalized in order to prevent the domination of the accuracy over the proportion of *matched_rules* and *total_rules*.

Step 4 of the algorithm groups triggers into several strata, then each stratum will be sorted according to the *Degree_of_Constraint* values. At a final step, each stratum has been returned with the priority value associated with it.

5 Related Work

The importance of integrating active behavior into the database systems has been recognized since the 1970s [12]. However, it was not until the late 1980s to early 1990s that the area of active databases has caught high interest among researchers [8, 10, 13, 22, 26]. At present most commercial database systems, such as Oracle, IBM DB2, Microsoft SQL Server, support the simple forms of triggers and also incorporate triggers into commercial object-oriented databases [7, 21]. The SQL standard [11, 14] has extensive coverage of triggers.

A simple concept of triggers has been successfully applied to solve problems in various domains [9]. For example, Sengupta et al [23] employ triggers to alert intrusion detection system administrator when a suspect event has been detected. In medical domain the employment of triggers to achieve active behavior is quite rare. Most of the proposed methods are for detecting static events such as the discovery of relationships that suggest risks of adverse events in patient records [20, 24], detection of dependency patterns of process sequences for curing brain stroke patients [18], the generation of rules to annotated protein data in medical database [16], or the exploration of environmental health data [6]. Our work differ from those appeared in the literature in that we propose a framework of employing knowledge discovery techniques to automatically create trigger rules and also prioritize rules in case of conflict. The utilization of our proposed method is to increase consistency in medical database. Any database modification events violating constraints will be alerted and undone. The system designed by Agrawal and Johnson [1] is also to support medical database but in a different aspect; they concentrate on security and privacy preservation of patients and other sensitive health data.

Although trigger is a powerful mechanism in active database systems, designing and writing correct trigger rules are not an easy and straightforward task. The difficulty is due to the complex and sometime unpredictable behavior of the triggers. Poorly designed triggers can activate each other indefinitely, which leads to the non-terminate execution. Several methods have been proposed [2, 3, 4, 5, 15, 25] to analyze trigger behavior at compile time and runtime. There exist some work on developing tools to aid trigger designing semantically [19] and visually [17]. Another problem regarding trigger behavior is the deterministic property of the triggers. Deterministic trigger processing guarantees the same order of execution when several trigger rules are activated simultaneously. We propose a mechanism to utilize domain-knowledge in choosing among activated triggered rules.

6 Conclusions

We present the design framework of active medical databases. Active behavior of the database system has been obtained through a set of trigger rules. Triggers are both created manually by database programmers and induced automatically via data mining techniques. We devise a method to induce trigger rules from existing database contents. We also propose a method to solve trigger conflict problem.

The trigger rule conflict occurs when an event activates several trigger rules simultaneously. To maintain the deterministic property of the active database processing, the database management system has to provide a conflict resolution policy. The common policy adopted by most systems is to assign rule priority. The rule prioritization is based on either the recent update or the complexity of the rule's condition. We propose a different scheme of prioritization by taking into account the knowledge regarding the database state. Moreover, we consider priority at the level of stratum, which may contain several related triggers. The concept of stratification preserves the termination property of trigger rule processing.

The design of medical active database framework and the trigger-conflict-resolution algorithm is the preliminary work toward the design and implementation of a set of tools to help database designer on designing and analyzing a complex set of triggers. A further investigation on a more practical active database with a larger trigger set is necessary.

Acknowledgments. The authors would like to thank all anonymous referees for their thorough reading and very helpful suggestion. This work has been fully supported by research fund from Suranaree University of Technology granted to the Data Engineering and Knowledge Discovery (DEKD) research unit. This research is also partly supported by grants from the National Research Council of Thailand (NRCT) and the Thailand Research Fund (TRF) under grant number RMU 5080026.

References

[1] Agrawal, R., Johnson, C.: Securing electronic health records without impeding the flow of information. Int. J. Medical Informatics 76, 471–479 (2007)

[2] Aiken, A., Hellerstein, J.M., Widom, J.: Static analysis techniques for predicting the behavior of active database rules. ACM Transactions on Database Systems 20(1), 3–41 (1995)

[3] Baralis, E., Ceri, S., Paraboschi, S.: Modularization techniques for active rule design. ACM Transactions on Database Systems 21(1), 1–29 (1996)

[4] Baralis, E., Ceri, S., Paraboschi, S.: Compile-time and runtime analysis of active behaviors. IEEE Transactions on Knowledge and Data Engineering 10(3), 353–370 (1998)

[5] Baralis, E., Widom, J.: An algebraic approach to rule analysis in expert database system. In: Proc. 20th VLDB, pp. 475–486 (1994)

[6] Bedard, Y., Gosselin, P., Rivest, S., et al.: Integrating GIS components with knowledge discovery technology for environmental health decision support. Int. J. Medical Informatics 70, 79–94 (2003)

[7] Bertino, E., Guerrini, G., Merlo, I.: Trigger inheritance and overriding in an active object database system. IEEE Transactions on Knowledge and Data Engineering 12(4), 588–608 (2000)

[8] Buchmann, A.: Current trends in active databases: Are we solving the right problems. In: Proc. Information Systems Design and Multimedia, pp. 121–133 (1994)

[9] Ceri, S., Cochrane, R.J., Widom, J.: Practical applications of triggers and constraints: Successes and lingering issues. In: Proc. 26th VLDB, pp. 254–262 (2000)

[10] Chakravarthy, S.: Rule management and evaluation: An active DBMS perspective. ACM SIGMOD Records 18(3), 20–28 (1989)

[11] Eisenberg, A., Melton, J.: SQL:1999, formerly known as SQL3. ACM SIGMOD Records 28(1), 131–138 (2000)

[12] Eswaran, K.P.: Specification, implementations and interactions of a trigger subsystem in an integrated database system. IBM Research Report RJ1820, San Jose, California (1976)

[13] Hanson, E.N., Widom, J.: An overview of production rules in database systems. Knowledge Engineering Review 8(2), 121–143 (1993)

[14] International Organization for Standardization, ISO/IEC 9075:2003 (2003)

[15] Karadimce, A.P., Urban, S.D.: Conditional term rewriting as a formal basis for analysis of active database rules. In: Proc. Research Issues in Data Engineering (RIDE), pp. 156–162 (1994)

[16] Kretschmann, E., Fleischmann, W., Apweiler, R.: Automatic rule generation for protein annotation with the C4.5 data mining algorithm applied on SWISS-PROT. Bioinformatics 17(10), 920–926 (2001)

[17] Lee, D., Mao, W., Chiu, H., et al.: Designing triggers with trigger-by-example. Knowledge and Information Systems 7, 110–134 (2005)

[18] Lin, F., Chou, S., Pan, S., et al.: Mining time dependency patterns in clinical pathways. Int. J. Medical Informatics 62(1), 11–25 (2001)

[19] Mota-Herranz, L., Celma-Gimenez, M.: Automatic generation of trigger rules for integrity enforcement in relational databases with view definition. In: Proc. 3rd Int. Conf. Flexible Query Answering Systems, pp. 286–297 (1998)

[20] Noren, G.N., Bate, A., Hopstadius, J., et al.: Temporal pattern discovery for trends and transient effects: Its application to patient records. In: Proc. KDD, pp. 963–971 (2008)

[21] Paton, N.: Active rules in database systems. Springer, Heidelberg (1999)

[22] Paton, N., Diaz, O.: Active database systems. ACM Computing Surveys 31(1), 63–103 (1999)

[23] Sengupta, S., Andriamanalimanana, B., Card, S.W., et al.: Towards data mining temporal patterns for anomaly intrusion detection systems. In: Proc. IEEE Workshop on Intelligent Data Acquisition and Advanced Computing Systems: Tech. and Applications, pp. 205–209 (2003)

[24] Silva, A., Cortez, P., Santos, M.F., et al.: Rating organ failure via adverse events using data mining in the intensive care unit. Artificial Intelligence in Medicine 43(3), 179–193 (2008)

[25] van der Voort, L., Siebes, A.: Termination and confluence of rule execution. In: Proc. 2nd Int. Conf. Information and Knowledge Management, pp. 245–255 (1993)

[26] Widom, J., Ceri, S.: Active database systems: Triggers and rules for advanced database processing. Morgan Kaufmann, San Francisco (1996)

Factorization of Sparse Bayesian Networks

Julio Michael Stern and Ernesto Coutinho Colla

Abstract. This paper shows how an efficient and parallel algorithm for inference in Bayesian Networks (BNs) can be built and implemented combining sparse matrix factorization methods with variable elimination algorithms for BNs. This entails a complete separation between a first symbolic phase, and a second numerical phase.

Keywords: Bayesian networks, Probabilistic Reasoning, Variable elimination, Elimination trees, Sparse Cholesky factorization, Sparse matrix factorizations.

1 Introduction

Bayesian Networks (BNs) are probabilistic graphical models used to represent and encode uncertain expert knowledge. BNs stand out for dealing with uncertainty in decision making and statistical inference, and many algorithms were described for inference in BNs, see Dechter (1996), Heckerman (1995), Jensen (1996), Lauritzen (1988), Pearl (1988), and Zang (1996). The parallel algorithm described in this paper is based on the sequential variable elimination algorithm of Cozman(2000), using algebraic operations on potentials. These algebraic schemata for inference in BNs are not only relatively simple to understand and to implement, but also allow us to use the techniques, heuristics and abstract combinatorial structures from the sparse matrix factorizations literature, see appendix, George (1993) and Stern (1994, 2008a,b).

The main goal of this paper is to show how variations of the variable elimination algorithm can be combined with sparse matrix factorization methods to implement a fast and efficient parallel algorithm for inference in BNs. This goal is achieved with the complete separation between a first symbolic phase, and a second numerical phase. In the symbolic phase the proposed algorithm explores the graphical

Julio Michael Stern and Ernesto Coutinho Colla
IME-USP, University of São Paulo, Brazil
e-mail: jmstern@hotmail.com

K. Nakamatsu et al. (Eds.): New Advan. in Intel. Decision Techno., SCI 199, pp. 275–285.
springerlink.com

structure of the model, without computing or even accessing probabilistic information. The second numerical phase can be fully vectorized and parallelized using static data structures previously defined in the first phase. This is done examining the decoupling or separation operators of sparse matrix factorization algorithms and BNs inference procedures from a unified combinatorial framework. This unified framework is the key for implementing efficiently this parallel algorithm.

2 Inference with Bayesian Network

A BN, see Jensen (1996), is a graphical model that efficiently encodes the joint probability distribution for a set (or list) of random variables, $X = \{X_1, X_2, ..., X_n\}$, each of them having a finite number of possible states. A BN consists of two components: (i) A Directed Acyclic Graph (DAG) defining the network structure and encoding the conditional dependence relations between the variables in X; (ii) A set of local probability densities associated with each variable. Each node, i, of the DAG represents a random variable, X_i. In order to make the notation lighter, we may write a node index, i, instead of its random variable, X_i, and vice versa. We also use the vectorized notation, X_S, for the subset $\{X_i\}, i \in S$.

The DAG representing the BN structure has an arc from node i to node j, that is, i is a *parent* of j, $i \in \text{pa}(j)$, if the probability distribution of variable X_j is directly dependent on variable X_i, and the strength of this influence is expressed by conditional probability distributions. In many specific statistical models an arc can be interpreted as a direct influence or causal effect of X_i on X_j, see Pearl (1988).

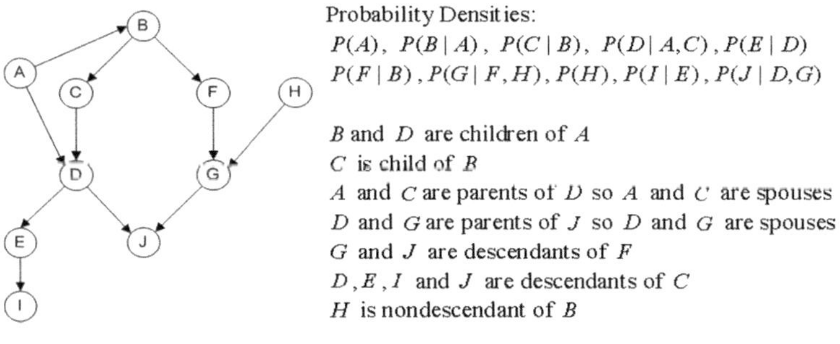

Fig. 1 Bayesian network example

The semantics of BNs implies a correspondence between the topology of a DAG and the network's probabilistic dependence relations, determined by the *Markov condition*: Every variable is independent of its nondescendants nonparents given its parents. Therefore, every X_i is associated with a local probability density, $P(X_i \mid X_{\text{pa}(i)})$, as showed in Fig 1. Based on this condition, a BN encodes a unique probability distribution: $P(X) = \prod_i P(X_i \mid X_{\text{pa}(i)})$.

Inference in BNs is based on queries, where the posterior marginal distribution for a set of *query variables*, X_Q, has to be computed given a set of observed variables, X_E. This set of observed variables is the *evidence* in the network and establishes the values of the variables in X_E. For example: $e = \{X_i = x_i, X_j = x_j\}$ establishes the values of X_i and X_j, so $E = \{i, j\}$.

The posterior probability of X_Q given e is:

$$P(X_Q \mid e) = \frac{P(X_Q, e)}{P(e)} = \frac{\sum_{X \setminus \{X_Q, X_E\}} P(X)}{\sum_{X \setminus X_E} P(X)} . \tag{1}$$

The expression $X \setminus Y$ indicates the set of all variables which belong to X but do not belong to Y, and the expression $\sum_X f(X,Y)$ indicates that all variables of X were *eliminated* or *marginalized out*, that is, were summed out from the function $f(X,Y)$.

Efficient computational algorithms rely on two important technical points:

(I) Given a BN over variables X, an evidence e and a query X_Q, not all variables of X may be required to compute $P(X_Q \mid e)$. If the local probability density $P(X_i \mid X_{\text{pa}(i)})$ is required to compute $P(X_Q \mid e)$, then X_i is a *requisite variable*, $i \in R$. Fortunately there are simple polynomial algorithms able to identify the set R. We have used Bayes-Ball algorithm, see Shachter (1998). It is important to realize that the requisite variables, X_R, can be identified exploring only the DAG topology, without any numerical information concerning probability distributions. Hence, in order to reduce the problem dimension, the identification of R should be done at the very first stage of inference calculation.

(II) At intermediate computations, it is not necessary to compute the normalization constants, that is, the denominator $P(e)$ of (1). We only need the numerator in (1),

$$P(X_Q \mid e) \propto P(X_Q, e) = \sum_{X_R \setminus \{X_Q, X_E\}} \left(\prod_{X_i \in X_R} P(X_i \mid X_{\text{pa}(i)}) \right) . \tag{2}$$

Hence, a basic rule for operation in BNs is: Compute the numerator $P(X_Q, e)$ and obtain normalization constant $P(e)$ only in the last stage. This rule means that we can perform the intermediate computations with un-normalized distributions, which are real-valued tables over a finite set of variables. These tables, ϕ, are called *potentials*, see Jensen (1996). A potential's domain, $\text{dom}(\phi)$, is its correspondent set of variables. In the following, we give some important properties of the algebra of potentials:

(1) A variable X_i can be *marginalized out* of a potential ϕ resulting in a new potential $\phi'_{X_i} = \sum_{X_i} \phi$ over the domain $\text{dom}(\phi'_{X_i}) = \text{dom}(\phi) \setminus \{X_i\}$. Marginalization follows:

(1a) the commutative law: $\sum_{X_i} \sum_{X_j} \phi = \sum_{X_j} \sum_{X_i} \phi$; and

(1b) the distributive law: if $X_i \notin \text{dom}(\phi_1)$, then $\sum_{X_i} \phi_1.\phi_2 = \phi_1.\sum_{X_i} \phi_2$.

(2) Two potentials can be *multiplied*, resulting in a new potential with $\text{dom}(\phi_1.\phi_2) = \text{dom}(\phi_1) \cup \text{dom}(\phi_2)$. Multiplication follows:

(2a) the commutative law: $\phi_1.\phi_2 = \phi_2.\phi_1$; and

(2b) the associative law: $(\phi_1.\phi_2).\phi_3 = \phi_1.(\phi_1.\phi_3)$.

As an example, consider the BN in Figure 1. The BN joint probability distribution can be rewritten as: $P(X) \propto \phi_A.\phi_B.\phi_C.\phi_D.\phi_E.\phi_F.\phi_G.\phi_H.\phi_I.\phi_J$, and the potentials specified for the network are: $P(A) \propto \phi_A(A), P(B \mid A) \propto \phi_B(B,A)$,

$P(C \mid B) \propto \phi_C(C,B), P(D \mid A,C) \propto \phi_D(D,A,C)$ and so on. Computing $P(I)$ can be accomplished by marginalizing out of $P(X)$ all the variables, except I.

$$P(I) = \sum_{A,B,C,D,E,F,G,H,J} P(X) . \tag{3}$$

BNs are particularly useful for calculating new probabilities when we acquire new information. However, in the preceding calculations no evidence was entered into the network. Now, assume information e has been acquired, stating that "$A = a_t$", where A is a variable and a_t is the t-th state of A. Let A have s states with probability distribution $P(A) = (x_1, ..., x_t, ..., x_s)$. This observed evidence e means that all states except tth one are impossible. So the new (un-normalized) probability distribution is $P(A,e) = (0, ..., 0, x_t, 0, ..., 0)$ which is the result of multiplying $P(A)$ with $\underline{e}_A = (0, ..., 0, 1, 0, ..., 0)$ in which only tth value is 1. The s-dimensional 0-1 potential $\underline{e}_A$ is called *finding*.

In the current example, assume that we have the evidence $A = a$, $H = h$ and $J = j$. This evidence e would be represented using three findings $\underline{e}_A$, $\underline{e}_H$ and $\underline{e}_J$. The posterior marginal $P(I \mid e)$ can be obtained normalizing $P(I,E)$:

$$P(I,e) = \sum_{A,B,C,D,E,F,G,H,J} P(X).\underline{e}_A.\underline{e}_H.\underline{e}_J . \tag{4}$$

To avoid calculating the product of all potentials, we use the distributive law:

$$P(I \mid e) = \sum_D \sum_E \phi_I(I,E).\phi_E(E,D) \sum_B \sum_C \phi_C(C,B).$$

$$\sum_A \phi_A(A).\phi_B(B,A).\phi_D(D,A,C).\underline{e}_A.$$

$$\sum_J \sum_G \phi_J(J,D,G).\underline{e}_J \sum_F \phi_F(F,B). \sum_H \phi_G(G,F,H).\phi_H(H).\underline{e}_H .$$

First, calculate $\phi_H^{'} = \sum_H \phi_G(G,F,H).\phi_H(H).\underline{e}_H$, then multiply $\phi_H^{'}(F,G)$ on $\phi_F(F,B)$ and calculate $\phi_F^{'} = \sum_F \phi_F(F,B).\phi_H^{'}(F,G)$. The later result is multiply on $\phi_J(J,D,G).\underline{e}_J$, to calculate $\phi_G^{'} = \sum_G \phi_J(J,D,G).\phi_F^{'}(B,G).\underline{e}_J$, and so forth. All the operations involved are represented in Figure 2.

Because marginalization is commutative it can be done in any order. In the preceding calculation the marginalization, also called variable elimination, was done in a particular order, namely $q = [H,F,G,E,A,C,J,B,D]$.

The diagram in Figure 2 also portrays the dependencies among potential operations to calculate $P(I \mid e)$. Notice that some operations could be done simultaneously. For example, at very first stage, we could perform the required operations on A, H, and I, calculating $\phi_A^{'} = \sum_A \phi_A(A).\phi_B(B,A).\phi_D(D,A,C).\underline{e}_A$, etc.

If a parallel computer is available, we can simultaneously execute all the marginalizations using already computed potentials. Hence, it is desirable to find:
(i) An efficient way to specify all dependencies among marginalization operations.
(ii) A way to specify an elimination order entailing a "simple" dependence structure, so that many operations can be done simultaneously.

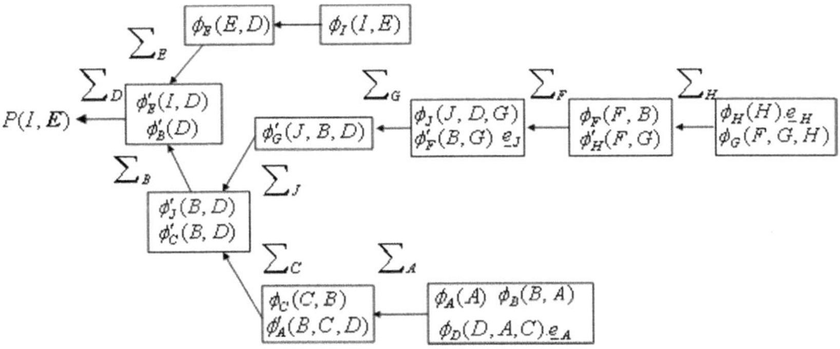

Fig. 2 The process of marginalizing down to I

The dependence structure of these operations is exactly the same as the dependence structure for "pivoting" operations appearing in numerical linear algebra, namely, in the Cholesky factorization of sparse matrices, see appendix, George (1993), Pissanetzky (1984), and Stern (1994, 2008b). We describe only the aspects pertinent to this paper.

An *Undirected Graph* (UG), $\mathcal{G} = (\mathcal{V}, \mathcal{E})$, has undirected edges, $\{i, j\} \in \mathcal{E}$, standing for pairs of opposite directed arcs, (i, j) and (j, i). The *Moral Graph* of a DAG, $\mathcal{G}$, is the UG with the same nodes as $\mathcal{G}$, and edges joining nodes i and j if they are immediate relatives in $\mathcal{G}$. The immediate relatives of a node in $\mathcal{G}$ include its parents, children and spouses (but not brothers or sisters). i is a spouse of j is they have a child in common, that is, $i \in \mathrm{sp}(j) \Leftrightarrow \exists k \,|\, i, j \in \mathrm{pa}(k)$.

The *Markov Blanket* of X_i, $X_{\mathrm{mb}(i)}$ is defined as the minimal set of variables that makes a variable X_i independent from all other variables in the BN. This means that the Markov Blanket of a variable "decouples" this variable from the rest of network: $P(X_i \,|\, X_{\mathrm{mb}(i)}, X_j) = P(X_i \,|\, X_{\mathrm{mb}(i)})$. It can be shown that the set of immediate relatives of node i is the Markov Blanket of node i. Figure 3a shows the Moral Graph of the BN in Figure 1. It is important to realize that if X_i and X_j are both in the same domain, of a variable X_k of the BN, then the edge $\{i, j\}$ is in the Moral Graph.

Given an UG, $\mathcal{G} = (\mathcal{V}, \mathcal{E})$, $\mathcal{V} = \{1, \ldots n\}$, and $q = [q(1), \ldots q(n)]$, an elimination order, we define the elimination process of its nodes as the sequence of *elimination graphs* $\mathcal{G}_k = (\mathcal{V}_k, \mathcal{E}_k)$, for $k = 1 \ldots n$, as follows: When eliminating node $q(k)$, we make its neighbors a *clique*, adding all missing edges between them.

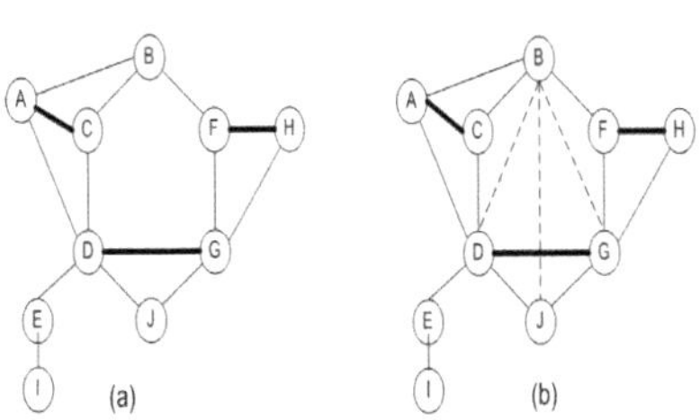

Fig. 3 (a) Moral Graph (b) Filled Graph

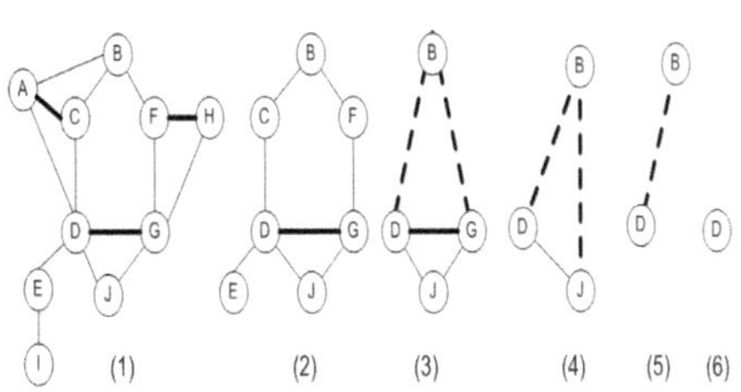

Fig. 4 Elimination graphs sequence

$$\mathscr{V}_k = \{q(k), q(k+1), \ldots q(n)\}, \quad \mathscr{E}_1 = \mathscr{E}, \text{ and, for } k > 1 ,$$

$$\{i,j\} \in \mathscr{E}_k \Leftrightarrow \begin{cases} \{i,j\} \in \mathscr{E}_{k-1} \text{ , or} \\ \{q(k-1), i\} \in \mathscr{E}_{k-1} \text{ and } \{q(k-1), j\} \in \mathscr{E}_{k-1} \text{ .} \end{cases}$$

The *Filled Graph* is the graph $(\mathscr{V}, \mathscr{F})$, where $\mathscr{F} = \cup_{k=1}^{n} \mathscr{E}_k$. The *original* edges and the *filled* edges in $\mathscr{F}$ are, respectively, the edges in $\mathscr{E}$ and in $\mathscr{F} \backslash \mathscr{E}$. There is a computationally more efficient form of obtaining the Filled Graph, known as *simplified elimination*: In the simplified version of the elimination graphs, $\mathscr{G}_k^*$, when eliminating vertex $q(k)$, we add only the clique edges incident to its neighbor, $q(l)$, that is next in the elimination order.

The marginalization of variable X_i out of $P(X)$ corresponds to the elimination of the correspondent node in the elimination sequence. In order to marginalize on X_i, we have first to multiply all the potentials having X_i in its domain, and than sum out X_i. The domain of the resulting potential includes all the neighbors of X_i. In the Elimination Graphs, the corresponding elimination of X_i forms a clique with all of X_i's neighbors. Figure 3b and 4 show the Filled Graph and a synthetic version of the eliminations graphs for the order $q = [H, F, G, E, A, C, J, B, D]$.

The *Elimination Tree*, see appendix, George (1993), Pissanetzky (1984), and Stern (1994, 2008b), portrays the dependencies among numeric operations on potentials, corresponding to dependencies in the node elimination process in the elimination graph. Hence, building the Elimination Tree for the corresponding Moral Graph makes it easy to see which variables can be eliminated simultaneously. Figure 5 shows the Elimination Tree for the order $q = [H, F, G, E, A, C, J, B, D]$.

According to the Figure 5, six steps would be enough to eliminate all nodes: Variables I, H and A could be eliminated at first step and variables E, F and C at second one. Note that: (i) The Elimination Tree has the same structure of the tree of operations portrayed in Figure 2; (ii) A serial elimination would require 10 steps.

Clearly the Elimination Tree depends on the chosen elimination order, and the sparse matrix literature has many heuristics designed for finding good elimination orders. In this paper we adopted an heuristic based on a nested dissections of the breadth-first tree rooted at a pseudo-peripheral vertex which, in turn, was found using the Gibbs heuristic, see Figures 6 and 7. These procedures are described in the appendix, see also George (1993), Pissanetzky (1984) and Stern (1994, 2008b).

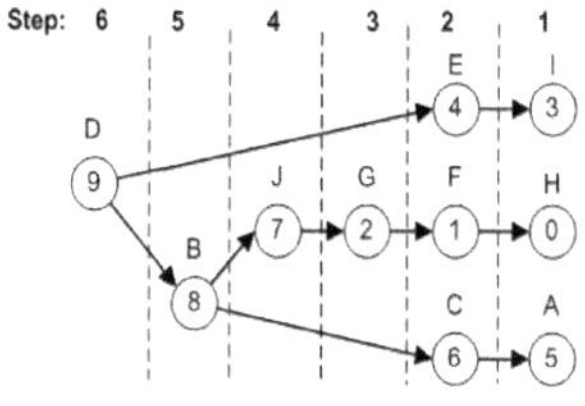

Fig. 5 Elimination Tree for order $q = [H, F, G, E, A, C, J, B, D]$

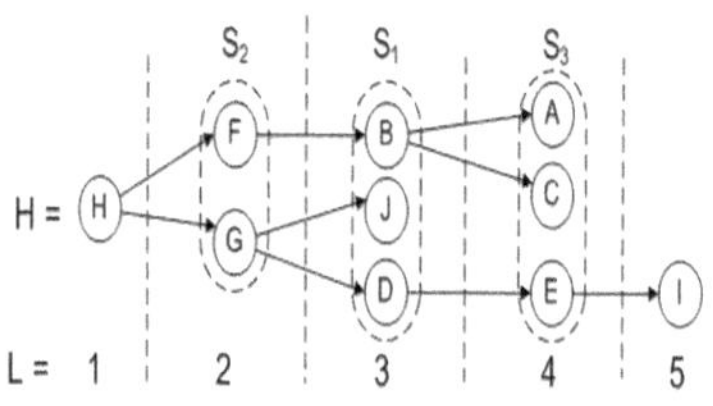

Fig. 6 Nested Dissection

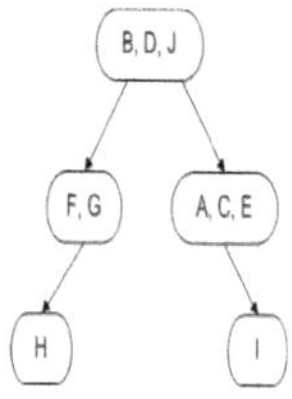

Fig. 7 Nested Dissection

3 Parallel Variable Elimination Algorithm

The sequence of operations described in the previous section for inference in BNs can be summarized in the *parallel variable elimination algorithm*:

1. Symbolic phase:

1.1* Define the requisite variables X_R (ex. using Bayes-Ball algorithm);

1.2 Build the Moral Graph (including only variables in X_R);

1.3 Choose a good elimination order using the Gibbs heuristics to find a pseudo-peripheral vertex used as a root for the Nested Dissection heuristic;

1.4 Symbolic Factorization: Execute the simplified elimination on the Moral Graph, and build the Elimination Tree;

1.5 Allocate the computation resources and prepare the data structures to execute the numeric operations.

2. Numeric phase: Using static data structures defined in the first phase:

2.1 While the root of the Elimination Tree was not executed: Based on the Elimination Tree hierarchy, trigger all threads executeing variable eliminations ready to be done, including its numeric operations of multiplication and marginalization;

2.2 Normalize the remaining potential at the root.

4 Results and Conclusions

The proposed parallel algorithm was implemented and its performance was compared with a serial implementation. Both implementations were done in C and use the same functions to execute the basic operations for: Load the network; Multiply and marginalize potentials; and define the elimination order. The only difference between the two implementations is that the parallel version builds the Elimination Tree and, if possible, eliminate two or more variables simultaneously. Following this strategy we hope to isolate the effect of parallelization.

Table 1 displays some illustrative results. These experiments were done in a bi-processed machine running Linux and consists of 100 inferences for 7 distinct queries using the Hailfinder25 network (55 variables). The set of experiments suggests that the parallel implementation is much faster than the serial one for larger experiments. Queries requiring more variables or with a branched structure in the Elimination Tree allow the simultaneous elimination of several variables, for example experiments 1 to 6. Models requiring less variables, or with a more linear structure in the elimination tree allow less parallelization of elimination operations. Consequently, in these examples, the serial implementation performed better due to the computational overheads imposed by the parallel version, namely, building of the Elimination Tree and the heavy context switch during execution. This was the case of experiment 7 in which the relations of dependence between the operations reduce the possibilities of parallelization.

Practitioners always want to solve larger models, most large models used in practice are sparse, and parallel or distributed computer are increasingly available. Hence, we see great potential for the parallel algorithm presented in this article.

Table 1 Query example, Numb. or requisite vars., Parallel time, Serial Time, Parallel context switches. Serial context switches

Q.E.	N.R.	P.T.	S.T.	P.C.S.	S.C.S.
1	44	174	812	6498	901
2	44	95	125	6514	142
3	45	71	155	6553	183
4	46	74	155	6681	164
5	46	104	126	6817	138
6	48	106	125	7067	152
7	22	98	66	2948	78

Appendix: Sparse Cholesky Factorization

In orther to highlight the analogy between sparse network and sparse matrix factorizations, this appendix presents a few (very summarized and condensed) examples of sparse Cholesky factorization. For a complete explanation, see Stern (1994, 2008b).

Figure 8 shows the positions filled in the Cholesky factorization of a matrix A, $A = LL'$, and in the Cholesky factorization of two symmetric permutation of the same matrix, $A(q,q)$. Initial Non Zero Elements, NZEs, are represented by x, initial zeros filled during the factorization are represented by 0, and initial zeros left unfilled are represented by blank spaces.

Fig. 8 Filled Positions in Cholesky Factorization

$$
\begin{matrix} 1 \\ 2 \\ 3 \\ 4 \\ 5 \\ 6 \end{matrix}
\begin{bmatrix} 1 & x & x & & & x \\ x & 2 & x & & & 0 \\ x & x & 3 & x & & 0 \\ & & x & 4 & & 0 \\ & & & & 5 & x \\ x & 0 & 0 & 0 & x & 6 \end{bmatrix}
\begin{bmatrix} 1 & x & x & x & & \\ x & 3 & 0 & x & x & \\ x & 0 & 6 & 0 & 0 & x \\ x & x & 0 & 2 & 0 & 0 \\ & x & 0 & 0 & 4 & 0 \\ & & x & 0 & 0 & 5 \end{bmatrix}
\begin{bmatrix} 5 & & & x & & \\ & 4 & & & x & \\ & & 2 & & x & x \\ x & & & 6 & & x \\ & x & x & & 3 & x \\ & & x & x & x & 1 \end{bmatrix}
$$

The Elimination Lemma, see Stern (1994, 2008b) states that, when eliminating the j-th column in the Cholesky factorization of matrix $A(q,q) = LL'$, we fill the positions in L corresponding to the filled edges in $\mathscr{F}$ at the elimination of vertex $q(j)$.

The *elimination tree*, $\mathscr{H}$, is defined by

$$
h(j) = \begin{cases} j, \text{ if } \operatorname{nze}(L^j) = \{j\}, \text{ or} \\ \min\{i > j \mid i \in \operatorname{nze}(L^j)\} \ , \text{ otherwise} \ ; \end{cases}
$$

where $h(j)$, the parent of j in $\mathscr{H}$, is the first (non diagonal) NZE in column j of L.

Figure 9 shows the elimination trees corresponding to the example A.1.

The elimination tree portrays the dependencies among the columns for the numeric factorization process. More exactly, we can eliminate column j of A, that is, compute all the multipliers in column j and update all the elements affected by these

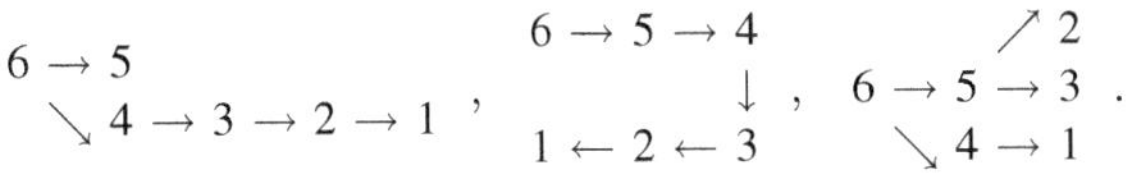

Fig. 9 Elimination Trees

multipliers, if and only if we have already eliminated all the descendents of j in the elimination tree. If we are able to perform parallel computations, we can simultaneously eliminate all the columns at a given level of the elimination tree, beginning with the leaves, and finishing at the root. For example, let us consider an elimination with the same pattern of the last permutation in 8, 9. This elimination tree has three levels that, from the leaves to the root, are: $\{1,3,2\}$, $\{4,5\}$, e $\{6\}$. Hence, the corresponding factorization requires only 2 steps, as illustrated in the following numerical example (in order to avoid taking square roots, we present the LU instead of the Cholesky factorization):

$$\begin{bmatrix} \mathbf{1} & & & 7 & & \\ & \mathbf{2} & & & 8 & \\ & & \mathbf{3} & & 6 & 9 \\ 7 & & & 53 & & 2 \\ & 8 & 6 & & 49 & 23 \\ & & 9 & 2 & 23 & 39 \end{bmatrix} \begin{bmatrix} 1 & & & 7 & & \\ & 2 & & & 8 & \\ & & 3 & & 6 & 9 \\ 7 & & & \mathbf{4} & & 2 \\ & \mathit{4} & \mathit{2} & & \mathbf{5} & 5 \\ & & \mathit{3} & 2 & 5 & 12 \end{bmatrix} \begin{bmatrix} 1 & & & 7 & & \\ & 2 & & & 8 & \\ & & 3 & & 6 & 9 \\ 7 & & & 4 & & 2 \\ & \mathit{4} & \mathit{2} & & 5 & 5 \\ & & \mathit{3} & \mathit{\frac{1}{2}} & \mathit{1} & \mathbf{6} \end{bmatrix}$$

The sparse matrix literature has many heuristics designed for finding good elimination orders. The example in Figures 10 and 11 show a good elimination order for a 13×13 sparse matrix.

The elimination order in Figure 10 was found using the Gibbs heuristic, described in Stern (1994, ch.6) or Pissanetzky (1984). The intuitive idea of Gibbs heuristic, see Figure 11, is as follows: 1- Starting from a 'peripheral' vertex, in our example, vertex 3; 2- Grow a breath-first tree $\mathscr{T}$ in $\mathscr{G}$. Notice that the vertices at a given level, l, of $\mathscr{T}$ form a separator, S_l, in the graph $\mathscr{G}$. 3- Chose a separator, S_l, that is 'small',

1	3	x											
2	x	8			x								
3			1		x							x	x
4				10	x							x	
5		x	x	x	**9**							0	
6						12				x			x
7							13			x			
8								2	x	x			
9								x	7	0			
10						x	x	x	0	**6**			0
11											11		x
12			x	x	0							**4**	0
13			x		0	x				0	x	0	**5**

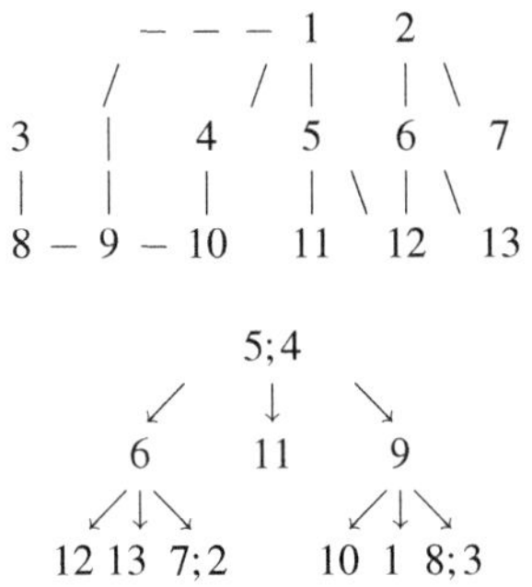

Fig. 10 Gibbs Heuristic's Elimination Order and Tree

$$
\begin{array}{cccccccccc}
 & & & 10 \rightarrow 4 & & 11 & & 13 & & \\
 & & & \nearrow & & \nearrow & & \nearrow & & \\
\mathscr{T} = & 3 \rightarrow & 8 \rightarrow & 9 \rightarrow & 1 \rightarrow & 5 \rightarrow & 12 \rightarrow & 6 \rightarrow & 2 \rightarrow & 7 \\
 & & & & & & & & & \\
l = & 1 & 2 & 3 & 4 & 5 & 6 & 7 & 8 & 9
\end{array}
$$

Fig. 11 Nested Dissection by Gibbs Heuristic

i.e. with few vertices, and 'central', i.e. dividing $\mathscr{G}$ in 'balanced' components. 4- Place in q, first the indices of each component separated by S_l, and, at last, the vertices in S_l. 5- Proceed recursively, separating each large component into smaller ones. In our example, we first use separator $S_5 = \{4,5\}$, dividing $\mathscr{G}$ in three components, $C_1 = \{3,8,1,10,9\}$ $C_2 = \{12,13,2,7,6\}$ $C_3 = \{11\}$. Next, we use separators $S_3 = \{9\}$ in C_1, and $S_7 = \{6\}$ in C_2.

References

Colla, E.C.: Aplicação de Técnicas de Fatoração de Matrizes Esparsas para Inferência em Redes Bayesianas. Ms.S. Thesis, MAC-IME-USP, Institute of Mathematics and Statistics, University of São Paulo (2007)

Cozman, F.G.: Generalizing variable elimination in Bayesian networks. In: IBERAMIA-SBIA, Workshop proceedings. São Paulo, Tec. Art, pp. 27–32 (2000)

Mandani, A., Heckerman, D., Wellman, M.P.: Real-world applications of Bayesian networks. Comm. of the ACM 38(3), 24–26 (1995)

Dechter, R.: Bucket elimination: An unifying framework for probabilistic inference. In: 12th UAI proceedings, pp. 211–219. Morgan Kaufmann Publishers, San Francisco (1996)

George, A., Gilbert, J.R., Liu, J.W.H. (eds.): Graph Theory and Sparse Matrix Computation. Springer, NY (1993)

Jensen, F.V.: An introduction to Bayesian networks. Springer, NY (1996)

Lauritzen, S.L., Spiegelhalter, D.J.: Local computations with probabilities on graphical structures and their application to expert systems. J. Royal Statistical Soc., B 50(2), 157–224 (1988)

Pearl, J.: Probabilistic reasoning in intelligent systems: Networks of plausive inference. Morgan Kaufmann, San Francisco (1988)

Pissanetzky, S.: Sparse matrix technology. Academic Press, New York (1984)

Shachter, R.: Bayes-ball: The rational pastime (for determining irrelevance and requisite information in belief networks and influence diagrams). In: 14th UAI proceedings, pp. 480–487. Morgan Kaufmann, San Francisco (1998)

Stern, J.M.: Simulated Annealing with a Temperature Dependent Penalty Function. ORSA Journal on Computing 4, 311–319 (1992)

Stern, J.M.: Esparsidade, Estrutura, Estabilidade e Escalonamento em Álgebra Linear Computacional. IX Escola de Computação. UFPE, Recife (1994)

Stern, J.M.: Decoupling, Sparsity, Randomization, and Objective Bayesian Inference. Cybernetics and Human Knowing 15(2), 49–68 (2008a)

Stern, J.M.: Cognitive Constructivism and the Epistemic Significance of Sharp Statistical Hypotheses. Tutorial book for MaxEnt, The 28th International Workshop on Bayesian Inference and Maximum Entropy Methods in Science and Engineering, July 06-11, Boracéia, São Paulo, Brazil (2008b)

Stern, J.M., Vavasis, S.A.: Nested Dissection for Sparse Nullspace Bases. SIAM Journal on Matrix Analysis and Applications 14(3), 766–775 (1993)

Stern, J.M., Vavasis, S.A.: Active Set Algorithms for Problems in Block Angular Form. Computational and Applied Mathemathics 12(3), 199–226 (1994)

van der Vorst, H.A., van Dooren, P. (eds.): Parallel Algorithms for Numerical Linear Algebra. North-Holland, Amsterdam (1990)

Zhang, N.L., Poole: Exploiting casual independence in Bayesian network inference. Journal of Artificial Intelligence Research, 301–328 (1996)

Incorporating Hopfield Neural Networks into Ant Colony System for Traveling Salesman Problem*

Yu-Lin Weng, Chou-Yuan Lee, and Zne-Jung Lee

Abstract. In this paper, the approach of incorporating Hopfield neural networks (HNN) into ant colony systems (ACS) is proposed and studied. In the proposed approach (HNNACS), HNN is used to find a plausibly good solution, which is then used in ACS as the currently best tour for the offline pheromone trail update. The idea is to deposit additional pheromone to ACS to enhance the search efficiency. From simulation results, the search efficiency of HNNACS is better than other existing algorithms.

1 Introduction

Traveling salesman problem (TSP) is a well-known combinatorial optimization problem. It is known as a classical NP-complete problem, which has an extremely large search space [2]. Some previous methods, such as simulated annealing (SA) [10] and GA [12] are also widely employed to solve TSP. Even though those approaches could find the best solution in those simulated cases, the search efficiency did not seem good enough. As a consequence, a novel approach that combines Hopfield neural networks (HNN) and ant colony system (ACS) is proposed and has been shown to have excellent search performance in this paper.

---

* This research was partly supported by the National Science Council, Taiwan, R.O.C. under grant NSC 97-2221-E-211 -017.

Yu-Lin Weng
Dept. of Information Management, ST. Mary's Medicine Nursing and Management College
e-mail: ylweng@smc.edu.tw

Chou-Yuan Lee
Dept. of Information Management, Lan Yang Institute of Technology
e-mail: yuan@orion.ee.ntust.edu.tw

Zne-Jung Lee
Dept. of Information Management, Huafan University
e-mail: johnlee@hfu.edu.tw

K. Nakamatsu et al. (Eds.): New Advan. in Intel. Decision Techno., SCI 199, pp. 287–294.
springerlink.com

Artificial ants are capable of mimicking the behavior of real ants [1]. In an ACS [2], ants are capable of exploring and exploiting possible solutions with the use of pheromone information left on the ground when they traversed. Each ant collects pheromone information and problem characteristics for building solutions individually and the pheromone information will also be modified to accumulate experience of this ant. With such mechanisms, this multi-agent (ant) algorithm is capable of solving optimization problems efficiently [3]. However, ACS may still suffer from inefficiency in various applications. HNN is a parallel network and can be used to solve optimization problems as shown in [6]-[9]. Nevertheless, HNN often fails to converge to valid solutions and may give solutions far from the known optimum [9] due to its greedy nature. Many researchers attempted to improve the performance of HNN based on linear stability analyses [6]. In this paper, we consider HNN as an initialization mechanism for ACS to improve ACS search efficiency. In the proposed approach, HNN can act as an initialization mechanism for ACS and such an algorithm is referred to as HNNACS. HNN is used to obtain a plausibly good solution, which is used for offline pheromone updating rules to deposit additional pheromone on the solution path. The experimental results indeed show significantly superior performance of those proposed approach over original ACS algorithms and other algorithms.

This paper is organized as follows. After this introduction section, Section 2 briefly describes how TSP is coded into HNN. In Section 3, the basics of ACS are introduced. In Section 4, the proposed approach to solve TSP is described. Section 5 shows the experimental results. Finally, Section 6 concludes the paper.

2 Formulation of Hopfield Neural Networks for Traveling Salesman Problem

An HNN is a dynamic system and can be written as [9].

$$\frac{du_i}{dt} = \sum_{j \neq i} w_{ij} s_j + I_i, \qquad s_i = f(u_i), \tag{1}$$

where $f(u_i)$ is the activation function of neuron u_i, s_i is the output of u_i and will be feedback to all neurons except its own, w_{ij} is the weight of connection between the i-th and j-th neurons and I_i is the external input [9]. Notice that the output of each neuron is fed back to the input of all the other neurons except its own input. The behavior of an HNN can be completely described by a Lyapunov function E which is called the Hopfield energy function [9]:

$$E = -\frac{1}{2} \sum_i \sum_{j \neq i} s_i s_j w_{ij} - \sum_i s_i I_i. \tag{2}$$

It is easy to verify that this is a Lyapunov function. It can also be found that $\frac{du_i}{dt} = -\frac{E}{ds_i}$. Assuming that the weight matrix is symmetric ($w_{ij}=w_{ji}$) and that the neurons are asynchronously updated, it can be proven that the Lyapunov function converges to a local minimum without oscillating between different states [9]. In other words, the state of an HNN converges and the found state corresponds to a local minimum of the Lyapunov function.

The basic approach of solving TSP with an HNN can be stated as follows. The output of all neurons represents a solution tour. The weights of the HNN are chosen so that the Hopfield energy function measures the length of a tour and the lowest energy state corresponds to the optimal tour [9].When the HNN evolves in time, the energy function decreases monotonically until a stable steady state is reached. This steady state corresponds to a locally optimal tour. In the following paragraphs, we shall describe how a tour is represented by one state of a set of neurons and how the weights of the network are determined.

An appropriate Hopfield energy function and the corresponding weights are defined. The following energy function is used in [9]:

$$E = \frac{V_1}{2}\sum_x\sum_i\sum_{j\neq i} s_{xi}s_{xj} + \frac{V_2}{2}\sum_i\sum_x\sum_{y\neq x} s_{xi}s_{yi} + \frac{V_3}{2}\left(\sum_x\sum_i s_{xi} - N\right)^2 + \frac{V_4}{2}\sum_x\sum_{y\neq x}\sum_i d_{xy}s_{xi}(s_{y,i+1}+s_{y,i-1}), \tag{3}$$

where V_1, V_2, V_3 and V_4 are positive coefficients [8] and all indices in Eq. (3) are defined in modulo N. For convenience, double-indices are used. For instance, s_{xi} denotes the output of the neuron corresponding to city x and tour position i. The distance between cities x and y is d_{xy}. The first three terms of the energy function correspond to the constraints of TSP and measure the degree of whether the solution is a valid path. Assume that the Hopfield net reaches a steady state with outputs only either '0' or '1'. The first term is zero if and only if each row has not more than one '1', which is the constraint that each city is allowed to be visited only once. The second term is zero if and only if each column has not more than one '1', which corresponds to the constraint that each position of the tour can only be occupied by one city. The third term is to meet the constraint that exactly N cities are visited on the tour. The final term penalizes tours with long distances. This term is directly proportional to the length of the tour [9]. Hence, the state with the lowest energy corresponds to the optimal tour.

Equation (3) can easily be transformed into the form of the Hopfield energy function as shown in Equation (2). Define

$$w_{xi,yj} = -V_1\delta_{xy}(1-\delta_{ij}) - V_2\delta_{ij}(1-\delta_{xy}) - V_3 - V_4 d_{xy}(\delta_{j,i+1}+\delta_{j,i-1}) \tag{4}$$

where δ_{ij} is 1 ($i=j$) or 0 (otherwise). Then, the energy function in Equation (3) can be rewritten as

$$E = -\frac{1}{2}\sum_{x}\sum_{y}\sum_{i}\sum_{j} w_{xi,yj} s_{xi,yj} - C\sum_{x}\sum_{i} Ns_{xi} + \frac{C}{2}N^2 \tag{5}$$

It can be shown that this is a Lyapunov equation [9]. This energy function is similar to that described in Equation (2) and then can easily be implemented as the Hopfield energy function for an HNN.

3 Ant Colony System

In the algorithms, three updating rules are required. They are the state transition rule, the local updating rule, and the global updating rule. The state transition rule is also referred to as the Ants generation and activity, the local updating rule is online updating rule, and the global updating rule is the offline updating rule. Considering an ant at time t positioned at node r, the state transition rule is to define the next node as s, which is governed by

$$s = \begin{cases} \arg\{\max\limits_{u=allowed_k(t)} [\tau_{ru}(t)\ \eta_{ru}^{\beta}]\} & when(q \le q_0) \\ S & otherwise \end{cases} \tag{6}$$

where $\tau_{ru}(t)$ is the pheromone trail at time t, $_{ru}$ is the problem-specific heuristic information, β is a parameter representing the importance of heuristic information, q is a random number uniformly distributed in [0,1], q_0 is a pre-specified parameter with $0 \le q_0 \le 1$, $allowed_k(t)$ is the set of feasible nodes currently not visited by ant k at time t, and S is an index of a node selected from $allowed_k(t)$ according to the probability distribution governed by

$$P_{rs}^{k}(t) = \begin{cases} \dfrac{\tau_{rs}(t)\ \eta_{rs}^{\beta}}{\sum\limits_{u \in allowed_k(t)} \tau_{ru}(t)\ \eta_{ru}^{\beta}} & \text{if } s \in allowed_k(t) \\ 0, & \text{otherwise;} \end{cases} \tag{7}$$

In this implementation, the heuristic information $_{ru}$ is set as the inverse of the distance between two cities.

While constructing feasible solutions, ants perform online step-by-step pheromone updates according to the following rule:

$$\tau_{ij}(t+1) \leftarrow (1-\psi)\tau_{ij}(t) + \psi\tau_0 \tag{8}$$

where ψ is a constant with $0<\psi\le 1$, $\tau_0 = (mF_{nn})^{-1}$ is the initial value of pheromone trails, m is the number of ants, and F_{nn} is the value of the tour length produced by nearest neighbor heuristic [5]. Equation (8) is the modification process of the pheromone trails. The pheromone value can either increase as ants deposit pheromone on the paths they traversed, or decrease because of pheromone evaporation for non-visited paths. In ACS, the offline updating rule is often used to collect useful global information by depositing additional pheromone. In this rule, only the paths in the currently best solution (i.e., the found shortest tour during the search up to now) are allowed to deposit pheromone. Some algorithms may activate a local search process to improve the current solutions before performing offline pheromone trail update. When the currently best solution is obtained, the offline pheromone trail update is performed for those paths in the solution as

$$\tau_{ij}(t+1) = (1-\rho)\,\tau_{ij}(t) + \rho\Delta\tau_{ij}(t) \tag{9}$$

Where $0<\rho \leq 1$ is a parameter governing the pheromone decay process, $\Delta\tau_{ij}(t)=1/C^{best}$, and C^{best} is the length of the currently best solution from the beginning of the search process. It is noted that there are also various modifications for ACS proposed in the literature. Nevertheless, this study intends to report the idea of incorporating HNN into ACS. Thus, in our implementation, we did not include those approaches so as to have a clear view on the effects of incorporating HNN into ACS.

4 The Proposed Algorithm

In the proposed approach, HNN is used to find a plausibly good solution (locally optimum), which is then used in Ant colony Systems (ACS) as the currently best tour for the offline pheromone trail update. The idea is to deposit additional pheromone to ACS to enhance the search efficiency and such an algorithm is referred to as HNNACS. The idea of HNNACS is to use HNN in finding plausibly good solutions and the found solutions are deposited with pheromone by the offline pheromone trail update rule. The proposed algorithm HNNACS is given in the following.

Procedure: The proposed approach HNNACS
Begin
Initialize s_{xi} at random and set parameters V_1, V_2, V_3, V_4 of HNN;
$t \leftarrow 0$;
While (not converge to generate feasible solution) **do**
    Conduct HNN to obtain a feasible solution ***H*** and calculate the length of the found solution as C^{hnn};
  If a feasible solution ***H*** is found
      Apply offline pheromone updating rule by using the found best solution;
   $t \leftarrow t+1$;
  End;
  While (ACS is not stopped) **do**
    While (not all ants have generated solutions) **do**
        Apply ant's generation and activity to generate solutions of each ant;
      Apply the online pheromone updating rule for each ant;
    End;
    Apply the offline pheromone updating rule;
  End;
  Find out the best path and minimum length of tour;
End;

In the first phase of HNNACS, a binary vector s_{xi} is randomly generated as the initial vector for an HNN and is evaluated via the energy function in Equation (5). Then, the HNN is in function to obtain a feasible solution. The obtained solution is deposited with pheromone by the offline pheromone trail update rule. Even though the selection of initial vectors may have influence on the search performance, all vectors used can have nice convergent behaviors in our experiment. We shall report the results in the next section. After the solution $\boldsymbol{H}$ for HNN is obtained and the length of $\boldsymbol{H}$ is calculated as C^{hnn}, the offline pheromone trail update is performed to provide clues for exploiting possibly good solutions in ACS. The solution $\boldsymbol{H}$ is used as the currently optimal solution for ACS and thus, is used to update the pheromone trail by Equation (9) with $\Delta\tau_{ij}(t) = 1/C^{\mathrm{hnn}}$. In the second phase, three main processes of ACS are performed in the sequence of ant's generation and activity, online updating rule, and offline updating rule. In other words, traditional ACS as stated in Section 3 is employed.

5 Experimental Results

In our implementation, the parameters of ACS used are ρ=0.1, ψ=0.1, q_0=0.7, β=2, the number of ants are 20 and the maximum number of ACS iterations=50000. The parameters in HNN are set as V_1=2500, V_2=500, V_3=200, V_4=0.1, and the HNN lasts iterations is 1000 with 10 sets initial vectors. The initial vector used is s_{11} to run HNN in HNNACS for all tested problems. Each HNN is provided with a random initial vector s_{xi} denoting the output of the neuron corresponding to the city x and tour position i. The TSP examples conducted here are obtained from the TSPSLIB website [4]. Several search algorithms are considered for comparison. The used search algorithms are Simulated Annealing (SA), Genetic Algorithm (GA), and traditional ACS. For fair comparison, the local search mechanism is also adopted in GA [13] and in ACS. Local search can explore the neighborhood in an attempt to enhance the fitness value of the solution in a local manner. Also, GA with 2-opt local search (GA-2-opt) [11][14] and ACS with 2-opt local search (ACS-2-opt) are also included in the comparison. Note that the results are all averaged over 10 trials to smooth out the randomness bearing in the algorithms conducted.

Since those algorithms are search algorithms, it is not easy to stop their search in a fair basis from the algorithm. In our study, those algorithms are simply stopped after a fixed time period of running. Experiments in C language were conducted on PCs with P4 processors, and were stopped after two hours of running. The results are listed in Table 1. From Table 1, it is easy to see that the proposed algorithms HNNACS have the best solution in those three tested instances. While performing offline pheromone trail update, the proposed algorithm becomes greedy to pass this useful global information by depositing additional pheromone. Consequently, the search efficiency of HNNACS can significantly better than other existing algorithms.

Table 1 Results of various search algorithms when experiments are stopped after two hours of running

Algorithms	kroA100 The best solution (21282)	Ch130 The best solution (6110)	kroA200 The best solution (29368)
SA	23038	6712	32412
GA	21941	6448	29744
GA with general local search	21764	6441	29644
GA-2-opt	21742	6436	29632
GA with HNN as local search	21733	6122	29385
ACS	21688	6131	29571
ACS with general local search	21426	6122	29561
ACS-2-opt	21404	6120	29546
ACS with SA as local search	21298	6118	29382
The proposed algorithm HNNACS	**21282**	**6110**	**29368**

6 Conclusions

In this paper, incorporating Hopfield neural networks into ACS for TSP is proposed. It is to use HNN to find a local optimal solution and to use it for pheromones trail updating once in ACS. The idea is simple, but the resultant performance is admirable. From our experiments, it is clearly evident that proposed algorithm can have superior performance over existing algorithms.

References

[1] Dorigo, M., Caro, G.D.: Ant colony optimization: A new meta-heuristic. In: Proceedings of the 1999 Congress on Evolutionary Computation, vol. 2, pp. 1470–1477 (1999)

[2] Gambardella, L.M., Dorigo, M.: Solving Symmetric and Asymmetric TSPs by Ant Colonies. In: Proceedings of 1997 IEEE International Conference on Evolutionary Computation (ECEC 1996), pp. 622–627. IEEE Press, Los Alamitos (1996)

[3] http://www.iwr.uni-heidelberg.de/groups/comopt/software/TSPLIB95/

[4] Stűtzle, T., Dorigo, M.: Ant Colony Optimization. MIT Press, Cambridge (2004)

[5] Wilson, G.V., Pawley, G.S.: On the stability of the traveling salesman problems algorithm of Hopfield and Tank. Biological Cybernetics 58, 63–70 (1988)

[6] Hopfield, J.J., Tank, D.W.: Neural computation of decisions in optimization problem. Biological Cybernetics 52, 141–152 (1985)

[7] Talavan, P.M., Yanez, J.: Parameter setting of the Hopfield network applied to TSP. Neural Networks 15, 363–373 (2002)

[8] Matsuda, S.: Optimal Hopfield network for combinatorial optimization with linear cost function. In: IEEE Int. Symp. Circuits Systems, pp. 2181–2184 (May 1989)

[9] Wang, L., Smith, K.: On chaotic simulated annealing. IEEE Trans. Neural Networks 9(4), 716–718 (1998)

[10] Aarts, E.H.L., Lenstra, J.K.: Local Search in Combinatorial Optimization. Wiley, New York (1997)

[11] Kolen, A., Pesch, E.: Genetic local search in combinatorial optimization. Discrete Applied mathematics and Combinatorial Operations Research and Computer Science 48, 273–284 (1994)

[12] Lee, Z.-J., Su, S.-F., Lee, C.-Y.: Efficiently Solving General Weapon-Target Assignment Problem by Genetic algorithms with Greedy Eugenics. IEEE Trans. Systems, Man and Cybernetics, Part B, 113–121 (2003)

[13] Goldberg, D., Lingle, R., Alleles: Loci and traveling salesman problem. In: Proceedings of the Second International Conference on Genetic Algorithms. Lawerence Eribaum Associated, Mahwah (1985)

Analysis and Diagnosis of Cardiovascular Diseases through the Paraconsistent Annotated Logic

João Inácio da Silva Filho, Gilberto A.T.A. Holms, Gabriel V. Hurtado, and Dorotéa V. Garcia

Abstract. All over the world cardiovascular diseases are responsible for a great number of clinical problems that result in death. Arterial Hypertension is considered one of the main agents of such diseases and, for that reason, the correct detection of Hypertension symptoms is very important for the medical area. In this work we use the Paraconsistent Logic (PL) as a method of treatment of uncertainties to help health professionals that act in this medical area. With base in the concepts of the Paraconsistent Logic we created a Expert System for the help in the analysis and diagnosis in cardiovascular diseases. This computational system is called "Para*Hyper* Analyzer" and it uses Algorithms of the Paraconsistent Logic, a non-classical kind of logic, for the treatment of contradictory information. The algorithms are obtained through the interpretation of the representative Lattice of Paraconsistent Annotated Logic with annotation of two values (PAL2v), where the certainty degree (D_C) and the contradiction degree (D_{ct}) are calculated, according to the logical methodology. The PAL2v algorithms are called Paraconsistent Analysis Nodes (PAN). The PAN allows the analysis of propositions which are related to changes in the blood pressure, so that the patient can be classified in a certain risk group. The Para*Hyper* Analyzer studies this information and presents a result which shows the possibility of developing a serious cardiovascular disease.

Keywords: cardiovascular diseases, expert systems, paraconsistent logic, hypertension diagnosis.

1 Introduction

According to the World Health Organization (WHO), Cardiovascular Diseases (CVD) cause a great number of morbidity, mortality and disability in the

João Inácio da Silva Filho, Gilberto A.T.A. Holms, Gabriel V. Hurtado and Dorotéa V. Garcia
UNISANTA-Santa Cecília University – Applied Paraconsistent Logic Group
Oswaldo Cruz street, 266,
11045-100, Santos, SP, Brazil
e-mail: `jinacsf@yahoo.com.br`

K. Nakamatsu et al. (Eds.): New Advan. in Intel. Decision Techno., SCI 199, pp. 295–303.
springerlink.com

industrialized countries. They are also bound to become, in the next decade, the major cause of mortality in developing countries. In Brazil, 27.4% of deaths occurred in 2003 were caused by Cardiovascular Diseases and this index can reach 37% when in these deaths we are not considering those in which the cause was violence [9]. The increase of CVD in developed regions is a result of the intensity and duration of the exposure to the risk situations. Urbanization changes have contributed to populace cardiovascular risk profile. A team of professionals of a variety of medical areas has considered the concept of "cardiovascular risks" as the main factor of heart disease incidents. Through the analysis of individual risk factors, it is possible to foresee events and introduce prevention programs, vigilance and health promoting actions for life improvement [8][9].

1.2 SCORE of Cardiovascular Risk Factors

Besides the genetic predisposition, environmental factors can contribute to the cardiovascular risk factors in families that keep poor health habit [6]. Among hypertensive individuals, this combination, can lead to physical inactivity, overweight, hyperglycemia and hyperlipidemia. As several studies, the Obesity is another strong risk factor for the cardiovascular diseases [9].

The Framingham Heart Study [9]quantifies the impact of measurable and changeable risk factors – such as age, gender, total cholesterol, HDL-C and blood pressure levels, smoking and presence of diabetes mellitus – on the development of coronary heart disease.

We present, in this work, an analysis model that can be used to generate an estimation of coronary heart disease whose information is very useful for developing an appropriate treatment and also motivate patients to change their behavior.

2 Paraconsistent Logic

In general, the systems of analysis, learning and recognition in the area of Artificial Intelligence, use the conventional logic, in which the world's description is considered by two states: False or True. This binary system does not deal with contradictory situations. However, the contradictions or inconsistencies are common when we describe parts of the real world and the Paraconsistent Logic has been created to find ways of dealing with those situations.

The studies of Paraconsistent Logic have presented results which permitted the consideration of inconsistencies [1][2][4]. Therefore, it is more appropriate to solve problems caused by contradictory situations when we work with the real world.

Based on theoretical concepts of the Paraconsistent Logic, with this work we present a Paraconsistent Expert System named Para*Hyper*. This Paraconsistent Expert System analyzes the possible contradictions among received information, being capable of informing the user about the risk of cardiovascular diseases.

2.1 The Paraconsistent Annotated Logic with Annotation of Two Values – PAL2v

The Paraconsistent Annotated Logic (PAL) [2][4] is a class of evidential logic that treats the signals represented by annotations. Each annotation, belongs to a finite lattice and attributes a value for the corresponding proposition *p*. A Paraconsistent Annotated Logic can be represented as a finite lattice of "four states" (figure 1a), where the propositional sentence will be followed by an Evidence degree. Therefore, an Annotated sentence associated to the Lattice of the Paraconsistent Annotated Logic can be read in the following way:

$P_{(T)}$ ==> the annotation or Evidence Degree T assigns a connotation of 'inconsistency' to the proposition *p*.

$P_{(t)}$ ==> the annotation or Evidence Degree *t* assigns a connotation of 'truth' to the proposition p.

$P_{(F)}$ ==> the annotation or Evidence Degree *F* assigns a connotation of 'falsehood' to the proposition *p*.

$P_{(\perp)}$ ==> the annotation or Evidence Degree ⊥ assigns a connotation of 'indefinity' to the proposition *p*.

The Paraconsistent Annotated Logic with annotation of two values (PAL2v) is an extension of PAL and it can be represented through a Lattice of four vertexes where we can determine some terminologies and conventions, as following:

Be $\tau = < |\tau|, \leq >$ a fixed finite lattice [2][3], where:

1. $|\tau| = [0, 1] \times [0, 1]$
2. $\leq = \{((\mu_1, \rho_1), (\mu_2, \rho_2)) \in ([0, 1] \times [0, 1])^2 \mid \mu_1 \leq \mu_2 \text{ e } \rho_1 \leq \rho_2\}$
   (where ≤ indicates the usual order of the real numbers).

The orderly pair's first element represents the Evidence Degree that is favorable to the proposition *P*, and the second element represents the Evidence Degree that is unfavorable or contrary to the same proposition. This way, the intuitive idea of the association of an annotation to a proposition $P_{(\mu, \lambda)}$ means that the evidence degrees favorable to *P* is μ, while the evidence degree unfavorable or contrary to *P* is λ.

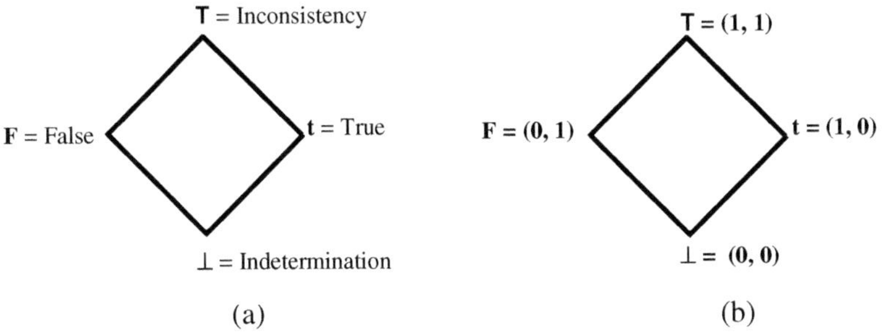

Fig. 1 Finite Lattice of the PAL2v four states with values

Through the methods presented in [4], a system using the Paraconsistent Annotated Logic receives signals of information in the form of degrees of evidence with values that vary from 0 to 1. These values can be placed in two representing axes of finite lattice (figure 2a) were are calculate the Certainty Degree D_C and the Contradicton Degree D_{ct} through the equations:

$$D_{ct} = \mu_1 + \lambda - 1 \quad (1)$$

and:

$$D_c = \mu_1 - \lambda \quad (2)$$

The Contradiction Degree value D_{ct} is an indicative of the inconsistency measure and the certainty degree value D_C is considered as the result of the analysis.

3 Paraconsistent Analysis Nodes – PAN

The element capable of treating a signal made up of one degree of favorable evidence and another of unfavorable evidence (μ_{1a}, μ_{2a}), and provide in its output a Resulting Evidence Degree, is called basic Paraconsistent Analysis Node (PANb). Figure 2(b) shows the representation of a PANb with two inputs of evidence degree:

μ_1 = favorable Evidence Degree of information source 1
λ = unfavorable Evidence Degree

Where:

$\lambda = 1 - \mu_2$
μ_2 is a favorable Evidence Degree of information source 2

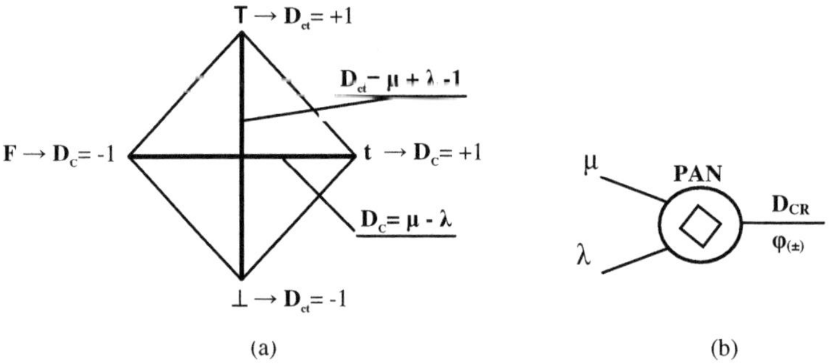

Fig. 2 Finite Lattice of PAL2v and Symbol of the Paraconsistent Analyzer Node PAN

A lattice description uses the values obtained by the equation results in the Paraconsistent Analyzer Node Algorithm [4] that can be written in a reduced form, as follows:

1. Enter with the input values

μ */ favorable evidence Degree $0 \leq \mu \leq 1$

λ */ unfavorable evidence Degree $0 \leq \lambda \leq 1$

2. Calculate the Contradiction Degree

$D_{ct} = (\mu + \lambda) - 1$

3. Calculate the Interval of Certainty

$\varphi = 1 - |D_{ct}|$

4. Calculate the Certainty Degree

$D_C = \mu - \lambda$

5. Calculate the distance *d* into Lattice

$$d = \sqrt{(1 - |D_C|)^2 + D_{ct}^2}$$

6. Compute the output signal

If $\varphi \leq 0{,}25$ *or* $d \geq 1$ *Then do* $S1 = 0.5$ *and* $S2 = \varphi$:

Indefinite logical state and

go to the steep 10

Or else go to the next step

7. Calculate the real Certainty Degree

If $D_C > 0$ $D_{CR} = (1 - d)$

If $D_C < 0$ $D_{CR} = (d - 1)$

8. Compute the signal of the Interval of Certainty

If $\mu + \lambda > 1$ Signal positive $\varphi_{(\pm)} = \varphi_{(+)}$

If $\mu + \lambda < 1$ Signal negative $\varphi_{(\pm)} = \varphi_{(-)}$

If $\mu + \lambda = 1$ Signal zero $\varphi_{(\pm)} = \varphi_{(0)}$

9. Present the outputs

Do $S1 = D_{CR}$ and $S2 = \varphi_{(\pm)}$

10. End

When PANs are interconnected in a network, the next lines are included in the Algorithm, in order to calculate the output Evidence Degree, as follows:

9.a Calculate the real Evidence Degree

$$\mu_{ER} = \frac{D_{CR} + 1}{2}$$

10.a Present the outputs

Do $S1 = \mu_{ER}$ and $S2 = \varphi_{(\pm)}$

11.a End

The Systems with the Paraconsistent Analysis Nodes (PAN) deal with the received signals through algorithms, showing the signals with a Certainty Degree Value and a Contradiction Degree Value in the output [4].

4 The Para*Hyper* Analyzer

Many works use the Expert Systems to study the correct diagnosis of cardiovascular diseases and this is an important goal for the Artificial Intelligence and Public Health fields. Within this researching area, the current work proposes an Expert

System that uses a new methodology based on non-classical logic, introducing new procedures for dealing with complex conflicting information, as it is required in Medical diagnosis. The Para*Hyper* Risk Rating attempts to allow a complete overview of the patient's state. For that, the Para*Hyper* generates a cardiovascular estimation disease risk, being an useful tool for the development of anticipated treatments.

The Para*Hyper* Expert System is fed by two types of information:

1 - Relevance to Risks;

2 - Blood pressure measurements.

To create the Expert System, the risk factors were classified into three groups. This classification is presented above (figure 3).

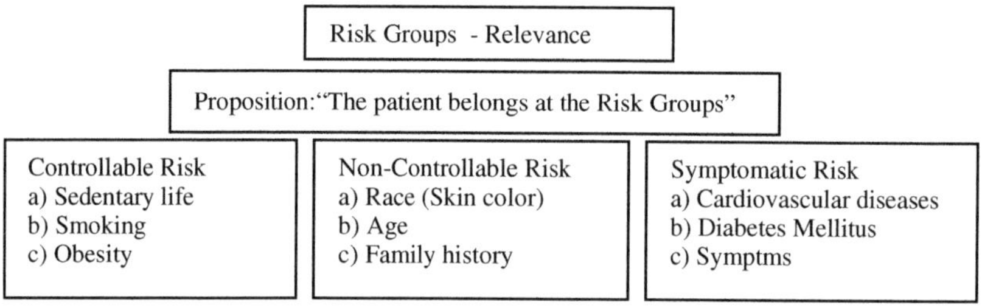

Fig. 3 Risk classification used in the System

The modeling of these values, the Proposition and the Universe of Discourse come from a knowledge basis. In this case, official data from the World Health Organization is used, which classifies obesity against BMI (Body Mass Index) as shown in Table 1 in figure 4. The Degrees of Evidence are obtained through modeling of risk-related sources that may cause cardiovascular diseases in patients and by blood pressure measurement. For instance, the variation of the Degrees of Evidence in contrast to the Body Mass Index in men is shown in figure 4.

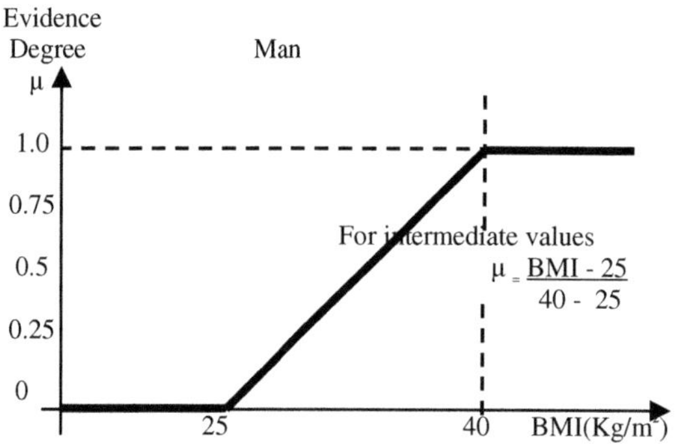

Table 1 - Classification of Obesity

	Male
Underweight	< 20
Normal	from 20 to 24.9
Light	from 25 to 29.9
Moderate	from 30 to 39.9
Serious	> 40

Fig. 4 Classification of Obesity for men and variation of the Evidence Degree against the Body Mass Index (BMI)

These values are inserted into the Para*Hyper* Expert System and analyzed by several PANs linked in a Paraconsistent network. Blood pressure measurement is modeled by establishing a Universe of Discourse and is divided into Diastolic and Systolic.

In figure 5 a part of the Paraconsistent network evaluates the evidences so as to provide a diagnosis related to the risks of cardiovascular diseases.

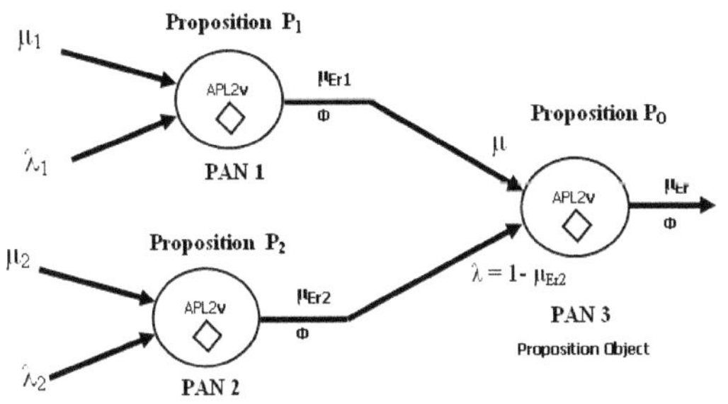

Fig. 5 Three interlinked PANs in a Paraconsistent network of analyses

The complete Para*Hyper* has about 30 PANs with several different configurations.

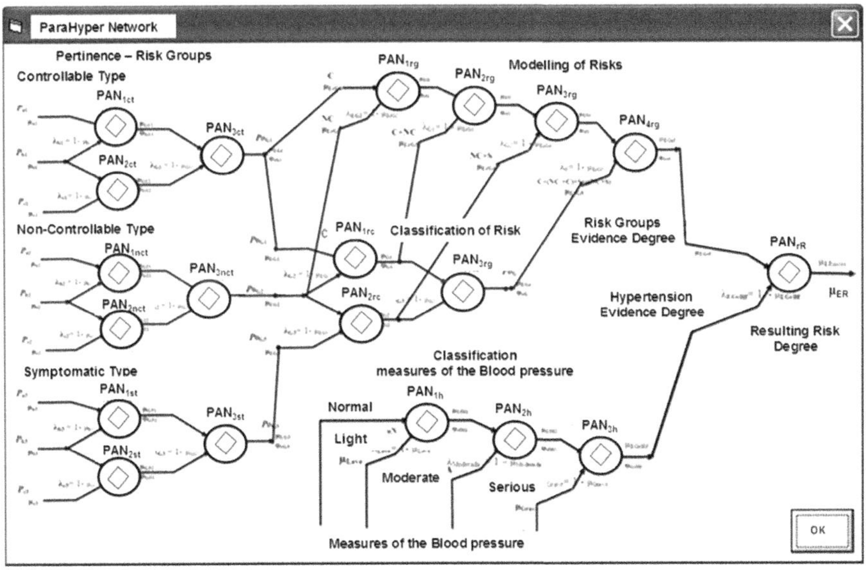

Fig. 6 Complete Network composites of the interlinked PANs

After the classification and logical analysis in the patient's data the Evidence Degree is extracted giving the information about the patient's inclusion in the Risk Groups.

The values of the blood pressure measurements are also analyzed and classified, we, then, obtain the Degree of Evidence about the symptoms of Hypertension. These two values are analyzed by the last algorithm that will give the Degree of final resulting Evidence.

5 Conclusions

Cardiovascular diseases are a concern to Public Health authorities all over the world due to the high number of people affected by them. Information about cardiovascular risk factors are uncertain and come in a combined form. It is very difficult to define a specific score to be used as a pattern in computational tools for support in diagnoses and treatments. A way to solve this problem is to treat uncertain information using techniques of Artificial Intelligence. For that reason, we created the Para*Hyper*, a software that enables the user to diagnose the propensity for the development of cardiovascular diseases through the use of clinical and laboratory data. The Para*Hyper* was built based on the Paraconsistent Logic, and this expert system can generate reliable scores even with contradictory information. This software easily accepts modifications and adjustments to be put to use in several situations, such as for children and adolescents. Despite being a first version, the initial tests have shown good results. After the conclusion of the software, the next phase will be the Para*Hyper* validation. Through the adjustment and normalization of inputs, as evidence degrees, the Para*Hyper* will be capable to provide answers through the score format, being able to determine risks for cardiovascular disease. Based on these first results, the Para*Hyper* is currently being improved so we can get its scores closer to the diagnosis made by specialists in this medical area.

References

1. Bobrie, G., Chatellier, G., Genes, N., et al.: Cardiovascular prognosis of "masked hypertension" detected by blood pressure self-measurement in elderly treated hypertensive patients. In: JAMA 2004, vol. 17, 291, pp. 1342–1349 (2004)
2. Da Costa, N.C.A., Abe, J.M., Subrahmanian, V.S.: Remarks on annotated logic. Zeitschrift f. math. Logik und Grundlagen d. Math. 37, 561–570 (1991)
3. Da Costa, N.C.A.: On Theory of Inconsistent Formal Systems. Notre Dame J. of formal Logic 15, 497–510 (1974)
4. Da Silva Filho, J.I.: Métodos de interpretação da Lógica Paraconsistente Anotada com anotação com dois valores LPA2v com construção de Algoritmo e implementação de Circuitos Eletrônicos., in Portuguese PhD thesis, University of São Paulo, POLI/USP - São Paulo (1999)
5. Foran, T.G., Sheahan, N.F., Cunningham, C., Feely, J.: Pseudo-hypertension and arterial stiffness: a review. Physiol. Meas. 25, R21–R33 (2004)

6. Knuiman, M.W., Divitini, M., Welborn, T.A., Bartholomew, H.C.: Familial correlations, cohabitation effects, and heritability for cardiovascular risk factors. Ann. Epidemiol. 6, 188–194 (1996)
7. Lewington, S.C., Qizilbash, N., et al.: Age-speci. c relevance of usual blood pressure to vascular mortality: a meta-analysis of individual data for one million adults in 61 prospective studies. Lancet. 360, 1903–1913 (2003)
8. Mancia, G., Rosei, A., DeBaker, G., et al.: European Society of Hypertension-European Society of Cardiology guidelines for the management of arterial hypertension. J. Hyperten. 21, 1011–1053 (2003)
9. O'Brien, E., Coats, A., Owens, P., et al.: Use and interpretation of ambulatory blood pressure monitoring: recommendations of the British Hypertension Society. BMJ 320, 1128–1134 (2000)

Hierarchical Forecasting with Polynomial Nets

M.S. Lauretto, F. Nakano, C.A.B. Pereira, and J.M. Stern

Abstract. This article presents a two level hierarchical forecasting model developed in a consulting project for a Brazilian magazine publishing company. The first level uses a VARMA model and considers econometric variables. The second level takes into account qualitative aspects of each publication issue, and is based on polynomial networks generated by Genetic Programming (GP).

Keywords: Genetic programming, Functional trees, Forecasting, Logistics, Meta-control, Polynomial networks.

1 Introduction

This article describes the authors' consulting project for a leading Brazilian magazine publishing company, nicknamed ABC, and its associated distributor company, nicknamed DE. One of the major logistic challenges of this business is the classic newsstand, newsvendor or newsboy problem, asking for optimal inventory levels. The standard operations research models for this problem assume fixed prices and random demand, see Hadley and Whitin (1963) and Denardo (1982). The inventory levels are then optimized in order to minimize the costs of being either over or under stocked.

The cost of overstock is captured by a well known Brazilian proverb stating that "a day-old newspaper is only good for wrapping fish". Unfortunately, old magazines do not even have that use. In most cases, only the cover page of unsold magazines are stripped and sent back some way along the distribution channel for control purposes, while the rest is recycled at the nearest paper factory. The immediate cost of under stock is lost sales. The long term costs of under stock include customer frustration, possibly leading to permanent fidelity or loyalty transfer to another magazine, low visibility, loss of mind and market share, etc.

M.S. Lauretto, F. Nakano, C.A.B. Pereira, and J.M. Stern
University of São Paulo, Brazil
e-mail: {lauretto,nakano}@ime.usp.br

K. Nakamatsu et al. (Eds.): New Advan. in Intel. Decision Techno., SCI 199, pp. 305–315.
springerlink.com

The distributor company, DE, deals with this problem at several hierarchical levels through the distribution channels, stocking and possibly restocking one or more times from large and small regional depots to individual newsstands. In this article the word newsstand is used as a generic name, encompassing point of sales ranging from street kiosks to supermarket or bookstore shelves.

The first step of the newsstand problem is to decide the print (or press) runs, that is, to determine the number of copies printed at each batch. Usually a magazine issue stays at the newsstand from one week to one month, and there is no time to reprint an issue.

Most magazines are printed in sections, typically of 16 pages, which are assembled within a cover and bounded. Sections containing articles and advertisement planned and written far ahead can be printed in advance, while the cover and other sections, containing articles referring to current events, are printed in a tight schedule. This process allows for substantial savings in the production costs, resulting in a complex operation that requires careful planning.

The optimization aspects of the problem are going to be reported elsewhere; at this article we focus on demand forecasting. The demand is generated by subscribers, newsstands and a small reserve for the back issue service. The number of subscribers is relatively stable over time, posing little challenge for accurate forecasting. In contrast, the newsstands demand is very sensitive to current events, specific aspects of individual issues, and current marketing efforts.

The forecasting tool developed at this consulting project uses a two level hierarchical approach. The first level uses a VARMA (vector auto-regressive moving average) model, see Brockwell and Davis (1991). The VARMA model is based on econometric variables like subscription and newsstand price, minimum, average or typical wage or income of the target populations, seasonal effects, number of days in the newsstand, delay between the release date of an issue and typical payday(s), etc. This first level gives good predictions for average sales, but can be improved to more accurately predict local fluctuations.

The second level of the hierarchical model is based on polynomial networks, an instance of general functional networks briefly described in section 2. This level takes into consideration the qualitative aspects specific to individual issues like the (quality of the) cover story, cover celebrity, cover photo, editorial content, point of sale advertising, national/regional marketing, promotional gifts, etc.

2 Functional Trees

This section presents an overview of some theoretical aspects of GP used in the synthesis of functional trees, including a few topics related to the authors' current research.

Functional tree methods are used for finding the specification of a complex function. This complex function must be composed recursively from a finite set of primitive functions or operators, $OP = \{op_1, op_2, \ldots op_p\}$, and from a set of atoms, $A = \{a_1, a_2, \ldots\}$. The k-th operator, op_k, takes a specific number, $r(k)$, of arguments, also known as the *arity* of op_k. We use three representations for (the value returned by) the operator op_k computed on the arguments $x_1, x_2, \ldots x_{r(k)}$:

$$op_k(x_1, \ldots x_{r(k)}) \ , \quad \begin{array}{c} op_k \\ / \quad\quad \backslash \\ x_1 \ \ldots \ x_{r(k)} \end{array} \ , \quad \left(op_k \, x_1 \ \ldots \ x_{r(k)}\right) \ .$$

The first is the usual form of representing a function in mathematics; the second is the tree representation which displays the operator with branches to its arguments; and the third is the prefix, preorder or LISP style representation, which is a compact form of the tree representation.

As a didactical example, let us consider the *Boolean network* specification problem, that is, the specification of a Boolean function of q variables, $f(x_1, \ldots x_q)$, to match a target table, $g(x_1, \ldots x_q)$, see Angeline (1996) and Banzhaf el al. (1998). The primitive set of operators and atoms for this problem are the standard ones used in classical logic:

$$OP = \{\sim, \wedge, \vee, \rightarrow, \odot, \otimes\} \ \text{and} \ A = \left\{x_1, \ldots x_q, 0, 1\right\} \ .$$

Notice that while the first operator, *not*, is unary, the last five, *and, or, imply, nand, xor*, are binary. Also, this set of Boolean operators is clearly redundant. Notice, for example, that all other operators can be synthesized using only the nand operator. This redundancy may, nevertheless, facilitate the search for the best configuration in the problem's functional space.

Figure 1 shows a target table, $g(a,b,c)$. As it is usual when the target function is an experimentally observed variable, the target function is *not* completely specified. Unspecified values in the target table are indicated by the don't-care symbol, $*$. The two solutions, f_1 and f_2, match the table in all specified cases. Solution f_1, however, is simpler and for that may be preferred (by some parsimony principle).

Starting from a given random tree, one can start a stochastic search in the problem's (topological) space. In Genetic Programming (GP) terminology, the individual's functional specification is called its *genotype*. the individual's expressed behavior, or computed solutions, is called its *phenotype*. Changing a genotype to a neighboring one is called a *mutation*. The quality of a phenotype, its performance, merit or adaptation, is measured by a *fitness* function. GP does not look at the evolution of a single individual, but rather at the evolution of a population. A time parameter, t, indexes the successive generations of the evolving population. In GP, individuals typically have short lives, surviving only a few generations before dying. Meanwhile, populations may evolve for a very long time.

In GP an individual may, during its ephemeral life, share information, that is, swap copies of its (partial) genome, with other individuals. This genomic sharing process is called *sex*. In GP an individual, called a *parent*, may also participate in the creation of a new individual, called its *child*, in a process called *reproduction*. In the reproduction process, an individual gives (partial) copies of its genotype to its offspring. Reproduction involving only one parent is called asexual, otherwise it is called sexual.

Sexual reproduction can be performed by crossover, with parents giving (partial) copies of their genome to their children. Figure 1b shows two parents and a child generated by a single crossover, for the Boolean problem considered in the last example. The tree representation indicates the crossover points by broken edges (=). Notice that in this example the child corresponds to a solution presented earlier. For further details see Stern (2008) and also Banzahf et al. (1998) and Goldberg (1989).

A	B	C	G	$F1$	$F2$
0	0	0	1	1	1
0	0	1	1	1	1
0	1	0	*	1	0
0	1	1	*	1	0
1	0	0	0	0	0
1	0	1	1	1	1
1	1	0	0	0	0
1	1	1	1	1	1

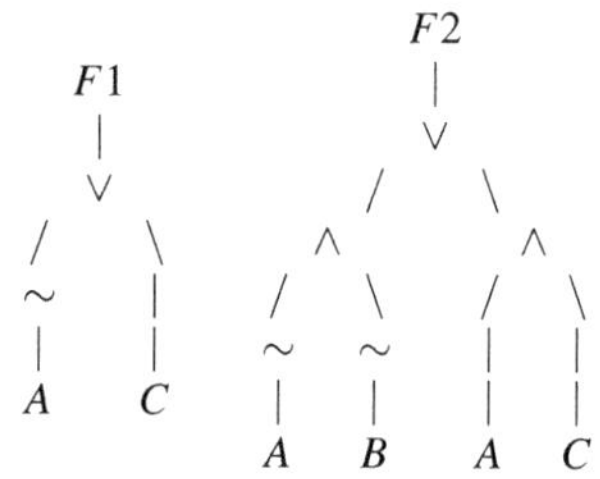

$$f_1 = (\sim a) \vee c\ ,\quad f_2 = (\sim a \wedge \sim b) \vee (a \wedge c)\ .$$

$$f_1 = (\vee\ (\sim a)\ c)\ ,\quad f_2 = (\vee\ (\wedge\ (\sim a)\ (\sim b))\ (a \wedge c))\ .$$

Fig. 1a Two Boolean functional trees for the target $g(a,b,c)$

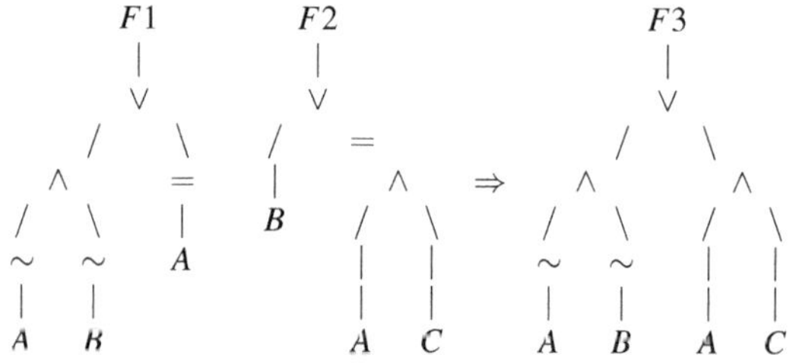

Fig. 1b Crossover between Boolean functional trees

Let us now consider the *polynomial network* specification problem in the consulting case at hand. These functional trees use as primitive operators *linear, quadratic* or *cubic* polynomials in one, two or three variables. Two auxiliary operators are defined as follows: A *normalizer* converts its input into an ouput of mean 0 and variance 1, and a *denormalizer* performs the inverse transformation.

Figure 2 displays a typical network for sales forecast used in the consulting project described in sections 1 and 3. Variable x_5 is the magazine's sales forecast obtained by a VARMA time series model. Variables x_1 to x_4 are qualitative variables, in a scale corresponding to approximate decile ranks of Bad (0-1), Weak (1-3), Average (3-7), Good (7-9) and Excellent (9-10). This scale is used by experts to assess the appeal or attractiveness of each individual issue of the magazine, according to:

(1) cover impact; (2) editorial content; (3) promotional items; and (4) point of sale marketing. Normalizers at the input edges and a denormalizer at the output edge of the polynomial network are not shown in the figure.

Of course, the optimization of a polynomial network is far more complex than the optimization of a Boolean network: Even having specified the network topology (identification problem), also the parameters $w_0, w_1, \ldots$ of the polynomial function have to be optimized (estimation problem). Parameter optimization can be based on recursive sub-tree regression; gradient, Partan or conjugate-gradient learning rules, etc. Topology optimization is based on GP algorithms.

Some aspects of GP relevant to the forecasting problem at hand and related to the authors' current research are briefly discussed in the following subsections.

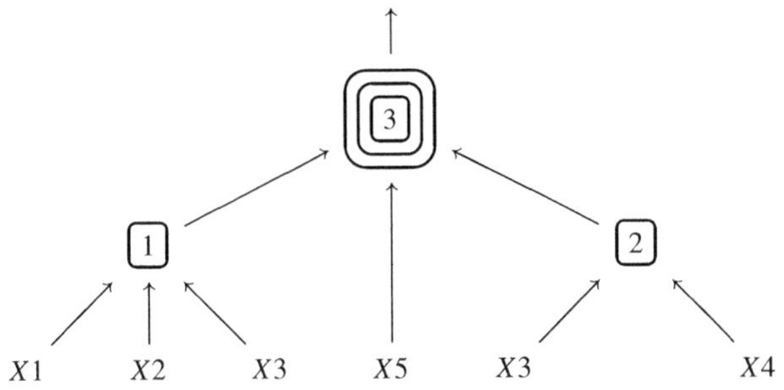

Fig. 2 Polynomial network. Rings on a node: 1- Linear; 2- (incomplete) Quadratic; 3- (incomplete) Cubic

2.1 Meta-control, Building Blocks and Modularity

P. Angeline (1996) noted that GP generated networks typically contain large segments of *extraneous* code, that is, code segments that, if removed, do not (significantly) alter the solution computed by the network. Trivial examples of extraneous code segments are $(+\,s\,0)$ and $(*\,s\,1)$, where s is a sub-expression. By their very definition, extraneous code segments cannot (significantly) contribute to an individual's fitness, and hence to its survival or mating probabilities. However, Angeline noticed that the presence of extraneous code could significantly contribute to the expected fitness of the individual's descendents! Apparently, the role of these (sometimes very large) patches of inert code is to isolate important blocks of working code, and to protect these *modules* or *building blocks* from being broken at recombination (destructive crossover).

In biological organisms, the genetic code of eukaryotes exhibits similar regions of code (DNA) that are or are not expressed in protein synthesis; these regions are called *exons* and *introns*, respectively. Introns do not directly code amino-acid sequences in proteins; nevertheless, they seem to have an important role in the meta-control of the genetic material expression and reproduction.

Iba and Sato (1992, p.548), in a pioneering work, proposed a meta-level strategy for GP based on a self-referential representation, where

> *"[a] self-referential representation maintains a meta-description, or meta-prescription, for crossover. This meta-genetic descriptions are allowed to co-evolve with the gene pool. Hence, genetic and meta-genetic code variations are jointly selected. How well the genetic code is adapted to the environment is translated by the merit or objective function which, in turn, is used for the immediate, short-term or individual selection process. How well the genetic and meta-genetic code are adapted to each other impacts on the system's evolvability, a characteristic of paramount importance in long-run survival of the species."*

Subsequent work of several authors (including these) tried to incorporate meta-control parameters to GP. Functional trees may, for example, incorporate edge annotations, interpreted as probability weights, linkage compatibility or affinity, etc. Such annotations are meta-parameters used to control the recombination of the sub-tree directly bellow a given edge. For example, weights may be used to specify the probability that recombination takes place at that edge, while linkage compatibility or affinity annotations may be used to identify homologous or compatible genes, specifying the possibility or probability of swapping two sub-trees. Other annotations, like context labels, semantic tags, variable type, etc., may provide additional information about the possibility or probability of recombination or crossover, the need of type-cast operations, etc.

Banzahf (1998, p.164), gives a simple example of functional tree annotation:

> *"Recently, we introduced the explicitly defined introns (EDI) into GP. An integer value is stored between every two nodes in the GP individual. This integer value is referred as the EDI value (EDIV). The crossover operator is changed so that the probability that crossover occurs between any two nodes in the GP program is proportional to the integer value between the nodes. That is, the EDIV integer value strongly influences the crossover sites chosen by the modified GP algorithm, Nordin et al. (1996).*
>
> *The idea behind EDIVs was to allow the EDIV vector to evolve during the GP run to identify the building blocks in the individual as an* emergent *phenomenon. Nature may have managed to identify genes and to protect them against crossover in a similar manner. Perhaps if we gave the GP algorithm the tools to do the same thing, GP, too, would learn how to identify and protect the building blocks. If so, we would predict that the EDIV values within a good building block should become low and, outside the good block, high."*

Let us finish this section presenting two interpretations for the role of modularity in genetic evolutionary processes. This interpretations are common in biology, computer science and engineering, an indication that they provide powerful insights. These two metaphors are commonly referred to as:

- New *technology dissemination* or *component design substitution*, and
- *Damage control* or *repair mechanism*.

The first interpretation is perhaps the more evident. In a modular system, a new design for an old component can be easily incorporated and, if successful, be rapidly disseminated. A classical example is the replacement of mechanical carburetors by electronic injection as the standard technology for this component of gasoline automotive engines. The large assortment of *upgrade kits* available in any automotive or computer store gives a strong evidence of how much these industries rely on modular design.

The second interpretation explains the possibility for the "continued evolution of germlines otherwise destined to extinction", see Michod and Levin (1988). A classic illustration related to the damage control and repair mechanisms offered by modular organization is given by the Hora and Tempus parable of Simon (1996), see also Growney (1998). The lessons learned from this parable may be captured by the following dicta of Herbert Simon:

> *"The time required for the evolution of a complex form from simple elements depends critically on the number and distribution of potential intermediate stable subassemblies."* Simon (1996, p.190).
> *"Hierarchy, I shall argue, is one of the central structural schemes that the architect of complexity uses."* Simon (1996, p.184).

2.2 Schemata and Parallelism

The *intrinsic parallelism* argument, first presented in Holland (1975), provides alternative insights into the concepts of building blocks and modularity. For a mathematical analysis of this argument, see Reeves (1993, Ch.4) or Stern (2008, H.2). According to Reeves,

> *"The underlying concept Holland used to develop a theoretical analysis of his GA [GP] was that of* schema. *The word comes from the past tense of the Greek verb* $\varepsilon\chi\omega$, echo, *to have, whence it came to mean shape or form; its plural is* schemata." (p.154)

Schemata are partially specified patterns in a program, like partially specified segments of prefix expressions, or partial code for functional sub-trees. The *length* and *order* of a schema are the *distance* between the first and last defined position on the schema, and the number of defined positions, respectively. The intrinsic parallelism theorem states that the number of schemata (of order l and length $2l$, in binary coded programs, in individuals of size n) present in a population of size m, is proportional to m^3. The crossover operator enriches the neighborhood of an individual with the schemata present in other individuals of the population. If, as suggested by the implicit parallelism theorem, the number of such schemata is large, GP is likely to be an effective strategy. Schaffer (1987, p.89), celebrates this theorem stating that:

> *"this [intrinsic parallelism] constitutes the only known example of combinatorial explosion working to advantage instead of disadvantage."*

Indeed, Schaffer has ample reasons to praise Holland's result. Nevertheless, we must analyze this important theorem carefully, in order to understand its consequences correctly. In particular, we should pay close attention to the unit, u, used to measure the population size, m. It so happens that this unit, $u = 2^l$, is itself exponential in the schemata order. Therefore, the combinatorial explosion works to our advantage as long as we use short schemata, relative to the log-size of the population. This situation is described by Reeves as:

> *"Thus the ideal situation for a GA [GP] are those where short, low-order schemata combine with each other to form better and better solutions. The assumption that this will work is called by Goldberg (1989) the* building-block hypothesis. *Empirical evidence is strong that this is a reasonable assumption in many problems."* (p.158)

One key question we must face in order to design a successful GP application is, therefore: How then can we organize our working space so that our programming effort can rely on short schemata?

The solution to this question is well known to computer scientists and software engineers: Organize the programs hierarchically (recursively) as self-contained (encapsulated) building-blocks (modules, functions, objects, sub-routines). As shown in the previous subsection, meta-contol is a key mechanism for modular organization in GP, promoting the spontaneous emergence of complex hierarchical systems.

For further details about the implementation of these concepts in the optimization of polynomial networks, see Nikolaev and Iba (2001, 2003, 2006). For further implications of these ideas in artificial intelligence and statistical modeling, see Stern (2008, ch.5) and the authors' forthcoming articles.

3 Example of Forecasting

In this section we return to our consulting project, and present a case study based on time series for a magazine published by ABC. Due to confidentiality and disclosure agreements, the time series given in this example has been de-trend and is presented in a relative percentage scale. Forecasts are always made three month ahead with the current past data. From the total of 39 months comprising the time series, the first 27 were used as training data, and the remaining 12 months as test data.

The first level of the hierarchical model consists of a VARMA model, implemented in order to capture trends, seasonal effects and market elasticities, built using automated variable selection procedures, see Brockwell and Davis (1991). The available explanatory variables include the dates of distribution and recall of each issue, its price for subscription and at the newsstand, minimum, average or typical wage or income of the target populations, typical payday schedules, etc.

Figure 3 presents the actual sales time series and the sales forecasts provided by the VARMA econometric model (top), as well as the models improved by qualitative data, using linear regression (center) and the polynomial network in Figure 2 (bottom). Table 1 shows the average error rates for VARMA, linear regression and polynomial network models. Notice that error rates provided by polynomial networks are smaller than in VARMA and linear regression models. The optimal polynomial network selection is guided by a regularization parameter, ρ, controlling the network complexity vs. training error. Its default value is $\rho = 1.0$. As expected, if ρ is too large, the network becomes too simple, resulting in an under-fitted model. On the other hand, if ρ is too small, the network becomes too complex, resulting in an

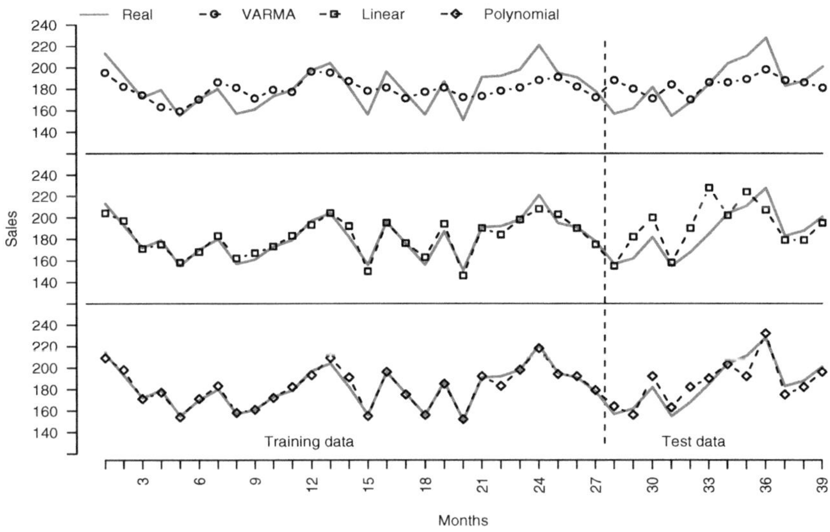

Fig. 3 Monthly sales and corresponding forecasts with VARMA models (top), linear regression (center) and polynomial network (bottom)

Table 1 Error averages using VARMA, Linear reg. and Polynomial networks

Model	Training dataset	Test dataset
VARMA alone	6.3%	8.6%
+ Linear regression	4.1%	7.4%
+ Polynomial network, $\rho = 0.5$	1.1%	5.1%
+ Polynomial network, $\rho = 1.0$	2.5%	4.2%
+ Polynomial network, $\rho = 2.0$	3.4%	4.6%

over-fitted model (over adjusted to the peculiarities of the training data). In either case, the network has low predictive (generalization) power.

4 Enterprise Integration

As in so many Operations Research projects, solving the mathematical and algorithmic aspects of the optimization and statistical models, and its computational implementation in a user friendly decision support tool, is just part of the entire consulting project. Training the corporate decision makers to use the tools and carefully explaining the concepts involved is also an essential part of the project. All those are prerequisites to the vital goal of integrating the new OR tools into the everyday life of the enterprise. Otherwise, the full benefits of the project are never achieved or, even worst, the new fancy tools are soon condemned to oblivion.

ABC is organized in business units according to major target populations, for example: children, including comic books; male teens; female teens; women, including arts, house and garden, gossip, etc.; men, including cars, computers, sports, swim suite, etc; business and economy; and general news.

ABC's business units were often evaluated by their total sales, market share, and other performance indices that do not take into account production and distribution costs. Meanwhile, DE and the printing plants were often evaluated by their operating costs, regardless of the global company performance. Needless to say, such evaluation metrics generated conflict and misunderstanding inside the company.

The primary objective of the statistical and optimization tools developed in this consulting project was to improve the quantitative fine tuning of the operation, and the project successfully accomplished this goal. However, the project could also make significant contributions to a secondary objective, namely, to improve the cooperation, integration, rational dialogue and mutual understanding concerning the different roles played by the several agents in such a complex operation. We hope that, in the future, it will also contribute for the development of more encompassive performance metrics, capable of harmonizing and integrating locally conflicting goals into global multi-objective functions.

References

Angeline, P.: Two Self-Adaptive Crossover Operators for Genetic Programming. In: Angeline, Kinnear (eds.) Advances in Genetic Programming, ch. 5, vol. 2, pp. 89–110. MIT, Cambridge (1996)

Banzahf, W., Nordin, P., Keller, R.E., Francone, F.D.: Genetic Algorithms (1998)

Brockwell, P.J., Davis, R.A.: Time Series: Theory and Methods, 2nd edn. Springer, Heidelberg (1991)

Denardo, E.: Dynamic Programming. Prentice-Hall, Englewood Cliffs (1982)

Farlow, S.J.: Self-Organizing Methods in Modeling. Marcel Dekker, New York (1984)

Goldberg, D.E.: Genetic Algorithms in Search, Optimization, and Machine Learning. Addison-Wesley, Reading (1989)

Growney, J.S.: Planning for Interruptions. Mathematics Magazine 55(4), 213–219 (1998)

Hadley, G., Whitin, H.M.: Analysis of Inventory Systems. Prentice-Hall, Englewood Cliffs (1963)

Holland, J.H.: Adaptation in Natural and Artificial Systems. University of Michigan Press (1975)

Iba, H., Sato, T.: Meta-Level Strategy for Genetic Algorithms Based on Structured Representation. In: Proc. of the Second Pacific Rim Int. Conf. on Artificial Intelligence, pp. 548–554 (1992)

Madala, H.R., Ivakhnenko, A.G.: Inductive Learning Algorithms for Complex Systems Modeling. CRC, Boca Raton (1994)

Michod, R.E., Levin, B.R.: The Evolution of Sex. An Examination of Current Ideas. Sinauer (1988)

Nikolaev, N.Y., Iba, H.: Regularization Approach to Inductive Genetic Programming. IEEE Transactions on Evolutionary Computation 5(4), 359–375 (2001)

Nikolaev, N.Y., Iba, H.: Learning Polynomial Fedforward Neural Networks by Genetic Programming and Backpropagation. IEEE Transactions on Neural Networks 14(2), 337–350 (2003)

Nikolaev, N.Y., Iba, H.: Adaptive Learning of Polynomial Networks. In: Genetic and Evolutionary Computation. Springer, Heidelberg (2006)

Reeves, C.R.: Modern Heuristics for Combinatorial Problems. Blackwell Scientific, Malden (1993)

Schaffer, J.D.: Some Effects of Selection Procedures on Hyperplane Sampling by Genetic Algorithms. In: Davis, L. (ed.) Genetic Algorithms and Simulated Annealing, Pittman, pp. 89–103 (1987)

Simon, H.A.: The Sciences of the Artificial. MIT Press, Cambridge (1996)

Stern, J.M.: Cognitive Constructivism and the Epistemic Significance of Sharp Statistical Hypotheses. In: MaxEnt 2008, The 28th International Workshop on Bayesian Inference and Maximum Entropy Methods in Science and Engineering, Boracéia, São Paulo, Brazil, July 6-11 (2008)

Autonomous Mobile Robot Emmy III

Claudio Rodrigo Torres, Jair Minoro Abe, Germano Lambert-Torres,
João Inácio Da Silva Filho, and Helga Gonzaga Martins

Abstract. This work presents some improvements regarding to the autonomous mobile robot Emmy based on Paraconsistent Annotated Evidential Logic Eτ. A discussion on navigation system is presented.

Keywords: Automation, paraconsistent logic, robotics, navigation system, logic controller.

1 Introduction

It is well known the use of non-classical logics in automation and robotics. In real applications, classical logic is inadequate for several reasons. The main point is that all concepts of real world encompass some imprecision degree. In order to overcome these limitations, several alternative systems were proposed. Maybe the most successful non-classical system is the so-called Fuzzy set theory [16]. In this work we employ another promising non-classical logic, namely the paraconsistent annotated systems. They've inspired applications in a variety of themes. Particularly in robotics, it was built some interesting autonomous mobile robots that can manipulate imprecise, inconsistent and paracomplete data. One of the robot series dubbed Emmy[1], based on a particular annotated system, namely, the paraconsistent annotated evidential logic Eτ [1], began with the 1st prototype studied in [2], [3]. Subsequently, some improvements were made in its 2nd prototype Emmy II [4] and in this paper we sketch the 3rd prototype discussing a navigation system.

Claudio Rodrigo Torres
Universidade Metodista de São Paulo, São Bernardo do Campo, SP, Brazil
e-mail: c.r.t@uol.com.br

Claudio Rodrigo Torres, Germano Lambert-Torres, and Helga Gonzaga Martins
UNIFEI – Federal University of Itajubá, Itajubá, GAIA - Artificial Intelligence Application Group, MG, Brazil

Jair Minoro Abe
Institute For Advanced Studies – University of São Paulo, SP, Brazil

João Inácio Da Silva Filho
Universidade Santa Cecília – UNISANTA, Santos, SP, Brazil

[1] The name Emmy is in homage to the mathematician Emmy Nöether (1882-1935). Such name was proposed by N.C.A. da Costa and communicated to J.M. Abe in 1999, University of Sao Paulo.

K. Nakamatsu et al. (Eds.): New Advan. in Intel. Decision Techno., SCI 199, pp. 317–327.
springerlink.com

2 Paraconsistent, Paracomplete, and Non-alethic Logics

In what follows, we sketch the non-classical logics discussed in the paper, establishing some conventions and definitions.

Let *T* be a theory whose underlying logic is *L*. *T* is called inconsistent when it contains theorems of the form *A* and ¬*A* (the negation of *A*). If *T* is not inconsistent, it is called *consistent*. *T* is said to be *trivial* if all formulas of the language of *T* are also theorems of *T*. Otherwise, *T* is called *non-trivial*. When *L* is classical logic (or one of several others, such as intuitionistic logic), *T* is inconsistent if *T* is trivial. So, in trivial theories the extensions of the concepts of formula and theorem coincide. Paraconsistent *logic* is a logic that can be used as the basis for inconsistent but non-trivial theories. A *theory* is called *paraconsistent* if its underlying logic is a paraconsistent logic. Issues such as those described above have been appreciated by many logicians. In 1910, the Russian logician Nikolaj A. Vasil'év (1880-1940) and the Polish logician Jan Łukasiewicz (1878-1956) independently glimpsed the possibility of developing such logics. Nevertheless, Stanislaw Jaśkowski (1996-1965) was in 1948 effectively the first logician to develop a paraconsistent system, at the propositional level [9]. His system is known as 'discussive propositional calculus'. Independently, some years later, the Brazilian logician Newton C.A. da Costa (1929-) constructed for the first time hierarchies of paraconsistent propositional calculi C_i, $1 \leq i \leq \omega$ of paraconsistent first-order predicate calculi (with and without equality), of paraconsistent description calculi, and paraconsistent higher-order logics (systems NF_i, $1 \leq i \leq \omega$). Also, independently of Da Costa [10], David Nelson (1918-2003) [11] has considered a paraconsistent logic as a version of his known as constructive logics with strong negation.

Nowadays, paraconsistent logic has established a distinctive position in a variety of fields of knowledge.

Another important class of non-classical logics are the paracomplete logics. A logical system is called *paracomplete* if it can function as the underlying logic of theories in which there are formulas such that these formulas and their negations are simultaneously false. Intuitionistic logic and several systems of many-valued logics are paracomplete in this sense (and the dual of intuitionistic logic, Brouwerian logic, is therefore paraconsistent).

As a consequence, paraconsistent theories do not satisfy the principle of non-contradiction, which can be stated as follows: of two contradictory propositions, i.e., one of which is the negation of the other, one must be false. And, paracomplete theories do not satisfy the principle of the excluded middle, formulated in the following form: of two contradictory propositions, one must be true.

Finally, logics which are simultaneously paraconsistent and paracomplete are called *non-alethic logics*.

3 Paraconsistent Annotated Evidential Logic Eτ

Annotated logics are a family of non-classical logics initially used in logic programming by [12]. An extensive study of annotated logics was made in [1]. Some applications are summarized in [13]. In view of the applicability of annotated logics to differing formalisms in computer science, it has become essential to study these logics more carefully, mainly from the foundational point of view.

In general, annotated logics are a kind of paraconsistent, paracomplete, and non-alethic logic. The latter systems are among the most original and imaginative systems of non-classical logic developed in the past century.

The atomic formulas of the logic $E\tau$ are of the type $p_{(\mu,\ \lambda)}$, where $(\mu, \lambda) \in [0, 1]^2$ and [0, 1] is the real unitary interval (p denotes a propositional variable). $p_{(\mu,\ \lambda)}$ can be intuitively read: "It is assumed that p's favorable evidence is μ and contrary evidence is λ." Thus, $p_{(1.0,\ 0.0)}$ can be read as a true proposition, $p_{(0.0,\ 1.0)}$ as false, $p_{(1.0,\ 1.0)}$ as inconsistent, $p_{(0.0,\ 0.0)}$ as paracomplete, and $p_{(0.5,\ 0.5)}$ as an indefinite proposition.

Also we introduce: Uncertainty Degree: $Gun(\mu, \lambda) = \mu + \lambda - 1$; Certainty Degree: $G_{ce}(\mu, \lambda) = \mu - \lambda$ $(0 \leq \mu, \lambda \leq 1)$; an order relation is defined on [0, 1]2: $(\mu1, \lambda1) \leq (\mu2, \lambda2) \Leftrightarrow \mu1 \leq \mu2$ and $\lambda1 \leq \lambda2$, constituting a lattice that will be symbolized by τ. With the uncertainty and certainty degrees we can get the following 12 output states: extreme state and non-extreme states, showed in the Table 1.

Table 1 Extreme and Non-extreme states

Extreme States	Symbol	Non-extreme states	Symbol
True	V	Quasi-true tending to Inconsistent	QV→T
False	F	Quasi-true tending to Paracomplete	QV→⊥
Inconsistent	T	Quasi-false tending to Inconsistent	QF→T
Paracomplete	⊥	Quasi-false tending to Paracomplete	QF→⊥
		Quasi-inconsistent tending to True	QT→V
		Quasi-inconsistent tending to False	QT→F
		Quasi-paracomplete tending to True	Q⊥→V
		Quasi-paracomplete tending to False	Q⊥→F

All states are represented in the next figure.

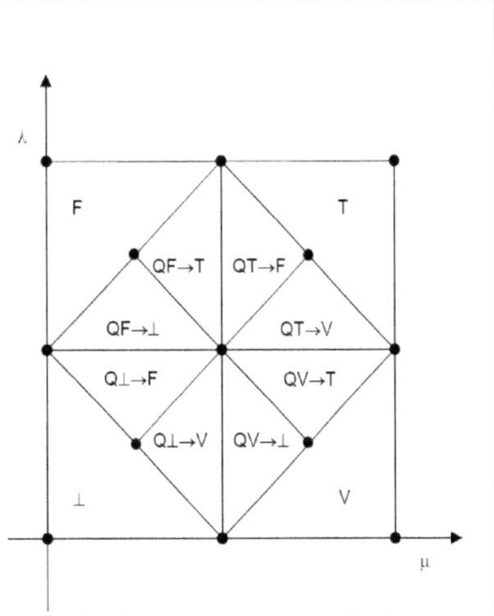

Fig. 1 Lattice τ

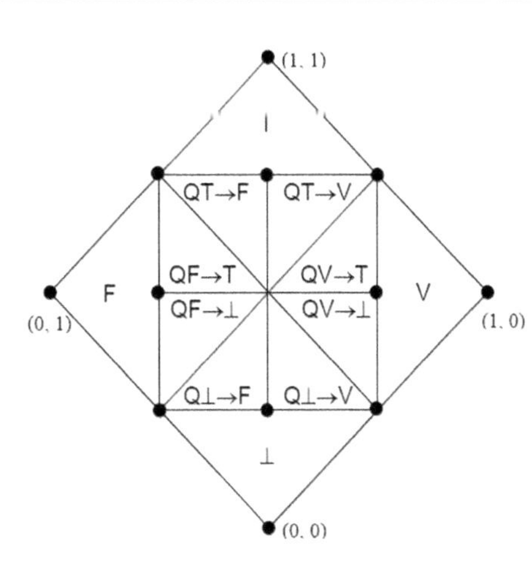

Fig. 2 Lattice τ - Hasse diagram

Some additional control values are:

V_{cic} = maximum value of uncertainty control = C_3
V_{cve} = maximum value of certainty control = C_1
V_{cpa} = minimum value of uncertainty control = C_4
V_{cfa} = minimum value of certainty control = C_2

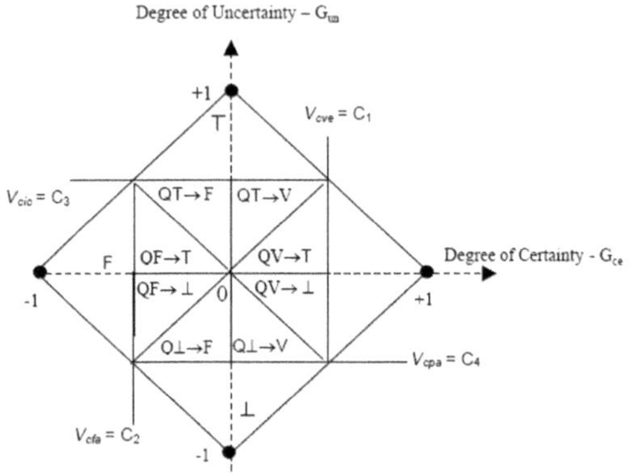

Fig. 3 Certainty and Uncertainty degrees

4 Paracontrol - Logical Controller

The Paracontrol [20] is an electronic materialization of the Para-analyzer algorithm [2], [15], which is basically an electronic circuitry, which treats logical signals in a context of logic Eτ. Such circuitry compares logical values and determines domains of a state lattice corresponding to output value. Favorable evidence and contrary evidence degrees are represented by voltage. Certainty and Uncertainty degrees are determined by analyze of operational amplifiers. The Paracontrol comprises both analogical and digital systems and it can be externally adjusted by applying positive and negative voltages. The Paracontrol was tested in real-life experiments with an autonomous mobile robot Emmy, whose favorable/contrary evidences coincide with the values of ultrasonic sensors and distances are represented by continuous values of voltage.

5 The Autonomous Mobile Robot Emmy

The controller Paracontrol was applied in this series of autonomous mobile robots. In some previous works [2], [19] is presented the autonomous mobile robot Emmy. The figure 4 shows the autonomous mobile robot Emmy. The Emmy robot consists of a circular mobile platform of aluminum 30 cm in diameter and 60 cm height. While moving in a non-structured environment the robot Emmy gets information about presence/absence of obstacles using the sonar system called Parasonic [3].

Fig. 4 The autonomous mobile robot Emmy

6 Robot Emmy II

Searching the Paracontrol, the robot Emmy controller, we perceived that the robot movements could be bettered by programming conveniently the no extreme logic state outs. This new Paracontrol version also is used to control an autonomous mobile robot named as Emmy II [4].

The platform used to assemble the Emmy II robot measures approximately 23cm of high and 25cm of diameter (circular format). The main components of Emmy II are a microcontroller from 8051 family, two ultrasonic sensors, and two DC motors. The figure 5 shows an Emmy II robot simplified block diagram.

The ultrasonic sensors are responsible for verifying whether there is any obstacle in front of the robot. The signals generated by the sensors are sent to the microcontroller. These signals are used to determine the favorable evidence degree value (μ) and the contrary evidence degree value (λ) on the proposition "The front of the robot is free".

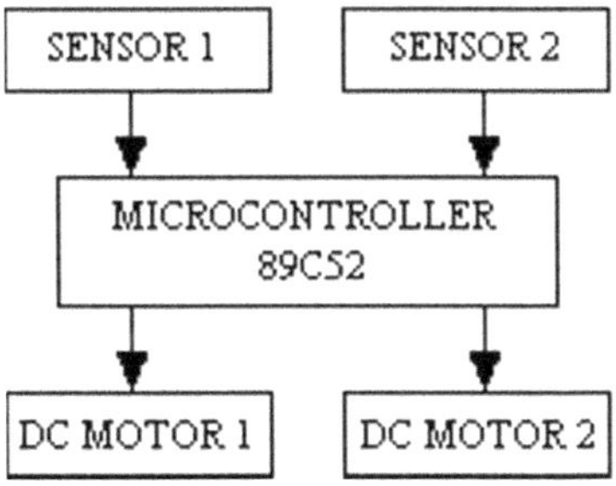

Fig. 5 Emmy II robot simplified block diagram

The Paracontrol, recorded in the internal memory of the microcontroller, uses the evidence degrees to determine the robot movements. The microcontroller is also responsible for applying power to the DC motors.

Figure 6 shows the Emmy II mechanical structure.

Fig. 6 Emmy II Mechanical Structure

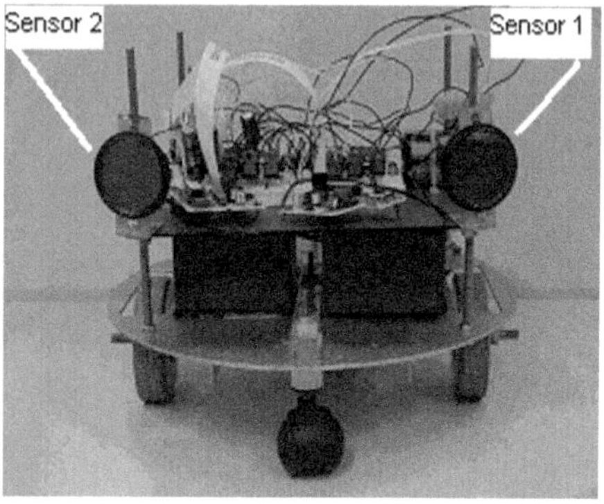

Figure 7 shows the decision state lattice that the Emmy II robot uses to determine the movement to perform.

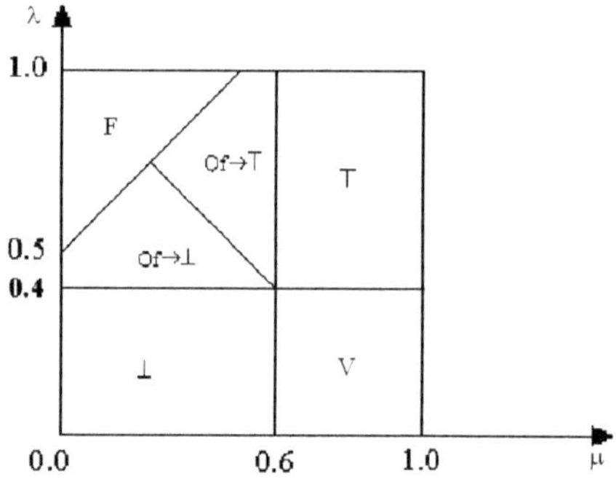

Fig. 7 Logical output lattice of Emmy II

Table 2 Logical states and action

Symbol	State	Action
V	True	Robot goes ahead
F	False	Robot goes back
⊥	Paracomplete	Robot turns right
T	Inconsistent	Robot turns left
QF→⊥	Quasi-false tending to paracomplete	Robot turns right
QF→T	Quasi-true tending to inconsistent	Robot turns left

Table 2 shows the actions related to each possible logic state. Each robot movement lasts approximately 0,4 seconds.

7 Autonomous Mobile Robot Emmy III

The aim of the Emmy III autonomous mobile robot is to be able to move from an origin point to an end point, both predetermined, in a non-structured environment. We will do it in steps. First, the robot must be able to move from a point to another in an environment without any obstacle. This environment is divided into cells [6] and a planning system gives the sequence of cells the robot must follow to reach the end cell. This idea was applied in [7], [8]. The second step is an evolving of the first step; the robot must be able to avoid cells that are supposed to have some obstacle in. A sensor system will detect the cells that have to be avoided. This sensor system will use Paraconsistent Annotated Logic to handle information captured by the sensors. The Emmy III structure is:

Sensing system. The robot's environment is composed of a set of cells. On the other hand, the sensing system has to determine the environment with enough precision, but the information captured by the sensors always has an inherent imprecision, which leads to an uncertainty regarding to the position actually the robot is in. In order to manipulate this kind of information, the sensing system is based on the Paraconsistent Annotated Evidential Logic $E\tau$, which captures the information generated by the sensors using favorable and contrary evidences degrees as seen in the logical controller Paracontrol.

Planning system. The objective is to build a planning system able to determine a path linking an initial point to an end point in a non-structured environment with some obstacles. For this, the environment is divided into cells and the planning system gives the sequence of cells that the robot starting from the initial point reaches successfully the end cell. The first step is to build a planning system for an environment without any obstacle, that is, an environment with all cells free. In the second step the sensing system informs the planning system the cells that have objects in.

Physical Construction. The Emmy III mechanical part must perform the schedule determined by the planning system. It must know the cell it is in, therefore, a monitoring position makes part of this construction. In the process, for each cell that the robot reaches, the possible error of position should be considered. In the items 8 and 9 it is described two Emmy III prototypes where a robot is able to follow a path determined by a planning system in an environment without any obstacle.

8 First Prototype of the Autonomous Mobile Robot Emmy III

The first prototype is composed of a planning system and a mechanical construction. The planning system considers an environment divided into cells. This first version considers all cells free. Then it asks for the initial point and the aimed point.

After that a sequence of movements is given on a screen. Also a sequence of pulses is sent to the step motors that are responsible for moving the physical platform of the robot. So, the robot moves from the initial point to the aimed point.

Figure 8 shows the planning system screen.

Fig. 8 Planning system screen

The physical construction of the first prototype of the Emmy III robot is basically composed of a circular platform of approximately 286 mm of diameter and two-step motors. The figure 9 shows the Emmy III first prototype. The planning system is recorded in a notebook. And the communication between the notebook and the physical construction is made through the parallel port. A potency driver is responsible to get the pulses from the notebook and send them to the step motors that are responsible for moving the robot.

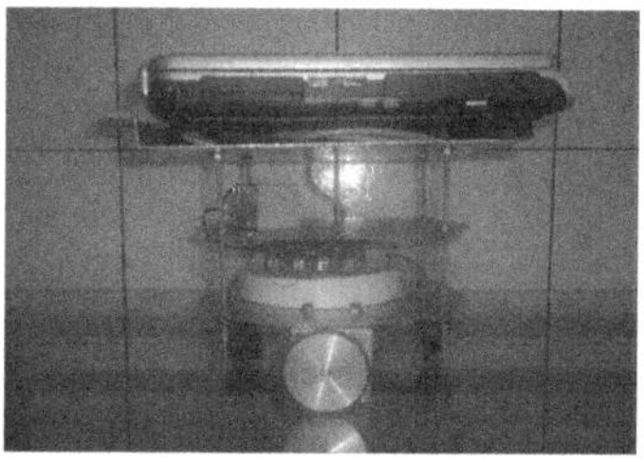

Fig. 9 The first prototype of Emmy III robot

9 Second Prototype of the Autonomous Mobile Robot Emmy III

Similarly to the first prototype, the second prototype of the autonomous mobile robot Emmy III is basically composed of a planning system and a mechanical structure. The planning system is recorded in any personal computer and the communication between the personal computer and the mechanical construction is done through a USB port. The planning system considers the environment around the robot divided into cells. So, it is necessary to inform the planning system the cell the robot is in and the aimed cell. The answer of the planning system is a sequence of cells that the robot must follow to go from the origin cell to the aimed cell.

The planning system considers all cells free. Figure 10 shows the screen of the planning system.

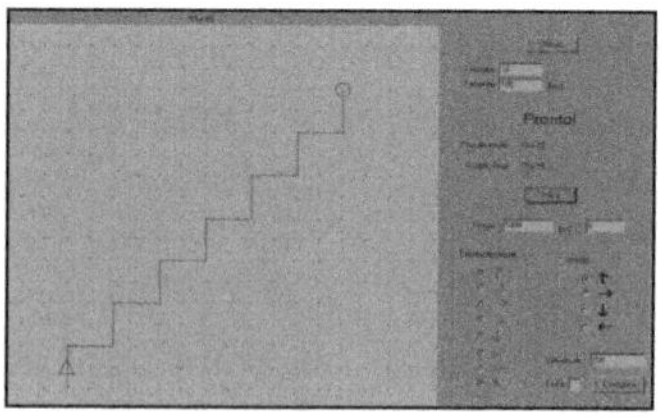

Fig. 10 The output of the planning system - EmmyIII

Figure 11 shows the mechanical structure of Emmy III second prototype.

Fig. 11 The mechanical structure of the Emmy III second prototype

The planning system considers all cells free. The mechanical construction is basically composed of a steel structure, two DC motors and three wheels. Each motor has a wheel fixed in its axis and there is a free wheel. There is an electronic circuitry on the steel structure. The main device of the electronic circuitry is the microcontroller PIC18F4550 that is responsible for receiving the schedule from the planning system and activates the DC motors. Also there is a potency driver between the microcontroller and the DC motors.

10 Conclusions

In this work we have studied the third prototype of the Emmy III autonomous mobile robot. The main concern is its navigating route planning, i.e. Emmy III is able to move from an origin point to an end point in a non-structured environment.

Two prototypes of this robot' version were built and tested. They are composed of a planning system and a mechanical structure and they were able to move from an origin point to an end point in an environment without any obstacle. Both of them had a satisfactory performance.

The next step is to build a mobile robot with the same characteristics of the described prototypes but adding a sensing system. So we expect that this new prototype will be able to move from an origin point to an end point in a non-structured environment with obstacles. The sensing system also is based on the context of Paraconsistent Annotated Evidential Logic Eτ. We hope to say more in forthcoming papers.

References

[1] Abe, J.M.: Fundamentos da Lógica Anotada (Foundations of Annotated Logics), in Portuguese, Ph. D. Thesis, University of São Paulo, São Paulo (1992)

[2] Da Silva Filho, J.I.: Métodos de Aplicações da Lógica Paraconsistente Anotada de Anotação com Dois Valores LPA2v com Construção de Algoritmo e Implementação de Circuitos Eletrônicos, in Portuguese, Ph. D. Thesis, University of São Paulo, São Paulo (1999)

[3] Abe, J.M., Da Silva Filho, J.I.: Manipulating Conflicts and Uncertainties in Robotics. Multiple-Valued Logic and Soft Computing 9, 147–169 (2003)

[4] Torres, C.R.: Sistema Inteligente Paraconsistente para Controle de Robôs Móveis Autônomos, in Portuguese, MSc Dissertation, Universidade Federal de Itajubá – UNIFEI, Itajubá (2004)

[5] Datasheet of Polaroid 6500 Series Sonar Ranging Module (1996)

[6] Elfes, A.: Using occupancy grids for mobile robot perception and navigation. Comp. Mag. 22(6), 46–57 (1989)

[7] Desiderato, J.M.G., De Oliveira, E.N.: Primeiro Protótipo do Robô Móvel Autônomo Emmy III, in Portuguese, Trabalho de Conclusão de Curso, Universidade Metodista de São Paulo, São Bernardo do Campo - SP (2006)

[8] Maran, L.H.C., Riba, P.A., Collett, R.G., De Souza, R.R.: Mapeamento de um Ambiente Não-Estruturado para Orientação de um Robô Móvel Autônomo Utilizando Redes Neurais Paraconsistente, in Portuguese, Trabalho de Conclusão de Curso, Universidade Metodista de São Paulo, São Bernardo do Campo - SP (2006)

[9] Jaskowski, S.: Um calcul des propositions pour les systems déductifs contradictoires. Studia Societatis Scientiarum Torunensis, Sect. A 1, 57–77 (1948)

[10] Da Costa, N.C.A.: On the theory of inconsistent formal systems. Notre Dame J. of Formal Logic 15, 497–510 (1974)

[11] Nelson, D.: Negation and separation of concepts in constructive systems. In: Heyting, A. (ed.) Constructivity in Mathematics, pp. 208–225. North-Holland, Amsterdam (1959)

[12] Subrahamanian, V.S.: On the semantics of quantitative Lógic programs. In: Proc. 4 th IEEE Symposium on Logic Programming. Computer Society Press, Washington D.C (1987)

[13] Abe, J.M.: Some Aspects of Paraconsistent Systems and Applications. Logique et Analyse 157, 83–96 (1997)

[14] Dacosta, N.C.A., Abe, J.M., Da Silva Filho, J.I., Murolo, A.C., Leite, C.F.S.: Lógica Paraconsistente Aplicada, Editôra Atlas, São Paulo, 214 p. (1999) ISBN 85-224-2218-4

[15] Da Silva Filho, J.I., Abe, J.M.: Paraconsistent analyzer module. International Journal of Computing Anticipatory Systems 9, 346–352 (2001)

[16] Zadeh, L.: Outline of a New Approach to the Analysis of Complex Systems and Decision Processes. IEEE Transaction on Systems, Mam and Cybernectics SMC-3(1), 28–44 (1973)

[17] Abe, J.M., Da Silva Filho, J.I.: Simulating Inconsistencies in a Paraconsistent Logic Controller. International Journal of Computing Anticipatory Systems 12, 315–323 (2002)

[18] Torres, C.R., Abe, J.M., Torres, G.L.: Sistema Inteligente Paraconsistente para Controle de Robôs Móveis Autônomos. In: Anais do I Workshop Universidade-Empresa em Automação, Energia e Materiais, November 5-6, 2004, Taubaté (SP), Brazil (2004)

[19] Da Silva Filho, J.I., Abe, J.M.: Emmy: a paraconsistent autonomous mobile robot, in Logic, Artificial Intelligence, and Robotics. In: Abe, J.M., Da Silva Filho, J.I. (eds.) Proc. 2nd Congress of Logic Applied to Technology – LAPTEC 2001, Ohmsha, Tokyo. Frontiers in Artificial Intelligence and Its Applications, pp. 53–61, 287 p. IOS Press, Amsterdam (2001)

[20] Da Silva Filho, J.I., Abe, J.M.: Para-Control: An Analyser Circuit Based On Algorithm For Treatment of Inconsistencies. In: Proc. of the World Multiconference on Systemics, Cybernetics and Informatics, ISAS, SCI 2001, vol. XVI, Cybernetics and Informatics: Concepts and Applications (Part I), Orlando, Florida, USA, pp. 199–203 (2001) ISBN 9800775560

[21] Abe, J.M., Da Silva Filho, J.I.: Simulating Inconsistencies in a Paraconsistent Logic Controller. International Journal of Computing Anticipatory Systems 12, 315–323 (2002)

Output Enhancement of Impulse Noise Filters by Edge Detection and Neuro-fuzzy Processing

Yakup Yuksel, Mustafa Alci, and M. Emin Yuksel

Abstract. A simple method for enhancing the output images of impulse noise filters for digital images is presented. The method is based on an edge detector and a simple 3-input 1-output neuro-fuzzy network. The internal parameters of the neuro-fuzzy network are adaptively optimized by training. The training is easily accomplished by using simple artificial images generated in a computer. The method can be used with any type of impulse noise filters since its operation is completely independent of the filter. The proposed method is applied to four representative impulse noise filters from the literature under different noise conditions and image properties. Results indicate that the proposed method may efficiently be used with any type of impulse noise filters to effectively reduce its distortion effects and enhance its output.

1 Introduction

Corruption of digital images by impulse noise is a frequently encountered problem in the acquisition and/or transmission of the images. In most applications, the noise has to be removed from the image data by using an appropriate filter before further processing of the image because the performances of subsequent processing stages usually depend on the success of the noise filtering stage.

A great majority of impulse noise removal methods are based on median filtering techniques, which utilize the rank order information of the pixels contained in the filtering window. The *standard median filter* [1], the *weighted median filter* [2] and the *center-weighted median filter* [3] are order statistic filters that attempt to remove impulse noise by replacing the center pixel of the filtering window with the median or a weighted median of the pixels in the window. A method for optimal design of these filters is presented in [4]. Although these filters provide a reasonable noise

Yakup Yuksel, Mustafa Alci, and M. Emin Yuksel
Department of Electrical and Electronics Engineering,
Erciyes University, Kayseri, 38039, Turkey
e-mail: yuksel@erciyes.edu.tr

K. Nakamatsu et al. (Eds.): New Advan. in Intel. Decision Techno., SCI 199, pp. 329–339.
springerlink.com

removal performance, they tend to remove thin lines, distort edges and blur fine details even at low noise densities since they are *spatially invariant* operators making no distinction between the corrupted and the uncorrupted pixels of the noisy input image.

The most straightforward way to avoid this problem is to use a *switching median filter* [5], which employs an *impulse detector* to check each pixel of the noisy input image prior to filtering to classify it as noisy or not. If the input pixel is detected as noisy, it is restored by the median filter. Otherwise, it is left unfiltered. Although this approach provides considerable improvement over the conventional median filtering, its performance naturally depends on the performance of the impulse detector. As a consequence, various impulse detection methods and switching median filter implementations have been proposed [5]-[13].

Different implementations of *mean filters* [14]-[16] are also used for impulse noise removal. These filters have been shown to exhibit better performance than some median based filters for both grey scale [14, 15] and color [16] images.

In addition to the methods discussed above, a number of nonlinear impulse noise filtering operators based on soft computing methodologies such as neural networks and fuzzy systems have been presented [17]-[21]. These filters offer relatively better noise removal and detail preservation performance than the median- and the mean-based operators at the cost of increased computational complexity. Methods based on soft computing techniques have also been adopted to improve the performances of traditional median- and the mean-based filtering methods [22]-[25]. Many different types of these approaches have been proposed for the removal of different kinds and mixtures of image noise including impulse noise.

All of the impulse noise filtering methods discussed so far more or less damage the useful data in the image during noise removal process by inevitably introducing some amount of distortion to thin lines, edges, fine details and texture in the image. Unfortunately, this inherent side effect of noise removal operators is highly undesirable in most image processing applications because preservation of useful features in the image during noise suppression is of vital importance regarding the success of subsequent image processing tasks such as boundary detection, pattern recognition, feature extraction, segmentation, object recognition, etc.

In the last few years, there has been a growing research interest in the applications of soft computing techniques, such as neural networks and fuzzy systems, to the problems in digital image processing [17]–[32]. Indeed, neuro-fuzzy (NF) systems offer the ability of neural networks to learn from examples and the capability of fuzzy systems to model the uncertainty which is inevitably encountered in noisy environments. Therefore, neuro-fuzzy systems may be utilized to efficiently reduce the distortion effects of impulse noise removal operators provided that appropriate network topologies and processing strategies are employed.

In this paper, we propose a novel method for efficiently reducing distortion effects of impulse noise filters for digital images. The method is based on a simple 3-input 1-output NF network. The internal parameters of the NF network are adaptively optimized by training. The training is easily accomplished by using simple artificial images that can be generated in a computer. The method can be used with any type of impulse noise removal operator to reduce its distortion effects. The operation of the method is completely independent of the noise removal operator and it does not interfere with the filtering behavior of the operator. The proposed method is applied to four popular impulse noise removal operators from the literature. Improvement in performance is evaluated for different test images and for different noise densities. Results indicate that the proposed method may efficiently be used with any type of impulse noise removal operator to significantly improve its filtering performance.

The rest of the paper is organized as follows: Section-2 explains the details of the proposed method and the structures of its building blocks. Section-3 discusses the application of the proposed method. Results of the experiments conducted to evaluate the performance of the proposed method for different impulse noise filters, for different noise conditions and for different image properties are reported in this section. Comparative discussion and interpretation of these results are also presented in this section. Section-4, which is the final section, presents the conclusions and remarks.

2 Method

Figure-1 shows the schematic representation of the proposed distortion reduction method. Here, the *noise filter* block represents the impulse noise removal operator whose distortion effects are to be reduced. There is no restriction on the type of the impulse noise removal operator. The proposed method can be used with any impulse noise removal operator since it does not interfere with the filtering behavior of the operator.

2.1 The Edge Detector

The *edge detector* block is used to generate the information related with the thin lines, edges, fine details and texture within the noisy input image. There are a number of different edge detectors in the literature [1]. The popular *Sobel* edge detector is used in this work for its simplicity and therefore it will be briefly reviewed here.

The Sobel edge detector employs the following convolution mask for edge detection.

$$M = \begin{bmatrix} -1 & -2 & -1 \\ 0 & 0 & 0 \\ 1 & 2 & 1 \end{bmatrix} \tag{1}$$

This mask is utilized for looking for the magnitude of the edge in the horizontal and vertical directions and then combining them into a single metric as follows:

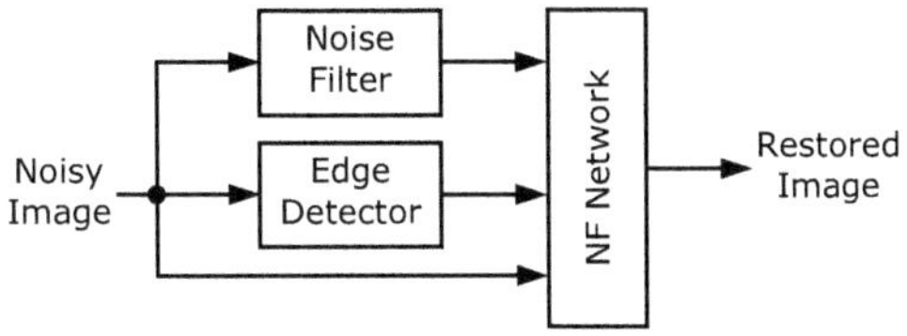

Fig. 1 The proposed setup for reducing distortion effects of an impulse noise filter

Let $x[r,c]$ represent the luminance value of a pixel at location (r,c) of an image. Here, r and c are the row and the column indices, respectively, with $1 \leq r \leq R$ and $1 \leq c \leq C$ for an image having a size of R-by-C pixels. The Sobel edge magnitude $e[r,c]$ of the pixel $x[r,c]$ of this image is given by

$$e[r,c] = \sqrt{m_H^2 + m_V^2} \tag{2}$$

where m_H and m_V denote the magnitude of the edge in the horizontal and the vertical directions, respectively, which are calculated by

$$m_H = \sum_{i=1}^{3}\sum_{j=1}^{3} x[r-2+i, c-2+j]\, M[i,j] \tag{3}$$

and

$$m_V = \sum_{i=1}^{3}\sum_{j=1}^{3} x[r-2+i, c-2+j]\, M[j,i] \tag{4}$$

2.2 *The Neuro-fuzzy Network*

The *NF network* block is the key element of the proposed method. It appropriately combines the information from the noisy input image, the output image of the noise filter and the output image of the edge detector to generate the final output image, which will hopefully have less distortion than the output image of the noise filter.

The NF network used in this work is a first order Sugeno type fuzzy system [33] with three inputs and one output. Each input has three *generalized bell* type membership functions whereas the output has a *linear* membership function. The input-output relationship of the NF network is as follows:

Let X_1, X_2, X_3 denote the inputs of the NF network and Y denote its output. Each possible combination of inputs and their associated membership functions is represented by a rule in the rule base of the NF network. Since the NF network has 3 inputs and each input has 3 membership functions, the rule base contains a total of 27 (3^3) rules, which are as follows:

1. if (X_1 is M_{11}) and (X_2 is M_{21}) and (X_3 is M_{31}), then $R_1 = F_1(X_1, X_2, X_3)$
2. if (X_1 is M_{11}) and (X_2 is M_{21}) and (X_3 is M_{32}), then $R_2 = F_2(X_1, X_2, X_3)$
3. if (X_1 is M_{11}) and (X_2 is M_{21}) and (X_3 is M_{33}), then $R_3 = F_3(X_1, X_2, X_3)$
4. if (X_1 is M_{11}) and (X_2 is M_{22}) and (X_3 is M_{31}), then $R_4 = F_4(X_1, X_2, X_3)$

$\vdots \qquad \vdots \qquad \vdots$

27. if (X_1 is M_{13}) and (X_2 is M_{23}) and (X_3 is M_{33}), then $R_{27} = F_{27}(X_1, X_2, X_3)$

where M_{ij} denotes the jth membership function of the ith input, R_k denotes the output of the kth rule, and F_k denotes the kth output membership function, with $i = 1,2,3$; $j = 1,2,3$; and $k = 1,2,\cdots,27$. The input membership functions are *generalized bell* type:

$$M_{ij}(u) = \frac{1}{1 + \left| \dfrac{u - a_{ij}}{b_{ij}} \right|^{2c_{ij}}} \tag{5}$$

and the output membership functions are *linear* type:

$$F_k(u_1, u_2, u_3) = d_{k1}u_1 + d_{k2}u_2 + d_{k3}u_3 + d_{k4} \tag{6}$$

Here the parameters a, b, c and d are constants that characterize the shape of the membership functions. The optimal values of these parameters are determined by training, which will be discussed in detail later on.

The output of the NF network is the weighted average of the individual rule outputs. The weighting factor, w_k, of each rule is calculated by evaluating the membership expressions in the antecedent of the rule. This is accomplished by first converting the input values to fuzzy membership values by utilizing the input membership functions and then applying the *and* operator to these membership values. The *and* operator corresponds to the multiplication of input membership values. Hence, the weighting factors of the rules are calculated as follows:

$$\begin{aligned}
w_1 &= M_{11}(X_1) \,.\, M_{21}(X_2) \,.\, M_{31}(X_3) \\
w_2 &= M_{11}(X_1) \,.\, M_{21}(X_2) \,.\, M_{32}(X_3) \\
w_3 &= M_{11}(X_1) \,.\, M_{21}(X_2) \,.\, M_{33}(X_3) \\
w_4 &= M_{11}(X_1) \,.\, M_{22}(X_2) \,.\, M_{31}(X_3) \\
&\vdots \\
w_{27} &= M_{13}(X_1) \,.\, M_{23}(X_2) \,.\, M_{33}(X_3)
\end{aligned} \tag{7}$$

Once the weighting factors are obtained, the output of the NF network can be found by calculating the weighted average of the individual rule outputs:

$$Y = \frac{\sum_{k=1}^{27} w_k R_k}{\sum_{k=1}^{27} w_k} \tag{8}$$

2.3 *Training of the Neuro-fuzzy Network*

The internal parameters of the NF network are optimized by training. Figure-2 represents the setup used for training. Here, the parameters of the NF network are iteratively optimized so that its output converges to the output of the *ideal noise filter* which, by definition, completely removes the noise from its input image. The ideal noise filter is *conceptual* only and does not necessarily exist in reality. It is only the output of the ideal noise filter that is necessary for training, and this is represented by the original (noise-free) training image.

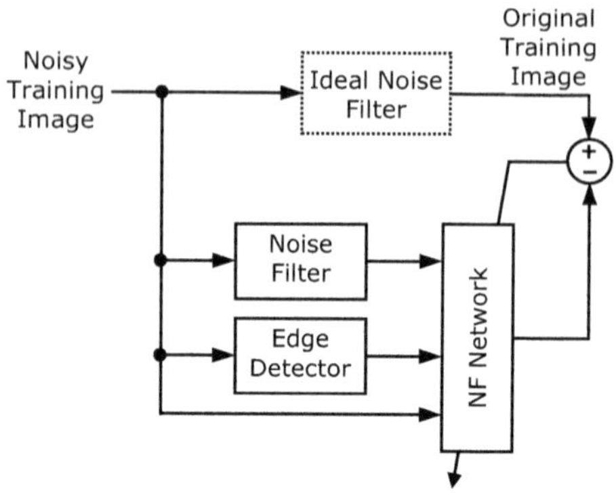

Fig. 2 Training of the neuro-fuzzy network

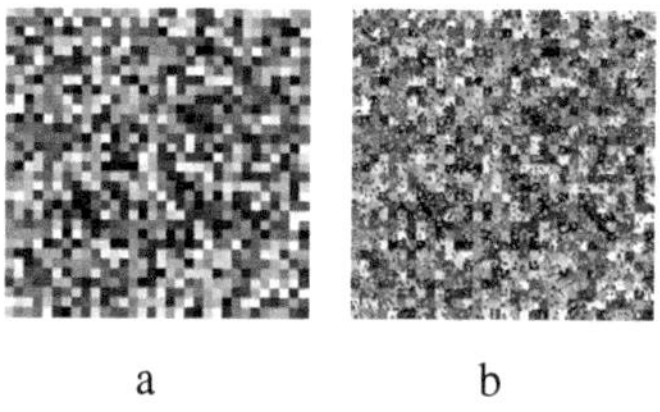

Fig. 3 Training images: (a) Original (b) Noisy

Figure-3 shows the images used for training. The image shown in Figure-3a is the *original training image*, which is a 128-by-128 pixel artificial image that can easily be generated in a computer. Each square box in this image has a size of 4-by-4 pixels and the 16 pixels contained within each box have the same luminance value, which is a random integer number uniformly distributed in [0, 255]. The image in Figure-3b is the *noisy training image* and is obtained by corrupting the original training image by impulse noise. The images in Figure-3b and Figure-3a are employed as the *input* and the *target (desired)* images during training, respectively. The parameters of the NF network are then iteratively tuned by using the Levenberg-Marquardt optimization algorithm [33] so as to minimize the learning error.

3 Results and Discussion

The proposed distortion reduction method is applied with four popular impulse noise filters. These are the standard median filter (MF) [1], the adaptive switching median filter utilizing center-weighted median filters for impulse detection (ACWMF) [7], the signal-dependent rank-ordered mean filter (SDROMF) [14] and the fuzzy filter (FF) [17]. These filters represent different approaches to impulse noise removal problem.

Several filtering experiments are performed to evaluate and compare the performances of the operators for the uses without and with the proposed method. The difference in performance is measured by employing the *mean squared error (MSE)* criterion, which is defined as:

$$\text{MSE} = \frac{1}{RC} \sum_{r=1}^{R} \sum_{c=1}^{C} \left(s[r,c] - y[r,c]\right)^2 \tag{9}$$

where $s[r,c]$ and $y[r,c]$ represent the luminance values of the pixels at location (r,c) of the original and the restored versions of a corrupted test image respectively.

The filtering experiments are performed on twelve popular test images available in the literature. These are the *Baboon*, *Boats*, *Bone*, *Bridge*, *Carpet*, *Goldhill*, *Lena*, *Pentagon*, *Peppers*, *Plant*, *Rock* and *Sail* images. These images are shown in Figure-4. All images are 8-bit gray level images having the same size of 256-by-256 pixels. The experimental images used in the simulations are generated by corrupting the original images by impulse noise with an appropriate noise density depending on the experiment.

The experiments are designed to evaluate the performance of the proposed distortion reduction method under varying noise conditions. The experimental procedure is as follows: The noise density is varied from 0% to 50% with 5% steps. For each noise density step, the twelve test images shown in Figure-4 are corrupted by impulse noise with that noise density. This produces twelve different experimental images each having the same noise density. These images are restored by using the setup shown in Figure-1. For each of the twelve noisy input images, two MSE values are calculated for the two restored images, one obtained at the output of the noise filter and the other at the output of the NF network, respectively. This gives two groups of MSE values with each group having twelve different MSE values. The twelve MSE values in each group are then averaged. Hence the average of the MSE values in the first group yields the representative MSE value of that filter for that noise density whereas the average of the MSE values in the second group yields the representative MSE value of the same filter when it is used together with the

Fig. 4 Test images: (a) Baboon, (b) Boats, (c) Bone, (d) Bridge, (e) Carpet, (f) Goldhill, (g) Lena, (h) Pentagon, (i) Peppers, (j) Plant, (k) Rock, (l) Sail

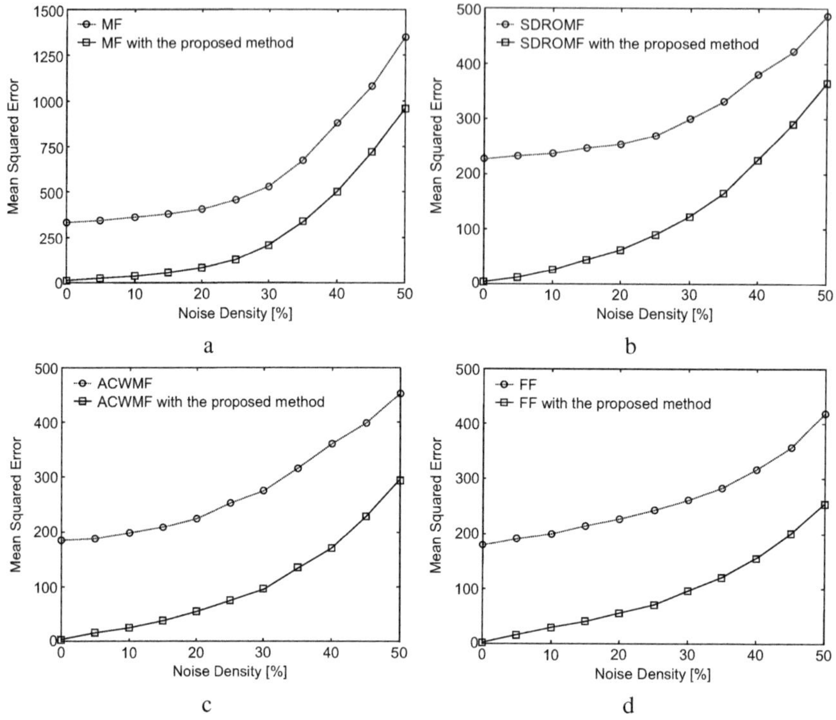

Fig. 5 Variation of the average MSE values of the four impulse noise filters as a function of noise density for the uses without and with the proposed method. (a) MF, (b) SDROMF, (c) ACWMF, (d) FF

proposed method. The difference between these two average MSE values indicates the reduction obtained by using the proposed method.

The above procedure is repeated for all noise densities to obtain the variation of the two average MSE values of the filter under experiment as a function of noise density. Finally, the overall experimental procedure is individually repeated for the remaining three impulse noise filters.

Figure-5 shows the variation of the average MSE values as a function of noise density. In each of the subfigures, the dotted curve represents the values calculated for the output image of the noise filter whereas the solid curve represents the values calculated for the distortion-reduced image obtained at the output of the NF network. It is obvious from these figures that the proposed method significantly decreases the MSE values of all operators for all noise densities.

For a visual evaluation of the distortion reduction obtained by using the proposed method, the output images of the four noise filters for the cases without and with the proposed method are shown in Figure-6 for the Baboon image corrupted by 25% impulse noise. The images show zoomed 128-by-128 pixel portions of the corresponding full size images for better visual comparison. Again, the undesirable

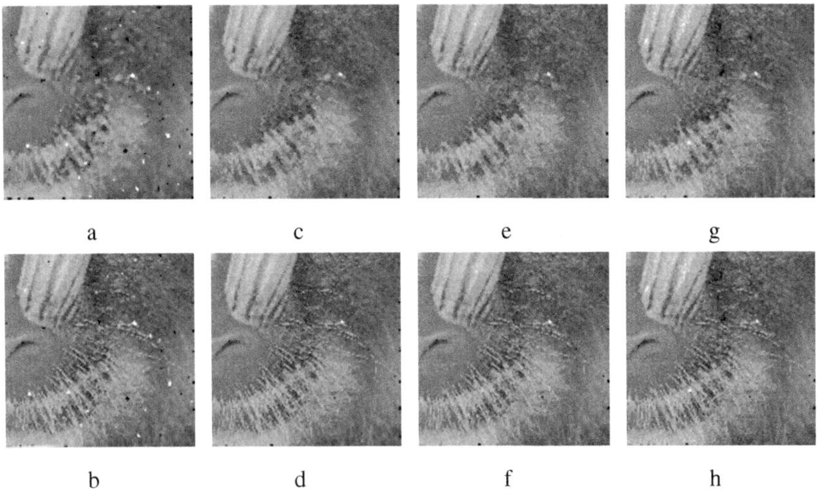

Fig. 6 Comparison of the direct and the distortion-reduced output images of the four impulse noise filters for the Baboon image corrupted by 25% impulse noise. The images show zoomed 128-by-128 pixel portions of the corresponding full size images. (a) MF, (b) MF with the proposed method, (c) SDROMF, (d) SDROMF with the proposed method, (e) ACWMF, (f) ACWMF with the proposed method, (g) FF, (h) FF with the proposed method

blurring distortions and the compensation of these distortions by the proposed method can easily be observed by carefully comparing the appearance of the hair around the mouth of the animal in all images in this figure.

Both numerical and visual results obtained in the simulation experiments presented in this section indicate that the proposed NF method effectively compensates for the distortions of the filters.

4 Conclusion

A novel method for reducing the undesirable distortion effects of impulse noise filters is presented. The fundamental advantages of the proposed method may be summarized as follows:

1. It is simple. It is applied by combining a given noise filter with an edge detector and a NF network. The structure of the NF network is also very simple. It is a first order Sugeno type fuzzy system with three inputs and one output.
2. The internal parameters of the NF network is determined by training. Moreover, training of the network is achieved by using artificial images which are conveniently generated in computer.
3. The method can be applied with virtually any type of noise filters since its operation is completely independent of the noise filter.

4. It does not interfere with the filtering behavior of the noise filter. It improves the overall filtering performance of the filter by reducing its distortion effects.

It is concluded that the proposed method can be used as a simple but powerful tool for efficiently reducing the distortion effects and improving the overall filtering performance of an impulse noise filter.

References

1. Umbaugh, S.E.: Computer Vision and Image Processing. Prentice-Hall International Inc., Englewood Cliffs (1998)
2. Yli-Harja, O., Astola, J., Neuvo, Y.: Analysis of the properties of median and weighted median filters using threshold logic and stack filter representation. IEEE Trans. on Signal Processing 39(2), 395–410 (1991)
3. Ko, S.-J., Lee, Y.H.: Center weighted median filters and their applications to image enhancement. IEEE Trans. on Circuit and Systems 38(9), 984–993 (1991)
4. Jeong, B., Lee, Y.H.: Design of weighted order statistic filters using the perceptron algorithm. IEEE Trans. on Signal Processing 42(11), 3264–3269 (1994)
5. Sun, T., Neuvo, Y.: Detail-preserving median based filters in image processing. Pattern Recognition Letters 15(4), 341–347 (1994)
6. Wang, Z., Zhang, D.: Progressive switching median filter for the removal of impulse noise from highly corrupted images. IEEE Trans. on Circuit and Systems 46(1), 78–80 (1999)
7. Chen, T., Wu, H.R.: Adaptive impulse detection using center-weighted median filters. IEEE Signal Proc. Letters 8(1), 1–3 (2001)
8. Chen, T., Wu, H.R.: Space variant median filters for the restoration of impulse noise corrupted images. IEEE Trans. on Circuit and Systems-II 48(8), 784–789 (2001)
9. Zhang, S., Karim, M.A.: A new impulse detector for switching median filters. IEEE Signal Proc. Letters 9(11), 360–363 (2002)
10. Yuksel, M.E., Besdok, E.: A simple neuro-fuzzy impulse detector for efficient blur reduction of impulse noise removal operators for digital images. IEEE Trans. on Fuzzy Systems 12(6), 854–865 (2004)
11. Yuksel, M.E., Basturk, A., Besdok, E.: Detail-preserving restoration of impulse noise corrupted images by a switching median filter guided by a simple neuro-fuzzy network. Applied Signal Processing 2004(16), 2451–2461 (2004)
12. Yuan, S.Q., Tan, Y.H.: Difference type noise detector for adaptive median filter. Electronics Letters 42(8), 454–455 (2006)
13. Dong, Y.H., Chan, R.H., Xu, S.F.: A detection statistic for random valued impulse noise. IEEE Trans. on Image Processing 16(4), 1112–1120 (2007)
14. Abreu, E., Lightstone, M., Mitra, S.K., Arakawa, K.: A new efficient approach for the removal of impulse noise from highly corrupted images. IEEE Trans. on Image Processing 5(6), 1012–1025 (1996)
15. Han, W.Y., Lin, J.C.: Minimum-maximum exclusive mean (MMEM) filter to remove impulse noise from highly corrupted images. Electronics Letters 33(2), 124–125 (1997)
16. Moore, M.S., Gabbouj, M., Mitra, S.K.: Vector SD-ROM filter for removal of impulse noise from color images. In: Proc. ECMCS 1999 EURASIP Conf. on DSP for Multimedia Communications and Services, Krakow (1999)
17. Russo, F., Ramponi, G.: A fuzzy filter for images corrupted by impulse noise. IEEE Signal Proc. Letters 3(6), 168–170 (1996)

18. Russo, F.: Noise removal from image data using recursive neurofuzzy filters. IEEE Trans. on Instrumentation Measurement 49(2), 307–314 (2000)
19. Yuksel, M.E., Basturk, A.: Efficient removal of impulse noise from highly corrupted digital images by a simple neuro-fuzzy operator. Int. J. Electron. Commun. 57(3), 214–219 (2003)
20. Yuksel, M.E., Basturk, A.: A simple generalized neuro-fuzzy operator for efficient removal of impulse noise from highly corrupted digital images. Int. J. Electron. Commun. 59(1), 1–7 (2005)
21. Own, C.M., Tsai, H.H., Yu, P.T., Lee, Y.J.: Adaptive type-2 fuzzy median filter design for removal of impulse noise. Imaging Science Journal 54(1), 3–18 (2006)
22. Kong, H., Guan, L.: Detection and removal of impulse noise by a neural network guided adaptive median filter. In: Proc. IEEE Int. Conf. on Neural Networks, vol. 2, pp. 845–849 (1995)
23. Lee, C.-S., Kuo, Y.-H., Yu, P.-T.: Weighted fuzzy mean filter for image processing. Fuzzy Sets and Systems 89(2), 157–180 (1997)
24. Lee, C.-S., Kuo, Y.-H.: The important properties and applications of the adaptive weighted fuzzy mean filter. Int. J. Intelligent Systems 14, 253–274 (1999)
25. Eng, H.-L., Ma, K.-K.: Noise adaptive soft-switching median filter. IEEE Trans. on Image Processing 10(2), 242–251 (2001)
26. Lin, C.-T., Lee, Y.-C., Pu, H.-C.: Satellite sensor image classification using cascaded architecture of neural fuzzy network. IEEE Trans. on Geoscience and Remote Sensing 38(2), 1033–1043 (2000)
27. Park, D.C., Woo, Y.J.: Weighted centroid neural network for edge preserving image compression. IEEE Trans. on Neural Networks 12(5), 1134–1146 (2001)
28. Boskowitz, V., Guterman, H.: An adaptive neuro-fuzzy system for automatic image segmentation and edge detection. IEEE Trans. on Fuzzy Systems 10(2), 247–262 (2002)
29. Lee, S.-J., Ouyang, C.-S., Du, S.-H.: A neuro-fuzzy approach for segmentation of human objects in image sequences. IEEE Trans. on Systems, Man and Cybernetics (Part B) 33(3), 420–437 (2003)
30. Qin, H., Yang, S.X.: Nonlinear noise cancellation for image with adaptive neuro-fuzzy inference systems. Electronics Letters 41(8), 474–475 (2005)
31. Pajares, G.: A Hopfield neural network for image change detection. IEEE Trans. on Neural Networks 17(5), 1250–1264 (2006)
32. Yuksel, M.E.: Edge detection in noisy images by neuro-fuzzy processing. Int. J. Electron. Commun. 61(2), 82–89 (2007)
33. Jang, J.-S.R., Sun, C.-T., Mizutani, E.: Neuro-fuzzy and soft computing. Prentice-Hall International Inc., Englewood Cliffs (1997)

An Algebraic Version of the Monadic System C_1

Jair Minoro Abe, Kazumi Nakamatsu, and Seiki Akama

Abstract. In this paper we present an algebraic version of the monadic system C_1* of Da Costa [6] by using the concept of Curry Algebra [4]. The algebraic structure obtained is called Curry Algebra C_1*. Some basic properties are discussed and presented.

Keywords: Curry algebras, system C_1, paraconsistent logic, algebraic logic, non-classical logic.

1 Introduction

Logic uses very important mathematical structures. That's true concerning with algebraic logic. Thus, we use in classical algebraic logical structures like semi-lattice, Boolean algebras, monadic algebras, cylindric algebras, besides topological space.

In general, the algebraic structures in classical logic are obtained by quotient passage: one has a logical structure, choose a suitable equivalence relation compatible with the logical operations, and by passing to the quotient we get the structure that algebraizes the system. By this process, one shows that Heyting algebras and Boolean algebras are respectively algebraization of the intuitionistic propositional calculus and classical propositional calculus.

Jair Minoro Abe
Graduate Program in Production Engineering, ICET - Paulista University
R. Dr. Bacelar, 1212, CEP 04026-002 São Paulo – SP – Brazil
Institute For Advanced Studies – University of São Paulo, Brazil
jairabe@uol.com.br

Jair Minoro Abe
School of Human Science and Environment/H.S.E. – University of Hyogo – Japan
nakamatu@shse.u-hyogo.ac.jp

Kazumi Nakamatsu
School of Human Science and Environment/H.S.E. – University of Hyogo – Japan
nakamatu@shse.u-hyogo.ac.jp

Seiki Akama
C-Republic Inc., 1-20-1, Higashi-Yurigaoka, Asao-ku, Kawasaki-shi, 215-0012, Japan
akama@jcom.home.ne.jp

K. Nakamatsu et al. (Eds.): New Advan. in Intel. Decision Techno., SCI 199, pp. 341–349.
springerlink.com

However, in some non-classical logics, sometimes we cannot use the method above. This happens when we don't have any significant congruence relation regarding to the logical system in question. Also, even in classical logic, sometimes the passage to the quotient can mask important facts. This succeeds, for instance, when we treat Smullyan tableaux and Hintikka set in the scope of classical propositional calculus.

In short, the use of pre-algebraic structure is relevant in situations like: 1) when one doesn't have (or it's unknown) a reasonable congruence relation, in particular when the equivalence relation chosen is not compatible with all logical operations; 2) In situations even it is available a congruence relation, we don't want to pass to quotient in order to not mask significant facts. In the first case we have structures that we call Curry algebras; in the second, we use pre-algebras in the common sense.

Generalization is an important concept in Mathematics. Curry systems can systematize a general theory of algebraization. Actually, all mathematical treatment of logical notions can be viewed as Curry systems. More than this, enriching or modifying the concepts of Curry system, we can obtain as particular cases, logical matrix, Kripke structures, theory of model, which are not directly coped with problem of algebraization. In a certain sense, we can say that logic reduces to the study of Curry systems. In what follows, we discuss the basic concepts of this paper [4].

2 Curry Systems

Definition 2.1. Suppose that in a non-empty set A is fixed an equivalence relation $\equiv$. We say that a n-ary operator φ on A is i-compatible with $\equiv$ if for any $x_1, \ldots, x_{i-1}, a, b, x_{i+1}, \ldots, x_n \in A$, if $a \equiv b$, implies $\varphi(x_1, \ldots, x_{i-1}, a, x_{i+1}, \ldots, x_n) \equiv \varphi(x_1, \ldots, x_{i-1}, b, x_{i+1}, \ldots, x_n)$. The operator is said to be compatible (or monotonic) with $\equiv$ if φ is i-compatible with $\equiv$ for all $i = 1, \ldots, n$. A relation R on A is said to be compatible with $\equiv$ if $(x_1, \ldots, x_n) \in R$ and $x_i \equiv x'_i$, $i = 1, \ldots, n$ then $(x'_1, \ldots, x'_n) \in R$.

Definition 2.2. A Curry system is a structure $<A, (\equiv)_{i \in I}, (S)_{j \in J}, (R)_{k \in K}, (\varphi)_{l \in L}, (C)_{m \in M}>$ such that $A \neq \varnothing$; $(\equiv)_{i \in I}$ is a collection of equivalence relations; $(S)_{j \in J}$ is a family of subsets of A; $(R)_{k \in K}$ is a finite collection of relations on A: $(\varphi)_{l \in L}$ is a family of operations on A; $(C)_{m \in M}$ is a finite collection of elements of A.

Normally we consider a unique equivalence relation $\equiv$ in a Curry system.

Definition 2.3. A Curry system is called a pre-algebra if the collection $(S)_{j \in J}$ is empty and all operations and relations regarding $\equiv$ are monotonic.

Definition 2.4. A Curry system is called a Curry algebra if there is at least one non-monotonic operation or relation on A.

Example 2.5. A system $<A, \equiv, \leq>$ is called a pre-ordered system if for all $x \in A$, $x \leq x$; for all $x, y, z \in A$, $x \leq y$ and $y \leq z$ imply $x \leq z$; fr all $x, y, x', y' \in A$, $x \leq y$, $x \equiv x'$, and $y \equiv y'$ imply $x' \leq y'$;

A pre-ordered system $<A, \equiv, \leq>$ is called a partially-ordered system if for all x, $y \in A$, $x \leq y$ and $y \leq x$ imply $x \equiv y$;

A partially-ordered system $<A, \equiv, \leq>$ is called a pre-lattice system if

For all x, $y \in A$, the set of $\sup\{x, y\} \neq \varnothing$ and the set of $\inf\{x, y\} \neq \varnothing$. We denote by $x \vee y$ one element of the set of $\sup\{x, y\}$and by $x \wedge y$ one element of the set of $\inf\{x, y\}$.

A system $<A, \equiv, \leq, \rightarrow>$ is called an implicative pre-lattice if $<A, \equiv, \leq>$ is a pre-lattice, and for all $x, y, z \in A$; $x \wedge (x \rightarrow y) \leq y$ and $x \wedge y \leq z$ iff $x \leq y \rightarrow z$.

$<A, \equiv, \leq, \rightarrow>$ is called classic implicative pre-lattice if it is an implicative pre-lattice and $(x \rightarrow y) \rightarrow x \leq x$ (Peirce's law).

With earlier example we can have an idea how we can extend the majority of algebraic systems to pre-algebraic systems considering an equivalence relation $\equiv$ instead of equality relation. In this way we can obtain Boolean pre-algebras, pre-filters, pre-lattices, etc.

Now we give a concrete example of a Curry algebra. It is obtained with the logical systems C_n $(1 \leq n < \omega)$ [6].

The primitive symbols of the language L of the calculi C_1 is are the following:

Propositional variables: a denumerable set of propositional variables; $\rightarrow$ (implication), $\wedge$ (conjunction), $\vee$ (disjunction), $\neg$ (negation), parentheses.

Formulas are defined in the usual manner. In L, we put:

Definition 2.6. Let A be any formula. Then A^{o} is shorthand for $\neg(A \wedge \neg A)$.

Definition 2.7. We write $A \leftrightarrow B$ for $(A \rightarrow B) \wedge (B \rightarrow A)$.

The postulates (axiom schemes and inference rules) of C_1 are: A, B, and C are formulas whatsoever.

(1) $A \rightarrow (B \rightarrow A)$

(2) $(A \rightarrow (B \rightarrow C)) \rightarrow ((A \rightarrow B) \rightarrow (A \rightarrow C))$

(3) $\dfrac{A,\ A \rightarrow B}{B}$

(4) $A \wedge B \rightarrow A$

(5) $A \wedge B \rightarrow B$

(6) $A \rightarrow (B \rightarrow (A \wedge B))$

(7) $A \rightarrow A \vee B$

(8) $B \rightarrow A \vee B$

(9) $(A \rightarrow C) \rightarrow ((B \rightarrow C) \rightarrow ((A \vee B) \rightarrow C))$

(10) $B^0 \rightarrow ((A \rightarrow B) \rightarrow ((A \rightarrow \neg B) \rightarrow \neg A)$

(11) $A^0 \wedge B^0 \rightarrow (A \wedge B)^0 \wedge (A \vee B)^0 \wedge (A \rightarrow B)^0$

(12) $\neg\neg A \rightarrow A$

(13) $A^0 \rightarrow (\neg A)^0$

Theorem 2.8. In C_1, all valid schemes and rules of classical positive propositional logic are true. In particular, the deduction theorem is valid in C_1.

In C_1, A^0 expresses intuitively that the formula A 'behaves' classically, so that the motivation of the postulates (10), (11), and (13) are clear. Furthermore, in this calculus, the set of all well-behaved formulas together with the connectives $\rightarrow$, $\wedge$, $\vee$, and $\neg$ have all the properties of classical implication, conjunction, disjunction, and negation, respectively. Therefore the classical propositional calculus is contained in C_1, though it constitutes a strict sub-calculus of the former. It is not difficult to prove that conditions a), b), c), and d) of the previous paragraph.

Theorem 2.9. In C_1, the following schemes are not valid (among others): $(A \wedge \neg A) \rightarrow B$; $(A \wedge \neg A) \rightarrow \neg B$; $A \rightarrow (\neg A \rightarrow B)$; $(A \leftrightarrow \neg A) \rightarrow B$; $(A \leftrightarrow \neg A) \rightarrow \neg B$; $A \rightarrow \neg\neg A$; $(\neg A \wedge (A \vee B)) \rightarrow B$; $(A \rightarrow B) \rightarrow (\neg B \rightarrow \neg A)$; $\neg(A \wedge \neg A)$

Theorem 2.10. C_1 is not trivial.
Let's define the operator $\neg^* A =_{\text{Def.}} \neg A \wedge A^0$ (i.e., the negation of a well-behaved formula A). Such operator $\neg^*$ is called strong negation and it is possible to proof in C_1 that with remaining connectives $\wedge$, $\vee$, and $\rightarrow$ they have all properties of classical connectives of negation, conjunction, disjunction, and implication, respectively. In short, the classical logic is contained in C_1.

3 The Curry Algebra C_1

Definition 3.1 A Curry algebra C_1 (or a C_1-algebra) is an implicative pre-lattice $<A, \equiv, \leq, \wedge, \vee, \rightarrow, '>$ with an maximum element 1 and operators $\wedge$, $\vee$, and $'$ satisfying the conditions below, where $x^0 =_{\text{Def.}} (x \wedge x')'$: $x \vee x' \equiv 1$; $x'' \leq x$; $y \leq (x \rightarrow y) \rightarrow ((x \rightarrow y') \rightarrow x')$; $x \wedge y \leq (x \rightarrow y)^0$; $x \wedge y \leq (x \wedge y)^0$; $x \wedge y \leq (x \vee y)^0$; $x \leq (x')^0$

Example 3.2. Let's consider the calculus C_1. A is the set of all formulas of C_1. Let's consider as operations, the logical connectives of conjunction, disjunction, implication, and negation. Let's define the relation on A.

$x \equiv y$ iff $\vdash x \leftrightarrow y$. It is easy to check that $\equiv$ is an equivalence relation on A.

$x \leq y$ iff $x \equiv x \wedge y$ and $y \leq x$ iff $y \equiv x \wedge y$. Also we take as 1 any fixed axiom instance.

The structure composed $<A, \equiv, \leq, \wedge, \vee, \rightarrow, '>$ is a C_1-algebra.

Theorem 3.3. Let's $<A, \equiv, \leq, \wedge, \vee, \rightarrow, '>$ be a C_1-algebra. Then the operator $'$ is non-monotone relatively $\equiv$.

Proof. See [10].

The earlier theorem says that every quotient algebra in the calculus C_1 is trivial. It's worthwhile to observe that [5] mention the existence of such non-monotone operators, but didn't give any concrete example.

Theorem 3.4. A C_1-algebra is distributive and has a greatest element, as well as a first element.

Definition 3.5. Let x be an element of a C_1-algebra. We put $x^* = x' \wedge x^o$.

Theorem 3.6. In a C_1-algebra, x^* is a Boolean complement of x; so $x \vee x^* \equiv 1$ and $x \wedge x^* \equiv 0$. Moreover, in a C_1-algebra, the structure composed by the underlying set and by operations ∧, ∨, and * is a (pre) Boolean algebra. If we pass to the quotient by the basic relation ≡, we obtain a Boolean algebra in the usual sense.

Definition 3.7. Let $<A, \equiv, \leq, \wedge, \vee, \rightarrow, '>$ be a C_1-algebra and $<A, \equiv, \leq, \wedge, \vee, \rightarrow, *>$ the Boolean algebra obtained as in the above theorem. Any Boolean algebra that is isomorphic to the quotient algebra of $<A, \equiv, \leq, \wedge, \vee, \rightarrow, *>$ by ≡ is called Boolean algebra *associated with the* C_1-algebra.

Hence, we have the following representation theorems for C_1-algebras.

Theorem 3.8. Any C_1-algebra is associated with a field of sets. Moreover, any C_1-algebra is associated with the field of sets simultaneously open and closed of a totally disconnected compact Hausdorff space.

One open problem concerning C_1-algebras remains. How many non-isomorphic associated with the C_1-algebra are there?

4 Some Generalizations

Now we show a chain of Curry algebras beginning with the C_1-algebra.
In Let $<A, \equiv, \leq, \wedge, \vee, \rightarrow, '>$ be a C_1-algebra. If $x \in A$, x^1 abbreviates x. x^n $(1 < n < \omega)$ abbreviates $x^{\overbrace{0...0}^{}}_{n-times}$. Also, $x^{(1)}$ abbreviates x^1. $x^{(n+1)}$ $(1 < n < \omega)$ abbreviates $x^{(n)} \wedge x^{n-1}$.

Definition 4.1. A C_n-algebra $(1 < n < \omega)$ is an implicative pre-lattice $<A, \equiv, \leq, \wedge, \vee, \rightarrow, '>$ with a first element 1 and operators ∧, ∨, and ' satisfying the conditions: $x \vee x' \equiv 1$; $x'' \leq x$; $y^{(n)} \leq (x \rightarrow y) \rightarrow ((x \rightarrow y') \rightarrow x')$; $x^{(n)} \wedge y^{(n)} \leq (x \rightarrow y)^{(n)}$; $x^{(n)} \wedge y^{(n)} \leq (x \wedge y)^{(n)}$; $x^{(n)} \wedge y^{(n)} \leq (x \vee y)^{(n)}$; $x^{(n)} \leq (x')^{(n)}$

Usual algebraic structural concepts like homomorphism, monomorphism, etc. can be introduced for Currry algebras without extensive comments.

Theorem 4.2. Every C_n-algebra is embedded in any C_{n-1}-algebra $(1 < n < \omega)$.

Corollary 4.2.1. Every C_n-algebra $(1 < n < \omega)$ is embedded in any C_1-algebra.

If we indicate a C_n-algebra by C_n, the embedding hierarchy can be represented as

$C_1 > C_2 > \ldots C_n > \ldots$

Definition 4.3. A C_ω-algebra is an implicative pre-lattice $<A, \equiv, \leq, \wedge, \vee, \rightarrow, '>$ with a first element 1 and operators ∧, ∨, and ' satisfying the conditions below:

1. $x \vee x' \equiv 1$
2. $x'' \leq x$

Example 4.4. The following structure $<\{1, 2, 3\}, \{1, 2\}, \rightarrow, \wedge, \vee, \neg>$is a C_ω-algebra:

Table 1 Operations

$A \to B$

A\B	1	2	3
1	1	1	3
2	1	1	3
3	1	1	1

$A \wedge B$

A\B	1	2	3
1	1	1	3
2	1	1	3
3	1	1	1

$A \vee B$

A\B	1	2	3
1	1	1	3
2	1	1	3
3	1	1	1

$\neg A$

A	¬A
1	3
2	2
3	1

The operator ¬ is not compatible with equivalence relation among formulas of the logic C_ω. In general the C_ω-algebras do not possesses a lower element but it can be extended to have a lower element [4]. If we have a C_ω-algebra $<A, \equiv, \leq, \wedge, \vee, \to, '>$, every Curry system isomorphic to the quotient algebra $<A, \equiv, \leq, \wedge, \vee, \to, '>$ by $\equiv$ is called an implicative pre-lattice associated to $<A, \equiv, \leq, \wedge, \vee, \to, '>$.

Thus, every implicative pre-lattice associated to $<A, \equiv, \leq, \wedge, \vee, \to, '>$ is a Heyting algebra or a maximal filter of a Heyting algebra (with or without first element)

There are not so many studies regarding algebraic versions of the systems C^*_n $(1 \leq n \leq \omega)$ (predicate calculi of the systems C_n). We propose in the sequence some extensions of the C_1-algebras.

5 C_1*-Monadic Algebras

In this section we introduce the monadic Curry algebra C_1*.

Definition 5.1. Let A be a C_1-algebra. Let $\exists$ (existential quantifier) and $\forall$ (universal quantifier) be operators on A. $(\exists, \forall)$ is called a quantifier on A if $\exists 0 \equiv 0$; $x \leq \exists x$; $\exists(x \vee y) \equiv \exists x \vee \exists y$; $\exists\exists x \equiv \exists x$; $\exists(\exists x)^* \equiv (\exists x)^*$; $\exists(x \wedge \exists y) \equiv \exists x \wedge \exists y$; $\forall 1 \equiv 1$; $\forall x \leq x$; $\forall(x \vee y) \equiv \forall x \vee \forall y$; $\forall\forall x \equiv \forall x$; $\forall(\forall x)^* \equiv (\forall x)^*$

We suppose in the above definition that, if $x \equiv y$, then $\exists x \equiv \exists y$ and $\forall x \equiv \forall y$. $\exists$ is called existential quantifier on A and $\forall$ is called universal quantifier on A. The pair $<A, (\exists, \forall)>$ is called a monadic Curry algebra C_1* or a C_1*-monadic algebra (or C_1*-algebra).

Given a Curry algebra C_1, let's assume that there is an universal quantifier defined on it, i.e., a structure $<A, \forall>$ such that conditions 7-11 above are satisfied. If we define $\exists_1 x =_{\text{Def.}} (\forall x^*)^*$, then $\exists_1$ is an existential quantifier (i.e. satisfying 1-6) and the structure composed by $<A, (\exists_1, \forall)>$ is a C_1*-monadic algebra. Also, we can get a monadic algebra considering an existential quantifier $\exists$ on a Curry algebra C_1 satisfying conditions 1-6 of above definition and defining a universal quantifier (i.e, satisfying 7-11) as $\forall_1 x =_{\text{Def.}} (\exists x^*)^*$. Then the structure composed by $<A, (\exists, \forall_1)>$ is a C_1*-monadic algebra. Given a Curry algebra C_1, in general, the algebras obtained $<A, (\exists_1, \forall)>$ and $<A, (\exists, \forall_1)>$ are not isomorphic. Also, given a C_1*-monadic algebra $<A, (\exists, \forall)>$, define the new quantifiers $\exists_1$ and $\forall_1$. In general, we have $\exists_1 \neq \exists$ and $\forall_1 \neq \forall$.

Let C be a C_1-algebra and $A = \{(x_1, x_2, \dots, x_n) | x_i \in C, i = 1, 2, \dots, n\}$. Let us suppose that if $x, y \in A$, then $x \vee y \in A$ and $\neg x \in A$, and if $(x_1, x_2, \dots, x_n), (y_1, y_2, \dots, y_n) \in A$ we define $(x_1, x_2, \dots, x_n) \equiv (y_1, y_2, \dots, y_n)$ iff $x_i \equiv y_i$, $i = 1, 2, \dots, n$. Also, we put $\overline{x} = \bigvee_{i=1}^{n} x_i$, $\underline{x} = \bigwedge_{i=1}^{n} x_i$, and we assume that $(\underbrace{\overline{x}, \dots, \overline{x}}_{n-times})$, $(\underbrace{\underline{x}, \dots, \underline{x}}_{n-times}) \in A$. If we define $\exists x = (\overline{x}, \dots, \overline{x})$ and $\forall x = (\underline{x}, \dots, \underline{x})$, then A is a C_1*-monadic algebra.

A more useful example of C_1*-monadic algebra is the following. Let C be a C_1-algebra, a set $K \neq \varnothing$, and C^K is the set of all functions of K into C. Let A be the set such that: (i) A is a C_1-algebra with respect the pointwise operations, and (ii) if $x \in A$, then the range of x has a supremum $\overline{x}$ and a infimum $\underline{x}$ in C, and the functions that take the value $\overline{x}$ at each point of K and $\underline{x}$ at each point of K are in A. If $\exists x$ and $\forall x$ are defined to be those functions, then A becomes a C_1*-monadic algebra. Every C_1*-monadic algebra obtained in this way is called a C-valued functional algebra with domain K.

Now we discuss some aspects of the algebraic structures originated in our discussion. In fact, the matter is very rich, but due limitations of this paper we'll concern only on some of them.

Theorem 5.2. In a C_1*-monadic algebra $<A, (\exists, \forall)>$, the structure composed by the underlying set and by operations $\wedge$, $\vee$, $*$, $\exists$, and $\forall$ is a (pre) monadic algebra. If we pass to the quotient by the basic relation $\equiv$, we obtain a monadic algebra in the usual sense [4].

Definition 5.3. Let $<A, (\exists, \forall)>$ be a C_1*-monadic algebra, and $<A, \equiv, \leq, \rightarrow, *, \exists, \forall>$ the monadic algebra obtained as in the above theorem. Any monadic algebra that is isomorphic to the quotient algebra of $<A, \equiv, \leq, \rightarrow, *, \exists, \forall>$ by $\equiv$ is called monadic algebra *associated with the* C_1*-monadic algebra.

Hence, we can establish the following representation theorems for C_1*-monadic algebras.

Theorem 5.4. If C is a C_1*-monadic algebra, then for its associated monadic algebra A, there exists a set X and there exits a Boolean algebra B, such that (i) A is isomorphic to a B-valued functional algebra A' with domain X, and (ii) for every element p of A' there exists a point x in X with $p(x) = \exists p(x)$.

Theorems 5.3 and 5.4 show us that C_1*-monadic algebras constitute interesting generalization of the concept of monadic algebras. Here, there is an open problem. How many non-isomorphic monadic algebras associated with a C_1*-monadic algebra are there?

6 Applications

Many important mathematical concepts can be accommodated with the concept of Curry system. For details, see [4]. Let's elaborate the concept of constructibility operator in Heyting algebras. For this, let's start with a pre-Boolean algebra <A, $\wedge$, $\vee$, ', 0, 1, ^{o}> in which is defined an operator ^{o}. The intuitive interpretation of this operator is 'if $x \in A$ represents a proposition, then x^{o} means that x is intuitioniscally true (or constructively true). x^{o} expresses that there exists an appropriate construction for x, i.e., x is true by construction. B^{o} indicates the range of B by ^{o} and $x \rightarrow_{o} y$ abbreviates $(x \rightarrow y)^{o}$.

The postulates are $x^{oo} \equiv x^{o}$; $x^{o} \leq x$; $(x \wedge y)^{o} \equiv x^{o} \wedge y^{o}$; If $x, y \in B^{o}$, then, $x \wedge y$, $x \vee y \in B^{o}$; If $x, y, z \in B^{o}$, then, $x \wedge z \leq y \Rightarrow z \leq x \rightarrow_{o} y$

The intuitive readings are clear; for instance 2 says that if x is constructively true, then x is classically true.

Let A be the Boolean algebra <A, $\wedge$, $\vee$, ', 0, 1, ^{o}> with the operator ^{o}. Then, the structure <A^{o}, $\wedge$, $\vee$, $\rightarrow$, 0°, 1°> where $0^{o} = 0$ and $1^{o} =_{\text{Def.}} x^{o} \rightarrow x^{o}$ is a Heyting algebra.

Theorem 6.1. Let <A, $\wedge$, $\vee$, ', 0, 1, ^{o}> be a pre-Boolean algebra with the operator ^{o}. If $x, y \in A$, $x \equiv y$ do not imply $x^{o} \equiv y^{o}$.

Proof. Let's consider the pre-Boolean algebra <{0, a, 1}, $\wedge$, $\vee$, ', 0, 1, ^{o}> with the basic equivalence relation $\equiv$ defined by $0 \equiv 0$, $a \equiv a$, $a \equiv 1$, $1 \equiv a$, $1 \equiv 1$ and the operations $\wedge$, $\vee$, ', and ^{o} defined by

Table 2 Operations

$x \wedge y$

$x\backslash y$	**0**	***a***	**1**
0	**0**	*a*	1
a	*a*	*a*	1
1	**1**	1	1

$x \vee y$

$x\backslash y$	**0**	***a***	**1**
0	**0**	*a*	1
a	*a*	*a*	1
1	**1**	1	1

x'; x^{0}

x	x'	x^{0}
0	1	0
a	0	0
1	0	1

We have $a \equiv 1$, but $a^{o} = 0$ and $1^{o} = 1$.

Theorem 6.2. Let <A, $\wedge$, $\vee$, ', 0, 1, ^{o}> be a pre-Boolean algebra with the operator ^{o}. The Peirce's law $(x \rightarrow y) \rightarrow x \leq x$ is not valid.

Theorem 6.4. Every pre-Boolean algebra with the constructibility operator is associated to a open sets algebra.

Let T be any mathematical theory. Certain propositions of T can be such that there is an effective method of computability for them. For instance, in the theory of differential equations, the existence of solutions is accompanied by a method that such solutions can be calculated with the approximation needed. If this happens, we say that such propositions x are computationally true and let's denote it by x^{*}. If we look for T as a pre-Boolean algebra, the operator * satisfies exactly the properties of the constructibility operator. In this direction, the pre-Boolean algebras with the operator ^{o} constitute a generalization of the concept of computability. The earlier considerations concerning the operator ^{o} also apply to generalize the idea of topological space. It also generalizes the concept of Boolean algebra with the interior operator.

The few observations above illustrate the enormous importance the study of Curry systems. The field is very large in horizon; in effect, it constitutes a promising area of researching with many interesting results to coming. We hope to say more in forthcoming papers.

References

1. Abe, J.M.: Curry algebras Pτ. Logique et Analyse 161, 5–15 (1998)
2. Abe, J.M.: Curry Algebras N1. Atti. Acc. Lincei. Rend. Fis. 7(9), 125–128 (1996)
3. Abe, J.M., Akama, S., Nakamatsu, K.: Monadic Curry Algebras Qτ. In: Apolloni, B., Howlett, R.J., Jain, L. (eds.) KES 2007, Part II. LNCS, vol. 4693, pp. 893–900. Springer, Heidelberg (2007)
4. Barros, C.M., da Costa, N.C.A., Abe, J.M.: Tópico de teoria dos sistemas ordenados: vol. II, sistemas de Curry, Coleção Documentos, Série Lógica e Teoria da Ciência, IEA-USP, 20, 132p. (1995)
5. Curry, H.B.: Foundations of Mathematical Logic. Dover, New York (1977)
6. Da Costa, N.C.A.: On the theory of inconsistent formal systems. Notre Dame J. of Formal Logic 15, 497–510 (1974)
7. Eytan, M.: Tableaux of Hintikka et Tout ça: un Point de Vue Algebrique. Math. Sci. Humaines 48, 21–27 (1975)
8. Halmos, P.R.: Algebraic Logic. Chelsea Publishing Co., New York (1962)
9. Kleene, S.C.: Introduction to Metamathematics. Van Nostrand, Princeton (1952)
10. Mortensen, C.: Every quotient algebra for C_1 is trivial. Notre Dame J. of Formal Logic 21, 694–700 (1977)

Some Three-Valued Temporal Logics for Future Contingents

Seiki Akama, Kazumi Nakamatsu, and Jair Minoro Abe

Abstract. To interpret the truth-value of future contingent events is of special importance for philosophy since the age of Aristotle. The subject is also interesting from the perspectives of linguistics and computer science. We compare two three-valued temporal logics for future contingents. We also suggest formalizing other types of useful three-valued temporal logics.

Keywords: future contingent, gap, glut, three-valued temporal logic.

1 Introduction

What is the truth-value of future contingent events? If future contingents are either true or false, namely if they have a truth-value, this contradicts our assumption that these propositions are contingent. Future contingents are closely related to *determinism*. If all sentences on the future are true or false, namely determined, then we have a conclusion that everything in the future is determined.

The problem of *future contingents* is of special importance for philosophy since the age of Aristotle. It is also interesting from the perspectives of linguistics and computer science. For linguists, the semantics for future contingent propositions challenges standard semantic theories. For computer scientists,

Seiki Akama
University of Tsukuba, Ibaraki, Japan
e-mail: akama@jcom.home.ne.jp

Kazumi Nakamatsu
University of Hyogo, Himeji, Japan
e-mail: nakamatus@shse.u-hyogo.ac.jp

Jair Minoro Abe
Paulista University, Sao Paulo, Brazil
e-mail: jairabe@uol.com.br

K. Nakamatsu et al. (Eds.): New Advan. in Intel. Decision Techno., SCI 199, pp. 351–361.
springerlink.com

there seem many future contingent events in various types of temporal-oriented computational systems.

One way to overcome the problem of future contingents is to develop a *three-valued temporal logic*. In the previous papers, we proposed two three-valued temporal logics, i.e. FCP in [2] and Q_t in [3], from different viewpoints. The purpose of this paper is to address the importance of three-valued temporal logics.

The rest of this paper is organized as follows. In section 2, we survey the problem of future contingents. In section 3, we give a three-valued temporal logic FCP. In section 4, we give a three-valued temporal logic Q_t. In section 5, we suggest formalizing other types of useful three-valued temporal logics. Section 6 gives our conclusions.

2 Future Contingents

To give the truth-value of future contingent events challenges philosophers for many years. If future contingents are either true or false, namely if they have a truth-value, this contradicts our assumption that these propositions are contingent. There is an intimate connection of future contingents and *determinism*. If all sentences on the future are true or false, namely determined, then we have a conclusion that everything in the future is determined. Namely, if something happens, it is necessary that it happens, and if something does not happen, it is impossible that it happens. Since determinism is controversial, it is very important to give a formal treatment of future contingents.

There are here two important concepts in investigating the nature of future contingents, namely *the law of excluded middle* (LEM) and *the principle of bivalence* (PB). However, they should be clearly distinguished. By (LEM), we mean the syntactic thesis of the form $A \vee \neg A$. (PB) is the semantic thesis that every proposition is either true or false. If (PB) holds, then these theses are equivalent. But if it does not hold, then there are two options whether (LEM) holds.

The study of future contingents has a long history. Aristotle [4] considered the issue in *De Interpretatione IX*. According to Aristotle, only propositions about the future which are either necessarily true, or necessarily false, or something determined have a determinate truth-value. In other words, Aristotle accepts (LEM), but rejects (PB) for future contingents. Consequently, we need to endorse Aristotle's argument that (PB) leads to fatalism (determinism) and seek to avoid the fatalist conclusion. There are several approaches to the solution.

Lukasiewicz [6] attempted to formalize Aristotle's idea by developing a *three-valued logic* in which the third truth-value reads "indeterminate". We know that Lukasiewicz presented the truth-value tables for negation, conjunction, disjunction and implication. For example, consider (1):

(1) I will go to Melbourne.

It seems correct to say that my going to Melbourne remains open. We can thus give (1) the third truth-value without any doubt. This suggests that a future contingent proposition lacks no determinate truth-values.

Unfortunately, many philosophers (or logicians) criticized Lukasiewicz's three-valued logic in that it is not successful as a logic for future contingents. There is a serious difficulty. In fact, the interpretation based on Lukasiewicz's three-valued logic does not work when we deal with the disjunctive proposition in which one disjunct is the negation of the other disjunct like (2).

(2) Either I will go to Melbourne or I will not go to Melbourne.

Obviously, (2) is *definitely* true. But, if we rely on Lukasiewicz's three-valued logic, the truth-value of (3) is indeterminate, and the result is not intuitively justified. (2) is an instance of (LEM), which of the form $A \vee \neg A$. A similar defect can be recognized in the treatment of (LNC), which of the form $\neg(A \wedge \neg A)$. These two principles have the truth-value "indeterminate" according to Lukasiewicz's three-valued logic.

We here claim that the interpretation of disjunction (conjunction) with indeterminate disjuncts (conjuncts) causes the trouble. In addition, Lukasiewicz's three-valued logic cannot formalize the above mentioned Aristotle's idea because Aristotle wants necessarily true propositions like $A \vee \neg A, \neg(A \wedge \neg A)$ to have the determinate truth-value, namely "true".

3 *FCP*

A three-valued temporal logic *FCP* was proposed by Akama, Nagata and Yamada in [2]. The logic is motivated by introducing new kind of negation into Kleene's strong three-valued logic.

Recall that Lukasiewicz's three-valued logic has the following truth-value tables:

$\sim$	
t	f
u	u
f	t

$\wedge$	t	u	f
t	t	u	f
u	u	u	f
f	f	f	f

$\vee$	t	u	f
t	t	t	t
u	t	u	u
f	t	u	f

$\rightarrow$	t	u	f
t	t	u	f
u	t	t	u
f	t	t	t

Here, t reads "true", f "false", and u "indeterminate", respectively. For the entries of determinate truth-values, the above agree with the classical truth-value tables. The gist of our revision is to change the truth-value table for negation, which causes several difficulties mentioned above. There seem no reasons to justify Lukasiewicz's interpretation of negation. The alternative interpretation is:

¬	
t	f
u	t
f	t

Here, we use ¬ for the new negation connective. This new interpretation is inspired by Prior's passage ([8], p. 136):

> Perhaps 'neither true nor false' is simply a possible way of describing the kind of falsehood which 'It will be that p' has, in Peircean logic, when the matter is undecided.

According to Prior's remark, we could identify "neither true nor false" as a subcase of "false". From this, the negation of "indeterminate" evaluates as "true". The proposed interpretation can also be justified as follows. If A is indeterminate, then it is not true that A. This gives rise to a possibility that something related to A is true. Thus, one can equate something related to A to the negation of A, supporting the new truth-value table. Those who are familiar with many-valued logic can recognize that the new negation ¬ is in fact the *external* negation due to Bochvar.

By our interpretation of "indeterminate", we should define the implication $A \supset B$ as $\neg(A \wedge \neg B)$, which has the truth-value table different from Lukasiewicz's:

$\supset$	t	u	f
t	t	f	f
u	t	t	t
f	t	t	t

Now, we are ready to formulate a three-valued temporal logic FCP. The language of FCP contains logical symbols: negation, conjunction, disjunction and implication, and temporal operators: F (it will be the case) and P (it has been the case). We can introduce other two temporal operators by definition.

$\mathrm{G} =_{\mathrm{def}} \neg\mathrm{F}\neg$ (it will always be the case)

$\mathrm{H} =_{\mathrm{def}} \neg\mathrm{P}\neg$ (it has always been the case)

We can also define the equivalence $\equiv$ as:

$$A \equiv B =_{\mathrm{def}} (A \supset B) \wedge (B \supset A).$$

The axiomatization of FCP, similar to that of the minimal temporal logic K_t, consists of axioms and rules of inference:

Three-Valued Temporal Logic FCP

Axioms

(A1) Axioms of Positive Classical Logic

(A2) $(\neg B \supset \neg A) \supset (A \supset B)$

(A3) $\mathrm{G}(A \supset B) \supset (\mathrm{G}A \supset \mathrm{G}B)$

(A4) $\mathrm{H}(A \supset B) \supset (\mathrm{H}A \supset \mathrm{H}B)$

(A5) $A \supset \mathrm{HF}A$

(A6) $A \supset \mathrm{GP}A$

Rules of Inference

(R1) $\vdash A,\ \vdash A \supset B \ \Rightarrow \vdash B$

(R2) $\vdash A \ \Rightarrow \vdash \mathrm{G}A$

(R3) $\vdash A \ \Rightarrow \vdash \mathrm{H}A$

Here, $\vdash A$ reads "A is provable in FCP". (R1) is *modus ponens*, and (R2) and (R3) are a temporal version of *necessitation.* The notion of a proof is defined as usual.

Surprisingly, FCP has classical flavors. For example, $A \supset A$ is true. We also note that $A \supset B$ is equivalent to $\neg B \supset \neg A$ or $\neg A \vee B$. This means that $A \supset B$ is interpreted as material implication. De Morgan and double negation laws hold. These facts can be well-understood by the interpretation of "indeterminate" sentences as a possible description of false sentences. This enables us to improve Lukasiewicz's approach. In FCP, we express (3) for (4).

(4) $\mathrm{F}A \vee \mathrm{F}\neg A$

which is equivalent to:

(5) $\mathrm{F}(A \vee \neg A)$

Even if A is indeterminate, we wish (5) to be true because $A \vee \neg A$ is a logical truth. In fact, this is possible in FCP as the semantics below shows.

A *FCP-model* is a triple $\langle T, \leq, V\rangle$, where T is a set of time points, $\leq$ is a binary relation on T, and V is a partial valuation function: $FOR \times T \rightarrow \{t, f, u\}$. Here, FOR is a set of formulas. $V(A, w) = t$ reads "A is true at $w \in T$" and $V(A, w) = f$ "A is false at $w \in T$", respectively. If neither $V(A, w) = t$ nor $V(A, w) = f$, $V(A, w) = u$, where u is "indeterminate". The valuation function V is extended for other connectives as follows:

$V(\neg A, w) = t$ iff $V(A, w) \neq t$
$V(\neg A, w) = f$ iff $V(A, w) = t$
$V(A \wedge B, w) = t$ iff $V(A, w) = V(B, w) = t$
$V(A \wedge B, w) = f$ iff $V(A, w) = f$ or $V(B, w) = f$
$V(A \vee B, w) = t$ iff $V(A, w) = t$ or $V(B, w) = t$
$V(A \vee B, w) = f$ iff $V(A, w) = V(B, w) = f$
$V(\mathrm{F}A, w) = t$ iff $\exists v(w \leq v$ and $V(A, v) = t)$
$V(\mathrm{F}A, w) = f$ iff $\forall v(w \leq v$ imply $V(A, v) = f)$
$V(\mathrm{F}A, w) = u$ iff $\forall v(w \leq v$ imply $V(A, v) = u)$
$V(\mathrm{P}A, w) = t$ iff $\exists v(v \leq w$ and $V(A, v) = t)$
$V(\mathrm{P}A, w) = f$ iff $\forall v(v \leq w$ imply $V(A, v) = f)$
$V(\mathrm{P}A, w) = u$ iff $\forall v(v \leq w$ imply $V(A, v) = u)$

A formula A is *valid*, written $\models A$, iff for every FCP-model and for every time point $w \in T, V(A, w) = t$.

By the tandem presentation of truth and falsity, conjunction and disjunction can be appropriately interpreted. For negation, the specification is trivial to simulate the truth-value table. Namely, negated sentences do not allow a truth-value gap. For temporal operators, we need to give three separate descriptions capable of interpreting future contingents.

Below we give some technical results about FCP. First, the interpretations of G and H can be described by the duality of F and P.

Theorem 1. The following hold.

$$V(\mathrm{G}A, w) = t \text{ iff } \forall v(w \leq v \text{ imply } V(A, v) = t)$$
$$V(\mathrm{G}A, w) = f \text{ iff } \exists v(w \leq v \text{ and } V(A, v) \neq t)$$
$$V(\mathrm{H}A, w) = t \text{ iff } \forall v(v \leq w \text{ imply } V(A, v) = t)$$
$$V(\mathrm{H}A, w) = f \text{ iff } \forall v(w \leq v \text{ and } V(A, v) \neq t)$$

Theorem 2. $\vdash A \ \Rightarrow \models A$.

Theorem 3. For future non-contingent (determinate) sentences A,

$$\models_{K_t} A \text{ iff } \models A.$$

Here, $\models_{K_t}$ denotes standard K_t validity.

Theorem 4 (completeness). $\vdash A$ iff $\models A$.

4 Q_t

A three-valued temporal logic Q_t was developed by Akama, Nagata and Yamada in citeany2 as a temporal version of Q in Prior [8]. A modal logic Q was proposed for the purpose of giving the formal account on a logic of *contingent beings*, in which one could intelligibly say that some beings are contingent and some are necessary. If we here temporarily interpret Prior's idea, a temporal version of Q can be obtained capable of addressing the problem of future contingents. In fact, Prior also suggests a three-valued temporal logic without detailed remarks.

Akama, Nagata and Yamada worked out Prior's suggestions with the formalization of Q_t. We denote the language of Q_t by $\mathcal{L}_{Q_t}$. The logical symbols of $\mathcal{L}_{Q_t}$ are $\neg$ (negation), $\wedge$ (conjunction), $\rightarrow$ (implication), F (future possibility), P (past possibility), $\mathrm{S_P}$ (past statability), and $\mathrm{S_F}$ (future statability). The formation rule of formulas is standard. Let A and B be formulas and $p, p_1, ..., p_n$ be propositional variables. We denote by FOR the set of all formulas.

Note that $\vee$ (disjunction) and $\leftrightarrow$ (equivalence) can be defined in the usual way. The formula $\mathrm{S_F}A$ ($\mathrm{S_P}A$) reads "A is statable at all future (past) time points", or "A has a truth-value at all future (past) time points". By using these two statability operators, we can define the statability operator S as follows:

$$\mathrm{S}A =_{\mathrm{def}} \mathrm{S_P}A \wedge A \wedge \mathrm{S_F}A$$

The formula SA reads "A is statable at all time points", or "A has a truth-value at all time points". Observe that for Q we have only one statability operator S, since all worlds are accessible from a world.

The temporal necessity operators G and H are not dual to temporal possibility operators in the ordinary sense, but they can be introduced in Q_t as follows.

$$\mathrm{G}A =_{\mathrm{def}} \mathrm{S_F}A \wedge \neg\mathrm{F}\neg A, \quad \mathrm{H}A =_{\mathrm{def}} \mathrm{S_P}A \wedge \neg\mathrm{P}\neg A$$

It will be clear why GA and HA respectively need the conjuncts $\mathrm{S_F}A$ and $\mathrm{S_P}A$ in addition to $\neg\mathrm{F}\neg A$ and $\neg\mathrm{P}\neg A$ when looking at the proposed semantics.

Now, we present a Hilbert system for Q_t whose axioms and rules of inference are as follows:

Hilbert System for Q_t

Axioms

(C) Axioms for Positive Classical Logic

(S1) $\mathrm{S}_*A \to \mathrm{S}_*p$, for any p in A

(S2) $(\mathrm{S}_*p_1 \wedge ... \wedge \mathrm{S}_*p_n) \to \mathrm{S}_*A$, where $p_1, ..., p_n$ are all the propositional variables in A

(S3) $\mathrm{PS_F}A \to \mathrm{S_F}A$

(S4) $\mathrm{FS_P}A \to \mathrm{S_P}A$

(T1) $(\mathrm{S_F}p_1 \wedge ... \wedge \mathrm{S_F}p_n \wedge \neg\mathrm{F}\neg(A \to B) \wedge \neg\mathrm{F}\neg A) \to \neg\mathrm{F}\neg B$, where $p_1, .., p_n$ are all the propositional variables in B that are not in A

(T2) $(\mathrm{S_P}p_1 \wedge ... \wedge \mathrm{S_P}p_n \wedge \neg\mathrm{P}\neg(A \to B) \wedge \neg\mathrm{P}\neg A) \to \neg\mathrm{P}\neg B$, where $p_1, .., p_n$ are all the propositional variables in B that are not in A

(T3) $A \to \neg\mathrm{F}\neg\mathrm{P}A$

(T4) $A \to \neg\mathrm{P}\neg\mathrm{F}A$

Rules of Inference

(MP) $\vdash A, \vdash A \to B \Rightarrow \vdash B$

(NF) $\vdash A \Rightarrow \vdash \neg\mathrm{F}\neg A$

(NP) $\vdash A \Rightarrow \vdash \neg\mathrm{P}\neg A$

Here, S_* denotes one of $\mathrm{S_F}, \mathrm{S_P}$, and S. $\vdash A$ reads "A is provable in Q". (MP) is called *modus ponens* and (NF) and (NP) are called *necessitation*, respectively.

There are some differences between the Hilbert systems of Q_t and standard temporal logics. In Q_t, as mentioned above, temporal necessity and possibility operators are not dual. Thus, (T3), (T4), (NF) and (NP) have different forms.

Due to the presence of truth-value gaps, (T1) and (T2), which correspond to the distribution axiom (K), need the additional conjuncts for statability.

We are now in a position to describe a Kripke type semantics for Q_t. Here, we use a three-valued valuation function. A *three-valued Kripke model* (Q_t-model, for short) for Q_t is denoted by $\langle T, <, stat, V\rangle$. Here, T is a set of time points, $<$ is an ordering relation on $T \times T$, $stat$ is the *statability relation*, and V is a three-valued valuation function, i.e. $FOR \times T \to \{1 \text{ (true)}, 0 \text{ (false)}, \frac{1}{2} \text{ (undefined)}\}$. Here, $V(p, s) = \frac{1}{2}$ reads "p is unstatable (undefined) at s".

$stat(A, s)$ is the statability relation on $FOR \times T$, meaning that A is *statable* at s. By definition, $stat(A, s)$ iff for every propositional variable p occurring in A, $V(p, s) \neq \frac{1}{2}$. This means that p has a truth-value (either 1 or 0) at s. The statability relation plays an important role to interpret a formula with gaps.

Then, the valuations V can be extended for other formulas as follows:

$$
\begin{array}{lll}
V(\neg A, s) = 1 & \text{iff} & stat(A, s) \text{ and } V(A, s) \neq 1 \\
V(\neg A, s) = 0 & \text{iff} & stat(A, s) \text{ and } V(A, s) \neq 0 \\
V(\neg A, s) = \frac{1}{2} & \text{otherwise} & \\
V(A \wedge B, s) = 1 & \text{iff} & V(A, s) = V(B, s) = 1 \\
V(A \wedge B, s) = 0 & \text{iff} & stat(A, s) \text{ and } stat(B, s) \\
 & & \quad \text{and } (V(A, s) = 0 \text{ or } V(B, s) = 0) \\
V(A \wedge B, s) = \frac{1}{2} & \text{otherwise} & \\
V(\mathrm{F}A, s) = 1 & \text{iff} & stat(A, s) \text{ and } \exists t(s < t \text{ and } V(A, t) = 1) \\
V(\mathrm{F}A, s) = 0 & \text{iff} & stat(A, s) \text{ and } \forall t(s < t \Rightarrow V(A, t) \neq 1) \\
V(\mathrm{F}A, s) = \frac{1}{2} & \text{otherwise} &
\end{array}
$$

$$
\begin{array}{lll}
V(\mathrm{P}A, s) = 1 & \text{iff} & stat(A, s) \text{ and } \exists r(r < s \text{ and } V(A, r) = 1) \\
V(\mathrm{P}A, s) = 0 & \text{iff} & stat(A, s) \text{ and } \forall r(r < s \Rightarrow V(A, r) \neq 1) \\
V(\mathrm{P}A, s) = \frac{1}{2} & \text{otherwise} & \\
V(\mathrm{S_F}A, s) = 1 & \text{iff} & stat(A, s) \text{ and } \forall t(s < t \Rightarrow stat(A, t)) \\
V(\mathrm{S_F}A, s) = 0 & \text{iff} & stat(A, s) \text{ and } \exists t(s < t \text{ and } not(stat(A, t))) \\
V(\mathrm{S_F}A, s) = \frac{1}{2} & \text{otherwise} & \\
V(\mathrm{S_P}A, s) = 1 & \text{iff} & stat(A, s) \text{ and } \forall r(r < s \Rightarrow stat(A, r)) \\
V(\mathrm{S_P}A, s) = 0 & \text{iff} & stat(A, s) \text{ and } \exists r(r < s \text{ and } not(stat(A, r))) \\
V(\mathrm{S_P}A, s) = \frac{1}{2} & \text{otherwise} &
\end{array}
$$

Upon our semantics, it is worth giving the following intuitive meanings of temporal operators.

FA is true at s iff A is statable at s and A is true at some time point t later than s.
PA is true at s iff A is statable at s and A is true at some time point r earlier than s.
GA is true at s iff A is statable at s and A is true at every time point t later than s.
HA is true at s iff A is statable at s and A is true at every time point r earlier than s.

Note that these interpretations of GA and HA, which are compatible with Prior's Q, should be distinguished from the following weaker interpretations.

GA is true at s iff A is statable at s and A is true at every time point t later than s at which A is statable.
HA is true at s iff A is statable at s and A is true at every time point r earlier than s at which A is statable.

The notion of validity can be defined as follows. We say that a formula A is *valid*, in symbol $\models A$, iff A is not false, i.e. $V(A, s) \neq 0$ for any $s \in T$ in every Q_t-model. The notion of validity in Q_t is different from that of standard temporal logic. It is needed to validate classical tautologies when the valuation of every propositional variable in them is undefined. As a result, all classical tautologies are shown to be valid in Q_t.

Consider the following two contingent sentences:

(3) Jan will read *Past, Present and Future.*
(4) Jan will not read *Past, Present and Future.*

(3) is expressed in Q_t as:

(5) FA

where A denotes the present tense proposition that Jan reads *Past, Present and Future.* Since 3() is a future contingent proposition, (5) evaluates as indeterminate. Similarly, we can represent (4) as (6) or (7):

(6) F$\neg A$
(7) $\neg$FA

(6) is indeterminate. Even if (7) is taken up, it is also indeterminate. Thus, there are no differences of (6) and (7) in Q_t. However, they should be distinguished unless truth-value gaps are admitted.

A controversial case for future contingents is like in (9):

(8) F$(A \vee \neg A)$

which is equivalent to (10):

(9) F$A \vee$ F$\neg A$

Both (8) and (9) are also shown to be indeterminate if A is indeterminate. Although Aristotle believed that (8) is a logical truth because of (LEM), it may be controversial that $A \vee \neg A$ holds for indeterminate A. Fortunately, our notion of validity in the semantics above ensures that $A \vee \neg A$ is also valid when A is indeterminate.

Now, we show some technical results whose proofs can be found in [3]. The first result states that the underlying logic of Q_t is Kleene's *weak three-valued logic*, which is equivalent to Bochvar's three-valued logic. In Kleene's weak three-valued logic, complex formulas receive the third truth-value if their compounds receive the third truth-value, as the following truth tables show.

$\rightarrow$	t	u	f	$\neg$
t	t	u	f	f
u	u	u	u	u
f	t	u	t	t

$\wedge$	t	u	f
t	t	u	f
u	u	u	u
f	f	u	f

$\vee$	t	u	f
t	t	u	t
u	u	u	u
f	t	u	f

Theorem 5. The logical connectives $\neg$ and $\wedge$ in Q_t are those in Kleene's weak three-valued logic.

Now, we define other connectives as

$$A \vee B =_{\text{def}} \neg(\neg A \wedge \neg B), \quad A \rightarrow B =_{\text{def}} \neg A \vee B.$$

Theorem 6. The fragment of $\mathcal{L}_{Q_t}$ with the logical connectives $\neg, \wedge, \vee, \rightarrow$ is Kleene's weak three-valued logic.

Theorem 7. $stat(A, s)$ iff $V(A, s) \neq \frac{1}{2}$.

Theorem 8. The future and past statability operators can be interpreted as follows:

$$V(\mathrm{S_F} A, s) = 1 \text{ iff } V(A, s) \neq \tfrac{1}{2} \text{ and } \forall t(s < t \Rightarrow V(A, t) \neq \tfrac{1}{2}),$$
$$V(\mathrm{S_P} A, s) = 1 \text{ iff } V(A, s) \neq \tfrac{1}{2} \text{ and } \forall r(r < s \Rightarrow V(A, r) \neq \tfrac{1}{2}).$$

Theorem 9 (Soundness)**.** $\vdash A \Rightarrow \models A$.

Theorem 10 (Completeness)**.** $\models A \Rightarrow \vdash A$.

5 Other Possibilities

There seem other possibilities of using of three-valued temporal logics. We here take up some of the interesting approaches, which will be worked out in other occasions. One can consider *intuitionistic tense logic* as an interesting candidate. Intuitionistic tense logic ITL was studied in detail by Ewald [5]. We find some similarities between Q_t and ITL, e.g. failures of PB and of duality of necessity and possibility operators. There are, however, differences. First, Q_t has a three-valued semantics, while ITL has no finite-valued semantics. Second, (LEM) holds in Q_t, but it does not hold in ITL. Although we cannot find practical applications of ITL, it is worth investigating future contingents within the framework of ITL from an intuitionistic point of view.

Although most approaches to future contingents rely on the notion of *gap*, i.e. neither true nor false, the notion of *glut*, i.e. both true and false, can be also used. This consideration gives rise to a *paraconsistent approach* in which a future contingent is interpreted as both true and false. The paraconsistent approach may be implicit in Priest's dialectical tense logic DTL in [7].

As is well known, Lukasiewicz proposed *L-modal systems*. This can be reformulated as a temporal logic. If this attempt is successful, Lukasiewicz's original idea to introduce his three-valued logic will be re-appreciated.

Three-valued temporal logics with *gap* can be semantically formalized by means of *supervaluation* of van Frrassen [9]. If the paraconsistent approach is philosophically defensible, three-valued temporal logics with *glut* can also be semantically formalized by means of *subvaluation*.

6 Conclusions

We have seen two three-valued temporal logics for future contingents and suggested other possibilities. These formalizations are based on different motivations, and it would be interesting to consider these philosophical aspects.

Three-valued temporal logics should be investigated for other issues. For instance, in computer science, we could study *model checking* in some three-valued logics. In philosophy, the so-called *Master Argument* is an interesting topic, which can be naturally formalized by three-valued modal temporal logic. In linguistics, the compositional semantics for future contingent propositions need a proper three-valued temporal logic. We hope to elaborate on these issues in other papers.

References

1. Akama, S., Nagata, Y.: On Prior's three-valued modal logic Q. In: Proc. of ISMVL 2005, pp. 14–19 (2005)
2. Akama, S., Nagata, Y., Yamada, C.: A three-valued temporal logic for future contingents. Logique et Analyse 198, 99–111 (2007)
3. Akama, S., Nagata, Y., Yamada, C.: Three-valued temporal logic Q_t and future contingents. Studia Logica 88, 215–231 (2008)
4. Aristotle: De Interpretatione. Translated by Edghill, E.M., Ross, W.D. (eds.) The Works of Aristotle. Oxford University Press, Oxford (1963)
5. Ewald, W.: Intuitionistic tense and modal logic. The Journal of Symbolic Logic 51, 166–179 (1986)
6. Lukasiewicz, J.: On 3-valued logic, 1920. In: McCall, S. (ed.) Polish Logic, pp. 16–18. Oxford University Press, Oxford (1967)
7. Priest, G.: To be and not to be: dialectical tense logic. Studia Logica 41, 63–76 (1982)
8. Prior, A.N.: Time and Modality. Oxford University Press, Oxford (1957)
9. van Fraassen, B.C.: Presuppositions, implications, and self-reference. Journal of Philosophy 65, 136–152 (1968)

Introduction to Plausible Reasoning Based on EVALPSN

Kazumi Nakamatsu, Toshiaki Imai, Jair M. Abe, and Seiki Akama

Abstract. In this paper, in order to deal with plausible (deontic) reasoning in the framework of EVALPSN as well as defeasible deontic one, we introduce translation rules from Billington's plausible theory into EVALPSN and show that the provability of the plausible theory can be preserved by the translation with a simple example.

Keywords: plausible logic, defeasible reasoning, paraconsistent annotated logic program, stable model, EVALPSN.

1 Introduction

In order to deal with defeasible deontic reasoning in the framework of a paraconsistent annotated logic program called EVALPSN by Nakamatsu, it has been shown that Billington's defeasible logic and Nute's defeasible deontic logic can be translated into EVALPSN with preserving their provabilities [4, 5]. Based on the translation, EVALPSN defeasible deontic reasoning has been applied to various intelligent control based on safety verification such as logical safety verification based pipeline control [6, 7].

Kazumi Nakamatsu
University of Hyogo, Himeji, Japan
e-mail: nakamatu@shse.u-hyogo.ac.jp

Toshiaki Imai
University of Hyogo, Himeji, Japan
e-mail: nc05a015@stshse.u-hyogo.ac.jp

Jair M. Abe
Paulista Univerity, Sao Paulo, Brazil
e-mail: jairabe@uol.com.br

Seiki Akama
University of Tsukuba, Tsukuba, Japan
e-mail: sub-akama@jcom.home.ne.jp

K. Nakamatsu et al. (Eds.): New Advan. in Intel. Decision Techno., SCI 199, pp. 363–372.
springerlink.com

Plausible reasoning is known as one of important reasoning ways in human being society such as judge's reasoning in court trial, and it has been formalized as a plausible logic by Billington [1].

In this paper, in order to deal with plausible (deontic) reasoning in the framework of EVALPSN as well as defeasible deontic one, we introduce translation rules from Billington's plausible theory into EVALPSN and show that the provability of the plausible theory can be preserved by the translation with a simple example, that is to say, the provability of plausible logic can be translated into the stable model satisfiability of EVALPSN.

This paper is organized in the following manner: firstly, EVALPSN is reviewed briefly; subsequently Billington's plausible logic and its translation into EVALPSN are introduced with a simple example; lastly, we conclude the future works for applications of EVALPSN plausible reasoning.

2 Evalpsn

We review EVALPSN briefly [5]. Generally, a truth value called an *annotation* is explicitly attached to each literal in annotated logic programs [2]. For example, let p be a literal, μ an annotation, then $p:\mu$ is called an *annotated literal*. The set of annotations constitutes a complete lattice. An annotation in EVALPSN has a form of $[(i,j),\mu]$ called an *extended vector annotation*. The first component (i,j) is called a *vector annotation* and the set of vector annotations constitutes the complete lattice,

$$\mathscr{T}_v(n) = \{\ (x,y)|0 \le x \le n, 0 \le y \le n, x, y \text{ and } n \text{ are integers }\}$$

in Fig.1. The ordering($\preceq_v$) of $\mathscr{T}_v(n)$ is defined as : let (x_1,y_1), $(x_2,y_2) \in \mathscr{T}_v(n)$,

$$(x_1,y_1) \preceq_v (x_2,y_2) \text{ iff } x_1 \le x_2 \text{ and } y_1 \le y_2.$$

For each extended vector annotated literal $p:[(i,j),\mu]$, the integer i denotes the amount of positive information to support the literal p and the integer j denotes that of negative one. The second component μ is an index of fact and deontic notions such as obligation, and the set of the second components constitutes the complete lattice,

$$\mathscr{T}_d = \{\bot, \alpha, \beta, \gamma, *_1, *_2, *_3, \top\}.$$

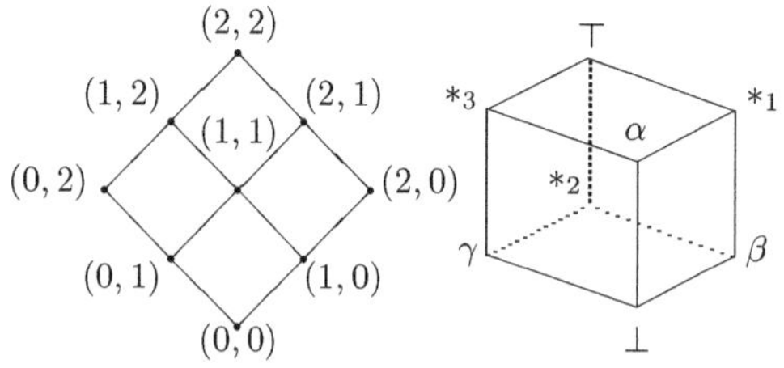

Fig. 1 Lattice $\mathscr{T}_v(2)$ and Lattice $\mathscr{T}_d$

The ordering($\preceq_d$) of $\mathcal{T}_d$ is described by the Hasse's diagram in **Fig.1**. The intuitive meaning of each member of $\mathcal{T}_d$ is $\perp$ (unknown), α (fact), β (obligation), γ (non-obligation), $*_1$ (fact and obligation), $*_2$ (obligation and non-obligation), $*_3$ (fact and non-obligation), and $\top$ (inconsistency). Then the complete lattice $\mathcal{T}_e(n)$ of extended vector annotations is defined as the product $\mathcal{T}_v(n) \times \mathcal{T}_d$. The ordering($\preceq_e$) of $\mathcal{T}_e(n)$ is defined as : let $[(i_1, j_1), \mu_1]$ and $[(i_2, j_2), \mu_2] \in \mathcal{T}_e$,

$$[(i_1, j_1), \mu_1] \preceq_e [(i_2, j_2), \mu_2] \quad \text{iff} \quad (i_1, j_1) \preceq_v (i_2, j_2) \quad \text{and} \quad \mu_1 \preceq_d \mu_2.$$

There are two kinds of *epistemic negation* ($\neg_1$ and $\neg_2$) in EVALPSN, which are defined as mappings over $\mathcal{T}_v(n)$ and $\mathcal{T}_d$, respectively.

Definition 1 (epistemic negations $\neg_1$ and $\neg_2$ in EVALPSN)

$$\begin{aligned}
&\neg_1([(i,j),\mu]) = [(j,i),\mu], \ \forall \mu \in \mathcal{T}_d,\\
&\neg_2([(i,j),\perp]) = [(i,j),\perp], \quad \neg_2([(i,j),\alpha]) = [(i,j),\alpha],\\
&\neg_2([(i,j),\beta]) = [(i,j),\gamma], \quad \neg_2([(i,j),\gamma]) = [(i,j),\beta],\\
&\neg_2([(i,j),*_1]) = [(i,j),*_3], \quad \neg_2([(i,j),*_2]) = [(i,j),*_2],\\
&\neg_2([(i,j),*_3]) = [(i,j),*_1], \quad \neg_2([(i,j),\top]) = [(i,j),\top].
\end{aligned}$$

If we regard the epistemic negations as syntactical operations, the epistemic negations followed by literals can be eliminated by the syntactical operations. For example, $\neg_1 p:[(2,0),\alpha] = p:[(0,2),\alpha]$ and $\neg_2 q:[(1,0),\beta] = p:[(1,0),\gamma]$.

There is another negation called *strong negation* ($\sim$) in EVALPSN, and it is treated as classical negation.

Definition 2 (strong negation $\sim$) [3]**.** Let F be any formula and $\neg$ be $\neg_1$ or $\neg_2$.

$$\sim F =_{def} F \rightarrow ((F \rightarrow F) \wedge \neg(F \rightarrow F)).$$

Definition 3 (well extended vector annotated literal)**.** Let p be a literal.

$$p:[(i,0),\mu] \quad \text{and} \quad p:[(0,j),\mu]$$

are called *weva(well extended vector annotated)-literals*, where $i, j \in \{1, 2, \cdots, n\}$, and $\mu \in \{\alpha, \beta, \gamma\}$.

Defintion 4 (EVALPSN)**.** If $L_0, \cdots, L_n$ are weva-literals,

$$L_1 \wedge \cdots \wedge L_i \wedge \sim L_{i+1} \wedge \cdots \wedge \sim L_n \rightarrow L_0$$

is called an *EVALPSN clause*. An *EVALPSN* is a finite set of EVALPSN clauses.

Fact and deontic notions, "obligation", "forbiddance" and "permission" are represented by extended vector annotations, $[(m,0),\alpha]$, $[(m,0),\beta]$, $[(0,m),\beta]$, and $[(0,m),\gamma]$, respectively, where m is a positive integer.

3 Plausible Logic

In this section, we simplify Billington's plausible logics in [1] and introduce the simplified version briefly. We note that some symbols are used duplicatedly as the previous section with different meanings.

First of all, as preliminary we introduce the basic notation and terminology in his logic.

Definition 5. Let Z be the set of all integers. If $m,n \in Z$ then $[m..n] = \{i \in Z \mid m \leq i \leq n\}$. The length of a sequence P is denoted by $|P|$. Let $P = (P(1),\ldots,P(|P|))$ be a finite sequence. If $i \in [1..|P|]$ then $P[1..i] = (P(1),\ldots,P(i))$, and if $i = 0$ then $P[1..i] = ()$, the empty sequence.

Definition 6 (Alphabet $\mathscr{A}$). Alphabet is the union of the following four pairwise disjoint sets of symbols:

- a non-empty countable set of atoms;
- the set $\{\neg,\wedge,\vee,\rightarrow,\Rightarrow,\rightsquigarrow\}$ of connectives ;
- the set $\{+,-,\Delta,\delta,\partial,\int\}$ of proof symbols ;
- and the set of punctuation marks consisting of the comma, braces and parentheses.

Definition 7

- If F is a non-empty set of formulas, $\bigwedge F(\bigvee F)$ denote the conjunction (disjunction) of all the formulas in F.
- A *clause* is the disjunction of a non-empty set of literals.
- The *complement*, $\sim f$, of a formula f and the *complement*, $\sim F$, of a set F of formulas are defined as follows: If f is an atom then $\sim f$ is $\neg f$; and $\sim \neg f$ is f. Otherwise $\sim F = \{\sim f | f \in F\}$, $\sim \bigwedge F = \bigvee \sim F$, and $\sim \bigvee F = \bigwedge \sim F$.

Definition 8 (Rules). A *rule* r has three parts:

- a finite set of formulas on the right called *antecedent* symbolized $A(r)$;
- an arrow in the middle
  (*strict arrow* $\rightarrow$, *plausible arrow* $\Rightarrow$, and *defeater arrow* $\rightsquigarrow$);
- a formula on the right called *consequent* symbolized $c(r)$.

$A \rightarrow f$ are called *strict rules*, whenever all the formulas in A are accepted, then f must be accepted; that is to say, Modus Ponens holds. $A \Rightarrow f$ are called *plausible rules*, if al the formulas in A are accepted then the formula f is accepted provided that all the evidence against f has been defeated. $A \rightsquigarrow f$ are called *defeater rules* or if A, *defeaters*, if all the formulas in A are accepted then the rule $A \rightsquigarrow f$ is an evidence against $\neg f$, but not for f. We can read $A \rightsquigarrow f$ as "if A then the negation of f is too risky". If f is a presumption then we can write it $\{\} \Rightarrow f$.

Definition 9 (Plausible Description). A *plausible description* over $\mathscr{A}$ of a situation is a triple

$$PD = (F, PR, DR)$$

such that F is a finite set of formulas, PR is a set of plausible rules, DR is a set of defeaters, and a finite conditions to be defined later. The formulas in F are called *facts*.

Definition 10. If f is a formula then $cnf(f)$ denotes a *conjunctive normal form* of f, and $dnf(f)$ denotes a *disjunctive normal form* of f. If $cnf(f) = \bigwedge\{\bigvee C_1, \dots, \bigvee C_m\}$ then let $Cl(f) = \{\bigvee C_1, \dots, \bigvee C_m\}$ and $Cll(f) = \{C_1, \dots, C_m\}$. If $dnf(f) = \bigvee\{\bigwedge D_1, \dots, \bigwedge D_m\}$ then let $Dcll(f) = \{D_1, \dots, \bigvee D_m\}$.

Definition 11 (Transformation into Simple Rules). The set of facts F is transformed into the set of simple strict rules,

$$Smpl(F) = \{\sim (C - \{q\}) \to q \mid q \in C, C \in Cll(f), f \in F\}.$$

The set of plausible rules PR is transformed into the set of simple plausible rules,

$$Smpl(PR) = \{B \cup \sim (C - \{q\}) \Rightarrow q \mid B \in Dcll(\bigwedge A(r)), \\ C \in Cll(c(r)), q \in C, r \in PR\}.$$

The set of defeaters DR is transformed into the set of simple defeaters,

$$Smpl(DR) = \{B \cup \sim (C - \{q\}) \rightsquigarrow q \mid B \in Dcll(\bigwedge A(r)), \\ C \in Cll(c(r)), q \in C, r \in DR\}.$$

The simple form of a plausible description PD is defined as

$$Smpl(PD) = Smpl(F) \cup Smpl(PR) \cup Smpl(DR).$$

Definition 12. If S is any set of rules then $S[q] = \{r \in S \mid q = c(r)\}$. the set of strict rules in S is denoted by S_s, the set of plausible rules in S by S_p and the set of defeater in S by S_d, moreover, $S_{pd} = S_p \cup S_d$ and $S_{sp} = S_s \cup S_p$.

Definition 13 (Non Simple Rules). Let S be a set of simple rules. non simple rule (strict rule)

$$ns(S_s[q]) = Cl(\bigvee\{\bigwedge A(r) \mid r \in S_s[q]\}) \to q,$$

non-simple rule (plausible rule)

$$ns(S_{sp}[q]) = Cl(\bigvee\{\bigwedge A(r) \mid r \in S_{sp}[q]\}) \Rightarrow q,$$

non-simple rule (defeater)

$$ns(S[q]) = Cl(\bigvee\{\bigwedge A(r) \mid r \in S_[q]\}) \rightsquigarrow q.$$

The set of non-simple rules generated from S,

$$Nsmpl(S) = \{ns(S_s[q]) \mid q \text{ is a literal}\} \cup \{ns(S_{sp}[q]) \mid q \text{ is a literal}\} \cup \\ \{ns(S[q]) \mid q \text{ is a literal}\}.$$

We define

$$\Theta(PD) = Smpl(PD) \cup Nsmpl(Smpl(PD)).$$

Definition 14 (Plausible Theory). A *plausible theory* over $\mathscr{A}$ is an ordered pair $(R, >)$ such that $R = \Theta(PD)$ for some plausible description PD over $\mathscr{A}$, and $>$ is a priority relation on R.

Definition 15 (Formal Proof). A *tag* is any element of $\{+\Delta, -\Delta, +\delta, -\delta, +\partial, -\partial, +\int, -\int\}$. A *formal proof* or *derivation*, P is a finite sequence $P = (P(1), \ldots, P(|P|))$ of ordered pairs $(T, \pm df)$ such that T is a plausible theory, $\pm \in \{+,-\}$, $d \in \{\Delta, \delta, \partial, \int\}$, f is a cnf-formula and for all i in $[0..|P|-1]$ the following inference conditions: $+\bigwedge$,$-\bigwedge$,$+\bigvee$,$-\bigvee$,$+\Delta$, $-\Delta$,$+\partial$,$-\partial$,$+\int$ and $-\int$ are all hold. The elements of a derivation is called *lines* of the derivation.

Definition 16 (Provability or Deducibility in Plausible Logic). Let T be a plausible theory and f a cnf-formula.

$$T \vdash f \iff (T, \pm df) \text{ is a line in some derivation.}$$

$(T, \pm df)$ is called *deducible* or *provable* or *d-provable*.

Definition 17. Define $\mathscr{P}^{-}(F) = \{G|G \subset F \text{ and } \{\} \neq G \text{ and } G \neq F\}$ to be the set of non-empty proper subsets of F, and $\mathscr{P}_{\geq 1}(F) = \{G|G \subset \text{ and } |G| \geq 1\}$ to be the set of non-empty subsets of F.

$+\bigwedge$) If $P(i+1) = (T, +d\bigwedge F)$ then $\forall f \in F, (T, +df) \in P[1..i]$.
$-\bigwedge$) If $P(i+1) = (T, -d\bigwedge F)$ then $\exists f \in F, (T, -df) \in P[1..i]$.

$+\bigvee$) If $P(i+1) = (T, +d\bigvee F)$ and F is a set of literals then
$\exists F' \in \mathscr{P}_{\geq 1}, \forall f \in F'$,
$((R \cup \{\{\} \rightarrow \sim q : q \in (F' - \{f\})\}, >), +df) \in P[1..i]$.
$-\bigvee$) If $P(i+1) = (T, -d\bigvee F)$ and F is a set of literals then
$forallF' \in \mathscr{P}_{\geq 1}, \exists f \in F'$,
$((R \cup \{\{\} \rightarrow \sim q : q \in (F' - \{f\})\}, >), -df) \in P[1..i]$.

In this paper, we will not introduce the translation of the above inference conditions $\pm\bigwedge$ and $\pm\bigvee$ to EVALPSN due to space restriction.

Definition 18. $+\Delta(T) = \{f|T \vdash +\Delta f\}$. The set of clauses in $+\Delta(T)$ is denoted by $Cl(+\Delta(T))$. The set of *prime* clauses in $+\Delta(T)$, $pr(+\Delta(T))$ is defined to be the set of clauses in $+\Delta(T)$ that have no subclause in $+\Delta(T)$, i.e., $pr(+\Delta(T)) = \{\bigvee C \in Cl(+\Delta(T))| \text{ if } D \subset C \text{ and } D \neq C \text{ then } \bigvee D \notin +\Delta(T)\}$. $Inc(T) = \{\sim C| \bigvee \in pr(+\Delta(T))\}$ is defined to be the minimal sets of literals that are *inconsistent with* T. $Inc(T,q) = \{I - \{q\}|q \in I \text{ and } I \in Inc(T)\}$.

$+\Delta$) If $P(i+1) = (T, +\Delta q)$ then
$\exists r \in R_s[q] \forall a \in A(r), (T, +\Delta a) \in P[1..i]$.
$-\Delta$) If $P(i+1) = (T, -\Delta q)$ then
$\forall r \in R_s[q] \exists a \in A(r), (T, -\Delta a) \in P[1..i]$.

$+\partial$) If $P(i+1) = (T, +\partial q)$ then either
 .1) $\exists r \in R_s[q] \forall a \in A(r), (T, +\partial a) \in P[1..i]$; or
 .2) $\exists r \in R_p[q]$ such that
  .1) $\forall a \in A(r), (T, +\partial a) \in P[1..i]$, and
  .2) $\forall C \in Inc(T,q), \exists c \in C \forall s \in R[c]$ either
   .1) $\exists a \in A(s), (T, -\partial a) \in P[1..i]$; or
   .2) $\exists t \in R_p[q]$ such that
    .1) $\forall a \in A(t), (T, +\partial a) \in P[1..i]$, and
    .2) $t > s$.

$-\partial$) If $P(i+1) = (T, -\partial q)$ then either
 .1) $\forall r \in R_s[q] \exists a \in A(r), (T, -\partial a) \in P[1..i]$, and
 .2) $\forall r \in R_p[q]$ either
  .1) $\exists a \in A(r), (T, -\partial a) \in P[1..i]$; or
  .2) $\exists C \in Inc(T,q), \forall c \in C \exists s \in R[c]$ such that
   .1) $\forall a \in A(s), (T, +\partial a) \in P[1..i]$, and
   .2) $\forall t \in R_p[q]$ either
    .1) $\exists a \in A(t), (T, -\partial a) \in P[1..i]$; or
    .2) $t \not> s$.

$+\delta$) If $P(i+1) = (T, +\delta q)$ then either
 .1) $\exists r \in R_s[q] \forall a \in A(r), (T, +\delta a) \in P[1..i]$; or
 .2) $\exists r \in R_p[q]$ such that
  .1) $\forall a \in A(r), (T, +\delta a) \in P[1..i]$, and
  .2) $\forall C \in Inc(T,q), \exists c \in C \forall s \in R[c]$ either
   .1) $\exists a \in A(s), (T, -\int a) \in P[1..i]$; or
   .2) $\exists t \in R_p[q]$ such that
    .1) $\forall a \in A(t), (T, +\delta a) \in P[1..i]$, and
    .2) $t > s$.

$-\delta$) If $P(i+1) = (T, -\delta q)$ then
 .1) $\forall r \in R_s[q] \exists a \in A(r), (T, -\delta a) \in P[1..i]$, and
 .2) $\forall r \in R_p[q]$ either
  .1) $\exists a \in A(r), (T, -\delta a) \in P[1..i]$; or
  .2) $\exists C \in Inc(T,q), \forall c \in C \exists s \in R[c]$ such that
   .1) $\forall a \in A(s), (T, +\int a) \in P[1..i]$, and
   .2) $\forall t \in R_p[q]$ either
    .1) $\exists a \in A(t), (T, -\delta a) \in P[1..i]$; or
    .2) $t \not> s$.

$+\int$) If $P(i+1) = (T, +\int q)$ then either
 .1) $\exists r \in R_s[q] \forall a \in A(r), (T, +\int a) \in P[1..i]$; or
 .2) $\exists r \in R_p[q]$ such that
  .1) $\forall a \in A(r), (T, +\int a) \in P[1..i]$, and
  .2) $\forall C \in Inc(T,q), \exists c \in C \forall s \in R[c]$ either
   .1) $\exists a \in A(s), (T, -\delta a) \in P[1..i]$; or
   .2) $t \not> s$.

$-\int)$ If $P(i+1) = (T, -\int q)$ then
 .1) $\forall r \in R_s[q] \exists a \in A(r), (T, -\int a) \in P[1..i]$, and
 .2) $\forall r \in R_p[q]$ either
  .1) $\exists a \in A(r), (T, -\int a) \in P[1..i]$; or
  .2) $\exists C \in Inc(T,q), \forall c \in C \exists s \in R[c]$ such that
   .1) $\forall a \in A(s), (T, +\delta a) \in P[1..i]$, and
   .2) $t > s$.

Definition 19 (Plausible Logic). A *plausible logic* consists of a plausible theory $(R, >)$, the twelve inference conditions $\{\pm\bigwedge, \pm\bigvee, \pm\Delta, \pm\delta, \pm\partial, \pm\int\}$. deducibility relation $\vdash$.

The following theorem holds with the definitions $+d(T) = \{f | T \vdash +df\}$ and $-d(T) = \{f | T \vdash -df\}$, where $d \in \{\Delta, \delta, \partial, \int\}$, and the proof is found in [1].

Theorem 1 [1]. Let T be a plausible theory.

(1) If $d \in \{\Delta, \delta, \partial, \int\}$ then $+d(T) \cap -d(T) = \{\}$.
(2) $+\Delta(T) \subset +\delta(T) \subset +\partial(T) \subset +\int(T)$.
(3) $-\int(T) \subset -\partial(T) \subset -\delta(T) \subset -\Delta(T)$.

4 Translation from Plausible Theory into EVALPSN

In this section, we consider to translate the plausible theory into EVALPSN, and assume the following correspondence between the provability of the plausible theory and satisfiability of EVALPSN stable model as well as [4].

$$\begin{aligned}
T \vdash +\Delta q &\longleftrightarrow M(Tran(T)) \models q:[(4,0),\alpha];\\
T \vdash +\delta q &\longleftrightarrow M(Tran(T)) \models q:[(3,0),\alpha];\\
T \vdash +\partial q &\longleftrightarrow M(Tran(T)) \models q:[(2,0),\alpha];\\
T \vdash +\int q &\longleftrightarrow M(Tran(T)) \models q:[(1,0),\alpha],
\end{aligned}$$

where $Tran(T)$ is the EVALPSN translated from the plausible theory T, and $M(Tran(T))$ is the stable model of $Tran(T)$. In the above assumption, the deontic annotation α is not essential.

Strict Rule
Based on the assumption we translate a strict rule $R : a_1, \ldots, a_k \rightarrow q$ into EVALPSN with considering the inference conditions $+\Delta)$, $+\partial)$, $+\delta)$ and $+\int)$. The strict rule R is translated into

$$Tran(R) = \{a_1:[(n,0),\alpha] \wedge \ldots \wedge a_k:[(n,0),\alpha] \rightarrow q:[(n,0),\alpha]\},\ n = 1, \ldots, 4$$

Plausible Rule
Suppose that there are plausible rules, $R_0 : a_1, \ldots, a_k \Rightarrow q$, and $R_1 : b_{11}, \ldots, b_{1k} \Rightarrow (\rightsquigarrow)\bar{q}_1$, $\cdots$, $R_m : b_{m1}, \ldots, b_{ml} \Rightarrow (\rightsquigarrow)\bar{q}_m$, where $R_0, \ldots, R_m \in R$, $k, l, m(\geq 1)$

are integers, and $\bar{q}_1, \ldots, \bar{q}_m \in Inc(T, q)$. Then, we obtain the following $Tran(R_0)$ for each plausible rule R_m in each provability,

$+\delta)$ for each rule R_m with the priority relation $(R_m > R_0)$, $Tran(R_0) =$

$$\{\ a_1 : [(3,0),\alpha] \wedge \ldots \wedge a_k : [(3,0),\alpha] \wedge \sim b_{m1} : [(1,0),\alpha] \rightarrow q : [(3,0),\alpha],$$

$$\vdots$$

$$a_1 : [(3,0),\alpha] \wedge \ldots \wedge a_k : [(3,0),\alpha] \wedge \sim b_{ml} : [(1,0),\alpha] \rightarrow q : [(3,0),\alpha]\ \},$$

otherwise, $Tran(R_0) = \{\ a_1 : [(3,0),\alpha] \wedge \ldots \wedge a_k : [(3,0),\alpha] \rightarrow q : [(3,0),\alpha]\ \}$.
$+\partial)$ for each rule R_m with the priority relation $(R_m > R_0)$, $Tran(R_0) =$

$$\{\ a_1 : [(2,0),\alpha] \wedge \ldots \wedge a_k : [(2,0),\alpha] \wedge \sim b_{m1} : [(2,0),\alpha] \rightarrow q : [(2,0),\alpha],$$

$$\vdots$$

$$a_1 : [(2,0),\alpha] \wedge \ldots \wedge a_k : [(2,0),\alpha] \wedge \sim b_{ml} : [(2,0),\alpha] \rightarrow q : [(2,0),\alpha]\ \},$$

otherwise, $Tran(R_0) = \{a_1 : [(2,0),\alpha] \wedge \ldots \wedge a_k : [(2,0),\alpha] \rightarrow q : [(2,0),\alpha]\}$.
$+\int)$ for each rule R_m with the priority relation $(R_m > R_0)$, $Tran(R_0) =$

$$\{\ a_1 : [(1,0),\alpha] \wedge \ldots \wedge a_k : [(1,0),\alpha] \wedge \sim b_{m1} : [(3,0),\alpha] \rightarrow q : [(1,0),\alpha],$$

$$\vdots$$

$$a_1 : [(1,0),\alpha] \wedge \ldots \wedge a_k : [(1,0),\alpha] \wedge \sim b_{ml} : [(3,0),\alpha] \rightarrow q : [(1,0),\alpha]\ \},$$

otherwise, $Tran(R_0) = \{a_1 : [(1,0),\alpha] \wedge \ldots \wedge a_k : [(1,0),\alpha] \rightarrow q : [(1,0),\alpha]\}$.

5 Translation Example

In this section, we present a simple example of the translation from a plausible theory to an EVALPSN.

Example 1 (see [1])
Suppose $T = (R, >)$ where $R = \{R_1 : \{\} \Rightarrow a,\ R_2 : \{\} \Rightarrow b,\ R_3 : \{\} \Rightarrow c,\ R_4 : a \Rightarrow e,\ R_5 : b \Rightarrow d,\ R_6 : c \Rightarrow \neg d\ R_7 : d \Rightarrow \neg e\}$ and $>$ is empty. Then, we have provabilities, $T \vdash +\delta a,\ +\delta b,\ +\delta c,\ -\partial d,\ +\int d,\ +\int \neg d,\ -\delta e,\ -\partial \neg e$, and $+\partial e,\ +\int \neg e$. The plausible theory T is translated into the EVALPSN $Tran(T) =$

$$\begin{aligned}
\{\ & a : [(3,0),\alpha], \quad b : [(3,0),\alpha], \quad c : [(3,0),\alpha],\\
& a : [(3,0),\alpha] \wedge \sim d : [(1,0),\alpha] \rightarrow e : [(3,0),\alpha], \quad a : [(1,0),\alpha] \rightarrow e : [(1,0),\alpha],\\
& a : [(2,0),\alpha] \wedge \sim d : [(2,0),\alpha] \rightarrow e : [(2,0),\alpha],\\
& d : [(3,0),\alpha] \wedge \sim a : [(1,0),\alpha] \rightarrow e : [(0,3),\alpha], \quad d : [(1,0),\alpha] \rightarrow e : [(0,1),\alpha],\\
& d : [(2,0),\alpha] \wedge \sim a : [(2,0),\alpha] \rightarrow e : [(0,2),\alpha],\\
& b : [(3,0),\alpha] \wedge \sim c : [(1,0),\alpha] \rightarrow d : [(3,0),\alpha], \quad b : [(1,0),\alpha] \rightarrow d : [(1,0),\alpha],
\end{aligned}$$

$$b:[(2,0),\alpha] \wedge \sim c:[(2,0),\alpha] \to d:[(2,0),\alpha],$$
$$c:[(3,0),\alpha] \wedge \sim b:[(1,0),\alpha] \to d:[(0,3),\alpha], \quad c:[(1,0),\alpha] \to d:[(0,1),\alpha],$$
$$c:[(2,0),\alpha] \wedge \sim b:[(2,0),\alpha] \to d:[(0,2),\alpha] \quad \}$$

Then we have the following stable model[4] of the EVALPSN $Tran(T)$,

$$M(Tran(T)) = \{ \ a:[(3,0),\alpha], \quad b:[(3,0),\alpha], \quad c:[(3,0),\alpha], \quad d:[(1,0),\alpha], \\ d:[(0,1),\alpha], \quad e:[(2,0),\alpha], \quad e:[(0,1),\alpha] \ \},$$

which corresponds to the provabilities of the original plausible theory T. The stable model shows that *a*, *b* and *c* are provable at δ-level that is represented by the vector annotation $(3,0)$, both d and $\neg d$ are just provable at $\int$-level that is represented by the vector annotation $(1,0)$, *e* is provable at ∂-level that is represented by the vector annotation $(2,0)$, and $\neg e$ is just provable at $\int$-level. Therefore, by **Theorem 1** it can be reasoned that neither d nor $\neg d$ are provable at both ∂ and δ-levels, *e* is unprovable at δ-level, and $\neg e$ is unprovable at both ∂ and δ-levels.

6 Conclusion and Future Work

In this paper, in order to deal with plausible reasoning in the flamework of EVALPSN as well as defeasible deontic one, we have introduced translation rules from Billington's plausible theory into EVALPSN and shown that the provability of the plausible logic can be translated into the stable model satisfiability of EVALPSN.

We will be planning to apply the EVALPSN plausible reasoning to some controls and analysing judge's reasoning in trial in our future work.

References

1. Billington, D.: Propositional Plausible Logic: Introduction and Implementation. Studia Logica 67, 243–269 (1989)
2. Blair, H.A., Subrahmanian, V.S.: Paraconsistent Logic Programming. Theoretical Computer Science 68, 135–154 (1989)
3. da Costa, N.C.A., Subrahmanian, V.S., Vago, C.: The Paraconsistent Logics P$\mathscr{T}$. Zeitschrift für Mathematische Logic und Grundlangen der Mathematik 37, 139–148 (1989)
4. Nakamatsu, K.: On the relation between vector annotated logic programs and defeasible theories. Logic and Logical Philosophy 8, 181–205 (2001)
5. Nakamatsu, K., Abe, J.M., Suzuki, A.: Annotated Semantics for Defeasible Deontic Reasoning. In: Ziarko, W.P., Yao, Y. (eds.) RSCTC 2000. LNCS (LNAI), vol. 2005, pp. 432–440. Springer, Heidelberg (2001)
6. Nakamatsu, K.: Pipeline Valve Control Based on EVALPSN Safety Verification. J. Advanced Computational Intelligence and Intelligent Informatics 10, 647–656 (2006)
7. Nakamatsu, K., Mita, Y., Shibata, T.: An Intelligent Action Control System Based on Extended Vector Annotated Logic Program and its Hardware Implementation. J. Intelligent Automation and Soft Computing 13, 289–304 (2007)

Rough Set Reducts Based Classification

Naohiro Ishii, Yongguang Bao, Yuta Hoki, and Hidekazu Tanaka

Abstract. Text classification aims to classify documents into categories or classes automatically based on their contents. While more and more textual information is available online, effective retrieval becomes difficult without good indexing and summarization of document contents. A rough set based reduct is a minimal subset of features, which has almost the same discernible power as the entire conditional features. Here, we propose a greedy algorithm to compute a set of rough set reducts which is followed by the k-nearest neighbor to classify documents. To improve the classification performance, reducts-kNN with confidence was developed. These proposed rough set reduct based classification methods for documents are compared by classification experiments. Experiments have been conducted on some benchmark datasets from the Reuters news data set.

Keywords: text classification, reducts of rough set , kNN, reducts with confidence.

1 Introduction

Rough sets theory firstly introduced by Pawlak[1,2,3,4] provides us a new approach to perform data analysis, practically. Rough set has been applied successfully and widely in machine learning and data mining[4,5,6,7]. An important task in rough set based data analysis is computation of the attribute or feature reduct for the classification. By Pawlak's rough set theory, a reduce is a minimal subset of features, which has the discernible power as using the entire features. Then, the reduct uses a minimum number of features and represents a minimal and complete

Naohiro Ishii and Yuta Hoki
Depart. of Information science, Aichi Institute of Technology
1247 Yachigusa , Yakusa-cho, Toyota, Aichi, 470-0392, Japan
e-mail: ishii@aitech.ac.jp

Yongguang Bao
Aichi Inormation System, Sumiyoshi-cho, Kariya, Aichi, 448-0852, Japan
e-mail: baoyg_860@hotmail.com

Hidekazu Tanaka
Depart. of Computer Science, Daido Institute of Technology
10-3 Takiharu-cho, Minamiku, Nagoya, Aichi, 457-8530, Japan
e-mail: hitanka@daido-it.ac.jp

K. Nakamatsu et al. (Eds.): New Advan. in Intel. Decision Techno., SCI 199, pp. 373–382.
springerlink.com

rues set to classify new objects. Finding all reducts of information system is combinatorial NP-hard computational problem. To classify unseen objects, it is optimal to use characteristics so that different reducts use different features as many as possible. Then, the union of these features in the reducts, is expected to include all indispensable features and the number of reducts used for classification is minimum. Here, a greedy algorithm to compute a set of reducts which considers these requirements, is presented. The proposed algorithm starts with the set of CORE features in the rough set. Through backtracking, multiple reducts are constructed using discernibility matrix. A reduct is computed by using forward stepwise selection and backward elimination based on the significance values of the features. The algorithm terminates when the features in the union of the reducts includes all the indispensable features in the database or the number of reducts is equal to the number given in advance. After multiple reducts construction, k-nearest neighbor(k-NN) approach is adopted for the data classification. Thus, we call here the method, Reducts-kNN, which combines multiple reducts with k-NN classifiers, respectively. To improve the classification performance[8,9], reducts-kNN with confidence was developed. These proposed rough set reduct based classification methods for documents are compared by classification experiments. Experiments have been conducted on some benchmark datasets from the Reuters news data set[10].

2 Rough Set and Its Reduct

2.1 Information Systems for Rough Set

An information system is composed of a 4-tuple as follows,

$$S = < U, Q, V, f >$$

where U is a closed universe, a finite nonempty set of N objects $(x_1, x_2, \ldots, x_N)$, Q is a finite nonempty set of n features $\{q_1, q_2, \ldots, q_n\}$, $V = \bigcup_{q \in Q} V_q$, where V_q is a domain (value) of the feature q, $f : U \times Q \rightarrow V$ is the total decision function called the information such that $f(x, q) \in V_q$, for every $q \in Q$, $x \in U$.

Any subset P of Q determines a binary relation on U, which will be called an indiscernibility relation denoted by $INP(P)$, and defined as follows: $xI_P y$ if and only if $f(x, a) = f(y, a)$ for every $a \in P$. Obviously $INP(P)$ is an equivalence relation. The family of all equivalence classes of $INP(P)$ will be denoted by $U / INP(P)$ or simply U / P; an equivalence class of $INP(P)$ containing x will be denoted by $P(x)$ or $[x]_p$.

2.2 *Reduct of Rough Set*

Reduct is a fundamental concept of rough set. A reduct is the essential part of an information system S that can discern all objects discernible by the original information system.

Let $q \in Q$. Feature q is dispensable in S ,if $IND(Q-q) = IND(Q)$;otherwise q is indispensable in S. The set $R \subseteq Q$ of feature will be called a reduct of Q, if $IND(R) = IND(Q)$ and all features of R are indispensable in S. We denote it as $RED(Q)$ or $RED(S)$. Feature reduct is the minimal subset of features Q with respect to decision features D, none of the features of any minimal subsets can be eliminated without affecting the essential information. These minimal subsets can discern classes with the same discriminating power as the entire features. The set of all indispensable from the set Q is called $CORE$ of Q and denoted by $CORE(Q)$:

$$CORE(Q) = \bigcap RED(Q)$$

2.3 *Discernibility Matrix*

One of basic notions in rough set is a discernibility matrix that helps us understand several properties to construct efficient algorithm to compute reducts. By $M(S)$, we denote an $n \times n$ matrix (c_{ij}), called the discernibility matrix of S, such as

$$c_{ij} = \{q \in Q : f(x_i, q) \neq f(x_j, q)\} \text{ for } i, j = 1, 2, ..., n .$$

Since $M(S)$ is symmetric and $c_{ii} = \phi$ for $i = 1, 2, ..., n$, we represent $M(S)$ only by elements in the lower triangle of $M(S)$, i.e. the c_{ij} is with $1 \leq j < i \leq n$.

2.4 *Feature Reductions*

Our algorithm to make reduct of data, starts with $CORE$ features. A reduct is computed by using forward stepwise selection and backward elimination based on the significance values of features as shown in Fig. 1. Through backtracking, multiple reducts are constructed using discernibility matrix. The algorithm terminates when the features in the union of reducts includes all the indispensable features in the database or the number of reducts is equal to the number determined by user. Let $COMP(B, ADL)$ denote the comparison procedure. Result of $COMP(B, ADL)$ becomes 1 if for each element c_j of ADL has $B \cap c_j \neq \phi$ otherwise 0, m be the parameter of reducts number.

```
Step 1 Create the discernibility matrix DM:[ Cij ];
  CORE = ∪{c ∈ DM : card(c) = 1};i = 1;
Step 2 While (i ≤ m) do begin
      REDU = CORE; DL = DM − REDU;
      /*forward selection*/
      While (DL≠φ) do begin
      Compute the frequency value for each feature q ∈ Q − ∪REDUi;
      Select the feature q with maximum frequency value and add it to REDU;
      Delete elements dl of DL which q ∈ dl from DL;
    End
    /*backward elimination*/
    N = card(REDU − CORE);
    For j = 0 to N − 1 do begin
      Remove aj ∈ REDU − CORE from REDU;
      If COMP(REDU, DM) = 0 Then add aj to REDU;
    End
    REDUi = REDU;  i = i + 1
    End
```

Fig. 1 Generation of multiple reducts

3 k-Nearest Neighbor Method

By using multiple reducts developed above, the data classification is needed. The basic k-nearest neighbor(kNN) is one of the simple methods for classification. It is intuitive and easy to understand its concept. Much attention is paid to combining a set of individual classifiers with the hope of improving the overall classification accuracy. In this paper, multiple reducts-kNN combined method is proposed to improve the classification accuracy. Then, these multiple reducts are expected to improve the performance of the k-nearest neighbor classifier.

3.1 k-Nearest Neighbor Classification with Multiple Reducts

The multiple reducts can be formulated precisely and in a unified way within the framework of rough set theory[Pawlak]. An attempt of combining multiple reducts followed by the k-nearest neighbor classification is performed here, which is shown in Fig.2. In Fig.2, we assume a set of m reducts $\{R_1, R_2, ..., R_m\}$. Each reduct R_j classifies features into a certain class by k-NN. Then, features extracted from data, are inputted first, to multiple reducts at the same time, which are determined in the above section. Then, each reduct followed by kNN, classifies feature class. The kNN adopted here, is interpreted in the following. For an unknown

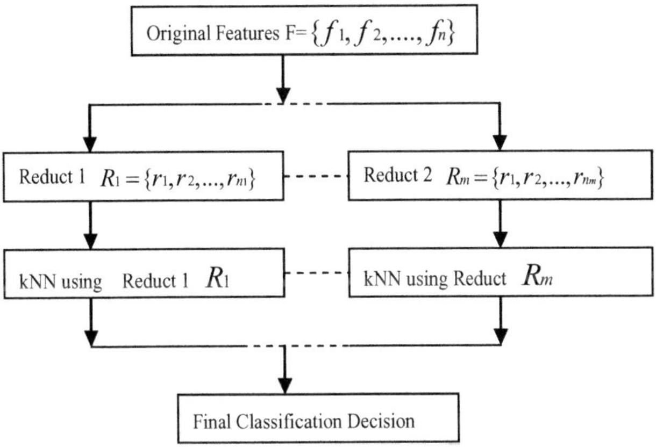

Fig. 2 Muliple reducts-kNN classifiers

object u, classification by kNN classifier can be performed as follows: calculating the distances between object and all training objects, choosing k-nearest neighbor objects and voting to class. Then, the algorithm by the reducts-kNN classification is given in the following.

1. Let the notation of training data set, be D and the total feature set be $F = \{f_1, f_2, ..., f_x\}$. From the data set D, multiple reducts $R_1, R_2, ..., R_m$ are computed.

2. Each kNN classifier computes rank of each class by using only features included in the given reduct . The distance $sim_i(d_a, d_b)$ between data d_a and d_b in the classifier kNNi corresponding to reduct R_i $(1 \le i \le m)$, is defined as follows.

$$sim_i(d_a, d_b) = \frac{\sum_{j=1}^{x}(count(R_i, f_j) \times x_{aj} \times x_{bj})}{\sqrt{\sum_{j=1}^{x}(count(R_i, f_j) \times (x_{aj})^2)} \times \sqrt{\sum_{j=1}^{x}(count(R_i, f_j) \times (x_{bj})^2)}}$$

where $count(R, f)$ takes value 1, when feature f is included in reduct R, while it takes value 0 when it is not included. Rank of the class c_j for data d_q, denoted by $rank_{cj,\,knni}(d_q)$, is given as

$$rank_{cj,\,knni}(d_q) = \frac{\sum_{j=n1}^{nk} sim(d_q, d_i) \times \delta(c_j, y_j)}{\sum_{j=n1}^{nk} sim(d_q, d_j)}, \text{ where } \delta(c_j, y_i) = \begin{cases} 1 : c_j = y_i \\ 0 : c_j \neq y_i \end{cases}$$

3. For integrating respective kNN classifiers, total rank score, $trank_{cj}(d_q)$ is computed as follows,

$$trank_{cj}(d_q) = \frac{\sum_{i=1}^{m}\{rank_{cj,\,knni}(d_q)\}}{m}$$

Finally, the classes that satisfy the condition of $trank_{cj}(d_q) \geq \theta$ are the final classification result.

3.2 Experimental Results

For evaluating the efficiency of the kNN classification with multiple reducts, experimental computations are carried out. The computer experiments were carried out by the well known news data called Reuters 21578, which consists of international politics and economical news documents. Documents in Reuters 21578, are represented in SGML and they are assigned to the class category. In this study, the documents are classified to 5 categories(cocoa, copper, cpi, gnp, rubber). To measure the classification accuracy in the class c_i, three indexes, $Recall_{ci}$, $Precision_{ci}$, and $Accuracy_{ci}$ are defined as follows,

$Recall_{ci}$ The ratio of documents classified to the class c_i within the total documents in the class c_i.

$Precision_{ci}$ The ratio of documents classified correctly in the class c_i within the documents assigned to the class c_i.

$Accuracy_{ci}$ The ratio of documents classified correctly to the class c_i and other than c_i among all the documents.

We assume here $\{c_i : i = 1 \sim m\}$ to be class set, where m is the number of class. The parameters of the classification accuracy, are defined as follows,

Let TP_i be the number of correctly classified documents as in c_i, which belong to the class c_i. Let FP_i be the number of incorrectly classified documents as in c_i, which do not belong to the class c_i. Let FN_i be the number of incorrectly classified documents as in not c_i, which are in the class c_i. Let TN_i be the number of correctly classified documents as in not c_i, which are not in the class c_i. These parameters are summarized in the contingency table as shown in Table 1.

Table 1 Contingency table

	Belong to c_i	Not belong to c_i
Classified to c_i	TP_i	FP_i
Not classified to c_i	FN_i	TN_i

The indexes, $Recall_{ci}$, $Precision_{ci}$, and $Accuracy_{ci}$ are for the class c_i as in Fig. 3. To measure and improve the inter-class accuracy, the indexes, micro-average, μ and macro-average, M are introduced. The former is the modified indexes for all classes $\{c_i\}$,while the latter is the average values of all classes.

Data weighting is an important pre-processing for vector space model of the classification of documents. Several data weighting methods are developed as word frequency occurrence, tf×idf , et al weighting[5,6]. Entropy weighting method[5,6] is applied here.

$Recall_{c_i} = \frac{TP_i}{TP_i + FN_i}$
$Precision_{c_i} = \frac{TP_i}{TP_i + FP_i}$
$Accuracy_{c_i} = \frac{TP_i + TN_i}{TP_i + FP_i + FN_i + TN_i}$

Fig. 3 Classification indexes

Three classification indexes are well known for the evaluation of classification. Accuracy is often used as a typical classification measure. To make clear how many number of reducts are applied to Reuters data analysis, classification indexes(recall, precision and accuracy) are computed for 2 class, 3 class,4-class and 5-class evaluation in micro average value, which is shown in Table 2.

Table 2 Evaluation for number of reducts by micro-average

		2- Class	3- Class	4-Class	5-Class
Number of reducts = 1	Recall	0.471	0.647	0.764	0.799
	Precision	0.500	0.667	1.000	0.786
	Accuracy	0.750	0.845	0.927	0.902
Number of reducs=5	Recall	0.716	0.868	0.812	0.830
	Precision	1.000	0.905	0.875	0.863
	Accuracy	0.859	0.925	0.909	0.928
Number of reducts=10	Recall	0.682	0.779	0.824	0.733
	Precision	0.909	0.824	0.918	0.857
	Accuracy	0.813	0.873	0.924	0.909

Similarly, those are computed for the evaluation in macro average value, which is shown in Table 3. In these experiments, the number of reducts are taken in case of to be 1 ,5 and 10.

From experimental evaluation results in Table 2 and Table 3, the number of reducts, 5 shows better results in recall, precision and accuracy values under the condition of the classification in 5-class, which is compared to those of the number of reducts , 1 and 10. In the dataset of Reuters data, the number of class will be 5- class in practice as shown in Table 3. We adopted 5 reducts in multiple reducts for Reuters data set. The first classification evaluation between kNN and the multiple reducts with 5 is compared in Table 4, which shows the micro-average case. Also, the second those evaluation is compared in Table 5 in macro-average case. From Table 4 and Table 5.

Table 3 Evaluation for number of reducts by macro-average

		2- Class	3- Class	4-Class	5-Class
Number of reducts = 1	Recall	0.471	0.647	0.764	0.799
	Precision	0.500	0.667	1.000	0.786
	Accuracy	0.750	0.845	0.927	0.902
Number of reducs=5	Recall	0.716	0.868	0.812	0.830
	Precision	1.000	0.905	0.875	0.863
	Accuracy	0.859	0.925	0.909	0.928
Number of reducts=10	Recall	0.682	0.779	0.824	0.733
	Precision	0.909	0.824	0.918	0.857
	Accuracy	0.813	0.873	0.924	0.909

Table 4 Comparison of classification evaluation by micro- average

kNN and multiple reducts		2- Class	3- Class	4-Class	5-Class
kNN	Recall	0.969	0.966	0.859	0.827
	Precision	0.721	0.700	0.725	0.741
	Accuracy	0.797	0.851	0.869	0.898
Reducts-kNN	Recall	0.719	0.879	0.815	0.827
	Precision	1.000	0.895	0.852	0.843
	Accuracy	0.859	0.925	0.909	0.928

The reducts-kNN method proposed here, shows better results compare to only kNN method. To improve further the performance of the reducts-kNN method, a confidence score[8] was introduced. In the algorithm by the reducts-kNN classification in section 3.1, a score $trank_{cj}(d_q)$, is introduced. Since the score is the average of all classification reducts-kNN, classifiers with the higher classification ability and other classifiers with the lower one, work equally for the classification determination.

Then, in case of classifiers with more than half lower ability, classification will make mistaken. By introducing confidence score to each reduct-kNN classifiers, it is expected to improve the classification evaluation for the reducts-kNN proposed

Table 5 Comparison of classification evaluation by macro- average

kNN and multiple reducts		2- Class	3- Class	4-Class	5-Class
kNN	Recall	0.971	0.961	0.860	0.811
	Precision	0.726	0.713	0.740	0.789
	Accuracy	0.797	0.851	0.869	0.898
Reducts-kNN	Recall	0.716	0.868	0.812	0.830
	Precision	1.000	0.905	0.875	0.863
	Accuracy	0.859	0.925	0.909	0.928

system. Thus, the confidence score was applied to the reducts-kNN classification. The experimental results with the confidence score are shown in Table 6, in which the improved multiple reduct-kNN is described in reducts-kNN with confidence.

In Table 6, 5-class data (cocoa, copper, cpi, gnp, rubber in Reuters) was experimented. The classification performance improves according to methods. The reducts-kNN shows better evaluation results than kNN only. The reducts-kNN with confidence shows better than the reducts-kNN without confidence.

Table 6 Comparison of classification evaluation by proposed methods

Methods and evaluation		cocoa	copper	cpi	gnp	rubber
kNN	Recall	0.867	0.823	0.615	1.000	0.750
	Precision	0.565	0.875	0.800	0.708	1.000
	Accuracy	0.872	0.947	0.851	0.851	0.968
Reducts-kNN	Recall	0.933	0.882	0.500	1.000	0.833
	Precision	0.823	1.000	0.867	0.791	0.833
	Accuracy	0.957	0.979	0.840	0.904	0.957
Reducts-kNN with Confidence	Recall	1.000	0.941	0.538	1.000	0.917
	Precision	0.882	1.000	0.933	0.810	1.000
	Accuracy	0.979	0.989	0.862	0.915	0.989

4 Conclusion

While more and more textual information is available online, effective retrieval becomes difficult without good indexing and summarization of document contents. A rough set based reduct is a minimal subset of features, which has almost the same discernible power as the entire conditional features. Here, we propose a greedy algorithm to compute a set of rough set reducts which is followed by the k-nearest neighbor to classify documents. Experimental studies are carried out by using Reuters news data for the number of reducts and classification classes. To improve the classification performance, reducts-kNN with confidence was developed. Classification by the reducts-kNN with confidence, shows better performance than the reducts-kNN and kNN methods. Relation between multiple reducts and confidence, is a remained problem in the future study.

References

1. Pawlak, Z.: Rough Sets. International Journal of Computer and Information Science 11, 315–329 (1982)
2. Pawlak, Z.: Rough Sets-Theoretical Aspects of Reasoning about Data. Kluwer Academic Publishers, Dordrecht (1991)

3. Pawlak, Z., Slowinski, R.: Rough Set Approach to Multi-attribute Decision Analysis. European Journal of Operations Research 72, 443–459 (1994)
4. Skowron, A., Polkowski, L.: Decision Algorithms, A Survey of Rough Set Theoretic Methods. Fundamenta Informaticae 30(3-4), 345–358 (1997)
5. Bao, Y., Aoyama, S., Du, X., Yamada, K., Ishii, N.: A Rough Set –Based Hybrid Method to Text Categorization. In: Proc. 2nd International Conference on Web Information Systems Engineering, pp. 254–261. IEEE Computer Society, Los Alamitos (2001)
6. Bao, Y., Tsuchiya, E., Ishii, N.: Classification by Instance-Based Learning. In: Gallagher, M., Hogan, J.P., Maire, F. (eds.) IDEAL 2005. LNCS, vol. 3578, pp. 133–140. Springer, Heidelberg (2005)
7. Momin, B.F., Mitra, S., Gupta, R.D.: Reduct Generation and Classification of Gene Expression Data. In: Proc. International Conference on Hybrid Information Technology ICHIT 2006, vol. I, pp. 699–708. IEEE Computer Society, Los Alamitos (2006)
8. Cheetham, W., Price, J.: Measures of solution accuracy in case-based reasoning systems. In: Funk, P., González Calero, P.A. (eds.) ECCBR 2004. LNCS, vol. 3155, pp. 106–118. Springer, Heidelberg (2004)
9. Delany, S.J., Cunningham, D.D., Zamolotskikh, A.: Generating Estimates of Classification Confidence for a Case-Based Spam Filter. In: Muñoz-Ávila, H., Ricci, F. (eds.) ICCBR 2005. LNCS, vol. 3620, pp. 177–190. Springer, Heidelberg (2005)
10. Reuters collection, `http://www.research.att.com/lewis/reuters21578.html`

Heterogeneous Multi-agents Learning Using Genetic Network Programming with Immune Adjustment Mechanism

Hirotaka Itoh, Naoki Ikeda, and Kenji Funahashi

Abstract. A heterogeneous multi-agent system is a system that involves two or more agents that cooperate in order to accomplish a certain task. Genetic Network Programming (GNP) is a technique to automatically build a multi-agent system. In the past, the authors proposed the use of the Immune evolved Genetic Network Programming (IGNP) technique for the automatic construction of multi-agent systems. In this paper, the authors propose the use of Genetic Network Programming with Immune Adjustment Mechanism (GNPIAM) as a technique to automatically build a heterogeneous multi-agent system. In this study, the authors carry out experiments using tile world to evaluate the validity of the proposed method and compare the three techniques—GNP, IGNP, and GNPIAM.

Keywords: GNP, GAIAM, GNPIAM, Heterogeneous multi-agent system.

1 Introduction

Recently, considerable research has been carried out on multi-agent systems. A multi-agent system is a system that solves a problem by using two or more agents. In the multi-agent system, the entire system is not collectively managed. The individual agent cooperates with other agents. The use of such a distributed management system has the advantage of being cost-effective. However, designing the behavior of two or more agents poses a serious problem, and it is difficult in the handwork.

Hirotaka Itoh
Information Technology Center, Nagoya Institute of Technology
Gokiso-cho, showa-ku, Nagoya, 466-8555, Japan
e-mail: ht-itoh@nitech.ac.jp

Naoki Ikeda
Graduate School of Information Science, Nagoya University
Furo-cho, Chikusa-ku, Nagoya, 464-8601, Japan
e-mail: icenine84@hotmail.com

Kenji Funahashi
Information Technology Center, Nagoya Institute of Technology
Gokiso-cho, showa-ku, Nagoya, 466-8555, Japan
e-mail: kenji@nitech.ac.jp

K. Nakamatsu et al. (Eds.): New Advan. in Intel. Decision Techno., SCI 199, pp. 383–391.
springerlink.com

Genetic Network Programming (GNP) [1, 2], which is an improvement of Genetic Programming (GP) [1], is a technique for automatically generating an agent's behavior. GNP uses a network structure and can make the agent's behavior correspond to a dynamic environment. The evolutionary algorithm of GNP is the same as the Genetic Algorithm (GA). The disadvantages of using GA are initial convergence and poor local search capability. In order to overcome these disadvantages, evolutionary adaptation algorithms based on the working of the immune system have been devised. Examples of such algorithms are Immune Algorithm (IA) [3] and the Genetic Algorithm with Immune Adjustment Mechanism (GAIAM) [4]. Previously, the author proposed the Immune evolved Genetic Network Programming (IGNP) [5] technique that used IA as the evolutionary algorithm of GNP. In this paper, the authors propose the use of Genetic Network Programming with Immune Adjustment Mechanism (GNPIAM) that uses GAIAM.

Previous researches that used GNP to study multi-agent systems [6, 7] aimed at increasing the efficiency of the system by using two or more agents for a task that could be accomplished by a single agent. However, the system proposed in this research is not one that seeks to increase the efficiency by employing two or more agents; rather, it is one that accomplishes a task by facilitating two or more agents. The task considered is one that cannot be accomplished by using a single agent. A heterogeneous multi-agent system is a system in which each individual agent has a separate operation program. The authors employed GNP, IGNP, and GNPIAM to study heterogeneous multi-agent systems and compared the three methods.

In this paper, GNPIAM is described in Section 2, and the heterogeneous multi-agent system is discussed in Section 3. The proposed experimental method is explained in Section 4 and conclusions are presented in Section 5.

2 GNPIAM

2.1 Structure of GNPIAM

The structure of GNPIAM is the same as that of GNP, and therefore, in this section, the structure of GNP is explained.

GNP [1, 2] is an automatic programming technique based on GP [2]. GP uses a tree structure, it returns to the root in GP after it changes from the root to the terminal. Therefore, GP is suitable for the study of a static environment. On the other hand, GNP uses a network structure. Since state transitions occur in the subnetwork, GNP is suitable for the study of a dynamic environment. Fig. 1 shows the structure of GNP. GNP consists of a start node, judgment node, and a processing node. The start node, which is displayed as a square in Fig. 1, is the point where processing begins; the processing node, which is displayed as an open circle in Fig. 1, refers to the stage where an agent initiates an operation. The judgment node, which is displayed as a diamond in Fig. 1, is the stage where the agent makes judgment.

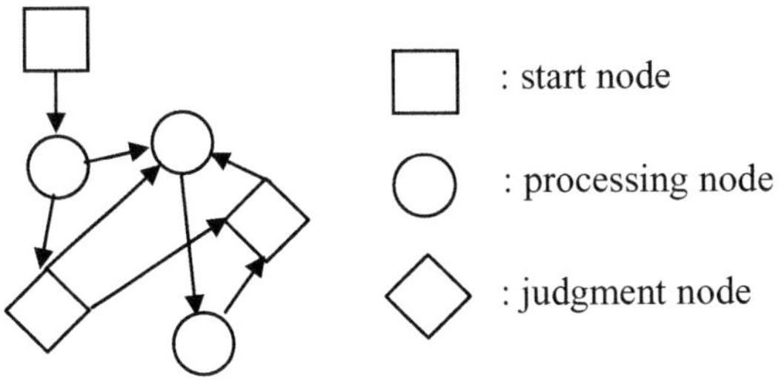

Fig. 1 Structure of the GNP

2.2 Genetic Operation of GNP

The genetic operation of GNP comprises crossover and mutation. During crossover, the nodes are replaced and connections are changed by replacing the specific area. During mutation, connections between the nodes are changed.

2.3 GNPIAM Algorithm

GNPIAM uses GAIAM [4] as the evolutionary algorithm. GAIAM is a GA that adopts two features of the immune system.

Various antibodies are present in the human body. As antigens invade the body, antibodies for these antigens are generated to eliminate the antigens. The immune system has the following two features:

[Feature 1] Capacity to adapt to mutations in antigens

It is difficult to produce antibodies for each and every antigen beforehand. In the absence of antibodies specific for an antigen, genes of the antibodies with the best specificity respond by mutating. Through repeated mutations, these genes produce antibodies that can adapt to the antigens.

[Feature 2] Mechanism to balance the generation of antibodies via other antibodies

The generation of antibodies for a given antigen is not a continuous process. Antibodies recognize each other on the basis of their structure, and when a given antibody is generated in excess, other antibodies that identify it as an antigen and are also generated to inhibit its growth, and a balance is maintained.

The GNPIAM algorithm is as follows:

Step 1: Generation of an initial group of antibodies

N antibodies are generated initially. These antibodies are similar to the individuals in GA and are helpful in solving optimization problems.

Step 2: Calculation of affinities

The affinities ax_i (i = 1,…, N) for antigens are calculated. ax_i is set in accordance with the problem being dealt with. The affinity for an antigen is similar to the concept of fitness in GNP.

Step 3: Calculation of expected values

The expected value e_i ($i = 1,\ldots, N$) of the number of antibodies that can survive into the next generation is calculated as

$$e_i = \frac{ax_i}{C_i} \tag{1}$$

Where C_i is the density of antibodies of type i ($i = 1,\ldots, N$).

$$C_i = \frac{1}{N}\sum_{j=1}^{N} ay_{i,j} \tag{2}$$

Here, $ay_{i,j}$ is the similarity between antibodies of type i and j ($i = 1,\ldots,N$, $j = 1,\ldots,N$) and is set in accordance with the problem. $N/2$ antibodies with low expected values are eliminated; however, among these, 10% of the antibodies with high affinities for antigens are excluded from elimination.

Step 4: Antibody generation

New antibodies are generated to replace the antibodies eliminated in Step 3. $N/2$ antibodies are selected from the surviving antibodies on the basis of the expected values. These selected antibodies are mutated, after which their affinities for antigens are calculated.

Step 5: Crossover and mutation

Antibodies are randomly selected (duplication permitted) from N antibodies, they undergo crossover depending on crossover probability Pc, thereby generating $N/2$ antibodies. The generated antibodies undergo mutation depending on mutation probability Pm, after which their affinities for antigens are calculated.

Step 6: Adjustment of antibodies

With respect to each antibody i of the $N/2$ antibodies generated in Step 5, an antibody j with the greatest affinity for i is sought from among the existing N antibodies. Among antibodies i and j, the one with the higher affinity for an antigen survives into the next generation, while the other is eliminated.

Step 7: Repetition of Steps 3 to 6 for a determined number of generations.

Step 4 models [Feature 1] of the immune system and Step 6 models [Feature 2] of the immune system. In GNPIAM, antibodies with low density and a high affinity for an antigen tend to survive in order to maintain diversity. Moreover, such antibodies are generated in GNPIAM; antibodies with a high affinity for an antigen are produced through mutation. In terms of GA, local search capability is improved. Moreover, narrowing of the search range is eliminated through Step 6. Thus, the narrowing of the search to the vicinity of a single local solution in GA is eliminated.

3 Heterogeneous Multi-agent System

In a multi-agent system, the entire system consists of two or more agents that operate individually and autonomously. If there is mutual cooperation between the agents, major tasks that are impossible to be performed by a single agent can be accomplished. The Multi-agent systems are classified into heterogeneous multi-agent [6] and homogeneous multi-agent systems. In the heterogeneous multi-agent system, each individual agent has a separate operation program. On the other hand, in a homogeneous multi-agent system, all the agents in the system have the same operation program. This study focuses on the heterogeneous multi-agent system. In this study, the learning method of the heterogeneous multi-agents is examined by using GNP, IGNP, and GAIAM.

4 Experiments

4.1 Tile World

The authors carried out experiments to evaluate the validity of the proposed methods; the experiments involved the use of tile world, which is a well-known example of a simulated dynamic environment. As shown in Fig. 2, tile world is a two-dimensional lattice plane wherein the agent, tile, floor, hole, and obstacles are arranged. The lattice plane is divided into cells, and the agent can move only by one cell at a time. In Fig. 2, the agent, tile, hole, and obstacles are denoted by "A," "T2," an open circle, and a full square, respectively. A blank cell in the lattice plane indicates the floor. In this experiment, weights are set to the tile for the co-operation of the agents. In Fig. 2, the number after "T" refers to the number of weights, i.e., "T2" is a tile with two weights; two agents are required for carrying a tile with two weights. In the tile world, the agents who cooperate with others and drop the tile into the hole are deemed fit. Two agents, after gathering in the cell consisting of a tile, grip the tile and move in tandem toward the cell containing the hole. When the agents release the tile, it falls into the hole.

The evaluation value, that is, the fitness of the agents in the tile world is calculated using the following equation:

$$
\begin{aligned}
fitness &= 45 \times DropTile \times TileWeight \\
&+ 5 \times \frac{InitialDist - LastDist}{InitialDist} \times TileWeight \\
&+ 10 \times GrabTile + 0.01 \times MoveArea
\end{aligned}
\tag{3}
$$

DropTile is the number of tiles dropped by the agents, *TileWeight* is the weight of the tile, *InitialDist* is the shortest distance between the tile and the hole before the start of simulation, *LastDist* is the shortest distance between the tile and the hole after the lapse of the time limit, *GrabTile* indicates whether or not the tile is still held after the lapse of the time limit, and *MoveArea* indicates the number of cells traversed by the agent. 45, 5, 10, and 0.01 are the weight of each paragraph. A fitness value exceeding 100 indicates that the agents can drop the tile into the hole and that exceeding 10 indicates that the agents are able to grip the tile. A fitness value exceeding 20 suggests that the agents can take the tile to the hole. Table 1 shows the values of the parameters in the experiment.

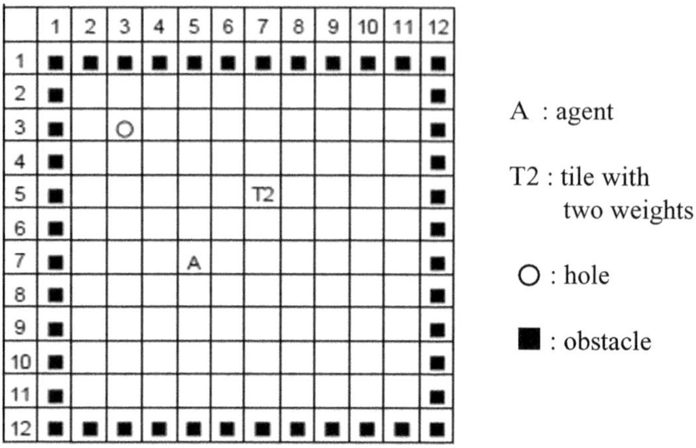

Fig. 2 Tile World

Table 1 Parameters in the experiment

Number of generations	500
Number of individuals	100
Limits time	2000
Mutation probability	0.01
Crossover probability	0.65
Number of trials	10

4.2 Experiment Using Tile World with One Tile

An experiment was performed with the initial arrangement shown in Fig. 2. One tile with two weights, one hole, and two agents are observed. The initial positions of the agents are also identical.

The learning result using GNP, IGNP, and GNPIAM is shown in Fig. 3, and the progress in learning during the ten trials is shown in Table 2.

In Fig. 3, "max" shows the transition of the number of times the maximum fitness was obtained from the ten trials, and "avg" is the average fitness of the ten trials. From Fig. 3, it can be seen that IGNP and GNPIAM are better than GNP. GNP is thought that it settles to the local solution, and the evaluation value did not go up. IGNP and GNPIAM result in better fitness because they have a wider search area. When IGNP and GNPIAM are compared using Table 2, it is evident that the learning of GNPIAM is better than that of IGNP. This is because the local search ability of GNPIAM is better (described in Section 2).

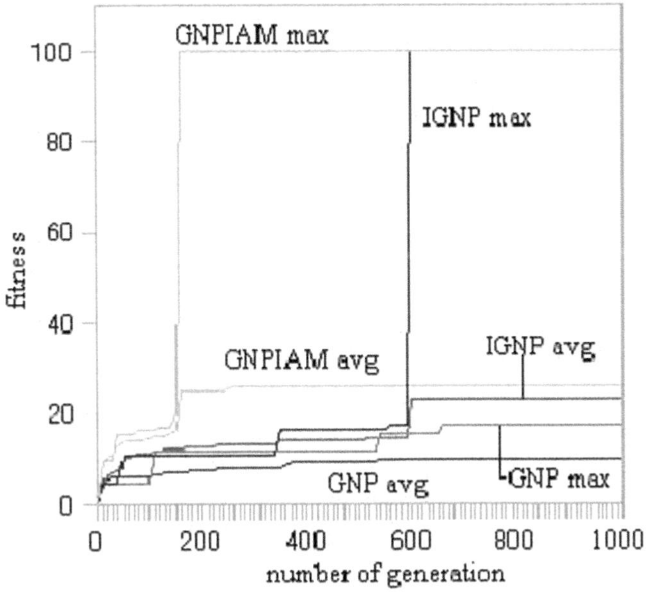

Fig. 3 Result of tile world with one tile

Table 2 Progress of the learning

	GNP	IGNP	GNPIAM
All the agents hold the tile	7	10	10
Agents move the tile	3	8	10
Agents reach the cell containing the hole	0	2	5
Agents drop the tile into the hole	0	1	1

4.3 Experiment Using Tile World with Two Tiles

This experiment was performed with the initial arrangement shown in Fig. 4. Two tiles with two weights each, one hole, and two agents are observed. The initial position of both the agents is the same.

In a tile world with two tiles, it is necessary to decide which tile to hold. Since both the agents are required to cooperate on selecting the same tile from the two tiles provided, the difficulty of this task is greater than that in a tile world with one tile.

The learning results when GNP, IGNP, and GNPIAM are used are shown in Fig. 5, and the progress in learning with each of the ten trials is shown in Table 3.

From Table 3 and Fig. 5, it is evident that IGNP and GNPIAM are better than GNP.

In GNP, the two agents could not hold the tile in three out of ten trials. In IGNP and GNPIAM, the two agents were able to hold the tile in all the trials. Only in

GNPIAM could the agents drop the tile into the hole, and this was observed in only one of the trials.

In this experiment, GNPIAM is the best.

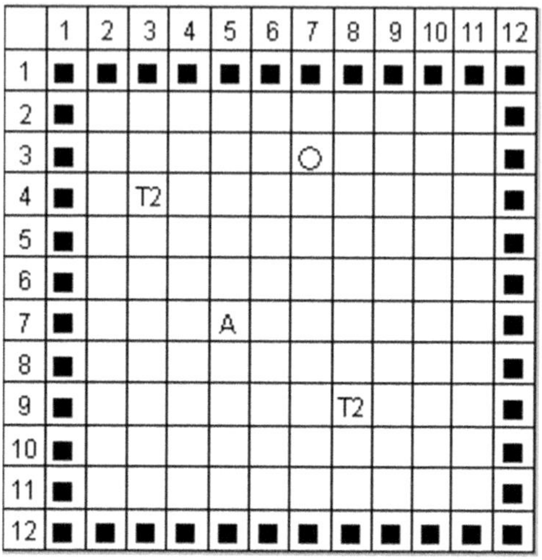

Fig. 4 Tile world with two tiles

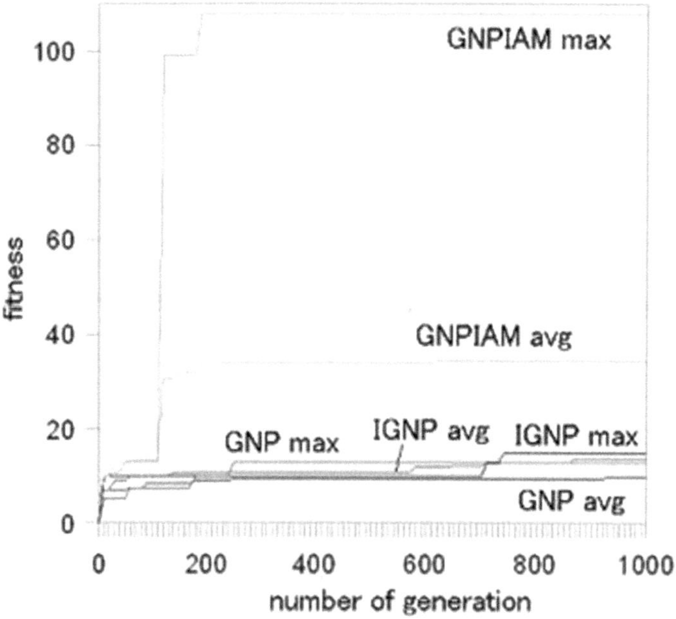

Fig. 5 Results for tile world with two tiles

Table 3 Progress of the learning

	GNP	IGNP	GNPIAM
All the agents hold the tile	7	10	10
Agents move the tile	3	9	10
Agents reach the cell of the hole	0	0	1
Agents drop the tile into the hole	0	0	1

5 Conclusion

Heterogeneous multi-agent systems were studied using GNP, IGNP, and GNPIAM. A task that cannot be accomplished by a single agent, but by multiple agents was chosen.

In this paper, the authors propose the use of GNPIAM for automatically generating an agent's behavior. To verify the effect of GNPIAM, the authors conducted two experiments. As a result, the learning of GNPIAM was found to be the best. GNPIAM has good local search ability and can search across a wide area.

In the future, the authors hope to improve the method and increase the number of trials in which the agents accomplish the task.

References

1. Hirasawa, K., Okubo, M., Katagiri, H., Hu, J., Murata, J.: Comparison between Genetic Network Programming and Genetic Programming Using Evolution of Ant's Behaviors. The Transactions of the Institute of Electrical Engineers of Japan 121(6), 1001–1009 (2001)
2. Murata, T., Nakamura, T.: Genetic Network Programming with Automatically Defined Groups for Assigning Proper Roles to Multiple Agents. In: Proceedings of Genetic and Evolutionary Computation, pp. 1705–1712 (2005)
3. Mori, K., Tsukiyama, M., Fukuda, T.: Immune Algorithm with Searching Diversity and its Application to Resource Allocation Problem. The Transactions of the Institute of Electrical Engineers of Japan 113(10), 872–878 (1997)
4. Itoh, H.: Genetic Algorithm with Immune Adjustment Mechanism. In: Proceedings of the 3rd IASTED Conference Computational Intelligence, pp. 79–84 (2007)
5. Itoh, H., Mase, T., Iwahori, Y.: Agent Learning using Immune Evolved Genetic Network Programming. Transactions of the Institute of Electrical Engineers of Japan 125(4), 537–544 (2005)
6. Hirasawa, K., Okubo, M., Hu, J., Murata, J., Matsuya, Y.: Co-evolution of Heteo-Multiagent Systems Using Genetic Network Programming. Transactions of the Institute of Electrical Engineers of Japan 123(3), 544–551 (2003)
7. Eguchi, T., Hirasawa, K., Furuzuki, T.: Construction of Symbiotic Evolutional Model in Multiagent Systems. IPSJ TOM 45(2), 144–156 (2004)

Analysis of Asymmetric Friendship among Students from Class Attendance Records

Nobuhiro Inuzuka, Toshitaka Kondo, and Shuhei Yamamoto

Abstract. In this paper we give an analysis of student friendship relation using class attendance records recorded by an attendance record system installed in Nagoya Institute of Technology (NIT). Our previous work[4] gives a prediction method of friendship relation among students using the data. This paper reveals further detailed analysis of friend and shows a possibility to have asymmetric relation among friendship from attendance records.

1 Introduction

Friendship is an underlying and basic relation to understand social structure among people[1, 3, 5]. Friendship and functionality of an organization are the two sides of a coin. Communication and activity among an organization impose friendship among people involved in the organization, and conversely functions are facilitated through friendship as their communication channel. Accordingly to understand friendship is effective to understand functionality of organizations.

Friendship is a base for comprehension of student behavior also in universities. At first friendship relation may be used to cluster students and the students' clusters may help to understand types of student behavior. For the second reason, it may give a method to find isolated students who are difficult in making communication ties and have problems to participate classes. For the third reason, we may count the importance of communication ability for students as their social skills and their total potential ability. The ability may also relate with progress of their study and to grasp their communication degree can be a monitor of their education environment.

Nobuhiro Inuzuka, Toshitaka Kondo, and Shuhei Yamamoto
Nagoya Institute of Technology,
Gokiso-cho Showa, Nagoya 466-8555, Japan
e-mail: inuzuka@nitech.ac.jp, kondo@phaser.elcom.nitech.ac.jp,
yamamoto@phaser.elcom.nitech.ac.jp

K. Nakamatsu et al. (Eds.): New Advan. in Intel. Decision Techno., SCI 199, pp. 393–403.
springerlink.com

In this paper we study on friendship using lecture class attendance records collected by a system which NIT installed in 2007. Our previous work[4] gave a prediction method of friendship under the assumption that friends have similar patterns of class attendance. The work focuses on the fact that differences of class attendance time between two students strongly depend on whether or not they are friends. In this paper we are interested in a type of asymmetricity of friendship. That is, we may observe for two friend students not only that they share time and activities but also that one of them proposes and leads their actions to another. We show a possibility that asymmetric commitments to their activities can be observed in the time differences.

The analysis of network connecting social entities, such as, persons, web-sites and topics, is a rapidly growing field in social studies and also in computer science[3, 6]. In special it gathers attention from data-mining communities under the situation that we can obtain online data appropriate for these studies. It aims to detect characteristics of network structure and to seek application in practical fields such as identification of activities in communities of people, web sites, and topics. Another topic is detection of roles of network nodes, such as leaders of a human community[2, 7]. Our interest shares the line of these researches and tries to understand a friendship network and to detect roles of students in student groups.

The following section explains the attendance record management system installed in NIT, and Section 3 reviews the friendship prediction method proposed in our previous work[4]. Section 4 gives our analysis of asymmetric relationship among students.

2 Class Attendance Record Management System and Friendship Relation

NIT installed a system to collect and manage class attendance records of students, which we call CARMS (Class Attendance Record Management System), in 2007. It aims to reduce tasks of instructors by collecting records automatically. The CARMS system consists of student ID-cards, card readers and a database management system (DBMS). A student ID card has a function of a wireless tag and keeps the information of the student ID of a card holder. A card reader has its own ID (reader-ID) and reads the information of an ID card when a holder places his/her card in the front of the reader. Each lecture room equips two or three readers near the entrances. Readers send the information as a tuple (SID, RID, ALT) to DBMS, where SID is a student ID, RID is a reader ID, and ALT is an attendance/leaving time, that is the time when the student puts close his/her card to the reader at his attendance to or leaving from a class. The DBMS collects and keeps all the information. An outline of the system is illustrated in Fig.1. CARMS gives lists of students who attended a class on a specified date. Although CARMS has also many other functions, we omit to describe the detail.

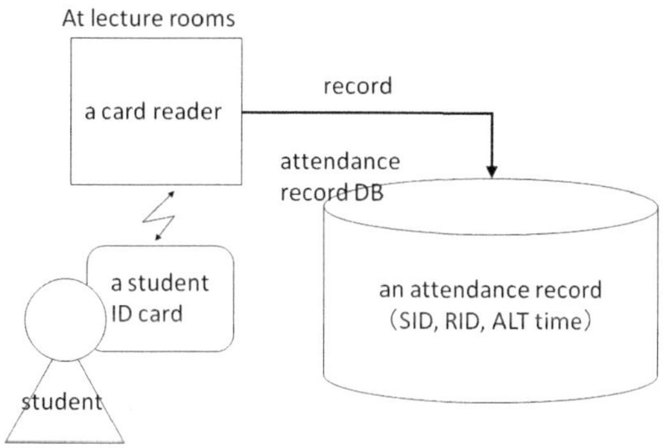

Fig. 1 The outline of CARMS system

Table 1 Summary of records collected by CARMS system

The period of records	2007.10.1 – 2008.3.31
# students recorded	4,403
of which first year students	942
of which second year students	936
of which third year students	929
of which fourth year students	287
# readers equipped	129
# records	864,882
of which first year students	295,700
of which second year students	242,001
of which third year students	165,588
of which fourth year students	18,463

Table 1 shows the basic data of the system and records collected and used for our experiments. Student IDs for fourth year students appeared in records and their attendance records are fewer than other years. This is because the most of fourth year students participate only graduate research projects or take a small number of lectures. Graduate course students are also in the similar situations. Accordingly we used only the data of 1st, 2nd and 3rd year undergraduate students.

Our previous work focuses on time difference of attendance records among students. A pair of friend students are expected to share a large part of time and places. It yields similar patterns in their attendance records.

For a pair of attendance records, (SID1, RID, ALT1) and (SID2, RID, ALT2), which share a reader ID RID, we call a signed time difference ALT2 – ALT1 of the two records an ALTD (attendance/leaving time difference) record from SID1 to SID2. When a student of SID1 attends a class at 9:00:00 am and another student of SID2 attends at 9:01:15, the ALTD from SID1 to SID2 is 00:01:15 and the ALTD from SID2 to SID1 is −00:01:15.

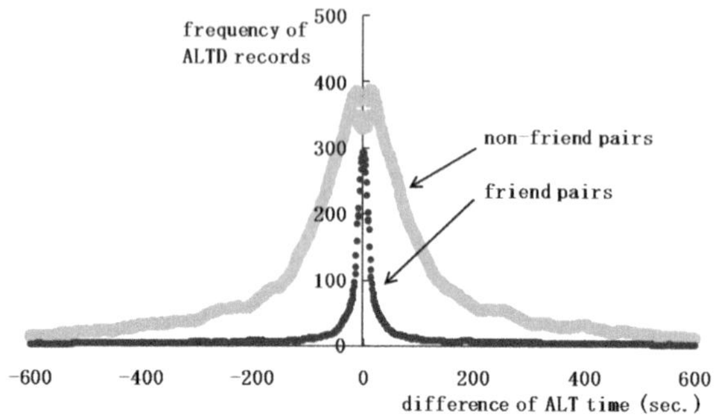

Fig. 2 Histograms of ALTD records for friend pairs and non-friend pairs

In order to see a causality among friendship and the ALTD distribution, our previous work observed histograms of ALTD records for friend student pairs and for non-friend pairs. This paper also reviewed it and got a histogram in Fig. 2. We obtained friend or non-friend pairs by a questionnaire. In the questionnaire, we asked students to choose their friends from a list. In order to avoid ambiguity of questions we asked to choose from choices (1) ones with whom you share your times not only in lecture rooms but also outside rooms, (2) ones with whom you have conversation in lecture rooms, (3) your acquaintances, whom you recognize their faces and names, and (4) others, and used ones selected for the first choice as their friends.

We can observe that friendship has large affection to the ALTD distribution. The ALTD distributions have peeks at a small time length. Ones for friend pairs the peeks move more near to zero and the peek becomes acute. Fig.3 shows the ratio of the number of ALTD records for friend pairs against all records. The range within one minute has larger ratio for friends pairs.

3 Prediction of Friendship

Noting on the character of ALTD our previous work gave a method to conjecture friendship from ALTD records between two students. Let $T = \{t_1, \ldots, t_n\}$ be a set of timestamps of all ALTD records from a student A to another student B during the period and try to have the probability $p(f|T)$, where f is the event than B is a friend of A. When we apply the Bayes' theorem to $p(f|T)$, we have the following equation.

$$p(f|T) = \frac{p(f)Ep(T|f)}{p(T)}$$

When we assume an ALTD record from A to B is independent to another ALTD record between the students, we can transform the equation as follows.

$$p(f|T) = p(f) \prod_{t \in T} \frac{p(t|f)}{p(t)} \tag{1}$$

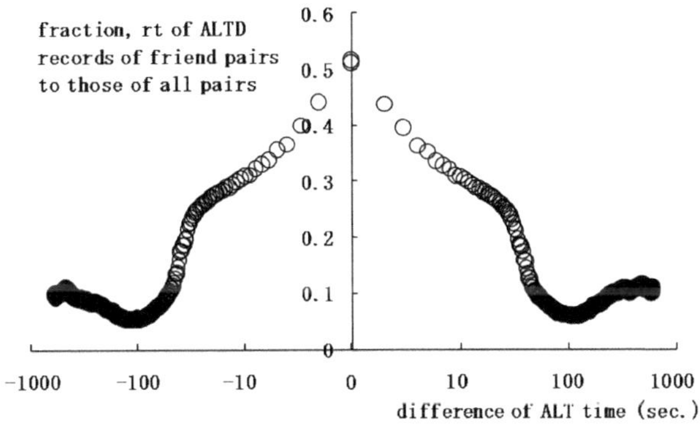

Fig. 3 The transition of fraction, r_t, of friend-pair ALTD records to all ALTD records

When a student meets one of his acquaintances in the morning, he may act through the day along with the person. Although if this situation happens frequently the independence assumption is not reasonable, we use the assumption here.

When we denote the fraction of ALTD records at t produced by friend pairs against all ALTD records at t by r_t, we can formulate it by,

$$r_t = \frac{\#\text{ALTD records for friend pairs at } t}{\#\text{ALTD records for all student pairs at } t} = \frac{X_f \cdot m_f \cdot p(t|f)}{X \cdot m \cdot p(t)},$$

where X is the number of all student pairs, X_f is the number of pairs A and B such that B is a friend of A, m is the expected number of ALTD records produced among randomly chosen two students, and m_f is the expected number of ALTD records among friend pairs.

Because $X_f = X \cdot p(f)$, we have

$$p(t|f) = \frac{X \cdot m \cdot p(t) \cdot r_t}{X_f \cdot m_f} = \frac{m \cdot p(t) \cdot r_t}{p(f) \cdot m_f}.$$

When we substitute this to Equation (1) we have the following equation,

$$p(f|T) = p(f) \prod_{t \in T} \frac{m \cdot r_t}{m_f \cdot p(f)} = p(f)^{1-n} \left(\frac{m}{m_f}\right)^n \prod_{t \in T} r_t$$

where n is the number of records in T.

The probability of the complement event $\bar{f}$ of non-friendship given is also derived similarly,

$$p(\bar{f}|T) = p(\bar{f})^{1-n} \left(\frac{m}{m_o}\right)^n \prod_{t \in T} (1 - r_t),$$

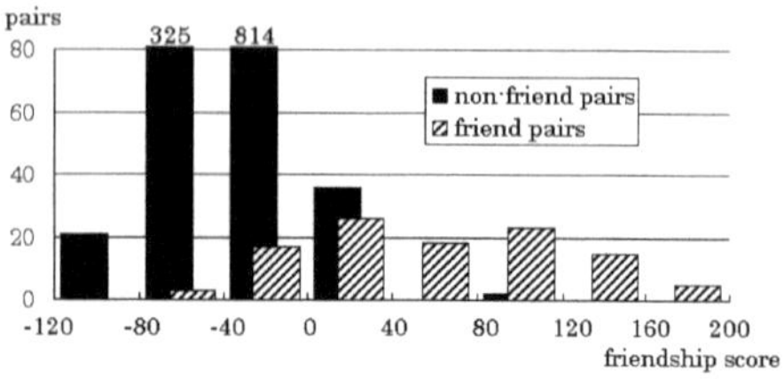

Fig. 4 The distribution of friendship scores

where m_o is the expected number of ALTD records among non-friend pairs. Then the friendship score from the student A to the student B can be given by the logit of $p(f|T)$ as follows,

$$\text{logit } p(f|T) = \log\Big(p(f|T)\Big) - \log\Big(p(\bar{f}|T)\Big) \tag{2}$$

When the score is positive we predict the pair as friends.

Fig.4 shows a result calculated the friendship score for a class. It shows the numbers of friend/non-friend pairs in ranges of friendship scores. Black bars show the numbers for non-friend pairs and shaded bars show ones for friend pairs. Large numbers of friend pairs take positive scores and large numbers of non-friend pairs take negative scores. The recall was 80% and the precision was 71% for this case.

4 Analysis of Asymmetric Friendship

By our previous work we confirmed that the friendship is closely associated with the ALTD distribution. We may expect to seek other clues to properties of students or of their groups in ALT records.

When two students are friends each other, it means not only that they are somehow close and share times for their everyday, but also that each of them has some role in their relationship. For example, a student A is reliable and then some student wants to be a friend of A and he/she expects to ask to have suggestions from A. For another example, two students are friends because they have many common interests and can enjoy conversation each other. The latter case we can understand the friendship is even and symmetric but the former show a asymmetric case. We call the former case *a asymmetric friendship*. The aim of this section is to seek evidence of asymmetric friendship in the ALT records.

A Student Pairs of Asymmetric ALTD Distribution. We have already see that ALTD distribution is symmetric even for friend pairs, which is shown in Fig. 2. This is trivial because when an ALTD record from A to B is t, the reverse records from B to A is $-t$. However, if we see only ALTD records for a pair of students, say A and B, the distribution must not be symmetric. Fig. 5 is a distribution of ALTD records for certain two students. The distribution is

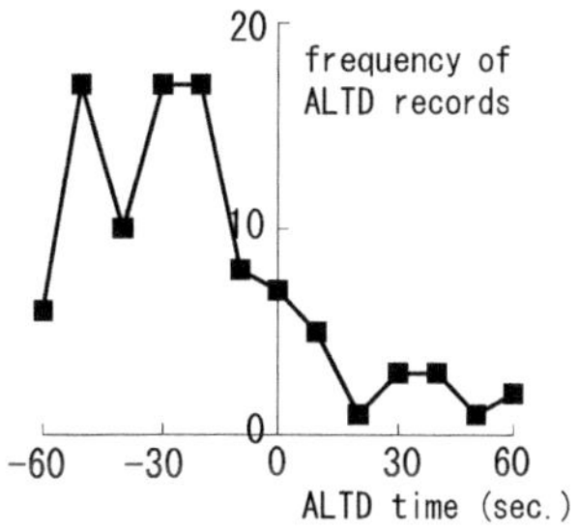

Fig. 5 An example of student pair which has an asymmetric ALTD distribution

largely shifted to minus. Although asymmetric distribution likely occurs when the number of records is small, it is notable that the two students shown in the figure have more than 100 records. We may guess that asymmetric friendship affect asymmetricity of distributions.

Asymmetric Friendship and Biased ALTD by Individual Behavior. The shift of ALTD distribution does not necessarily mean asymmetric friendship. Students have their own characters and styles. A student may attend lecture rooms considerable time before lectures begin and another student may attend on time or in late. That is ALT times are biased by their own nature and also ALTD distribution is affected by the relative difference of two students' characters.

Nevertheless we have a hypothesis that if two students have asymmetric roles or friendship their ALTD records are largely influenced by their asymmetricity, independently of their characters. We do not refute that their ALTDs are affected by the difference of their characters but we maintain that the affection is only for average behaviour of them and that their ALTD records for chances in which they behave as friends include an evidence of their asymmetric relationship.

In order to confirm this hypothesis we have done an analysis. Before we are going to the analysis we prepare some words.

When an ALT record (SID, RID, ALT) is collected for a lecture class that begins at t a modified record (SID, RID, ALT$-t$) is called a *relative* ALT. For example when a student recorded his ALT record at 8:47 for a lecture starting at 8:50, its relative ALT has time -3 minutes. For both ALT records for arrival for a lecture and records for leaving the previous lecture we take the same origin for their relative ALT. We consider only relative ALT within -15 munutes to $+5$ minutes, that is, from 5 minutes before ending the previous lecture to 5 minutes after starting the lecture, where NIT has 10 minutes for an interval of lectures. For a student we call the average of his/her relative ALT *an average relative* ALT. The average relative ALT of a student reflects tendency for ALT from his/her character and life style. Further we consider the difference of average relative ALT for a pair of students A and B, that is, (the average relative ALT of B) $-$ (the average relative ALT of A), which

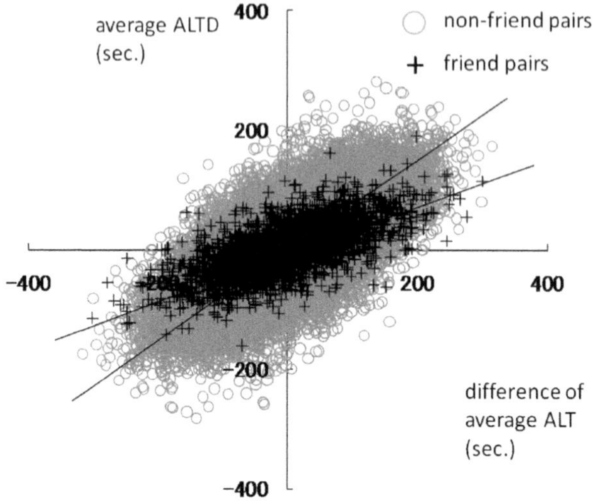

Fig. 6 Differences of average ALT and average ALTDs

reflects the average difference come from their personal tendency. We call it *a difference of average* ALT from A to B.

We define another measure which also reflects difference of two students. We call the average of all ALTD from student A to student B *an average* ALTD from A to B. It may affect the relationship of the two students. A difference of average ALT counts all ALT records of two students, the records which is for lectures that may not be attended by both students. On the other hand, an average ALTD counts only records for lectures attended by both students. Furthermore for the later discussion we consider average ALTDs by taking only ALTD records which has time difference less than a certain time in order to restrict records into ones in which two students can behave as friends if they are friends.

Let us see the correlation between the two difference, the differences of average ALT and the average ALTDs. We examine the correlation by the records of third year students. If friendship does not affect asymmetricity of ALTD records, we can observe a strong correlation between the two measures. If average ALTDs are affected by the form of friendship between students independently from their personal tendency, the correlation may be low. Fig.6 is a scatter diagram in which a student pair is plotted by their differences of average ALT on the horizontal axis and their average ALTD on the vertical axis. Points for friend pairs are plotted by black crosses and points for non-friend pairs are by gray circles. Pairs are classified to friends and non-friends using the friendship score.

The diagram shows ovals of which the major axis is right up. The values for the horizontal axis, differences of average ALT, are differences of personal tendency of students. The values for the vertical axis are the same but restricted

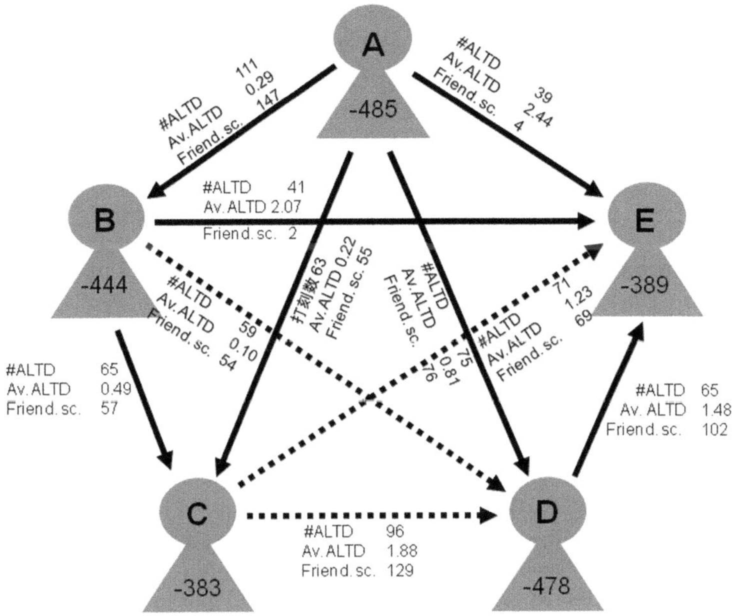

Fig. 7 Asymmetric relationship among a friend group

for ALT of the same lectures and short differences. If ALTs of each student are distributed around its average (personal tendency) independently of the other student, the both value will have strong correlation. We can see in the diagram that the correlation is not strong and it is weaker for friend pairs. When we fit a line to the ovals, the slope is 0.35 for friend pairs and 0.54 for non-friend pairs. For the case of friend pairs the dependency of average ALTD to the differences of average ALT seems weaker than that for non-friend pairs. It may be guessed that there are other factors for friend pairs which decide characters of friendship relations.

A Friend Group with Asymmetric Friendship. Fig.7 shows an example of a friend group in which some friend pairs seem to have asymmetric friendship. The group was chosen by the friendship scores. In the figure, the shapes with symbols A to E means students, a value in a shapes means the average relative ALT of the student, i.e. his/her personal tendency for arriving or leaving lectures. When a student A has an earlier average relative ALT than another B, A acts earlier than B in average. We also see average ALTD for student pairs. In order to restrict ALTD records into the cases in which two students are supposed to act as friends, we only count the ALTD records within 10 seconds. The figure draws arrows from a student A to another B when the average ALTD from A to B is positive. Values above the arrows are the average ALTD for records within 10 seconds differences between students. That is, the value on an arrow from A to B shows how early the student A acts than B when two students act as friends. Arrows have also friendship scores for the students and the numbers of ALTD records within 10 seconds.

An interesting point is that we can find student pairs who have an arrow from a late average ALT to a early average ALT, that is the direction of average ALTD is opposite to that of average ALT. Such arrows are drawn in broken arrows. Students connected by a broken arrow act in different ways when they meet in a lecture. We may take this observation as an evidence that such two students have asymmetric friendship relation and act in a proper way as friends from other situations where each of them act as an individual. We can also expect to see a broken of transitivity in the arrows but we could not observe it in collected records.

5 Conclusions

We focused on asymmetric friendship relations among students and tried to observe and analyze the relations in ALT records recorded in CARMS system. We observed that some student pairs have distinguished asymmetric distributions in their ALTD records and it does not necessarily relate to their difference of average ALT. We discussed the reason of this observation as an expression of asymmetric friendship among students in ALT records. Because a distribution of ALTD was symmetric and was different from the difference of average ALT we seek other factor to our hypothesis, asymmetric friendship. We execute a questionnaire survey which asked students if he/she is tend to propose to friends or to obey proposals from friends when he/she passes over time. However, we can not find clear causality between results of questionnaire and our analysis. We continue to see objective and semantical evidence to our observations. We also need detailed statistical analysis for our observation to maintain the hypothesis.

Acknowledgements. This work was supported in part by a grant from the learning field project of Nagoya Institute of Technology, a research grant by Ministry of Education, Culture, Sports, Science and Technology, Japan. Authors thank Information Technology Center of Nagoya Institute of Technology, which allows us to use the data collected by CARMS system.

References

1. Allan, G.: Friendship, Sociology and Social Structure. J. Social and Personal Relationships 15 (1998)
2. Du, N., Wu, B., Pei, X., Wang, B., Xu, L.: Community Detection in Large-Scale Social Networks. In: Workshop on Web Mining and Social Network Analysis, held in conjunction with KDD 2007, vol. (8) (2007)
3. Gonzalez, M.C., Herrmann, H.J., Kertesz, J., Vicsek, T.: Community structure and ethnic preferences in school friendship networks (2006) arXiv:physics/0611268v1 [physics.soc-ph]

4. Inuzuka, N., Nakano, T., Shimomura, K.: Friendship analysis using attendance records to university lecture classes. In: Proc. IASK International Conference on Teaching and Learninig 2008, pp. 478–486 (2008)
5. Shen, D., Sun, J.-T., Yang, Q., Chen, Z.: Latent Friend Mining from Blog Data. In: ICDM 2006. IEEE Computer Society, Los Alamitos (2006)
6. Scott, J.: Social Network Analysis, 2nd edn. Sage Publications, Thousand Oaks (2000)
7. Scripps, J., Tan, P.-N., Esfahanian, A.-H.: Node Roles and Community Structure in Networks. In: Workshop on Web Mining and Social Network Analysis, held in conjunction with KDD 2007, vol. 10 (2007)

Updating Background Image for Motion Tracking Using Particle Filter

Yuji Iwahori, Wataru Kurahashi, Shinji Fukui, and Robert J. Woodham

Abstract. Particle filtering based motion tracking needs the extraction of the region of moving object to assign particles for moving object. To extract the moving object, a background subtraction is often used. However, it is difficult to extract the moving object when illumination changes during motion. Updating background image using RANSAC has been proposed to solve this problem, but it is still difficult for RANSAC to update the background image with high accuracy when many exception values are included in the data for updating background. In addition, another constraint includes such that the first background image is necessary for updating background image to extract the moving object. This paper proposes an extended new approach to update the background image with high accuracy using the data which excepts the exception values based on the tracking result with particle filtering. PSA (Pixel State Analysis) is further introduced to distribute particles before updating the first background, which can assign particles without preparing background image in advance.

Keywords: Updating of Background Image, Particle Filter, RANSAC, Pixel State Analysis.

Yuji Iwahori and Wataru Kurahashi
Dept. of Computer Science, Chubu University
Matsumoto-cho 1200, Kasugai 487-8501, Japan
e-mail: `iwahori,kurahasi@cs.chubu.ac.jp`
`http://www.cvl.cs.chubu.ac.jp/`

Shinji Fukui
Faculty of Education, Aichi University of Education
Hirosawa, Igaya-cho, Kariya 448-8542, Japan
e-mail: `sfukui@auecc.aichi-edu.ac.jp`

Robert J. Woodham
Dept. of Computer Science, University of British Columbia
Vancouver, B.C. Canada V6T 1Z4
e-mail: `woodham@cs.ubc.ca`

K. Nakamatsu et al. (Eds.): New Advan. in Intel. Decision Techno., SCI 199, pp. 405–414.
springerlink.com

1 Introduction

As the tracking approach for the sequential frames, particle filter [1] is a good probabilistic approach for the robust tracking to both noise and occlusion. The initial setting for particle filter requires the extraction of moving object. The background subtraction is often used to extract the moving object. The previous approaches using background subtraction has some problems to extract the moving object under illumination changes and to prepare a background image in advance. The low processing cost and high accuracy are desired for the robust tracking. This requires to update the optimal background image sequentially for corresponding to the illumination changes.

Some previous approaches [2]-[7] have been proposed to update background images. However, there are still some problems such that the update interval should be set appropriate value [2], background images under the various illuminating conditions are necessary in advance [4], it is difficult to be applied to the environment inside room [5], it is difficult to correspond to the illumination change [6], the exception values are included to the updated background image when the background model has not been constructed [7].

This paper proposes a new approach using both RANSAC and PSA (Pixel State Analysis) [8] to estimate the background image, which is robust to the illumination changes without preparing the background image in advance. The approach updates the background image using data which remove the exception values for the moving object using both particle filter and RANSAC. This can improve the accuracy to estimate background. The method can also obtain the initial background image with high accuracy without preparing background image which does not include moving object. The method also can correspond to the illumination change based on updating the background image quickly. The approach are evaluated and demonstrated with computer experiments.

2 Updating Background Image

The line detection has been proposed for the detection and tracking of moving object in our system [9]. The system assigns the detection line to the possible points where a moving object enters or leaves, and tracks the detected moving object with particle filter. The system [9] detects the moving object with the background subtraction inside the region to track the moving object with the detection line. It is sufficient that this approach updates the background image inside the detection line.

The procedure of the background subtraction by this approach is shown as follows.

STEP1. Obtaining the data for updating background image
STEP2. Removal of exception value using the estimated value obtained with RANSAC

STEP3. Calculation of intensity value by the estimation based on the LMS criteria

2.1 Obtaining Data for Updating Background Image

n image data is used to update the background image with RANSAC. Data for updating background image sometimes include the moving object which becomes the exception value. The update of background image with RANSAC can remove the exception values when the number of exception values is less than $n/2$, however, the exception values with more than $n/2$ are included when the time interval to obtain the background image is taken to be short. Therefore, the previous approach [2] has the difficulty to update the background image with high accuracy without taking the appropriate time interval.

This approach tries to solve this problem of time interval to obtain the background image by using the detected result of moving object with the detection line of [9]. This system tracks the moving object with particle filter after detecting the moving object on the detection line. This means that the existing region of moving object is assumed to be known. When the data for updating background image are obtained, the regions which are tracked with particle filter and detected on the detection lines are removed. This enables to update the background image using data which do not include the exception value, that is, it is not necessary to set the time interval of data for updating background.

2.2 Updating Background Using RANSAC

RANSAC is a method to extract sample data among all data based on random sampling and to obtain the near optimal solution by repeating the fitting the sample data to the LMS estimation or LMedS estimation. When the exception data are not included in the extracted samples, the estimation result gives higher accuracy, and when the number of exception values is less than that of all data, more estimation values are included within the range of estimated errors. From this observation, when most measured values are included in the range, the estimation is regarded as correct result.

Here, a method to update background image using RANSAC is explained. Let the intensity value at the pixel (u, v) in the image be $f_{u,v}$, and it is prepared that n frames of image data are taken. First, select a sample $f^i_{u,v}$ among n data of (u, v). Here, $f^i_{u,v}$ is the intensity value selected using random sampling at time i. Next, M samples $f^j_{u,v}(j = 1, \cdots, M)$ are randomly selected from n image data which exclude $f^i_{u,v}$. The residual $e^j_{u,v}$ is obtained using selected samples as follows.

$$e^j_{u,v} = f^i_{u,v} - f^j_{u,v} \tag{1}$$

Square value of residual, $(e^j_{u,v})^2$, is taken and the median value of $(e^i_{u,v})^2$ is taken. These processes are iterated for N times. The iteration times N is determined by the following equation from the reliability for the requested solution.

$$P = 1 - \{1 - (1 - \epsilon)^F\}^N \tag{2}$$

where P represents the probability which includes at least one non-exception value among $f^i_{u,v}$, ϵ represents the ratio of exception values among all measurement data, and F represents the number of parameters.

After N iteration processes, the minumum value $(e^{min}_{u,v})^2$ is obtained among $(e^i_{u,v})^2 (i = 1, \cdots, N)$, and $f^{min}_{u,v}$ is assumed to be a temporary background and $f^{bg}_{u,v}$ is assumed to be a pixel value. Swinging of estimated values is calculated with the standard derivation $\sigma_{u,v}$, where $\sigma_{u,v}$ is given using $(e^{min}_{u,v})^2$ as

$$\sigma_{u,v} = 1.4826(1 + \frac{1}{n-1})\sqrt{(e^{min}_{u,v})^2} \tag{3}$$

where the constant 1.4826 is used for the modification to normalize the distribution.

Data among (u, v) beyond the range of $f^i_{u,v} \pm 2.5\sigma_{u,v}$ are removed as the exception values. The remained data are used for the LMS estimation, then the pixel value $f^{BG}_{u,v}$ is estimated.

$$f^{BG}_{u,v} = \frac{1}{m_{u,v}} \sum_{k=1}^{n} f^k_{u,v} \cdot \phi(f^k_{u,v}, f^{bg}_{u,v}, 2.5\sigma_{u,v}) \tag{4}$$

$$m_{u,v} = \sum_{k=1}^{n} \phi(f^k_{u,v}, f^{bg}_{u,v}, 2.5\sigma_{u,v}) \tag{5}$$

where $\phi(x, a, b)$ is taken to be 1 when $a - b < x < a + b$, 0 for other cases.

These processes are applied to whole pixels and the background image is made based on the estimated values. These processes will have the strong robustness for the case that the exception values are included.

3 Initial Arrangement of Particles by PSA and Acquiring Initial Background Image

This approach requires a background image before updating background is done for the first time because the exception values are removed using the tracking result with background subtraction and particle filter. It may be considered that a background image without moving object is prepared in advance, but such a system has the problem for basic conditions. It is also possible to make the background image without using the results of tracking with particle filter and the detection line, however there are some cases that the background image cannot be acquired with high accuracy.

So this paper proposes a new approach to set the initial arrangement of particle filter and to acquire the initial background image. The proposed approach makes it possible to track the moving object and to acquire the initial background image without preparing the background image in advance.

3.1 Initial Arrangement of Particles by PSA

PSA is an approach to discriminate the status of each pixel into three states of moving, stationary, and background using the data of all frames during the present frame to the previous $k + l$ frames. PSA estimates the status for each of the present frame to the previous k frames. Pixel inside the moving object region is discriminated into moving status or stationary status, that is, the moving object is recognized as moving status and the stopping object is recognized as stationary status.

This approach assumes that moving object exists inside the region recognized as moving status, and particles are initially arranged for the moving region. The discrimination for moving status does not need the background image. This enables the extraction of moving object region without preparing the background image, and the initial arrangement of particle filter can be done without decreasing the accuracy. Here, the initial arrangement of particles is done for the k frames before the present frame, since PSA can discriminate the status of the previous k frames not the present status.

The background image is acquired by updating the background of section 2 when the image data for n frames are acquired, then the particle filter is arranged for the present frame. After acquiring the background image, the background subtraction is applied to detect the object on the detection line.

3.2 State Transition of Stationary to Background

The status with PSA is discriminated as either the stationary or the background. Discriminating the stationary or the background status needs the background image but this becomes a constraint. This approach discriminates the status using the first frame image of the video sequences as a temporary background image.

When the temporary background image is used or when the object which is stopping at the first frame and begins to move on the way, the stopping object region is discriminated as the stationary status. If the discrimination of such a region into the appropriate status fails, the exception values will be included into the data for updating background image. The solution of the proposed approach to this problem is when the status is the background then the status is transited to the appropriate status by confirming if the region discriminated as the stationary status is true or not.

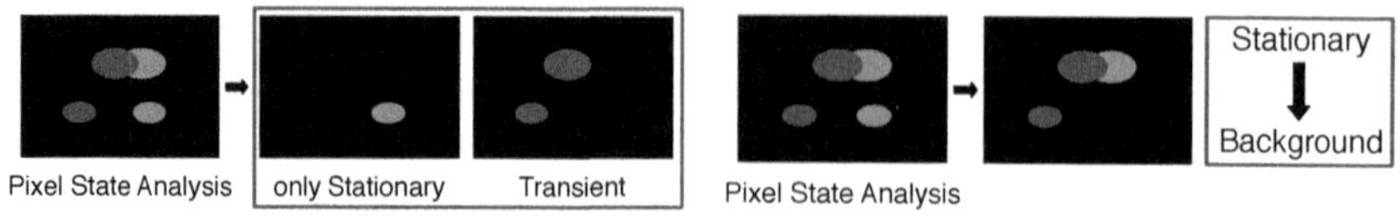

(a) Extraction of Stationary Status and Moving Status

(b) Removal of Stationary Status with Status Transition

Fig. 1 Transition of Stationary Status into Background Status

When the region of stationary status appears, the region is either of that after the stationary object moved or that the stationary object exists in. Discriminating to which region that belongs, and discriminating the status becomes stationary after the object included in the temporary background moved, the status will be transited from stationary status to background status. The procedures are as follows.

STEP1. The region of stationary status where the moving status is not connected, and the region of moving status are extracted (See Fig.1-(a)).
STEP2. Each histogram is obtained for each region of extracted stationary status and moving status.
STEP3. Similarity between the region of moving status and the region of stationary status is obtained using histogram.
STEP4. In case that there is the region with the higher value of similarity, it is judged that the object in the region of stationary status moved to the region of moving status. In this case, the status will be transited from the stationary status into the background status (See Fig.1-(b)).

When the status is transited from the stationary status to the background status, it is considered that the object had existed in the region and the data for updating background may include the object. From this situation, the data for updating background obtained should be removed once, and the background is updated with only the data obtained after the status transition. This enables to update the background with high accuracy including the case when the object is stopping at the first frame.

3.3 Automatic Determination of Threshold Value Used in PSA

PSA discriminates the status of each pixel using the change value T of the image intensity at each pixel and the stable degree S. The status is discriminated into the moving status when the change value T is greater than the threshold value th_{PSA}, otherwise the status is discriminated into the stationary or background status. The appropriate threshold value is used to discriminate the status.

This approach determines the threshold value as follows. The range for the variance of image intensity is obtained for the background with noise, and if the threshold value exceeds the range, the threshold value is taken to become the change of image intensity affected by the moving object. The procedure is as follows.

STEP1. The absolute difference value of each RGB component is calculated from the first frame and the second frame of video sequence.
STEP2. The standard derivation $\sigma^c(c = R, G, B)$ of absolute value of the difference is calculated using pixels whose absolute values of the difference calculated in STEP1 are less than some threshold value.
STEP3. The normalized distribution with mean O and standard derivation σ^c is assumed from the assumption that noise included in each pixel is independent each other. Under this assumption, the threshold value th_{PSA} is taken to be $6\sigma^c$.

The reason for that pixels with less value than some threshold are used in STEP2 is that the pixel values inside the region of moving object should not be used for the calculation. This could obtain the range of variation of image intensity value of background with only the pixel values in the background region.

4 Robust Approach to Illumination Change

When the sudden illumination change occurs, the image intensity value is changedmuch for whole of input image. The background subtraction may extract not only the object region but also the background region. When the illumination changes with large values, background is updated with terminating line detection temporarily.

It is recognized that the sudden illumination change occurred when the object is detected with more than 80% pixels on the detection line. In that case, the following processing is given.

STEP1. The detection of object is temporarily terminated on the detection line, while the tracking with particle filter is continued.
STEP2. The sequential n frames are taken with data for updating background from video sequence under the condition that the region where particles exist is removed.
STEP3. Background image is updated with obtained n image data for updating background.
STEP4. Object detection on the detection line restarts using updated background image.

Thus, the approach can update the background image with high accuracy even when the illumination change suddenly occurs, and can detect the object on the detection line.

5 Experiments

The computer experiment was done with the video sequence images to confirm the efficiency of the proposed approach. Digital video camera was used to take video. The image size is 720×480 and the specification of PC is Core2Duo 2.4GHz with 2 GB main memory. A total of 30 frames background data is used for updating background, and the interval for updating background was taken to be 2. The number of frames of image data used in PSA was 10.

Part of images used in the experiment is shown in Fig.2, and the background image taken by the proposed approach is shown in Fig.3-(a). The illumination change occurs at 94th frame. It was shown that the approach

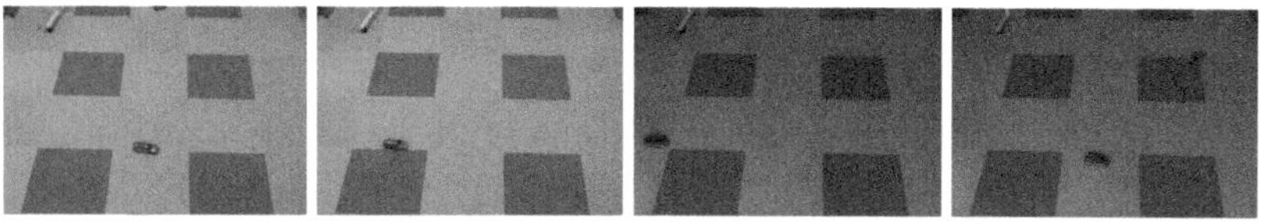

1st frame 60th frame 120th frame 180th frame

Fig. 2 Input Image Examples

(a)
(b)
(c)

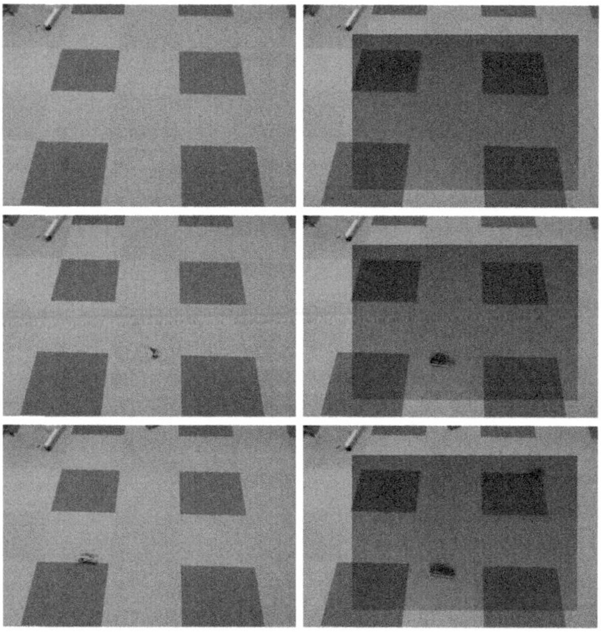

Before Illumination Change After Illumination Change

Fig. 3 Obtained Background Image: (a) This Approach (b) Approach [2] (c) Approach [7]

Table 1 Mean Error of Image Intensities

	Before Illumination Change	After Illumination Change
Proposed Approach	1.693	1.946
Approach [2]	3.608	4.962
Approach [7]	6.023	7.327

could obtain the background image without being affected by the moving object even when the illumination suddenly changed.

It took 501.71 msec to update the background image for each time. It is necessary to make the processing speed up to take the background image in real time based on this approach.

Next, the approach was evaluated in comparison with the approach [2] and the dynamic construction of background model with mixed Gaussian distribution [7]. The result by [2] is shown in Fig.3-(b) while that by [7] is shown in Fig.3-(c). The mean error of intensity values inside the detection line between the result image and the correct image is shown in Table 1. The proposed method gives the better result from this comparison. The approach [2] needs to set the appropriate time interval to take data, which fails to remove the exception values with such a short time interval like this condition. As a result, the moving object may be included in the background image. The approach [7] needs time to construct the background model when the illumination changes, or the moving object may be included in the background image when the moving object suddenly stops.

This approach has an advantage to take data without the exception values, and can generate the background image with high accuracy even when the time interval to take data is short.

6 Conclusion

This paper proposed a method of using particle filter for the initial arrangement of particles, RANSAC and PSA to obtain and update the background image.

The approach removes the exception values using the detection result of moving object on the detection line with the particle filter, then removes the exception values using the estimated value obtained by RANSAC and obtains the background image. The approach makes it possible to update the background image with high accuracy even when the time interval to take data for updating background is short.

PSA is used to make the appropriate transition of status without preparing a background image which excludes the moving object in advance. This could arrange the initial setting of particle filter without preparing the background image.

Further subjects include the real time implementation based on the approach for updating background and PSA processing.

Acknowledgements. The authors would like to thank Takuya Fujikawa for his help of experimental implementation. Iwahori's research is supported by JSPS Grant-in-Aid for Scientific Research (C)(20500168), Chubu University Grant, Nitto Foundation. Wooodham's research is supported by the Natural Sciences and Engineering Research Council (NSERC).

References

1. Isard, M., Blake, A.: CONDENSATION - conditional density propagation for visual tracking. Intl. J. of Computer Vision 29(1), 5–28 (1998)
2. Hori, T., Nami, M., Iijima, T.: A Technique of Moving Object Detection and Tracking in Real Enviroments, Hokkaido Industrial Research Institute Report, No. 305, pp. 9–15 (2006)
3. Takeuchi, K., Kaneko, S., Igarashi, S., Satoh, Y., Hane, T.: Image Analysis for Pedestrian Behavior based on Robust Subtraction and Segmentation. IIEEJ 31(2), 193–201 (2002)
4. Tsuchida, M., Kawanishi, T., Murase, H., Takagi, S.: Sequential Monte-Carlo Estimation of Background Image for Background Subtraction under Changing Illumination. IEICE D-II J87(5), 1062–1070 (2004)
5. Yoshimura, H., Iwai, Y., Yachida, M.: Sequential estimation of background components in outdoor environments. CVIM 2005(38), 61–68 (2005)
6. Ken, M., Hitoshi, H., Takashi, M.: Dynamic Background Modeling using Linear Dynamical System. CVIM 2006(51), 61–68 (2006)
7. Shimada, A., Arita, D., Taniguchi, R.-i.: Fast Dynamic Control of Adaptive Mixture-of-Gaussian Background Models. IEICE, D J90(9), 2606–2614 (2007)
8. Fujiyoshi, H., Kanade, T.: Layered Detection for Multiple Overlapping Objects. IEICE Trans. on Info. and Sys E87-D(12), 2821–2827 (2004)
9. Fujikawa, T., Iwahori, Y., Fukui, S.: Tracking of Entering and Leaving Object with Particle Filtering. In: Proceedings of 2007 Tokai-Section Conference on Electrical and Related Engineering, vol. O-227 (2007)

Driver Assistance Systems to Rate Drowsiness: A Preliminary Study

Md. Shoaib Bhuiyan

Abstract. This paper attempts to present a comprehensive survey of what is being done to automate the drowsiness ratings to be employed within a vehicle. The paper analyses the evidences for the usefulness of the measures currently used in drowsiness detection devices, which are not invasive and is based solely on eye activity. Their relationships with drowsiness and performance are described, and general problems and pitfalls associated with their practical use in passenger vehicles are identified. It also simulates a non-intrusive drowsiness detection system that is the core detection technique of several devices under review to understand how all the components of the system respond in real-time. A rating table to aid in automating the drowsiness rating in future is also included based upon analysis of drowsiness observed from recorded video.

Keyword: Drowsy, sleepy, detection, rating, scale, non-invasive.

1 Neurobiological Factors Affecting Drowsiness

Drowsiness, also referred to as sleepiness, is a state which is the result of both the circadian rhythm and the need to sleep. Sleep is a neurobiological need with predictable patterns of sleepiness and wakefulness. It is being recognized that neurobiologically based sleepiness contributes to human error in a variety of settings, and driving is no exception (Åkerstedt, 1995a, 1995b; Dinges, 1995; Horne, 1988; Sharpley, 1996; Martikainen, 1992). Drowsiness contributes to an estimated 76,000 to 100,000 crashes each year in the United States, resulting in an estimated 1,500 deaths and thousands of injuries (Knipling and Wang 1995; Wang, Knipling and Goodman 1996). Most crash database statistics based on police-papered crashes suggest that 2% to 4% of vehicle crashes nationwide involve a drowsy driver. These statistics, however, likely underestimate the role of drowsiness in crashes because it is difficult to identify and because drivers are unlikely to admit they were drowsy after being involved in a crash. A recent in-depth naturalistic study (Klauer et al 2006) found that drowsiness affected crashes in which the study sample was involved at much higher rates than would have been predicted

Md. Shoaib Bhuiyan
Department of Medical Information Science
Suzuka University of Medical Science
1001-1 Kishioka, Suzuka, Mie, Japan 510-0293
e-mail: bhuiyan *at* suzuka-u.ac.jp

K. Nakamatsu et al. (Eds.): New Advan. in Intel. Decision Techno., SCI 199, pp. 415–425.
springerlink.com

based on existing databases. That study found drowsiness to be a contributing factor in 20% of all crashes and 16% of near-crashes in the study sample.

For the purposes of this paper, we shall use the term drowsiness interchangeably with sleepiness. The terms "fatigue" and "inattention" also carry unique meanings (Brown, 1994). Fatigue is the consequence of physical labor or a prolonged experience and is defined as a disinclination to continue the task at hand. In regard to driving, a psychologically based conflict occurs between the disinclination to drive and the need to drive. One result can be a progressive withdrawal of attention to the tasks required for safe driving. Inattention can result from fatigue, but the crash literature also identifies preoccupation, distractions inside the vehicle, and other behaviors as inattention (Treat et al., 1979).

Literature related to the driving made little mention of sleepiness before 1985 and instead focused on the prevention of inattention and fatigue; traffic crash forms did not have a category for papering sleepiness as a crash cause. In the more recent surveys of noncommercial crashes, investigators have begun to collect and analyze data for instances in which the driver may have fallen asleep.

Sleepiness can certainly contribute to fatigue and inattention, and given the lack of objective tests or uniform papering requirements to distinguish these different crash causes, misclassification and inconsistencies in the primary data and the literature can be expected. Some, but not all, recent studies and reviews make an explicit assumption that given the uncertainty in crash papers, all crashes in the fatigue and inattention categories should be attributed to sleepiness. US Department of Transportation (DOT) suspects that sleepiness-related crashes are still very often papered in the categories of fatigue and inattention, and it reached consensus that sleepiness is an under-recognized feature of noncommercial automobile crashes. DOT expert panel concluded that the data on fatigue and inattention provide less support for defining risk factors and high-risk groups than the data on sleepiness or drowsiness. In addition, sleepiness is identifiable, predictable, and preventable (Rau P.S. 2005).

Swedish National Road and Transport Research Institute carried out a study to identify the most relevant parameters for the prediction and detection of fatigue and drowsiness in drivers (Kircher A et al 2002). It focused on methods to automatically detect drowsiness. The main variables of interest were: time-to-line crossing, lateral position of the car and related measures, steering wheel measures. Other measurement variables of possible interest were: longitudinal position deviation, eye blink frequency, physiological variables, and driving performance measures (other than lateral position). The survey clearly indicated that no single indicator could be used to detect drowsy driving. A combination of different measures was recommended e.g. analysis of lateral control performance and eye blink pattern. Furthermore, it was noted that so far there is no commercial system available that provides a sufficiently reliable method to detect a drowsy driver. Their analysis of experimental data did not reveal any clear answer to what indicators are the most prominent with respect to detecting drowsy driving behavior. Further investigations and analyses of driving behavior data are needed. However, the data analysis conforms in large to the findings in the literature survey.

2 Methodology

To identify potential devices and associated technical information, the keywords used for searching were: 'driver fatigue'; 'driver alert'; 'driver distraction'; 'drowsy driver'. This was repeated using 'sleepiness', 'sleep', 'drowsiness'. Other keywords were 'eye blinking and sleepiness'.

The databases and internet search engines used for the information search were 'IEEE Xplore', 'ACM digital library', 'Google'. The 'US patents database' was also searched as part of the Google search. More than a hundred peer-reviewed journal and conference papers and a large number of technical reports and websites were reviewed.

2.1 *Fatigue or Drowsiness Detection Methods*

A review (by Johanes, 2006) of existing fatigue detection and monitoring systems shows that the various techniques available can be classified into four groups:

A. Fit-for-duty and readiness-to-performance evaluation
These technologies attempt to assess the vigilance capacity of an operator before the work is actually performed.

B. Mathematical models for predicting alertness dynamics
These technologies use mathematical models to predict operator alertness and performance at different times based on interactions of sleep, circadian and related antecedents of fatigue.

C. Vehicle-based performance technologies
These technologies detect the behavior of the driver by monitoring the transportation hardware systems under the control of the driver, such as driver's steering wheel movements, acceleration, braking, gear changing, lane deviation etc.

D. In-vehicle online operator status monitoring technologies
These technologies seek to offer real time monitoring and recording of some bio behavioral dimensions of an operator, such as features of the eyes, face, head, brain activity, reaction time etc. during driving.

2.2 *Organization of this Survey Paper*

The focus of this paper will be on analyzing and evaluating the different techniques available under **Group D**. We exclude all intrusive, head or eye mounted devices and psycho-physiological measures like EEG. We shall also look into different drowsiness rating scales and what is being done to automate these ratings.

3 Drowsiness Detection Algorithms

3.1 Eye Localization and Tracking – for Detecting Micro Sleeps

This method of fatigue detection is a vision-based approach, making use of an on-board video camera and microprocessor, to enable non-intrusive, automatic diagnosis. The system receives continuous sequence of images input from the camera, analyze the eyes in each image and make comparison between images. This real time system generally consists of four steps:

a) Localization of the eyes (in the first frame)
b) Tracking the eyes in the subsequent frame
c) Detection of success or failure in tracking
d) Detection of possible drowsiness, based on pre-defined algorithm logic.

Localization involves looking at the entire image of the face and determining the eye envelopes. This can be done using the Template-based matching technique – which uses knowledge about the shape of an object and searches the image for that particular object, and/or Feature-based matching technique – which involves the use of geometrical constraints such as the relative locations of facial features to approximate locations. The search method used had several purposes; first is to compute the rough regional location of the eyes, using the fact that eyes are regions of rapidly changing intensity and that the entire eye regions are darker than their surroundings, so as to reduce the search space and extract the relevant regions at a reduced computational cost. Secondly, it is to identify the exact location of the eye, by relying on the fact that the eyes correspond to intensity valleys in the image, and finally obtain an estimation of the position of the iris by applying a simple pre-defined template.

3.2 Non-intrusive System Design

The algorithm extracts the movie to single frames. For every frame, it applies the Sobel edge detector on the image frame. Edge detector threshold is given as an input from the user. Also for each frame, it runs Circular Hough Transform algorithm on the binary image (output image of the edge detector). Then it draws the circles that were detected on each original image frame. If the number of circles detected were less than 2 then it marks that image frame as 'eyes shut'. Otherwise, it marks the image frame in question as 'open eye'. If eyes were detected as shut in 8 consecutive frames then it produces an actionable output, which in this case is a 'beep' sound.

The Circular Hough Transform detects circles in an image by letting each edge point of the image to "vote" for all circles (with given radius) it may belong to. Circles with maximum votes "win". The circular Hough transform maps a two-space (x,y) to a three-space of circles (x,y,r).

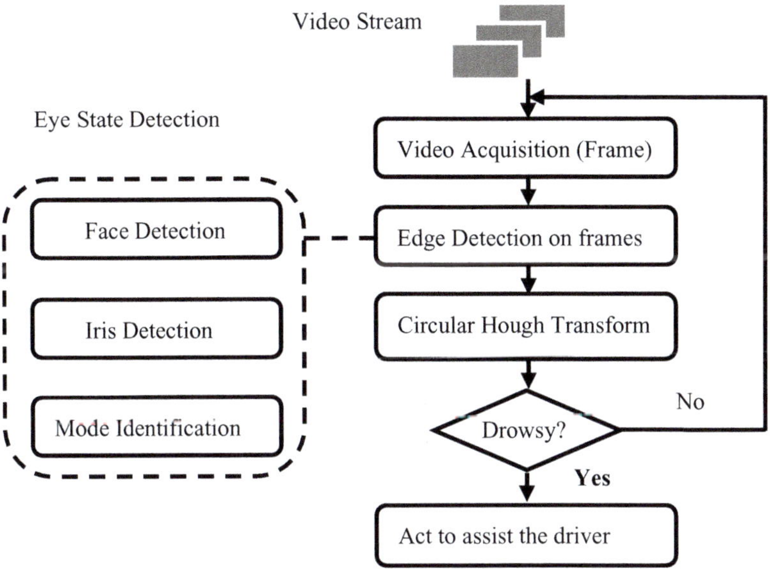

Fig. 1 Flow chart of driver's drowsiness detection system

3.3 Drowsiness Detection Devices Based on Non-invasive, Eye Activity

Many drowsiness detection devices are based upon indices of eye activity that are correlated with sleepiness, including blinking behavior, saccadic eye movement patterns (or changes in fixation point), pupil size, eye point of regard and eye closure for brief periods. Table 1 lists the detection devices we consider practical.

The main technologies that have been used to measure eye activity for the purposes of sleepiness detection are infrared reflectance devices and the analysis of video imagery. The infrared technique typically uses a transmitter and receiver that are mounted on a pair of spectacles worn by the driver, and directs a beam of infrared light at the eyelids of the driver (Webster and Leder, 1997). The method may allow the driver to wear prescription glasses or sunglasses if they are suitably modified. Image analysis of eye activity is achieved using a dashboard-mounted camera pointed at the driver's face, and uses image processing techniques to locate the driver's head and eye positions, although operational problems may be encountered with large head movements. They do however enable eye point of regard to be measured through the combined use of head and eye position; eye point of regard can provide an indication of the driver's focus of visual attention, and therefore may provide information additional to the prediction of sleepiness. Both technologies can have difficulty coping with the wide range of illumination levels typically encountered while driving, such as bright sunlight, darkness and rainy conditions. Therefore a careful assessment of the practical issues associated with measuring eye activity to detect sleepiness is required.

Table 1 Non-invasive, eye-based drowsiness detection devices (Wright NA. et al, 2007)

Name of device *Physiological, physical, behavioral or model-based operation*	**Operational, validation, market and assessment status** *For example, in routine use, tested in lab, validated in operational setting, available commercially*	**Intrusiveness and likely operator acceptance** *Limitations on use, for example environmental noise or light; relevance to specified occupational settings*	**Format of warning or alert** *For example, auditory, visual or other*
PERCLOS Detects eye closure using infrared, retinal reflectance device. Measures duration of blinks and eye closures, and proportion of time eyes closed over a specified time interval.	Marketed by Attention Technologies Inc. through the Co-Pilot system. Is being evaluated by a trucking company along with Sleep Watch and SafeTrac.	Non-intrusive. Has problems with detecting eye closure during night driving	Provides continuous green-amber-red warning light to indicate sleepiness level, accompanied by auditory tone when critical level detected (quoted as 3-4s eye closure)
CoPilot CoPilot is based. upon the PERCLOS system It detects percentage of time eyes are closed over a specified time interval (PERCLOS system) via infrared camera system	A spin-off from Carnegie Mellon University, it is being evaluated in a variety of research settings including simulator and on-road scenarios. US Patent obtained.	Non-intrusive. Current version limited to night-time use. Developer is currently extending system to operate under daylight conditions.	Auditory and visual stimuli. Different levels of warning depending on PERCLOS measures. Current system has only audible tone
ETS-PC Eye Tracking System University of South Dakota spin-off. Detects eye closure via a camera	Available commercially and currently being used in research environment by car manufacturers	Non-intrusive. Manufacturer states that the system adapts to different lighting levels (operates in daylight and darkness)	No warning stimuli currently included
Toyota Driver Drowsiness Detection and Warning System Detects eyelid movement using camera mounted on rear-view mirror.	Prototype system under research to determine if a driver's eyes are properly open	Non-intrusive	Acoustic warning. Provides warning through an in-vehicle navigation system
faceLAB 4.5 Spin-off from Australia National University. Measures eye-gaze and eye closure Uses PERCLOS fatigue assessment scale	faceLAB 4.5 system is now commercially available and is currently being used as a research tool. But not as an alertness device	Non-intrusive. Is OK with day or night-time, and copes with both. Also OK with (sun) glasses. Copes with partial head occlusion.	No warnings in place
Smart Eye Pro 5.0 Detects head position and point of gaze via image processing	Commercially available. Company in Gothenburg, Sweden	Non-intrusive	No warnings described

4 Drowsiness Ratings Scales

The literature on sleepiness often cites specific sleepiness scales such as Karolinska Sleepiness Scale (KSS), Stanford Sleepiness Scale (SSS), and the Epworth Sleepiness Scale, (ESS). Each scale measures different characteristics. Other more general scales are the Visual Analogue Scale (VAS) and the category ratio scale (CR-10 scale). A short description of each follows. Evaluation of the following different scales is left for future work.

1. Karolinska Sleepiness Scale, KSS

The Karolinska Sleepiness Scale has been used widely (Åkerstedt T et al, 1990; Gillberg M et al, 1994). This scale uses nine steps of categories to cover the entire wakeful/sleepy continuum, making it a bipolar scale. Thus, only a few categories and scale values on the upper end of these scales are available for ratings in sleepy subjects. To study long distance truck drivers KSS was used to rate subjective sleepiness. Night drivers rated their sleepiness higher and had increased alpha and theta activity (Kecklund G et al 1993). There was also an individual correlation between KSS ratings and the alpha activity.

2. Stanford Sleepiness Scale, SSS

This rating scale consists of seven descriptive statements to describe the feelings a person may feel at the time of scale administration (Hoddes E et al 1973). The statements on the scale are bipolar, ranging from "feeling active and vital, alert, wide awake". SSS has been used to measure sleepiness among drivers. In a study with professional truck drivers under different types of driving regimes, SSS was used to evaluate the drivers' level of fatigue (Williamson AM et al 1996).

3. Epworth Sleepiness Scale, ESS

ESS is a simple questionnaire that measures the subject's general level of daytime sleepiness (Johns MW, 1991). ESS consists of eight different common situations in which the subjects rate their chances that they would doze off or fall asleep. One can refer to it as measuring average sleep propensity according to definition set by Johns. It correlates with MSLT and during overnight polysomnography.

4. Visual Analogue Scale, VAS

The VAS scale is useful for measuring several subjective phenomena and is most commonly used in the assessment of pain. It is usually a 10 cm line with end anchors labeled with the extremes of the phenomena to be measured. Subjects can indicate their level of the phenomena being measured by making a mark on the line (Kushida C, 2005). It can be used either as a unipolar scale, from not sleepy to very much sleepy, or as a bipolar scale, from very alert to very sleepy. The unipolar scale version was used in a study of young adult drivers investigating the relationship between perceived sleepiness while driving and a sleep model.

5. Category ratio scale, CR-10

The CR-10, also called the Borg-scale, combines level-anchored judgment aspects of a category scale with the growth rate determination possibilities of a ratio scale by using verbal labels on an open-ended number scale. CR-10 is a general scale that can be applied to various situations. Because it has a greater discrimination than KSS and SSS, it is easier to differentiate the rated symptoms, particularly at the lower end of the scale but also at a higher level of intensity.

5 PERCLOS Based Drowsiness Rating Scale

PERCLOS is defined as "PERcent of the time a driver's eyelids are CLOSed". The advantage of PERCLOS and measures derived from the driving performance data is that drowsiness rating process could be automated (and it is validated in the case of PERCLOS). However multiple video images and availability of a trained observer enable arguably more reliable method of assessing drowsiness.

Table 2 Observer drowsiness scale based on the video analysis (Zilberg E et al 2007)

Drowsiness Level	Drowsiness State	Video image indicators
0	Alert	Normal fast eye blinks, often reasonably regular;
		Apparent focus on driving with occasional fast sideway glances;
		Normal facial tone;
		Occasional head, arm and body movements.
1	Slightly drowsy	Increase in duration of eye blinks;
		Possible increase in the rate of eye blinks;
		Increase in duration and frequency of sideway glances;
		Appearance of "glazed-eye" look;
		Appearance of abrupt irregular movements – rubbing face/eyes, moving restlessly on the seat;
		Abnormally large body movements following drowsiness episodes
		Occasional yawning.
2	Moderately drowsy	Occasional disruption of eye focus;
		Significant increase in the eye blink duration;
		Disappearance of eye blink patterns observed during the alert state;
		Reduction on the degree of eye opening;
		Occasional disappearance of facial tone;
		Episodes without any body movements.
3	Significantly drowsy	Discernible episodes of almost complete eye closure, eyes are never fully open;
		Significant disruption of eye focus;
		Periods without any body movements (longer than for level 2) and facial tone followed by abrupt large body movements.
4	Extremely drowsy	Significant increase in duration of the eye closure episodes;
		Longer durations of episodes with no body movements followed by large isolated "correction" movements

One potential disadvantage of PERCLOS is that it is based on 3 minute running average with subsequent delay as well as being based on a single physiological indicator (thus ignoring multitude of other observable signs). Also there is a risk in relying on the automated estimate of PERCLOS produced by the FaceLAB system where the estimated PERCLOS clearly did not match the video images. The potential disadvantages of using measures derived from the driving performance data are that they were only derived to estimate PERCLOS and appear to be less directly associated with variation in drowsiness than analysis of video or physiological variables. These measures were also clearly associated with the road bends and other changes in driving scenery with potential of confounding associations with the states of drowsiness.

The above drowsiness rating scale included 5 levels similar to the scale developed by Wierwille and Ellsworth – from alert to extremely drowsy – and was based on observing a number of indicators including duration and rate of eye blinks and other eye lid movements, degree and duration of eye closures, direction and focus of eye gaze, patterns of facial, hand and other body movements, yawning etc (Zilberg E et al 2007). Guidelines for determining the drowsiness states in this scale are presented in the Table 2. The drowsiness level was estimated for 10 seconds intervals. This enabled responsive evaluation of drowsiness that could be potentially associated with a number of physiological indicators for development of an optimum drowsiness detection algorithm. If an observer was not confident which one of the two consecutive drowsiness levels to allocate then (Zilberg E et al 2007) suggests use of a new fractional level between those two levels. We are leaving the non-PERCLOS based evaluation methods for a future study.

6 Summary and Conclusions

We wanted to know what has already been done and what is being done to automate the drowsiness ratings to be employed within a vehicle. This paper attempts to present a comprehensive survey. It is meant to be a resource of useful notes and references to guide the reader. We simulated the drowsiness detection part with non-intrusive video processing that is at the core of most of the vision systems employed to assist drivers. In order for the system to detect drowsiness successfully, a set of parameters had to be given to the system manually, and they might vary from movie to movie (threshold, radius). We have also learned that no automatic system exists to classify drowsiness, yet.

Also wearing glasses of any kind caused the system to fail. For some special cases of drivers such as the ones who will not close his/her eyes when drowsy or the ones wearing sunglasses, the above research should combine with other detection methods to improve the accuracy and robustness of the drowsiness detection.

Acknowldgements. The author is grateful to Professor Koji Oguri, Faculty of Information Science and Technology, Aichi Prefectural University for discussing related topics and to Professor Yuji Iwahori, Computer Science, Chubu University for his patience and encouragement during the manuscript preparation. Thanks are due to the two anonymous reviewers for suggesting ways to improve the presentation of this paper.

References

Åkerstedt, T.: Work hours, sleepiness and accidents: introduction and summary. J. Sleep. Res. 4(2), 1–3 (1995a)

Åkerstedt, T.: Work hours, sleepiness and the underlying mechanisms. J. Sleep Res. 4(2), 15–22 (1995b)

Brown, I.: Driver Fatigue. Hum Factors 36(2), 298–314 (1994)

Dinges, D.: An overview of sleepiness and accidents. J. Sleep Res. 4(2), 4–14 (1995)

Horne, J.: Why we sleep, pp. 1–12. Oxford University Press, Oxford (1988)

van den Berg, J.: Indicators and Predictors of Sleepiness, PhD Dissertation, Umeå University, Umeå Sweden (2006)

Johns, M.W.: A new method for measuring daytime sleepiness: the Epworth sleepiness scale. Sleep 14(6), 540–545 (1991)

Kushida, C. (ed.): Sleep deprivation. Clinical issues, pharmacology and sleep loss effect. Marcell Dekker, New York (2005)

Rau, P.S.: Drowsy Driver Detection and Warning System for Commercial Vehicle Drivers: Field Operational Test Design, Data Analyses, and Progress

`http://www-nrd.nhtsa.dot.gov/pdf/nrd-01/esv/esv19/05-0192-W.pdf` (Accessed November 20, 2008)

Sharpley, A.: Impact of daytime sleepiness underrated. Lancet. 338, 71 (1996)

Åkerstedt, T., Gillberg, M.: Subjective and objective sleepiness in the active individual. Int. J. Neurosci. 52(1-2), 29–37 (1990)

Gillberg, M., Kecklund, G., Åkerstedt, T.: Relations between performance and subjective ratings of sleepiness during a night awake. Sleep 17(3), 236–241 (1994)

Hoddes, E., Zarcone, V., Smythe, H., Phillips, R., Dement, W.C.: Quantification of sleepiness: a new approach. Psychophysiology 10(4), 431–436 (1973)

Kircher, A., Uddman, M., Sandin, J.: Vehicle Control and Drowsiness. Swedish. National Road and Transport Research Institute (May 2002)

Klauer, S.G., Dingus, T.A., Neale, V.L., Sudweeks, J.D., Ramsey, D.J.: The impact of driver inattention on near crash/crash risk: An analysis using the 100 Car Naturalistic Driving Study Data (Contract No. DTNH22-00-C-07007, Task order 23). National Highway Traffic Safety Administration, Washington (2006)

Knipling, R.R., Wang, J.S.: Revised estimates of the US drowsy driver crash problem size based on general estimates system case reviews. In: Proceedings of the 39th Annual Association for the Advancement of Automotive Medicine, Chicago, IL, pp. 451–466 (1995)

Kecklund, G., Åkerstedt, T.: Sleepiness in long distance truck driving: an ambulatory EEG study of night driving. Ergonomics 36(9), 1007–1017 (1993)

Martikainen, K., Hasan, J., Urponen, H., Vuori, I., Partinen, M.: Daytime sleepiness: a risk factor in community life. Acta Neurol. Scand. 86(4), 337–341 (1992)

Ji, Q., Zhu, Z., Lan, P.: Real-Time Non-intrusive Monitoring and Prediction of Driver Fatigue. IEEE Transactions on Vehicular Technology 53(4), 1052–1068 (2004)

Reyner, L.A., Horne, J.A.: Evaluation: "in-car" countermeasures to sleepiness: cold air and radio. Sleep 21(1), 46–50 (1998)

Treat, J.R., Tumbas, N.S., McDonald, S.T., Shinar, R.D., Mayer, R.E., Sansifer, R.L., Castellan, N.J.:Tri-Level Study Of The Causes of Traffic Accidents. Executive Summary, Indiana University, DOT HS 805 099 (May 1979)

Wang, J.S., Knipling, R.R., Goodman, M.J.: The role of driver inattention in crashes: New statistics from the 1995 Crashworthiness Data System. In: The 40th Annual Proceedings of the Association for the Advancement of Automotive Medicine, pp. 377–392 (1996)

Webster, J.G., Leder, R.: Tiny device in eye glasses could help keep employees awake and safe while on the job. College of Engineering 1997 Annual Report Engineering Ideas for Tomorrow (Accessed November 13, 2008), http://www.engr.wise.edu/news/ar/1997

Wierwille, W.W., Ellsworth, L.A., Wreggit, S.S., Fairbanks, R.J., Kirn, C.L.: Research on Vehicle-Based Driver Status/Performance Monitoring: Development, Validation, and Refinement of Algorithms for Detection of Driver Drowsiness. National Highway Traffic Safety Administration Final Paper: DOT HS 808 247, 1994, Publication No. FHWA-MCRT-98-006) (October 1998)

Wierwille, W.W., Ellsworth, L.A.: Evaluation of Driver Drowsiness by Trained Raters. Accident Analysis and Prevention 26, 571–581 (1994)

Williamson, A.M., Feyer, A.M., Friswell, R.: The impact of work practices on fatigue in long distance truck drivers. Accid. Anal. Prev. 28(6), 709–719 (1996)

Wright, N.A., Stone, B.M., Horberry, T.J., Reed, N.: A review of in-vehicle sleepiness detection devices (Published Project Paper 157). Berkshire, UK Transport Research Laboratory (2007)

Zilberg, E., Xu, Z.M., Burton, D., Karrar, M., Lal, S.: Methodology and initial analysis results for development of non-invasive and hybrid driver drowsiness detection systems. In: Proc. 2nd Int'l Conf. Wireless Broadband and Ultra Wideband Communications, AusWireless 2007. IEEE, Los Alamitos (2007)

Anomaly Foreground Detection through Background Learning in Video Surveillance

Cheng-Yuan Tang, Yi-Leh Wu, Shih-Pin Chao, Wen-Chao Chen, and Pan-Lan Chen

Abstract. We present a new set of rapid detection of background subtraction algorithms using codebooks to established Background Model (BG Model) and the concept of Color Model originally proposed by [6]. The proposed methods do not require prior learning, as in [6], and can create an instant BG Model detection and training with instant learning mechanism. Our proposed methods can also turn the latter coming but stationary foreground objects gradually as background, which are more adaptive to the actual environments. The proposed methods can also use instant learning to absorb sudden camera movements caused by the environments. We show that the proposed methods are effective and efficient in video surveillance applications.

Keywords: foreground detection, codebook, background model.

1 Introduction

In the study of visual surveillance, effective background detection has been a very important issue. A common approach to capture moving foreground is to use back-ground subtraction with the pure background and the current image. Traditional methods require a precise static background and thus cannot be used properly in a dynamic environment where the background changes dramatically. Some re-searchers propose to establish a background model by a series of statistical color images, by the time sequence of each color pixel to establish a single distribution model [1-2,7-8]. However, the background model of a single color

Cheng-Yuan Tang and Pan-Lan Chen
Department of Information Management, Huafan Uniersity, Taiwan

Yi-Leh Wu
Department of Computer Science and Information Engineering, National Taiwan University of Science and Technology, Taiwan

Shih-Pin Chao and Wen-Chao Chen
Electronics and Optoelectronics Research Lab., ITRI, Hsinchu, Taiwan
e-mail: `cytang@cc.hfu.edu.tw`

K. Nakamatsu et al. (Eds.): New Advan. in Intel. Decision Techno., SCI 199, pp. 427–435.
springerlink.com

distribution model cannot record and describe the dynamic background such as shaking trees. A complex and non-statistical background model is generally established by mixture of Gaussians (MOG) [3,4]. But when the background changes dramatically, it has been shown that a small amount of Gaussians fail to accurately describe the changes in the background, which will lead to an over-sensitive problem for detection. The pixel-based method [6], which is based on the assumption that the time series of each pixel is independent, employs the codebook (CB) for background subtraction.

Until recently, researches in background detection in video surveillance use a training process to create the background model for detection. Those methods will fail when the background changes or the detection time lasts too long. Under these circumstances, the background scene may be far different from the training one because of changes of objects or light sources.

In this work, we explore how to simplify the characteristics of saving various features and establish algorithms which could detect and update the background models on the fly based on the codebook algorithm proposed in [6].

2 Background Model

The general codebook algorithm uses quantify/clustering techniques [5] to determine whether a cluster is background according to pixel sampling. If the background is quantified into a set of code words, one can also compile background on the pixel basis.

2.1 Construction of Codebook

Let ***P*** be a sequence of the same pixel of ***N*** RGB consecutive images: $P=\{p_1, p_2 \dots p_N\}$. Let ***C*** as Codebook: $C=\{c_1, c_2 \dots c_L\}$. Each pixel has different quantities of Codeword. ***L*** is the amount of codeword, c_i includes both, $v_i=(\bar{R},\bar{G},\bar{B})$, $aux_i=<\hat{I},\check{I}>$, $\hat{I}$ and $\check{I}$ are the max and min brightness, TempColor (TC) is the temporary storage of p_i color: $TC-\{tc_1, tc_2 \dots tc_J\}$, ***J*** is the amount of TempColor. During run time, for each pixel in ***t*** sampling time we compare the codeword in ***C***. If matched, the pixel is stored in the temporary memory ***TC***. If the size of ***TC*** exceeds the preset threshold value, ***TC*** will produce a set of codeword and store in ***C***. Following is the detailed algorithm.

Ⅰ.L $\leftarrow 0$, C $\leftarrow \phi$

Ⅱ.For t=1 to N

① $P_t=(R,G,B)$, $I \leftarrow R+G+B$

② find Codeword c_m from $C=\{c_i \mid 1<=i<=L\}$

(1) $colordist(P_t, v_m) <= \varepsilon$

(2) $brightness(I, <\hat{I}, \check{I}>) = true$

(3) $BGS(x)\begin{cases} foreground, \text{if colordist and brightness no match} \\ background, \text{otherwise} \end{cases}$

③ If $colordist(x_t, v_m) <= \varepsilon_1$ and $brightness(I, <\hat{I}, \check{I}>) = true$

matching: $TC \leftarrow <P_t, I>$, $TC = \{tc_i \mid 1 <= i <= J\}$

④ If $C = \psi$

(1) $C \leftarrow <x_t, I, I>$

⑤ If $J >= MaxTCNum$, $TC = \{tc_i \mid 1 <= i <= J\}$

(1) $(\bar{R}, \bar{G}, \bar{B}) = mean(TC)$, $v_m \leftarrow (\bar{R}, \bar{G}, \bar{B})$

(2) $\hat{I} = \min(TC(I))$, $\check{I} = \max(TC(I))$

(3) $C \leftarrow <v_m, \hat{I}, \check{I}>$

Fig. 1 Detailed algorithm

Step (1) and (2) will be detailed in the following sections. When $\boldsymbol{x_t}$ and $\boldsymbol{v_m}$ are of similar colors and their brightness are below the threshold, the value of color $\boldsymbol{x_t}$ and the brightness value $\boldsymbol{I}$ will be stored into the $\boldsymbol{TC}$. ε is the sampling threshold. When the size of $\boldsymbol{TC}$ exceeds threshold MaxTCNum, all of color values in $\boldsymbol{TC}$ will be averaged to $(\bar{R}, \bar{G}, \bar{B})$ and the minimum and maximum value of $\boldsymbol{I}$ will be computed. When the current frame matches the features in codebook, we update the codebook so that the background model fit more accurately with future object characteristics and light sources.

2.2 *Color and Brightness Match*

Changing light in environment is often a serious problem. If we only rely on RGB color, it cannot be detected accurately. We thus segment each pixel into color and brightness separately to compute similarity.

3 Detection and Training

This section shows step by step our logical deduction with our initial detection results, problems and the ways of improvements.

Fig. 2 Initial detection results

When we set up the mechanisms to detect and study, we start by implementing the algorithm in [6]. Fig. 3.1 shows the initial detection results used the algorithm discussed in Section 2.1. In Fig. 2, when the brightness satisfies the detection conditions, the corresponding pixel is set to red. When the color distance satisfies the detection conditions, the corresponding pixel is set to green. When brightness and color distance both satisfy the detection conditions, the corresponding pixel is set to set to yellow (red + green). Black part represents that brightness and color distance both fail the detection conditions.

After running the initial detection method for some time, we gradually find some blurs from detection results. The part originally regarded as background starts to be detected as foreground by mistakes as shown in Fig. 3.

According to detection results, there are many detailed sections very similar to the background color of foreground object. It makes these detailed sections be mistakenly judged as the background. After its mistaken judge, those sections would immediately be learned to the codebook. The codebook will be stored with wrong background information instead of original correct information due to the pulling out of those wrong foreground colors. The result is as shown in Fig. 4.

The original algorithm uses color and brightness as the basis to classify the fore-ground or the background. But when the foreground is too close to the back-ground, this detection results will be completely wrong to classify and thus we can not rely on the detection results of the background are indeed the background.

The algorithm in [6] separates training background and detection function. In training, all the similar background parts are promised to be correct. But when in the actual detection, the color match the background does not mean it is the back-ground.

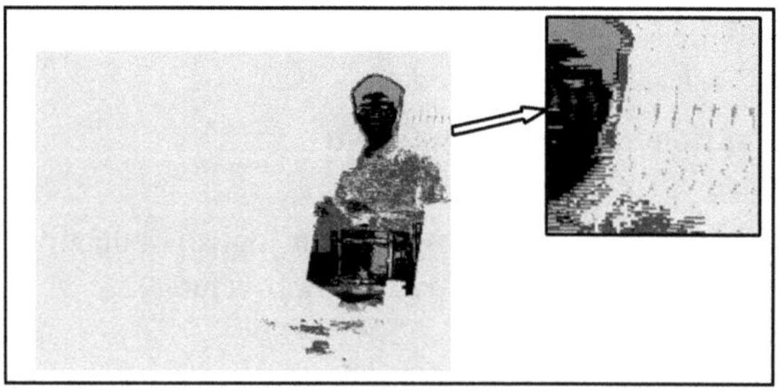

Fig. 3 Detection errors: false-foreground

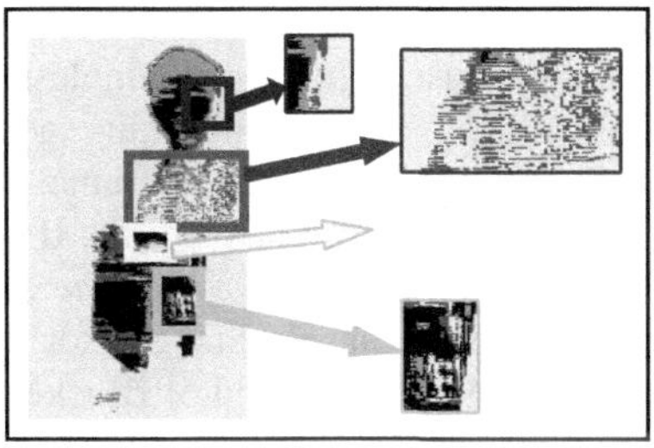

Fig. 4 Detection errors: false background

The proposed solution is to keep learning the background information even during the detection phase. To accomplish this goal, we need to establish a rapid and effective mechanism to locate the image regions that require learning. We propose to segment each image frame into fixed blocks of the same size and compute the percentage of non-match pixels in each block. If the percentage of non-match pixels of the block is less than the threshold, this entire block will be treated as background. If the percentage of non-match pixels is higher than the threshold then the entire block will be treated as foreground. The threshold is typically set to 96% to 98%. The filtering result is as shown in Fig. 5.

In Fig. 5, the input MatchMap output a boolean map that separates background and foreground by the block filtering method and the result is the BlockMap in Fig. 5. We can use this map to cut and show the result of detection, and also use this map to decide where the real background is, and to learn the detected background into the codebook.

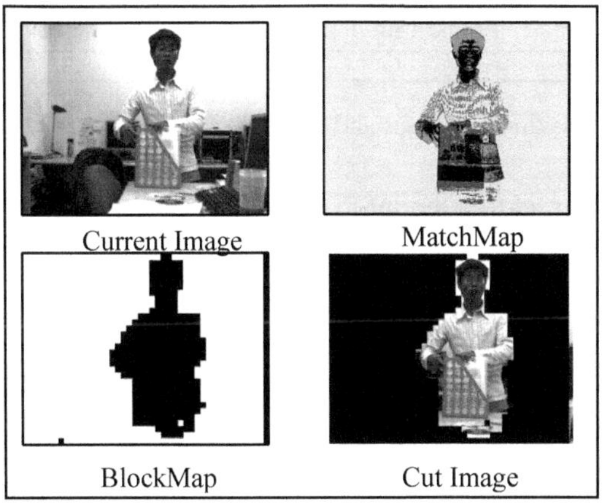

Fig. 5 Results after filtering

In the original algorithm, the foreground is the part which is not learned at the be-ginning. In practical visual surveillance system, we have assumed foreground that is invading objects at the scope of surveillance, and the object is displayed. How-ever, not all invading objects at scope of surveillance are monitored. It must be detection and identification of objects as meaningless and no longer continue to move, such as a person takes a bookcase into the scope of surveillance and leaves.

In [6], that stationary bookcase will always be regarded as the foreground. But in reality, the bookcase is just a meaningless background. The propose algorithm includes new detect and update training functions. Our approach could record BlockMap generated from each frame, and compare their similarity with the BlockMap generated from next frame. If the two are exactly the same, it will add one to the cumulative record. When the cumulative value is over the threshold, it means that the image hasn't changed for a long time. The proposed algorithm can learn those detected foreground of BlockMap. The cumulative threshold value will be corrected depending on speeds of every monitoring system's camera shots.

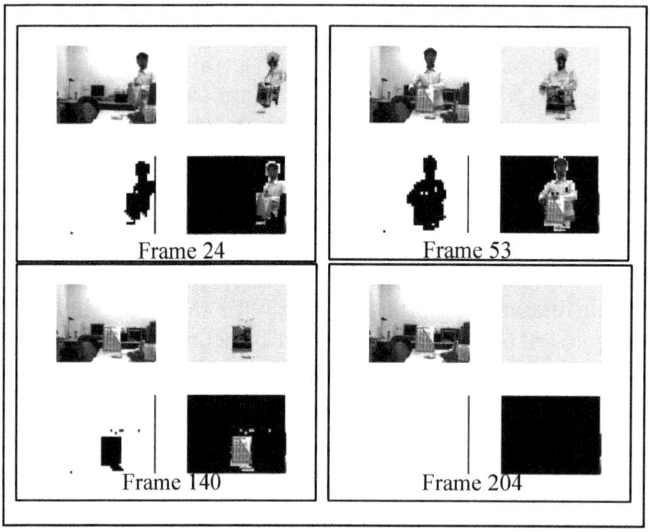

Fig. 6 Stationary foreground detection

After adding the mechanism, the stationary foreground is decided as the background through time, as shown in Fig. 6. If the monitoring system is moving, this algorithm needs only a small number of image frames to correct the back ground model to normal condition.

The proposed method is detailed as follows:

(1) Read the current image pixel and check with the Codebook. If the Codebook has the same value, then store pixel's RGB value to CodeBook and return to (1); otherwise goto (2).
(2) Use the each pixel's Codebook and RGB Color of image to produce a MatchMap by matching the color distance and brightness as discussed in Section 2.2.
(3) Translate the MatchMap to BlockMap. And if the previous BlockMap exist, then goto (4); otherwise goto (5).

(4) Compare the current BlockMap with the previous BlockMap, if exactly the same, then add 1 to the CumLastBlockMap; otherwise, set the CumLastBlockMap to 0,
(5) Test if the CumLastBlockMap is greater than or equal to the MaxBlockMap (a preset threshold); if so, store each pixel's RGB of image to Codebook; otherwise, store to the TempColor.
(6) Test if the TempColor's size exceed the MaxTempColorNum (a default threshold); if so, store the average of TempColor's all Value into the Code-Book.
(7) Test if the CodeBook's size exceed the MaxCodeBookNum (a default thresh-old); if so, remove the oldest of Codeword from CodeBook.
(8) Goto (1).

Through time, the proposed algorithm gradually saves many codewords. To reducc thc burdcn of mcmory, thc algorithm additionally creates a mechanism to delete the older codewords from the codebook. The algorithm needs to set the max amount of codewords. If the amount of codewords is above the threshold value, it will remove the oldest codeword. An example of updating the codebook is illustrated in Fig. 7.

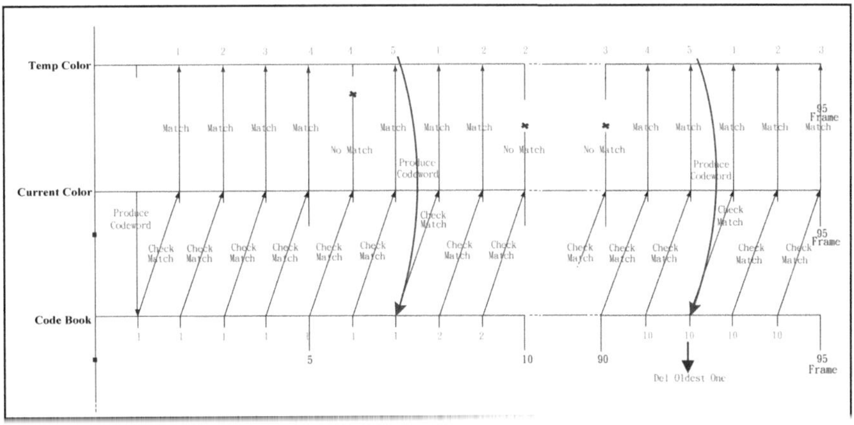

Fig. 7 Codebook update example

4 Experiments

In the following experiment, we compare our algorithm with background model algorithm in [6]. We will use the same sequence of images for the two algorithms. Before the detection phase, we allow [6] to train 10 seconds (300 frames) of no-changing background, and then we start the detection phase at the same time for both algorithms.

In Fig. 8, we show frame numbered 240 of the detection results of both algorithms with an intruding object and the results are comparable. We also find little difference between the results of frame numbered 300 of both algorithms. After frame numbered 300, the bookcase was put in a stationary position. We show the detection results of frame numbered 900, because it has been over the threshold of pre-background stay, the proposed method will train the foreground bookcase which is stationary for too long to become background. The bookcase will be shown as background afterwards.

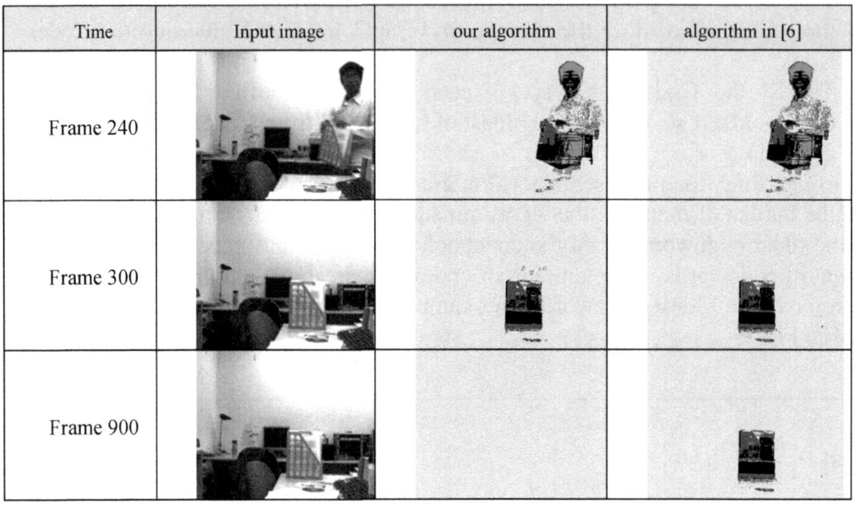

Fig. 8 Detection results

A comparison chart of the proposed algorithm and [6] is shown in Table 4.1. The proposed algorithm is marginally less accurate than [6]. But the proposed algorithm does not require pre-training to establish the background model. And during the detection phase, the proposed algorithm require additional training process and thus is marginally slower than [6]. However, the proposed algorithm eliminates the background model problem in [6] that after a long period of detection the detection accuracy will decrease. The proposed algorithm also can train during the detection phase and train those stationary foregrounds into background model. By adding the training process to the detection process, the proposed algorithm holds a distinct advantage against fixed background learning algorithms such as [6].

Table 1 Comparison of the proposed algorithm and [6]

	algorithm in [6]	proposed algorithm
detection accuracy	72%	68%
training + detection	X	O
training stationary foreground	X	O
Compensation of ca-mera movement	X	O

5 Conclusion and Future Work

This study proposes a background detection algorithm that can detect and train at the same time. By combining the training and detection, the proposed algorithm can make the established background model more accurate to the environment through time. The proposed algorithm can also learn stationary foreground objects as background through time. This same feature can be extended to compensate camera movements. Our preliminary experiment results with indoor videos suggest that the proposed algorithm is very promising in real world video surveillance applications. One of our future directions is to extend the current algorithm to the outdoor environments with significant changes in light sources.

References

1. Wren, C.R., Azarbayejani, A., Darrell, T., Pentland, A.: Pefinder: Real-time tracking of the human body. IEEE Transactions on PAMI 19(7), 780–785 (1997)
2. Horprasert, T., Harwood, D., Davis, L.S.: A statistical approach for real-time robust background subtraction and shadow detection. In: IEEE Frame-Rate Applications Workshop, Kerkyra, Greece (1999)
3. Stauffer, C., Grimson, W.E.L.: Adaptive background mixture models for real-time tracking. In: Int. Conf. Computer Vision and Pattern Recognition, vol. 2, pp. 246–252 (1999)
4. Elgammal, A., Harwood, D., Davis, L.S.: Non-parametric model for background subtraction. In: Vernon, D. (ed.) ECCV 2000. LNCS, vol. 1843, pp. 751–767. Springer, Heidelberg (2000)
5. Kohonen, T.: Learning vector quantization. Neural Networks 1, 3–16 (1988)
6. Kim, K., Chalidabhongse, T.H., Harwood, D., Davis, L.: Background Modeling and Subtraction by Codebook Construction. In: IEEE International Conference on Image Processing (ICIP) (2004)
7. Chalidabhongse, T.H., Kim, K., Harwood, D., Davis, L.: A Perturbation Method for Evaluating Background Subtraction Algorithms. In: Joint IEEE International Workshop on Visual Surveillance and Performance Evaluation of Tracking and Surveillance (VS-PETS 2003), Nice, France, October 11–12 (2003)
8. Horprasert, T., Harwood, D., Davis, L.S.: A Statistical Approach for Real-time Robust Background Subtraction and Shadow Detection. IEEE International Conference on Computer Vision (ICCV), pp. 1–19 (1999)

Adaptive Alarm Filtering by Causal Correlation Consideration in Intrusion Detection*

Heng-Sheng Lin, Hsing-Kuo Pao, Ching-Hao Mao, Hahn-Ming Lee, Tsuhan Chen, and Yuh-Jye Lee

Abstract. One of the main difficulties in most modern Intrusion Detection Systems is the problem of massive alarms generated by the systems. The alarms may either be false alarms which are wrongly classified by a sensitive model, or duplicated alarms which may be issued by various intrusion detectors or be issued at different time for the same attack. We focus on learning-based alarm filtering system. The system takes alarms as the input which may include the alarms from several intrusion detectors, or the alarms issued in different time such as for multi-step attacks. The goal is to filter those alarms with high accuracy and enough representative capability so that the number of false alarms and duplicated alarms can be reduced and the efforts from alarm analysts can be significantly saved. To achieve that, we consider the causal correlation between relevant alarms in the temporal domain to re-label the alarm either to be a false alarm, a duplicated alarm, or a representative true alarm. To be more specific, recognizing the importance of causal correlation can also help us to find novel attacks. As another feature of our system, our system can deal with the frequent changes of network environment. The framework gives the judgment of attacks adaptively. An ensemble of classifiers is

---

* This work was partially supported by the iCAST project sponsored by the National Science Council, Taiwan, under the Grant No. NSC97-2745-P-001-001.

Heng-Sheng Lin
Trade-Van Information Services Co., Taipei, Taiwan
e-mail: ed.lin@tradevan.com.tw

Hsing-Kuo Pao, Ching-Hao Mao, Hahn-Ming Lee, and Yuh-Jye Lee
National Taiwan University of Science and Technology, Taipei, Taiwan
e-mail: pao@mail.ntust.edu.tw, d9415004@mail.ntust.edu.tw, yuh-jye@mail.ntust.edu.tw

Hahn-Ming Lee
Academia Sinica, Taipei, Taiwan
e-mail: hmlee@mail.ntust.edu.tw

Tsuhan Chen
Carnegie Mellon University, Pittsburgh, PA, USA
e-mail: tsuhan@cmu.edu

K. Nakamatsu et al. (Eds.): New Advan. in Intel. Decision Techno., SCI 199, pp. 437–447.
springerlink.com

adopted for the purpose. Accordingly, we propose a system mainly consisting of two components: one is for alarm filtering to reduce the number of false alarms and duplicated alarms; and one is the ensemble-based adaptive learner which is capable of adapting to environment changes through automatic tuning given the expertise feedback. Two datasets are evaluated.

Keywords: Intrusion detection, alarm filtering, false alarm, adaptive learning, ensemble.

1 Introduction

Over the last couple of decades, the intrusion methods are getting sophisticated and diversified. Variety of rootkits and exploit codes are easily obtained for the hackers to attack the systems. Therefore, individual data can be illegally read or overwritten by intruders. Many different Intrusion Detection Systems (IDSs) have been provided to detect those malicious attacks. However, one of the weakest points with those IDSs is the problem of massive false alarms (or false positives). As revealed in several reports, IDS usually generates nearly 99% of false alarms in the detection [1, 5]. On the other hand, various alarms, which could be issued by naïve decision rules may come from the same unique attack. For instance, several minor alarms may suggest a multi-step attack. A clever detection system should be able to single out the reason for further automatic or non-automatic analysis. Overall, false alarms or duplicated alarms can waste significant time from human analyzers. Without considering those issues, an IDS can be virtually useless. The problem is more serious when no enough human analyzers can be assigned for further analysis of the generated alarms. E.g., a personal IDS will not afford such overhead. In this work, we propose an alarm filtering (AF) framework which can significantly reduce these two kinds of alarms: false alarms and duplicated alarms. The framework considers causal correlation between alarms and alarms will then be issued with high accuracy and no redundancy. As another important feature, to apply our system to real network, we would like to make the final decision of alarm classification adaptively to different periods and to different environment. An ensemble of classifiers called ensemble-based adaptive learner (EAL) will be adopted to adjust the prediction precision and sensitivity for the system according to network conditions. The feedback from alarm analysts will be used to tune the setting periodically.

To reduce the false alarms and duplicated alarms, our AF system considers causal correlation between several alarms when they are either temporally correlated or associated with a single attack. Alarm correlation [12, 13] has been used in discovering the intentions or root cause of the attackers [3] and how they achieve their goals [8], i.e. the attack methods. Based on our observations, single minor alarm in small scale may be confusing, but minor alarms collected as a whole may indicate a serious attack. When lacking of considerations in large scale, some alarms may be mislabeled, so called the *false alarm problem*. On the other hand, multi-step attacks often trigger a bunch of alarms in a sensitive system to downgrade the performance of the system, so called the *duplicated alarm problem*. To deal with these two problems, one has to consider causal correlated alarms instead of a single alarm. Note that reduction of false alarms may lower the detection sensitivity (also known as recall). A trusted system must still be sensitive enough to detect serious attack and at the same time only a small number of false alarms are generated. As a challenging but an important extreme, our AF system will have the ability to identify novel alarms or alarms related to *novel attacks*. Causal correlation is considered for those alarms which may be associated with some anomaly behaviors and the final judgment can then be given with high confidence. We need to emphasize that the single alarm is usually issued from some naïve decision rules or signature alignment, to deal with various special cases or to solve some particular problems. Such rules may be created by a simple-minded consideration without too much rigorous efficiency analysis of the whole system. On the other hand, some rules may be created with a global view and lack of ability to fit into special environment, e.g., the period when new attacks just being released, or the "normal" period with low number of "background alarms". Our system offers a solution for that. To consider the causal correlation for a set of alarms, we are possible to relate the alarms to true attack or the attack of high risk, including novel attacks.

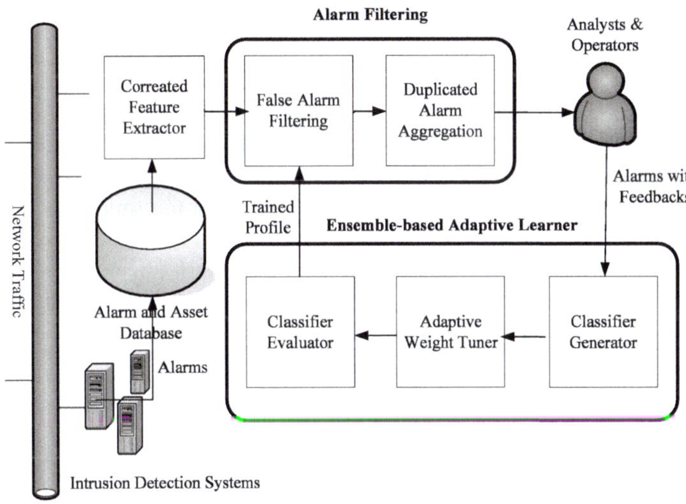

Fig. 1 Architecture of Adaptive Alarm Filtering System

The consideration of causal correlation can reduce both of the false alarms and duplicated alarms, and at the same time can deal with novel attacks. However, the network environment is usually not stable. As a result, operators have to tune and confirm the setting of IDSs frequently for the changes. That creates a burden for the operators. Due to the changes of network environment including devices, services or attack approaches, the pre-trained classifier will be getting to lose its accuracy on prediction after a certain period. This phenomenon is called *concept drift* [11] and happens very often in the real world. To address the problem, we consider automatically tuning the system so that the intrusion detection can adapt to the environment. An *adaptive* learning system filters false and duplicated alarms after IDSs and adaptively learns from responses of experts recurrently [6, 9]. That helps system operators relieved from laboring works on parameter tuning. Different from previous research, our proposed framework gets more focused on practical issues of changing network environment over different periods or even different sites. Using the proposed *ensemble*-based adaptive learner (called EAL) is eligible to give robust performance on prediction as time goes by or for different network environment. It is understood that the network data are large-scale stream data, and usually highly unbalanced between attacks and normal data. Besides, most attacks happen in a continuous fashion in a very short period of time. To deal with those concerns, our EAL is proposed based on entropy computation, also inspired by AdaBoost [10]. Also, some aging effect is added to the system. By our approach, the rare attacks are not to be overlooked and can contribute some to our system. Other than adapting to time, we can also make our system adaptive to different commercial organizations. We will take the risk of assets as the input for further system improvement. Overall, we aim at designing a system which can be applied to real environment.

2 Feature Extraction, Alarm Filtering and Adaptation

In this section, we discuss our proposed system in full details, as illustrated in Fig. 1. Different from most IDSs, we focus on the reduction of false alarms and duplicated alarms after IDSs issue alarms. To deal with real network data, we also consider an adaptive system where the attack call may depend on the time information. Our system can take alarms from several sources, e.g., distributed IDSs, as the input. Basically, the system can be separated into three parts: Feature Extraction Unit, Alarm Filtering Unit, and Ensemble-based Adaptive Learning Unit. We proceed to give details for those different units.

2.1 Feature Extraction

Many factors combined together to decide which alarm comes from a true attack. They include causal correlated alarms, unusual changes of frequency in alarm issuing, and asset information, etc. That is, the feature extraction set is beyond the common attributes of intrusion alarms, such as packet size, signature names, IP addresses, port numbers and so on. We need to know that, in some cases, the IP address may limit the generalization ability of the model [3]. Opposed to that, the properties of hosts are more strongly relevant to attacks. Therefore, an IP address is replaced by its corresponding asset information. Below, we illustrate those features one by one.

Causal Alarm Features. Our idea is to correlate the alarm in the previous step and the alarm at this moment as causal correlation features. As we can imagine, the features will help us to detect a multi-step attack. More than that, the features also provide more information on recognizing alarms which may be from novel attacks. For instance, if an alarm never happened in historical data, it is still possible to be classified correctly according to the conditional probability of the alarm after observing the pre-step alarms. To be more specific, those features consist of the combination of present alarm tag and the alarm tag in the previous stage (or pre-step alarm), under three different conditions, pre-step alarm with the same source, or same target as the current alarm, or both being the same.

Abnormal Frequency Value. Several triggered alarms are like "background noises" which happen all the time and may not suggest any meaningful information. The true alarm usually has an instant change that is out of the range of its normal frequency. We can compute mean μ, and variance σ^2, of each individual alarm a, on daily basis from historical data. The attribute, being able to real-time measures the degree of anomaly on individual alarm, is formulated as follows:

$$\mathrm{AFV}(a,x)=\begin{cases}\ln\left(\dfrac{(x-\mu_a)^2}{2\sigma_a^{\ 2}}\right) & x>\mu_a\\ 0 & \text{others}\end{cases} \tag{1}$$

where x is an accumulated amount of corresponding alarms of the present day, a refers to the alarm identity of x, μ_a is the average number of alarms correspondent to alarm a, and σ_a^2 is the correspondent variance.

2.2 Alarm Filtering

Alarm Filtering (AF) is to reduce the amount of alarms. To provide succinct alarm report to users, we first classify alarms to a class of relevant or irrelevant to an attack, and then aggregate duplicated alarms as an alarm group on behalf of a high level event for users. We consider causal correlation of alarms to reduce the false alarms, as described previously. An ensemble-based classifier, combined with basic learners is adopted as false alarm filtering. In order to adapt to changing network condition, the filter employs ensemble-based adaptive learner to keep updating (discussed later) as time goes by.

2.3 Ensemble-Based Adaptive Learner

Inspired by AdaBoost, the proposed EAL algorithm is a meta-learning method aimed to combine multi-classifiers when data are incrementally grown with time. Different from many related works [2, 4], the proposed algorithm is specifically focused on the characteristics of computer network and practical requirement of SOC.

In our ensemble of classifiers, there are two types of weights should be optimized. They are example weight for each instance and classifier weight, so called voting weight for each weak classifier. To learn from feedbacks, alarm log is separated by day as $D_J = D_{J-1}\bigcup d_J$ and $D_0=\phi$ where $d_j=\{(\mathbf{x}_{ji}, w_{ji}, z_{ji})_{i=1}^{n(j)}\}$, $n(j)$

denotes the amounts of alarms in the j-th day, $\boldsymbol{x}_{ji}$ is an alarm feature vector of the i-th example, $w_{ji}\in[0, 1]$ denotes the corresponding example weight, and $z_{ji}\in Z=\{-1,1\}$ indicates the corresponding real class in the j-th day. We set different example weights according to its class in our experiments. Borrowing the concept of entropy from information theory, we can deal with the problem of unbalanced data. Moreover, sample re-weighting, like AdaBoost, and complementary learning from previous wrongly predicted examples, strongly enhance the robustness of our system. The function is listed as follows:

$$h_{final}(x) = \arg\max_{z\in Z} \sum_{k=1}^{m} \varphi(a_k, \lambda, \tau) \cdot v_k \cdot h_k(x, z) \tag{2}$$

Where $h_{final}(\circ)$ is the finial hypothesis of committee decision with m member classifiers, $h_k(\circ)$ represents the hypothesis of the k-th day, $v_k = (1+entropy(P(d_k)))\cdot\log((1-\varepsilon_k)/\varepsilon_k)$ is the corresponding voting weight decided by its error rate ε, and the entropy defined by, $entropy(P) = -P\ln(P)-(1-P)\ln(1-P)$, indicating the information of the k-th training data point for the distribution. The $P(d_k)$ is the portion of true alarms in a training set. Finally, in order to being adaptive to changes, the Memory Decline Ratio (MDR) listed as follows will be used:

$$\varphi_k(a_k, \lambda, \tau) = \frac{\exp(-\lambda(a_k-\tau))}{1+\exp(-\lambda(a_k-\tau))}, \tag{3}$$

which is inspired by aging-forgetting mechanism and modified by sigmoid function. It is employed to tune the voting weight through the time in each individual classifier. The forgetting slope λ, is set for how fast to drop out a useless classifier with tolerating time τ. Setting the pair of parameters will be discussed in experiment and, actually, depends on the degree of concept drift or change in each dataset.

3 Experiments

We have built a prototype system to demonstrate the proposed approach that is able to filter the alarms of both kinds, the false alarms and the duplicated alarms. Below, we discuss different measurement on the system of alarms filtering such as False Positive (FP), which analysts have to pay extra effort with, and True Negative (TN), which means that filtered alarms are indeed not correlated to attacks. Of course, filtering out the true alarms associated to attacks is more serious than anything else, e.g. achieving high TP rate. Moreover, the ability to identify novel alarms is also taken into account. There are two experiments designed for demonstration. One is to evaluate that the proposed feature set including the causal correlation features is able to enhance the performance of intrusion detection. The second experiment is to compare with different learning schemes to support that our approach is able to filter out false or duplicated alarms and effectively adapt to network change especially when novel alarms happen.

Novel alarms make operators tune the setting of IDSs from time to time. The novel alarms also make pre-trained model useless after a while. Hence, we especially discuss the ability of our framework to identify novel alarms, meaning that our AF is able to give the correct predicted class on an unseen alarm under an acceptable level of false positive rate. By means of Receiver Operating Characteristic (ROC) curve, the False Positive (FP) rate referring to cost and the True Positive (TP) rate referring to detection ability allow analysts to know the trade off between detection rate and cost.

3.1 Datasets

The system was validated by two datasets. The first one is made by a popular benchmark, DARPA 1999 [7], the other is a real world private alarm dataset, which is provided from an SOC operated in Taiwan, called A-SOC 2007. The center offers a service of security surveillance to their clients, including many organizations, government departments and companies. Both of the alarms of DARPA and A-SOC are manually labeled for evaluation according to its official report and warning tickets to monitored client, respectively.

Data Distribution. Their data distributions are shown in Table 1.

Table 1 Distribution of each dataset for Experiments. Novel alarms represent that an alarm is never seen before the day

Dataset	Duration	Distinct IP	Distribution of Alarm Label		
			Total (Novel)	True (Novel)	False (Novel)
DARPA	1999.3.1 ~ 1999.4.10	546	55,473 (3,693)	19,109 (2,191)	36,364 (1,502)
A-SOC	2007.8.30 ~ 2007.9.7	5,368	308,063 (56,984)	6,743 (2,978)	301,320 (54,006)

3.2 Performance Measurement

Receiver Operating Characteristic (ROC) curve is adopted as the main performance measurement. The unbalanced problem makes the performance hard to be evaluated. Because the number difference between true alarms and false alarms is large, the enhancement of performance on identifying rare true alarms is easy to be overlooked if using Accuracy as a measurement. ROC curve, which consists of TP rate and FP rate, is suitable to be a performance measurement for this. In the viewpoint of system security operators, they want to know how much cost (FP rate) they have to pay if keeping a level of sensitivity of recognizing rare attacks. ROC curve can serve this purpose because false positive rate is like a cost we have to pay if we want to reach a level of true positive rate (alarm detection rate).

Table 2 Performance Test with Different Feature Combinations. The performance comparison is TP rate (detection rate) vs. FP rate (cost). The detection rate of novel alarms is specifically demonstrated for revealing the ability to detect novel alarms with different feature combinations. Basic feature set is the original attributes generating from Snort IDS but excluding source and destination IP addresses

Dataset	Feature Combination	Cost FP rate	All Alarms TP rate	Novel Alarms TP rate	Correctly Filtered
DARPA	Basic	18.83%	82.45%	85.76%	89.8%
	Causal	5.26%	95.08%	87.36%	97.34%
A-SOC	Basic	37.19%	64.01%	87.58%	98.73%
	Causal	28.03%	62.36%	88.68%	98.84%

Note: Our experiment takes all alarms as the input. Correctly Filtered Alarms represent the re-labeled false alarms that are indeed not associated to attacks and can be filtered out appropriately

3.3 *Results and Discussion*

Feature Set Evaluation. The first experiment is to demonstrate that the causal correlation is helpful on alarms filtering without sacrificing the detection rate. To achieve that, we test different feature combinations with DARPA and A-SOC datasets and analyze the performance result. As shown in Table 2, the proposed causal feature set including basic alarm information, asset, causal correlation and variance frequency has the best performance, higher detection rate and lower FP rate, especially on detecting novel alarms. Moreover, adopting the causal correlation features to classify SOC dataset causes that the false positive rate is greatly reduced about 10%. The reason is that the SOC dataset gathered in 2007 has more sophisticated multi-step attacks than DARPA 1999. Therefore, the causal correlation feature set has greater enhancement on SOC dataset than on DARPA 1999.

Cost and Detection Rate. To evaluate our approach on the tradeoff of cost and detection rate, our proposed EAL are compared with two other generic schemes in the second experiment. The first controlled scheme is that the decision model only keeps the last classifier in alarm ensemble classifiers for prediction. The second one is to keep all previous trained classifiers and combine them with the same weight in alarm ensemble classifiers for prediction. The comparison of ROC curve is shown in Fig. 2, which demonstrates that our approach receives the largest area-under-curve (AUC) values in both of the DARPA and A-SOC datasets. The ROC curve provides analysts the view of how much the cost has to pay if the model can identify alarms including novel alarms, as shown in Fig 2(a) and 2(b). The cost implicitly means that analysts have to spend their time to pick out the false alarms. Obviously, the high cost is unpractical when the system is operated on real environment. All comparison results are illustrated in Table 3. Our proposed EAL also performs very well on correctly filtering alarms without

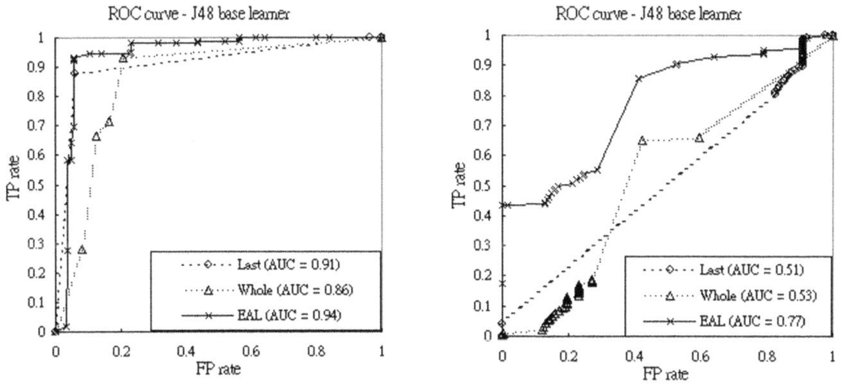

(a) ROC curve of DARPA dataset

(b) ROC curve of A-SOC dataset

Fig. 2 Capability of detecting false alarms - ROC curves of different incremental classifier schemes using a base learner, J48 from Weka

Table 3 Performance comparison, TP rate (detection rate) vs. FP rate (cost), with different learning schemes. The scheme of only last classifier is to predict alarms by the classifier trained from the previous day. And, whole classifier means the alarm ensemble classifiers combining all of previous trained classifiers for prediction. The last scheme is our proposed approach EAL, for prediction

Dataset	Learning scheme	Cost FP rate	All Alarms TP rate	Novel Alarms TP rate	Correctly Filtered Alarms TN / (TN + FN)
DARPA	Only Last Classifier	5.92%	87.96%	81.79%	93.7% (34,210 / 36,511)
	Whole Classifier	16.46%	71.52%	20.58%	84.81% (30,378 / 35,820)
	EAL	5.7%	93.22%	87.17%	96.36% (34,292 / 35,588)
A-SOC	Only Last Classifier	6.43%	48.63%	56.28%	98.79% (281,938 / 285,402)
	Whole Classifier	30.08%	49.35%	80.86%	98.4% (210,690 / 214,105)
	EAL	28.03%	62.36%	88.68%	98.84% (216,854 / 219,392)

sacrificing much to deal with novel or rare true alarms. In traditional methods on false alarm reduction, assessing risk through an asset table or setting a correlation rule to recognize true alarms is possible to ignore novel alarms. Therefore, they can only aware of known attack and lack of ability to defense from a new threat. However, in our experiment, it actually shows the ability to find out novel alarms.

Aggregation Duplicated Alarm. There are 55,473 individual alarms in DARPA dataset grouped into 11,197 groups, including only two impure groups, which include true and false alarms in the same group. The reduced amount significantly relieves about 75% of overall alarms. On the other hand, the 308,063 individual alarms in A-SOC dataset are grouped into 25,662 groups with 27 impure groups but still helpful for analysts. As a result, analysts just need to confirm the succinct alarm groups with predicted true class of alarms instead of a large number of individual alarms. About 92% labor is saved in A-SOC dataset by means of our approach. Actually, our grouping approach does not only save work for analysts but also make analysts easily giving feedback for enhancing the ability of the alarm ensemble classifiers on further prediction.

4 Conclusions

We proposed a system for adaptive alarm filtering. Our goal is to enhance the performance of IDS through reducing the number of false alarms and duplicated alarms. Other than that, our system can be operated in an adaptive fashion. The proposed learning-based alarm filtering system does not only classify alarms with high confidence but also adaptively change with time goes by according to feedback from experts. Moreover, through our feature set including the causal correlation features, the system also makes identifying novel alarms possible. After evaluation of experiments on DARPA and A-SOC dataset, all individual alarms are aggregated as groups, which reduces size to about 25% and 8% from original alarms respectively. In the mean while, with at least 87% novel alarm detection rate, about 96% to 98% of the false alarms have been correctly filtered out in DARPA and A-SOC dataset. After false alarms significantly identified by proposed approaches, analysts can actually pay more attentions on the main courses such as intrusion analysis and related responses.

References

1. Alharbt, A., Imai, H.: IDS False Alarm Reduction Using Continuous and Discontinuous Patterns. In: Proc. of the 3th International conf. on Applied Cryptography and Network Security (ACNS 2005), pp. 192–205 (2005)
2. Fern, A., Givan, R.: Online ensemble learning: An empirical study. Machine Learning 53(1), 71–109 (2003)
3. Julisch, K.: Clustering Intrusion Detection Alarms to Support Root Cause Analysis. ACM Trans. on Information and System Security (TISSEC) 6(4), 443–471 (2003)
4. Kidera, T., Ozawa, S., Abe, S.: An Incremental Learning Algorithm of Ensemble Classifier Systems. In: Proc. of the International Joint Conf. on Neural Networks (IJCNN 2006), BC, Canada, pp. 3421–3427 (2006)
5. Law, K.H., Kwok, L.F.: IDS False Alarm Filtering Using KNN Classifier. In: Lim, C.H., Yung, M. (eds.) WISA 2004. LNCS, vol. 3325, pp. 114–121. Springer, Heidelberg (2005)
6. Liaw, K.-K., Wu, Y.-L.: False Alarm Filtering Using SVM and Sliding Window. In: The 3rd Joint Workshop on Information Security (JWIS 2008), Seoul, Korea (July 2008)
7. Mahoney, M.V., Chan, P.K.: An Analysis of the 1999 DARPA/Lincoln Laboratory Evaluation Data for Network Anomaly Detection. In: Vigna, G., Krügel, C., Jonsson, E. (eds.) RAID 2003. LNCS, vol. 2820, pp. 220–237. Springer, Heidelberg (2003)

8. Ning, P., Cui, Y., Reeves, D.S., Xu, D.: Techniques and tools for analyzing intrusion alerts. ACM Trans. on Information and System Security (TISSEC) 7(2), 274–318 (2004)
9. Pietraszek, T.: Using adaptive alert classification to reduce false positives in intrusion detection. In: Jonsson, E., Valdes, A., Almgren, M. (eds.) RAID 2004. LNCS, vol. 3224, pp. 102–124. Springer, Heidelberg (2004)
10. Schapire, R., Freund, Y., Bartlett, P., Lee, W.S.: Boosting the margins: A new explanation for the effectiveness of voting methods. The Annals of Statistics 26(5), 1651–1686 (1998)
11. Widmer, G., Kubat, M.: Learning in the presence of concept drift and hidden contexts. Machine Learning 23(1), 69–101 (1996)
12. Valeur, F., Vigna, G., Kruegel, C., Kemmerer, R.A.: Comprehensive approach to intrusion detection alert correlation. IEEE Trans. on Dependable and Secure Computing 1(3), 146–169 (2004)
13. Zhu, B., Ghorbani, A.A.: Alert Correlation for Extracting Attack Strategies. International Journal of Network Security 3(3), 224–258 (2006)

Anomaly Detection via Over-Sampling Principal Component Analysis

Yi-Ren Yeh, Zheng-Yi Lee, and Yuh-Jye Lee

Abstract. Outlier detection is an important issue in data mining and has been studied in different research areas. It can be used for detecting the small amount of deviated data. In this article, we use "Leave One Out" procedure to check each individual point the "with or without" effect on the variation of principal directions. Based on this idea, an over-sampling principal component analysis outlier detection method is proposed for emphasizing the influence of an abnormal instance (or an outlier). Except for identifying the suspicious outliers, we also design an on-line anomaly detection to detect the new arriving anomaly. In addition, we also study the quick updating of the principal directions for the effective computation and satisfying the on-line detecting demand. Numerical experiments show that our proposed method is effective in computation time and anomaly detection.

1 Introduction

Due to the reasons that only very few labeled data are available in real applications and the events that people are interested in are extremely rare or do not happen before, the outlier detection is getting people's attention more and more [3, 4, 7, 8, 9, 11]. Outlier detection can be used in many application domains such as homeland security, credit card fraud detection, intrusion and insider threat detection in cyber-security, fault detection and malignant diagnosis etc. [8, 11, 12, 13]. Thus, the outlier detection methods are designed for finding the rare instances or the deviated data. In other words, an outlier detection method can be applied to deal with extremely unbalanced data distribution problems, such as capturing the anomaly which exists in a small proportion of network traffic.

Yi-Ren Yeh, Zheng-Yi Lee, and Yuh-Jye Lee
Computer Science and Information Engineering, National Taiwan University of Science and Technology, No.43, Sec.4, Keelung Rd., Taipei, Taiwan 10607
e-mail: {D9515009,M9615018,yuh-jye}@mail.ntust.edu.tw

K. Nakamatsu et al. (Eds.): New Advan. in Intel. Decision Techno., SCI 199, pp. 449–458.
springerlink.com

In the past, many outlier detection methods have been proposed [3, 7, 9]. One of the most popular outlier methods is using the density-based local outlier factor (LOF) to measure the outlierness for each instance [3]. The LOF uses the density of each individual instance's neighbors to define the degree of outlierness and concludes a suspicious ranking for all instances. The most important property of the LOF is considering the local data structure for estimating the density. This property makes the LOF discover the outliers which are sheltered under a global data structure. Besides, an angle-based outlier detection (ABOD) method has also been proposed recently [9]. The main concept of ABOD is using the variation of the angles between the each target instance and the rest instances. An outlier or deviated instance will generate a smaller variance among its associated angles. Based on this observation, the ABOD considers all the variance of the angles between the target instance and any pair of instances to detect outliers. However, the time complexity of ABOD is too high to deal with large datasets. In [9], the authors also proposed the fast ABOD which is an approximation of the original ABOD. The difference is that fast ABOD only considers the variance of the angles between the target instance and any pair of instances of target instance's k nearest neighbors. Even though, these methods mentioned above can not be scaled up to massive datasets because of the very expensive computational cost.

In this paper, we observe that removing (or adding) an abnormal instance (or outlier) will cause a lager effect on principal directions than removing (or adding) a normal one. From this observation, we apply the "Leave One Out" (LOO) procedure to check each individual point the "with or without" effect on the variation of principal directions. This will help us to remove the suspicious outliers in the dataset. Thus, it can be used for the data cleaning purpose. Once we have a clean dataset, we can extract the leading principal directions from it and use these directions to characterize the normal profile for the dataset. Similarly, we can evaluate the "with or without" effect of new arriving data point. That defines a suspicious score for the new arriving data point. If the score is greater than a certain threshold, we regard this point as an outlier. Based on this mechanism, we proposed an on-line anomaly detection method. Intuitively, the "with or without" effect on the principal direction will be diminished for a single data point even it is an outlier when the dataset is large. To overcome this problem, we employ the "over-sampling" scheme that will amplify the "with or without" influence made by an outlier. We also are aware of computation issues in the whole process. How to compute the principle directions efficiently when the mean and covariance matrix are changed slightly is also a key issue and the tricks for matrix computation will be included in this work as well.

2 Over-Sampling Principal Component Analysis

In this section, we first introduce the classical dimension reduction method PCA briefly. The study on the influence of the variation of principal directions via LOO procedure is also be exhibited. Finally, we introduce the over-sampling scheme in PCA to emphasize the influence of an abnormal instance. In addition, an effective computation for computing the covariance matrix and estimating principal directions in LOO procedure is also proposed.

2.1 *Principal Component Analysis*

PCA is an unsupervised dimension reduction method. It can retain those characteristics of the data set that contribute most to its variance by keeping lower-order principal components. These few components often contain the "most important" aspects of the data. Let $A \in \mathbb{R}^{p \times n}$ be the data matrix and each column, $x_i \in \mathbb{R}^p$, represents an instance. PCA involves the eigenvalue decomposition in the covariance matrix of the data. Its formulation is solving an eigenvalue problem as follows:

$$\Sigma_A \Gamma - \lambda \Gamma, \tag{1}$$

where $\Sigma_A = \frac{1}{n}\sum_{i=i}^{n}(x_i - \mu)(x_i - \mu)^\top$ is the covariance matrix, μ is the grand mean, and the resulting Γ is the eigenvector set. In practical, some eigenvalues have little contribution to variance and can be discarded. It means that we only need to keep few components to represent the data. In addition, PCA explains variance and is sensitive to outliers. A few points distant from the center would have a large influence on variance and its principal directions. In other words, these first few principal directions will be influenced seriously if our data contain some outliers.

2.2 *The Influence of an Outlier on Principal Directions*

Based on the concept that we mentioned in the previous section, PCA is sensitive to outliers and we only need few principal components to represent the main data structure. That is, an outlier or a deviated instance will cause a larger effect on these principal directions. Hence, we explore the variation of principal directions when removing or adding an instance. This concept is illustrated in Fig. 1 where the clustered blue circles represent the normal data, the red square represents an outlier, and the green arrow is the first principal direction. From the right panel to the left panel in Fig. 1, we can see that the first principal direction is affected when we remove an outlier. The first principal direction is changed and forms a larger angle between the old one and itself. In this case, the first principal direction will not be affected and only form an extremely small angle between the old first principal direction and the new one if we remove a normal instance. Via this observation, we use LOO procedure to check each individual point the "with or without" effect. On the other hand, we might have the pure normal data in hand. In this case, we use the same concept in LOO setting but with incremental strategy. That is, adding an instance to see the variation of the principal directions. Similarly, adding a normal data point will create a smaller angle between the old one and itself while it will form a larger angle with adding an outlier (from the left panel to the right panel in Fig. 1).

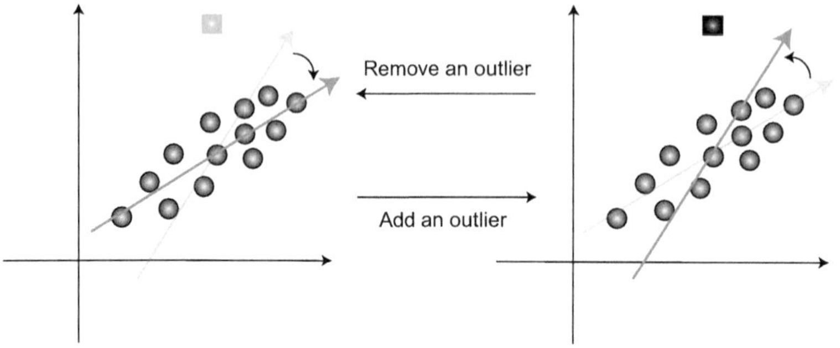

Fig. 1 The illustration for the effect of an outlier on the first principal direction

We check the variation of the principal directions for each new arriving instance and regard it as an outlier if the variation of the principal directions is significant.

In summary, we find that the principal directions will be affected with removing an outlier while the variation of the principal direction will be smaller with removing a normal instance. This concept can be used for identifying the anomaly or outliers in our data. On the contrary, adding an outlier will also cause a larger influence on the principal directions while the variation of the principal directions will be smaller with adding a normal one. It means that we can use the incremental strategy to detect the new arriving abnormal data or outliers. In other words, we explore the variation of the principal directions with removing or adding a data point and use this information to identify outliers and detect new arriving deviated data.

2.3 Over-Sampling Principal Components Analysis

As we mentioned in Section 2.2, we identify outliers in our data and detect the new arriving outliers through the variation of the principal directions. However, the effect of "with or without" a particular data may be diminished when the size of the data is large. On the other hand, the computation in estimating the principal directions will be heavy because we need to recompute the principal directions many times in LOO scenario.

In order to overcome the first problem, we employ "over-sampling" scheme to amplify the outlierness on each data point. For identifying an outlier via LOO strategy, we duplicate the target instance instead of removing it. That is, we duplicate the target instance many times (10% of the whole data in our experiments) and observe how much variation do the principal directions vary. With this over-sampling scheme, the principal directions and mean of the data will only be affected slightly if the target instance is a normal data point (see Fig. 2(a)). On the contrary, the variation will be enlarged if we duplicate an outlier (see Fig. 2(b)). On the other hand, we also can apply over-sampling scheme in the LOO procedure with incremental case. The main idea is to enlarge the difference of the effect between a normal data

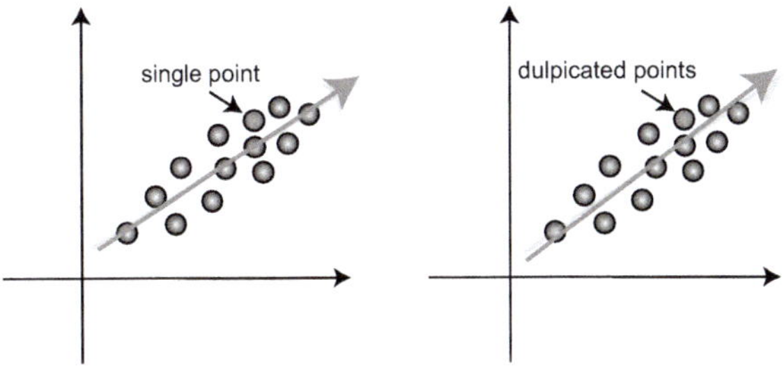

(a) The effect of over-sampling on normal data

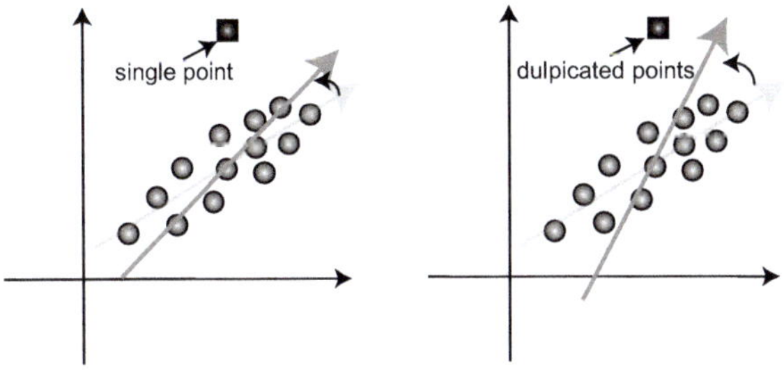

(b) The effect of over-sampling on an outlier

Fig. 2 The effect of over-sampling on an outlier and a normal instance

point and an outlier. Based on the over-sampling PCA, we make the idea discussed in Section 2.2 more practical.

For computation issue, we need to recompute the principal directions many times in the LOO scenario. In order to avoid this heavy loading, we also proposed two strategies to accelerate the procedure in estimating principal directions. The first one is the fast updating for the covariance matrix. The another one is the solving the eigenvalue problem via the power method [6]. As (1) shows, the formulation of PCA is solving an eigenvalue decomposition on the covariance matrix of the data. However, it is unnecessary to completely re-compute the covariance matrix in the LOO procedure. The difference of covariance matrix can be easily adjusted while we only duplicate one instance. Hence, we consider a light updating of covariance matrix for fast computation [5]. Let $Q = \frac{AA^\top}{n}$ be the pre-computed scaled outer-product matrix. We use the following updating for the adjusted mean vector $\tilde{\mu}$ and covariance matrix $\tilde{\Sigma}$:

$$\tilde{\mu} = \frac{\mu + r \cdot x_t}{1+r} \tag{2}$$

and

$$\tilde{\Sigma} = \frac{1}{1+r} Q + \frac{r}{1+r} x_t x_t^\top - \tilde{\mu}\tilde{\mu}^\top, \tag{3}$$

where $A \in \mathbb{R}^{p \times n}$ is the data matrix, x_t is the target instance and r is the parameter of the proportion of the whole data in duplicating x_t. From (3), it shows that we

Algorithm 1. Over-sampling Principal Component Analysis Outlier Detection for Data Cleaning

Input: a data matrix $A \in \mathbb{R}^{p \times n}$ and the ratio r
Output: the suspicious outlier ranking for the data
1. Compute outer-product $Q = \frac{AA^\top}{n}$, the mean μ, and the first principal direction v
2. Using LOO strategy to duplicate the target instance x_t and compute the adjusted mean vector $\tilde{\mu}$ and covariance matrix $\tilde{\Sigma}$:
    $\tilde{\mu} = \frac{\mu + r \cdot x_t}{1+r}$
    $\tilde{\Sigma} = \frac{1}{1+r}Q + \frac{r}{1+r}x_t x_t^\top - \tilde{\mu}\tilde{\mu}^\top$
3. Extract the adjusted first principal direction $\tilde{v}$ and compute the cosine similarity of v and $\tilde{v}$
4. Repeat step 2 and 3 until scanning all the data
5. Ranking all instances according to their suspicious outlier scores ($1-$ | cosine similarity |)

can keep the matrix Q in advance and need not to recompute it completely in LOO procedure.

In extracting the first principal direction, we also apply the power method for fast computation. Power method [6] is an eigenvalue algorithm for computing the greatest eigenvalue and the corresponding eigenvector. Given a matrix M, this method starts with an initial normalized vector u_0, which could be an approximation to the dominant eigenvector or a nonzero random vector, then iteratively computes the u_{k+1} as follows:

$$u_{k+1} = \frac{Mu_k}{\|Mu_k\|}. \tag{4}$$

The sequence $\{u_k\}$ converges on the assumption that there exists an largest eigenvalue of M in absolute value. From (4), we can see that power method does not compute a matrix decomposition but only uses the matrix multiplication. Based on this property, the power method can converge rapidly and make our LOO procedure faster. On the other hand, if we want to find the remaining eigenvectors, we could use deflation process [6]. Note that we only use the first principal component in our experiments so we only apply the power method in estimating the first principal direction.

3 Data Cleaning and On-Line Anomaly Detection

In this section, we present the framework of our data analysis. There are two phases in our framework, data cleaning and on-line anomaly detection. In the data cleaning phase, the goal is to identify the suspicious outliers. First, we over-sample each instance with LOO strategy to see the variation of the first principal direction. Here we use the absolute value of cosine similarity to measure the difference of the first principal direction and define "one minus the absolute value of cosine similarity" as the suspicious outlier scores. A higher suspicious outlier score implies the higher probability of being an outlier. Once we have the suspicious outlier scores for each instance, we can rank the instances and filter out the outliers in the given data

Algorithm 2. Over-sampling Principal Component Analysis for On-line Anomaly Detection

Input: the scaled outer-product matrix $Q = \frac{AA^\top}{n}$, the mean vector μ and the first principal direction v of the normal data, the ratio r, and threshold h, and the new arriving instance x
Output: x is an outlier or not
1. Compute the updated mean vector $\tilde{\mu}$ and covariance matrix $\tilde{\Sigma}$:
$\tilde{\mu} = \frac{\mu + r \cdot x}{1+r}$
$\tilde{\Sigma} = \frac{1}{1+r}Q + \frac{r}{1+r}xx^\top - \tilde{\mu}\tilde{\mu}^\top$
2. Extract the updated first principal direction $\tilde{v}$ and compute the cosine similarity of v and $\tilde{v}$
3. Check the cosine similarity of v and $\tilde{v}$ and see if it is higher than the specified threshold h

according to the ranking. The over-sampling principal component analysis outlier detection algorithm (OPCAOD) for data cleaning is described in Algorithm 1.

After filtering the suspicious points, we can get the pure normal data and apply the on-line anomaly detection which is not suitable for LOF and ABOD. Nevertheless, the quick updating of the principal directions in our proposed method can satisfy the on-line detecting demand. In this phase, the goal is to identify the new arriving abnormal instance. Similarly, we also apply over-sampling PCA for the new arriving instance to check the variation of the principal directions. However, how to determine the threshold for identifying an abnormal instance is a problem. In order to overcome this problem, we use some statistics to set the threshold. The idea is calculating the mean and standard deviation of the suspicious scores which are computed from all normal data points. Once we have the mean and standard deviation, a new arriving instance will be marked if its suspicious score is higher than the mean plus a specified multiple of the standard deviation. The over-sampling principal component analysis for on-line anomaly detection (OPCAAD) is also described in Algorithm 2.

4 Experimental Results

In our experiments, we evaluate our methods in three datasets. For the outlier detection, we first generate a 2-D synthetic data for testing our method. The synthetic data is consisting of 200 normal instances (blue circle in Fig. 3) and 10 deviated instances (red stars in Fig. 3). The normal data points are generated for the normal distribution with zero mean and standard deviation 1. On the other hand, the deviated data points are generated from the normal distribution with zero mean and standard deviation 15. Note that the clustering algorithm is not useful here because the outliers do not belong to a certain cluster. In this 2-D synthetic data, we apply our over-sampling principal component analysis outlier detection (OPCAOD) on it and filter 5% of the whole data (10 points) as outliers (with black crosses). The result is shown in the Fig. 3 and we can see the effectiveness of our proposed because of catching all the outliers in this synthetic contaminated data. Except for the 2-D synthetic data, we also evaluate our outlier detection method on `pendigits` dataset which can be

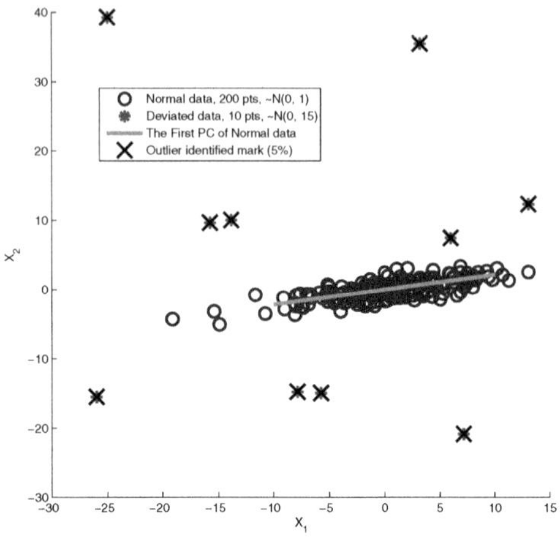

Fig. 3 The result of identifying outliers in the 2-D synthetic data

Table 1 The AUC scores of PCAOD, OPCAOD, LOF, and fast ABOD in 9 different combinations of `pendigits` dataset

Scenario	PCAOD	OPCAOD ($r = 0.1$)	LOF ($k = 100$)	Fast ABOD ($k = 40$)
`0 vs. 1`	0.9098	0.9994	0.9942	0.9541
`0 vs. 2`	0.9235	0.9999	0.9962	0.9502
`0 vs. 3`	0.4677	0.9978	0.9972	0.9232
`0 vs. 4`	0.8765	0.9533	0.9859	0.9201
`0 vs. 5`	0.8421	0.9515	0.9981	0.9472
`0 vs. 6`	0.9865	0.9939	0.9785	0.9386
`0 vs. 7`	0.9227	0.9984	0.9966	0.9297
`0 vs. 8`	0.8343	0.9556	0.9947	0.9246
`0 vs. 9`	0.7881	0.9985	0.9944	0.9641

obtained from UCI Repository of machine learning data archive [1]. We fixed the digit "0" as the normal data (780 instances) and set up 9 different combination via other digits "1" to "9" (20 data points for each) to evaluate our method in outlier detection. In this dataset, we compare our methods PCAOD (only removing one instance in LOO) and OPCAOD with LOF and fast ABOD. We use the area under the ROC curve (AUC) [2] to evaluate the suspicious outlier ranking. The results are shown in Table 1 and Table 2. Here r is the ratio of the duplicated points relative to the whole data and k is the number of nearest neighbors which is needed to be given in LOF and fast ABOD. In our experiments, we have tried several parameters for these methods and used the best parameter for each method respectively. These results show that our method is comparable with LOF and fast ABOD in detecting the outliers. Nevertheless, our method is faster than LOF and fast ABOD. On the other hand, we also can see the effect of over-sampling strategy from Table 1. The AUC score of over-sampling PCA is much better than that without over-sampling.

Table 2 The average cpu times of PCAOD, OPCAOD, LOF, and fast ABOD in `pendigits` dataset

	PCAOD	OPCAOD ($r = 0.1$)	LOF ($k = 100$)	Fast ABOD ($k = 40$)
cpu time (sec.)	0.2671	0.2878	3.221	18.772

Table 3 The true positive (TP) rate, false positive (FP) rate, and error rate of KDD Cup 99 dataset. TP rate is the percentage of attacks detected; FP rate is the percentage of normal connections falsely classified as attacks

Attack	Testing data size		TP	FP	Error
type	normal	attack	Rate	Rate	Rate
Dos	2000	100	0.940	0.073	0.073
Probe	2000	100	0.980	0.022	0.023
R2L	2000	100	0.900	0.071	0.072
U2R	2000	49	0.816	0.038	0.038

For the on-line anomaly detection phase, we evaluate our method with KDD cup 99 dataset [10]. In our experiments, we focus on the 10% training subset under the tcp protocol. We extract 2000 normal instances points as the training set and also re-sample another 2000 normal instances and different size of attacks as our testing set. The details are recorded in Table 3. In the beginning, we apply the data cleaning phase to filter 100 points (5%) in the normal data to avoid the deviated data. After that, we extract the normal pattern (the first principal direction) and use the on-line anomaly detection to detect the new arriving attack. Note that we use the training set and re-sample other attacks to determine the threshold. The results are shown in Table 3. From Table 3, we can see the good performance of our proposed method because of the high true positive rates and low false negative rates. On the other hand, our proposed method also work well in detecting the rare attacks, like *U2R*. It shows that an outlier detection method is suitable for the extremely unbalanced data distribution. In summary, our proposed method and framework not only can detect the outliers in the given data but also can be applied to predict the abnormal behavior.

5 Conclusion and Future Work

We have explored the variation of principal directions in the leave one out scenario. From the experimental results, we demonstrated that the variation of principal directions caused by outliers indeed can help us to detect the anomaly. We also proposed the over-sampling PCA to enlarge the outlierness of an outlier. In addition, an effective computation for computing the covariance matrix and estimating principal directions in LOO is also proposed for reducing the computational loading and satisfying the on-line detecting demand which is not suitable for LOF and ABOD. On the other hand, our proposed PCA based anomaly detection is suitable for the extremely unbalanced data distribution (such as network security problems). In the future, we will also study how to speed up the procedure via online learning techniques (ie., develop a quick adjusting for the principal directions directly).

References

1. Asuncion, A., Newman, D.J.: UCI repository of machine learning databases (2007), `http://www.ics.uci.edu/ mlearn/mlrepository.html`
2. Bradley, A.P.: The use of the area under the ROC curve in the evaluation of machine learning algorithms. Pattern Recognition 30, 1145–1159 (1997)
3. Breunig, M.M., Kriegel, H.-P., Ng, R., Sander, J.: LOF: Identifying density-based local outliers. In: Proc. of the 2000 ACM SIGMOD Int. Conf. on Management of Data, Dallas, Texas (2000)
4. Chandola, V., Banerjee, A., Kumar, V.: Anomaly detection: a survey. ACM Computing Surveys (2009)
5. Erdogmus, D., Rao, Y., Peddaneni, H., Hegde, A., Principe, J.C.: Recursive principal components analysis using eigenvector matrix perturbation. Journal of Applied Signal Process 13, 2034–2041 (2004)
6. Golub, G.H., Van Loan, C.F.: Matrix Computations. Johns Hopkins University Press, Baltimore (1983)
7. Hawkins, D.: Identification of Outliers. Chapman and Hall, London (1980)
8. Huang, L., Nguyen, X., Garofalakis, M., Jordan, M.I., Joseph, A., Taft, N.: In-network pca and anomaly detection. In: Advances in Neural Information Processing Systems, vol. 19, pp. 617–624. MIT Press, Cambridge (2007)
9. Kriegel, H.-P., Schubert, M., Zimek, A.: Angle-based outlier detection. In: Proc. of 14th ACM SIGKDD Int. Conf. on Knowledge Discovery and Data Mining, Las Vegas, NV (2008)
10. KDD Cup 1999 Data (August 2003), `http://kdd.ics.uci.edu/databases/kddcup99/kddcup99.html`
11. Lazarevic, A., Ertoz, L., Kumar, V., Ozgur, A., Srivastava, J.: A comparative study of anomaly detection schemes in network intrusion detection. In: Proc. of the Third SIAM Conference on Data Mining (2003)
12. Rawat, S., Gulati, V.P., Pujari, A.K.: On the use of singular value decomposition for a fast intrusion detection system. Electronic Notes in Theoretical Computer Science 142, 215–228 (2006)
13. Wang, W., Guan, X., Zhang, X.: A novel intrusion detection method based on principal component analysis in computer security. In: Proceedings of the International Symposium on Neural Networks, Dalian, China, pp. 657–662 (2004)

Towards a New Medical Decision Support System with Bio-inspired Interpretive Structural Modelling

Ikno Kim and Junzo Watada

Abstract. Interpretive structural modelling (ISM) is a useful method employed in decision making in industrial and systems engineering fields. Moreover, ISM often plays an important role in structuralising particular issues or problems related to medical issues. A small number of elements can be straightforwardly calculated using ISM, but it is difficult to structuralise the problem with a large number of elements using electronic computers in polynomial time. In the real world, medical decision support systems (MDSS) are basically composed of electronic computer-based systems. Therefore, in this paper, we show results on the basis of using a bio-inspired ISM that measures the efficiency of combining a computer-based decision support system towards the creation of a new MDSS, using an example of a rehabilitation centre selection problem.

1 Introduction

In structuralising medical selection problems or processes, complicated contextual relations are often selected as a possible statement of relationship. In this case, interpretive structural modelling (ISM) constructs a hierarchic structure to minimise risk and uncertainty. ISM was developed by Warfield [6] and is a popular decision support method for structural decision making. A set of element sets is basically related to a contextual relation. For example, in selecting hospitals or rehabilitation centres, the different contexts of medical issues often come from different areas of the organisations, and those given medical issues, problems, affairs, questions, etc. can be a set of elements.

In order to construct a hierarchic structure that is composed of medical issues, a large number of elements are intricately interconnected to be handled properly for understanding their contextual relations. A number of elements within the given medical issues are often needed to be process in the real world, meaning that those

Ikno Kim and Junzo Watada
Graduate School of Information, Production and Systems,
Waseda University
2-7 Hibikino, Wakamatsu-ku, Kitakyushu 808-0135, Japan
e-mail: octoberkim@akane.waseda.jp, watada@waseda.jp

K. Nakamatsu et al. (Eds.): New Advan. in Intel. Decision Techno., SCI 199, pp. 459–466.
springerlink.com

elements with medical issues exponentially increase. Most computer applications process only numerical elements instead of a group of issues [5]. To address this problem, a bio-inspired evolutionary computation is adapted to ISM to construct a hierarchic structure showing contextual relations of medical issues.

In this paper, the authors expand the capability of bio-inspired evolutionary computation and create a new bio-inspired ISM, based on innovative bioscience technologies [7] as an approach to a new MDSS. This system will measure the efficiency of bioscience technologies, and, combined with a computer-based method, create a better tool for decision-making.

Table 1 List of the given element; information cited from Sonoda's Rehabilitation Square [3]

Element	Contents of Elements
Element-1	Time Period of Training Programmes
Element-2	Facilities of Rehabilitation Wards
Element-3	Obtaining Sight of Rehabilitation Centres in the Neighbourhood
Element-4	Basis of Centre Functional Examination Outcomes
Element-5	Existence of Medical Rehabilitation Specialists
Element-6	Distance Between Rehabilitation Centre and House
Element-7	Quality of Training Programmes
Element-8	Easy to be Admitted to Rehabilitation Centre

	v_1	v_2	v_3	v_4	v_5	v_6	v_7	v_8
v_1	0	1	1	0	1	0	0	0
v_2	0	0	1	0	1	0	0	0
v_3	0	1	0	0	1	0	0	0
v_4	0	0	0	0	0	0	1	0
v_5	0	0	0	0	0	0	0	0
v_6	0	0	0	0	0	0	0	1
v_7	0	0	1	1	0	0	0	0
v_8	1	0	0	0	0	1	0	0

Fig. 1 Binary adjacency matrix of the given elements

2 Interpretive Structural Modelling

In an ISM process, the complex issues of a contextual relation among an element set are exploited to efficiently construct a directed graph, or network representation based on graph theory. In this section, the process of a conventional ISM method is briefly described with an example of selecting a rehabilitation centre, which is shown in Table 1.

2.1 Model Selection Problem

Table 1 shows an example of elements required for selecting a medical rehabilitation centre. The elements in Table 1 can be transformed to a digraph. The digraph with 8 related selection ideas can be also expressed in a binary adjacency matrix, as shown in Fig. 1. This matrix is constructed by setting $t_{i,j} = 1$ wherever there is an arc, whereas by setting $t_{i,j} = 0$ elsewhere.

2.2 Process of ISM

Conventional mathematical methodology has been used in various works [4] to construct a hierarchic structure. The binary adjacency matrix **B** shows a reachability example in a path of length 1 as shown in Fig. 1. To obtain a reachability matrix, the binary adjacency matrix **B** is calculated by adding the identity matrix **I**. In the reachability matrix, the labelled rows and columns are denoted by $r_{i,j}$. Hence, each element node v_i in the reachability matrix **R** is defined as follows:

$$R_s(i) = \{v_j \in V \mid r_{i,j} = 1\} \quad (1)$$

$$A_s(i) = \{v_j \in V \mid r_{j,i} = 1\} \quad (2)$$

where $R_s(i)$ is a reachability set of element nodes that should be reachable from v_i, and $A_s(i)$ is an antecedent set of element nodes that should be reachable to v_i. A set of the intersections of $R_s(i)$ and $A_s(i)$ is defined as follows:

$$Z = \{v_i \in V \mid R_s(i) \cap A_s(i) = R_s(i)\} \quad (3)$$

In a conventional ISM method, it is necessary to examine all element nodes whether they are partitioned or not.

For the conventional ISM procedure, the steps of constructing hierarchic structure(s) are as follows:

- *Step 1*: Enumerate ideas representing element nodes using the KJ method or other idea generation support methods.
- *Step 2*: Based on the opinions of individual and group members that are shown in binary numbers, a digraph can be created.
- *Step 3*: A binary adjacency matrix can be created by transforming the digraph that is composed of element nodes and arcs. A set of element nodes that belongs to this intersection set Z is shown in (3).
- *Step 4*: Examine all of the element nodes in the set Z whether they are composed of a single digraph or partitioned in different digraphs.

- *Step 5*: All of the element nodes that satisfy (3) are selected, and all of these selected nodes are denoted as level 1 element nodes.
- *Step 6*: The selected element nodes at level 1 should be removed, and a repetition for selecting element nodes that satisfy (3) should be performed; these selected element nodes are denoted as level 2 element nodes.
- *Step 7*: Similarly, the removal of selected element nodes should be performed, and element nodes should be selected again, unless all of the element nodes have disappeared.
- *Step 8*: A structural matrix and digraph can be constructed by representing each level of all divided element nodes.
- *Step 9*: Transform the structural matrix and digraph to construct a reachable condensation matrix, which can be determined by reducing the specific element nodes that contain the exact same characters and are adjacent to each other.
- *Step 10*: Transform the reachable condensation matrix to construct a reachable skelton matrix, which can be determined by calculating the reachable condensation matrix with Boolean algebra until no new entries are obtained.
- *Step 11*: Construct hierarchic structure(s), which is/are represented by the reachable skelton matrix.

3 Bio-inspired Evolutionary Computation

Bio-inspired computation is basically executed through DNA oligonucleotides on the basis of biological computation. Therefore, in this section, the authors briefly introduce biological computation and DNA oligonucleotides.

3.1 Biological Computation

DNA molecules as information storage media, and the functions of biological and chemical reactions are caused by the effects of various kinds of enzymes and proteins. To make the best use of DNA molecular structures, characters, and functions at the nano level, a molecular form of computation can be created based on calculation structures and architectures. This computation is called molecular computation, also known as biological computation that was discovered by Adleman [1].

3.2 Watson-Crick Complementarity

DNA [2] consists of polymer chains called DNA strands that are composed of nucleotides. These nucleotides, partially comprised of adenine (A), guanine (G), cytosine (C), and thymine (T), are bonded in covalent and hydrogen bonds. DNA calculations are executed using these scientific bonding reactions. Adenine always bonds with thymine, while guanine always bonds with cytosine. This phenomenon is called Watson-Crick complementarity.

4 Bio-inspired ISM

In bio-inspired ISM, there are several kinds of biochemical techniques using the encoded DNA sequences that can be employed to implement the DNA algorithm to construct a hierarchic structure. Thus, in this section, the authors describe a method of bio-inspired ISM.

4.1 *Process of Bio-inspired ISM*

For the bio-inspired ISM procedure, the method for constructing a hierarchic structure(s) is shown in the following steps:

- *Step 1*: The proposed element nodes with ideas were enumerated by the KJ method or other idea generation support methods.
- *Step 2*: The relations between elements are evaluated by a pair-wise comparison. In that case, the opinions of individual and group members are shown as binary numbers, and a digraph can be created.
- *Step 3*: A binary adjacency matrix can be created by transforming the digraph, and the DNA sequences for the DNA algorithm can be encoded based on the binary adjacency matrix.
- *Step 4*: Biochemical techniques including hybridization and ligation, gel electrophoresis, polymerase chain reaction, and affinity separation are employed to distinguish the encoded DNA sequences, measure the length of the DNA strands, and separate the DNA strands into each hierarchical level in each hierarchic structure.
- *Step 5*: Hierarchic structure(s) can be constructed based on the selected single-stranded DNAs that have been classified into each level by the experiment.

4.2 *Encoding Process*

First, type 1 represents the specific two element nodes for type 1 in which there is a direction of the arrow that indicates the direction from the element node v_i to the element node v_j. For type 1, all of the row and column labels are denoted by i and j for determining types 1-1 and 1-2. Each of the different element nodes is encoded as an oligonucleotide consisting of two unique sites. For type 1-1, a subset of arcs in a set A is denoted by $A_{1\text{-}1}$, which is defined as

$$A_{1-1} = \left\{ \overrightarrow{(v_y, v_q)} \in A \mid y \in Y, q \in Q, \text{and } y \neq q \right\} \tag{4}$$

where the element node v_y is directed to the element node v_q. Hence, for type 1-1, a set of each element node v_y in this subset, where y is transformed to $D(y)$, which is expressed as

$$D(y) = \{ v_q \in V \mid t_{p,x} = t_{y,q} = 1 \} \tag{5}$$

where both p and y are the row labels, both x and q are the column labels, and y = 1, 2,..., n. Type 1-2 represents directed cycles and should be encoded with either type 3 or type 4, and a set of each element node v_y in this subset, where y is transformed to $G(y)$.

Second, type 2 is a double-encoded substring of two element nodes in a one way direction, except for the two or more element nodes that are already satisfied in either type 1-1 or 1-2. For type 2, a subset of arcs in a set A is denoted by A_2 for type 2, which is defined as

$$A_2 = \left\{ \overrightarrow{(v_i, v_j)} \in A \mid i \text{ and } j = 1, 2, \ldots, n \text{ and } i \neq j \right\} \tag{6}$$

Hence, there are two specific element nodes that are encoded as an oligonucleotide consisting of two unique sites.

Third, type 3 represents a double-encoded substring of the element nodes in which the element node v_i is directed to the two or more element nodes included in either the subset $D(y)$ or $G(y)$ and is satisfied in either type 1-1 or 1-2. For type 3, a subset of arcs in a set A is denoted by A_3, which is defined as

$$A_3 = \left\{ \overrightarrow{(v_i, v_s)} \in A \mid i = 1, 2, \ldots, n, \, s \in D(y) \text{ or } G(y), \text{ and } i \neq s \right\} \tag{7}$$

where the element node v_i is directed to the element node v_s. Hence, the single element node and the subset of the element nodes are encoded together as an oligonucleotide consisting of two unique sites.

Fourth, type 4 is also basically encoded with either type 1-1 or 1-2 in a digraph. Type 4 also represents a double-encoded substring of the element nodes in the same manner as type 3, but the element node v_j is directed from the two or more element nodes included in either the subset $D(y)$ or $G(y)$. For type 4, a subset of arcs in a set A is denoted by A_4, which is defined as

$$A_4 = \left\{ \overrightarrow{(v_s, v_j)} \in A \mid s \in D(y) \text{ or } G(y), \, j = 1, 2, \ldots, n, \text{ and } s \neq j \right\}. \tag{8}$$

where the element node v_s is directed to the element node v_j. Hence, the subset of the element nodes and the single element node are encoded together as an oligonucleotide consisting of two unique sites.

Fifth, type 5 represents a complementary site between only type 3 and 4 that are sequenced in a 5' to 3' direction. Each subset of these complementary sites is encoded. Here, a subset of the element nodes is encoded as an oligonucleotide consisting of two complementary sites sequentially.

Finally, type 6 is temporarily created for attaching each of the different double-encoded substrings to determine whether circular DNA fragment(s) correspond to type 1-1 or 1-2. Each element node of these complementary sites is encoded. So, each element node except the element nodes included in either the subset $D(y)$ or $G(y)$ is encoded as an oligonucleotide consisting of two complementary sequential sites.

4.3 Simulated Experimental Results

For this selection problem, a hierarchic structure is created based on the proposed bio-inspired ISM and shown in Fig. 2. It was constructed by distinguishing the complementary sites of the element nodes, and by determining the longest length of DNA strands from the simulated gel electrophoresis apparatus. As shown in Fig. 2, the hierarchic structure is mainly composed of the 4-level digraph.

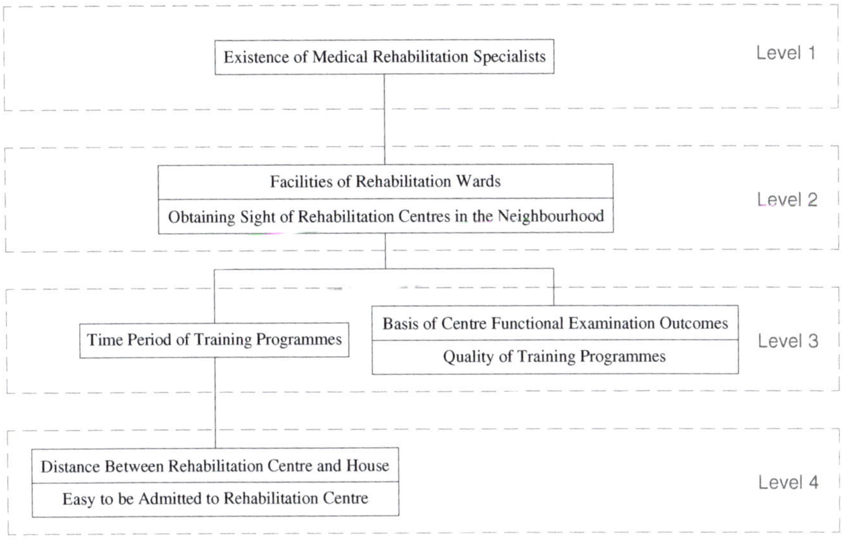

Fig. 2 Results of the hierarchically restructured selection problem

5 Conclusions

The advantage of using bio-inspired computation to analyse complex situations with a large number of elements is the capability of structuralising the medical selection problem. In addition, the authors have proposed a new method of bio-inspired ISM that can be applied to a variety of medical problems to create a new MDSS.

In future work, the proposed bio-inspired ISM for encoding element nodes will be applied to other models that are composed of a digraph, or other graphs related to medical interpretive issues or problems.

References

1. Leonard, A.: Molecular computation of solutions to combinatorial problems. Science 266, 1021–1024 (1994)
2. Daniel, H., Elizabeth, J.: Essential genetics: A genomics perspective, 3rd edn., pp. 210–242. Jones and Bartlett Publishers, Inc. (2005)
3. `http://www.mars.dti.ne.jp/~shigerus/` (November 11, 2008)
4. Japan Industrial Management Association, Handbook of industrial management. Maruzen Co., Ltd., pp. 639–650 (1994) (in Japanese)

5. Donna, M.L.: Structured decision making with interpretive structural modelling: Implementing the core of interactive management, pp. 2--11. Sorach Inc. (2007)
6. John, N.W.: Structuring complex systems. Battelle Monograph, Columbus, Ohio 4 (1974)
7. Junzo, W.: DNA computing and its application. In: Fulcher, J., Jain, L.C. (eds.) Computational Intelligence: A Compendium, pp. 1065–1086. Springer, Heidelberg (2008)

Construct an Approximation Decision Model of Medical Record by Neural Networks - The Ophthalmology Department as an Example

Yaw-Jen Lin, Chin-Dr Fan, Cheng-Chin Huang, and Cheng-Yu Fan

The most traditional formative education of doctors is to have students to simulate diagnosis and treatment modes of their teachers, or carry handbooks such as "clinical summary" and "suit the remedy to the case for clinical illness" with them as the guideline. However, the temporary approach by finding answers in the books is not only time-consuming, but also affects the confidence of the patients to the doctor. This research aims at the above-said clinical practice in clinical emergency call and doctor formative education. We take the ophthalmology as an example; utilize the back-propagation algorithm of the artificial neural networks, to construct an "approximation decision model of medical record" for clinical diagnosis guideline. The doctor can input information such as chief complaint, other complaint and diagnosis etc. into the decision model and then correct ophthalmologic approximation medical records are outputted as the reference of diagnosis and treatment to improve the quality of medical treatment and medical care.

1 Introduction

When a fresh doctor undertake resident physician, clinic doctor or emergency treatment doctor, he will begin contacting with the patients and conduct the medical service such as inquiry, examination, diagnosis, treatment and medication. This is easy to the rich-experienced and professional chief physician or attending doctor. However, it is common for a fresh doctor to be perplexed in how to deal with some relatively troublesome patients or symptoms. This will not only influence the mood and judgment of the doctor for his diagnosis and treatment, but also have negative effect on the quality of medical treatment and medical care to the patients. Accordingly, the clinical medicine formative education for training doctors is still a task for related organizations to breakthrough at present.

Yaw-Jen Lin
Central Taiwan University of Science and Technology

Chin-Dr Fan
Chutung Veterans Hospital,VAC

Chin-Dr Fan, Cheng-Chin Huang, and Cheng-Yu Fan
Yu Da College of Business

K. Nakamatsu et al. (Eds.): New Advan. in Intel. Decision Techno., SCI 199, pp. 467–480.
springerlink.com

1.1 Research Background and Intention

Most hospitals have equipped the information system to improve their service quality. As the rise of global network application, doctors often readily obtain a great deal of information over the computer and networks to solve problems arising in medical treatment. A British medical journal reported that, according to investigation [1], when a doctor meets with some patients who suffer peculiar diseases, he can seek suggestions on the Google, and the correctness for answerback is about 58 percent.

Although the above-mentioned method can be utilized, facing with imperative "clinical treatment" task, there are also some shortcomings to overcome in the practice:

1. The method to take reference from clinical medicine guidance handbook is not only time-consuming, but also affects the confidence of the patients to the doctor.
2. Seeking solutions from the Internet, though the answerback correctness of diagnosis is up to 58%, this percentage is still on the low side and there is still a long way to go from the practical application.
3. Medical science is an important field for the application of the artificial neural network, but on the basis of the understanding from published research literatures, most of them are only limited to the research of diagnosis of single disease.

1.2 Purpose of Research

Since the middle period of 1980's, because the development of the artificial neural network theory and modes has achieved critical breakthrough, the research on artificial neural network has become hotter and hotter and papers about disease diagnosis have been presented for many times. Medical science is an important field for the application of the artificial neural network, the application scopes in the published papers include: diagnosis of epilepsy [2], dermatosis [3], cephalalgia diagnosis [4], acute coronary occlusion [5], and heart attack [6], cardiogram assortment, assessment of the bladder cancer cell in the urine and pulmonary function etc., altogether more than 20 categories.

This research is to apply back propagation algorithm of the artificial neural networks to construct the approximation medical record decision model by taking ophthalmology as an example to achieve the following purposes:

1. To verify back-propagation algorithm and the feasibility of applying it to the diagnosis of multi-diseases of ophthalmology.
2. The experience for construction of the ophthalmology approximation medical record decision model can be used as the reference for subsequent construction of decision models in other departments.

3. After adding a user's interface, it can be used for the assessment of the clinical application and misdiagnosis rate of the doctors.

2 Materials and Methods

This research adapts the back-propagation algorithm of the artificial neural networks and utilizes PCNeuron V4.0 software tools to conduct the experiment of ophthalmologic "approximation decision model of medical record". The main procedures for constructing the decision model are: data collection, establishment of rule library, coding of output and input variables, and operation of neural networks. The configuration is shown in Figure 2.1.

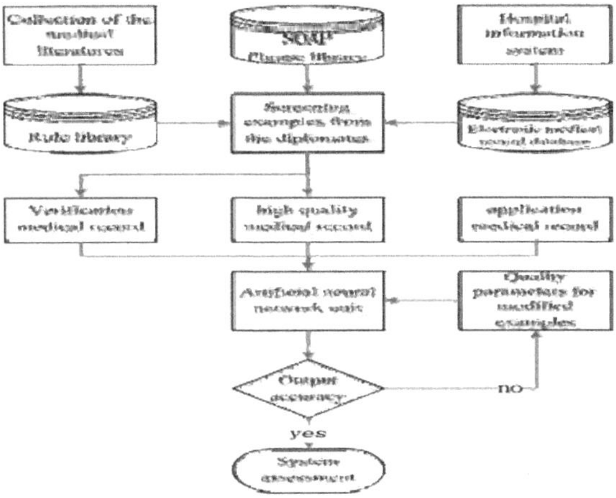

Fig. 2.1 Configuration of approximation decision model of clinical record

2.1 Data Collection

The data collection includes: (1) to collect medical literatures related to ophthalmologic symptoms and diseases so as to establish the rule library; (2) to collect the original ophthalmologic clinical record from the medical record database of the hospital information system (HIS), and consult the doctor's common phrase library and rule library to create high quality medical record and verify the medical record; and (3) to establish data such as application medical record, etc.

1. Rule library establishment

The "rule library" of this research is the summary of the consulted medical literatures and widely collected data related to ophthalmologic symptoms and diseases, for example, "Handbook of prescribing specific medicine for clinical diseases", the 4th Edition [7] compiled by Doctor Huang Qixun, "Self diagnosis for ophthalmologic diseases" [8] compiled by Doctor Chen Mushi and the consultation and guidance from the diplomates in ophthalmologic department of Zhudong Rongmin Hospital, as shown in Table 1. The ophthalmologic diseases include:

hordeolum, chalazion, eyelash insertion inward, cataract, glaucoma, xerophthalma, conjunctivitis, adult epiphora and traumas etc. Among them, the conjunctivitis can be divided into three classes: bacteroidal, viral, and allergic conjunctivitis. Because the adult epiphora is mostly caused by other diseases, it is not solely listed here. And also, the portion of traumas belongs in injured disease, so it is not listed in the research scope.

Table 1 Association rules for common ophthalmologic diseases and symptoms (extract)

Symptom \ disease	Conjunctivitis			Hordeolum	Chalazion	Trichiasis	Cataract	Glaucoma	Dry eye syndrome
	Bacterial	Viral	Allergic						
ICD_9_CM	+	+	+	+	+	+	+	+	+
Poor eyesight	-	-	-	-	+	-	+	++	-
Pain	+	+	-	+	-	-	-	+	+
Photophobia	+	+/-	+/-	-	-	-	+	-	-
Foreign body sensation	+	+/-	+/-	-	-	+	-	-	-
Itching	+	+	++	-	-	-	-	-	-
Lachrymation	+	++	+	-	-	+	-	-	++
Secretion	++	+	-	-	-	-	-	-	++
Pathologic change of front lymph of ears	-	+	-	-	-	-	-	-	-
Conjunctival congestion	++	++	++	-	-	-	-	+	-
Intraocular	-	-	-	-	-	-	-	+	-

2. High quality medical record (**HQMR**)

There shall be a great deal of representative data for the training of artificial neural networks. The "Original medical record" of ophthalmology required by this research is taken from the electronic medical record database in past years of Chutung Veterans Hospital. In the original medical records, the privacy data of the patients have to be concealed and substituted by serial numbers to guarantee the rights and interests of the patients. Then the ophthalmologic diplomates of the hospital are invited to filter the data of "doctor's advice", which will delete the obsolete or incomplete data and rank those in accordance with the contents of the "phrase library" and "rule library" as the high quality medical record. Altogether 120 medical records have been screened out. The network training adopts data format of high quality medical record (**HQMR**), including data about symptoms (input variables) and diseases (output variables).

3. Verification medical records (**VMR**)

The data format, the material source and the screening method for verification medical record (**VMR**) are identical to that of **HQMR**. There are twenty **VMR** used in this research for verification of the accuracy of recall operation of the artificial neural network.

4. Application medical records (**AMR**)

The data of application medical record (**AMR**) only has input variables but does not have target variables, so not all the symptom data are fully in conformity with the regulation in the "rule library". The main contents of **AMR** are to emulate the clinic examination data of the doctors. There are ten **AMR** used in this research for validation of the accuracy of recall operation of the artificial neural networks.

2.2 Coding of the Rule Library

Input and output variable data for network training only uses numerical mode, hence the rule library shall be firstly be coded and used as the basis of high quality medical record, verification medical record and application medical record. According to the nature of the data, the input variables in this research adopt "numeral variable, logic variable and ordered classification variable" respectively, and the output variables adopt "multi-neuron of disordered classification variable" coding, as shown in Table 2.

Table 2 Variable normalization of the rule library (extract)

Symptom \ disease	Conjunctivitis			Hordeolum	Chalazion	Trichiasis	Cataract	Glaucoma	Dry eye syndrome
	Bacterial	Viral	Allergic						
ICD_9_CM	372.0	372.1	372.14	373.11	373.2	374.05	366.19	365.9	375.15
Poor eyesight	0	0	0	0	1	0	1	2	0
Pain	1	1	0	1	0	0	0	1	1
Photophobia	1	0.5	0.5	0	0	0	1	0	0
Foreign body sensation	1	0.5	0.5	0	0	1	0	0	0
Itching	1	1	2	0	0	0	0	0	0
Lachrymation	1	2	1	0	0	1	0	0	2
Secretion	2	1	0	0	0	0	0	0	2
Pathologic change of front lymph of ears	0	1	0	0	0	0	0	0	0
Conjunctival congestion	2	2	2	0	0	0	0	1	0
Intraocular pressure	0	0	0	0	0	0	0	1	0
Eyelid	0	0	0	1	0	0	0	0	0

2.3 Network Training

Network training is to load the **HQMR** into the network, then conduct pre-processing, parameter setting, execution, post processing and analysis. If the network convergence and the misjudgment rate are in accordance with the expectancy, this means that the network learning has completed. The procedures are as follows:

1. Execute the experiment tool software PCNeuron4.0.
2. Meanwhile, load the **HQMR**, **VMR**, and **AMR** files.
3. Execute pre-processing: to generate variable statistic files, training example files, and measurement example files. The sampling of examples adopts random sampling method.
4. Set network parameter: to generate network parameter files and network listing files, recommended values in the reference books or error-trial method can be adopted.

5. Execute network training: the system outputs convergence iteration files and generates result output files and weighting matrix files.

Execute post processing: analysis of the convergence iteration procedure chart, calculation of confusion matrix and reverse scaling and misjudgment rate, connection of weighted value histogram and sensitivity etc. If the network convergence and misjudgment rate are not desirable, Step 2 can be re-executed to change the quality and quantity of the high quality medical record, or Step 4 and Step 5 are repeatedly executed until the network behaves well [9].

2.4 Network Verification

Verification mode is to load the input variables of the **VMR** into the trained network, then compare the forecasted output variables with the target output variables of the **VMR**. If both are very approximate, that means that the reliability of the decision is good. The procedures are as follows:

1. Execute pre-processing: to select **VMR** processing.
2. Set network parameter: (1) select the network mode of the original item; (2) Change the measurement example number into the actual number of the **VMR**; (3) Change the network execution mode into the verification mode.
3. Execute network recall: the system will generate new result output files.
4. Execute post processing: to output the confusion matrix chart, reverse scaling and misjudgment rate. If the network result is not as good as the expectancy, the training mode has to be re-executed [9].

2.5 Network Application

The application mode is to load the input variables of the **AMR** into the trained artificial neural network, then validate its output variables. If the output result is approximate to that of artificial judgment, which means the reliability of the decision mode is good.

The procedures are analogous to those of network verification mode, the differences are: (1) unknown data processing shall be selected for pre-processing operation; (2) the measurement example number shall be changed into the **AMR** file number, and the application mode should be selected for the execution mode.

2.6 Network Decision Rules

The decision rules of the input and output variables are as follows (extract one as example)

Input variables:

1. ICD-9-CM: the diseases range is not limited.

2. Eyesight: (0) not affected by diseases; (1) blurred vision; (2) having a strong impact on the eyesight.
3. Pain: (0) negative; (1) positive.
4. Photophobia: (0) negative; (0.5) negative or positive; (1) positive.
5. Intraocular pressure: (0) negative; (1) positive;
6. Firm mass: (0) negative; (1) positive.
7. Ghost: (0) negative; (1) positive.

Output variables:

1. Chalazion diseases
2. Cataract diseases
3. Glaucoma discases

The rule is that f the preliminary diagnosis of the patient ICD9 =range of eye diseases, and following parameters (eyesight=1, pain=0, photophobia=0, intraocular pressure=0, firm mass=0, and ghost=1) then the decision of the diseases=in conformity to chalazion.

3 Experimental Results

We Take 120 **HQMR** from the diagnosis, treatment and prescription of the ophthalmologic diplomats, using back-propagation network, the "approximation decision model of medical record" is constructed. According to the association rule between the diseases and symptoms, the input and output variables of the **HQMR** are normalized and coded, whose numerical data are shown in Table 3, in which 1 to 27 on the left side is the input variable coding, and 1 to 10 on the right side is the output variable coding.

Table 3 Truth table for input and output variables (extract)

no	Symptom									disease					
	ICD_9_CM	Age	Poor eyesight	Pain	Photophobia	Foreign body sensation	Itching	Lachrymation	Secretion	Bacterial conjunctivitis	Viral conjunctivitis	Allergic conjunctivitis	Hordeolum	Chalazion	Trichiasis
1	372.0	87	0	1	1	1	1	1	2	1	0	0	0	0	0
2	372.05	71	0	1	1	1	1	1	2	1	0	0	0	0	0
3	372.0	45	0	1	1	1	1	1	2	1	0	0	0	0	0
4	372.00	44	0	1	1	1	1	1	2	1	0	0	0	0	0
5	372.0	87	0	1	1	1	1	1	2	1	0	0	0	0	0
6	372.05	71	0	1	1	1	1	1	2	1	0	0	0	0	0
7	372.0	45	0	1	1	1	1	1	2	1	0	0	0	0	0
8	372.00	44	0	1	1	1	1	1	2	1	0	0	0	0	0
9	372.0	87	0	1	1	1	1	1	2	1	0	0	0	0	0
10	372.05	71	0	1	1	1	1	1	2	1	0	0	0	0	0
11	372.0	45	0	1	1	1	1	1	2	1	0	0	0	0	0
12	372.00	44	0	1	1	1	1	1	2	1	0	0	0	0	0
13	077.3	84	0	1	0.5	0.5	1	2	1	0	1	0	0	0	0
14	077.0	56	0	1	0.5	0.5	1	2	1	0	1	0	0	0	0
15	077.0	22	0	1	0.5	0.5	1	2	1	0	1	0	0	0	0
16	077.1	56	0	1	0.5	0.5	1	2	1	0	1	0	0	0	0
17	077.3	84	0	1	0.5	0.5	1	2	1	0	1	0	0	0	0
18	077.0	56	0	1	0.5	0.5	1	2	1	0	1	0	0	0	0

3.1 Analyze the Network Training Result

Load the **HQMR** into the network, after the pre-processing, parameter setting (Table 4), network execution and post processing, the results are as follows:

Table 4 Network parameter setting

Parameter	Value
Input (1-hidden<50,2-hidden<40)	27
Hidden 1(1-hidden<50,2-hidden<40)	8
Hidden 2(1-hidden<50,2-hidden<40)	8
Output(1-hidden<50,2-hidden<40)	10
Number of Train Examples	72
Number of Test Examples	48
Train Cycles (1-30000, usually=1000)	100
Test Period(usually = Number of train Cycles/100=10)	1
Using batch learn (yes=1,no=0, usually 0)	0
Using learned weights (yes=1,no=0, usually 0)	0
Range of weight (0.1~0.5, usually 0.3)	0.3
Random seed (0.1~0.9)	0.456
Learn rate (0.1~10.0 usually 1.0)	1
Learn rate reduced factor (0.9~1.0 usually 0.95)	0.95
Learn rate minimum bound (0.01~1.0 usually 0.1)	0.1
Momentum Factor (0.0~0.8 usually 0.5)	0.5
Momentum Factor reduced factor (0.9~1.0 usually 0.95)	0.95
Momentum Factor minimum bound(0.0~0.1 usually 0.1)	0.1
Run Model(train=1,verify or Apply=2 usually 1)	1

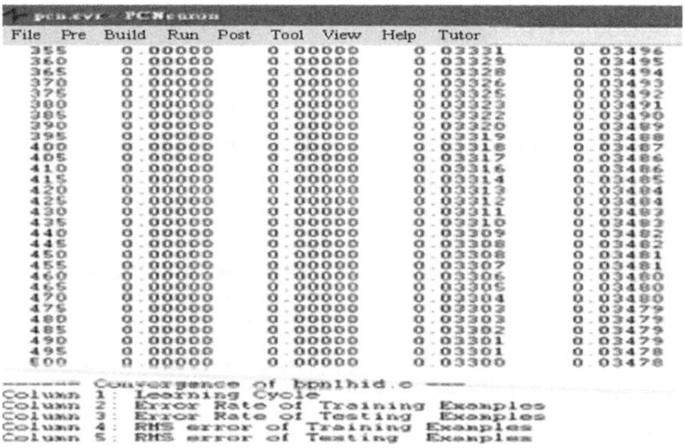

Fig. 3.1 Convergence and iteration files

1. Status of convergence and iteration: If the misjudgment rate of the 2nd behavior "training example" equals to 0.00000 and that of the 3rd behavior "training example" also equals to 0.00000 in the observation of Figure 3.1, this shows the effect of network training is satisfactory.

Measurement output status of network training: as shown in Table 5, the left figure is the target values of the measurement examples, the results show that the target values are identical to the forecasted output values, and the right figure shows the operation output values of the measurement examples of the network training, so the misjudgment rate is 0.

Table 5 The output values of the measurement examples (extract)

The target values of the train examples	The output values of the train examples
0.800 0.200 0.200 0.200 0.200 0.200 0.200 0.200 0.200 0.200	0.774 0.196 0.226 0.224 0.238 0.249 0.229 0.235 0.240 0.003
0.800 0.200 0.200 0.200 0.200 0.200 0.200 0.200 0.200 0.200	0.774 0.196 0.226 0.224 0.238 0.249 0.229 0.235 0.240 0.003
0.800 0.200 0.200 0.200 0.200 0.200 0.200 0.200 0.200 0.200	0.774 0.196 0.226 0.224 0.238 0.249 0.229 0.235 0.240 0.003
0.800 0.200 0.200 0.200 0.200 0.200 0.200 0.200 0.200 0.200	0.774 0.196 0.226 0.224 0.238 0.249 0.229 0.235 0.240 0.003
0.800 0.200 0.200 0.200 0.200 0.200 0.200 0.200 0.200 0.200	0.774 0.196 0.226 0.224 0.238 0.249 0.229 0.235 0.240 0.003
0.200 0.800 0.200 0.200 0.200 0.200 0.200 0.200 0.200 0.200	0.125 0.788 0.207 0.208 0.211 0.213 0.207 0.210 0.213 0.165
0.200 0.800 0.200 0.200 0.200 0.200 0.200 0.200 0.200 0.200	0.124 0.787 0.207 0.208 0.211 0.213 0.208 0.210 0.213 0.165
0.200 0.800 0.200 0.200 0.200 0.200 0.200 0.200 0.200 0.200	0.124 0.787 0.207 0.208 0.211 0.213 0.208 0.210 0.213 0.165
0.200 0.800 0.200 0.200 0.200 0.200 0.200 0.200 0.200 0.200	0.124 0.787 0.207 0.208 0.211 0.213 0.208 0.210 0.213 0.165
0.200 0.800 0.200 0.200 0.200 0.200 0.200 0.200 0.200 0.200	0.124 0.787 0.207 0.208 0.211 0.213 0.208 0.209 0.212 0.166
0.200 0.200 0.800 0.200 0.200 0.200 0.200 0.200 0.200 0.200	0.208 0.204 0.794 0.196 0.191 0.189 0.195 0.192 0.191 0.202
0.200 0.200 0.800 0.200 0.200 0.200 0.200 0.200 0.200 0.200	0.208 0.204 0.794 0.196 0.191 0.189 0.195 0.192 0.191 0.202
0.200 0.200 0.800 0.200 0.200 0.200 0.200 0.200 0.200 0.200	0.209 0.204 0.794 0.196 0.191 0.189 0.195 0.192 0.191 0.201

2. Produce procedure chart for network convergence and iteration: the results for the execution of convergence procedure chart are shown in Figure 3.2. The x coordinate is 0~495, and y coordinate is 0~0.5. The symbol "■" stands for the misjudgment rate of the training, and the symbol "▲" stands for the misjudgment rate of the measurement example. The symbol "+" stands for the RMS value of the training examples, and the symbol "X" stands for the RMS value of the measurement examples. The former two are suitable for the application of the assortment type. Because the misjudgment rate approaches to 0, the misjudgment rate curve is also 0.
3. Produce of the confusion matrix: the total misjudgment rate of the confusion matrix can be used for the judgment of the effect of the network learning for the application of assortment problem. The execution results of the confusion matrix in this research show that the elements on the diagonals are large and those on the non-diagonals are small, which means the emulation results are good, as shown in Figure 3.3.

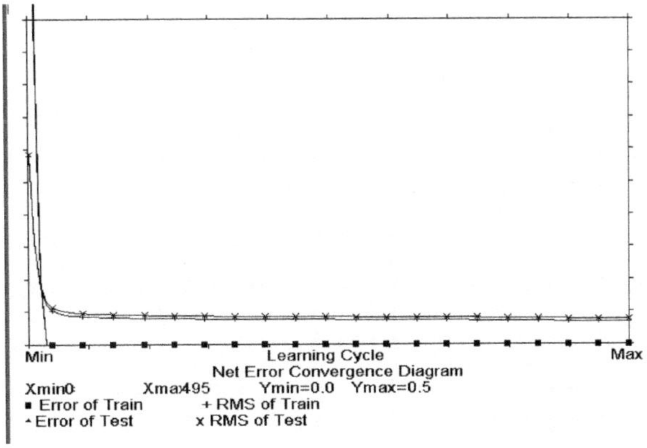

Fig. 3.2 Convergence procedure chart

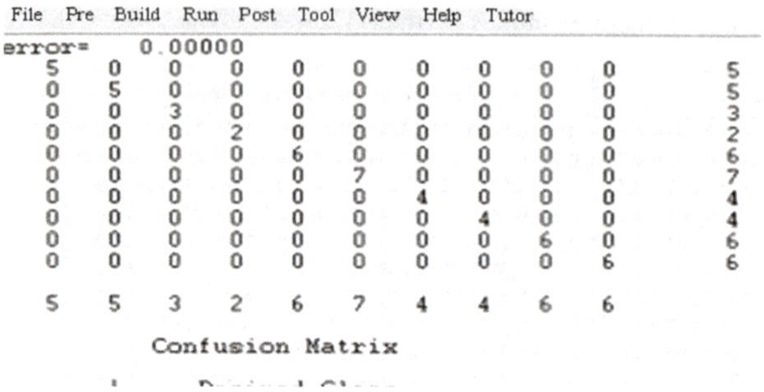

File Pre Build Run Post Tool View Help Tutor

error= 0.00000

5	0	0	0	0	0	0	0	0	0	5
0	5	0	0	0	0	0	0	0	0	5
0	0	3	0	0	0	0	0	0	0	3
0	0	0	2	0	0	0	0	0	0	2
0	0	0	0	6	0	0	0	0	0	6
0	0	0	0	0	7	0	0	0	0	7
0	0	0	0	0	0	4	0	0	0	4
0	0	0	0	0	0	0	4	0	0	4
0	0	0	0	0	0	0	0	6	0	6
0	0	0	0	0	0	0	0	0	6	6
5	5	3	2	6	7	4	4	6	6	

Confusion Matrix

Fig. 3.3 The confusion matrix chart

4. Reverse scaling and misjudgment rate in final execution: there are 10 research targets and 48 measurement examples in this research, after executing "reverse scaling and misjudgment rate"; the individual misjudgment rate among the 10 kinds of diseases is 0.00000, as shown in Figure 3.4.

Fig. 3.4 Reverse scaling and misjudgment rate

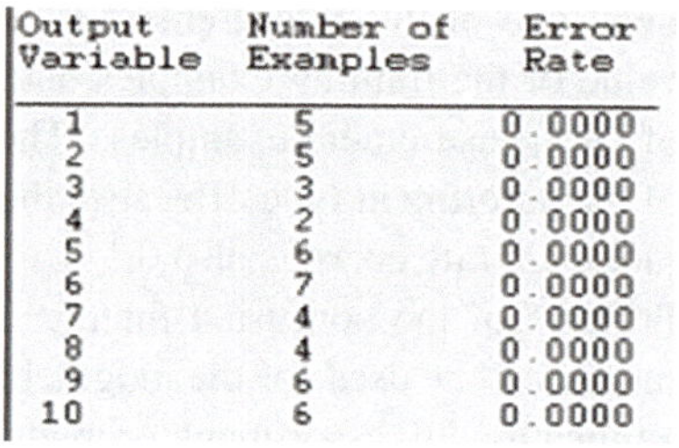

Output Variable	Number of Examples	Error Rate
1	5	0.0000
2	5	0.0000
3	3	0.0000
4	2	0.0000
5	6	0.0000
6	7	0.0000
7	4	0.0000
8	4	0.0000
9	6	0.0000
10	6	0.0000

5. Then using the **VMR** whose target values are known to verify the reliability of the trained network. Selecting two medical records each from the ten ophthalmologic diseases, twenty medical records in total are used for the recalled operation verification, and the target values are in conformity, as shown in Table 6, which means that the network is reliable.
6. Produce of the confusion matrix: the execution results of the confusion matrix show that the elements on the diagonals are large and those on the non-diagonals are small, which means the verification results of the recall are good, as shown in Figure 3.5. The overall misjudgment rate is 0.00000.

The execution of reverse scaling and the misjudgment rate: the number of verification examples is twenty, after executing "reverse scaling and the misjudgment rate"; the individual misjudgment rate is 0.00000.

7. Then using the **AMR** whose "target values" are unknown to validate the reliability of the trained network. Selecting one record each from the ten ophthalmologic diseases, ten records in total are used for the recalled operation verification. Because the **AMR** do not have the target values, and the recall operation output value is in conformity with the artificial judgment, as shown in Table 7, which means that the network is reliable.

Table 6 Recalled operation output values of the verification medical records

The target values of the verify examples	The output values of the verify examples
0.800 0.200 0.200 0.200 0.200 0.200 0.200 0.200 0.200 0.200	0.774 0.197 0.225 0.224 0.238 0.249 0.229 0.236 0.240 0.003
0.800 0.200 0.200 0.200 0.200 0.200 0.200 0.200 0.200 0.200	0.774 0.197 0.226 0.224 0.238 0.249 0.229 0.235 0.240 0.003
0.200 0.800 0.200 0.200 0.200 0.200 0.200 0.200 0.200 0.200	0.125 0.788 0.207 0.208 0.211 0.213 0.207 0.210 0.213 0.165
0.200 0.800 0.200 0.200 0.200 0.200 0.200 0.200 0.200 0.200	0.125 0.788 0.207 0.208 0.211 0.213 0.207 0.209 0.213 0.166
0.200 0.200 0.800 0.200 0.200 0.200 0.200 0.200 0.200 0.200	0.209 0.204 0.794 0.196 0.191 0.189 0.195 0.192 0.191 0.201
0.200 0.200 0.800 0.200 0.200 0.200 0.200 0.200 0.200 0.200	0.208 0.204 0.794 0.196 0.191 0.189 0.195 0.192 0.191 0.202
0.200 0.200 0.200 0.800 0.200 0.200 0.200 0.200 0.200 0.200	0.210 0.201 0.195 0.795 0.194 0.191 0.194 0.194 0.194 0.205
0.200 0.200 0.200 0.800 0.200 0.200 0.200 0.200 0.200 0.200	0.209 0.201 0.196 0.795 0.194 0.191 0.195 0.194 0.193 0.206
0.200 0.200 0.200 0.200 0.800 0.200 0.200 0.200 0.200 0.200	0.222 0.203 0.186 0.186 0.782 0.170 0.183 0.182 0.176 0.213
0.200 0.200 0.200 0.200 0.800 0.200 0.200 0.200 0.200 0.200	0.222 0.203 0.186 0.186 0.782 0.170 0.183 0.182 0.176 0.214
0.200 0.200 0.200 0.200 0.800 0.200 0.200 0.200 0.200 0.200	0.228 0.203 0.180 0.174 0.168 0.763 0.170 0.166 0.158 0.213
0.200 0.200 0.200 0.200 0.200 0.800 0.200 0.200 0.200 0.200	0.227 0.203 0.180 0.174 0.168 0.764 0.170 0.166 0.157 0.213

Fig. 3.5 The confusion matrix chart

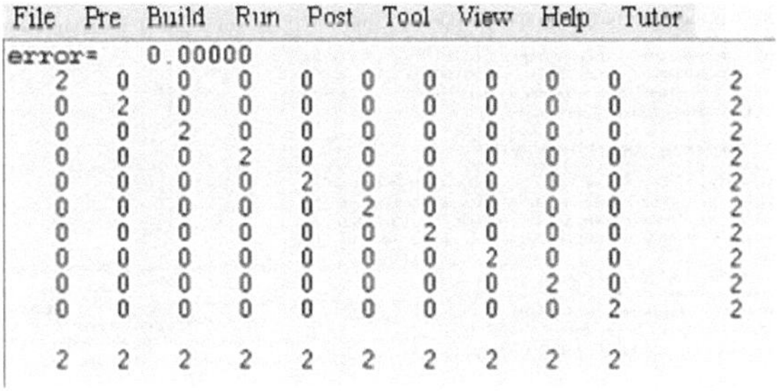

Table 7 Recalled operation output values of the Validation medical records

The target values of the validate examples	The output values of the validate examples
0.000 0.000 0.000 0.000 0.000 0.000 0.000 0.000 0.000 0.000	0.774 0.197 0.225 0.224 0.238 0.249 0.229 0.236 0.240 0.003
0.000 0.000 0.000 0.000 0.000 0.000 0.000 0.000 0.000 0.000	0.125 0.788 0.207 0.208 0.211 0.213 0.207 0.209 0.213 0.166
0.000 0.000 0.000 0.000 0.000 0.000 0.000 0.000 0.000 0.000	0.209 0.204 0.794 0.196 0.191 0.189 0.195 0.192 0.191 0.201
0.000 0.000 0.000 0.000 0.000 0.000 0.000 0.000 0.000 0.000	0.209 0.201 0.196 0.795 0.194 0.191 0.195 0.194 0.193 0.206
0.000 0.000 0.000 0.000 0.000 0.000 0.000 0.000 0.000 0.000	0.222 0.203 0.186 0.186 0.782 0.170 0.183 0.182 0.176 0.213
0.000 0.000 0.000 0.000 0.000 0.000 0.000 0.000 0.000 0.000	0.227 0.203 0.180 0.174 0.168 0.764 0.170 0.166 0.157 0.213
0.000 0.000 0.000 0.000 0.000 0.000 0.000 0.000 0.000 0.000	0.213 0.201 0.190 0.192 0.187 0.183 0.790 0.188 0.186 0.205
0.000 0.000 0.000 0.000 0.000 0.000 0.000 0.000 0.000 0.000	0.212 0.204 0.189 0.191 0.183 0.178 0.187 0.785 0.184 0.207
0.000 0.000 0.000 0.000 0.000 0.000 0.000 0.000 0.000 0.000	0.225 0.204 0.182 0.197 0.172 0.160 0.101 0.174 0.777 0.209
0.000 0.000 0.000 0.000 0.000 0.000 0.000 0.000 0.000 0.000	0.060 0.209 0.247 0.237 0.262 0.274 0.259 0.259 0.261 0.786

4 Discussion

There are 27 input variables, 10 output variables, altogether 37 input and output variables. The coding mode of the input and output variables adopt numerical variables, logic variables, ordered classification variables and the multi-neurons of disordered classification variables, therefore the complexity is on the high side. Besides consulting the advanced experience in the experiment, the "error trial method" is adopted to try modifying parameters such as the number of concealed nodes, learning cycle number, learning rate and practicability factor, etc., so the expected effect can be reached. For instance:

1. Range of weighted values: it is generally suggested that, the relatively bigger one should be selected as the input variable and output variable, if the value is less than 10, the weighted value will be 0.3; and 0.15 for 10 to 20. For this research whose input variable is 27 and output variable is 10, the weighted value is generally 0.1, but the experience shows that 0.3 is appropriate. If 0.1 is taken, the convergence state is inversely not good, as shown in left of Fig. 7. The misjudgment rate of the training examples is fairly good, but the misjudgment for measurement example is 0.50000, which is on the high side. Meanwhile, observed from the convergence procedure chart, the RMS values of the measurement examples are on the high side, as shown in right of Figure 4.1.

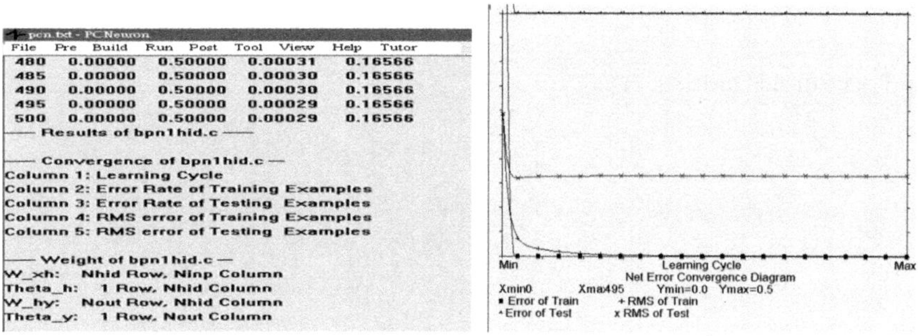

Fig. 4.1 Convergence iteration status and procedure chart

Table 8 Sensitivity matrix data

Symptom	Disease								
1	0.036	-0.352	0.349	-0.087	-0.478	0.621	-0.199	-0.071	-0.026	0.139
2	0.016	-0.003	-0.002	0.008	-0.007	-0.004	-0.016	0.006	0.013	-0.027
3	2.948	-0.901	-0.362	-0.517	1.713	-1.499	0.731	1.018	-0.844	-4.883
4	3.218	1.625	-3.073	1.463	-0.217	-0.371	-1.173	1.784	-0.309	-6.677
5	3.430	-0.653	-1.210	-0.757	-0.627	-0.251	2.892	0.578	-0.874	-5.904
6	4.144	-1.208	-0.961	0.198	0.409	1.419	-0.498	0.018	-1.141	-6.727
7	-0.178	-0.453	4.823	-0.579	-0.917	-1.391	-1.221	0.375	0.167	0.306
8	-0.769	1.410	0.816	-2.309	-1.091	0.691	-0.964	-0.405	1.835	0.756
9	4.566	-0.908	-3.405	-0.117	0.837	-0.105	0.289	0.130	1.238	-7.541
10	-5.000	6.301	-2.371	-0.401	-0.209	0.484	1.124	-0.828	-1.064	5.640
11	1.127	0.783	3.381	-0.301	-1.191	-0.629	-1.621	1.338	-1.022	-2.701
12	0.602	0.430	0.321	-1.069	-1.568	0.298	-0.151	1.931	-0.238	-1.375
13	0.220	-0.968	-0.061	4.025	-0.437	-0.618	-1.324	0.660	-0.802	-0.067
14	0.561	-0.232	0.886	4.041	-1.542	-0.181	-0.863	0.270	-1.521	-1.176
15	2.119	-0.540	-2.435	-0.868	6.811	-1.548	-0.967	-0.667	-0.898	-3.241
16	-0.047	-0.347	-1.047	-0.584	-1.028	-0.366	5.104	-1.176	-0.395	0.173
17	-0.076	0.639	-0.345	0.177	-1.501	2.287	-1.125	0.283	-0.686	-0.139
18	0.103	-0.016	-0.928	-1.114	-1.416	3.145	0.265	-0.361	-0.575	0.004
19	0.750	-0.624	-0.889	-0.478	-1.272	1.907	-0.353	0.803	-0.111	-0.903
20	-1.125	0.242	0.684	-0.762	-0.673	-0.460	-0.453	1.366	0.469	1.856

2. Initial value of the learning rate is generally 0.1 to 1, the empirical value is 0.5.
3. The cell number of concealed layer is generally suggested as the input cell number plus output cell number, and it is desirable to take 8 in this experiment.

4. It can be found from the data of the sensitivity matrix that, the first (diagnosis code) and second symptoms (age) do not behave so evidently to the diseases, hence there is no evident difference in the data obtained by network operation, as shown in Table 8.

5 Conclusion

This paper is different from the conventional research that only aimed at multi-symptoms for one disease to judge if it is really that disease. It attempts to aim at the association rule for 27 symptoms of 10 diseases. And it applies the back-propagation algorithm of the artificial neural network to construct ophthalmologic approximation medical record decision model. It computes the input data of the unknown diseases to acquire best approximated reference medical record, taking the best approximated medical records as the reference of the clinic treatment and prescription for the doctors, and provide the ophthalmologic approximation decision modular of the clinic treatment guidance system to facilitate the subsequent construction of the whole system. The experimental results have proved that it has achieved the expected results.

During the experiment, it is found that the collection and establishment of the "rule library" and "coding" of the input and output variables are of great importance, which are directly related to the success or failure of the network training. If there is no approach to distinctly differentiate the relation between the clustering of the symptoms and the diseases, it will be very difficult to achieve network convergence. Facing with the finer and finer medical specialties, more and newer discoveries will be made about the relation between the symptoms and diseases with the efforts and advancement of the medical field. This information must be considered when more disease diagnosis decision models for other medical departments or practical clinic diagnosis and treatment guidance system need constructing in the future. In other words, for either a decision model or the complete system, both the expandability and compatibility should be taken into account so as to bring the due effect into play and adapt to the development of medical science in future.

References

[1] Tang, H., Ng, J.H.K.: Googling for a diagnosis-use of Google as a diagnostic aid: internet based study. BMJ, November 10 (2006)

[2] Apolloni, B., et al.: Diagnosis of epilepsy via back-propagation. In: IJCNN 1990-Wash, vol. II, pp. 571–574 (1990)

[3] Yoon, Y., et al.: A connectionist expert system for dermatology diagnosis. Expert Systems, 23–31 (Winter 1990)

[4] Saito, K., Nakano, R.: Medical diagnostic expert system based on PDP model. In: ICNN 1987, vol. I, pp. 255–262 (1987)

[5] Baxt, W.G.: Use an artificial neural network for data analysis in clinical decision-making: the diagnosis for acute coronary occlusion. Neural Computation 2, 480–489 (1990)
[6] Harrison, R.F., Marshall, S.J., Kennedy, R.L.: The early diagnosis of heart attacks: a neuro computational approach. In: IJCNN 1991, vol. I, pp. 1–5 (1991)
[7] Qixun, H.: Handbook of prescribing specific medicine for clinical diseases, 4th edn. Jiazhou Publishing House, Taiwan (2000)
[8] Mushi, C.: Self diagnosis for ophthalmologic diseases, 1st edn. Huacheng Publishing Co. Ltd., Taiwan (2004)
[9] Yaomao, C., Chunyuan, Y.: Instruction Manual for Artificial Neural Network PCNeuron, 1st edn. Dingmao Publishing Co. Ltd., Taiwan (2004)
[10] Yicheng, Y.: The Application of Artificial Neural Networks, 3rd edn. Rulin Publishing Company, Taiwan (2004)
[11] Yicheng, Y.: The Application and practical construction of the Artificial Neural Networks, 8th edn. Rulin Publishing Company, Taiwan (2004)

A Multi Model Voting Enhancement for Newborn Screening Healthcare Information System

Sung-Huai Hsieh, Po-Hsun Cheng, Sheau-Ling Hsieh, Po-Hao Chen, Yung-Ching Weng, Yin-Hsiu Chien, Zhenyu Wang, and Feipei Lai

Abstract. The clinical symptoms of metabolic disorders during neonatal period are often not apparent. If not treated early, irreversible damages such as mental retardation may occur, even death. Therefore, practicing newborn screening is essential, imperative to prevent neonatal from these damages. In the paper, we establish a newborn screening model that utilizes Support Vector Machines (SVM) techniques and enhancements to evaluate, interpret the Methylmalonic Acidemia (MMA) metabolic disorders. The model encompasses the Feature Selections, Grid Search, Cross Validations as well as multi model Voting Mechanism. In the model, the predicting accuracy, sensitivity and specificity of MMA can be improved dramatically. The model will be able to apply to other metabolic diseases as well.

Sung-Huai Hsieh, Sheau-Ling Hsieh, Yung-Ching Weng, and Feipei Lai
Information Systems Office, National Taiwan University Hospital, Taipei, Taiwan

Sung-Huai Hsieh, Po-Hao Chen, Yung-Ching Weng, and Feipei Lai
Department of Computer Science and Information Engineering National Taiwan University, Taipei, Taiwan

Feipei Lai
Department of Electrical Engineering, National Taiwan University, Taipei, Taiwan
Graduate Institute of Biomedical Electronics and Bioinformatics National Taiwan University, Taiwan

Sheau-Ling Hsieh
Network and Computer Centre, National Chiao Tung University, Hsin Chu, Taiwan

Department of Information Management, National Taiwan University, Taipei, Taiwan

Yin-Hsiu Chien
Department of Medical Genetics National Taiwan University Hospital

Po-Hsun Cheng
National Kaohsiung Normal University, Taiwan

Zhenyu Wang
Computing Laboratory,Oxford University, Oxford, UK

K. Nakamatsu et al. (Eds.): New Advan. in Intel. Decision Techno., SCI 199, pp. 481–492.
springerlink.com

Keywords: Newborn Screening, Tandem Mass Spectrometry, Support Vector Machines, Methylmalonic Acidemia.

1 Introduction

Tandem Mass Spectrometry (MS/MS) has been used for years to identify and measure inborn errors of metabolism [1][2]. In National Taiwan University Hospital (NTUH), Taiwan, MS/MS is used to quantify concentrations of up to 35 metabolites simultaneously from a single blood spot. It leads high dimension data of each newborn. The primary markers are selected by clinical experience and then tested against patient and control groups to maximize diagnostic accuracy. According to the most important feature, cut-off value is applied to be a threshold of concentration value of metabolism [3]. The newborn's data, which exceed the cut-off values in the primary tests, are submitted to the confirmatory tests with the same cards. If the result is successively positive, the case will be diagnosed of suspect positive. For some diseases, it has already been demonstrated that sensitivity and specificity can be improved by taking into account combinational features rather than single feature alone. For example, for methylmalonic acidemia (MMA) disease, the features of C3 and C4DC have been proven useful.

Machine learning has been widely and successfully applied to many real world classification problems such as text categorization, face detection, protein secondary structure prediction, etc. In the most cases, the generalization performance of machine learning is outstanding. The basic idea of machine learning, for example, a Support Vector Machine (SVM) which is a classifier that the set of binary labeled training data vectors can be separated by a hyperplane. In the simplest case of a linear hyperplane there may exist many possible separating hyperplanes. There are many possible linear classifiers that can separate the data, but among them, the SVM classifier seeks the separating hyperplane that produces the largest separation margin (maximizes the distance between it and the nearest data point of each class). This linear classifier is termed to the optimal separating hyperplane. The hyperplane with maximal margin is the ultimate learning goal in statistical learning theory, and will probably perform well in classifying the new data.

Finding good features of the data for the learning algorithms [10], in the form of suitable features, is known to have high influence on the performance of the resulting classifiers. This task is addressed by feature selection and feature cross validation techniques, which are directed at optimizing the data representation for subsequent data-driven learning algorithms.

In the following sections of the paper, we first elaborate the method of our proposed adaptive feature selection and SVM. The experiment results as well as comparison with original method in NTUH and published results will be illustrated in Section 3. Finally, the paper concludes in Section 4.

2 Methodology

We propose a proper supervised classification [4] data flow to enhance the accuracy and sensitivity of the Newborn Screening process, as depicted in Figure 1. In the diagram, the Train Dataset undergoes learning to produce the SVM prediction model; the New Dataset processes the same methods to obtain the prediction result according to the trained model. Before training or predicting, the dataset is preprocessed by the MS/MS machine. The Feature Selection part will generate the most relevant features by a Pearson-like formula [5]. The Scaling method is used to avoid biasing and to improve computing efficiency. The SVM machine learning with Grid Search then generates the prediction model for the New Dataset.

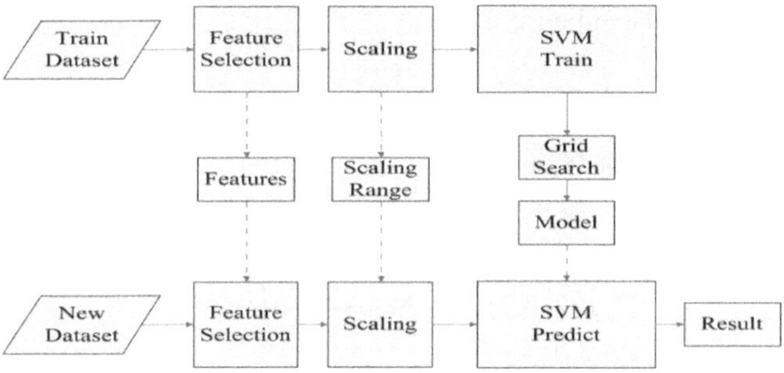

Fig. 1 Data Flow Model for Newborn Screening

A. Preprocessing

Tandem mass spectrometry (MS/MS) has been used for years to identify and measure carnitine ester concentrations in blood and urine of children suspected of having inborn errors of metabolism. MS/MS permits very rapid, sensitive and, with internal standards, accurate quantitative measurement of many different types of metabolites by conversion of raw mass spectra into clinically meaningful results.

B. Feature Selection

We apply a Pearson-like Correlation coefficient for the feature selection method. After ranking the features' coefficients, we choose the most relevant features for the continued machine learning model and the new predicted datasets. For a series of n measurements of X and Y, written as x_i and y_i where i = 1..., n, the Correlation coefficient formula as the form:

$$r_{xy} = \frac{\sum(x_i - \bar{x})(y_i - \bar{y})}{(n-1)s_x s_y}, \quad (1)$$

Where $\bar{x}$ and $\bar{y}$ are the sample means of X and Y, s_x and s_y are the sample standard deviations of X and Y and the sum is from $i = 1$ to n.

C. Scaling

The range of the feature's raw data is not always the same, and can be either too large or too small. Therefore, all features must be scaled to a proper range. The training and predicting processes will be faster, and they can avoid certain large value features so as not to induce a learning model bias. The most common range is [0, 1] or [-1, 1].

D. SVM Training

The central idea of SVM classification [8] is to use a linear separating hyperplane to create a classifier. The vector which can effect the separation is called a "support vector".

Given a training set of instance-label pairs:

$$(x_i, y_i), i = 1,\dots, l, x_i \in R^n, y \in \{1, -1\}^l \tag{2}$$

The hyper-plane can be expressed as bellow:

$$< w^T \cdot x > + b = 0 \tag{3}$$

Where

$$w = [\mathrm{w}_1, \mathrm{w}_2, \dots, \mathrm{w}_n]^{\mathrm{T}}, \quad x = [\mathrm{x}_1, \mathrm{x}_2, \dots, \mathrm{x}_n]^{\mathrm{T}}$$

Then the definition of a decision function is

$$f(\mathbf{x}) = sign(w^T \phi(x) + b) \tag{4}$$

with the largest possible margin, which apart from being an intuitive idea has been shown to provide theoretical guarantees in terms of generalization ability.

If the data is distributed in a highly nonlinear way, employing only a linear function causes many training instances to be on the wrong side of the hyperplane, which results in underfitting occurs and the decision function does not perform well. Thus SVM non-linearly transforms the original input into a higher dimensional feature space. More precisely, the training data x is mapped into a (possibly infinite) vector

$$\Phi(x) = (\phi_1(x), \phi_2(x) \cdots \cdots, \phi_i(x), \cdots)$$

In this higher dimensional space, it is more possible that data can be linearly separated. We therefore try to find a linear separating plane in a higher dimensional space.

$$\begin{aligned} \min_{w,b,\xi} \quad & \frac{1}{2} w^T w + C \sum_{i=1}^{l} \xi_i \\ \text{subject to} \quad & y_i \left(w^T x_i + b \right) \geq 1 - \xi_i \\ & \xi_i \geq 0 \end{aligned} \tag{5}$$

The parameters ξ_i are called the slack variables, and they ensure that the problem has a solution in case the data is not linear separable. The constraints in (5) contain a penalty term, $C \cdot \sum_{i=1}^{L} \xi_l$, where C > 0, and the parameter is chosen by the user to assign a penalty to errors. Usually this problem is called a primal problem. After the data are mapped into a higher dimensional space, the number of variables (*w*, *b*) becomes very large or even infinite. We handle this difficulty by solving the dual problem [6][7].

A Kernel for a nonlinear SVM projects the samples to a feature space of higher dimension via a nonlinear mapping function. Among the nonlinear kernels, the radial-based function (RBF) is defined as

$$K(x_i, x_j) = \exp\left(-\gamma \|x_i - x_j\|^2\right), \gamma > 0 \quad (6)$$

Where γ is a kernel parameter.

E. Parameter Optimization and Cross Validation

There are two parameters while using RBF kernels: a penalty (C) and a gamma (γ). It is not known beforehand which C and γ are best for a given problem; consequently, some kind of model selection (parameter search) must be performed. The goal is to identify a good set of (C, γ) such that the classifier can accurately predict unknown data (i.e., testing data). Note that it may not be useful to achieve a high training accuracy (i.e., the classifiers accurately predict training data whose class labels are already known). Therefore, a common method is to separate the training data into two parts where one part is considered unknown in training the classifier. Then, the prediction accuracy on this set can more precisely reflect the performance in classifying the unknown data. An improved version of this procedure is a k-fold cross-validation.

3 Experiment Results

In the experiment, we intend to build a robust model, as depicted in Figure 2, to classifier the MMA cases by using the data mining system we proposed. In addition, the model will be implemented in the decision support system to help the NTUH staff of Screening Department identify the MMA cases.

A. Data Set

During the experiment, we collected 350 samples for SVM training. The samples were divided into 5 parts equally; in a part, we randomly selected 35 sick samples and 35 normal samples, i.e., the total data comprising (35+35)*5 = 350 samples. We utilized the data to build 5 models and to tune the model weights. Afterwards,

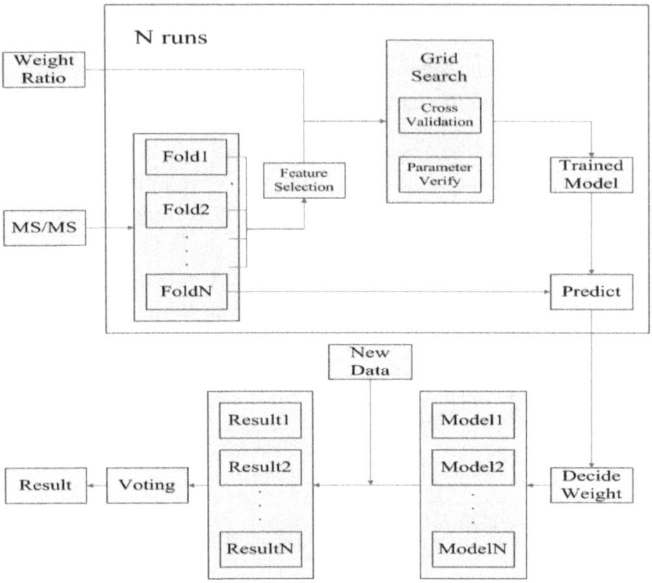

Fig. 2 Experimental Processes in Robust Model

we took 60 sick samples and 60 normal samples from the origin MS/MS data set to be the New Data, and we used the new data to evaluate our voting mechanism.

B. Experiment Steps

At beginning, we decide weight ratios and adjust different misprediction penalties. While a positive case being predicted as a negative one is a very serious problem which should be avoided. However, a negative case being erroneously predicted as a positive one is not imperative. Under the current NTUH Newborn Screening Diagnoses MMA rule, it allows 1.6% false positive ratios [9]. In here, we require a higher specificity and a more accurate classification model.

Secondly, we randomly separate the samples into N parts (N Folds), with each part possessing the same number of sick and normal cases (currently N = 5). During a Cross Validation, we choose N-1 parts as the training data and the remaining part (i.e., FoldN) as the prediction data. In the Grid Search, we select the different feature number, C and γ values for iteration in order to generate the weight ratio parameter. In the Search, all parameters are using the default values initially. After the N-1 parts are trained to generate the Train Model, we then verify the parameters. We utilize the N part to be the prediction fold and obtain the weight ratio. After N iterations, N weight ratios are generated to decide the individual model weight for voting mechanism later.

In the third step, we use all models to predict a New Data set and create Results. Finally, based on the Results and the model weights, we apply the voting rule to decide whether the case is positive or negative. The voting algorithm will be illustrated later.

C. Weight ratio

We randomly chose 235 sick samples and 235 normal samples in order to tune the weight ratio. This was different from the data set which we used to train the models. Using an equal number of sick and normal samples can avoid biasing the learning model. We used 5-part cross validation for the weight ratio result.

A cross validation was performed for these experiments. We had 235 abnormal sample data points, so we divided the samples into 5 parts. Each part included the screening data of the same case, and 200 randomly chosen normal cases were added to each part. One part was used for the prediction, and the others were used as the training data. (i.e., part 1 was used to predict, and parts 2-5 were used to train). We performed the experiment five times, and took the average of the results.

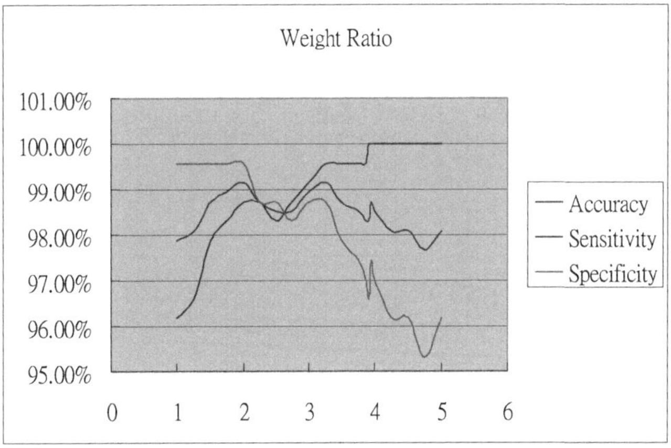

Fig. 3 Weight ratio results

Since we wanted the model to identify the sick cases as best as possible, we needed to check the sensitivity, and we were not overly concerned about the accuracy or specificity. Figure 5-4 shows that weight ratios of 3.2 and 3.9 performed well in classifying the sick samples. In this work, we decided to use 3.9 as the weight ratio result.

D. Grid Search with Feature Numbers

Next, a grid search was conducted on the features selected as indicated in red rectangular of Figure 4. In the diagram, by the definition of correlation [ ], a

Fig. 4 Features Ranking and Selection

0.707452319	F32
0.700834545	F15
0.635640171	F18
0.593935525	F13
0.552183402	F8
0.520412774	F5
0.498286848	F4
0.452625167	F10
0.428105964	F21
0.420345616	F19
0.366240882	F12
0.334937529	F29
.	.
.	.
.	.

coefficient greater than 0.8 means strong correlation; between 0.6 to 0.8 means high correlation; and between 0.4 to 0.6 means intermediate correlation. Therefore, we set the Features Numbers Ranking threshold equal to 0.5. In the experiment, we decide to use the most relevant features to build the model. All 6 features selected are greater than 0.5.

After determined 6 features, we plan to find the best model in each iteration repeatedly by tuning the parameters: Feature, C and Gamma. For example, we can

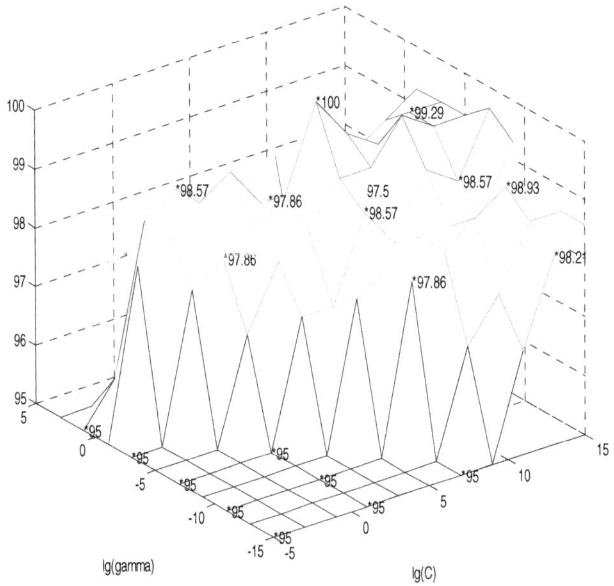

Fig. 5 The accuracy from Grid Search with 5 Features in model1 to obtain the best Trained Model1

obtain the best Trained Model1 via the highest accuracy of Grid Search involving the top 5 correlation coefficient as the selected Features. In Figure 5, it indicates the best Trained Model of model having the best parameters C = 7, γ = -3 and Feature Number = 5. In addition, in each Grid Search, the Cross Validation is performed to prevent over fitting problem.

From Figure 5, we can fix one axis and run through the other axis values. We can then analyze the trend of the log(C) and log(γ) for 5 feature numbers. We present this figure to show that the best parameter set (denoted by 100%), which is the model parameter chosen in model 1.

Fig. 6 comparison of default C & γ parameters # (default) with the best Trained Model ones @(model)

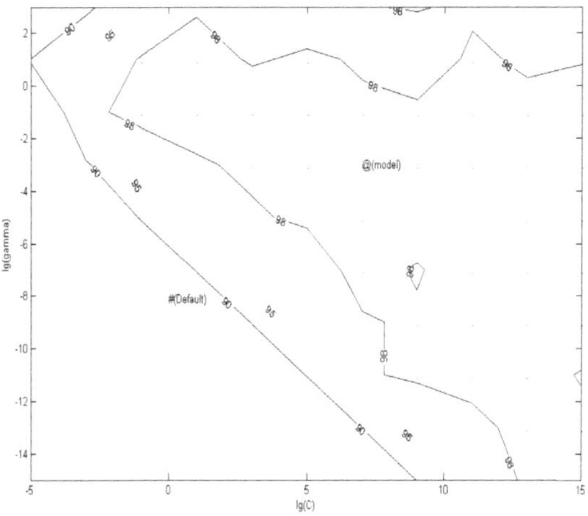

Figure 6 shows that in the log(C) and log(γ) space, point accuracy results above 98% are concentrated on the right side, which is the good region [Asymptotic Behaviors of Support vector machines with Gaussian Kernal / S. Sathiya Keerthi and Chih-Jen Lin]. @(model) denotes that the point is the model1 parameters we chose, and #(default) denotes the default parameter of the model. The default parameters gave a poor result, with the accuracy lower than 90%.

E. Model parameters model weight

In Figure 7, the above graph shows the results for the 5 models with their parameters respectively, and the table below shows the predicted results. We decided the weighting of the model by the weight results. Models 4 and 5 yielded better results, so they were given higher weights of 3. Model 3 gave the worst result, and were given the lowest weight of 1. Models 1 and 2 were between model 3 and models 4 and 5 in terms of accuracy, so the weight value was chosen to be between the weight values of model 3 and models 4 and 5.

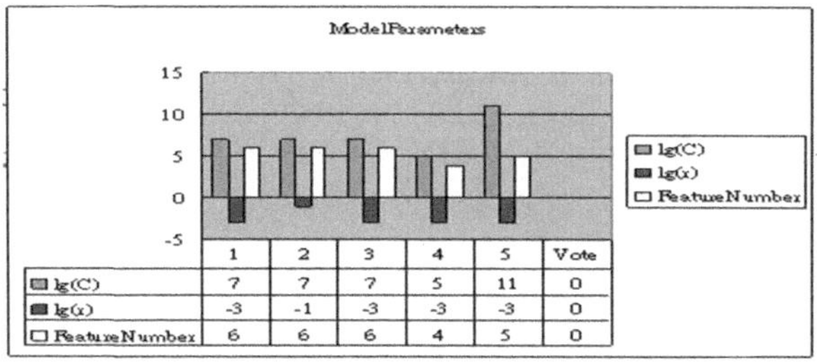

Model	1	2	3	4	5
Accuracy	97.14 (%)	97.14 (%)	95.71 (%)	100 (%)	100 (%)
Weight	2	2	1	3	3

Fig. 7 Model Parameters and model prediction results

F. Voting

The last step of the experiment was voting. We proposed a voting algorithm, which is implemented in our system and was introduced below. When the model weights are determined, we follow the algorithm to decide the voting results based on a sample having 60 sick ones and 60 normal ones.

Figure 8 show that the voting result has a better performance than the original models. 99.17% was the highest accuracy, although the specificity of the vote was poor for models 1 and 3. The sensitivities of model 1 and model 3, however, were poor. The two models missed some sick samples, so that the vote system was still better. The predicted results of the 5 models are shown below.

The algorithm is only biased for sick samples when Ss = Ns, i.e., when the sick score equals the normal score, and we judge the voting results as sick. In our experiment, there are 5 different models.

Consider that we want to predict a sample Q.
Model set M = {M1, M2,…Mn};
Wi = weight of Mi;
Ri = predicted results of Mi;
Ss is the sick score with an initial value of 0;
Ns is the normal score with an initial value of 0;
For i = 1 to n
If Ri = sick, then Ss = Ss + weight * 1;
If Ri = normal, then Ns = Ns + weight * 1;
End of for
If Ss >=Ns then vote result of Q = sick
else vote result of Q = normal;

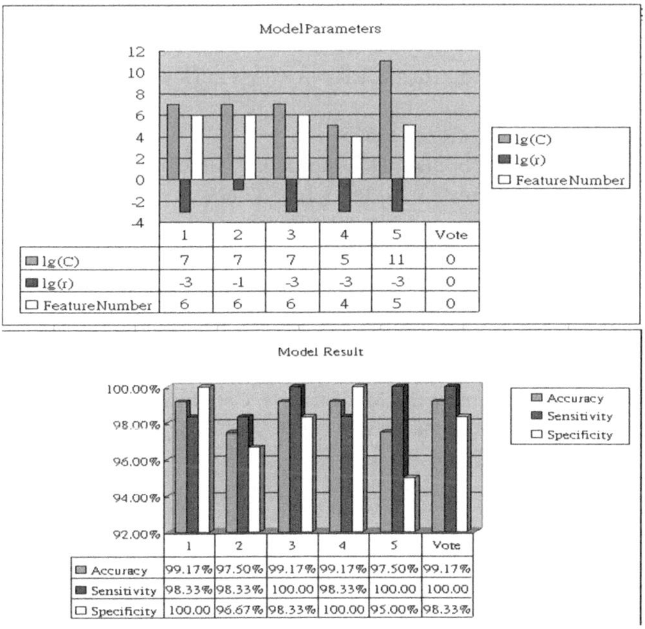

	1	2	3	4	5	Vote
lg(C)	7	7	7	5	11	0
lg(r)	-3	-1	-3	-3	-3	0
FeatureNumber	6	6	6	4	5	0

	1	2	3	4	5	Vote
Accuracy	99.17%	97.50%	99.17%	99.17%	97.50%	99.17%
Sensitivity	98.33%	98.33%	100.00	98.33%	100.00	100.00
Specificity	100.00	96.67%	98.33%	100.00	95.00%	98.33%

Fig. 8 The voting results with Accuracy, Sensitivity and Specificity

4 Conclusion

In the paper, we design and develop a model for Newborn Screening System. The model provides Support Vector Machines (SVM) classification techniques and enhancements to evaluate, interpret, and determine whether a newborn has MMA metabolic disorders. The model encompasses the Feature Selections, Grid Search, Cross Validations and multi model Voting Mechanism. In the model, the predicting accuracy, sensitivity and specificity of MMA can be improved dramatically. The model will be able to apply to other metabolic diseases as well.

References

[1] Chace, D.H., Kalas, T.A., Naylor, E.W.: Use of Tandem Mass Spectrometry for Multianalyte Screening of Dried Blood Specimens from Newborns. Clin. Chem. 49, 1797–1817 (2003)

[2] Pinheiro, M., Oliveira, J.L., Santos, M.A.S., Rocha, H., Cardoso, M.L., Vilarinho, L.: NeoScreen: A software application for MS/MS newborn screening analysis. In: Barreiro, J.M., Martín-Sánchez, F., Maojo, V., Sanz, F. (eds.) ISBMDA 2004. LNCS, vol. 3337, pp. 450–457. Springer, Heidelberg (2004)

[3] Tu, C.-M., Chang, H.-Y., Tang, M.-Y., Lai, F., et al.: The Design and Implementation of a Next Generation Information System for Newborn Screening. In: HEALTHCOM 2007 (June 2007)

[4] Michie, D., Spiegelhalter, D.J., Taylor, C.C., Campbell, J.: Machine learning, neural and statistical classification (1995)

[5] Forthofer, N., Sul Lee, E., Hernandez, M.: Biostatistics: A Guide to Design, Analysis and Discovery, 2nd edn. (2006)

[6] Cortes, C., Vapnik, V.: Support-vector network (1995)

[7] Ward, J.J., McGuffin, L.J., Buxton, B.F., Jones, D.T.: Secondary Structure Prediction with Support Vector Machine. Bioinformatics 19, 1650–1655 (2003)

[8] Chen, P.H., Fan, R.E., Lin, C.J.: A Study on SMO-type Decomposition Methods for Support Vector Machines (January 2005)

[9] Tu, C.-M.: The New Generation of Information System for Newborn Screening – A Case Study of National Taiwan University Hospital., Dept. of Computer Science and Information Engineering, National Taiwan University, Taiwan, Master Thesis (June 2007)

[10] Baumgartner, C., Böhm, C., Baumgartner, D.: Modelling of classification rules on metabolic patterns including machine learning and expert knowledge. Journal of Biomedical Informatics 38(2), 89–98 (2005)

Design and Implementation of Mobile Electronic Medication Administration Record

Sung-Huai Hsieh, I-Ching Hou, Ching-Ting Tan, Po-Chao Shen, Hui-Chu Yu, Sheau-Ling Hsieh, Po-Hsun Cheng, and Feipei Lai

Abstract. Patients' safety is the most essential, critical issue, however, errors can hardly prevent, especially for human faults. In order to reduce the errors caused by human, we construct Electronic Health Records (EHR) in the Health Information System (HIS) to facilitate patients' safety and to improve the quality of medical care. During the medical care processing, all the tasks are based upon physicians' orders. In National Taiwan University Hospital (NTUH), the Electronic Health Record committee proposed a localized standard of flows of orders. There are objectives of the standard: first, to enhance medical procedures and enforce hospital policies; secondly, to improve the quality of medical care; third, to collect sufficient, adequate data for EHR in the near future. Among the proposed procedures, NTUH decides to establish a Mobile Electronic Medication Administration Record (ME-MAR) System. According to researches, it indicates that medication errors are highly proportion to total medical faults. Therefore, we expect the ME-MAR system can reduce medication errors. In addition, we predict ME-MAR can assist nurses or healthcare practitioners to administer, manage medication properly.

Keywords: patients' safety, quality of medical care, Electronic Health Records, Mobile Electronic Medication Administration Record.

Sung-Huai Hsieh, I-Ching Hou, Ching-Ting Tan, Hui-Chu Yu, Sheau-Ling Hsieh, and Feipei Lai
Information Systems Office, National Taiwan University Hospital, Taipei, Taiwan

Sung-Huai Hsieh, Po-Chao Shen, and Feipei Lai
Department of Computer Science and Information Engineering National Taiwan University, Taipei, Taiwan

Feipei Lai
Department of Electrical Engineering, National Taiwan University, Taipei, Taiwan
Graduate Institute of Biomedical Electronics and Bioinformatics National Taiwan University, Taiwan

Sheau-Ling Hsieh
Network and Computer Centre, National Chiao Tung University, Hsin Chu, Taiwan

Po-Hsun Cheng
National Kaohsiung Normal University, Taiwan

K. Nakamatsu et al. (Eds.): New Advan. in Intel. Decision Techno., SCI 199, pp. 493–507.
springerlink.com

1 Introduction

1.1 The Trend of Health Information System

In last two decades, many hospitals have applied information system to aid the medical, nursing, and administrative staff for daily routine operations. It is not only for the convenience of the hospital staffs, but also for better medical quality of patient cares. However, the aspects of the information system related to the hospitals in the past may be different from those at current. The goal of health information systems was and is as simple as relevant: to contribute a high-quality, efficient patient care [1]. Therefore, it is a patient-centered approach towards medical and nursing cares. In addition, the administrative and management tasks need to support those cares [2].

As mentioned above, because of the transition, from traditional paper-based to computer-based, the Electronic Health Record (EHR) has become the most popular topic among medical informatics. By developing EHR, we can manage and retrieve patients' health data easily, reduce paper usages. Health records are the critical assets of patients. Once all medical records are translated or transformed into EHR, the patients can easily carry their own data. The essence of developing HIS is in fact to improve safety over the health care institutions. Publication of the U.S. Institute of Medicine's 1999 reported: "To Err is Human" [3]. By implementing the safety rule sets (for both medical, nursing staffs and patients) into HIS, computer can always notify the staffs without inducing errors. Furthermore, once medical errors inevitably occur, HIS has the ability to collect those errors, then the researches can analyze them and stored in the HIS database. Later, the system can provide solutions, issue new rules and strategies, as well as indicate further modifications to make the system safer and more reliable.

1.2 NTUH Present Status

National Taiwan University Hospital (NTUH) has adopted healthcare information system for more than twenty-five years, and developed a new generation of HIS since four years ago. However, the health record is still paper-based due to ease of usage and legal reasons. In fact, currently we are still in a transition phase. Most patients' data have been installed into the database, but portions of the medical processes have not been implemented in the NTUH HIS. Thus, not all legal health records are available in the system. Now, the EHR committee is taking charge of organizing standard processes for physicians prescribed data in NTUH. The committee decides what should be modified among the existing HIS subsystems and which subsystem should be constructed with additional, specific functionalities in order to support EHR. Moreover, all the computerized data, collected during the processes, will be formalized into EHR in the near future to replace the current paper-based charts.

1.3 Motivation and Objective

As mentioned above, the EHR committee has sketched a complete plan of the flows of EHR orders. We plan to modify the existing systems or implement new ones to cover the entire process. The orders are prescribed, generated by the physicians initially, and the orders are received by the nurses. After the nurses review the orders, the nurses initiate and perform those orders. However, the nurses execute the orders written on papers. Thus, we need to develop new systems according to the EHR committee standards to assist the nurses' operations electronically and with mobility. Before the NTUH electronic medical administration record system establishes, the orders are transcribed by hands or hard copies. The accuracy of the transcribed orders, records may induce errors or doubtable issues that need to prevent by the nursing staffs. Our objective is to construct the Mobile Electronic Medication Administration Record system (ME-MAR) to complete the closed-loop medication administration system, and then to achieve the entire process computerized. To reduce human involvements can decrease human errors, so as to improve the medication safety.

In the following sections of the paper, we first elaborate the related work and background at NTUH in Section 2. Detailed design principles and features of ME-MAR are illustrated in Section 3. In Section 4, the preliminary system assessment is provided. Finally, the paper concludes in Section 5.

2 Backgrounds

2.1 NTUH Health Information System

NTUH is an education, medical center in Taiwan with large scale. There are approximately 8000 outpatients per day, and the hospital has 2200 beds for inpatients. The number of staffs is more than 5000. As numerous patients and staffs in the hospital, we need to construct a reliable, effective information system to provide high quality of health care as well as to assist staffs' daily routines. The legacy NTUH information system, developed by venders mainly, has been incompatible with current technologies. It is difficult to manage or control by NTUH. In addition, the cost of maintenance, provided by venders, is a great burden for NTUH. Therefore, the department of information in the hospital has begun to develop our own HIS five years ago; the processing is still carrying on. The NTUH HIS encompasses many subsystems, i.e., clinics, medical report, pharmacy, pathology, etc. as well as vender developed subsystems. Those subsystems will be replaced by our own newly, implemented ones in the near future.

The new HIS system developed by the department of information are deploying under advanced technologies to enhance flexibility, efficiency, and ease of maintenance, using Service Oriented Architecture (SOA) over web services in a wired, wireless environment. The over all architecture of the NTUH HIS with multi-tier is shown in Figure 1. The system separates user interfaces, subsystems and database accessing in layers to ensure data security and integrity. In the diagram, the Web Services Servers represent the subsystems. Moreover, the Single-Sign-On mechanism, navigating via the subsystems, is included, by controlling the accessibility of every single web page for every user. The healthcare certification IC card for validation during the login processes is also enclosed in the HIS.

In the NTUH HIS, the most important system is Health Level Seven (HL7) [5] Central. It takes the responsibilities to exchange data among the subsystems, i.e., Web Services Servers. As long as the subsystem communicates according to the HL7 standards, despite the platform based upon, it can exchange data correctly in the HIS [6]. In Figure 1, it also indicates that the HIS is heavily dependent on web accessing by either wireless or wired front-ends, users. To expedite the processes, an efficient web adaption technique is essential. The issue will be addressed in the design approaches later.

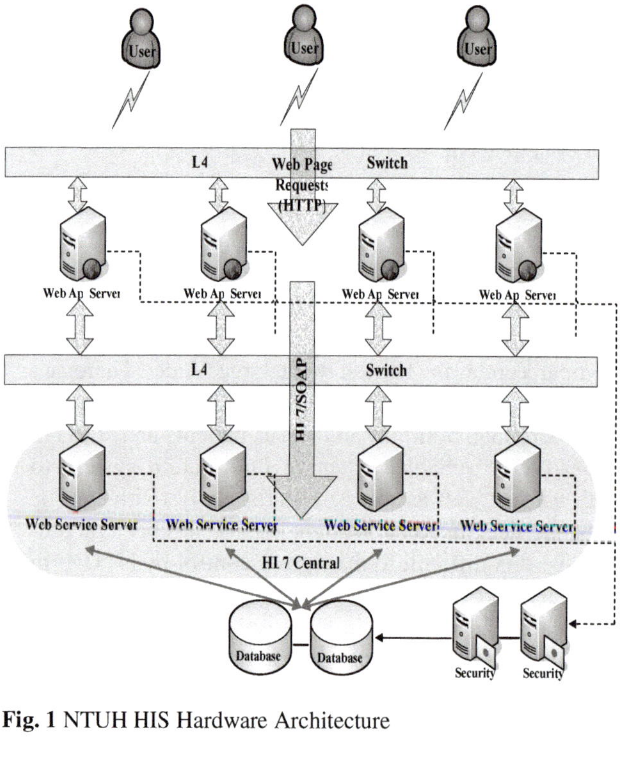

Fig. 1 NTUH HIS Hardware Architecture

2.2 *Related Systems with Mobile Medication Administration in NTUH*

Through the process of health care, the orders prescribed by the physicians are the initial bases. Any examination, inspection, and dispensing are made after the

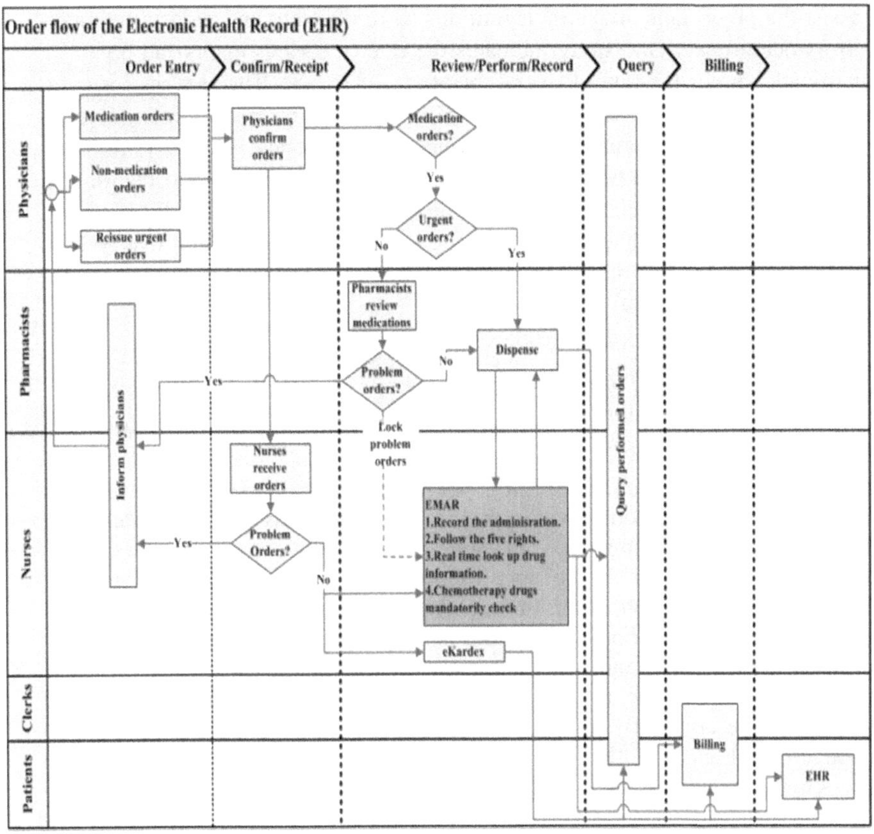

Fig. 2 NTUH order flow of EHR

orders generated. Therefore, to establish the medical records is the process after receiving the orders. The EHR committee proposed a flowchart of the prescribed orders in NTUH, shown in Figure 2. There are five steps included and five characters or roles involved in the workflow. The five steps are: order entry, confirm/receipt, review/perform/record, query, as well as billing. The five roles are: physicians, pharmacists, nurses, clerks, and patients. In the diagram, physicians generate, enter the orders. After confirmed the correctness of the orders, the pharmacists dispense the prescriptions with care. In the meanwhile, the ME-MAR system, planned to develop in the HIS, provides services: recording the administration flows electronically; looking up or searching prescribed drugs or doses dynamically; verifying chemotherapy drugs mandatorily in real time. The nurses distribute the medicine in wireless, mobile nursing carts to the associated patients. The clerks calculate the billing; the patients pay the corresponding bills and can receive their own EHR.

Currently, only the Computerized Provider Order Entry (CPOE) has been implemented. Both the out-patient and the in-patient systems use the CPOE for diagnosing, prescribing, and other medical acts embedded in ME-MAR. The CPOE can assist the physicians, make their routines more efficient and reduce the medical errors occurrence. One study indicates the rate of serious errors fall 55% [7], another study shows the rate of overall errors fall 83% [8]. The reasons for computerizing of prescribing orders can improve safety are: first, all orders are formatted, so that they must include a dose, route, and frequency; secondly, they are legible and the orders can be identified under all instances; third, information can be provided to the physicians during the prescription; and fourth, all orders are checked for hidden problems including allergies, drug interactions, overly high dose, drug-laboratory problems, as well as whether the dose is appropriate for liver or kidney functions [9].

2.3 Introduce New Technique to Improve the Web Development of NTUH HIS

Because NTUH adopts web service, it is very important to avoid the web inherent shortcomings. Asynchronous JavaScript and XML (Ajax) is a popular technique in recent years, and adopting Ajax can significantly improve front-end users [10].

By introducing Ajax technique can eliminate the start-stop-start-stop nature interactive style of web application. From software architecture view, using Ajax is to add intermediary layer between the user and the server, and it seems making the Ajax application slower and less responsive, but actually it does not. When it starts to load an Ajax page, the server only initializes few data, and the most logic of the Ajax page is directly downloaded to the client, so the user would feel the Ajax page responses instantly. When the Ajax page is completely downloaded, the "Ajax" engine, it actually is written in JavaScript, communicates with the server, renders the interface, and prepares all the data needed by the Ajax page. The Ajax engine allows any interaction of the user happening asynchronously rather than communicating with the server via every single action. In this way, the client side can handle almost any logic of a task, such as simple data validation, editing data in memory, etc. The server side only takes responsible for data exchanging between database and other complex data processing. In fact, according to our experience in NTUH, the users do not aware of the page "refresh" more than waiting. More precisely speaking, waiting cannot be totally eliminated. But by applying Ajax, refresh can be totally eliminated unless users want to do that. Although waiting cannot be eliminated, the waiting time can be reduced.

3 Mobile Electronic Medication Administration Record

3.1 Architecture Overview

The ME-MAR System is one of the subsystems of NTUH HIS as mentioned in Section 2. We separate the access of database from user interfaces and the application logics. Moreover, in order to control the accessibility, the system architecture also includes session services. Figure 3 shows the three components of the system architecture: ME-MAR (web application server), session services, and HL7 Central. In the diagram, the session services implement the login validation including authentication and authorization. The functionalities of the ME-MAR subsystem have been described clearly in Section 2.2. The HL7 Central glues the ME-MAR subsystem with all other HIS subsystem via HL7 communication standards. The front-end users are mainly designed for nurses moving mobile nursing carts and performing the prescribed orders for daily operation or routines as mentioned in the previous Sections. The wireless techniques are based upon WLAN 802.11a; the client browsers reside in Notebooks.

3.2 The Design Principle of ME-MAR

As mentioned in Section 2, we introduced Ajax to improve the development of web applications. When the requested page is downloaded, the JavaScript codes, contained in the page, will render user interfaces and process data at background, thus the users only see the phenomenon of "flash" while the page is being loaded. Figure 4 depicts the principle of our ME-MAR Ajax approach.

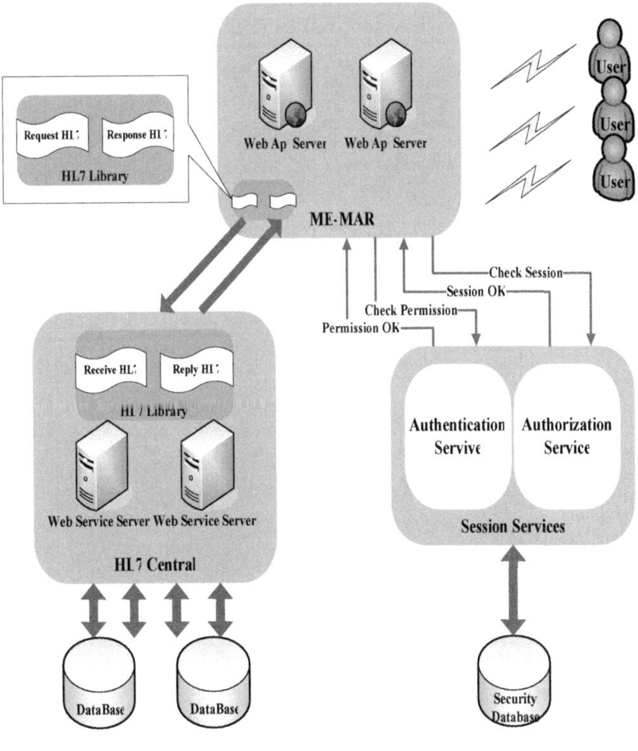

Fig. 3 NTUH Electronic Medication Administration Record System Architecture

The Ajax technique can also reduce the users' waiting time, for example, our EMAR needs to prepare data for three days, and we can query data of the selected date first, and then query the data of the other two days in background. If we use the traditional approach for our program, the users need to wait for the data of three days, and it must be longer than wait for data of one day. The Ajax technology can provide the flow, feature for ME-MAR.

3.3 Medication Administration

Nurses are recognized as the "last line of defense" in preventing medication errors made by pharmacists and physicians [11]. The main reasons of medication errors are: wrong route, wrong dose, wrong time, wrong drug, wrong frequency, wrong patient, without administering, ignore cosigning administration, prepare wrong diluted solution, wrong setting of devices, not comply with act (i.e. administer the drug which is only able to be administered by physician) [7]. The medication administration procedure that has been taught for many years is focused on the five rights: right patient, right drug, right dose, right route, and right time [12]. The ME-MAR system must support the five rights as possible, and comply with NTUH local policies: one sign of one case which means the nurses administer one patient at a time. Previous paper-based sheet has a biggest shortcoming, because every station prints the sheets of patients once a day, but in real clinical situations,

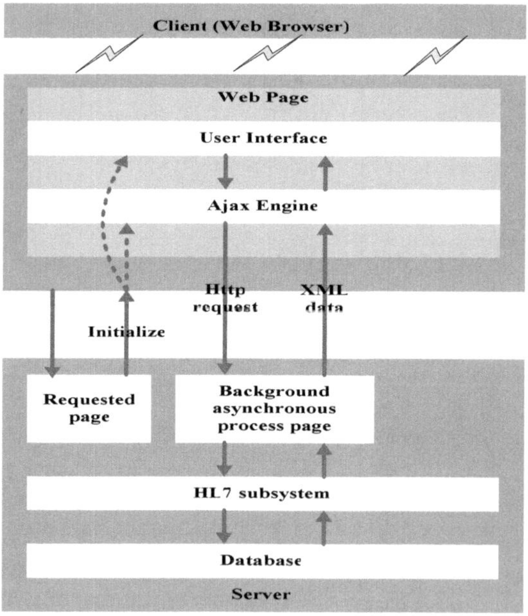

Fig. 4 The principle of ME-MAR Ajax approach

orders might be deleted, and modified after the sheets were printed, so the nurses must transcript new orders by hands, and this action is very error-prone, besides transcription is much more time consuming than computers directly transcript orders to ME-MAR system, moreover sometimes the nurses might waste more time to figure out what illegible handwriting was supposed to. In order to reduce human errors, during the workflow of medical care manual steps must be reduced, so we want to construct a comprehensive computerized closed-loop medication administration system. Until now, NTUH HIS still lacks ME-MAR system to complete the closed-loop medication administration system.

3.4 Detailed Features and Functionalities of ME-MAR System in NTUH

3.4.1 Record the Administration

The main purpose of ME-MAR is to record the data of patients' medication administration, the nurses need to record the administration which has already been given to the patients. In our design, the system will present the default timeslot by filling the buttons with blue according to the route code and frequency. In most cases, the nurses will administer medications to patients in the default timeslot, but in some other cases, the nurses might administer medications later or earlier, because of many other possible clinical situations.

3.4.2 Follow the Five Rights of Medication

Our system provides nurses with patient's personal data includes patient's name, drug name includes generic name and trade name, dose for single usage, route code, and system can calculate default time of administration according to order frequency; so our system is fully following the five rights of medication.

3.4.3 Highly Warning Drugs Remind

For those highly warning drugs, our system displays the generic name with red to notify nurse they must carefully administer this medication to avoid medical errors or damage to patients; highly warning drug such as "Potassium Solution".

3.4.4 Provide Special Orders

Sometimes physicians would give additional direction of the drug they prescribed, for example, before administering blood glucose drugs, nurses should administer appropriate dose according to glucose value level. Besides, there are some special order frequencies, e.g. ASORDER, when physicians prescribe such frequency, they will give additional direction about the time of administration.

3.4.5 New and Deleted Orders Remind

In real clinical situation, physicians may prescribe new orders or modify some active orders anytime as needed, so when the nurses review patients' order lists, they must be able to know that there are new orders need to be done or no more perform those old orders. So we implement the mechanism to highlight the new orders, this makes sure that the nurses will note the new orders and perform them immediately if necessary.

3.4.6 Easily Display the Details of the Administration

If the nurses would like to know the details of a medication administration of particular some timeslot, they do not need to double click the button to show edit panel and then view all of the information, instead the nurses just need to move the cursor over the button, and all of the information will show under the cursor, including dose, dose unit, the reason if did not give drugs to the patients, the name of the nurse who signed for the record, name of the nurse who cosigned for the record before giving drugs, time of cosigning for the record before giving drugs, name of the nurse who cosigned for the record after giving drugs, time of cosigning for the record after giving drugs, and ME-MAR.

3.4.7 Real Time Lookup Drug Information

Unlike the previous paper records, the online system can provide real-time lookup drug information, this functionality can significantly improve medication safety, because all the details of a drug are available if the nurses are not familiar with any drugs, they can look up for it, and avoid give wrong drugs to the patients, at the same time, the nurses can increase their knowledge about the drug.

3.4.8 Sort Orders by Route or Frequency

In order to facilitate works, the nurses would view orders sorted by specific fields on the order list, they would like to check the orders with the same frequency or same route listed together, so that they can prepare the medications which should be administered at the same time or same route. This convenient functionality can be achieved easily by programs, but it is impossible for paper sheets.

3.4.9 Cosign before and after Administering

In order to enhance medication administration safety, when the nurses are going to administer medications to the patients, sometimes there are other nurses except the nurse who administers the medication to check orders and drugs, then give the drugs to the patients. The system does not allow the administering nurse cosigns for his or her own administration, i.e. the cosigner and administering nurse must be different.

3.4.10 Chemotherapy Drugs Mandatory Check before Administering

The usage of chemotherapy drugs is very critical for patients' safety, any miss using will cause great damage to patients, so we implement a mechanism to enforce administering chemotherapy drugs to follow three steps: chemotherapy drug order review, chemotherapy drug administration review, and chemotherapy drug administration.

3.4.11 Check Single Administration Dose and Daily Dose Limitation

We implement the mechanism to limit the administration dose, nurses cannot administer dose that exceed order dose in single administration, because they should not violate physicians' orders, if necessary, physicians should modify or prescribe new orders rather than nurses sign an over dose administration.

3.4.12 Support to Prepare Drugs

Before the nurses administer medications to the patients, they will prepare all drugs of one patient into his or her drug box, and in order to let the nurses know what drugs have already been prepared, we implement the mechanism to record those drugs have been prepared, but this is just auxiliary functionality, the preparation records are not formal clinical data.

3.4.13 Medication Administration Record Summary Print

Because of policy and legal reason, we still print all the Mobile electronic medication administration records into paper sheets for patients' health records at present. We are now in a phase transition, and it is hard to dispose paper-based record suddenly. Although we still need to print the records into paper sheets, but the quantities are greatly reduced, because previous paper-based records are printed every day, and it has to reserve much space for nurses administering, writing. After adopting ME-MAR, the paper sheets of medication administration records are printed only when the patients transfer wards or discharge, and the sheets leave no space for writing, so that we can reduce paper usage of medication administration records.

4 Preliminary System Assessments

4.1 Criteria

There are three factors to evaluate the ME-MAR proposed by the Medication Administration System — Nurses Assessment of Satisfaction (MAS-NAS) Scale. The three factors are: (a) efficacy, dependable and effective system: efficiency, user-friendliness, readily, available supplies, error prevention/reduction, and adequate turnaround time; (b) safety, system components assure the nurse that it is correct to administer the medication: pharmacist check, physician/pharmacist agreement, ease of checking, drug alert feature, following the five rights, and communication; and (c) access, having necessary information and medications immediately at hand: ease of finding information about drugs, managing medication reactions and knowledge of intended and side effects, access to needed systems, and knowing where to find medications [13].

4.2 Assessment

4.2.1 Efficacy

The inherent shortcoming of web page is the phenomenon of "flash", but we introduce new technique Ajax to avoid this shortcoming to achieve user-friendliness. We also test the system correctness for two months to make sure ME-MAR is able to record the administration correctly. We implement ME-MAR in a simple and straightforward way; we believe that nurses are able to operate this system after a class of education. There are descriptions of the colors which represent the status of administration. Besides, we implement sorted orders by route and frequency, and support to prepare drugs.

4.2.2 Safety

We design ME-MAR system following the five rights of medication to enhance patients' safety, and other features and functionalities to facilitate safety: highly warning drugs remind, special orders, new and deleted orders remind, cosign before and after administering, chemotherapy drugs mandatory check before administering, and check single administration dose and daily dose limitation.

4.2.3 Access

The ME-MAR system is an independent page pop up from original web page, so if the nurses need to look for other information in NTUH HIS, they can continue their medication administration at the same time, besides the real time lookup drug information functionality is also a new pop up window provided by department of pharmacy, and nurses are able to look for drug information and keep their medication administration works. Table 1 concludes the features and functionalities of ME-MAR and corresponding criteria.

The online progress of ME-MAR system in NTUH is divided into four phases, and the first phase started on April 21st, 2008. In this phase, there are few units deploying ME-MAR system, and we also collect data and advices from the clinical users to modify our system. Table 2 outlines the main differences between the paper-based MAR and the ME-MAR system.

There were five adverse drug events reported from January 1st, 2008 to April 20th, 2008, and there was one case reported from April 21st, 2008 to July 18th, 2008. Although we are not sure that the adverse drug events reduced by ME-MAR, but our ME-MAR does not increase the adverse drug events. Although we

Table 1 The features and functionalities of ME-MAR and corresponding criteria

Criterion	Features and Functionalities
Safety	Follow the five rights of medication
Safety	Highly warning drugs remind
Safety	Provide special orders
Safety	New and deleted orders remind
Efficacy	Easily display the details of the administration
Access	Real time lookup drug information
Efficacy	Sort orders by route or frequency
Safety	Cosign before and after administering
Safety	Chemotherapy drugs mandatory check before administering
Safety	Check single administration dose and daily dose limitation
Efficacy	Support to prepare drugs

Table 2 Main differences between the paper-based MAR and the ME-MAR system

Action/Functionality	Before	After
New or modified orders are transcribed to the MAR	Y	Y
Sort orders by route and frequency		Y
Follow the NTUH local medication policy "one sign of one case"		Y
Real time look up drug information		Y
Chemotherapy drugs mandatory check before administering		Y
Automatically check single administration dose and daily dose limitation		Y
Light shading is enhanced by implement ME-MAR; white is for user-friendly; dark shading is new functionalities to promote safety.		

believe the ME-MAR system can promote the quality of health care safety, but successful adoption requires acceptance by nursing staff [11]. If users perceive the system's inefficiencies, then the system may not promote safety, instead the system becomes the reasons for nursing staff to circumvent hospital procedure or protocol to administer medication to the patients. The procedural features and functionalities of ME-MAR system must be acceptable to nurses to support them in this nursing role. There is a study showing that medication administration consumes up to 40% of nurses' time [14], so if nurses are satisfied with the ME-MAR system, they may less likely to develop "workaround" to save their time and then more likely to spend their time and energy focusing on the professional aspect of medication administration [11]. In order to evaluate the satisfaction of our ME-MAR, we decide to adopt The Clinical Information System Implementation Evaluation Scale (CISIES) [16] for evaluation. The CISIES contains 37 items and is administered in the post implementation period of a CIS implementation. It can be used for formative purpose, to adjust on-going implementations, or summative purpose, to assist in judging the relative success of an implementation. The proposes recommend administering the CISIES for the first time between 45 days (after system turn-on) and 60 days, and we administer the CISIES on June 5th, actually after system turned-on 45 days. We select three units which started to use ME-MAR on April 21st, and we sent out 40 questionnaires but received only 27 questionnaires. Our average score of satisfaction is 0.1 and according to the Administration, Scoring and Interpretation Guidelines for the Clinical Information Systems Implementation Evaluation Scale (CISIES), this score means: in general the respondents are neutral to the implementation of the system. Neutral can mean several things—that they are satisfied with some aspects of the implementation and not others or that some respondents are satisfied overall and some are not satisfied overall. Especially in this range, it is best to examine responses to individual items and individuals mean scores [16].

5 Conclusion and Discussion

In the paper, we have developed the Mobile Electronic Medication Administration Record System to aid the nursing staff for administering medication. We must emphasize that the main purpose the ME-MAR system is to promote patients' safety. In addition, there is a further benefit of the system, that is, NTUH started to establish the electronic health record recently, and the medication administration record is an essential part of health record. Therefore, after we constructed the ME-MAR system, all of the patients' medication administration data are stored in NTUH HIS. Furthermore, once the electronic health record is installed, the medication administration data is available automatically. The evaluation results of the nurses' satisfaction indicate that, in general, respondents are neutral regarding to the ME-MAR system. In other words, the system has spaces to improve or enhance. We will collect the users' suggestions to modify the current ME-MAR in order to coordinate with their requirements. Moreover, we can combine with the RFID mechanism to further improve the system, because using machines to scan patients' tags is conducive to identify patients correctly. The system can automatically search for patients' ME-MAR. As mentioned in Section 2, our goal is to implement all procedures in the order flow of HER. We have already begun to develop advanced CPOE currently; the advanced CPOE basically depends on ME-MAR, but it deals with variations of orders. ME-MAR handles medication orders, and the advanced CPOE controls non-medication orders. The two systems cover the order flow of performing and recording; these are the essences of nurses' daily operations. Another important procedure is pharmacists review medication orders, because pharmacists are experts of medication. If all medication orders are reviewed by them, when they discover problems, they can directly lock the problematic orders, and nurses will not be able to propagate the errors. The mechanism provides higher safety to patients. The systems we developed, implemented can facilitate entire medical staff for their operations in NTUH, and improve patients' safety as well as the quality of medical cares.

References

[1] Haux, R., Winter, A., Ammenwerth, E., Brigl, B.: Strategic Information Management in Hospitals. In: An Introduction to Hospital Information Systems. Springer, New York (2004)
[2] Haux, R.: Health Information System — past, present, future. Int. J. Med. Inf. 75, 268–281 (2006)
[3] Kohn, L., Corrigan, J., Donaldson, M. (eds.): To Err is Human. National Academy Press for the U.S. Institute of Medicine, Washington (1999)
[4] Yang, T.H., Cheng, P.H., et al.: A Scalable Multi-tier Architecture for the National Taiwan University Hospital Information System based on HL7 Standard (unpublished)
[5] Health Level Seven (HL7), HL7 Standard v.2.5 (2003)
[6] Ko, L.-F., Lin, J.-C., et al.: HL7 Middleware Framework for Healthcare Information System. In: HEALTHCOM 2006 (August 2006)
[7] Bates, D.W., Leape, L.L., Cullen, D.J., Laird, N., Petersen, L.A., Teich, J.M., et al.: Effect of computerized physician order entry and a team intervention on prevention of serious medication errors. JAMA 280, 1311–1316 (1998)
[8] Bates, D.W., Teich, J., Lee, J., Seger, D., Kuperman, G.J., Boyle, D., et al.: The impact of computerized physician order entry on medication error prevention. J. Am. Med. Informatics Assoc. 6, 313–321 (1999)
[9] Bates, D.W.: Using information technology to reduce rates of medication errors in hospital. BMJ 320, 788–791 (2000)
[10] Garrett, J.J.: Ajax: A New Approach to Web Application (February 2005), `http://www.adaptivepath.com/ideas/essays/archives/000385.php`
[11] Hurley, A.C., Bane, A., Fotakis, S., Duffy, M.E., Sevigny, A., Poon, E.G., Gandhi, T.K.: Nurses' Satisfaction With Medication Administration Point-of-Care Technology. JONA 37, 343–349 (2007)
[12] Eisenhauer, L.A., Hurley, A.C., Dolan, N.: Nurses' Reported Thinking During Medication Administration. Journal of Nursing Scholarship 39, 82–87 (2007)
[13] Hurley, A.C., Lancaster, D., Hayes, J., Wilson-Chase, C., Bane, A., Griffin, M., Warden, V., Duffy, M.E., Poon, E.G., Gandhi, T.K.: The Medication Administration System—Nurses Assessment of Satisfaction (MAS-NAS) Scale. Journal of Nursing Scholarship 38, 298–300 (2006)
[14] Armitage, G., Knapman, H.: Adverse events in drug administration: A literature review. Journal of Nursing Management 11, 130–140 (2003)
[15] Staggers, N., Kobus, D., Brown, C.: Nurses Evaluations of a Novel Design for an Electronic Medication Administration Record. Computers, Informatics, Nursing 25, 67–75 (2007)
[16] Maranda, M.J.: PhD, Brian Gugerty, DNS, MS, RN. CISIES: An Informatics Measurement Instrument, `http://cisevaluation.com/`
[17] DeLone, W.H., McLean, E.R.: The DeLone and McLean model of information systems success: A ten-year update. J. Manage. Inform. Syst. 19, 9–30 (2003)
[18] Rocheleau, B.: Evaluating public sector information systems: Satisfaction versus impact. Eval. Program Plann. 16, 119–129 (1993)
[19] Ives, B., Olson, M., Baroudi, J.J.: The measurement of user information satisfaction. Commun. ACM 26, 785–793 (1983)
[20] Davis, F.D.: Perceived usefulness, perceived ease of use, and user acceptance of information technology. MIS Q 13, 319–339 (1989)
[21] Venkatesh, V.: Determinants of perceived ease of use: Integrating control, intrinsic motivation, and emotion into the technology acceptance model. Inform. Syst. Res. 11, 342–365 (2000)
[22] Lee, Y., Kozar, K., Larsen, K.: The technology acceptance model: Past, present, and future. Commun. AIS 12, 752–780 (2003)

Customer Relationship Management in Healthcare Service – An Integrated DSS Framework for Patient Loyalty

Chi-I Hsu, Pei-Lun Hsu, and Chaochang Chiu

Abstract. Patient loyalty is a critical criterion for healthcare customer relationship management (CRM). An integrated framework with a case-based prediction model and a constraint-based optimization model is proposed to support the decision making of healthcare providers. This research first adopts a case-based prediction mechanism to forecast the possible loyalty level. We also proposes a constraint-based optimization approach as a subsequent mechanism to determine the optimum values of case features that may lead to the optimal patient loyalty. The potential use of this framework helps a decision maker allocate resources to increase the loyalty level for the given target patient segmentation.

1 Introduction

CRM is an important issue in healthcare service especially when there is potential for healthcare providers to build ongoing relationships with patients. In the competitive healthcare environment, more healthcare providers focus on CRM with the assistant of information systems to deliver value over price. The purpose of CRM is to create and maintain good and long-term relationships with customers. Customer loyalty is a critical criterion for CRM. Oliver [8] suggested that loyalty refers to the repeated use of certain products or services by customers and even changes in business scenarios will not affect the purchase patterns or willingness to continue to use those products or services. Therefore, the core of the CRM in healthcare service is to satisfy patient interest and needs to increase the patient loyalty level.

Chi-I Hsu
Department of Information Management, Kai Nan University, Taiwan, R.O.C.
e-mail: imchsu@mail.knu.edu.tw

Pei-Lun Hsu
Department of Electronic Engineering, Ching Yun University, Taiwan, R.O.C.

Chaochang Chiu
Department of Information Management, Yuan Ze University, Taiwan, R.O.C.

K. Nakamatsu et al. (Eds.): New Advan. in Intel. Decision Techno., SCI 199, pp. 509–517.
springerlink.com

Effective CRM requires comprehensive data and applications to support analysis and to provide consistent customer services. Case-based reasoning (CBR) is a machine reasoning technique that adapts previous similar cases to infer further similarity of the new input case. When developing a CBR system, a set of useful case features must first be determined to differentiate one case from the others. The case features are usually observable with concrete meanings. By retrieving and adapting previous similar cases, this research adopts a case-based prediction mechanism to predict the overall level of patient loyalty for a target patient segmentation. The prediction result enables administrators of healthcare providers to assess the effectiveness of policies and programs and whether these are adapting to changing patient needs.

Based on the prediction model, this research further proposes a constraint-based optimization mechanism to determine the optimum values of case features that best approximate the goal of patient loyalty. The optimization model is useful to make suggestions regarding the increase or decrease levels of adjustable case features for a target patient segmentation. To upgrade the loyalty level, the tracking result provides information regarding the areas for improvement and prompts healthcare providers to take appropriate actions.

A questionnaire is developed to collect the patient cases of seeking medical care. From a regional hospital in Taiwan, two hundred eighty-four real patient cases are collected in the case base for the experiment. The integrated system with both the case-based prediction model and the constraint-based optimization model can support the decision making of healthcare providers by the capability to predict the possible loyalty level for a target patient segmentation as well as to provide optimal suggestions leading to the goal of loyalty level.

2 Healthcare CRM

2.1 Healthcare in Taiwan

With the implementation of the National Health Insurance (NHI) program in 1995, through risk pooling, the public in Taiwan has obtained comprehensive medical care such as health prevention, clinical care, hospitalization, resident care and social rehabilitation [1]. It is a change in payment methods for healthcare providers and does not affect the medical rights of the public. When seeking medical treatments using their NHI cards, patients are not only focusing on the medical expenses, but also the quality of healthcare services. Healthcare CRM concepts and solutions was initially addressed by private medical institutions [4]. To respond to a changing environment, both public and private healthcare providers are moving towards creating and maintaining good relationships with their patients.

In addition to the implementation of NHI program, other reasons for Taiwan healthcare providers to emphasize CRM include (1) the competitive environment of health care, (2) the increase of patients' knowledge level, and (3) the open of mass communication media [3]. Regarding the last point, the Internet is a new potential media to build relationship with patients and to implement e-healthcare business on the net. Growing number of healthcare providers in Taiwan are using the Internet to provide on-line information, appointments, and market services to patients.

2.2 Patient Loyalty

Loyalty refers to a long-term commitment that a customer makes to the target products or companies. Loyalty is a manifestation of both attitudes and actual purchasing patterns towards specific products and services. Loyal customers continue to favor their preferred products or product mixes [8] despite they receive information regarding other similar products or services.

Chaudhuri and Holbrook [2] discussed the issues of loyalty from two perspectives: the levels of loyalty to the brands and willingness to purchase despite the brand premiums. Loyal customers tend to have the following characteristics: purchases frequently and repeatedly, purchase other categories of products and services from the same company, actively promote, through word-of-mouth, for the company, and indifferent to competitors' promotions [9]. In the context of medical service, a loyal patient often choose to go to the preferred healthcare provider to seek medical treatments for the same or the other syndromes, and recommend to others his or her preferred healthcare provider despite the promotions provided by other healthcare providers.

3 Research Framework

As shown in figure 1, the integrated DSS framework for patient loyalty consists of the human-computer interface, a patient loyalty case base, and both prediction and optimization models. For a target patient segment, users can issue queries to inquire the possible loyalty level for a target patient segmentation as well as to wait for optimal suggestions leading to the goal of loyalty level.

3.1 Case-Based Prediction Model

CBR is a problem-solving method that is similar to the analogical reasoning process. Several researches have applied CBR for prediction in different business domains, including the early cost prediction of structural systems [5], the predicting high risk software components [6] and etc. Our previous research proposed a case-based prediction mechanism to predict IS outsourcing success using genetic algorithms (GA) to support the retrieval of similar cases [7].

The general process of GA-CBR prediction is shown in Figure 2. A description of the prediction problem as the input case is feed into the system. The similarity is computed between the input case and retrieved cases. The higher the similarity the more likely the retrieved case matches the input case. The GA is employed to determine the optimal weights for the circulations of the similarity degree. Based on the similarity degree, the system retrieves similar cases from the case base. The adapting process aims to apply the outcome features of the most similar cases to

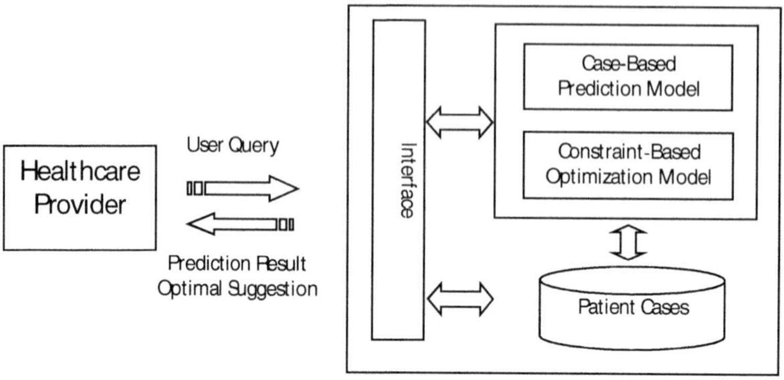

Fig. 1 Integrated DSS Framework for Patient Loyalty

Fig. 2 Case-based Prediction

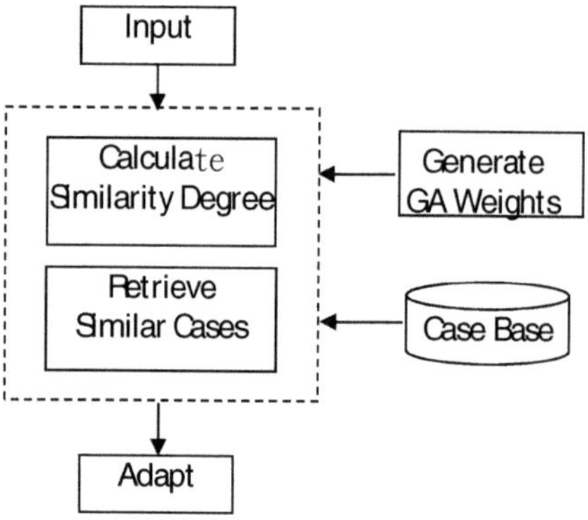

generate a solution for the current problem, that is, the expected outcome features for a certain input case. The mean absolute error (MAE) is applied to evaluate the results. The fitness function is defined as the equation (Eq.1).

Minimize

$$Y = \frac{\sum_{t=1}^{p} \left| O_t^{'} - O_t \right|}{p} \tag{1}$$

where $t=1$ *to* p, p is the number of training cases. O_t' and O_t are the expected and real composite outcome feature of the t_{th} training case, respectively. Y is MAE.

3.2 Constraint-Based Optimization Model

A goal-seeking problem is to find out the most appropriate level for the adjustable inputs under the condition of a setting goal. The search space for the goal-seeking problem could become very huge as the number of case features increases. To reduce the search space, a feasible approach is to use constraints to define the set of adjustable case features and adjust ranges. In general, the GA can deal with soft and hard constraints. Soft constraints refer to integrating a penalty into the fitness function. Hard constraints refer to abandoning invalid chromosomes after a population is generated. Instead of soft and hard constraints, this research integrates the constraint-based reasoning into the GA process as a filter to assure that each chromosome is generated satisfying the constraints. This approach excludes invalid chromosomes being generated and therefore considerably reduces the search space compared with the general GA approach.

The proposed constraint-based optimization consists of four modules. As shown in Figure 3, these modules include GA initialization, chromosome filtering, fitness evaluation, and GA operation.

In the GA initialization module, chromosomes are encoded as the values of the adjustable case features and the potential chromosome candidates for the first population are generated.

The chromosome filtering module applies the constraint-based reasoning to ensure each chromosome satisfies the requirements of constraints. In this research, the constraints can be the valid adjust ranges of the case features. Figure 4 illustrates the detail chromosome filtering process. The filtering process adopts a constraint-based reasoning technique called the local propagation. As shown in Figure 3, the local propagation uses the constraint information regarding each valid gene range (G_j domain). If the gene value is inconsistent with the constraints, the value of gene (g_{ij}) is replaced by a value randomly selected from the satisfactory universe set (S_{Gj}). By continuously executing the validation process on chromosome $C_i(g_{i1}, g_{i2}, \ldots g_{ij}, \ldots, g_{im})$ in sequence by the loop ($j=j+1$) under the condition ($j<=m$), the new chromosome C'_i is made to satisfy the constraints. By checking the defined gene range, the local propagation excludes invalid chromosomes being made and assures that invalid chromosomes will not further be processed by the

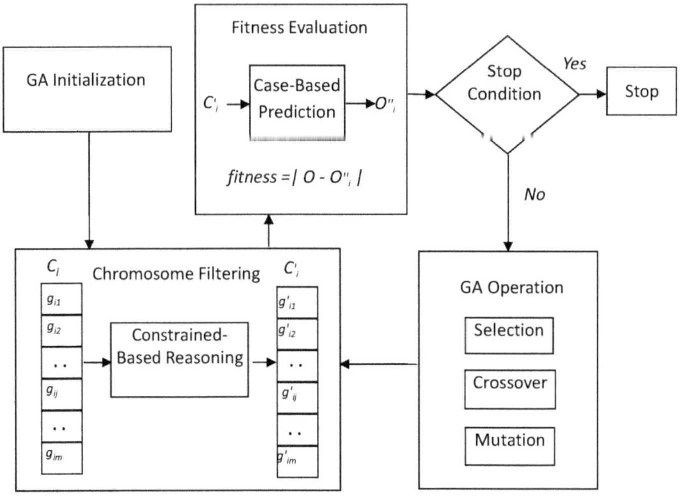

Fig. 3 Constraint-Based Optimization

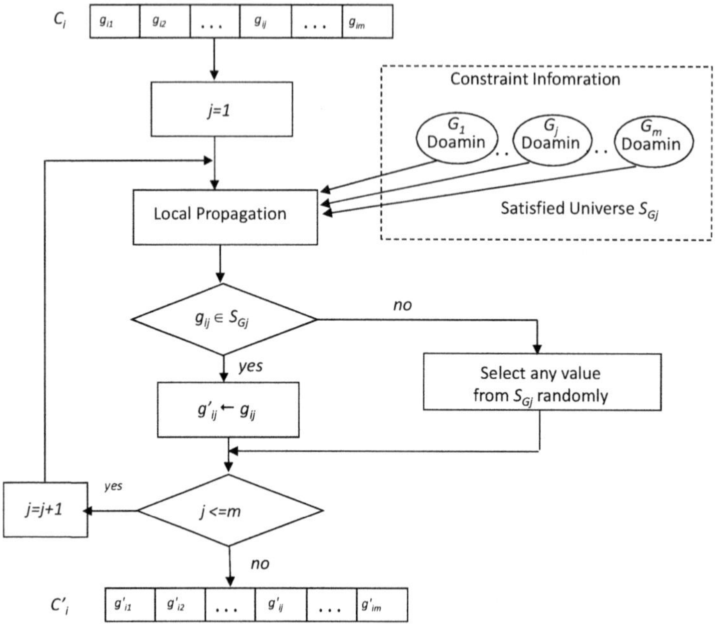

Fig. 4 Chromosome Filtering

Fitness Evaluation module and the GA operation module. This approach offers a way to reduce the search space of the GA and help the GA search for the optimal solution efficiently.

In the fitness evaluation module, the case-based prediction denoted as *CBR( )* is used to determine the predicted loyal level (O''_i) for each chromosome (C'_i) in the population. The fitness function is defined as the equation 2 which is to minimize the difference between the goal of loyalty level (O) and the predicted outcome (O''_i).

Minimize

$$fitness = | \; O - O''_i \; | \quad where\ O\ is\ the\ goal\ and\ O''_i = CBR\,(\,C'_i\,) \tag{2}$$

4 The Experiments and Results

4.1 Case Description

A case is represented with features tied to a context. In the healthcare service domain, this research proposed four dimensions for patient loyalty as shown in Table 1. These dimensions focus on general concerns of patient loyalty with reference to the literatures of loyalty.

In addition to some socio-demographic characteristics of the patients (gender, age, marriage status, profession, education level, income level, and the department of receiving medical treatment), this research proposed features for a patient case as shown in table 2. These are perceived features when a patient seeks a medical treatment from a hospital.

4.2 Case Collection

Based on the above case description, a questionnaire was developed for collecting patient cases from a regional hospital in Taiwan. Ten patients were pre-tested with the entire questionnaire to ensure the face validity. The final questionnaires were distributed to patients at all departments in the hospital. Each patient was asked to rate on a scale of 1-5 about his or her perception with the questionnaire items. Two hundred eighty-six patient cases were collected in the case base.

4.3 GA Control Parameters

Chromosomes are designed for encoding the weights which use a number of 0~1 to represent a weight. The key defined GA parameters consists of the population size, crossover rate, and mutation rate. The following values were adopted in this research. The population size was 100, crossover rate was 0.6, and mutation rate was 0.01. The entire learning process stopped after 500 generations.

4.4 Prediction Results

The prediction results are illustrated in Table 3 and Figure 5. All methods were tested with a 5-fold cross-validation method. The equal weights approach set feature weights equally without the support of GA. Our approach denoted as GA-CBR demonstrated more accurate results than the equal weights approach. The GA-CBR results were the best with average MAE equal to 0.282 for training and 0.293 for testing.

Table 1 Outcome Feature — Patient Loyalty

Dimension	Definition
Continued Treatments	The willingness to seek medical treatments for the same syndrome in the healthcare provider.
Extended Treatments	The willingness to seek medical treatments for the other syndromes in the healthcare provider.
Recommendation	The willingness to recommend the healthcare provider to others.
Resistance Against competitor's promotions	The willingness to seek medical treatments in the healthcare provider though other health providers are on promotions

Table 2 Case Features

Features	Definition
SPecialty (SP)	Perceived medical specialty of attending medical personnel
Service Attitude (SA)	Perceived service attitude of attending medical personnel
Service Flow (SF)	Perceived efficiency of medical service flow
ENvironment (EN)	Space and facilities of receiving medical treatment
Traffic Time (TT)	Traffic time to hospital
REputation (RE)	Perceived reputation of attending medical personnel and hospital
Time Consumed (TC)	Average time consumed of receiving one medical treatment

Table 3 MAE of Different Methods

	GA-CBR		Equal Weights	
Fold	Training	Testing	Training	Testing
1	0.284	0.28	0.407	0.426
2	0.281	0.314	0.412	0.418
3	0.283	0.301	0.424	0.367
4	0.287	0.283	0.418	0.38
5	0.275	0.287	0.393	0.41
Average	0.282	0.293	0.411	0.4

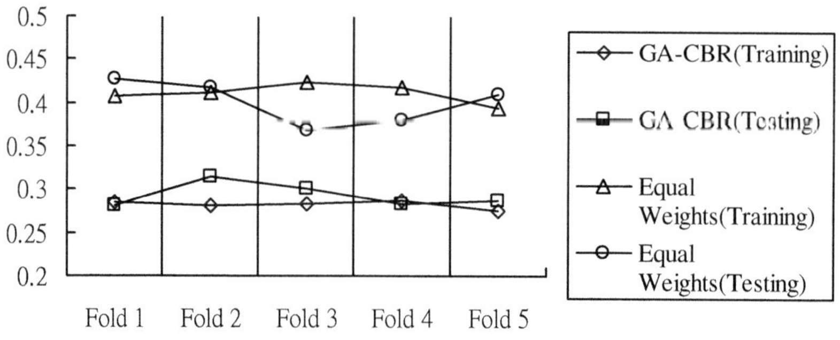

Fig. 5 Five-fold MAE results

4.5 Optimization Example

The following draws on an example to illustrate the potential use the constraint-based optimization model. A target patient segment can be defined as specify the socio-demographic characteristics of the patients including gender, age, marriage status, profession, education level, income level, and the department of receiving medical treatment. To increase the overall patient loyalty for a target patient segmentation, for instance, to achieve the goal of loyalty level four in average, the healthcare provider would like to explore the appropriate levels for case features 'service flow efficiency' (SF) and 'average time consumed of receiving a medical treatment' (TC). Chromosomes are designed for encoding the values of SF and TC, which use a number of -1 ~ 1 to represent the adjust direction and percentage (-100% ~ +100%). Constraints are defined as the feasible adjust range for each adjustable case feature. The constraint-based optimization model can generate the optimal levels for SF and TC which suggest the increase or decrease levels for the service flow efficiency and the average time consumed of receiving a medical treatment. The results can be used for the reference of healthcare providers to improve their resource allocation and CRM plans.

5 Conclusion

In the domain of healthcare service, this research proposes a case-based prediction model to predict patient loyalty level and a constraint-based optimization model to determine the optimal values of the case features to achieve the setting goal of patient loyalty. This research suggests that appropriate input features can be found that best approximate the goal of patient loyalty. The result information is with highly strategic values as CRM becomes a strategic important practice to organizations. The proposed research framework can help administrators of healthcare providers establish a proactive customer relationship management as well as provide better healthcare services.

References

1. Bureau of National Health Insurance (BNHI) (accessed August 5, 2008), http://www.nhi.gov.tw/english/
2. Chaudhuri, A., Holbrook, M.B.: The chain of effects from brand trust and brand affect to brand performance: the role of brand loyalty. Journal of Marketing 65(2), 81–93 (2001)
3. Chen, C.: Hospital Organization and Management. HonHan Publication, Taipei (1998)
4. Chun, G.: The influence of taiwan's medical policy on the national health insurance (nhi) planning. Journal of National Policy 103 (1995) (Times Review)
5. Dogan, S.Z., Arditi, D., Günaydin, H.M.: Determining attribute weights in a cbr model for early cost prediction of structural systems. Journal of Construction Engineering and Management 132(10), 1092–1098 (2006)
6. Emam, K., Benlarbi, S., Goel, N., Rai, S.: Comparing case-based classifiers for predicting high risk software components. Journal of Systems and Software, 301–320 (2001)
7. Hsu, C., Chiu, C., Hsu, P.L.: Predicting information systems outsourcing success using a hierarchical design of case-based reasoning. Expert Systems with Applications 26(3), 435–441 (2004)
8. Oliver, R.L.: Satisfaction: a Behavioral Perspective on the Consumer. McGraw-Hill, Irwin (1997)
9. Stum, D.L., Thiry, A.: Building customer loyalty. Training & Development Journal 45(4), 34 (1991)

Applying Fuzzy TOPSIS Approach for Evaluating RFID System Suppliers in Healthcare Industry

Tien-Chin Wang, Hsien-Da Lee, and Po-Hsun Cheng

Abstract. RFID is an emerging technology intends to replace traditional barcode. Compared to barcode, RFID has the practical advantages of improving total product traceability and increasing accuracy. Especially healthcare service is life-critical, so that any careless mistakes may result in reversible loss. Due to the huge demand for automatic identification, many corporations and researchers devoted their efforts in RFID approaches for better achieving their goals. However, since there are various RFID solutions on the markets, it is critical to select the optimal solution to fit the desired scenarios. RFID deployment and implementation require not only technology concern, but also financial investment and user involvement. In this paper, we applied fuzzy TOPSIS to effectively evaluate suitable RFID solution providers. The proposed framework intends to help decision makers better evaluate the important factors affecting RFID implementation, forecasting the probability of a successful RFID project, as well as identifying the actions necessary before implementation. In addition, an experimental case study is presented to illustrate the application of the proposed approach.

Keywords: RFID; MCDM; Fuzzy TOPSIS; Health care system.

1 Introduction

RFID stands for radio frequency identification, which is a wireless communication technology which is used to identity tagged objects. The RFID technology has

Tien-Chin Wang
Department of Information Management, I-Shou University, Kaohsiung, Taiwan
e-mail: tcwang@isu.edu.tw

Hsien-Da Lee
Department of Information Engineering, I-Shou University, Kaohsiung, Taiwan
e-mail: leesd@center.fotech.edu.tw

Po-Hsun Cheng
Department of Software Engineering, National Kaohsiung Normal University, Kaohsiung, Taiwan
e-mail: cph@nknu.edu.tw

K. Nakamatsu et al. (Eds.): New Advan. in Intel. Decision Techno., SCI 199, pp. 519–526.
springerlink.com

recently attracted much attention for application in many domains including product tracking, point-of-sale applications, supply chain tracking, safety and access control and medical service. After the outbreak of SARS, Taiwan government invests a large sum of fund to enhance patient safety. RFID can provide capability to promote visibility in healthcare management to ensure patient safety. Since RFID technology is still developing, many standards and devices have not yet acquired wide recognition. Most importantly, various scenarios may need to adopt different RFID solutions. In the selection process of evaluating RFID suppliers, many criteria need to take into consideration. The selection of suitable RFID suppliers can be viewed as a multi-criteria decision making (MCDM) problem. Among many famous MCDM methods, Technique for Order Performance by Similarity to Ideal Solution (TOPSIS) is a practical and useful technique for ranking and selection of a number of possible alternatives through measuring Euclidean distances. TOPSIS was first developed by Hwang and Yoon [2]. It bases upon the concept that the chosen alternative should have the shortest distance from the Positive Ideal Solution (PIS), i.e., the solution that maximizes the benefit criteria and minimizes the cost criteria; and the farthest from the Negative Ideal Solution (NIS), i.e., the solution that maximizes the cost criteria and minimizes the benefit criteria.

In this paper, we adopt fuzzy TOPSIS approach to model RFID supplier selection process in healthcare industry. The rest of the paper is organized as follows. In the next section, the literature reviews some RFID application systems in Taiwan healthcare institutions and theoretic background of TOPSIS. Section 3 will focus on the proposed group TOPSIS model in a step by step fashion. In section 4, an illustrated example is presented to confirm the feasibility of the proposed approach. In the final section, conclusions are drawn.

2 Literature Review

2.1 RFID Applications in Healthcare Industry

In recently years, many countries have invested a large sum of budget and have initiated some pilot projects for the RFID deployment in healthcare industry. RFID technology can help medical institutions enhance the accuracy of identifying patients without physical contact. RFID also can speed up the tracking process due to its active radio frequency feature. Deploying RFID technology could provide advantages identifying, tracking and tracing objects to provide high medical quality service and promote value-added operation efficiency. In USA, the emergency unit of Washington Hospital Center in Washington, D.C. dispatched over 20 readers and 100 tags to its patients since October, 2004. The well-known pharmaceutical company Pfizer declared it would place RFID tag on each product bottle from 2005. According to ID TechEx, the RFID market in USA healthcare industry will be worth more than 8.6 billion US dollar. Deployment of RFID technology in medical institutions has become a cost-effective trend of management and operations.

Many researchers present their studies about the critical factors of deploying RFID technology in medical institutions. Kuo and Chen [1] classify four dimensions of deployment critical issues: technological, economical, industrial and social dimension. Benfatto and Vecchio [4] examine some of the key considerations and opportunities for SCM solutions in the field of health specific goods and focus the benefits of an integrated management of the processes and the main actors involved. Sun et al. [5] stress the patient safety issue as the most important priority while implementing a RFID system. Fisher and Monahan [6] applied qualitative methods for tracking the social dimensions of RFID systems in hospitals and found that hospital staff, especially nurses, expressed concerns about increasing job load by operating newly implementing RFID systems. However, deploying RFID in healthcare industry for promoting patient safety and avoiding human errors is a complicated decision making process. This can viewed as MCDM problems. In this research, we proposed a fuzzy TOPSIS approach for a medical institution to select optimal RFID supplier from several alternatives.

3 Proposed Method

Assume a MCDM problem which has m alternatives $A_i(i=1,2,...,m)$ and n criteria $C_j(j=1,2,...,n)$. The decision matrix with respect to each criterion and alternative denoted as follow:

$$D = \begin{matrix} & \begin{matrix} C_1 & C_2 & \cdots & C_n \end{matrix} \\ \begin{matrix} A_1 \\ A_2 \\ \vdots \\ A_m \end{matrix} & \begin{bmatrix} X_{11} & X_{12} & \cdots & X_{1n} \\ X_{21} & X_{22} & \cdots & X_{2n} \\ \vdots & \vdots & \ddots & \vdots \\ X_{m1} & X_{m2} & \cdots & X_{mn} \end{bmatrix} \end{matrix}$$

The fuzzy TOPSIS approach can be summarized as follows:

Step 1: Calculate weighted decision matrix: Subjective assessments are to be made by DM to determine the weighting vector $W=(w_1,w_2,...,w_n)$. The weighting vector W represents the relative importance of n selection criteria $C_j(j=1,2,...,n)$ for the problem.

$$\tilde{W}_j = \frac{1}{n}\left(\sum_{j=1}^{n} w_j^e\right), j=1,2,\cdots,n \tag{1}$$

Assume in a fuzzy environment, ratings are described using triangular fuzzy number. The importance weights of the *kth* decision maker are $\tilde{R}_k=(a_k,b_k,c_k)$, then the aggregated fuzzy rating can be expressed as $\tilde{R}=(a,b,c)$ and the aggregated fuzzy weights $\tilde{w}_j=(w_{j1},w_{j2},w_{j3})$

$$w_{j1} = \min_k \{w_{jk1}\}, w_{j2} = \frac{1}{K}\sum_{k=1}^{K} w_{jk2}, w_{j3} = \max\{w_{jk3}\} \tag{2}$$

Step 2: Normalize the decision matrix so that each criterion is comparable. The initial data with respect to each criterion will be normalized by dividing the sum of criterion values. For fuzzy data denoted by triangular fuzzy number as (a_{ij}, b_{ij}, c_{ij}), the normalized values for benefit-related criteria and cost-related criteria are calculated as follows.

$$\tilde{r}_{ij} = \left\{\left(\frac{a_{ij}}{c_{ij}^{+}}, \frac{b_{ij}}{c_{ij}^{+}}, \frac{c_{ij}}{c_{ij}^{+}}\right), j \in B \tag{3}$$

$$\tilde{r}_{ij} = \left\{\left(\frac{a_{ij}}{c_{ij}^{-}}, \frac{b_{ij}}{c_{ij}^{-}}, \frac{c_{ij}}{c_{ij}^{-}}\right), j \in C \tag{4}$$

Step 3: Determine the positive ideal solution A^{+} and the negative ideal solution A^{-}. Sort the weighted normalized values for each criterion in descending order.

$$A^{+} = \left(\tilde{v}_1^{+}, \tilde{v}_2^{+}, \cdots, \tilde{v}_k^{+}\right) \tag{5}$$

$$A^{-} = \left(\tilde{v}_1^{-}, \tilde{v}_2^{-}, \cdots, \tilde{v}_k^{-}\right) \tag{6}$$

Step 4: Calculate the distance from the positive ideal solution and the negative ideal solution for each alternative. The distance between two triangular fuzzy numbers $A_1 = (a_1, b_1, c_1)$ and $A_2 = (a_2, b_2, c_2)$ is calculated as

$$d(A_1, A_2) = \sqrt{\frac{1}{3}\left[(a_1 - a_2)^2 + (b_1 - b_2)^2 + (c_1 - c_2)^2\right]} \tag{7}$$

$$d_i^{+} = \sum_{j=1}^{k} d\left(\tilde{v}_{ij}, \tilde{v}_j^{+}\right), \; i = 1,2,\cdots,m \tag{8}$$

$$d_i^{-} = \sum_{j=1}^{k} d\left(\tilde{v}_{ij}, \tilde{v}_j^{-}\right), \; i = 1,2,\cdots,m \tag{9}$$

Step 5: Calculate the closeness coefficient (CC). And rank each CC of each alternative in descending order. The alternative with the highest CC value will be the best choice.

$$CC_i = \frac{d_i^-}{d_i^+ + d_i^-}, \quad i = 1,2,\cdots,m \tag{10}$$

4 Illustrative Example

This section is related with a medical institution which intends to deployment RFID technology to its emergency unit. We apply Wang and Lee's fuzzy TOPSIS approach [3] to evaluate the RFID suppliers as to select the optimal one. Since the emergency operations need quick response, high accuracy and protection for staff from contamination. The medical institution desires to find an optimal RFID solution. First of all, a committee is formed including three decision makers D_1, D_2 , D_3 . Then the criteria are determined as function accuracy C_1, implement cost C_2, supplier service quality C_3 and organization acceptance C_4. After preliminary screening, three alternatives A_1, A_2, A_3 have remained in the candidate list.

4.1 Application with Fuzzy TOPSIS Method

Firstly, three decision makers evaluate the importance of each criterion by using linguistic variables in Table 1. The importance weights of each criterion determined by decision makers are shown in Table 2.

Table 1 Fuzzy linguistic terms for importance and fuzzy numbers for each criterion

Importance	Abbreviation	Fuzzy Number
Very Low	VL	(0, 0, 0.2)
Low	L	(0.1, 0.2, 0.3)
Medium Low	ML	(0.2, 0.4, 0.5)
Medium	M	(0.4, 0.5, 0.6)
Medium High	MH	(0.5, 0.7, 0.8)
High	H	(0.7, 0.8, 0.9)
Very High	VH	(0.8, 1, 1)

Table 2 Importance weight of each criterion assigned by decision maker

DM C	D1	D2	D3
C1	H	VH	H
C2	VH	H	VH
C3	MH	VH	M
C4	H	MH	MH

The aggregated fuzzy weights of each criterion are shown in Table 3.

Table 3 Aggregated fuzzy weights

Criterion	Fuzzy number
C1	(0.7,0.87,1)
C2	(0.7,0.93,1)
C3	(0.4,0.67,0.9)
C4	(0.5,0.73,0.9)

The fuzzy linguistic terms and correspondent fuzzy numbers for each alternative listed in Table 4.

Table 4 Fuzzy linguistic terms and correspondent fuzzy numbers for each alternative

Performance	Abbreviation	Fuzzy Number
Very Poor	VP	(0, 0, 0.2)
Poor	P	(0.05, 0.2, 0.35)
Medium Poor	MP	(0.2, 0.35, 0.5)
Fair	F	(0.35, 0.5, 0.65)
Medium Good	MG	(0.5, 0.65, 0.8)
Good	G	(0.65, 0.8, 0.95)
Very Good	VG	(0.8, 1, 1)

The original DM rating table obtained from DMs and normalized fuzzy decision matrix can be calculated. The results are shown in Table 5 and Table 6, respectively.

Table 5 The initial DM rating table

Criteria	Alternatives	D1	D2	D3
C1	A1	F	G	G
	A2	P	F	P
	A3	G	MG	G
C2	A1	MP	P	G
	A2	VP	P	F
	A3	F	F	P
C3	A1	G	VG	F
	A2	G	MG	G
	A3	G	VG	G
C4	A1	F	F	MP
	A2	P	F	VG
	A3	F	G	G

Table 6 Normalized decision matrix

	C_1	C_2	C_3	C_4
A_1	(0.35,0.70,0.95)	(0.05,0.45,0.65)	(0.35,0.77,1.00)	(0.20,0.45,0.65)
A_2	(0.05,0.30,0.65)	(0.00,0.23,0.65)	(0.50,0.75,0.95)	(0.05,0.57,1.00)
A_3	(0.50,0.75,0.95)	(0.05,0.40,0.65)	(0.65,0.87,1.00)	(0.35,0.70,0.95)

The overall weighted normalized ratings of each alternative in related with each criterion are shown in Table 7.

Table 7 Overall weighted normalized ratings for each alternative

	C_1	C_2	C_3	C_4
A_1	(0.25,0.61,0.95)	(0.04,0.42,0.65)	(0.14,0.51,0.90)	(0.10,0.33,0.59)
A_2	(0.04,0.42,0.65)	(0.00,0.22,0.65)	(0.20,0.50,0.86)	(0.03,0.42,0.90)
A_3	(0.35,0.65,0.95)	(0.04,0.37,0.65)	(0.26,0.58,0.90)	(0.18,0.51,0.86)

According to Table 7, fuzzy positive ideal solution (PIS) and fuzzy negative ideal solution are obtained as follows:

$$A^{+} = [(0.95,0.95,0.95),(0.65,0.65,0.65),(0.90,0.90,0.90),(0.90,0.90,0.90)]$$
$$A^{-} = [(0.04,0.04,0.04),(0,0,0),(0.14,0.14,0.14),(0.03,0.03,0.03)]$$

Finally, the PIS and NIS for each alternative can be calculated and closeness coefficients are used to rank the optimal alternative. The result is shown in Table 8.

Table 8 Closeness coefficient table

	d^{+}	d^{-}	CC	Ranking
A1	1.92	1.94	0.50	2
A2	2.18	1.80	0.45	3
A3	1.67	2.17	0.57	1

From above, the order of rating among those alternatives is $A_3 \succ A_1 \succ A_2$, The best alternative would be A_3 .

5 Conclusion

Deploying RFID in healthcare industry for promoting patient safety and avoiding human errors is a complicated decision making process. Evaluating RFID alternatives is not only about technologic and financial concerns, but also it may related to organizational and management issues. Motivated by a need to systematically evaluate the RFID solutions, the paper presents a model based on fuzzy TOPSIS method. The model allows decision makers to utilize fuzzy linguistic terms to rate criterion importance and alternative performance. An experimental example also confirms the model's feasibility. The proposed model can provide decision makers a systematic framework to select the optimal solution.

References

1. Kuo, C.H., Chen, H.G.: The Critical Issues about Deploying RFID in Healthcare Industry by Service Perspective. In: Proceedings of the 41st Annual Hawaii International Conference on System Sciences, Hawaii, USA, January 7-10 (2008)
2. Hwang, C.L., Yoon, K.S.: Multiple attribute decision making: methods and applications. Springer, Berlin (1981)
3. Wang, T.C., Lee, H.D.: Developing a fuzzy TOPSIS approach based on subjective weights and objective weights. Expert Systems with Applications (December 2008)
4. Benfatto, M.C., Del Vecchio, C.: Organizational Impact of Technological Innovation on the Supply Chain Management in the Healthcare Organizations. In: Interdisciplinary Aspects of Information Systems Studies, pp. 71–77. Physica-Verlag (2008)
5. Sun, P.R., Wang, B.H., Wu, F.: A New Method to Guard Inpatient Medication Safety by the Implementation of RFID. Journal of Medical Systems 32(4), 327–332 (2008)
6. Fisher, J.A., Monahan, T.: Tracking the social dimensions of RFID systems in hospitals. International Journal of Medical Informatics 77, 176–183 (2008)

An Aspect of Decision Making in Rough Non-deterministic Information Analysis

Hiroshi Sakai, Kohei Hayashi, Hiroshi Kimura, and Michinori Nakata

Abstract. We have been proposing a framework Rough Non-deterministic Information Analysis (*RNIA*), which handles rough sets based concepts in not only Deterministic Information Systems (*DISs*) but also Non-deterministic Information Systems (*NISs*). We have recently developed some algorithms and software tools for rule generation from *NISs*. Obtained rules characterize the tendencies in *NISs*, and they are often applied to decision making. However, if the condition parts in such rules are not satisfied, obtained rules are not applied to decision making. In this case, we need to examine each data in *NISs*, directly. In this paper, we add a question-answering with criterion values to *RNIA*. This addition enhances the aspect of decision making in *RNIA*.

1 Introduction

Rough set theory offers a mathematical approach to vagueness and uncertainty, and the rough sets based concepts have been recognized to be very useful [1-4]. This theory usually handles tables with deterministic information, which we call *Deterministic Information Systems* (*DISs*). Many applications of this theory to information analysis, data mining, rule generation, machine learning and knowledge discovery have been investigated [1-4].

Non-deterministic Information Systems (*NISs*) and Incomplete Information Systems have been proposed for handling information incompleteness in *DISs*

Hiroshi Sakai, Kohei Hayashi, and Hiroshi Kimura
Department of Basic Sciences, Faculty of Engineering, Kyushu Institute of Technology, Tobata, Kitakyushu 804-8550, Japan
e-mail: sakai@mns.kyutech.ac.jp, kimura@mns.kyutech.ac.jp

Michinori Nakata
Faculty of Management and Information Science, Josai International University, Gumyo, Togane, Chiba 283, Japan
e-mail: nakatam@ieee.org

K. Nakamatsu et al. (Eds.): New Advan. in Intel. Decision Techno., SCI 199, pp. 527–536.
springerlink.com

[5-8]. *NISs* have been recognized to be the most important framework for handling information incompleteness in tables, and several theoretical works has been reported. We follow this robust framework, and we have been developing algorithms and software tools, which can handle rough sets based concepts in *NISs*. We are simply calling this work *Rough Non-deterministic Information Analysis* (*RNIA*).

In this paper, we briefly refer to rule generation in *RNIA*, and develop the aspect of decision making in *RNIA*. Throughout this paper, we employ examples instead of giving the details of definitions. The details of definitions are in [11-17].

2 Issues on Rough Non-deterministic Information Analysis

Let us consider Table 1 and Table 2. Table 1 and Table 2 are typical a *DIS* and a *NIS*, respectively.

Table 1 A Deterministic Information System

Patient	*Headache*	*Temperature*	*Flu*
p_1	*no*	*very_high*	*yes*
p_2	*yes*	*very_high*	*yes*
p_3	*no*	*normal*	*no*

Table 2 A Non-deterministic Information System

Patient	*Headache*	*Temperature*	*Flu*
p_1	$\{no\}$	$\{very\_high\}$	$\{yes\}$
p_2	$\{yes, no\}$	$\{high, very\_high\}$	$\{yes\}$
p_3	$\{no\}$	$\{normal, high\}$	$\{yes, no\}$

In a *DIS*, each attribute value is fixed. However, in a *NIS* each attribute value is given as a set. We interpret this set as that the actual value exists in this set but it is unknown due to the information incompleteness. *NISs* were proposed by Pawlak, Orlowska and Lipski in order to handle information incompleteness in *DISs* [5-8].

In *NISs*, the concept of an *extension* has been employed. Namely, it is possible to generate a *DIS* by means of replacing each set with a value in the set. We name such *DISs derived DISs* from a *NIS*. In Table 2, there are 16 derived *DISs*, and Table 1 is a derived *DIS* from a *NIS* in Table 2. The following two modalities, *certainty* and *possibility*, are introduced into *NISs*.

(Certainty). If a formula F holds in each derived *DIS* from a *NIS*, F also holds in the unknown real *DIS*.

(Possibility). If a formula F holds in some derived *DISs* from a *NIS*, there exist such a possibility that F holds in the unknown real *DIS*.

We have coped with several issues related to these two modalities, for example, the definability of a set in *NISs* [9], the consistency of an object in *NISs* [9,10], data dependency in *NISs* [10,11], rules and discernibility functions in *NISs* [12,14], reduction of attributes in *NISs* [11-13], Apriori algorithm in *NISs* [15-17], the manipulation of numerical values in *NISs* [18], etc. The most important problem is how to compute two modalities depending upon all derived *DISs* from a *NIS*. The number of all derived *DISs* increases in exponential order, therefore a simple method, such that every definition is sequentially computed in all derived *DISs* from a *NIS*, is not suitable. We have proposed some rough sets based methods for solving this problem.

3 Rule Generation in Rough Non-deterministic Information Analysis

This section surveys rules and algorithms in rule generation. Then, we show the real execution by an implemented software tool.

3.1 Definitions of Rules in *NISs*

In Table 1, we can pick up a consistent implication $\tau_1 : [Temperature, very\_high] \Rightarrow [Flu, yes]$ from p_1 and p_2. In Table 2, we can also pick up $\tau_2 : [Temperature, high] \Rightarrow [Flu, yes]$ from p_2. The implication τ_1 from p_1 appears in all 16 derived *DISs*, and we say this implication is *definite* in this case. This τ_1 from p_1 is not influenced by the information incompleteness at all, and τ_1 from p_1 is consistent in each derived *DIS*. This seems so characteristic property that we define rules according to this property. On the other hand, τ_1 and τ_2 from p_2 appear in the half of 16 derived *DISs*, and we say these implications are *indefinite*. Here, τ_1 from p_2 is consistent in all 8 derived *DISs*, and τ_2 from p_2 is consistent in 6 derived *DISs*.

Remark 1. For evaluating the validity of τ^x from an object x, we consider a set of derived *DISs* with τ^x. We do not consider any derived *DISs* without τ^x. In the above τ_1^1 from p_1, we consider 16 derived *DISs*, however we consider the half of 16 derived *DISs* for τ_1^2 from p_2. We may employ all derived *DISs* for evaluating τ^x, however we miss some useful implications like τ_1^2 from p_2 in this case. Therefore, we employ a set of derived *DISs* with τ^x for evaluating the validity of τ^x.

In rule generation, we first defined the consistency based rules in the following, and realized software tools.

(Consistency based rules from a *NIS***)**

(1) A *consistency based definite certain rule* is defined by a definite implication τ from a *NIS*, which is consistent in each derived *DIS* from a *NIS*.
(2) A *consistency based indefinite certain rule* is defined by an indefinite implication from a *NIS*, which is consistent in each derived *DIS* with τ from a *NIS*.
(3) A *consistency based possible rule* is defined by a definite or an indefinite implication τ from a *NIS*, which is consistent in some derived *DISs* from a *NIS*.

Then, we also defined the criterion based rules, which depend upon the criteria *support* and *accuracy*. In a *DIS*, a rule is often defined by an implication τ satisfying $support(\tau) \geq \alpha$ and $accuracy(\tau) \geq \beta$ $(0 < \alpha, \beta \leq 1.0)$. We extended this definition to *NISs*, and defined the following.

(Criterion (α,β) based rules from a *NIS*)

(1) A *criterion* (α,β) *based definite certain rule* is defined by a definite implication τ from a *NIS*, which satisfies both $support(\tau) \geq \alpha$ and $accuracy(\tau) \geq \beta$ in each derived *DIS* from a *NIS*.
(2) A *criterion* (α,β) *based indefinite certain rule* is defined by an indefinite implication τ from a *NIS*, which satisfies both $support(\tau) \geq \alpha$ and $accuracy(\tau) \geq \beta$ in each derived *DIS* with τ from a *NIS*.
(3) A *criterion* (α,β) *based possible rule* is defined by a definite or an indefinite implication τ from a *NIS*, which satisfies both $support(\tau) \geq \alpha$ and $accuracy(\tau) \geq \beta$ in some derived *DISs* from a *NIS*.

We have the next proposition, which connects the consistency based rules with the criterion (α,β) based rules.

Proposition 1
(1) A consistency based definite certain rule is a criterion (α,1.0) based definite certain rule.
(2) A consistency based indefinite certain rule is a criterion (α,1.0) based indefinite certain rule.
(3) A consistency based possible rule is a criterion (α,1.0) based possible rule.

In both consistency based rules and criterion based rules, definite certain rules are the most reliable. Possible rules specify implications which may be rules, and any implication except possible rules does not become a rule.

3.2 Some Properties in *NISs*

We have already proved some important properties, and we apply them to criterion (α,β) based rule generation.

Proposition 2
(1) For each implication τ^x from object x, there is a derived DIS_{worst}, where both $support(\tau^x)$ and $accuracy(\tau^x)$ are minimum. Furthermore, there exist formulas calculating these minimum values (minsupp and minacc), and this calculation does not depend upon the number of derived DISs.
(2) For each implication τ^x from object x, there is a derived DIS_{best}, where both $support(\tau^x)$ and $accuracy(\tau^x)$ are maximum. Furthermore, there exist formulas calculating these maximum values (maxsupp and maxacc), and this calculation also does not depend upon the number of derived DISs.

According to Proposition 2, the definition of criterion (α,β) based rules is equivalently translated to the following.

(Criterion (α,β) based rules from a *NIS*): Revised equivalently

(1) A *criterion* (α,β) *based definite certain rule* is a definite implication τ from a *NIS*, which satisfies both $minsupp(\tau) \geq \alpha$ and $minacc(\tau) \geq \beta$.
(2) A *criterion* (α,β) *based indefinite certain rule* is an indefinite implication τ from a *NIS*, which satisfies both $minsupp(\tau) \geq \alpha$ and $minacc(\tau) \geq \beta$.
(3) A *criterion* (α,β) *based possible rule* is a definite or an indefinite implication from a *NIS*, which satisfies both $maxsupp(\tau) \geq \alpha$ and $maxacc(\tau) \geq \beta$.

In this revision, criterion (α,β) based rule generation does not depend upon the number of all derived *DISs*. Like this, we can escape from the computational issue.

Now, we briefly refer to the method to calculate *minsupp*, *minacc*, *maxsupp* and *maxacc*. We employ the following two sets for attributes A_i and attribute values $\zeta_{i,j}$.
$Descinf([A_i,\zeta_{i,j}])$: A set of objects, whose attribute value is definite and $\zeta_{i,j}$.
$Descinf(\wedge_i[A_i,\zeta_{i,j}])=\cap_i Descinf([A_i,\zeta_{i,j}])$.
$Descsup([A_i,\zeta_i])$: A set of objects, which include a value $\zeta_{i,j}$.
$Descsup(\wedge_i[A_i,\zeta_i])=\cap_i Descsup([A_i,\zeta_{i,j}])$.
Descinf and *Descsup* are the minimum and the maximum sets for an equivalence class, respectively. Clearly, $Descinf(\wedge_i[A_i,\zeta_{i,j}]) \subseteq Descsup(\wedge_i[A_i,\zeta_{i,j}])$ holds. The unknown real equivalence class by $\wedge_i[A_i,\zeta_{i,j}]$ is between $Descinf(\wedge_i[A_i,\zeta_{i,j}])$ and $Descsup(\wedge_i[A_i,\zeta_{i,j}])$.

According to the above two sets $Descinf([A_i,\zeta_{i,j}])$ and $Descsup([A_i,\zeta_{i,j}])$, we can calculate most of definitions. For example, we can obtain the minimum and the maximum criterion values of an implication τ.
If $\tau : [CON,\zeta] \Rightarrow [DEC,\eta]$ is definite,
$minsupp(\tau)=|Descinf([CON,\zeta]) \cap Descinf([DEC,\eta])|/|OB|$,
$minacc(\tau)=\frac{|Descinf([CON,\zeta]) \cap Descinf([DEC,\eta])|}{|Descinf([CON,\zeta])|+|OUTACC|}$,
$(OUTACC=[Descsup([CON,\zeta]) - Descinf([CON,\zeta])] - Descinf([DEC,\eta])$,
$maxsupp(\tau)=|Descsup([CON,\zeta]) \cap Descsup([DEC,\eta])|/|OB|$,
$maxacc(\tau)=\frac{|Descinf([CON,\zeta]) \cap Descsup([DEC,\eta])|+|INACC|}{|Descinf([CON,\zeta])|+|INACC|}$.
$(INACC=[Descsup([CON,\zeta]) - Descinf([CON,\zeta])] \cap Descsup([DEC,\eta]))$.

If τ is indefinite, four adjusted formulas can be similarly derived. As for the definite τ_1 from object p_1 in Table 2, the following holds.
$Descinf([Temperature, very\_high])=\{p_1\}$,
$Descsup([Temperature, very\_high])=\{p_1,p_2\}$,
$Descinf([Flu, yes])=\{p_1,p_2\}$, $Descsup([Flu, yes])=\{p_1,p_2,p_3\}$,
$minsupp(\tau_1)=(|\{p_1\} \cap \{p_1,p_2\}|)/3=1/3$,
$OUTACC=[\{p_1,p_2\} - \{p_1\}] - \{p_1,p_2\}=\{\}$,
$minacc(\tau_1)=(|\{p_1\} \cap \{p_1,p_2\}|)/(|\{p_1\}|+|\{\}|)=1.0$,
$maxsupp(\tau_1)=(|\{p_1,p_2\} \cap \{p_1,p_2,p_3\}|)/3=2/3$,

$INACC=[\{p_1,p_2\}-\{p_1\}]\cap\{p_1,p_2,p_3\}=\{p_2\}$,
$maxacc(\tau_1)=(|\{p_1\}\cap\{p_1,p_2,p_3\}|+|\{p_2\}|)/(|\{p_1\}|+|\{p_2\}|)=1.0$.
Since $minacc(\tau_1)=1.0$ holds, τ_1^1 is a consistency based definite certain rule, i.e., τ_1^1 from object p_1 appears in all derived *DISs* and τ_1^1 is consistent in each derived *DIS*.

We have to remark that the definitions of four criterion values depend upon all derived *DISs*, whose number increases in exponential order. However, each formula does not depend upon the number of all derived *DISs*.

3.3 An Algorithm for Criterion (α,β) Based Rule Generation

We follow *Apriori* algorithm in transaction data [19,20], and extend it to an algorithm in *NISs*. *Apriori* algorithm employs large item sets, which correspond to equivalence classes for descriptors, and *support* and *accuracy* values are calculated in order to discriminate rules from lots of implications. On the other hand, we employ two sets for each descriptor, i.e., *Descinf*, *Descsup* and the criterion values like *minsupp* and *minacc* are calculated. Since the calculation of criterion values does not depend upon the number of derived *DISs*, the complexity of the extended algorithm is almost the same as *Apriori* algorithm.

An Overview of Criterion (α,β) Based Certain Rule Generation

(1) For the condition $minsupp(\tau)=|SET|/|OB|\geq\alpha$, obtain the number NUM of elements in SET satisfying this condition.
(2) Generate conjunctions of descriptors.
(3) For every conjunction of descriptors $[CON,\zeta]\wedge[DEC,\eta]$ $(\tau:[CON,\zeta]\Rightarrow[DEC,\eta])$, if $minsupp(\tau)\geq\alpha$ and $minacc(\tau)\geq\beta$, τ is a certain rule. Remove such τ. Otherwise, pick up candidates of conjunctions satisfying $minsupp\geq\alpha$, and repeat (2) and (3) until there is no conjunctions.

The following is the real execution of criterion (0.3,0.8) based certain rule generation in Table 2.

```
===== Lower Approximation Strategy =====
CAN(1)={[Head,no],[Head,yes],[Temp,high],[Temp,normal],
  [Temp,very_high],[Flu,no],[Flu,yes]}(7)
CAN(2)={[Head,no][Flu,no](<DEF>0.000,<INDEF>0.333),
  [Temp,high][Flu,no](<DEF>0.000,<INDEF>0.500),
  [Temp,normal][Flu,no](<DEF>0.000,<INDEF>1.000),
  [Head,no][Flu,yes](<DEF>0.500,<INDEF>1.000),
          :      :      :      :
  [Temp,very_high][Flu,yes](<DEF>1.000,<INDEF>1.000)}(8)
  ---------- OBTAINED RULE ----------
  [Temp,normal]=>[Flu,no]
    minsupp<DEF>=0.000,minsupp<INDEF>=0.333,minacc<DEF>=0.000,
    minacc<INDEF>=1.000,(<DEF>from )(<INDEF>from 3)
  [Head,no]=>[Flu,yes]
    minsupp<DEF>=0.333,minsupp<INDEF>=0.667,minacc<DEF>=0.500,
    minacc<INDEF>=1.000,(<DEF>from )(<INDEF>from 2,3)
```

```
[Temp,high]=>[Flu,yes]
  minsupp<DEF>=0.000,minsupp<INDEF>=0.333,minacc<DEF>=0.000,
  minacc<INDEF>=1.000)(<DEF>from )(<INDEF>from 2,3)
      :     :     :     :
[Temp,very_high]=>[Flu,yes]
  minsupp<DEF>=0.333,minsupp<INDEF>=0.667,minacc<DEF>=1.000,
  minacc<INDEF>=1.000,(<DEF>from 1)(<INDEF>from 2)
CAN(3)={[Head,no][Temp,high][Flu,no](<DEF>0.000,<INDEF>0.500)}(1)
EXEC_TIME=0.000(sec)
```

In the above execution, an indefinite implication $\tau : [Temp, normal] \Rightarrow [Flu, no]$ from p_3 satisfies $minsupp(\tau) = 0.333 \geq 0.3$ and $minacc(\tau) = 1.000 \geq 0.8$. Therefore, τ is a criterion (0.3,0.8) based indefinite certain rule. On the other hand, $[Temperature, very\_high] \Rightarrow [Flu, yes]$ from p_1 is a criterion (0.3,0.8) based definite certain rule.

4 Decision Making in Rough Non-deterministic Information Analysis

Obtained rules characterize the tendencies of data, and they are often applied to decision making. In reality, we applied our software tools to a revised mammographic data (150 objects, 5 attributes:$\{Age, Shape, Margin, Density, Severity\}$, Decision attribute: *Severity*, *support*=0.3, *accuracy*=0.5) in UCI machine learning repository, and we obtained the following implications.
(1) Criterion (0.3,0.5) based definite certain rules:

```
[Margin,1]⇒[Severity,0].
```

(2) Criterion (0.3,0.5) based possible rules:

```
[Density,3]⇒[Severity,0],[Density,3]⇒[Severity,1].
```

Even though two possible rules are inconsistent, this may occur. Because, there exists a derived *DIS*, where two criterion values satisfy the condition.

These implications are very characteristic, however if the condition parts in such rules are not satisfied, obtained rules are not applied to decision making. In reality, we do not know that what does $[Shape, 1]$ conclude. For handling such cases, we newly added a *question-answering* with *criterion values* to *RNIA*. We specify the conditions, and we directly obtain the decision attribute values with criterion values. According to the criterion values of each decision, we see the validity of the decision. Let us consider some real cases.

(CASE 1)

```
Condition:[Shape,1]
Decision 1:[Severity,1]
  minsupp=9/150=0.060,minacc=9/48=0.181
  maxsupp=13/150=0.087,maxacc=13/48=0.270
Decision 2:[Severity,0]
  minsupp=35/150=0.233,minacc=35/48=0.729
  maxsupp=39/150=0.260,maxacc=39/48=0.813
```

There are two decisions, [*Severity*, 1] and [*Severity*, 0]. In this case, [*Severity*, 0] is probably concluded from the criterion values. Because, every value in Decision 2 is uniquely better than that in Decision 1. In reality, an implication [*Shape*, 1] $\Rightarrow$ [*Severity*, 0] is a criterion (0.2,0.7) based certain rule. If we employ lower values of α and β for generating rules, the question-answering with criterion values may not be necessary. However, lots of improper rules may exist in this case.

```
(CASE 2)
  Condition:[Density,3]
  Decision 1:[Severity,1]
    minsupp=45/150=0.300,minacc=45/115=0.391
    maxsupp=55/150=0.367,maxacc=55/88=0.625
  Decision 2:[Severity,0]
    minsupp=33/150=0.220,minacc=33/88=0.375
    maxsupp=70/150=0.467,maxacc=70/115=0.609
```

In CASE 2, it may be difficult to decide the decision according to the criterion values. In this case, 33 percent of the 150 objects are Imp_1 : [*Density*, ?] $\Rightarrow$ [*Severity*, 1] or Imp_2 : [*Density*, ?] $\Rightarrow$ [*Severity*, 0]. If we replace ? in Imp_1 with 3 and ? in Imp_2 with 2, [*Severity*, 1] will be concluded. Because, both *support* and *accuracy* values of Imp_1 are much better than those of Imp_2. On the other hand, if we replace ? in Imp_1 with 2 and ? in Imp_2 with 3, [*Severity*, 0] will be concluded. Like this, CASE 2 shows that the conclusion strongly depends upon the information incompleteness.

```
(CASE 3)
  Condition:[Shape,2]∧[Margin,2]
  Decision 1:[Severity,1]
    minsupp=0/150=0.000,minacc=0/6=0.000
    maxsupp=2/150=0.013,maxacc=2/2=1.000
  Decision 2:[Severity,0]
    minsupp=0/150=0.000,minacc=0/2=0.000
    maxsupp=6/150=0.040,maxacc=6/6=1.000
```

In CASE 3, it may also be difficult to decide the decision according to the criterion values. There exist less objects with the condition [*Shape*, 2] $\wedge$ [*Margin*, 2]. In both cases, *maxacc*=1.0 holds, but the *support* values are too low.

According to three cases, we showed the aspect of decision making in *RNIA*. The following is a real execution for a condition [*Margin*, 1] in 961 total objects. The *support* and *accuracy* values of [*Severity*, 0] are much better than those of [*Severity*, 1].

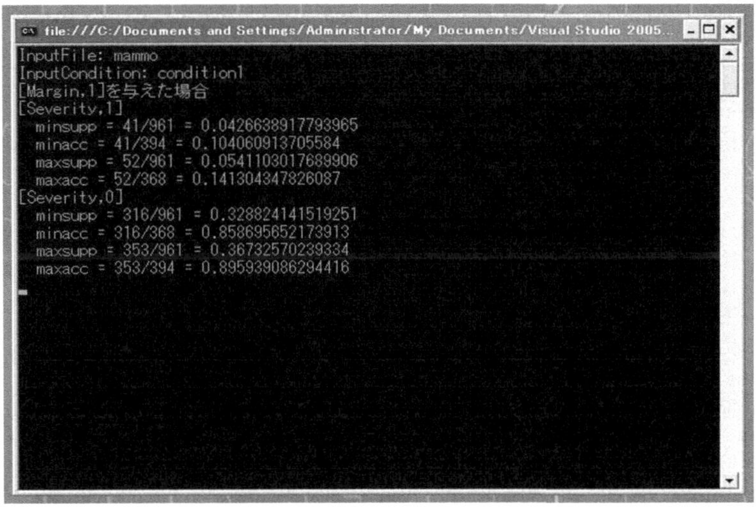

Fig. 1 Real execution for the condition [Margin,1]

5 Concluding Remarks

This paper briefly surveyed the framework of *RNIA*, and clarified the aspects of decision making in *RNIA*. Decision making in *RNIA* is depending upon all derived *DISs* from a *NISs*, and we see that each derived *DIS* satisfying some constraints is an *evidence* for a decision. According to a set of evidences, we make a decision. If we employ a sequential method depending upon each derived *DIS*, this method depends upon the number of derived *DISs*. Therefore, a computational problem occurs, however we employ a method of granular computing, i.e., the manipulation of *Descinf* and *Descsup*, and solved this computational problem.

We have just started this *evidence based decision making* in *RNIA*, which will be different from statistical decision making nor fuzzy decision making. This will be a new framework for decision making from incomplete information.

References

1. Pawlak, Z.: Rough Sets. Kluwer Academic Publishers, Dordrecht (1991)
2. Pawlak, Z.: Some Issues on Rough Sets. Transactions on Rough Sets 1, 1–58 (2004)
3. Komorowski, J., Pawlak, Z., Polkowski, L., Skowron, A.: Rough Sets: a tutorial. Rough Fuzzy Hybridization, 3–98 (1999)
4. Rough Set Software. Bulletin of Int'l. Rough Set Society 2, 15–46 (1998)
5. Orlowska, E., Pawlak, Z.: Representation of Nondeterministic Information. Theoretical Computer Science 29, 27–39 (1984)
6. Orlowska, E. (ed.): Incomplete Information: Rough Set Analysis. Physica-Verlag (1998)
7. Lipski, W.: On Semantic Issues Connected with Incomplete Information Data Base. ACM Transaction on DBS 4, 269–296 (1979)

8. Lipski, W.: On Databases with Incomplete Information. Journal of the ACM 28, 41–70 (1981)
9. Sakai, H.: Effective Procedures for Handling Possible Equivalence Relations in Non-deterministic Information Systems. Fundamenta Informaticae 48, 343–362 (2001)
10. Sakai, H.: Effective Procedures for Data Dependencies in Information Systems. In: Rough Set Theory and Granular Computing. Studies in Fuzziness and Soft Computing, vol. 125, pp. 167–176. Springer, Heidelberg (2003)
11. Sakai, H., Okuma, A.: Basic Algorithms and Tools for Rough Non-deterministic Information Analysis. Transactions on Rough Sets 1, 209–231 (2004)
12. Sakai, H., Nakata, M.: Rough sets based Minimal Certain Rule Generation in Non-deterministic Information Systems: An Overview. Frontiers in Artificial Intelligence and Applications, vol. 132, pp. 256–263. IOS Press, Amsterdam (2005)
13. Sakai, H.: On a rough sets based data mining tool in prolog: An overview. In: Umeda, M., Wolf, A., Bartenstein, O., Geske, U., Seipel, D., Takata, O. (eds.) INAP 2005. LNCS, vol. 4369, pp. 48–65. Springer, Heidelberg (2006)
14. Sakai, H., Nakata, M.: An Application of Discernibility Functions to Generating Minimal Rules in Non-deterministic Information Systems. Journal of Advanced Computational Intelligence and Intelligent Informatics 10, 695–702 (2006)
15. Sakai, H., Ishibashi, R., Koba, K., Nakata, M.: On Possible Rules and Apriori Algorithm in Non-deterministic Information Systems: Part 2. In: An, A., Stefanowski, J., Ramanna, S., Butz, C.J., Pedrycz, W., Wang, G. (eds.) RSFDGrC 2007. LNCS, vol. 4482, pp. 280–288. Springer, Heidelberg (2007)
16. Sakai, H., Ishibashi, R., Nakata, M.: Lower and Upper Approximations of Rules in Non-deterministic Information Systems. In: Chan, C.-C., Grzymala-Busse, J.W., Ziarko, W.P. (eds.) RSCTC 2008. LNCS, vol. 5306, pp. 299–309. Springer, Heidelberg (2008)
17. Sakai, H., Ishibashi, R., Koba, K., Nakata, M.: Rules and Apriori Algorithm in Non-deterministic Information Systems. Transactions on Rough Sets 9 (accepted)
18. Sakai, H., Koba, K., Nakata, M.: Rough Sets Based Rule Generation from Data with Categorical and Numerical Data. Journal of Advanced Computational Intelligence and Intelligent Informatics 12(5), 426–434 (2008)
19. Agrawal, R., Srikant, R.: Fast Algorithms for Mining Association Rules. In: Proc. 20th Very Large Data Base, pp. 487–499 (1994)
20. Agrawal, R., Mannila, H., Srikant, R., Toivonen, H., Verkamo, A.: Fast Discovery of Association Rules. In: Advances in Knowledge Discovery and Data Mining, pp. 307–328. AAAI/MIT Press (1996)

A Value for Multi-alternative Games with Restricted Coalitions under the Equally Divided Spoils Assumption

Satoshi Masuya, Masahiro Inuiguchi, and Teruhisa Nakai

Abstract. This paper deals with cooperative games with n players and r alternatives which are called multi-alternative games with restricted choice situations. In these games, a value based on marginal contributions has been proposed. Many well-known values such as the Shapley value and the Banzhaf value are based on marginal contributions. On the other hand, some values such as the Deegan-Packel value are based on equally divided payoffs. Then, in this paper, we investigate a value based on equally divided payoffs for multi-alternative games with restricted choice situations.

Keywords: game theory, cooperative game, multi-alternative game, restricted game.

1 Introduction

The cooperative game theory provides useful tools to analyze cost allocation, voting power, and so on. The problems analyzed by the cooperative game theory include n entities called players and are usually expressed by characteristic functions called games which map each subset of players to a real number. The real number shows a profit or a cost when all players in the subset cooperate. The solutions to the problems are given by value functions which assign a real number to each player. The real number called a value can show the cost borne by the player, power of

Satoshi Masuya and Masahiro Inuiguchi
Graduate School of Engineering Science, Osaka University
1-3 Machikaneyama, Toyonaka, Osaka 560-8531, Japan
e-mail: masuya@inulab.sys.es.osaka-u.ac.jp,
inuiguti@sys.es.osaka-u.ac.jp

Teruhisa Nakai
Faculty of Environmental and Urban Engineering, Kansai University
3-3-35 Yamate, Suita, Osaka 564-8680, Japan
e-mail: nakai@iecs.kansai-u.ac.jp

K. Nakamatsu et al. (Eds.): New Advan. in Intel. Decision Techno., SCI 199, pp. 537–546.
springerlink.com

influence, and so on depending on the problem setting. Several value functions have been proposed. As representative examples of value functions, the Shapley value [8] and the Banzhaf value [1, 5] are well-known. Reasonable axiom systems characterizing those values uniquely are shown in literature [8, 5].

In the conventional cooperative games, each player can take one from two options: cooperate and non-cooperate. However, in the real world problems, we may face a decision problem to choose one from several options. From this point of view, it is worthwhile to treat cooperative games in which each player has r options. Then multi-alternative games also called games with r alternatives have been proposed by Bolger [2]. A multi-alternative game is expressed by a generalized characteristic function which maps an arrangement showing all players' choices to an r-dimensional real vector. Bolger [2] proposed a generalized value function which maps a multi-alternative game to an n-dimensional real vector whose i-th component shows the value of player i. This function is a generalization of the Shapley function. On the other hand, Ono [7] proposed a multi-alternative Banzhaf value (an MBZ value) as a generalization of the Banzhaf value.

The value functions/generalized value functions described above are considered under the assumption that all coalitions/arrangements are formed with equal possibilities. In the real world, there are many cases when this assumption does not hold. For example, when each player has his/her own ideology, he/she would be difficult to cooperate with players having totally different ideologies. Moreover, when a certain license is necessary to choose an option in a multi-alternative game, players without the licenses cannot choose it and then some arrangements cannot be realized.

Considering such restrictions on coalition forming, Masuya and Inuiguchi [6] investigated restricted multi-alternative games, more specifically, multi-alternative games with *restricted choice situations*. In the situation, some alternatives cannot be chosen by certain players. They [6] proposed a value based on marginal contributions for these games and succeeded in giving an axiom system which characterizes it uniquely.

While many well-known values such as the Shapley value and the Banzhaf value are based on marginal contributions, some other values such as the Deegan-Packel value [4] are not. The Deegan-Packel value is proposed under three fundamental assumptions with respect to simple games. The most characteristic assumption for the Deegan-Packel value is the equally divided spoils assumption. The equally divided spoils assumption is supported by some experimental evidence [3, 10].

In this paper, we propose a value for multi-alternative games with restricted choice situations under the equally divided spoils assumption. Further, we give an axiom system for the proposed value. The system is composed of four axioms concerning zero players, additivity, symmetry, and sharing the expected payoff of a coalition.

In Section 2, we introduce the multi-alternative games with restricted choice situations and a value based on marginal contributions which are proposed by Masuya and Inuiguchi [6]. In Section 3, we propose an explict function as a value for the restricted multi-alternative games under equally divided spoils assumption. Further, some concepts for characterizing the proposed value are given. In Section 4, we show an axiom system characterizing the proposed value uniquely. In Section 5, we give a numerical example and compare the proposed value and the value based on marginal contributions.

2 Multi-alternative Games with Restricted Choice Situations and a Value for These Games

In this section, we introduce the multi-alternative games with restricted choice situations and a value for these games proposed by Masuya and Inuiguchi [6].

First, we provide the conventional multi-alternative games (games with r alternatives) proposed originally by Bolger [2] and extended by Tsurumi et al. [9].

Let $N = \{1,\ldots,n\}$ be the set of players and $R = \{1,\ldots,r\}$ the set of alternatives. Let Γ_j be the set of players who have chosen the alternative $j \in R$. A finite sequence of subsets of players, $\Gamma = (\Gamma_1,\ldots,\Gamma_r)$, is called an arrangement. Each arrangement Γ satisfies $\Gamma_1 \cup \cdots \cup \Gamma_r \subseteq N$ and $\Gamma_k \cap \Gamma_l = \emptyset$ ($\forall k \neq l$). Let Γ_0 be a subset of players who have chosen none of alternatives. In the original multi-alternative games treated by Bolger [2], Γ_0 is assumed to be empty. Then we have $\Gamma_0 = N - \bigcup_{k\in R} \Gamma_k$. For the sake of convenience, we define $R_0 = \{0,1,\ldots,r\}$. We denote $\exists k \in R$ such that $S = \Gamma_k$ by $S \in \Gamma$. For any $S \in \Gamma$, we call (S,Γ) an embedded coalition (ECL). Let $E(N,R)$ be the set of ECLs and $A(N,R)$ the set of arrangements on N and R. Then a function $v : E(N,R) \to \mathbb{R}$ satisfying $v(\emptyset,\Gamma) = 0$ is called an extended multi-alternative game on N with r alternatives. Let $MG(N,R)$ be the set of extended multi-alternative games on N and R.

Example 1 (Job Selection Game)
Three students A, B and C are considering to work part-time. There are two jobs 1 and 2 but students cannot take both. Then each student can take one job or nothing. They can take the same job. If only two students would take different jobs, the remaining student would not get any payoff but the students taking jobs would get some payoffs independently. The payoff does not depend on the job taken but on the student taking a job. The payoffs of students A, B and C would be 8, 6 and 4 units, respectively. If student A would work alone while students B and C would make the same choice, independent of the job taken by A, student A would get 5 units as a payoff. If students B and C would work together while student A would not work with them, independent of the job taken by them, students B and C would get 18 units as the total payoff. If student B would work alone while students A and C would make the same choice, independent of the job taken by B, student B would get 3 units as a payoff. If students A and C would work together while student B would not work with them, independent of the job taken by them, students A and C would get 25 units as the total payoff. If student C would work alone while students A and B would make the same choice, independent of the job taken by C, student C would get 1 unit as a payoff. If students A and B would work together while student C would not work with them, independent of the job taken by them, students A and

B would get 30 units as the total payoff. If all students A, B and C would work together, independent of the job taken, they would get 50 units as the total payoff.

This game can be represented by the following multi-alternative game v with $N = \{A,B,C\}$ and $R = \{1,2\}$:

$$
\begin{aligned}
&v(\{A\},\Gamma) = 8, \text{ for } \Gamma \text{ such that } \{A\} \in \Gamma \text{ and } |\Gamma_i| = 1, i = 1,2,\\
&v(\{B\},\Gamma) = 6, \text{ for } \Gamma \text{ such that } \{B\} \in \Gamma \text{ and } |\Gamma_i| = 1, i = 1,2,\\
&v(\{C\},\Gamma) = 4, \text{ for } \Gamma \text{ such that } \{C\} \in \Gamma \text{ and } |\Gamma_i| = 1, i = 1,2,\\
&v(\{A\},\Gamma) = 5, \text{ for } \Gamma \text{ such that } \{A\} \in \Gamma \text{ and } (\{B,C\} \in \Gamma \text{ or } \emptyset \in \Gamma),\\
&v(\{B\},\Gamma) = 3, \text{ for } \Gamma \text{ such that } \{B\} \in \Gamma \text{ and } \{A,C\} \in \Gamma \text{ or } \emptyset \in \Gamma),\\
&v(\{C\},\Gamma) = 1, \text{ for } \Gamma \text{ such that } \{C\} \in \Gamma \text{ and } \{A,B\} \in \Gamma \text{ or } \emptyset \in \Gamma),\\
&v(\{A,B\},\Gamma) = 30, \text{ for } \Gamma \text{ such that } \{A,B\} \in \Gamma,\\
&v(\{A,C\},\Gamma) = 25, \text{ for } \Gamma \text{ such that } \{A,C\} \in \Gamma,\\
&v(\{B,C\},\Gamma) = 18, \text{ for } \Gamma \text{ such that } \{B,C\} \in \Gamma,\\
&v(N,\Gamma) = 50, \text{ for } \Gamma \text{ such that } N \in \Gamma,\\
&v(\emptyset,\Gamma) = 0, \text{ for } \Gamma \text{ such that } \emptyset \in \Gamma,
\end{aligned}
$$

where $|\Gamma_i|$ is the cardinality of Γ_i.

Now we introduce the restrictions on choice of alternatives. In the conventional multi-alternative games, each player can choose any alternative from a given set of alternatives. However, in the real world, there exists a situation where some alternatives cannot be chosen by all players. For example, in Job Selection Game described above, some students cannot take some job due to their inabilities or conflicts with regular lessons. In order to treat such situations, restricted games with r alternatives (restricted multi-alternative games) have been proposed.

Let W be the set of arrangements which can be formed. Namely, W is the set of feasible arrangements. Let R_i be the set of alternatives which player $i \in N$ can choose. Obviously, we have $R_i \subseteq R_0$ and $R_i \ni 0, \ \forall i \in N$. Especially, $R_i = R_0$ holds if player i can choose any alternatives and $R_i = \{0\}$ holds if player i can choose none of alternatives. Then W can be regarded as the sequence of $(R_1,\ldots,R_n)$. We call the set W of feasible arrangements *a restricted choice situation*. Let $AR(N,R)$ be the set of restricted choice situations. We define a multi-alternative game with a restricted choice situation as a pair (v,W) where $v \in MG(N,R)$ and $W \in AR(N,R)$. An extended multi-alternative game can be represented by $(v,A(N,R))$.

In order to describe a value based on marginal contributions for multi-alternative games with restricted choice situations, we define a set of arrangements.

Definition 1. Given $W \in AR(N,R)$, we define $W_{i,j}$ by

$$W_{i,j} = \{\Gamma \in W \mid i \in \Gamma_j\}.$$

The value based on marginal contributions is defined by the functions g^j : $MG(N,R) \rightarrow (\mathbb{R}^n)^{AR(N,R)}$, $j = 1,\ldots,r$ with the following i-th component:

$$g_i^j(v)(W) = \begin{cases} \sum_{\substack{\Gamma \in W, \\ \Gamma_j \ni i}} \sum_{\substack{k \in R_0 - \{j\}, \\ \Gamma^{i \to k} \in W}} \frac{1}{|W|}[v(\Gamma_j, \Gamma) - v(\Gamma_j - \{i\}, \Gamma^{i \to k})], & \text{if } W_{i,j} \neq \emptyset, \\ 0, & \text{otherwise,} \end{cases} \tag{1}$$

where $\Gamma^{i \to k}$ is the arrangement obtained by moving $i \in \Gamma_j$ from Γ_j to Γ_k ($k \in R_0 - \{j\}$).

The function defined by (1) is interpreted as follows. The term $v(\Gamma_j, \Gamma) - v(\Gamma_j - \{i\}, \Gamma^{i \to k})$ can be interpreted as the marginal contribution of player i to Γ_j. $|W|$ shows the number of feasible arrangements. Therefore, the weight $\frac{1}{|W|}$ means that each feasible arrangement is formed with equal probability. Then $g_i^j(v)(W)$ is the expected value of the marginal contribution of player i to alternative j in restricted game (v, W).

For the comparison with the axiom system with respect to the proposed value in this paper, we describe the axiom system with respect to the value defined by (1). To this end, we introduce a few notions in multi-alternative games with restricted choice situations.

Definition 2. Given $v, w \in MG(N, R)$, we define the sum game $v + w$ by

$$(v + w)(\Gamma_j, \Gamma) = v(\Gamma_j, \Gamma) + w(\Gamma_j, \Gamma), \quad \forall (\Gamma_j, \Gamma) \in E(N, R).$$

Definition 3 (*j-null player* for restricted multi-alternative games [6]). Let $v \in MG(N, R)$, $W \in AR(N, R)$, $i \in N$ and $j \in R$. Player i is called *a j-null player* on (v, W) if and only if the following holds:

$$\text{if } W_{i,j} \neq \emptyset \text{ then } v(\Gamma_j, \Gamma) - v(\Gamma_j - \{i\}, \Gamma^{i \to k}) = 0, \quad \forall \Gamma \in W_{i,j}, k \in R_0 - \{j\}, \Gamma^{i \to k} \in W.$$

Note that if $W_{i,j} = \emptyset$, Player i is a j-null player.

Definition 4. Let $v \in MG(N, R)$, $W \in AR(N, R)$ and $i \in N$. Then player i is called an *unrelated player* if $R_i = \{0\}$.

Unrelated players are the players who cannot choose any alternatives.

An axiom system of the value g^j defined by (1) is given by Masuya and Inuiguchi [6]. The system is composed of the four axioms described below. The first two axioms are employed to those of the Bolger value [2] and the MBZ value [7].

Let π^j be a vector function from $MG(N, R)$ into $(\mathbb{R}^n)^{AR(N,R)}$. The i-th component of π^j is denoted by π_i^j. Note that for any $v, w \in MG(N, R)$, $v + w \in MG(N, R)$ holds.

Axiom 1 (j-null player). *Given $v \in MG(N, R)$, $i \in N$, $j \in R$ and $W \in AR(N, R)$, the following holds:*

$$\pi_i^j(v)(W) = 0 \Leftrightarrow i \text{ is a } j\text{-null player on } (v, W)$$

Axiom 2 (Linearity). *Given $v_1, v_2 \in MG(N, R)$ $c_1, c_2 \in \mathbb{R}$ and $W \in AR(N, R)$, the following holds:*

$$\pi^j(c_1 v_1 + c_2 v_2)(W) = c_1 \pi^j(v_1)(W) + c_2 \pi^j(v_2)(W),\ j = 1, \dots, r$$

Axiom 3 (Independence from unrelated players). *Let $v \in MG(N,R)$ and $W \in AR(N,R)$. Let us add an unrelated player $n+1$ to the set of players N, and we denote v' the $(n+1)$-person game. Then, the following holds:*

$$\pi_i^j(v', W) = \pi_i^j(v, W),\quad \forall i \in N,\quad \forall j \in R.$$

Axiom 4 described in what follows is a property with respect to voting games [2]. In a voting game, $v(\Gamma_j, \Gamma)$ takes 0 or 1. Γ_j such that $v(\Gamma_j, \Gamma) = 1$ is called a winning coalition while Γ_j such that $v(\Gamma_j, \Gamma) = 0$ is called a losing coalition.

If Γ_j is changed from a winning coalition to a losing coalition by player i's moving from Γ_j to Γ_k $(k \in R_0 - \{j\})$, the movement is called a negatively influential movement under arrangement Γ. On the contrary, if Γ_j changed from a losing coalition to a winning coalition by player i's moving from Γ_j to Γ_k $(k \in R_0 - \{j\})$, the movement is called a positively influential movement under arrangement Γ. Under an arrangement Γ, let $M_{i,j}^-(\Gamma)$ be the number of negatively influential movements of player i from Γ_j and $M_{i,j}^+(\Gamma)$ the number of positively influential movements of player i from Γ_j. Using $M_{i,j}^-(\Gamma)$ and $M_{i,j}^+(\Gamma)$, we define $M_{i,j} = \sum_{\Gamma \in W}(M_{i,j}^-(\Gamma) - M_{i,j}^+(\Gamma))$. Then Axiom 4 is given as follows.

Axiom 4 (Property of voting games). *Let $W \in AR(N,R)$, $\Gamma \in W$, $i,k \in N$ and $j \in R$, and let v be a voting game with r alternatives. If $M_{i,j} - M_{k,j} = l$, we have*

$$\pi_i^j(v)(W) = \pi_k^j(v)(W) + \frac{l}{|W|}.$$

Then we have the following theorem which shows value g^j defined by (1) is the unique function satisfying all those four axioms.

Theorem 1 (Masuya and Inuiguchi [6]). *Function g^j, $j = 1, \dots, r$ defined by (1) is the unique function which satisfies Axiom 1 through Axiom 4.*

3 The Proposed Value

In this section, we propose a value for multi-alternative games with restricted choice situations under equally divided spoils assumption. Moreover, we introduce some notions used for characterizing the proposed value.

We define a function $f^j : MG(N,R) \to (\mathbb{R}_+^n)^{AR(N,R)}$ $(j = 1, \dots, r)$ by its i-th component,

$$f_i^j(v)(W) = \begin{cases} \displaystyle\sum_{\substack{\Gamma \in W, \\ \Gamma_j \ni i}} \frac{1}{|W|} \cdot \frac{v(\Gamma_j, \Gamma)}{|\Gamma_j|}, & \text{if } W_{i,j} \neq \emptyset, \\ 0, & \text{otherwise.} \end{cases} \tag{2}$$

The interpretation of this function is given as follows. The term $\frac{v(\Gamma_j,\Gamma)}{|\Gamma_j|}$ means that the coalition Γ_j divides their common payoff equally. $|W|$ shows the number of feasible arrangements. Therefore, the weight $\frac{1}{|W|}$ means that each feasible arrangement is formed with equal probability. Then $f_i^j(v)(W)$ is the expected value of equally divided spoils that player i may obtain by the selection of alternative j in restricted game (v,W).

Definition 5. Given $W \in AR(N,R)$, $v \in MG(N,R)$ and $j \in R$, we define $\bar{v}_j(W)$ as follows:

$$\bar{v}_j(W) = \frac{1}{|W|} \sum_{\Gamma \in W} v(\Gamma_j,\Gamma). \tag{3}$$

$\bar{v}_j(W)$ is understood as the expected payoff which the coalition choosing the jth alternative gets.

Definition 6 (*j-zero player* for restricted multi-alternative games). Let $v \in MG(N,R), W \in AR(N,R)$ satisfying $W_{i,j} \neq \emptyset$ for a player $i \in N$ and an alternative $j \in R$. If $v(\Gamma_j,\Gamma) = 0 \ \ \forall \Gamma \in W$ satisfying $\Gamma_j \ni i$ holds , i is called a *j-zero player* on (v,W). Note that if $W_{i,j} = \emptyset$, Player i is a j-zero player.

This concept is a generalization of *zero players* defined by Deegan and Packel [4]. The j-zero players for restricted multi-alternative games coincide with zero players when $r = 1$ and $R_i = \{0,1\} \ \ \forall i \in N$, i.e., there is no restriction on player's choice.

The j-null players defined in the previous section are the players who have "no marginal contributions" in all coalitions while the j-zero players are the players who can participate only to zero value coalitions. Therefore, j-null players would be more influential to their coalitions than j-zero players.

Definition 7 (symmetry). Let $W \in AR(N,R)$ and two players $i, \ k \in N$ satisfying $R_i = R_k$. Interchanging i with k in any $\Gamma \in W$, we make the new arrangement $\Gamma' \in W$. Two players i and k are called *symmetric* in the game (v,W) for alternative j if and only if

$$v(\Gamma_j,\Gamma) = v(\Gamma'_j,\Gamma') \ \ \forall \Gamma \in W. \tag{4}$$

Definition 8 (monotonicity). (v,W) is said to be monotone if and only if v satisfies

$$v(\Gamma_j,\Gamma) \geq v(\Gamma_j - \{i\},\Gamma^{i\to k}), \forall i \in \Gamma_j, \forall k \text{ such that } \Gamma^{i\to k} \in W, \forall j \in R_0, \forall \Gamma \in W. \tag{5}$$

If (v,W) is monotone, all j-zero players are j-null players and j-null players could contribute more than j-zero players.

4 An Axiomatic Approach

In this section, we give four axioms other than the axioms described in Section 2. The axioms given in this section are also regarded as reasonable for a value function to multi-alternative games with restricted choice situations. The four axioms are concerning zero players, additivity, symmetry, and sharing the expected payoff of a coalition.

Let π^j be a vector function from $MG(N,R)$ into $(\mathbb{R}^n_+)^{AR(N,R)}$. The i-th component of π^j is denoted by π^j_i.

Axiom 5 (j-zero player). *Given $v \in MG(N,R)$, $i \in N$, $j \in R$ and $W \in AR(N,R)$, the following holds:*

$$\pi^j_i(v)(W) = 0 \Leftrightarrow i \text{ is a } j\text{-zero player on } (v,W)$$

Axiom 5 is similar to Axiom 1. The j-null player in Axiom 1 is replaced with the j-zero player in Axiom 5. If (v,W) is monotone, Axiom 5 is a weaker requirement than Axiom 1 because all j-zero players are j-null players. By the difference between j-zero players and j-null players, even if we assume Axiom 5, j-null players may take non-zero values.

Axiom 6 (Additivity). *Given $v_1,\ v_2 \in MG(N,R)$ and $W \in AR(N,R)$, the following holds:*

$$\pi^j(v_1+v_2)(W) = \pi^j(v_1)(W) + \pi^j(v_2)(W),\ j = 1,\dots,r,$$

where v_1+v_2 is the sum game of v_1 and v_2.

Axiom 6 is weaker requirement than Axiom 2. Let $c_1 = c_2 = 1$ in Axiom 2. Then we obtain Axiom 6.

Axiom 7 (Symmetry). *Let $v \in MG(N,R), W \in AR(N,R)$, $i,k \in N$ $j \in R$. If i and k are symmetric in the game (v,W) with respect to alternative j, the following holds:*

$$\pi^j_i(v)(W) = \pi^j_k(v)(W).$$

The axiom with respect to the symmetry is often employed in the values of cooperative games such as Shapley value, Banzhaf value, Bolger value, and so on. Moreover, value g^j defined in Section 2 satisfies Axiom 7.

Axiom 8 (Sharing the expected payoff of a coalition). *Given $v \in MG(N,R), W \in AR(N,R)$, and $j \in R$ the following holds:*

$$\sum_{i\in N} \pi^j_i(v)(W) = \bar{v}_j(W). \tag{6}$$

Axiom 8 requires that the sum of all players' values with respect to each alternative equals to the expected payoff of the coalition choosing the alternative. This axiom is resemble to j-efficiency in Bolger value [2]. Namely, the sum of all players' value is required to equal to the payoff of a grand coalition (all players choose same alternative). Bolger assumes that a grand coalition is formed even if the formation process is different. On the other hand, Axiom 8 assumes that any arrangement (coalition situation) can be formed with equal probability. Considering those four axioms, we obtain the following theorem.

Theorem 2. *Function f^j, $j = 1,\dots,r$ defined by (2) is the unique function which satisfies Axiom 5 through 8.*

5 Numerical Example

We calculate the proposed value (f^j) in Job Selection Game described in Example 1 and compare it with the value (g^j) based on marginal contributions. Because the sum of f^j over players is different from that of g^j, we compare them by their normalized values $\bar{f}^j$ and $\bar{g}^j$. We consider two different situations: a situation when all students can choose all jobs and a situation when student A cannot choose job 2.

The results of computations are shown in Tables 1 and 2. Table 1 shows the values when all students can choose all jobs while Table 2 shows the values when student A cannot choose job 2. Because v is symmetric with respect to jobs, the values are same independent of the jobs students choose when all students can choose all jobs. Let us compare $\bar{f}^j$-values with $\bar{g}^j$-values when all students can choose all jobs. The range of $\bar{f}^j$-values are smaller than that of $\bar{g}^j$-values. Since this Job Selection Game is monotone, this result may be obtained from the fact that Axiom 5 is weaker than Axiom 1. In other words, by the weakness, $\bar{f}^j$ might share the profit more impartially among players than $\bar{g}^j$.

As shown in Table 2, $\bar{f}^2$-value as well as $\bar{g}^2$-value are zero for student A. This would be natural from the restriction that A cannot take Job 2. A remarkable difference between $\bar{f}^j$-values and $\bar{g}^2$-values is observed in the changes of those values

Table 1 The normalized values when all students can choose all jobs

	Job 1		Job 2	
Player	$\bar{f}^1$	$\bar{g}^1$ $\bar{f}^2$	$\bar{g}^2$	
A	0.390	0.413	0.390	0.413
B	0.330	0.327	0.330	0.327
C	0.278	0.254	0.278	0.254
Total	1	1	1	1

Table 2 The normalized values when student A cannot choose job 2

	Job 1		Job 2	
Player	$\bar{f}^1$	$\bar{g}^1$ $\bar{f}^2$	$\bar{g}^2$	
A	0.448	0.309	0	0
B	0.296	0.387	0.562	0.569
C	0.255	0.303	0.437	0.430
Total	1	1	1	1

of student A with $j = 1$ by the restriction (Job 2 cannot be taken by student A). From Tables 1 and 2, we know that $\bar{f}^1$-value increases by the restriction while $\bar{g}^1$-value decreases. This would be a reflection of Axiom 4, i.e., $\bar{g}^j$-value increases as the student has more selectable alternatives. On the other hand, $\bar{f}^j$ does not satisfy Axiom 4.

6 Conclusion

We have proposed the value for multi-alternative games with *restricted choice situations* under equally divided spoils assumption. A system of axioms have been given to characterize the value uniquely. In numerical example, we have compared the proposed value with the value based on marginal contributions. We observed big differences between those two values. The proposed value tends to share payoffs more impartially among players while the other value tends to decrease as selectable alternatives decreases. A further investigations on properties of those values as well as comparisons are future topics of our research.

References

1. Banzhaf, J.F.: Weighted Voting Doesn't Work: a Mathematical Analysis. Rutgers Law Review 19, 317–343 (1965)
2. Bolger, E.M.: A Value for Games with *n* Players and *r* Alternatives. International Journal of Game Theory 22, 319–334 (1993)
3. Caplow, T.: A Theory of Coalitions in the Triad. American Sociological Review 21, 489–493 (1956)
4. Deegan, J., Packel, E.W.: A New Index of Power for Simple *n*-Person Games. International Journal of Game Theory 7, 113–123 (1978)
5. Lehrer, E.: An Axiomatization of the Banzhaf value. International Journal of Game Theory 17, 89–99 (1988)
6. Masuya, S., Inuiguchi, M.: A value based on marginal contributions for multi-alternative games with restricted coalitions. In: Proceedings of the 11th Czech-Japan Seminar, pp. 15–20 (2008)
7. Ono, R.: Values for Multialternative Games and Multilinear Extensions. In: Holler, M., Owen, G. (eds.) Power indices and coalition formation, pp. 63–86. Kluwer Academic Publishers, Dordrecht (2001)
8. Shapley, L.S.: A Value for *n*-person Games. In: Kuhn, H., Tucker, A. (eds.) Contributions to the theory of games II, Princeton, pp. 307–317 (1953)
9. Tsurumi, M., Inuiguchi, M., Tanino, T.: A Solution for Fuzzy Generalized Multi-alternative Games. In: The 2006 NOLTA Proceedings, pp. 95–98 (2006)
10. Vinacke, W.E., Arkoff, A.: Experimental Study of Coalitions in the Triad. American Sociological Review 22, 406–415 (1957)

Variable Accessibility Models for Modal Logic on Topological Spaces

Tetsuya Murai, Seiki Ubukata, and Yasuo Kudo

Abstract. In this paper, we described a relationship between topological spaces and neighborhood frames in modal logic. Then we introduce variable accessibility models for modal logic by, for each world, selecting a neighborhood adequate under a given context and/or time.

Keywords: Topological spaces, Modal logic, Kripke frames, Neighborhood frames, Variable Accessibility, Agent control.

1 Introduction

Topological spaces[5] are very important in agent control in artificial intelligence. Logical properties of topological spaces has been investigated in the tradition of modal logics. McKinsey el al.[4] showed that normal system S4(=*KT4*) corresponds to topological systems. On the other hand, Kripke semantics (cf. [1, 2, 3]) shows that reflexive and transitive frames determines S4, but a Kripke frame generated from a neighborhood system is too strong in the sense that a generated accessibility relation is just the equality relation, which corresponds to S5. In this paper, given a topological space, we propose a variable accessibility model as a kind of Kripke models by selecting one neighborhood from a given neighborhood system under a given context and show some properties.

2 Preliminaries

This section is arranged based on several textbooks[1, 2, 3] on modal logics.

Tetsuya Murai and Seiki Ubukata
Hokkaido University, Kita 14, Nishi 9, Kita-ku, Sapporo 060-0814, Japan
e-mail: {murahiko,ubukata}@main.ist.hokudai.ac.jp

Yasuo Kudo
Muroran Institute of Technology, 27-1 Mizumoto-cho, Muroran 050-8585, Japan
e-mail: kudo@csse.muroran-it.ac.jp

K. Nakamatsu et al. (Eds.): New Advan. in Intel. Decision Techno., SCI 199, pp. 547–553.
springerlink.com

2.1 Kripke Frames and Models

Given a set of atomic sentences $\mathscr{P}$, a language $\mathscr{L}_{\mathrm{ML}}(\mathscr{P})$ for modal logic is formed from $\mathscr{P}$ using logical operators $\top$ (the truth constant), $\bot$ (the falsity constant), $\neg$ (negation), $\wedge$ (conjunction), $\vee$ (disjunction), $\rightarrow$ (material implication), $\leftrightarrow$ (equivalence), $\Box$ (necessity), and $\Diamond$ (possibility) as the least set of sentences generated by the following formation rules:

(1) $\mathsf{p} \in \mathscr{P} \Rightarrow (\mathsf{p}) \in \mathscr{L}_{\mathrm{ML}}(\mathscr{P})$
(2) $(\top), (\bot) \in \mathscr{L}_{\mathrm{ML}}(\mathscr{P})$
(3) $p \in \mathscr{L}_{\mathrm{ML}}(\mathscr{P}) \Rightarrow (\neg p), (\Box p), (\Diamond p) \in \mathscr{L}_{\mathrm{ML}}(\mathscr{P})$
(4) $p, q \in \mathscr{L}_{\mathrm{ML}}(\mathscr{P}) \Rightarrow (p \wedge q), (p \vee q), (p \rightarrow q), (p \leftrightarrow q) \in \mathscr{L}_{\mathrm{ML}}(\mathscr{P})$.

A *Kripke frame* $\mathscr{F}$ is a structure

$$\langle U, R \rangle,$$

where U is a universe (set of possible worlds) and R is a binary relation on U. Given a Kripke frame $\mathscr{F}$, a *Kripke model* $\mathscr{M}$ on $\mathscr{F}$ is a structure

$$\langle U, R, V \rangle,$$

where V is a valuation which assigns truth values to every atomic sentence at each world. We define two subsets in U by

$$\|p\|^{\mathscr{M}} = \{x \in U \mid \mathscr{M}, x \models p\},$$
$$U^R(x) = \{y \in U \mid xRy\}.$$

Then, the truth conditions for modal sentences are given by

$$\mathscr{M}, x \models \Box p \Leftrightarrow U^R(x) \subseteq \|p\|^{\mathscr{M}},$$
$$\mathscr{M}, x \models \Diamond p \Leftrightarrow U^R(x) \cap \|p\|^{\mathscr{M}} \neq \emptyset.$$

It is well-known that different properties of binary relations validate different axiom schemata. Thus they play an essential role of soundness and completeness of systems of modal logic with respect to corresponding classes of Kripke frames. In particular, the correspondences shown in Table 1 are well-known in literature: For example, the second line shows that the smallest normal system K is sound and complete with respect to any class of Kripke models.

2.2 Topological Spaces and S4

The relationship between *topological spaces* and normal system S4 was examined McKinsey and Tarski[4] in 1944. Here we take a definition of topological space as

$$\langle U, N \rangle,$$

where N is a *neighborhood system*, that is,

Table 1 Modal systems and properties of a relation R

System	Relation R
K	a binary relation (no conditions)
D(=KD)	serial
T(=KT)	reflexive
B(=KTB)	reflexive and symmetric
S4(=$KT4$)	reflexive and transitive
S5(=$KT5$)	an equivalence relation

$$N : U \rightarrow 2^{2^U}$$

is a function that satisfies the following conditions.

(N_1) $U \in N(x)$
(N_2) $X \in N(x) \Rightarrow x \in X$
(N_3) $X_1, X_2 \in N(x) \Rightarrow X_1 \cap X_2 \in N(x)$
(N_4) $(X \in N(x) \text{ and } X \subseteq Y) \Rightarrow Y \in N(x)$
(N_5) $X \in N(x) \Rightarrow \exists Y \in N(x)[Y \subseteq X \text{ and } \forall y(y \in Y \Rightarrow X \in N(y))]$

On the other hand, S4 is usually axiomatized as $KT4$ by the following rules of inference and axiom schemata

RN. From p infer $\Box p$
Df$\Diamond$. $\Diamond p \leftrightarrow \neg\Box\neg p$
K. $\Box(p \rightarrow q) \rightarrow (\Box p \rightarrow \Box q)$
T. $\Box p \rightarrow p$
4. $\Box p \rightarrow \Box\Box p$

and their corresponding conditions in a Kripke model are reflexivity and transitivity of the relation in the model. Thus we cannot find deep relationship between neighborhood systems and modal logic S4 in Kripke frames. It is, however, possible when we consider the relationship in neighborhood frames.

2.3 Neighborhood Frames and Models

The standard neighborhood system in a topological space must satisfy the five axioms (N_1),...,(N_5) in the previous section. When we do not assume any axiom on N, we obtain a *neighborhood frame* $\mathscr{F}$ as a structure

$$\langle U, N \rangle,$$

where N is just a function

$$N : U \longrightarrow 2^{2^U}.$$

A neighborhood frame is also called a *Scott-Montague frame*. Conversely, when a neighborhood frame satisfies the following conditions (N_1),...,(N_5) in the previous section, it is just a neighborhood system in the sense of the standard topology.

Given a neighborhood frame $\mathcal{F}$, a *neighborhood model* $\mathcal{M}$ on $\mathcal{F}$ is a structure

$$\langle U, N, V \rangle,$$

where V is a valuation.

The truth conditions of modal sentences are given by

$$\mathcal{M}, x \models \Box p \Leftrightarrow \|p\|^{\mathcal{M}} \in N(x),$$
$$\mathcal{M}, x \models \Diamond p \Leftrightarrow (\|p\|^{\mathcal{M}})^{\mathrm{C}} \notin N(x).$$

Normal system S4 is proved to be both sound and complete with respect to the class of topological spaces.

2.4 From Kripke to Neighborhood Models

Following to Chellas[1], given a Kripke model, we can generate a neighborhood model, which is equivalent to the given Kripke one. First recall the truth condition of necessity sentence in a Kripke model $\mathcal{M}$:

$$\mathcal{M}, x \models \Box p \Leftrightarrow U^R(x) \subseteq \|p\|^{\mathcal{M}}$$

Then we can define a function $N_R : U \to 2^{2^U}$ by

$$N_R(x) = \{X \in 2^U \mid U^R(x) \subseteq X\}$$

Then the truth condition can be rewritten into an equivalent expression

$$\mathcal{M}, x \models \Box p \Leftrightarrow \|p\|^{\mathcal{M}} \in N_R(x).$$

Thus the Kripke model has another equivalent expression

$$\langle U, N_R, V \rangle,$$

which is nothing but a neighborhood model.

The above N_R has the following properties:

[m] $(X \in N_R(x) \text{ and } X \subseteq X') \Rightarrow X' \in N_R(x)$
[c] $X, X' \in N_R(x) \Rightarrow X \cap X' \in N_R(x)$
[n] $W \in N_R(x)$
[a] $\cap N_R(x) (= \cap_{X \in N_R(x)} X) \in N_R(x)$

The former three conditions validate the following three schemata

M. $\Box(p \wedge p') \to (\Box p \wedge \Box p')$
C. $(\Box p \wedge \Box p') \to \Box(p \wedge p')$
N. $\Box\top$.

All of them are valid in any Kripke model. This means that Kripke semantics cannot distinguish these schemata in their framework. On the other hand, neighborhood semantics can do, because it has their corresponding properties. If you want some axiom, say C, invalidate, we can by removing, say, [c]. Note that the axiom schemata are not necessarily independent of each other.

The last condition [a] has a very important role in the sense that, by [a], we have the following relationship between the set of worlds from x and $N_R(x)$:

$$U^R(x) = \cap N_R(x).$$

2.5 *From Neighborhood to Kripke Models*

A neighborhood frame $\langle U, N \rangle$ is called a *filter* when its function N satisfies the three condition [m], [c], and [n]. By a filter we also mean a neighborhood model on a filter if confusion does not arise. The smallest normal system K is sound and complete with respect to a class of filters. Note that the class of filters is not equal to the class of Kripke models.

A filter is said to be *augmented* when it satisfies [a]. Then, it has the least element in every $N(x)$, that is, $\cap N(x)$, by which we can define an accessibility relation R_N by

$$xR_N y \Leftrightarrow y \in \cap N(x).$$

Thus an augmented neighborhood models is equivalent to its corresponding Kripke models by the above formula. Hence we can say the smallest normal system K is sound and complete with respect to the class of augmented neighborhood models.

We show various properties of a function N and their corresponding axiom schemata in Table 2.

3 Topological Spaces and S4 Revisited

Following Chellas[1], there are other ways of axiomatization for S4 as classical modal system $EMCNT4$:

RE. From $p \leftrightarrow p'$ infer $\Box p \leftrightarrow \Box p'$
M. $\Box(p \wedge p') \to (\Box p \wedge \Box p')$
C. $(\Box p \wedge \Box p') \to \Box(p \wedge p')$
N. $\Box\top$
T. $\Box p \to p$
4. $\Box p \to \Box\Box p$

Table 2 Properties of a function N and axiom schemata

Properties of N	Axiom Schemata
[m] $(X \in N(x) \text{ and } X \subseteq X') \Rightarrow X' \in N(x)$	**M.** $\Box(p \wedge p') \rightarrow (\Box p \wedge \Box p')$
[c] $X, X' \in N(x) \Rightarrow X \cap X' \in N(x)$	**C.** $(\Box p \wedge \Box p') \rightarrow \Box(p \wedge p')$
[n] $U \in N(x)$	**N.** $\Box\top$
[p] $\emptyset \notin N(x)$	**P.** $\neg\Diamond\bot$
[f] $X \cup X' \in N(x) \Rightarrow (X \in N(x) \text{ or } X' \in N(x))$	**F.** $\Box(p \vee p') \rightarrow (\Box p \vee \Box p')$
[k] $(X^{\mathrm{C}} \cup X' \in N(x) \text{ and } X' \in N(x)) \Rightarrow X' \in N(x)$	**K.** $\Box(p \rightarrow p') \rightarrow (\Box p \rightarrow \Box p')$
[d] $\text{Not } (X \in N(x) \text{ and } X^{\mathrm{C}} \in N(x))$	**D.** $\Box p \rightarrow \Diamond p$
[d_{C}] $X \in N(x) \text{ or } X^{\mathrm{C}} \in N(x)$	**$\mathbf{D}_{\mathrm{C}}$.** $\Diamond p \rightarrow \Box p$
[t] $X \in N(x) \Rightarrow x \in X$	**T.** $\Box p \rightarrow p$
[b] $x \in X \Rightarrow \{y \in W \mid X^{\mathrm{C}} \notin N(y)\} \in N(x)$	**B.** $p \rightarrow \Box\Diamond p$
[4] $X \in N(x) \Rightarrow \{y \in W \mid X \in N(y)\} \in N(x)$	**4.** $\Box p \rightarrow \Box\Box p$
[5] $X \notin N(x) \Rightarrow \{y \in W \mid X \notin N(y)\} \in N(x)$	**T.** $\Diamond p \rightarrow \Box\Diamond p$

Such systems whose code starts from E are called classical systems of modal logic[1]. The properties of a function N corresponding to S4 is

$$
\begin{array}{ll}
[\mathrm{m}] & (X \in N(x) \text{ and } X \subseteq X') \Rightarrow X' \in N(x) \\
[\mathrm{c}] & X, X' \in N(x) \Rightarrow X \cap X' \in N(x) \\
[\mathrm{n}] & U \in N(x) \\
[\mathrm{t}] & X \in N(x) \Rightarrow x \in X \\
[4] & X \in N(x) \Rightarrow \{y \in W \mid X \in N(y)\} \in N(x)
\end{array}
$$

Thus we have the following correspondence

$$
\begin{array}{l}
(\mathrm{N}_1) \Leftrightarrow [\mathrm{n}] \\
(\mathrm{N}_2) \Leftrightarrow [\mathrm{t}] \\
(\mathrm{N}_3) \Leftrightarrow [\mathrm{c}] \\
(\mathrm{N}_4) \Leftrightarrow [\mathrm{m}]
\end{array}
$$

Here condition (N_5) seems to be different from condition [4], but we can easily show that

$$[m] \Rightarrow ((\mathrm{N}_5) \Leftrightarrow [4]),$$

thus the two formulations are equivalent to each other. Now we can explicitly understand a deep relationship between neighborhood systems and modal logic S4.

4 Variable Accessibility Models

Topological spaces, particularly distance spaces, or moreover, Euclidean spaces, are very important in agent control in artificial intelligence. Any topological spaces $\langle U, N \rangle$ is an augmented filter, but, unfortunately, we have

$$\cap N(x) = \{x\},$$

for $x \in U$, so its generating relation R_N is just equality, which means neighborhood systems have too much freedom.

Hence we need select just one neighborhood from $N(x)$ at each world x. Such selection of neighborhoods is, in general, done under a given context. In some cases, it depends on time when we must consider time sequences like agent control. Let U_x^C be such a neighborhood selected from $N(x)$ under context C. Then we have a Kripke frame

$$\mathscr{F}_C = \langle U, R_C \rangle$$

by defining an accessibility relation

$$xR_C y \Leftrightarrow y \in U_x^C.$$

When such frame depends on time t, we denote $\mathscr{F}_t$ and R_t instead of $\mathscr{F}_C$ and R_C, respectively.

By condition (N_2), $\mathscr{F}$ is apparently a reflexive frame, so Kripke models on such frames are *KT*-models. Further, by condition (N_5), if we choose neighborhoods in a well-defined way, then $\mathscr{F}$ may also be transitive, thus we may have *KT4*-models.

5 Concluding Remarks

In this paper, we described a relationship between topological spaces and neighborhood frames in modal logic. Then we introduce variable accessibility models for modal logic by, for each world, selecting a neighborhood adequate under a given context and/or time. Some experimental results will be shown at the conference site.

Acknowledgements. This work was partially supported by Grant-in-Aid No.19300074 for Scientific Research(B) and Grant-in-Aid No.19650046 for Exploratory Research of the Japan Society for the Promotion of Science.

References

1. Chellas, B.F.: Modal Logic: an Introduction. Cambridge University Press, Cambridge (1980)
2. Hughes, G.E., Cresswell, M.J.: A New Introduction to Modal Logic. Routledge (1996)
3. Goldblatt, R.: Logics of Time and Computation. CSLI (1987)
4. McKinsey, J.C.C., Tarski, A.: The Algebra of Topology. Annals of Mathematics 45, 141–191 (1944)
5. Sierpiński, W.: General Topology. Univ. of Toronto Press (1956)

A Heuristic Algorithm for Selective Calculation of a Better Relative Reduct in Rough Set Theory

Yasuo Kudo and Tetsuya Murai

Abstract. In this paper, we consider a heuristic method to partially calculate relative reducts with better evaluation by the evaluation criterion proposed by the authors. By using the average of certainty and coverage of decision rules constructed from each condition attribute, we introduce an evaluation criterion of condition attributes, and consider a heuristic method for calculating a relative reduct with better evaluation.

1 Introduction

Rough set theory [4, 5] provides mathematical foundations of set-theoretical approximation of concepts and reasoning about data. In the aspect of reasoning about data by rough set, generating relative reducts and decision rules from a given decision table have been central topics, and there are various studies about this topics. Skowron and Rauszer [8] have proposed an algorithm to calculate all relative reducts by using the concept of the discernibility matrix. However, they have also proved that computational complexity of calculation of all relative reducts in the given decision table is NP-hard [8]. Thus, there are many proposals of approximate algorithms to partially calculate relative reducts instead of calculating all relative reducts [1, 2, 6, 9, 10, 11].

The authors have proposed an evaluation criterion of relative reducts based on partitions of objects that are generated from the relative reducts [3]. However, the evaluation criterion needs to construct partitions from generated relative reducts, it

Yasuo Kudo
Department of Computer Science and Systems Engineering, Muroran Institute of Technology
27-1 Mizumoto, Muroran 050-8585, Japan
e-mail: kudo@csse.muroran-it.ac.jp

Tetsuya Murai
Graduate School of Information Science and Technology, Hokkaido University
Kita 14, Nishi 9, Kita-ku, Sapporo 060-0814, Japan
e-mail: murahiko@main.ist.hokudai.ac.jp

K. Nakamatsu et al. (Eds.): New Advan. in Intel. Decision Techno., SCI 199, pp. 555–564.
springerlink.com

is difficult to use the evaluation criterion for approximate algorithms to partially calculate relative reducts.

Thus, in this paper, we consider a heuristic method to calculate a relative reduct with better evaluation by the evaluation criterion proposed by the authors.

2 Rough Sets

In this section, we review rough set theory and the evaluation criterion of relative reducts proposed by the authros. Note that contents of this section are based on [3, 7].

2.1 Decision Tables and Lower and Upper Approximations

Generally, subjects of data analysis by rough sets are illustrated by decision tables. Formally, a *decision table* is the following triple:

$$DT = (U, C, d), \tag{1}$$

where U is the set of objects, C is the set of condition attributes such that each attribute $a \in C$ is a function $a : U \rightarrow V_a$ from U to the set V_a of values of the attribute a, and d is a function $d : U \rightarrow V_d$ called the decision attribute.

Classification of elements in decision tables are done based on *indiscernibility relations*. For any set of attributes $A \subseteq C \cup \{d\}$, the indiscernibility relations R_A is the following binary relation on U:

$$R_A = \{(x,y) \mid a(x) = a(y), \forall a \in A\}. \tag{2}$$

If a pair (x,y) is in R_A, then two elements x and y are indiscernible with respect to all attributes in A. It is well-known that any indiscernibility relation is an equivalence relation, and equivalence classes by an equivalence relation consists of a partition on the domain of the equivalence relation. In particular, the indiscernibility relation R_d based on the decision attribute d provides a partition $\mathscr{D} = \{D_1, \cdots, D_k\}$, and each element $D_i \in \mathscr{D}$ is called a *decision class*.

Classifying elements with respect to condition attributes provides approximation of decision classes. Formally, for any set $B \subseteq C$ of condition attributes and any decision class $D_i \in \mathscr{D}$, we let

$$\underline{B}(D_i) = \{x \in U \mid [x]_B \subseteq D_i\}, \tag{3}$$

$$\overline{B}(D_i) = \{x \in U \mid [x]_B \cap D_i \neq \emptyset\}, \tag{4}$$

where the set $[x]_B$ is the equivalence class of x by the indiscernibility relation R_B. The set $\underline{B}(D_i)$ and the set $\overline{B}(D_i)$ are called the *lower approximation* and the *upper approximation* of the decision class D_i with respect to B, respectively. Note that the lower approximation $\underline{B}(D_i)$ illustrates the set of elements which are correctly

Table 1 A decision table

U	c_1	c_2	c_3	c_4	c_5	c_6	d
x_1	1	0	0	0	0	1	1
x_2	0	1	0	0	0	1	1
x_3	0	2	1	0	1	0	2
x_4	0	1	1	1	0	0	2
x_5	0	1	2	0	0	1	1
x_6	0	1	0	0	1	1	3

classified to the decision class D_i by checking all attributes in B. A pair $(\underline{B}(D_i), \overline{B}(D_i))$ is called a *rough set*.

Example 1. Table 1 illustrates a decision table which consists of the set of objects $U = \{x_1, \cdots, x_6\}$, the set of condition attributes $C = \{c_1, \cdots, c_6\}$ and the decision attribute d. For example, an attribute c_2 is a function $c_2 : U \rightarrow \{0, 1, 2\}$, and the value of an object $x_3 \in U$ at c_2 is 2, that is, $c_2(x_3) = 2$. Moreover, the decision attributed d provides the following three decision classes, $D_1 = \{x_1, x_2, x_5\}$, $D_2 = \{x_3, x_4\}$ and $D_3 = \{x_6\}$.

2.2 Decision Rules

In this paper, we denote a decision rule constructed from a subset $B \subseteq C$ of condition attribute, the decision attribute d and an object $x \in U$ by $(B,x) \rightarrow (d,x)$. The concepts of certainty and coverage are well-known criteria for evaluating decision rules. For any decision rule $(B,x) \rightarrow (d,x)$, the degree $Cer(\cdot)$ of certainty and the degree $Cov(\cdot)$ of coverage are defined as follows, respectively:

$$Cer((B,x) \rightarrow (d,x)) = \frac{|[x]_B \cap D_i|}{|[x]_B|}, \quad (5)$$

$$Cov((B,x) \rightarrow (d,x)) = \frac{|[x]_B \cap D_i|}{|D_i|}, \quad (6)$$

where the set D_i is the decision class such that $x \in D_i$, and the notation $|X|$ is the cardinality of the set X.

For example, in Table 1, a decision rule $(B,x_1) \rightarrow (d,x_1)$ constructed from a set $B = \{c_2, c_3\}$, the decision attribute d and an object $x_1 \in U$ has actually the following form:

$$(c_2 = 0) \wedge (c_3 = 0) \rightarrow (d = 1),$$

and its certainty is 1, and the coverage is $\frac{1}{3}$.

2.3 Relative Reducts

By checking values of all condition attributes, we can classify all discernible elements in a given decision table to those correct decision classes. However, not all condition attributes may need to be checked in the sense that some condition attributes are essential to classify, and the other attributes are redundant. A minimal set of condition attributes to classify all discernible elements to correct decision classes is called a *relative reduct* of the decision table.

To introduce the concept of relative reducts, for any subset $B \subseteq C$ of condition attributes in a decision table DT, we let

$$POS_B(\mathscr{D}) = \bigcup_{D_i \in \mathscr{D}} \underline{B}(D_i). \tag{7}$$

The set $POS_B(\mathscr{D})$ is called the positive region of $\mathscr{D}$ by B. All elements $x \in POS_B(\mathscr{D})$ are classified to correct decision classes by checking all attributes in B. In particular, the set $POS_C(\mathscr{D})$ is the set of all discernible elements in DT.

Here, we define relative reducts formally. A set $A \subseteq C$ is called a *relative reduct* of the decision table DT if the set A satisfies the following conditions:

1. $\mathrm{POS}_A(\mathscr{D}) = \mathrm{POS}_C(\mathscr{D})$.
2. $\mathrm{POS}_B(\mathscr{D}) \neq \mathrm{POS}_C(\mathscr{D})$ for any proper subset $B \subset A$.

Note that, in general, there are plural relative reducts in a decision table. Common part of all relative reducts are called the *core* of the decision table.

For example, there are the following three relative reducts in Table 1: $\{c_3, c_5\}$, $\{c_5, c_6\}$, and $\{c_2, c_4, c_5\}$. The condition attribute c_5 appears in all of the relative reducts in Table 1, and therefore the core of Table 1 is $\{c_5\}$.

The discernibility matrix is one of the most popular method to calculate all relative reducts in the decision table. Let DT be a decision table with $|U|$ elements. The *discernibility matrix DM* of DT is a symmetric $|U| \times |U|$ matrix whose element at i-th row and j-th column is the following set of condition attributes to discern between two elements x_i and x_j:

$$\delta_{ij} = \begin{cases} \{a \in C \mid a(x_i) \neq a(x_j)\}, & \text{if } d(x_i) \neq d(x_j), \\ \emptyset, & \text{otherwise.} \end{cases} \tag{8}$$

Each attribute $a \in \delta_{ij}$ in the non-empty element δ_{ij} represents that x_i and x_j are discernible by checking the value of a.

Using the discernibility matrix, we get all relative reducts of the decision table as follows:

1. Construct the following logical formula $L(\delta_{ij})$ from each non-empty set $\delta_{ij} = \{a_{k1}, \cdots, a_{kl}\}$ ($i > j$ and $l \geq 1$) in the discernibility matrix:

$$L(\delta_{ij}) : a_{k1} \vee \cdots \vee a_{kl}. \tag{9}$$

2. Construct a conjunctive normal form $\bigwedge_{i>j} L(\delta_{ij})$.

Table 2 The discernibility matrix of Table 1

	x_1	x_2	x_3	x_4	x_5	x_6
x_1	$\emptyset$					
x_2	$\emptyset$	$\emptyset$				
x_3	$\{c_1,c_2,c_3,c_5,c_6\}$	$\{c_2,c_3,c_5,c_6\}$	$\emptyset$			
x_4	$\{c_1,c_2,c_3,c_4,c_6\}$	$\{c_3,c_4,c_6\}$	$\emptyset$	$\emptyset$		
x_5	$\emptyset$	$\emptyset$	$\{c_2,c_3,c_5,c_6\}$	$\{c_3,c_4,c_6\}$	$\emptyset$	
x_6	$\{c_1,c_2,c_5\}$	$\{c_5\}$	$\{c_2,c_3,c_6\}$	$\{c_3,c_4,c_5,c_6\}$	$\{c_3,c_5\}$	$\emptyset$

3. Transform the conjunctive normal form to the minimal disjunctive normal form:

$$\bigwedge_{i>j} L(\delta_{ij}) \equiv \bigvee_{p=1}^{s} \bigwedge_{q=1}^{t_p} a_{pq} \tag{10}$$

4. For each conjunction $a_{p1} \wedge \cdots \wedge a_{pt_p}$ $(1 \leq p \leq s)$ in the minimal disjunctive normal form, construct a relative reduct $\{a_{p1}, \cdots, a_{pt_p}\}$.

Example 2. Table 2 illustrates the discernibility matrix of the decision table by Table 1. Each non-empty set that appears in the matrix represents the set of condition attributes that we should check to discern the corresponding objects. For example, the set $\delta_{65} = \{c_3, c_5\}$ represents that we can discern the objects x_6 and x_5 by either of values of these objects at the condition attribute c_3 and c_5. Note that we omit upper triangular components of the discernibility matrix in Table 2 because the discernibility matrix is symmetric by the definition. We construct a conjunctive normal form by connecting logical formulas based on non-empty elements in Table 2, and transform the conjunctive normal form to the minimal disjunctive normal form as follows;

$$\begin{aligned}&(c_1 \vee c_2 \vee c_3 \vee c_5 \vee c_6) \wedge (c_2 \vee c_3 \vee c_5 \vee c_6) \wedge \cdots \wedge (c_3 \vee c_5)\\ \equiv\ &(c_3 \wedge c_5) \vee (c_5 \wedge c_6) \vee (c_2 \wedge c_4 \wedge c_5).\end{aligned}$$

Consequently, from this minimal disjunctive normal form, we have the three relative reducts $\{c_3, c_5\}$, $\{c_5, c_6\}$, and $\{c_2, c_4, c_5\}$.

2.4 *Evaluation of Relative Reducts Based on Partitions*

In this subsection, we review an evaluation method for relative reducts by using partitions constructed from the relative reducts [3].

This evaluation method is based on the following idea; We consider that the more rough partition a relative reduct constructs, the better evaluation the relative reduct has. From the viewpoint of rule generation, such rough partitions constructed from relative reducts provide decision rules with higher degrees of coverage rather than degrees of coverage of decision rules based on fine partitions. Thus, we consider to evaluate relative reducts by using criteria of decision rules constructed from relative reducts. Suppose we fix a non-empty subset of condition attributes $B \subseteq C$. For any equivalence class $[x]_B \in U/R_B$, we define the following set:

$$Dec([x]_B) \stackrel{\text{def}}{=} \{D_i \in \mathscr{D} \mid D_i \cap [x]_B \neq \emptyset\}. \tag{11}$$

The set $Dec([x]_B)$ corresponds to the set of conclusions of decision rules with the formula (B,x) as the antecedent. Thus, the degree defined as follows:

$$N_R \stackrel{\text{def}}{=} \sum_{[x]_B \in U/R_B} |Dec([x]_B)| \tag{12}$$

is the total sum of the number of all decision rules constructed from B.

We have shown the following theorem about averages of certainty and coverage of all decision rules constructed from B [3].

Theorem 1. *For any non-empty subset $B \subseteq C$ of condition attributes, the average of certainty $ACer(B)$ of all decision rules $(B,x) \to (d,x)$ $(\forall x \in U)$ constructed from B is calculated by the following equation:*

$$ACer(B) = \frac{|U/R_B|}{N_R}. \tag{13}$$

Similarly, the average of coverage $ACov(B)$ is calculated as follows:

$$ACov(B) = \frac{|\mathscr{D}|}{N_R}. \tag{14}$$

In Theorem 1, if we use relative reducts as subsets of condition attributes in any consistent decision table, the average values of certainty of decision rules constructed from the relative reduct are equal to 1 for any relative reducts. On the other hand, the smaller the number of equivalence classes constructed from the relative reduct is, that is, the more "rough" the partition is, the higher the average of coverage of decision rules constructed from the relative reduct is. Therefore, in this sense, we can use the average values of coverage of decision rules constructed from relative reducts as an evaluation criterion for relative reducts.

Example 3. We can construct the following five decision rules from a relative reduct $\{c_3, c_5\}$ of Table 1:

- $(c_3 = 0) \wedge (c_5 = 0) \to (d = 1)$, Certainty= 1, Coverage= $\frac{2}{3}$,
- $(c_3 = 2) \wedge (c_5 = 0) \to (d = 1)$, Certainty= 1, Coverage= $\frac{1}{3}$,
- $(c_3 = 1) \wedge (c_5 = 0) \to (d = 2)$, Certainty= 1, Coverage= $\frac{1}{2}$,
- $(c_3 = 1) \wedge (c_5 = 1) \to (d = 2)$, Certainty= 1, Coverage= $\frac{1}{2}$,
- $(c_3 = 0) \wedge (c_5 = 1) \to (d = 3)$, Certainty= 1, Coverage= 1.

The average of certainty of these rules is $(1+1+1+1+1)/5 = 1$, and this value is equal to the degree "the number of equivalence classes / the number of decision

Table 3 Average of certainty and coverage calculated from relative reducts

Relative reducts	Avg. of certainty	Avg. of coverage
$\{c_3, c_5\}$	1	0.6
$\{c_5, c_6\}$	1	0.75
$\{c_2, c_4, c_5\}$	1	0.6

rules" by Theorem 1. On the other hand, the average of coverage is $(\frac{2}{3} + \frac{1}{3} + \frac{1}{2} + \frac{1}{2} + 1)/5 = \frac{3}{5}$, and it is also equal to the degree "the number of decision classes / the number of decision rules" by Theorem 1.

Thus, we have the evaluation degree $\frac{3}{5}$ of the relative reduct $\{c_3, c_5\}$. Table 3 shows the average values of certainty and coverage by each relative reduct. By this result, we regard the relative reduct $\{c_5, c_6\}$ as the best relative reduct that provides the most rough and correct approximations of decision classes.

3 A Heuristic Algorithm for Selective Calculation of a Better Relative Reduct

In this section, we propose a heuristic algorithm for calculating a better relative reduct of a given decision table. Here, we intend to construct a relative reduct with higher evaluation degree by the evaluation criterion in Section 2.4.

As we have reviewed, the evaluation of a relative reduct is based on the number of generated decision rules from the relative reduct, and therefore, the evaluation relies essentially on a partition constructed from the relative reduct. Because the constructed partition is represented by combination of partitions constructed from each condition attribute in the relative reduct, we consider that, for each condition attribute, we use some evaluation of partitions constructed from each condition attribute for selective calculation of a better relative reduct.

Generally, decision rules constructed from a condition attribute are not correct, and we need to consider not only the average of coverage of decision rules, but also the average of certainty. Thus, we introduce the following evaluation criterion of condition attributes.

Definition 1. Let $DT = (U, C, d)$ be a decision table. For each condition attribute $a \in C$, an evaluation degree $Eval(a)$ of a is defined as follows:

$$Eval(a) \stackrel{\text{def}}{=} \frac{1}{2}(ACer(\{a\}) + ACov(\{a\})). \tag{15}$$

The evaluation degree $Eval(a)$ of a condition attribute a is simply the average of $ACer(\{a\})$ defined by (13) and $ACov(\{a\})$ defined by (14). We consider that the higher the evaluation $Eval(a)$ is, the more suitable for the better relative reduct the attribute a is.

However, in some cases, evaluation of condition attributes using the average of certainty and coverage may not be sufficient for selecting condition attributes used

Table 4 Avg. of certainty and coverage, evaluation of each condition attribute

Attributes	Avg. of certainty	Avg. of coverage	Evaluation	Core
c_1	0.5	0.75	0.625	×
c_2	0.4	0.6	0.5	×
c_3	0.75	0.75	0.75	×
c_4	0.5	0.75	0.625	×
c_5	0.5	0.75	0.625	○
c_6	0.67	1	0.83	×

in the better relative reduct. This is because the proposed evaluation of condition attributes does not consider the existence of core of decision table, and therefore condition attributes in the core may have low evaluation even though the selectively calculated relative reduct has to contain such condition attributes.

Thus, we need to consider whether the core of decision table exists. The following property enables us considering the core of decision table without calculating all relative reducts.

Proposition 1. *Let $DM = \{\delta_{ij} \mid 1 \leq i,j \leq |U|\}$ be a discernibility matrix of a decision table $DT = (U,C,d)$. For each condition attribute $a \in C$, a is in the core of DT if and only if a satisfies the following property:*

$$\min_{a \in \delta_{ij}} \left|\delta_{ij}\right| = 1. \tag{16}$$

This proposition exhibits that we can detect the core of decision table by checking the size of elements in the discernibility matrix. Table 4 represents the average of certainty, the average of coverage and the evaluation of each condition attribute in Table 1, and whether each condition attribute is in the core of Table 1. The symbol ○ represents that the condition attribute c_5 is in the core because the element $\delta_{62} = \{c_5\}$ in Table 2 is the singleton such that $c_5 \in \delta_{62}$.

By using the evaluation criterion of condition attributes defined by (15) and the property of core by (16), we consider the following heuristic algorithm for calculating a relative reduct with better evaluation:

1. Calculate the discernibility matrix DM, and let
$$n = \left|\{\delta_{ij} \in DM \mid \delta_{ij} \neq \emptyset\}\right|.$$
Moreover, let $m = 0$.
2. If there exist condition attributes $a_1, \cdots, a_i$ that satisfy the equation (16), let $S = \{a_1, \cdots, a_i\}$ and remove $a_1, \cdots, a_i$ from the list of condition attributes; Otherwise, let $S = \emptyset$.
3. Sort the list of condition attributes by descending order of the evaluation degrees of attributes.
4. From the current list of condition attributes, select condition attributes $a_j, \cdots, a_k$ with the highest evaluation in the current list. Update the set S as follows:

$$S := S \cup \{a_j, \cdots, a_k\},$$

and remove $a_j, \cdots, a_k$ from the current list.

5. Update the number m as follows:

$$m := \left| \{\delta_{ij} \in DM \mid \delta_{ij} \cap S \neq \emptyset\} \right|.$$

6. If $m = n$, output S as the result of selective calculation, and finish; Otherwise, go back to Step 4.

Example 4. We calculate a better relative reduct of Table 1 by the proposed algorithm. In Step 1, we have 11 non-empty elements in the discernibility matrix by Table 2, and let $n = 11$.

In Step 2, from Table 4, we know that the condition attribute c_5 is in the core of Table 1, and therefore we have the set $S = \{c_5\}$ and remove c_5 from the list of condition attributes.

In Step 3, the list of current condition attributes c_1, c_2, c_3, c_4, c_6 are sorted as follows; 1. c_6, 2. c_1, c_3 and c_4, and 3. c_2.

In Step 4, we select the attribute c_6, update the set S as $S = \{c_5, c_6\}$, and remove c_6 from the current list of condition attributes.

In Step 5, we count the number m of elements δ_{ij} in Table 2 such that $\delta_{ij} \cap S \neq \emptyset$, and consequently, we have $m = n$. Therefore, in Step 6, we get $S = \{c_5, c_6\}$ as the result of selective calculation of a better relative reduct. From Table 3, the set $S = \{c_5, c_6\}$ is actually the relative reduct of Table 1 with the highest evaluation.

4 Conclusion

In this paper, we considered a heuristic method for partial calculation of relative reducts with better evaluation by the evaluation criterion proposed by the authors. By using the average of certainty and coverage of decision rules constructed from each condition attribute, we introduced an evaluation criterion of condition attributes, and proposed a heuristic method for calculating a relative reduct with better evaluation. The proposed method was demonstrated by a small example. More refinement of the proposed method and experiments using more large data are future issues.

References

1. Hedar, A.H., Wang, J., Fukushima, M.: Tabu search for attribute reduction in rough set theory. Soft Couping 12(9), 909–918 (2008)
2. Hu, F., Wang, G., Feng, L.: Fast Knowledge Reduction Algorithms Based on Quick Sort. In: Wang, G., Li, T., Grzymala-Busse, J.W., Miao, D., Skowron, A., Yao, Y. (eds.) RSKT 2008. LNCS (LNAI), vol. 5009, pp. 72–79. Springer, Heidelberg (2008)
3. Kudo, Y.: An Evaluation Method of Relative Reducts Based on Roughness of Partitions (Extended Abstract). In: The 6th International Conferences on Rough Sets and Current Trends in Computing (RSCTC 2008): Programs & Abstracts, pp. 26–29 (2008)

4. Pawlak, Z.: Rough Sets. International Journal of Computer and Information Science 11, 341–356 (1982)
5. Pawlak, Z.: Rough Sets: Theoretical Aspects of Reasoning about Data. Kluwer Academic Publishers, Dordrecht (1991)
6. Pawlak, Z., Słowiński, R.: Rough Set Approach to Multi-Attribute Decision Analysis. European Journal of Operation Research 74, 443–459 (1994)
7. Polkowski, L.: Rough Sets: Mathematical Foundations. Advances in Soft Computing. Physica-Verlag (2002)
8. Skowron, A., Rauszer, C.M.: The discernibility matrix and functions in information systems. In: Słowiński, R. (ed.) Intelligent Decision Support: Handbook of Application and Advance of the Rough Set Theory, pp. 331–362. Kluwer Academic Publishers, Dordrecht (1992)
9. Ślęzak, D.: Approximate Entropy Reducts. Fundamenta Informaticae 53(3–4), 365–387 (2002)
10. Xu, J., Sun, L.: New Reduction Algorithm Based on Decision Power of Decision Table. In: Wang, G., Li, T., Grzymala-Busse, J.W., Miao, D., Skowron, A., Yao, Y. (eds.) RSKT 2008. LNCS (LNAI), vol. 5009, pp. 180–188. Springer, Heidelberg (2008)
11. Xu, Z., Zhang, C., Zhang, S., Song, W., Yang, B.: Efficient Attribute Reduction Based on Discernibility Matrix. In: Yao, J., Lingras, P., Wu, W.-Z., Szczuka, M.S., Cercone, N.J., Ślęzak, D. (eds.) RSKT 2007. LNCS (LNAI), vol. 4481, pp. 13–21. Springer, Heidelberg (2007)

Rule Induction for Decision Tables with Ordered Classes

Yoshifumi Kusunoki, Masanori Inoue, and Masahiro Inuiguchi

Abstract. In this paper, we study rule induction based on the rough set theory. In the rough set theory, we induce minimal rules from a decision table, which is a data set composed of objects. Each object is described by condition attributes and classified by a decision attribute. When the decision attribute of a given decision table is ordinal, we may induce rules w.r.t. upward/downward unions of decision classes. This approach would be better in simplicity of obtained rules than inducing rules w.r.t. decision classes directly. However, because of independent applications of rule induction methods, inclusion relations among upward/downward unions in the conclusions of obtained rules are not inherited to the premises of those. This non-inheritance may debase the quality of obtained rules. In this paper, we propose two approaches to inherit the implication relations among the conclusions of obtained rules to the premises of those. Moreover, we propose an approach to classification unseen objects using rules w.r.t. upward/downward unions of decision classes. The performances of the proposed approaches are examined by numerical experiments.

Keywords: Rule induction; Rough sets; Decision rule.

1 Introduction

Rough set theory [8] provides useful tools for analysis of decision tables. Rule induction is one of major applications of rough set theory. Methods for inducing minimal rules from a given decision table have been proposed. Let us mention some methods related to our research. Grzymala-Busse [1] proposed LEM2 as a rule induction algorithm under a given decision table with categorical attributes.

Yoshifumi Kusunoki, Masanori Inoue, and Masahiro Inuiguchi
Graduate School of Engineering Science, Osaka University
1-3, Machikaneyama, Toyonaka, Osaka 560-8531, Japan
e-mail: {kusunoki,inoue}@inulab.sys.es.osaka-u.ac.jp, inuiguti@sys.es.osaka-u.ac.jp

K. Nakamatsu et al. (Eds.): New Advan. in Intel. Decision Techno., SCI 199, pp. 565–574.
springerlink.com

This method does not induce all decision rules in a decision table but a minimal set of rules which can explain all consistent objects in the decision table. Grzymala-Busse [2] proposed MLEM2 as an extension of LEM2 for accommodating numerical/ordinal attributes in a decision table. Greco et al. [3] proposed DRSA (dominance-based rough set approach) assuming the monotonicity between condition attributes and a decision attribute. Greco et al. [4] proposed the rule induction algorithm DOMLEM as another extension of LEM2 based on DRSA. DOMLEM induces rules showing monotonic relations between condition and decision attributes.

We consider decision tables with ordinal decision attributes. In this case, even if we do not assume the monotonicity between condition and decision attributes, we may induce rules with respect to upward and downward unions of decision classes as DOMLEM does. This approach would be better in the simplicity of induced rules than the approach of inducing rules from decision classes directly. However, any LEM2-based algorithms described above are independently applied to the induction of rules inferring the membership to each upward/downward union of decision classes. By the nature of upward and downward unions, we have inclusion relations between two upward unions as well as between two downward unions. However, because of the independent applications of LEM2-based algorithms, the premises of induced rules corresponding to two upward/downward unions do not reflect the inclusion relation of the upward/downward unions. This non-reflection may debase the quality of obtained rules. In this paper, we propose a few approaches to reflect the inclusion relation between two upward/downward unions to premises of the corresponding rules. The performances of the proposed approaches as well as the conventional approaches are examined and compared by numerical experiments.

This paper is organized as follows. In next section, we briefly review rule induction and rough sets. Then the approaches to induce rules whose premises reflect the implication relations between conclusions are described in Sect. 3. In Sect. 4, the proposed combined approaches and applications of induced rules to unseen object classification are delivered. In Sect. 5, numerical experiments are explained and results of experiments are shown and interpreted. Finally, in Sect. 6, concluding remarks are given.

2 Rule Induction Based on Rough Sets

2.1 Decision Tables, Decision Rules and Rough Sets

In rough set theory, decision tables showing object features are analyzed. A decision table is formally characterized by a quadruple $\langle U, C \cup \{d\}, V, f \rangle$, where U is a finite set of objects, C is a finite set of condition attributes, $d \notin C$ is a unique decision attribute, $V = \bigcup_{a \in C \cup \{d\}} V_a$ is a set of attribute values, V_a is a set of values with respect to attribute a and $f : U \times C \cup \{d\} \to V$ is a total function called an information function. Objects in a decision table are classified by their decision attribute values into decision classes Cl_i, $i = 1, 2, \ldots, p$. We assume that each attribute is nominal or ordinal. Values for a nominal attribute are considered labels and used to distinguish objects. Therefore, there is neither an order relation between nominal attribute values nor degrees of differences. On the other hand, values of an ordinal attribute are assumed to be totally ordered, however degrees of differences are meaningless.

A decision rule is an if...then... rule composed of a premise and a conclusion. It is described as "**IF** *premise* **THEN** *conclusion*". The premise of a decision rule is expressed as a conjunction of elementary conditions. The elementary condition is represented as '$f(a,x) = v$' ($v \in V_a$) for a nominal condition attribute a while it is represented as '$f(a,x) \geq v^{\mathrm{L}}$' or '$f(a,x) < v^{\mathrm{R}}$' ($v^{\mathrm{L}}, v^{\mathrm{R}} \in V_a$) for an ordinal/numerical condition attribute a. The conclusion is represented as '$x \in Cl_i$' or '$x \in \bigcup_i Cl_i$' in this paper. Decision rules are induced based on objects in the given table by generalizing their conditions. A decision rule is simply called a rule if there is no confusion.

When a given decision table includes at least two objects which share the same condition attribute values but take different decision attribute values, the table is inconsistent. In this case, we cannot induce exact rules corresponding to those objects. Then we apply the rough set theory [8]. In the rough set theory, the inconsistency is treated well by lower and upper approximations of decision classes.

Now let us introduce lower and upper approximations. First, we define an equivalence class of an object x with respect to C by $[x]_C = \{y \in U \mid f(y,a) = f(x,a), \forall a \in C\}$. Then lower and upper approximations of Cl_i are defined by $C_*(Cl_i) = \{x \in U \mid [x]_C \subseteq Cl_i\}$ and $C^*(Cl_i) = \{x \in U \mid [x]_C \cap Cl_i \neq \emptyset\}$. If $C_*(Cl_i) = C^*(Cl_i)$, the decision class Cl_i is rough; otherwise, Cl_i is crisp.

The lower approximation of a decision class includes only consistent objects while the upper approximation includes possible objects. Objects in lower approximations as well as those in upper approximations are consistent in the sense that there is no conflict between members and nonmembers.

When a given decision table includes inconsistent objects, we induce rules with respect to Cl_i (resp. $\bigcup_i Cl_i$) based on objects in its lower approximation $C_*(Cl_i)$ (resp. $\bigcup_i Cl_i$) or those in its upper approximation $C^*(Cl_i)$ (resp. $\bigcup_i Cl_i$). Rules induced based on objects in lower approximations are called certain rules while rules induced based on objects in upper approximations are called possible rules. In this paper, we focus on certain rules.

2.2 Rule Induction Algorithm

In this paper, we study methods for inducing rules whose premises reflect the implication relations between conclusions. We use MLEM2 proposed by Grzymala-Busse [2] as the fundamental rule induction algorithm.

The algorithm of MLEM2 is shown in Fig. 1. First, a target decision class or a set of decision classes is approximated by the manner of rough set theory. In this algorithm, B is the input and it is defined by a lower approximation or a upper approximation of a certain class or a union of classes. The algorithm outputs $\mathbf{T}$, a minimal set of minimal rules. The minimal set of decision rules means that there exists no redundant rule in itself. On the other hand, the minimal rule means there

Input: a set B
Output: a single local covering $\mathbf{T}$ of B
1: $\mathbf{T} := \emptyset$; $G := B$;
2: **while** $G \neq \emptyset$ **do**
3: $T := \emptyset$; $T(G) := \{t \mid [t] \cap G \neq \emptyset\}$;
4: **while** $T = \emptyset$ **or** $[T] \not\subseteq B$ **do**
5: † select $t \in T(G)$ with the highest priority,
if a tie occurs, select $t \in T(G)$ such that $|[t] \cap G|$ is maximum;
if another tie occurs, select $t \in T(G)$ with smallest cardinality of $[t]$;
if a further tie occurs, select a first one;
6: $T := T \cup \{t\}$;
7: $G := [t] \cap G$;
8: $T(G) := \{t \mid [t] \cap G \neq \emptyset\}$; $T(G) := T(G) - T$;
9: **end while**
10: **for each** t in T **do**
11: **if** $[T - \{t\}] \subseteq B$ **then** $T := T - \{t\}$; **end if**
12: **end for**
13: $\mathbf{T} := \mathbf{T} \cup \{T\}$;
14: $G := B - \bigcup_{T \in \mathbf{T}} [T]$;
15: **end while**
16: **for each** T in $\mathbf{T}$ **do**
17: **if** $\bigcup_{S \in \mathbf{T} - \{T\}} [S] = B$ **then** $\mathbf{T} := \mathbf{T} - \{T\}$; **end if**
18: **end for**

Fig. 1 Algorithm of MLEM2

exists no redundant condition in its premise. The algorithm is based on the sequential covering method [7] equipped with the general to specific search [7].

The algorithm iteratively adds the best condition in a given greedy criterion until objects satisfying all conditions belong to B, and removes those objects from target set G. Redundant conditions are removed at first screening process at lines 10–12. Taking a conjunction of remaining conditions, a rule having the conjunction as its premise is induced. This iteration written from lines 2–15 of Fig. 1 repeats until G becomes empty. The greedy criterion is lexicographical as specified at line 5. From the first criterion, a condition satisfied with the most objects in target set G is selected. The lines 16–18 are the second screening process which removes redundant rules.

3 Rule Induction for Ordered Classes

In the real world, we may have decision tables whose decision attribute is ordinal. As examples of such an ordinal decision attribute, we may have the evaluation of candidates for the employment in a company, the environment assessment of cities, and so on. In this paper, we focus on such decision tables, and assume $Cl_1 \prec \ldots \prec$

Cl_p, i.e., Cl_1 and Cl_p are the worst and best, respectively, and the other decision classes are linearly ordered.

When decision classes are ordered, we may induce rules inferring memberships of upward and downward unions of decision classes. The upward and downward unions of decision classes are respectively defined by $Cl_i^{\geq} = \bigcup_{j \geq i} Cl_j$ and $Cl_i^{\leq} = \bigcup_{j \leq i} Cl_j$. The induction of those rules would be more advantageous than the induction of rules inferring memberships of decision classes concerning to the simplicity of induced rules. Then we consider the induction of rules inferring memberships to upward/downward unions of decision classes.

In the conventional application, we use a rule induction algorithm independently for each upward/downward union of decision classes. Due to the independent applications, the two sets of induced rules with respect to upward unions $Cl_i^{\geq}$ and $Cl_j^{\geq}$ $(i > j)$ may have no relation despite the fact that $Cl_i^{\geq} \subseteq Cl_j^{\geq}$. Considering inclusion relation $Cl_i^{\geq} \subseteq Cl_j^{\geq}$, it is natural that there exists an induced rule concluding $Cl_j^{\geq}$ having a weaker premise than the premise of any induced rule concluding $Cl_i^{\geq}$. However, such implication relations are not guaranteed in the conventional application of a rule induction algorithm. Similar unfavorable results may also happen for rules with respect to downward unions. Under such unfavorable results, we may have a strange consequence that an unseen object is in $Cl_i^{\geq}$ whereas it is not in $Cl_j^{\geq}$ in the utilization of induced rules.

In this paper, we consider approaches to rule induction which guarantee the implication relations described above between any two sets of induced rules with respect to upward/downward unions. In [6], we have proposed possible approaches to this subject using MLEM2 as the rule induction algorithm. Two approaches used in this paper are explained as follows:

Refining Approach. Let X and Y be conclusions of rules and X implies Y. This approach is based on the idea that the premise of rules concluding X should be stronger than the premise of rules concluding Y. To crystallize this idea, after inducing all rules having conclusion Y, rules having conclusion X are induced by refining the premises of rules having conclusion Y. In MLEM2, we may realize this approach by the following changes: (1) we replace "$T := \emptyset$;" by "$T := T'$; $G := [T] \cap G$;" at line 3 of Fig. 1, where T' is the premise of a rule concluding Y. (2) Before this command, we should select T' from premises of all rules concluding Y by the same criterion described at line 5. (3) Moreover we must modify the screening of T so that we maintain the relation that T implies T'.

Alleviating Approach. In contrast to the refining method, this method is based on the idea that the premise of rules concluding Y should be weaker than the premise of rules concluding X when X implies Y. To crystallize this idea, after inducing all rules having conclusion X, rules having conclusion Y are induced by alleviating the premises of rules having conclusion X. In MLEM2, we may realize this approach by the following changes: (1) we replace "$T(G) := \{t \mid [t] \cap G \neq \emptyset\}$" by "$T(G) := \{t \mid [t] \cap G \neq \emptyset, T' \text{ implies } t\}$" at lines 3 and 8 of Fig. 1, where T' is the premise of a rule concluding X. (2) Moreover we must modify the screening of **T** so that we maintain the relation that for each rule concluding

Table 1 Five possible combinations

Combination	Upward unions	Downward unions
CA1	Refining approach	Refining approach
CA2	Refining approach	Alleviating approach
CA3	Alleviating approach	Refining approach
CA4	Alleviating approach	Alleviating approach

X, there exists at least one rule concluding Y whose premise is implied by the premise of it.

Other than the two approaches described above, we can have more sophisticated approaches. For example, one possible approach is as follows: (0) we select $t \in \{1,\ldots,p\}$, appropriately. (1) we induce rules with respect to $Cl_t^{\geq}$ using MLEM2, (2) we induce rules with respect to $Cl_{t+1}^{\geq},\ldots,Cl_n^{\geq}$ by the refining approach, and (3) we induce rules with respect to $Cl_{t-1}^{\geq},\ldots,Cl_1^{\geq}$ using the alleviating approach. This approach is the mixture of the refining and alleviating approaches. However, in this paper, we restrict ourselves within the two basic approaches described above.

4 The Proposed Combined Approaches and Rule Application

4.1 Combinations of Approaches for Upward and Downward Unions

In the previous section, we have described rule induction approaches to upward or downward unions. However, we have not yet mentioned about the rule induction approaches to obtain both rules with respect to upward unions and rules with respect to downward unions. Unless both rules with respect to upward unions and rules with respect to downward unions are induced, we would not infer the membership to decision class from condition attribute values of an unseen object with a sufficient accuracy. In this paper, we consider four combinations of two approaches described in the previous section. The four combinations are shown in Table 1.

As shown in Table 1, the first combined approach, CA1 uses the refining approach to both rule induction with respect to upward and downward unions. The second combined approach, CA2 uses the refining approach to rule induction with respect to upward unions and the alleviating approach to rule induction with respect to downward unions. The third combined approach, CA3 is the opposite to CA2 and uses the alleviating approach to rule induction with respect to upward unions and the refining approach to rule induction with respect to downward unions. The fourth combined approach, CA4 uses the alleviating approach to both rule induction with respect to upward and downward unions.

4.2 Application of Induced Rules to Unseen Objects

Once we obtain rules concluding upward unions as well as those concluding downward unions, we can use those rules to classify an unseen object into a decision class. However, due to the lack of the comprehensiveness and total consistency of data in a given decision table, induced rules are often imperfect. They would have some conflicts, inapplicabilities and indecisiveness. Therefore, we have proposed a resolving method to those difficulties in [6]. The method consists of two steps. First, we decide whether an unseen object u is classified into $Cl_i^{\geq}$ or $Cl_{i-1}^{\leq}$ for each $i = 2,\ldots,p$ using the method described in [2]. The method in [2] classifies u based on factors pertaining to degrees of support and lengths of rules whose premises are completely or partially matched by u. Second, we aggregate results in the first step. We adopt a majority voting, namely, if u is classified into $Cl_i^{\geq}$ (and not into $Cl_{i-1}^{\leq}$) in the first step, then it votes for $Cl_i, Cl_{i+1}, \ldots, Cl_p$, and u is finally classified into a decision class with the most votes. A tie is broken by selecting arbitrarily. For example, there exist four classes Cl_1, Cl_2, Cl_3 and Cl_4, and an unseen object u is classified into $Cl_2^{\geq}$, $Cl_3^{\geq}$ and $Cl_4^{\leq}$ in the first step, then votes of Cl_1, Cl_2, Cl_3 and Cl_4 become 1, 2, 3 and 2, respectively. Consequently, u is classified into Cl_3.

5 Numerical Experiments

5.1 Explanation of the Experiments

In order to examine the performances of the proposed approaches, we executed numerical experiments. We compare the proposed approaches with the conventional approach in MLEM2, i.e., independent applications of MLEM2 to each of upward and downward unions.

We apply the proposed approaches as well as the conventional approach to a set of training data. We then evaluate the performance of the obtained set of rules by a set of checking data. The performance is measured by classification accuracy, i.e., the rate of objects correctly classified, to all checking data.

To execute the experiments, we need data suitable for proposed approaches. However, unfortunately, it is not very easy to obtain such data set from the public domain, even though we often to come across them in the real world. Then we artificially generate data which has nested structures. To obtain the data, we randomly generated hyper rectangles $H_1 \supseteq \cdots \supseteq H_p$, where p is a given number of decision class and H_l, $l = 1,\ldots,p$ are subsets of the Cartesian product of attribute value sets to assumed condition attributes. We then regard a point in the Cartesian product as an object and the vector representing the point as condition attribute values. The decision class Cl_l is composed of points in $H_l \setminus H_{l+1}$, where, for convenience, we define $H_{p+1} = \emptyset$. Several sequences of hyper rectangles $H_{j1} \supseteq \cdots \supseteq H_{jp}$, $j = 1,\ldots,q$ are independently generated. Then some hyper rectangles of different sequences may have intersections. In this case, the decision attribute value of an

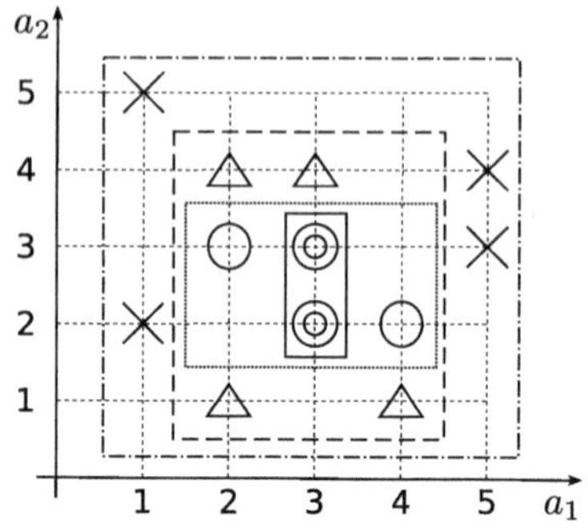

Fig. 2 The figure is an example of the nested structure with two condition attributes each of which takes five ordered values. A decision attribute has four values, ⊚, ○, △ and × which are ordered as ⊚ ≻ ○ ≻ △ ≻ ×

object x in an intersection is determined by the largest value l among H_{jl}'s such that $x \in H_{jl}$. Figure 2 shows an example of the nested structure.

We generate all objects in H_{j1}, $j = 1,\ldots,q$ and collect pairs of condition attribute values and the decision attribute value of them as data. The set of training data are composed of sampled data and the checking data are composed of the remaining data.

5.2 Results and Discussion

Results in four decision tables which vary in numbers of condition attributes and numbers of its values are shown in Tables 2–5. All of the decision tables have four ordered decision classes and three sequences of hyper rectangles. All condition attributes in each decision table are ordered. We executed the experiments with several different sampling rates for training data. The sampling rates are 1%, 2%, ..., 8%, 9%. In each table, row 'Rate' shows the sampling rates and row 'ML' indicates the results of the conventional approach based on MLEM2, i.e., the independent applications to upward/downward unions. Rows 'CA1', 'CA2', 'CA3' and 'CA4' show results in the proposed approaches of Table 1.

All degrees of the classification accuracy are calculated using checking data. We prepare 100 different sets of training data for each sampling rate. We conducted 100 experiments for each proposed approach at a sampling rate. Each entry of Tables shows the average *ave* and the standard deviation *dev* in the form of *ave* ± *dev*. The mark * means the average of the accuracy of the proposed approach is not significantly different from that of MLEM2 by the paired t-test with significance level $\alpha = 0.05$.

From Table 2–5, CA2 is significantly better than ML for all sampling rates. However, CA1, CA3 and CA4 are worse than or equal to ML, especially CA3 is significantly worse than ML. CA2 is better than CA1 and CA4 is better than CA3 at averages of accuracy. Note that CA2 and CA4 use the alleviating approach to downward unions while CA1 and CA3 apply the refining approach. Moreover, CA2 is better than CA4 and CA1 is better than CA3. CA2 and CA1 use refining approach to upward unions. Combining this observation with the data generation in the experiments, alleviating approach seems to be good for reverse-hill/reverse-valley

Table 2 Classification accuracy in a decision table with five ordinal condition attributes which take four values, four ordered classes and three sequences of hyper rectangles

Rate	1%	2%	3%	4%	5%	6%	7%	8%	9%
ML	34.2 ±7.3	43.3 ±7.7	51.2 ±6.4	55.6 ±7.5	61.0 ±7.2	65.7 ±5.8	67.1 ±6.1	70.2 ±6.1	73.1 ±6.4
CA1	32.9 ±7.2	42.3* ±7.2	50.5* ±7.3	54.0 ±7.5	59.1 ±6.6	63.1 ±6.1	65.6 ±6.7	68.6 ±6.2	71.7 ±6.3
CA2	34.0* ±8.0	43.8* ±8.1	53.6 ±7.5	56.6* ±7.4	61.7* ±6.3	66.4* ±6.1	68.2 ±6.5	70.4* ±5.8	73.5* ±5.9
CA3	32.4 ±6.6	41.3 ±7.0	47.5 ±6.9	51.4 ±6.8	56.2 ±6.9	61.4 ±6.6	64.1 ±6.9	67.3 ±6.8	69.3 ±6.9
CA4	33.5* ±6.9	41.4 ±6.5	48.9 ±7.0	52.6 ±6.5	57.9 ±6.8	62.4 ±6.8	65.6 ±7.1	68.3 ±6.8	70.3 ±6.9

Table 3 Classification accuracy in a decision table with five ordinal condition attributes which take five values, four ordered classes and three sequences of hyper rectangles

Rate	1%	2%	3%	4%	5%	6%	7%	8%	9%
ML	48.1 ±8.3	64.1 ±8.1	76.7 ±6.5	83.2 ±5.8	88.2 ±4.9	92.7 ±4.3	95.6 ±3.0	96.6 ±2.6	97.3 ±2.5
CA1	47.7* ±9.0	64.1* ±10.0	76.3* ±8.4	83.3* ±6.8	89.4 ±5.2	93.1* ±4.5	95.9* ±3.4	97.0* ±2.5	97.9 ±2.4
CA2	49.4* ±9.7	67.5 ±9.0	80.5 ±8.3	86.9 ±6.4	92.1 ±5.5	96.0 ±3.9	97.8 ±2.9	98.6 ±2.1	98.9 ±1.8
CA3	46.7 ±9.5	61.5 ±10.0	73.8 ±8.5	81.4 ±7.1	87.6* ±5.7	91.9 ±5.1	95.2* ±3.6	96.6* ±3.1	97.6* ±2.9
CA4	47.2* ±9.5	64.5* ±10.2	77.3* ±7.8	84.6 ±7.0	90.2 ±5.5	94.3 ±4.6	96.9 ±3.5	98.1 ±2.6	98.6 ±2.2

Table 4 Classification accuracy in a decision table with six ordinal condition attributes which take four values, four ordered classes and three sequences of hyper rectangles

Rate	1%	2%	3%	4%	5%	6%	7%	8%	9%
ML	47.1 ±6.8	65.2 ±6.9	76.6 ±5.9	85.7 ±4.9	90.6 ±4.0	93.6 ±4.0	96.0 ±2.6	97.9 ±2.1	98.4 ±1.7
CA1	46.9* ±7.3	62.9 ±7.6	74.3 ±7.5	83.9 ±5.3	88.8 ±4.8	92.0 ±3.7	94.6 ±2.8	96.6 ±2.8	97.7 ±2.2
CA2	51.2 ±8.8	68.4 ±8.2	81.6 ±7.3	90.5 ±5.7	94.1 ±4.1	96.7 ±3.1	98.1 ±2.1	99.2 ±1.2	99.5 ±0.9
CA3	43.7 ±6.9	58.6 ±7.6	69.7 ±6.7	81.2 ±6.6	86.8 ±5.7	90.8 ±5.1	93.7 ±3.4	96.4 ±3.1	97.3 ±2.4
CA4	46.8* ±7.9	62.7 ±8.1	75.7* ±7.0	86.4* ±6.9	91.6 ±5.2	94.9 ±4.5	97.1 ±3.3	98.9 ±1.7	99.1 ±1.5

Table 5 Classification accuracy in a decision table with six ordinal condition attributes which take five values, four ordered classes and three sequences of hyper rectangles

Rate	1%	2%	3%	4%	5%	6%	7%	8%	9%
ML	72.1 ±4.9	86.8 ±3.6	92.7 ±2.3	95.5 ±1.6	96.9 ±1.2	97.8 ±0.9	98.6 ±0.7	98.8 ±0.6	99.1 ±0.6
CA1	70.7 ±5.8	85.5 ±3.7	91.8 ±2.7	94.9 ±1.8	96.6 ±1.5	97.6* ±1.2	98.3 ±0.9	98.6 ±0.8	98.9 ±0.7
CA2	74.5 ±6.2	88.1 ±3.2	93.8 ±2.2	96.1 ±1.6	97.5 ±1.3	98.3 ±1.0	99.0 ±0.6	99.2 ±0.6	99.4 ±0.5
CA3	65.1 ±5.3	82.1 ±4.2	89.6 ±2.8	93.4 ±2.0	95.2 ±1.6	96.5 ±1.3	97.7 ±1.0	98.1 ±0.8	98.5 ±0.8
CA4	68.1 ±5.4	84.0 ±3.8	91.2 ±2.5	94.5 ±1.8	95.9 ±1.5	97.1 ±1.2	98.3 ±0.8	98.6 ±0.7	98.8 ±0.6

structures, where a reverse-hill (resp. reverse-valley) structure is a structure composed of complements of upward (resp. downward) unions when upward (resp. downward) unions form a hill (resp. valley) structure. Moreover, refining approach seems to be good for hill/valley structures.

6 Concluding Remarks

In this paper, we have investigated the rule induction approaches when the decision attribute values are ordered. We have studied the refining approach, the alleviating approach to induce rules whose premises reflect implication relations among conclusions. Since we may apply different methods to upward unions and downward unions, we consider four combinations of the rule induction approaches. In order to evaluate the performances of the proposed combined approaches, we executed numerical experiments. We compared the four combined approaches and the conventional approach in the classification accuracy of the induced set of rules. As the result, we conclude that the best combined approach adopts the refining approach for hill/valley structures and adopts the alleviating approach for reverse-hill/reverse-valley structures.

We may construct a new rule induction approach by hybridization of refining and alleviating approaches as described in Sect. 3. The investigation of the utilization of the new rule induction approach would be a future topic.

Acknowledgements. This work was partially supported by the Grant-in-Aid for Scientific Research (B) No.17310098.

References

1. Grzymala-Busse, J.W.: LERS-A system for learning from examples based on rough sets. In: Słowiński, R. (ed.) Intelligent Decision Support.Handbook of Application and Advances of the Rough Set Theory, pp. 3–18. Kluwer Academic Publishers, Dordrecht (1992)
2. Grzymala-Busse, J.W.: MLEM2-Discretization During Rule Induction. In: Proceedings of the IIPWM, pp. 499–508. Springer, Heidelberg (2003)
3. Greco, S., Matarazzo, B., Słowiński, R.: Rough sets theory for multicriteria decision analysis. Eur. J. Oper. Res. 129, 1–47 (2001)
4. Greco, S., Matarazzo, B., Słowiński, R.: Rough approximation by dominance relations. Int. J. Intell. Syst. 17, 153–171 (2002)
5. Inuiguchi, M.: Generalizations of Rough Sets: Rough sets under similarity, fuzzy and dominance relation. J. Jpn. Soc. Fuzzy Theory Syst. 13(6), 562–570 (2001)
6. Kusunoki, Y., Inoue, M., Inuiguchi, M.: Considering about inclusion of conclusion part of induction rules. In: RIMS Kôkyûroku 1584, Numerical Optimization Methods, Theory and Applications, pp. 59–71 (2008) (in Japanese)
7. Mitchell, T.M.: Machine learning. McGraw-Hill, New York (1997)
8. Pawlak, Z.: Rough sets. Int. J. Inf. Comp. Sci. 11(5), 341–356 (1982)

Dynamic Index Fund Optimization by a Heuristic GA Method Based on Correlation Coefficients

Yukiko Orito, Hisashi Yamamoto, Yasuhiro Tsujimura, and Yasushi Kambayashi

Abstract. The portfolio optimizations are generally to determine the proportion of funds in the portfolio consisting of the static assets. Then, it is hard to determine the proportion-weighted combination for the optimal portfolio consisting of the static large number of assets. In order to avoid this problem, we propose a Heuristic GA Method that optimizes the portfolio that consists of not only the given static assets but also the dynamically selected assets in this paper. In order to demonstrate the effectiveness of our method, we apply the method to creating an index fund based on correlation coefficients for the Tokyo Stock Exchange. This fund is one of the passively managed portfolios. The results show that our method works well for a dynamic index fund optimization.

1 Introduction

Portfolio optimization problems, based on the Markowitz's modern portfolio theory [1], are to select assets for a certain portfolio in advance and to optimize the proportion of funds in these static assets in order to achieve investment targets. Recently, many researchers have expanded the Markowitz's mean-variance model into

Yukiko Orito
Ashikaga Institute of Technology, 268-1, Ohmae-cho, Ashikaga, Tochigi 326-8558, Japan
e-mail: orito@ashitech.ac.jp

Hisashi Yamamoto
Tokyo Metropolitan University, 6-6, Asahigaoka, Hino, Tokyo 191-0065, Japan
e-mail: yamamoto@sd.tmu.ac.jp

Yasuhiro Tsujimura
Nippon Institute of Technology, 4-1, Gakuendai, Miyashiro-cho, Saitama 345-8501, Japan
e-mail: tujimr@nit.ac.jp

Yasushi Kambayashi
Nippon Institute of Technology, 4-1, Gakuendai, Miyashiro-cho, Saitama 345-8501, Japan
e-mail: yasushi@nit.ac.jp

K. Nakamatsu et al. (Eds.): New Advan. in Intel. Decision Techno., SCI 199, pp. 575–584.
springerlink.com

the practical applications by using evolutional methods [2, 3, 4, 5]. However, the static assets included in their portfolios have already been determined before applying optimization methods. The proportion-weighted combination of these static assets are represented as a real-valued array and they apply their own method to these assets. Therefore it is hard to determine the proportion-weighted combination for the optimal portfolio consisting of the static large number of assets.

On the other hand, Orito et al. [6] proposed a method to construct a portfolio by selecting a subset of the given assets. Aranha and Iba [7] also proposed a similar but different method. Both methods employ the binary digit 0 or 1 as the representations of "not-selected" or "selected" assets. However, when the number of selected assets for the portfolios increases, the number of proportion-weighted combinations also increases. Therefore it is hard to determine the proportion-weighted combination for the optimal portfolio consisting of the static large number of assets.

In order to avoid these problems, we propose a Heuristic GA Method that makes portfolios that consists of not only the given static assets but also the dynamically selected assets in this paper. Our method alternately repeats to move the assets expected to have good influence on the objective function into the portfolio and to remove the assets expected to have not-so-good influence on the objective function from the portfolio. The portfolio is optimized through a GA whenever the assets are dynamically selected. Hence, our method has the following advantages. First, even the assets not included in the portfolio in the first place can be selected to the portfolio by the repetition of procedure of our method. Second, we can optimize the portfolio that consists of only the valuable assets expected to have good influence on the objective function.

2 Portfolio Optimization Problem

In this section, we describe the portfolio optimization problem. First, we define the following notations.

i: Asset i, $i = 1, \cdots, N$.
A: the set consisting of all the assets. That is $\mathbf{A} = \{1, \cdots, N\}$.
X: the set consisting of assets. That is $\mathbf{X} = \{1, \cdots, \#\mathbf{X}\}$ whose $\#\mathbf{X}$ represents the number of elements in the set **X**.
t: time basis, dates data $t = 1, \cdots, T$.
P_{index}: the value of the given benchmark index at t.
$\mathbf{P}_{index}$: the rates of changes of benchmark index. That is the vector $\mathbf{P}_{index} = (P_{index}(1), \cdots, P_{index}(T))$ whose $P_{index}(t)$ is defined by $P_{index}(t) = (P_{index}(t+1) - P_{index}(t)) / P_{index}(t)$.
$F_i(t)$: the price of Asset i at t.
$\mathbf{P}_i$: the rates of return of Asset i. That is the vector $\mathbf{P}_i = (P_i(1), \cdots, P_i(T))$ whose $P_i(t)$ is defined by $P_i(t) = (F_i(t+1) - F_i(t))/F_i(t)$.
w_i: the weight (the proportion of funds) of Asset i included in a portfolio. That is $0 \le w_i \le 1$. Note that we do not discuss the short sale in this paper.

$\mathbf{G_X}$: the portfolio for the set $\mathbf{X}$. That is the vector $\mathbf{G_X} = (w_1, \cdots, w_{\#\mathbf{X}})$ such that $\sum_{i=1}^{\#\mathbf{X}} w_i = 1$.

$\mathbf{P_{G_X}}$: the return rates of portfolio $\mathbf{G_X}$. That is the vector $\mathbf{P_{G_X}} = (P_{\mathbf{G_X}}(1), \cdots, P_{\mathbf{G_X}}(T))$ whose $P_{\mathbf{G_X}}(t)$ is defined by $P_{\mathbf{G_X}}(t) = \sum_{i=1}^{\#\mathbf{X}} w_i \cdot P_i(t)$.

In this paper, we apply the Heuristic GA Method to optimize the index funds. Index funds are popular passively managed portfolios and are constructed to mimic the performance of the given benchmark index, such as the S&P 500 in New York. As an objective function to evaluate the relation between the fund's price and the benchmark index, Orito et al. [8] adopted a coefficient of determination and Oh et al. [9] adopted a fitness measure based on the beta. The coefficient of determination or the beta is one of the major functions in order to evaluate correlation between the fund and the overall market. In this paper, we focus on the index fund optimization problems based on Orito et al. [8]. However a coefficient of determination has an important problem that it is expressed by the square of a correlation coefficient. It means that the coefficient of determination might approach 1 even though the correlation coefficient approaches -1. In order to avoid this problem, we adopt the correlation coefficient between the fund's return rates and the rates of changes of the benchmark index as our objective function.

In this paper, we define the portfolio optimization problem based on the coefficient of determination between the return rates of portfolio $\mathbf{P_{G_X}}$ and the rates of changes of benchmark index $\mathbf{P}_{index}$ as

$$\max_{\mathbf{X} \in \mathbf{A}} \quad R_{\mathbf{G_X}} = \frac{cov(\mathbf{P}_{index}, \mathbf{P_{G_X}})}{\sqrt{var(\mathbf{P}_{index}) \cdot var(\mathbf{P_{G_X}})}}, \qquad (1)$$

$$\text{s.t.} \sum_{i=1}^{\#\mathbf{X}} w_i = 1 \ (i \in \mathbf{X})$$

where $cov(\mathbf{P}_{index}, \mathbf{P_{G_X}})$ is covariance of $\mathbf{P}_{index}$ and $\mathbf{P_{G_X}}$, $var(\mathbf{P}_{index})$ is variance of $\mathbf{P}_{index}$ and $var(\mathbf{P_{G_X}})$ is variance of $\mathbf{P_{G_X}}$, respectively.

3 Heuristic GA Method

Suppose that the initial set consisting of N assets is defined as a set $\mathbf{A} = \{1, \cdots, N\}$. Based on the set $\mathbf{A}$, two subsets are defined as $\mathbf{H}$ $(\subseteq \mathbf{A})$ and $\mathbf{L}$ $(\subseteq \mathbf{A})$. We note that the subset $\mathbf{H}$ is a subset consisting of assets selected for the portfolio and the subset $\mathbf{L}$ is a subset consisting of assets not selected for the portfolio. Hence, as the relation between the subsets $\mathbf{H}$ and $\mathbf{L}$, note that $\mathbf{A} = \mathbf{H} \cup \mathbf{L}$ and $\mathbf{H} \cap \mathbf{L} = \emptyset$ hold. In this paper, we call the subsets $\mathbf{H}$ and $\mathbf{L}$ "Portfolio Set" and "Non-portfolio Set", respectively.

In order to maximize the correlation coefficient defined by Eq. (1), we propose a Heuristic GA Method that alternately repeats to move the "valuable assets" from

Non-portfolio Set **L** into Portfolio Set **H** and to remove the "less-valuable assets" from Portfolio Set **H** to Non-portfolio Set **L**. The valuable asset means the asset expected to have good influence on the correlation coefficient. On the other hand, the less-valuable asset means the asset expected to have not-so-good influence on the correlation coefficient. These valuable and less-valuable assets are selected through a GA.

We describe a GA in Sect. 3.1. In addition, the selection criteria of the valuable and less-valuable assets are given by a property of correlation coefficient. We describe this property in Sect. 3.2 and describe the procedure of our Heuristic GA Method in Sect. 3.3.

3.1 GA

In Heuristic GA Method, we apply the following GA to Portfolio Set **H** and Non-portfolio Set **L**.

For the genetic representation, a gene represents the weight of Asset i, and is denoted by w_i $(0 \leq w_i \leq 1)$. A chromosome represents a portfolio for the set **X**, and is denoted by $\mathbf{G_X} = (w_1, \cdots, w_{\#\mathbf{X}})$. The fitness value of GA is to maximize the correlation coefficient given by $R_{\mathbf{G_X}}$.

For the genetic operations, the GA randomly generates S_{pop} chromosomes for the initial population. In the above genetic representation, the chromosome $\mathbf{G_X}$ consists of the dynamically selected genes. We apply the uniform crossover and the uniform mutation. After making a new offspring, the GA repairs the new genes via renormalization. On each generation, the GA selects S_{pop} chromosomes by using the elitism selection in the order of high fitness values of GA. The GA repeats these operations until the terminate criterion, the final K-th generation, is satisfied. On the K-th generation, we choose one chromosome with the highest fitness value of GA in the population. This chromosome is expressed as $\mathbf{G_X}*$ for the subset **X**.

3.2 Property of Correlation Coefficient

The Portfolio Set **H** is defined by the union set of all the valuable assets selected by the repetition of procedure of the Heuristic GA Method. Therefore, in our method, it is assumed that the correlation coefficient becomes high by the union set of valuable assets.

We are now ready to present the following property that the correlation coefficient $R_{\mathbf{G}_{\mathbf{X}_a \cup \mathbf{X}_b}}$, which is obtained by the portfolio consisting of the union set of $\mathbf{X}_a$ and $\mathbf{X}_b$, is greater than or equal to the minimum of coefficients $R_{\mathbf{G}_{\mathbf{X}_a}}$ and $R_{\mathbf{G}_{\mathbf{X}_b}}$.

Property 1. For the two sets $\mathbf{X}_a$ and $\mathbf{X}_b$, let w $(0 \leq w \leq 1)$ be the weight of the portfolio for the set $\mathbf{X}_a$. If $R_{\mathbf{G}_{\mathbf{X}_a}} \geq 0$ and $R_{\mathbf{G}_{\mathbf{X}_b}} \geq 0$ are satisfied, the relation of correlation coefficients obtained by the portfolios is defined as follows:

$$R_{\mathbf{G}_{\mathbf{X}_a \cup \mathbf{X}_b}} \geq \min\left(R_{\mathbf{G}_{\mathbf{X}_a}}, R_{\mathbf{G}_{\mathbf{X}_b}}\right) \tag{2}$$

Proof. Using Eq. (1), the correlation coefficient obtained by the union of set $\mathbf{X}_a \cup \mathbf{X}_b$ is defined by

$$R_{\mathbf{G}_{\mathbf{X}_a \cup \mathbf{X}_b}} = \frac{cov(\mathbf{P}_{index}, \mathbf{P}_{\mathbf{G}_{\mathbf{X}_a \cup \mathbf{X}_b}})}{\sqrt{var(\mathbf{P}_{index}) \cdot var(\mathbf{P}_{\mathbf{G}_{\mathbf{X}_a \cup \mathbf{X}_b}})}}. \tag{3}$$

Since Eq. (3) holds the properties of $\mathbf{P}_{\mathbf{G}_{\mathbf{X}_a \cup \mathbf{X}_b}} = w\mathbf{P}_{\mathbf{G}_{\mathbf{X}_a}} + (1-w)\mathbf{P}_{\mathbf{G}_{\mathbf{X}_b}}$, it is represented by

$$R_{\mathbf{G}_{\mathbf{X}_a \cup \mathbf{X}_b}} = \frac{w \cdot R_{\mathbf{G}_{\mathbf{X}_a}} \sqrt{var(\mathbf{P}_{\mathbf{G}_{\mathbf{X}_a}})} + (1-w) \cdot R_{\mathbf{G}_{\mathbf{X}_b}} \sqrt{var(\mathbf{P}_{\mathbf{G}_{\mathbf{X}_b}})}}{w \cdot \sqrt{var(\mathbf{P}_{\mathbf{G}_{\mathbf{X}_a}})} + (1-w) \cdot \sqrt{var(\mathbf{P}_{\mathbf{G}_{\mathbf{X}_b}})}}$$
$$\times \frac{w \cdot \sqrt{var(\mathbf{P}_{\mathbf{G}_{\mathbf{X}_a}})} + (1-w) \cdot \sqrt{var(\mathbf{P}_{\mathbf{G}_{\mathbf{X}_b}})}}{\sqrt{w^2 var(\mathbf{P}_{\mathbf{G}_{\mathbf{X}_a}}) + (1-w)^2 var(\mathbf{P}_{\mathbf{G}_{\mathbf{X}_b}}) + 2w(1-w) cov(\mathbf{P}_{\mathbf{G}_{\mathbf{X}_a}}, \mathbf{P}_{\mathbf{G}_{\mathbf{X}_b}})}}.$$

From the Cauchy-Schwarz inequality, that is $\sqrt{var(\mathbf{P}_{\mathbf{G}_{\mathbf{X}_a}})}\sqrt{var(\mathbf{P}_{\mathbf{G}_{\mathbf{X}_b}})} \geq cov(\mathbf{P}_{\mathbf{G}_{\mathbf{X}_a}}, \mathbf{P}_{\mathbf{G}_{\mathbf{X}_b}})$, we have the following equation for $0 \leq w \leq 1$,

$$\frac{w \cdot \sqrt{var(\mathbf{P}_{\mathbf{G}_{\mathbf{X}_a}})} + (1-w) \cdot \sqrt{var(\mathbf{P}_{\mathbf{G}_{\mathbf{X}_b}})}}{\sqrt{w^2 var(\mathbf{P}_{\mathbf{G}_{\mathbf{X}_a}}) + (1-w)^2 var(\mathbf{P}_{\mathbf{G}_{\mathbf{X}_b}}) + 2w(1-w) cov(\mathbf{P}_{\mathbf{G}_{\mathbf{X}_a}}, \mathbf{P}_{\mathbf{G}_{\mathbf{X}_b}})}} \geq 1.$$

Therefore, if $0 \leq w \leq 1$, $R_{\mathbf{G}_{\mathbf{X}_a}} \geq 0$ and $R_{\mathbf{G}_{\mathbf{X}_b}} \geq 0$ are satisfied,

$$\begin{aligned} R_{\mathbf{G}_{\mathbf{X}_a \cup \mathbf{X}_b}} &\geq \frac{w\sqrt{var(\mathbf{P}_{\mathbf{G}_{\mathbf{X}_a}})}}{w\sqrt{var(\mathbf{P}_{\mathbf{G}_{\mathbf{X}_a}})} + (1-w)\sqrt{var(\mathbf{P}_{\mathbf{G}_{\mathbf{X}_b}})}} R_{\mathbf{G}_{\mathbf{X}_a}} \\ &\quad + \frac{(1-w)\sqrt{var(\mathbf{P}_{\mathbf{G}_{\mathbf{X}_b}})}}{w\sqrt{var(\mathbf{P}_{\mathbf{G}_{\mathbf{X}_a}})} + (1-w)\sqrt{var(\mathbf{P}_{\mathbf{G}_{\mathbf{X}_b}})}} R_{\mathbf{G}_{\mathbf{X}_b}} \\ &\geq \min\left(R_{\mathbf{G}_{\mathbf{X}_a}}, R_{\mathbf{G}_{\mathbf{X}_b}}\right). \end{aligned} \tag{4}$$

From Eq. (4), the correlation coefficient obtained by the portfolio consisting of the union set of $\mathbf{X}_a$ and $\mathbf{X}_b$ is greater than or equal to the minimum of coefficients for $\mathbf{X}_a$ or $\mathbf{X}_b$. □

3.3 Procedure of Heuristic GA Method

The Heuristic GA Method alternately repeats to move the valuable assets from Non-portfolio Set **L** into Portfolio Set **H** and to remove the less-valuable assets from

Portfolio Set **H** to Non-portfolio Set **L**. In this paper, we call the movement operation of valuable assets Step A and the removal operation of less-valuable assets Step B, respectively.

Let s_A be the repetition number of Step A and s_B be the repetition number of Step B. For example, the first process in Step A is expressed as 1_A. The Portfolio Set **H** and the Non-portfolio Set **L** on the s-th process from $s = 1$ to $s = 3$ are shown in Fig. 1. We describe the procedure of Steps A and B in Sects. 3.3.1 and 3.3.2, respectively.

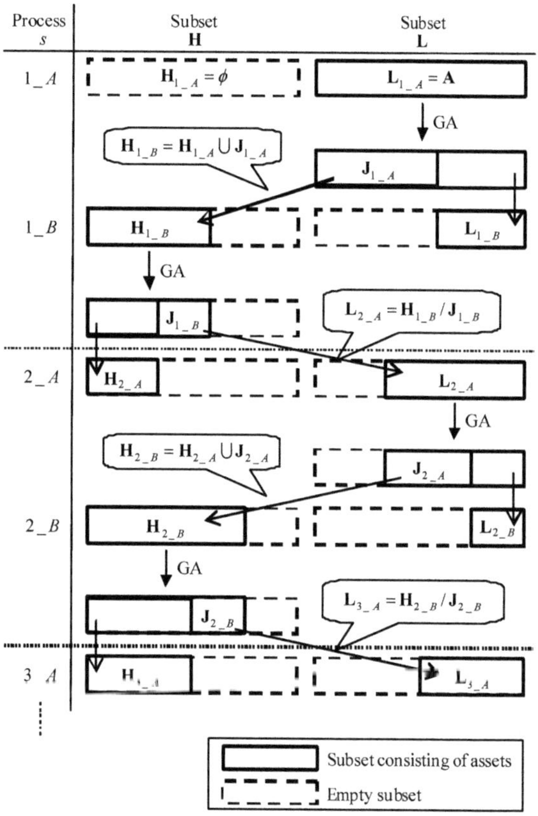

Fig. 1 Steps A and B in Heuristic GA Method

3.3.1 Step A

Step A in our method is to move the valuable assets expected to have good influence on the correlation coefficient from Non-portfolio Set $\mathbf{L}_{s\_A}$ into Portfolio Set $\mathbf{H}_{s\_B}$. First, the subsets $\mathbf{L}_{1\_A}$ and $\mathbf{H}_{1\_A}$ are initialized as $\mathbf{L}_{1\_A} = \mathbf{A}$ and $\mathbf{H}_{1\_A} = \emptyset$. On the s-th process, we apply the above GA to the $\mathbf{L}_{s\_A}$ and then we choose the best chromosome $\mathbf{G}_{\mathbf{L}_{s\_A}}*$ on the K-th generation. Our method moves the group of valuable assets, defined by $\mathbf{J}_{s\_A}$, whose weight belongs to the chromosome $\mathbf{G}_{\mathbf{L}_{s\_A}}*$ from the subset $\mathbf{L}_{s\_A}$ to the subset $\mathbf{H}_{s\_B}$ by using the heuristic rule defined as

$$\begin{aligned}\mathbf{J}_{s\_A} &= \left\{i | w_i \geq A_{bound}, i \in \mathbf{L}_{s\_A}, w_i \in \mathbf{G}_{\mathbf{L}_{s\_A}}*\right\},\\ \mathbf{H}_{s\_B} &= \mathbf{H}_{s\_A} \cup \mathbf{J}_{s\_A},\\ \mathbf{L}_{s\_B} &= \mathbf{L}_{s\_A} \setminus \mathbf{J}_{s\_A},\end{aligned} \tag{5}$$

where the boundary parameter A_{bound} is given in advance.

Since $\mathbf{H}_{s\_A}$ and $\mathbf{J}_{s\_A}$ in Eq. (5) hold the same properties of $\mathbf{X}_a$ and $\mathbf{X}_b$ in the property 1, the relation $R_{\mathbf{G}_{\mathbf{H}_{s\_A} \cup \mathbf{J}_{s\_A}}} \geq \min\left(R_{\mathbf{G}_{\mathbf{H}_{s\_A}}}, R_{\mathbf{G}_{\mathbf{J}_{s\_A}}}\right)$ holds. Hence, the correlation coefficient obtained by the union set $\mathbf{H}_{s\_B}$ is greater than or equal to the minimum of coefficients obtained by $\mathbf{H}_{s\_A}$ and $\mathbf{J}_{s\_A}$.

In addition, it is desired that $\mathbf{J}_{s\_A}$ consists of more valuable assets to the Portfolio Set $\mathbf{H}_{s\_B}$ than the assets included in $\mathbf{L}_{s\_B}$. We define the relation of $\mathbf{J}_{s\_A}$ and $\mathbf{L}_{s\_B}$ as

$$R_{\mathbf{G}_{\mathbf{J}_{s\_A} \cup \mathbf{H}_{s\_A}}} \geq R_{\mathbf{G}_{\mathbf{L}_{s\_B} \cup \mathbf{H}_{s\_A}}}. \tag{6}$$

However, the property 1 is not strong enough to select $\mathbf{J}_{s\_A}$ consisting of valuable assets that holds Eq. (6). Therefore, we give the following constraints to select $\mathbf{J}_{s\_A}$ that holds Eq. (6).

$$\begin{cases} cov\left(\mathbf{P}_{\mathbf{G}_{\mathbf{L}_{s\_B}}}, \mathbf{P}_{\mathbf{G}_{\mathbf{H}_{s\_A}}}\right) > 0 \\ cov\left(\mathbf{P}_{\mathbf{G}_{\mathbf{J}_{s\_A}}}, \mathbf{P}_{\mathbf{G}_{\mathbf{H}_{s\_A}}}\right) > 0 \\ R_{\mathbf{G}_{\mathbf{J}_{s\_A}}} \geq R_{\mathbf{G}_{\mathbf{L}_{s\_B}}} \geq 0 \end{cases} \tag{7}$$

Hence, Step A in our method moves the valuable assets to the Portfolio Set $\mathbf{H}_{s\_B}$ that holds Eq. (5) if the set $\mathbf{J}_{s\_A}$ holds Eq. (7).

3.3.2 Step B

Step B in our method is to remove the less-valuable assets expected to have not-so-good influence on the correlation coefficient from Portfolio Set $\mathbf{H}_{s\_B}$ into Non-portfolio Set $\mathbf{L}_{s+1\_A}$. The subsets $\mathbf{H}_{s\_B}$ and $\mathbf{L}_{s\_B}$ are given by Step A. On the s-th process, we apply the above GA to the $\mathbf{H}_{s\_B}$ and then we choose the best chromosome $\mathbf{G}_{\mathbf{H}_{s\_B}}*$ on the K-th generation. Our method removes the group of less-valuable assets, defined by $\mathbf{J}_{s\_B}$, whose weight belongs to the chromosome $\mathbf{G}_{\mathbf{H}_{s\_B}}*$ from the subset $\mathbf{H}_{s\_B}$ to the subset $\mathbf{L}_{s+1\_A}$ by using the heuristic rule defined as

$$\begin{aligned}\mathbf{J}_{s\_B} &= \{i | w_i \leq B_{bound}, i \in \mathbf{H}_{s\_B}, w_i \in \mathbf{G}_{\mathbf{H}_{s\_B}}*\},\\ \mathbf{L}_{s+1\_A} &= \mathbf{L}_{s\_B} \cup \mathbf{J}_{s\_B},\\ \mathbf{H}_{s+1\_A} &= \mathbf{H}_{s\_B} \setminus \mathbf{J}_{s\_B},\end{aligned} \tag{8}$$

where the boundary parameter B_{bound} is given in advance. As we did in Sect. 3.3.1 for Step A, we give the following similar constraints to select $\mathbf{J}_{s\_B}$ in Step B.

$$\begin{cases} cov\left(\mathbf{P}_{\mathbf{G}_{\mathbf{J}_{s\_B}}}, \mathbf{P}_{\mathbf{G}_{\mathbf{L}_{s\_B}}}\right) > 0 \\ cov\left(\mathbf{P}_{\mathbf{G}_{\mathbf{H}_{s+1\_A}}}, \mathbf{P}_{\mathbf{G}_{\mathbf{L}_{s\_B}}}\right) > 0 \\ R_{\mathbf{G}_{\mathbf{H}_{s+1\_A}}} \geq R_{\mathbf{G}_{\mathbf{J}_{s\_B}}} \geq 0 \end{cases} \tag{9}$$

Hence, Step B in our method removes the less-valuable assets from the Portfolio Set $\mathbf{H}_{s\_B}$ that holds Eq. (8) if the set $\mathbf{J}_{s\_B}$ holds Eq. (9).

3.3.3 Terminate Criterion

Repeating the processes of Steps A and B, we can define the $\mathbf{G}_{\mathbf{H}_{s\_B}}*$ as our portfolio. However, we cannot obtain the accurate number of assets that should move to the portfolio only by the property 1. Hence, we define Eq. (10) as the terminate criterion of procedure of Steps A and B.

$$R_{\mathbf{G}_{\mathbf{H}_{s+1\_B}}*} \leq R_{\mathbf{G}_{\mathbf{H}_{s\_B}}*} \tag{10}$$

When the terminate criterion is satisfied, we get the chromosome $\mathbf{G}_{\mathbf{H}_{s\_B}}*$ as our portfolio obtained by our Heuristic GA Method.

4 Numerical Experiments

We have applied our method to each of 12 data periods of the First Section of Tokyo Stock Exchange from Jan. 6, 1997 to Oct. 2, 2006. Each data period consists of 100 days, and is shifted every 200 days. The dataset is a subset of the TOPIX (Tokyo Stock Price Index). The TOPIX is a well known benchmark index and represents the increase or decrease in stock values of all assets on the market. The parameters used in our method are as follows: Total number of assets: $N = 1000$; Population size: S_{pop}=100; Crossover rate: 0.9; Mutation rate: 0.1; Generation size: $K = 100$; Method run: 20; Boundary parameter in Eq. (5): $A_{bound} = \alpha \cdot 1/\#\mathbf{L}_{s\_A}$ ($\alpha = 1.5, 2.0, 2.5, 3.0$); Boundary parameter in Eq. (8): $B_{bound} = \beta \cdot 1/\#\mathbf{H}_{s\_B}$ ($\beta = 0.3, 0.5, 0.7$).

In order to demonstrate the efficiency of our method, we compare the coefficients and the numbers of assets obtained by our method with those of obtained by a method that applies the GA to optimize the portfolio consisting of the static 1000 ($= N$) assets. The parameters and operations of GA for the compared method are the same of our GA except generation size. The generation size for the compared method is set to $K = 500$.

For each period, the average of correlation coefficients and the average of number of selected assets to the portfolio obtained by our method with $\alpha = 2.0$ and $\beta = 0.5$ and those of obtained by compared method are shown in Table 1.

We can observe that the correlation coefficients obtained by our method are higher than those of obtained by the compared method and our portfolio consists of the small number of assets. Hence, the results suggest that our method works well for the selecting and optimizing the index funds based on correlation coefficients.

Table 1 The correlation coefficient and the number of selected assets

Data	Our method		Compared method	
Period	Coefficient	Number of assets	Coefficient	Number of assets
1	0.999922	323.65	0.998377	1000
2	0.999879	319.55	0.998228	1000
3	0.999952	336.60	0.998929	1000
4	0.999886	381.65	0.997674	1000
5	0.999019	207.40	0.989667	1000
6	0.999791	311.45	0.995740	1000
7	0.999888	318.85	0.997331	1000
8	0.999906	315.35	0.998267	1000
9	0.999852	296.25	0.998399	1000
10	0.999970	335.35	0.999368	1000
11	0.999961	376.30	0.998173	1000
12	0.999971	400.00	0.999480	1000

For our portfolios, however, the number of selected assets depends on the boundary parameters A_{bound} in Step A and B_{bound} in Step B. This is our future subject.

5 Conclusions

In this paper, we have proposed a Heuristic GA Method for the dynamic index fund optimization based on correlation coefficients.

The numerical experiments demonstrate that our method works well for the optimization problem that makes the portfolios consisting of the small number of dynamically selected assets.

However, the number of selected assets depends on the boundary parameters in our method. Our next objectives are to improve these parameters in the method and to stabilize the number of dynamically selected assets.

Acknowledgements. This research was partially supported by Grant #20710119, Grant-in-Aid for Young Scientists (B) from JSPS, (2008-).

References

1. Markowitz, H.: Portfolio Selection. Journal of Finance 7, 77–91 (1952)
2. Xia, Y., Liu, B., Wang, S., Lai, K.K.: A Model for Portfolio Selection with Order of Expected Returns. Computers & Operations Research 27, 409–422 (2000)
3. Chang, T.-J., Meade, N., Beasley, J.E., Sharaiha, Y.M.: Heuristics for Cardinality Constrained Portfolio Optimization. Computers & Operations Research 27, 1271–1302 (2000)
4. Lin, C.C., Liu, Y.T.: Genetic Algorithms for Portfolio Selection Problems with Minimum Transaction Lots. European Journal of Operational Research 185(1), 393–404 (2008)

5. Streichert, F., Tanaka-Yamawaki, M.: The Effect of Local Search on the Constrained Portfolio Selection Problem. In: Proceedings of 2006 IEEE Congress on Evolutionary Computation, pp. 2368–2374 (2006)
6. Orito, Y., Yamamoto, H., Yamazaki, G.: Index Fund Selections with Genetic Algorithms and Heuristic Classifications. Journal of Computers and Industrial Engineering 45, 97–109 (2003)
7. Aranha, C., Iba, H.: Portfolio Management by Genetic Algorithms with Error Modeling. In: JCIS Online Proceedings of International Conference on Computational Intelligence in Economics & Finance (2007)
8. Orito, Y., Takeda, M., Iimura, K., Yamazaki, G.: Evaluating the Efficiency of Index Fund Selections Over the Fund's Future Period. Computational Intelligence in Economics and Finance 2, 157–168 (2007)
9. Oh, K.J., Kim, T.Y., Min, S.: Using Genetic Algorithm to Support Portfolio Optimization for Index Fund Management. Expert Systems with Applications 28, 371–379 (2005)

Simulation Modeling of Emergence-of-Money Phenomenon by Doubly Structural Network

Masato Kobayashi, Masaaki Kunigami, Satoru Yamadera, Takashi Yamada, and Takao Terano

Abstract. This paper describes simulation studies in order to examine the validity of our previous predictions of the emergence of money using Doubly Structural Network Model (DSN model). DSN model consists of two levels of networks: the one of inner-agent model to represent their beliefs or knowledge about the world and the other of inter-agent model to represent a social network among agents. Using DSN model, we have explained how the concepts of money as a exchangeable media emerges through agent interaction. DSN Model is congenial to agent-based simulation. In this paper, using large scale intensive computer experiments, we investigate the bifurcation analysis derived from dynamics of DSN model. We also show new emergent phenomena on various types of social networks.

Masato Kobayashi
Tokyo Institute of Technology, 4259 Nagatsuta-cho, Midori-ku, Yokohama 226-8502, Japan
e-mail: masato.gssm@gmail.com

Masaaki Kunigami
Ministry of Defense, 5-1 Ichigaya Honmuracho, Shinjuku-ku, Tokyo, 162-8801 Japan
e-mail: mkunigami@gakushikai.jp

Satoru Yamadera
Tokyo Institute of Technology
e-mail: satoru_yamadera@ybb.ne.jp

Takashi Yamada
Tokyo Institute of Technology
e-mail: tyamada@trn.dis.titech.ac.jp

Takao Terano
Tokyo Institute of Technology
e-mail: terano@dis.titech.ac.jp

K. Nakamatsu et al. (Eds.): New Advan. in Intel. Decision Techno., SCI 199, pp. 585–594.
springerlink.com

1 Introduction

A kind of goods called "money" plays an unique role as a "medium for exchange" in economy and society, even though that has little practical advantage. In this paper, we executed the Agent-Based Simulation for the emergence of money to verify our previous work's analytical results.

In our previous work, we proposed a new social model "Doubly Structural Network Model" [7][8]. The "Doubly Structural Network Model" is describing "the emergence of the money from the barter economy". Especially [8] shows the following bifurcation analysis results. a) The dynamics shows that even a commodity with no distinctive properties can become a medium of exchange. b) The dynamics enables bifurcation analysis, which shows that the degree of connection of the social network strongly affects the number of media of exchange.

1.1 Problem of the Origin/Emergence of Money

In economics, "money" is usually defined by the following functions [3, 4, 10];

- a medium of exchange,
- an unit of value,
- storage of value.

Amongst these, many economists maintain that money is essentially a "medium of exchange" [3, 5, 4, 10, 12]. For emergence of "medium of exchange", almost all the agents within the society must recognize that "a particular commodity is exchangeable with almost all others". This nature is called "general acceptability". In this paper, commodities that have general acceptability as a medium of exchange are defined as "proto-money" [8], the most primitive form of money. The emergence of money, therefore, means the emergence of "proto-money" from a barter economy.

1.2 Mathematical Models for the Emergence of Money

The emergence of money is studied in not only economics but also mathematical models [2, 6, 11, 9, 14].

A simulation model of exchanging commodities [17] shows that by adding the Maxim 'Accept what others accept!' (Menger's 'salability'), a commodity-money emerges when the "threshold of exchange" in the "view vector" of the Maxim exceeds a certain level. Another agent simulation [16] in lattice space shows a commodity becomes money based on the "trust" from agents.

In [17] and [14], they point out that the general acceptability of a certain commodity can be represented as a star-shaped network of acceptability or exchangeability around the commodity. We focus on that the networks of exchangeability are essentially important for bottom up approach. In the next section, we introduce our "Doubly Structural Network Model". This model can describe a emerging process of the general acceptability via coherent growth of individual star-shaped networks of recognition.

From another point of view, these researches suggest that emergence of money needs some structural change (change of the "threshold of exchange" in the "view vector" [17], establishing common recognition and a large disturbance [5], establishing trust [16]) in the society. Following sections illustrate that our model is useful to describe an emergence mechanism based on a social structure.

1.3 Why We Need Simulation Studies

In [7, 8], they have investigated the theoretical analyses of the emergence of money phenomena with doubly structural network model. However, the analyses were based on the simple assumptions about the network structures, because we must simplify the model to apply isocline analysis techniques in nonlinear systems. To employ the more realistic situation in to the model, we should move the theoretical analyses to agent-based computer simulation. The objectives of the simulation are summarized as follows:

1) Validate the simulation results compared with the theoretical ones applied to the regular network environment.
2) Investigate the hub effects extended from the regular network environment. The Environment is too complex to examine with the isoclines method.
3) Examine what will happen in the much more complex network environments in order to investigate the emergence mechanism of the money concept and the behaviors of the processes.

In section 2, we explain the principles of doubly structured network model and the simulation model will follow in section 3.

2 The Doubly Structural Network Model of the Emergence of Money

2.1 Doubly Structural Network Model

In this subsection, we review the model introduced in our previous work [8]. This model is unique from other related work since it has double structure of inter-agent social network and inner-agent recognition networks. This double structure of networks enables us to describe and to analyze the emergence of common knowledge or organized/collective recognitions in the society.

We proposed the Doubly Structural Network Model that handles the propagation of knowledge and recognition between agents in society. The structure of this model is illustrated as in Fig.1, and is defined by formula (1).

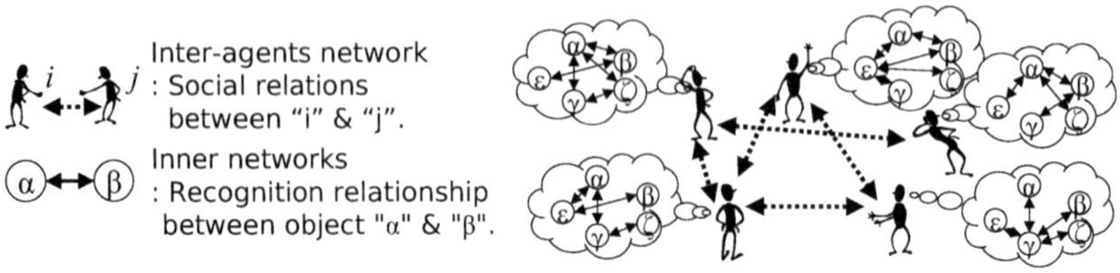

Fig. 1 A doubly structural network of society

$$\begin{array}{ll} G^S \equiv (V^S, E^S), V^S \equiv \{v_i^S | i = 1 \dots N\}, E^S \subseteq V^S \times V^S & \text{(a)} \\ G_i^I \equiv (V^I, E_i^I), V^I \equiv \{v_\alpha^I | \alpha = 1 \dots M\}, E_i^I \subseteq V^I \times V^I & \text{(b)} \\ G^D \equiv \{\{v_i^S, G_i^I | i = 1 \dots N\}, E^S\} & \text{(c)} \\ G_{t+dt}^D \equiv F(t, G_t^D) & \text{(d)} \end{array} \quad (1)$$

- In formula (1a), "social (inter-agent) network" G^S represents the social structure composed of N agents. The node (vertex) v_i^S represents the i-th agent. The edge set E^S represents connection or disconnection between these agents.
- In formula (1b), "internal (recognition) network" G_i^I represents the internal landscape or recognition of the i-th agent on certain objects $(\alpha, \beta, \dots)$. The node (vertex) v_α^I represents the object α. The edge set E_i^I represents connection or disconnection between those objects in the i-th agent's recognition.
  (Whichever directed/undirected graph is available for social or internal network.)
- Formula (1c) shows that "doubly structural network" G^D is created by attaching (/ mounting) each internal / recognition network $G_i^I (i = 1, 2, \dots, N)$ onto the corresponding node i (i-th agent) of the social network G^S.
- Formula (1d) shows that a propagation / learning model of the doubly structural network is defined through providing change of state (edges connect / disconnect) in the internal network via interaction of social network nodes (agents) for this network. At this time, we call as the doubly structural network model of static society if the social network does not change autonomously. On the other hand, we call it as the model of dynamic society if its social network changes autonomously.

In the next section the doubly structural network model is applied into a classical problem of economics (the emergence of money) and we employ an analytical method suited to this feature.

2.2 Emergence Model

Here, we implement a specific mechanism to describe the emergence of money in our model. On the implementation, the social network reflects the topology of economical/social relationship between agents (indicated as $i, j = 1 \dots N$). The agents' inner networks show their own recognition on the exchangeability between commodities (indicated by $\alpha, \beta, \gamma = 1 \dots M$). Each element of adjacent matrix is defined by $e^{(i)}_{\alpha\beta} = e^{(i)}_{\beta\alpha} = 1$ if "α and β are exchangeable", and $e^{(i)}_{\alpha\beta} = e^{(i)}_{\beta\alpha} = 0$ if not.

Among possible stage for emergence of money, this paper focuses on the emergence of proto-money in which a certain commodity achieves "general acceptability" in the society. In our doubly structural network model, the emergence of proto-money α is represented as a self-organizing process in which almost inner-networks become similar star-shaped networks with common hub α. (Fig.2)

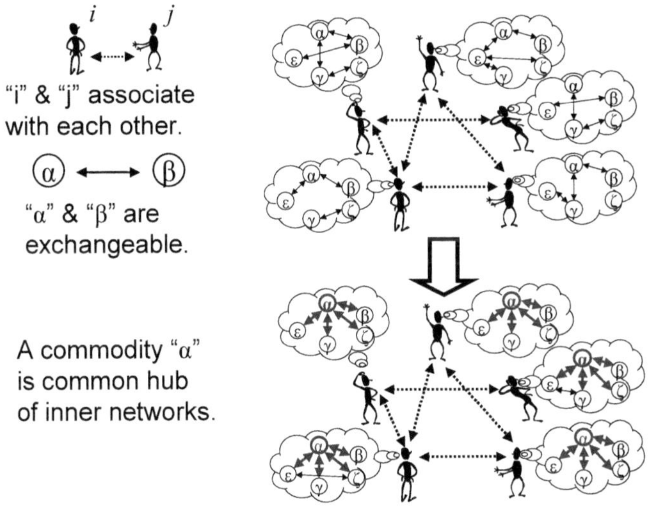

Fig. 2 Emergence of a proto-money: common hub represents general acceptability

Agents in our model interact each other by the following manners in each time step.

1. Exchange: In the social network, neighboring agents i and j exchange commodities α and β with probability P_E, if both of them recognize that α and β are exchangeable.
2. Learning: Learning process of agents consists of the following four ways.
   - Imitation: If an agent i's neighbor j succeeded in exchanging $\alpha - \beta$, then i imitate j (i.e. $e^{(i)}_{\alpha\beta} \to 1$) with the probability P_I.
   - Trimming: If an agent i has cycle recognition of exchangeability, then the agent i will trim its inner-network by cutting randomly one of these cyclic edges with the probability P_T. (Zirkulartausch [13])
   - Conceiving: Even if an agent i has no recognition of $\alpha - \beta$ exchangeability, it will happen to conceive that with the probability P_C.

- Forgetting: Vice versa, even if an agent i has recognition of $\alpha - \beta$ exchangeability, it will happen to forget this with the probability P_F.

3 Agent-Based Simulation

This section shows the main outcomes of this paper. Here we use an agent based simulation model to examine the following research issues.

a) To verify the analytical results [8] found by the mean field approximation on the homogeneous (regular) networks.
b) To verify that heterogeneity (Hub) of networks encourages the emergence.
c) To verify that complexity of networks (Small World / Scale Free) encourages the emergence.

The agent-based simulation is based on the emergence model shown in section 2.2. We executed the simulation 200 times, changing the degree of social network, and observed the number of the proto-money emerging for each network degree and their corresponding percentage. Table 1. shows the parameters in the simulation.

Table 1 Parameters and corresponding values for simulation

Parameters		Values
Probability of Imitation	(P_I)	0.2
Probability of Trimming	(P_T)	0.1
Probability of Cenceiving	(P_C)	0.01
Probability of Forgetting	(P_F)	0.01
Num of Agents		250
Num of Goods		32

3.1 Regular Network

To verify the prediction of mathematical analysis, we executed a simulation using Regular Network as a social network and we compare the experimental result with the prediction depended on Mean-field Dynamics shown in [8].

Fig.3(a) is the experimental result that used the regular network as a social network. A horizontal axis expresses the network degree and the vertical axis expresses the percentage of different the numbers of the emerged proto-money.

Fig.3(a) shows that in the beginning, no proto-money emerged till the degree was increased to 5 when single emergence phenomenon was observed. As the degree was further increased, multiple emergences phenomenon began to rise till degree up to 15. This result is in agreement with the result shown by the mean-field approximation in [8]. However, when the network degree was increased further, the percentage of a multiple emergences decreased and the percentage of a single emergence increased.

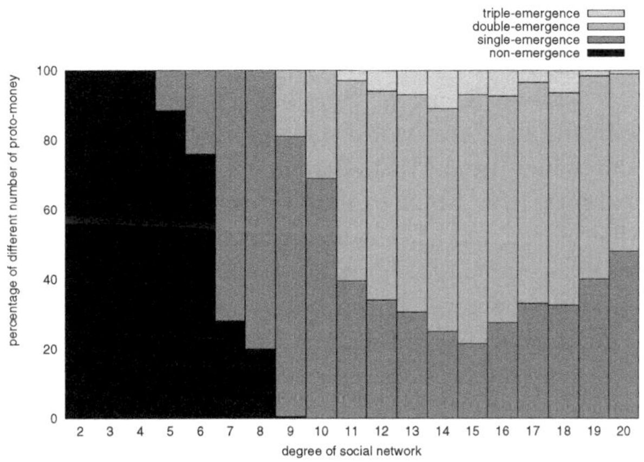

(a) Regular Network

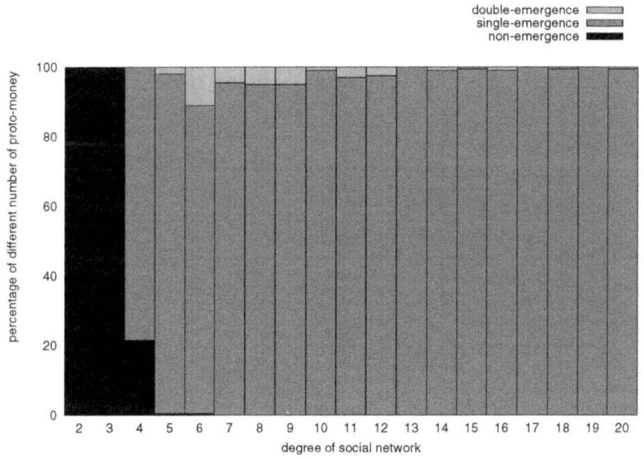

(b) Regular Network with Hub

Fig. 3 The Simulation Result of Regular Network

3.2 Regular Network with Hub

Fig.3(b) shows the result when hub agents were added to the regular network. The social network was made by replacing 5% of the agents in a regular network with hub agents. The horizontal axis represents the average degree. The simulation result verified the results shown in [8]. Compared with Fig.3(a), the emergence started earlier at lower degree when a hub agents were added to the social network. Unlike regular network without hub, it can be seen in Fig.3(b) that multiple emergence of proto-money are very rare and difficult to achieve. However similar to regular network without hub, the percentage of single emergence roe as the degreee of the social network was increased.

3.3 Complex Network

Next, we alter the social network into the network reflecting reality. We executed the same simulation as section 3.1, using "Small-World network" and "Scale-Free network" as the example of real-world's social network. Fig.4(a) shows the experimental result which used Watts-Strogatz (WS) model [15] network with re-wiring probability 10%. Fig.4(b) illustrates the result of the simulation used Barabasi-Albert (BA) model [1] which is Scale-Free network. As for the graph of BA model, the value of a horizontal axis is increased by two units. These graphs show that emergence of proto-money has occurred at a lower network degree as compared to the regular network with hub shown in Fig.3(b) and there are almost no multiple emergences.

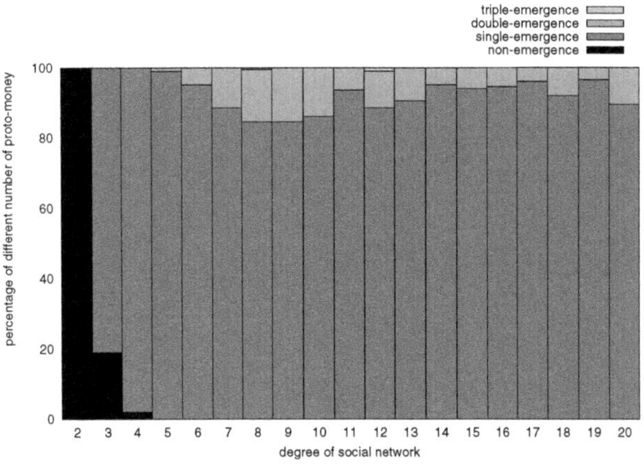

(a) Small-World network

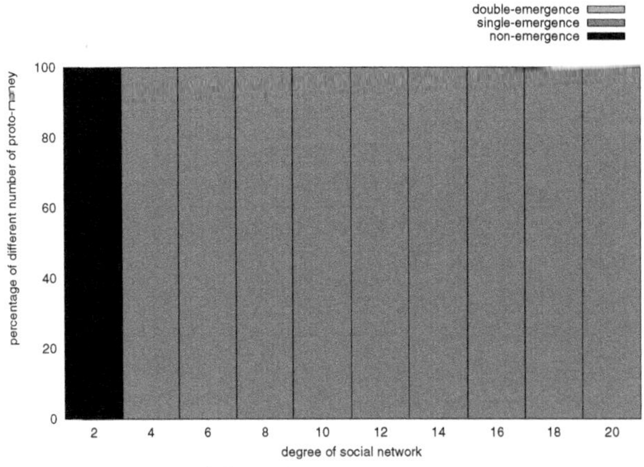

(b) Scale-Free network

Fig. 4 The Simulation Result of Complex Network

4 Conclusion

In this paper, we executed the Agent-Based Simulation for the emergence of money using "doubly structural network model". From the intensive simulation experiments, we confirmed the result of mathematical analysis, and verified the effect which a social network structure has on the emergence. As a result, we are successful in the illustration of the emergence of money (establishment of general acceptability). It also illustrated that a nature (degree of social network: k) of social structure plays an important role in emergence.

When k was large, the experimental results differed from the mathematical analysis. An elucidation of this phenomenon needs further study. When network structures are changed even if the network degree is the same, emergence speed will vary. The variance of degree has also considerable effect on phenomenon of emergence.

Insight gained by this analysis on emergence of money suggests that the doubly structural network model can be a valid analytical method when applied to other fields by looking at money as a communication medium in different transactions.

The proposed model and simulation results are not able to be validated in real cases, because we have not had the data on what really happened in the history of money. However, recently, we have experienced much about the use of regional money in various areas in the real cities, virtual money in cyber world such as the Second-Life, and mileage points systems in the markets. To examine what would happen in such money systems, we would utilize our theoretical and simulation methods: whether 1) they would have enough degrees of connections among users, 2) the hub effect would work, and 3) the exchange system would survive against the competitive environments with the other systems. We would not examine such situations without the proposed agent-based model. Future studies include the application of the proposed model to such realistic situations.

References

1. Barabasi, A., Albert, R.: Emergence of scaling in random networks. Science 286, 509–512 (1999)
2. Duffy, J.: Learning to speculate: Experiments with artificial and real agents. Journal of Economic Dynamics & Control 25, 295–319 (2001)
3. Hayek, F.: Denationalization of Money -The Argument Refined-ch.10/12. The Institute of Economic Affairs (1976)
4. Hicks, J.: Critical Essays in Monetary Theory, ch. 1. Clarendon Press (1967)
5. Iwai, K.: The bootstrap theory of money: A search-theoretic foundation of monetary economics. Structural Change and Economic Dynamics 7, 451–477 (1996)
6. Kiyotaki, N., Wright, R.: On money as a medium of exchange. Journal of Political Economy 97, 927–954 (1989)

7. Kunigami, M., Kobayashi, M., Yamadera, S., Terano, T.: On emergence of money in self-organizing micro-macro network model. In: Proceedings of ESSA 2007, pp. 417–425 (2007)
8. Kunigami, M., Kobayashi, M., Yamadera, S., Yamada, T., Terano, T.: A doubly structural network model and analysis on emergence of money. In: World Congress on Social Simulation 2008 (WCSS 2008) (2008)
9. Luo, G.Y.: The evolution of money as a medium of exchange. Journal of Economic Dynamicsand Control 23, 415–458 (1999)
10. Mankiw, N.: Macroeconomics, 4th edn., vol. 5. Worth Pub. (1999)
11. Marimon, R., McGrattan, E., Sargent, T.: Money as a medium of exchange in an economy with artificially intelligent agents. Journal of Economic Dynamics and Control 14(2), 329–373 (1990)
12. Menger, C.: Grandsatze der Vorkswwirtschaftslehre, ch. 8 (1871)
13. Menger, C.: Grandsatze der Vorkswwirtschaftslehre (Zweite Aufage), ch. 9 (1923)
14. Starr, R.: Why is there money? endogenous derivation of 'money' as the most liquid asset: a class of examples. Economic Theory 21, 455–474 (2003)
15. Watts, D., Strogatz, S.: Collective dynamics of small-world networks. Nature 393, 1302–1305 (1998)
16. Yamadera, S., Terano, T.: Examining the myth of money with agent-based modeling. Social Simulation - Technologies, Advances, and New Discoveries. Information Science Reference, Hershey, pp. 252–262 (2007)
17. Yasutomi, A.: The emergence and collapse of money. Physica D 82, 180–194 (1995)

Detecting Environmental Changes through High-Resolution Data of Financial Markets

Aki-Hiro Sato

Abstract. This article proposes methods to detect states of financial markets both comprehensively and with a high-resolution. In order to quantify trading patterns several mathematical methods are proposed based on frequencies of quotations/transactions estimated from high-resolution data of financial markets. The empirical results (graphical network representation and quantification of states of market participants) for the foreign exchange market are shown. It is concluded that synchronous behavior associated with a large population of market participants may be a candidate of precursory signs leading to an environmental change.

1 Introduction

Recent world economy tightly couples due to both globalization of trading and development of communication technology. Globally spreading of digital highways from the end of the last century enables us to have launched electronic commerce systems through digital computer networks due to development of Information and Communication Technology (ICT). As a result one can collect massive data about human activities and analyze our society from a physical perspective based on such data [1, 2].

Since the electronic commerce systems possess several advantages to human-based commerce systems they have been covered with our globe since the end of the last century. At the same time electronic trading systems in financial markets have launched across the globe. As a result one can directly access financial markets through computer terminals on one's desktop. Furthermore a vast amount of data about trading (order book data and trades and quotes data) has been accumulated on server computers of such electronic trading systems.

Aki-Hiro Sato
Department of Applied Mathematics and Physics, Graduate School of Informatics, Kyoto University
e-mail: aki@i.kyoto-u.ac.jp

K. Nakamatsu et al. (Eds.): New Advan. in Intel. Decision Techno., SCI 199, pp. 595–603.
springerlink.com

Moreover market participants developed automated trading systems based on techniques of algorithmic trading or financial robots and administrators of markets introduced high-performance matching systems which can operate a massive amount of quotations and transactions in order to avoid systemic risks for increases of the number of quotations.

However technological development may be increasing difficulties that nobody can catch up their information generating speed. Therefore it is a challenging subject to understand states of financial markets based on a larger amount of data than human cognitive capability.

Several researchers in fields of financial engineering, econophysics, and computational finance have paid significant attention to this challenging issue [3, 4, 5, 6, 7, 8, 9, 10]. Specifically ICT is also helpful for us to collect and handle a massive amount of data and to extract information from them.

Detecting environmental states both comprehensively and with a high-resolution seems to provide us with meaningful insights for us to act in our incomplete world. If one can learn spatio-temporal correlation patterns for environment, then one can predict a post-environmental change from a pre-environmental change, and/or one can infer the other environmental change which one has not observed yet from an environmental change which one has observed. Therefore detecting environmental changes in detail and learning patterns of such changes bring us predictability.

In this article visualization techniques and quantification methods to detect environmental changes through high-resolution financial data are proposed. Furthermore by using high-resolution data of the foreign exchange market empirical investigation is conducted.

This article is organized as follows. In Sec. 2 high-resolution data of the foreign exchange market is briefly explained. In Sec. 3 mathematical methods to visualize and quantify states of market participants are established and several empirical results obtained from data of the foreign exchange market by means of the proposal methods are shown. Finally Sec. 4 is devoted to concluding remarks.

2 Data

The foreign exchange market is a network consisting of bank traders, investors, and brokers. They gather together in order to exchange currencies. The trades between currencies occur based on quotations including ask/bid rates which market participants want to sell/buy currencies.

Before 1990s all the quotations were transfered via voice through the international telephone system and were matched via human brokers. At the beginning of 1990s several brokerage institutions (Reuters, EBS) have been launched computer-based brokerage systems, which consists of terminal computers that traders use and server computers that conduct matching operation automatically via digital highway across the globe.

Since the computer-based brokerage systems have advantages to human-based brokerage system because of cost-effectiveness, high-speed, and fairness, computer-based brokerage systems have been dominant in the foreign exchange market. Over 90% turnovers have been reported to be traded via computer-based brokerage system as of 2006.

As a result massive data about quotations and transactions have been accumulated on the matching engine and it becomes possible to analyze states of the foreign exchange market by using such data.

In this analysis we use ICAP EBS Data Mine 1.0 [11]. This data for a period from 1st June to 30th September 2008 includes exchangeable currency pairs consisting of 23 currencies and 4 commodities. In this data we found that 40 kinds of currency pairs [1]. From the records one can confirm arrivals of quotations and occurrences of transactions with 1-second resolution.

3 Methods and Results

Recently several researchers have paid remarkable attention to constituents flows on topological networks [12]. In general topological networks are extracted from relationship between two elements which exchange constituents. Therefore it seems to be important to consider processes of constituents as well as their network structure. Recently several studies focus on conveyance process of constituents on networks [13, 14, 15, 16].

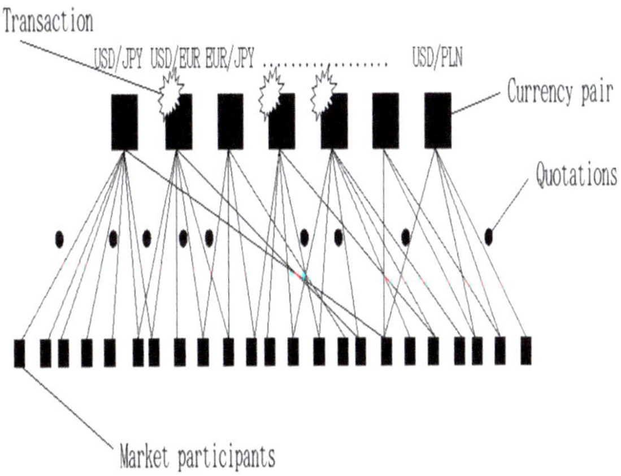

Fig. 1 Conceptual illustration of the foreign exchange market as a bipartite graph. Large filled boxes represent currency pairs, and small filled boxes market participants. If a market participant enters a quotation for a currency pair then the market participant is regarded to be connected with the currency pair. Transactions occur when two quotations which two market participants entered for the same currency pair are matched

[1] The analyzed data include quotations or transactions for AUD/JPY, AUD/NZD, AUD/USD, CHF/JPY, EUR/CHF, EUR/CZK, EUR/DKK, EUR/GBP, EUR/HUF, EUR/ISK, EUR/JPY, EUR/NOK, EUR/PLN, EUR/SEK, EUR/SKK ,EUR/USD, EUR/ZAR, GBP/CHF, GBP/JPY, GBP/USD, NZD/USD, USD/CAD, USD/CHF, USD/HKD, USD/JPY, USD/MXN, USD/PLN, USD/RUB, USD/SGD, USD/TRY, USD/ZAR, XAG/USD, XAU/USD, XPD/USD, XPT/USD, USD/THB, CAD/JPY, NZD/JPY, EUR/RUB, ZAR/JPY

The foreign exchange market can be described as bipartite graph consisting of N market participants and $M(M-1)$ currency pairs with M currencies (See. Fig. 1). Define $f_{jk}(t) \quad (t = 0,\ldots,T-1)$ as the number of quotations/transactions for the currency pair (j/k) for a period in $[t\Delta,(t+1)\Delta] \quad (\Delta > 0)$.

Ask/bid rates of the currency pair j/k means that selling/buying prices at the second currency k for a unit of the first currency j. Since trading of currencies is symmetric, buying/selling currency j with currency i is equivalent to selling/buying currency i with currency j. Therefore we assume that the number of quotations/transactions is symmetric $f_{jk}(t) = f_{kj}(t)$. From the definition it obviously satisfies $f_{jj}(t) = 0$.

Fig. 2 shows quotation/transaction flows at the foreign exchange market from 04:57 to 05:00 on 7th July, 2008 (UTC).

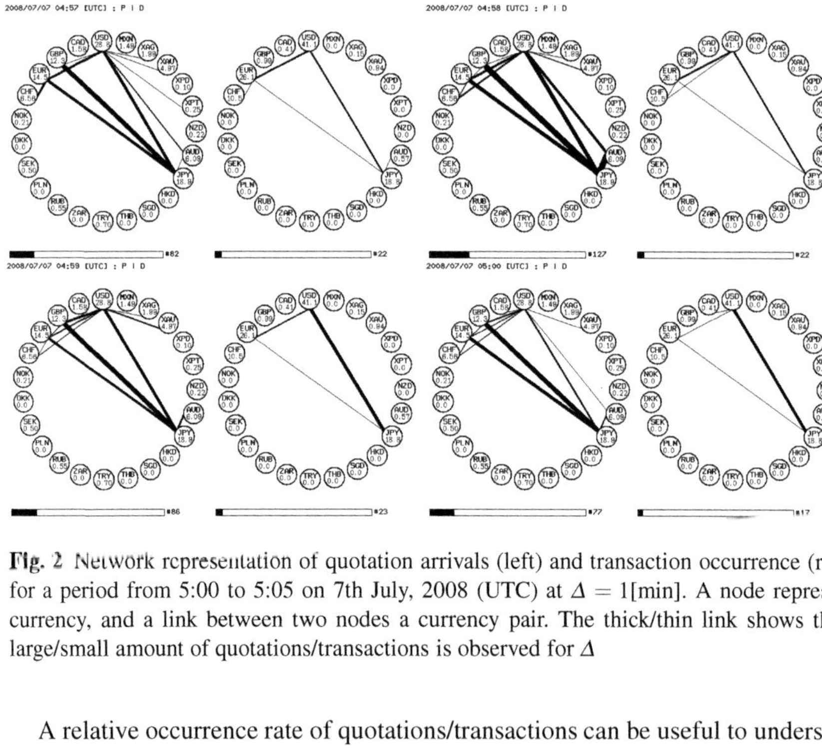

Fig. 2 Network representation of quotation arrivals (left) and transaction occurrence (right) for a period from 5:00 to 5:05 on 7th July, 2008 (UTC) at $\Delta = 1$[min]. A node represents currency, and a link between two nodes a currency pair. The thick/thin link shows that a large/small amount of quotations/transactions is observed for Δ

A relative occurrence rate of quotations/transactions can be useful to understand comprehensive behavior of market participants. The relative occurrence rate for the currency pair j/k is estimated from $f_{jk}(t)$ as

$$p_{jk} = \frac{\sum_{t=0}^{T-1} f_{jk}(t)}{\sum_{j=1}^{M}\sum_{k=1}^{M}\sum_{t=0}^{T-1} f_{jk}(t)}. \tag{1}$$

Furthermore the occurrence rate of quotations/transactions associated with the currency k is defined as

$$q_k = \sum_{j=1}^{M} p_{jk}. \tag{2}$$

Note that a certain kind of uncertainty relationship between temporal resolutions and estimation accuracies of relative occurrence rates exists. Namely if $T\Delta$ is small (large) then environmental changes can be detected with a high (low) temporal resolution with low (high) estimation accuracy of the relative occurrence rates.

From the bipartite graph description p_{jk} (q_k) are related with a population of market participants who want to quote/trade the currency pair j/k (the currency k). Therefore the shape of p_{jk} (q_k) may quantify a degree of monopolization in the foreign exchange market. In order to quantify the degree of monopolization we employ the normalized Shannon entropy for p_{jk} (q_k),

$$H_{cp} = -\frac{\sum_{j=1}^{M}\sum_{k=1}^{M} p_{jk}\log p_{jk}}{\log M(M-1)}, \tag{3}$$

$$H_c = -\frac{\sum_{k=1}^{M} q_k \log q_k}{\log M}, \tag{4}$$

where $0\log 0 = 0$. From this definition H_{cp} (H_c) takes values from 0 to 1. If all the currency pairs or currencies are equally traded then H_{cp} (H_c) takes the maximum value 1.

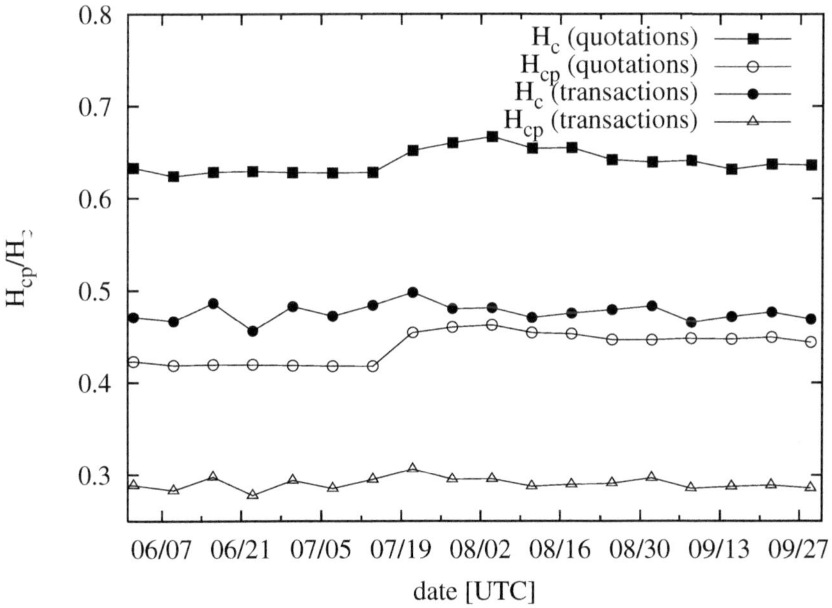

Fig. 3 The normalized Shannon entropy of relative frequencies for currency pairs and currencies. These values are estimated by using relative frequencies of quotations/transaction for a week ($\Delta T = 10,080$ [min])

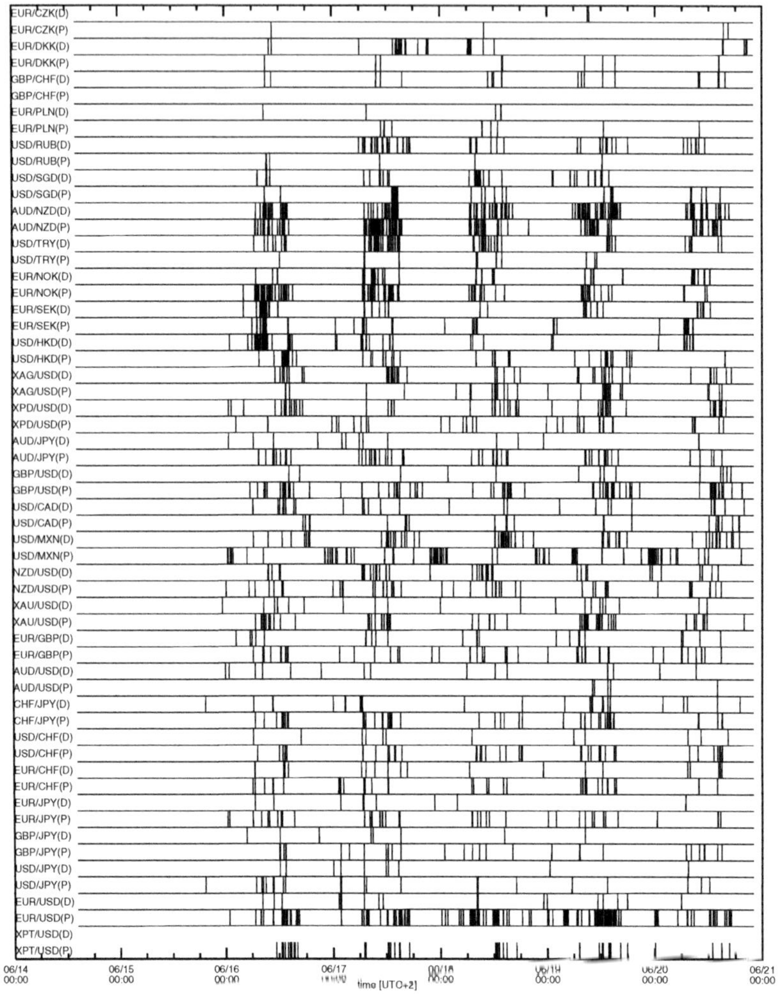

Fig. 4 Time series of impulsive occurrence of quotations/transactions for currency pairs. Lines represents time when the impulsive occurrence is detected. The horizontal axis represents time (UTC), and the vertical axis currency pairs

Fig. 3 shows H_{cp} and H_c estimated from relative frequencies of quotations and transactions for $T\Delta = 7200$[min]. We can find that the degree of monopolization temporally varies. Specifically it is found that the trading patterns at the foreign exchange market may drastically change around the middle of July 2008.

Sometimes synchronized behavior of human actions are found. For example hit products in merchandise sales [17], bubbles and crashes of financial markets [18], business cycles alternating between boom and bust [19, 20] have been considered in successive studies.

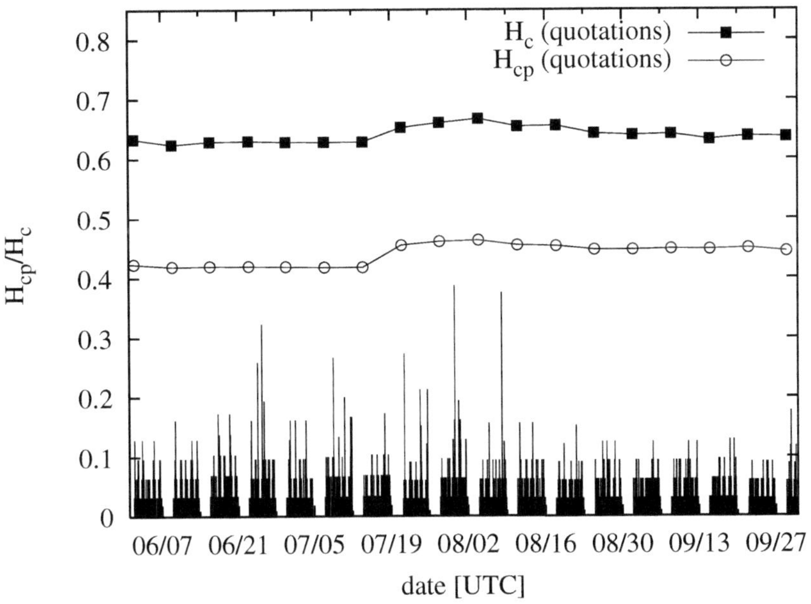

Fig. 5 The normalized Shannon entropy of relative frequencies of quotations for currency pairs and currencies and synchronous index computed at $D = 5.0$

These phenomena commonly happen when participants synchronously behave (active/inactive, buying/selling, and so on). Therefore it seems to be important to quantify and to understand synchronous states of participants. Here we propose a method to visualize and quantify the synchronous states of market participants from $f_{jk}(t)$. If $f_{jk}(t)$ instantaneously becomes larger than before then one may regard that many participants make the same decision at the same time. Therefore it is meaningful to detect an impulsive change of $f_{jk}(t)$. In order to detect the impulsive variation we employ a median filter, which is one of rank-order filters. Let filtered time series $\tilde{f}_{jk}(t)$ be defined as

$$\tilde{f}_{jk}(t) = \text{Median}_{t' \in [t-K,t+K]}\Big(f_{jk}(t')\Big), \tag{5}$$

where $\text{Median}_{t' \in [t-K,t+K]}\Big(x(t')\Big)$ is denoted as a median of $x(t')$ in a range of $[t - K, t + K]$. Defining $\hat{f}_{jk}(t) = |f_{jk}(t) - \tilde{f}_{jk}(t)|$ one can determine that an impulsive change happens at t if $\hat{f}_{jk}(t) \geq \theta$ for any $\theta > 0$. Although a selection of θ is arbitrary, θ is assumed as $\theta = m + D\sigma$, where m denotes a mean value averaged over time, σ a standard deviation, and D a parameter to determine a resolution. We set $D = 5.0$ and calculated time when an impulsive occurrence of quotations/transactions happened. If the impulsive occurrence of quotations/transactions is detected/undetected then we assign $y_{jk}(t) = 1/0$.

As shown in Fig. 4 it is found that several impulsive events are detected and that sometimes some currency pairs show synchronous big jumps. In order to quantify synchronous states in the foreign exchange market we define a synchronization index as

$$Q(t) = \frac{1}{M(M-1)} \sum_{j,k=1;j\neq k}^{M} y_{jk}(t). \tag{6}$$

Fig. 5 shows temporal development of H_{cp} (H_c) and $Q(t)$. Before H_{cp} (H_c) becomes larger than previous values around the middle of July 2008 $Q(t)$ sharply took large values around 21st June 2008. This synchronous behavior associated with a large population of market participants may be considered as precursory changes leading to the environmental change.

4 Conclusions

In order to detect environmental changes through high-resolution financial data both comprehensively and circumstantially mathematical methods to visualize and quantify states of market participants from quotation flows at the financial market were proposed. The normalized Shannon entropy for relative frequencies of quotations/transactions and the synchronous index by means of median filtering for quotation flows and transaction occurrences were established. By using high-resolution data of the foreign exchange market empirical investigation was conducted. From comparative analysis with results obtained from these proposed methods it was found that synchronous behavior associated with a large population of market participants may be a premonitory phenomenon leading to environmental changes.

In order to conclude that the synchronous behavior of market participants can be the precursors of environmental changes both persistent and comprehensive investigation is needed. At this moment the proposal methods become one of candidates to provide us with some useful information on market situations and to capture environmental changes at least since to visualize situations is to understand them.

Both comprehensive and precise visualization of states of financial markets allow us to improve our detectability for environmental changes. Such kinds of investigation with high-resolution financial data will be helpful for us to understand our social environment.

Acknowledgements. This work is partially supported by Global COE program "Informatics Center for the Development of Knowledge Society Infrastructure" in Kyoto University from the Japanese Ministry of Education, Culture, Sports, Science and Technology. Furthermore the author is grateful for the financial support from Ken Millennium, Corporation.

References

1. Carbone, A., Kaniadakis, G., Scarfone, A.M.: Tails and Ties, Editorial — Topical Issue on Physics in Society. The European Physical Journal B 57, 121–125 (2007)
2. Lambiotte, R., Ausloos, M., Thelwall, M.: Word statistics in Blogs and RSS feeds: Towards empirical universal evidence. Journal of Informatics 1, 277–286 (2007)
3. Mantegna, R.N., Stanley, H.E.: An Introduction to Econophyscis: Correlations and complexity in finance. Cambridge University Press, Cambridge (1999)
4. Souma, W., Fujiwara, Y., Aoyama, H.: Random matrix approach to shareholding networks. Physica A 344, 73–76 (2004)
5. Mizuno, T., Takayasu, H., Takayasu, M.: Correlation networks among currencies. Physica A 364, 336–342 (2006)
6. Hayashi, K., Kaizoji, T., Pichl, L.: Correlation patterns of NIKKEI index constituents: Towards a mean-field model. Physica A 383, 16–21 (2007)
7. Naylor, M.J., Rose, L.C., Moyle, B.J.: Topology of foreign exchange markets using hierarchical structure methods. Physica A 382, 199–208 (2007)
8. Sato, A.-H., Hołyst, J.A.: Characteristic periodicities of collective behavior at the foreign exchange market. The European Physical Journal B 62, 373–380 (2008)
9. Heimo, T., Tibély, G., Saramäki, J., Kaski, K., Kertész, J.: Spectral methods and cluster structure in correlation-based networks. Physica A 387, 5930–5945 (2008)
10. Sato, A.-H.: Comprehensive high-resolution analysis on behavior of market participants in the foreign exchange market (in preparation)
11. ICAP PLC, http://www.icap.com
12. Albert, R., Barabási, A.-L.: Statistical mechanics of complex networks. Reviews of Modern Physics 74, 47–97 (2002)
13. Noh, J.D., Rieger, H.: Random walks on complex networks. Physical Review Letters 92, 118701 (2004)
14. Menezes, M.A., Barabáshi, A.-L.: Separating internal and external dynamics of complex systems. Physical Review Letters 93, 068701 (2004)
15. Eisler, Z., Kertész, J.: Scaling theory of temporal correlations and size-dependent fluctuations in the traded value of stocks. Physical Review E 73, 046109 (2006)
16. Meloni, S., Gómez-Gradeñes, J., Latora, V., Moreno, Y.: Scaling breakdown in flow fluctuations on complex networks. Physical Review Letters 100, 208701 (2008)
17. Sornette, D., Deschâtres, F., Gilbert, T., Ageon, Y.: Endogenous versus exogenous shocks in complex networks — an empirical test using book sale rankings. Physical Review Letters 93, 228701 (2004)
18. Lux, T.: Herd behavior, bullbes and crashes. Economic Journal 105, 881–896 (1995)
19. Schumpeter, J.A.: Theory of Economic Development: An inquiry into profits, capital, credit, interest and the business cycle. Oxford University Press, Oxford (1961) (Originally published in 1912)
20. Honggang, L., Gao, Y.: A GDP fluctuation model based on interacting firms. Physica A 387, 5225–5230 (2008)

A Stochastic Model for Pareto's Law and the Log-Normal Distribution under the Detailed Balance and Extended-Gibrat's Law

Shouji Fujimoto, Masashi Tomoyose, and Atushi Ishikawa

Abstract. We verify that Takayasu–Sato–Takayasu (TST) model satisfies not only Pareto's law but also the detailed balance under Gibrat's law, by using numerical simulation. We employ a tent-shaped function as multiplicative noise. We also numerically confirm that the reflection law is equivalent to the equation which gives the Pareto index μ in TST model. We extend the model modifying the stochastic coefficient under a Non-Gibrat's law, and also numerically observe the detailed balance. The obtained pdf is power-law in the large scale region, and is the log-normal distribution in the middle scale one. We also study the dependence of Pareto index on the average of the additive noise.

1 Introduction

In the large scale region of a company size such as wealth, sales, profits, the number of employees and so forth (x), the probability distribution function (pdf) $P(x)$ obeys Pareto's law [1]:

$$P(x) \propto x^{-(\mu+1)} \quad \text{for} \quad x > x_0 \, . \tag{1}$$

Recently, Fujiwara et al. [2] find that the Pareto's law is derived from the law of detailed balance and Gibrat's law [3]. The detailed balance is time-reversal symmetry ($x_1 \leftrightarrow x_2$) observed in a stable economy

Shouji Fujimoto
Kanazawa Gakuin University, 920-1392, Japan
e-mail: fujimoto@kanazawa-gu.ac.jp

Masashi Tomoyose
University of the Ryukyus, 903-0213, Japan
e-mail: mtomo@weiss.sci.u-ryukyu.ac.jp

Atushi Ishikawa
Kanazawa Gakuin University, 920-1392, Japan
e-mail: ishikawa@kanazawa-gu.ac.jp

K. Nakamatsu et al. (Eds.): New Advan. in Intel. Decision Techno., SCI 199, pp. 605–614.
springerlink.com

$$P_{12}(x_1, x_2) = P_{12}(x_2, x_1) \,. \tag{2}$$

Here, x_1, x_2 are two successive company sizes, and $P_{12}(x_1, x_2)$ is the joint pdf. Also, we define a growth rate as $R = x_2/x_1$. Gibrat's law means that the conditional pdf $Q(R|x_1)$ of the growth rate R is independent of the past value x_1 in large scale region

$$Q(R|x_1) = Q(R) \quad \text{for} \quad x > x_0 \,. \tag{3}$$

This derivation is, however, only valid in the large scale region. The Pareto's law (1) is not observed below some threshold x_0 [3, 4] because the Gibrat's law (3) is not satisfied in the middle scale region [5, 6, 7]. In Ref. [8], we show that the log-normal distribution in the middle scale region:

$$P(x_1) = C x_1{}^{-(\mu+1)} \, e^{-\alpha \ln^2 \frac{x_1}{x_0}} \quad \text{for} \quad x_{\min} < x_1 < x_0 \tag{4}$$

is derived from a Non-Gibrat's law

$$Q(R|x_1) = Const. \; R^{\mp t_{\pm}(x_1) - 1} \quad \text{for } R \gtrless 1 \,, \tag{5}$$

$$t_{\pm}(x_1) = t_{\pm}(x_0) \pm \alpha \, \ln \frac{x_1}{x_0} \tag{6}$$

under the detailed balance (2). The Non-Gibrat's law means that a statistical dependence of $Q(R|x_1)$ on the past value x_1 exists in the middle scale region. In the derivation, under the detailed balance (2), we show that the expression of $t_{\pm}(x_1)$ (6) is unique if the pdf of growth rate distribution is approximated by tent-shaped function (5) observed in profits data of Japanese firms[9]. We observe that the distribution (4) fits with the above data in the large scale Pareto's law region and those in the middle scale log-normal region consistently. The parameters are estimated as follows: $\alpha \sim 0$ for $x_1 > x_0$, $\alpha \sim 0.14$ for $x_{\min} < x_1 < x_0$, $x_0 \sim 63{,}000$ thousand yen and $x_{\min} \sim 1{,}600$ thousand yen [8].

All these analyses are, however, concerned with non-negative profits data. The exhaustiveness of negative data is untrustworthy and reliable statistical analysis is difficult because the number of negative profits data is much less than that of non-negative data, in the database. Also, these data analyses are restricted in the finite period and category, because it is not easy to obtain complete data.

On the other hand, negative company size distributions are thought to be deeply related to default probabilities which are important when one considers investments, bond rating, insurance, and so forth. Especially negative profits must be taken into account to derive company's default. In this study, we propose a stochastic model based on the detailed balance and (Non-)Gibrat's law, and confirm the consistency of the model. Pareto index is calculated for the case in which x can be negative, as well as the case that x is non-negative. We also examine the dependence of Pareto index on the functional form of the noise which is included in the model. These numerical analyses aim at calculating default probabilities under restricted data in the context of decision-making in environmental changes.

2 Simulation in the Large Scale Region

There is a model which leads the Pareto's law by using multiplicative stochastic process. It is called Takayasu–Sato–Takayasu(TST) model[10]. TST model is given by the Langevin equation with a multiplicative stochastic noise $b(t)$ and an additive stochastic noise $f(t)$:

$$x(t+1) = b(t)x(t) + f(t) \,. \tag{7}$$

They find that the necessary and sufficient conditions for power-law (1) are as follows:

$$\langle \ln b(t) \rangle < 0 \,, \quad \langle b(t)^2 \rangle > 1 \,. \tag{8}$$

They also show that the index μ of power-law (1) is derived from the equation[1]

$$\langle b(t)^{\mu} \rangle = 1 \,. \tag{9}$$

Here $\langle \cdots \rangle$ denotes an average over realizations. In order to clarify that this model is consistent with the empirical observation, the detailed balance and Gibrat's law must be satisfied. As known from Eq. (7), Gibrat's law holds in the region where a noise $f(t)$ is negligible. The result in Ref. [2], therefore, suggests that the detailed balance exists in TST model which leads Pareto's law.

In the following two sections, we utilize a tent-shaped distribution: Eqs. (5)–(6) with $\alpha = 0$ and Weibull distribution:

$$\frac{k}{\lambda}(\frac{x}{\lambda})^{k-1} \exp[-(\frac{x}{\lambda})^k] \tag{10}$$

as distribution of $b(t)$ and $f(t)$ respectively. In this case, the multiplicative noise b and the additive noise f are non-negative. Consequently, x should also be non-negative. Of course, profits, income of firms and so forth can be negative; however, the simulation in the following two sections is concerned with the only non-negative quantity x.

Figure 1 shows the typical scatter plot of the simulation in which we choose the following parameters: $t_+(x_0) = 2.5$, $t_-(x_0) = 1.5$, $x_0 = 10^0$, $k = 0.5$ and $\lambda = 30$. To clarify the validity of the detailed balance, we apply the one-dimensional Kolmogorov-Smirnov (K-S) test. We compare the distribution sample for $P(x_1 \in [10^{3+0.2(n-1)}, 10^{3+0.2n}], x_2)$ with another sample for $P(x_1, x_2 \in [10^{3+0.2(n-1)}, 10^{3+0.2n}])$ with $n = 1, 2, \cdots, 20$ by making the null hypothesis that these two samples are taken from a same parent distribution. Each p value is shown in Fig. 2. The null hypothesis is not rejected in 5% significance level in the region where the noise $f(t)$ is negligible. The detailed balance is observed in the region. On the other hand, in the region where the detailed balance is not observed, the additive noise f is not

[1] Using Eq. (11) reduced from this equation, the range of μ corresponding to Eqs. (8) is clarified to be $0 < \mu < 2$.

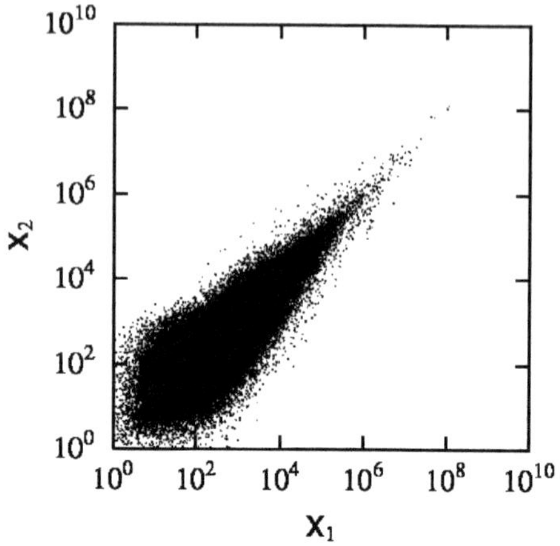

Fig. 1 The scatter plot of data points under Gibrat's law, the number of which is "500,000 ". We employ the following parameters: $t_+(x_0) = 2.5$, $t_-(x_0) = 1.5$, $x_0 = 10^0$, $k = 0.5$ and $\lambda = 30$

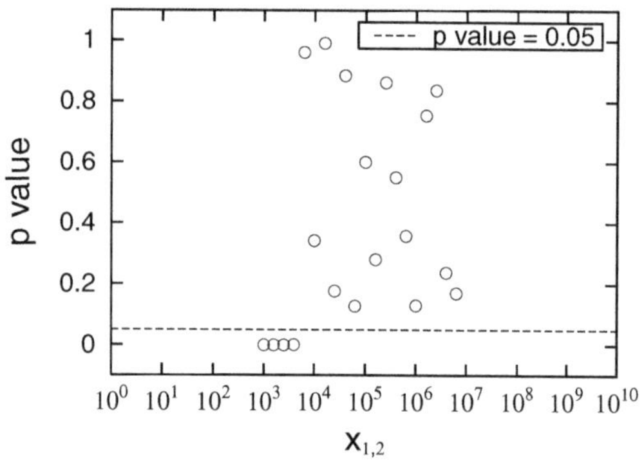

Fig. 2 Each p value of the one-dimensional Kolmogorov-Smirnov test for Fig. 1. The null hypothesis is not rejected in 5% significance level in the region where the noise $f(t)$ is negligible

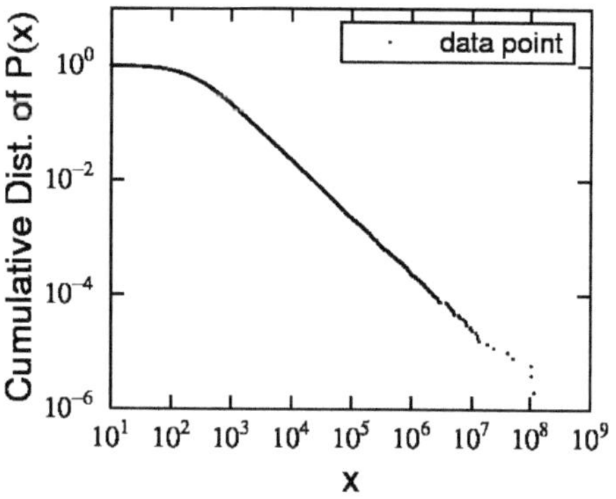

Fig. 3 The cumulative distribution of $P(x)$ shows that the Pareto's law is observed above $x_{min} \sim 10^4$. The Pareto index μ is estimated by nearly 1

negligible and Gibrat's law is not satisfied. The boundary between the two regions is denoted by x_{min}. In the upper region, the additive noise f is negligible, and Pareto's law is observed (Fig. 3).

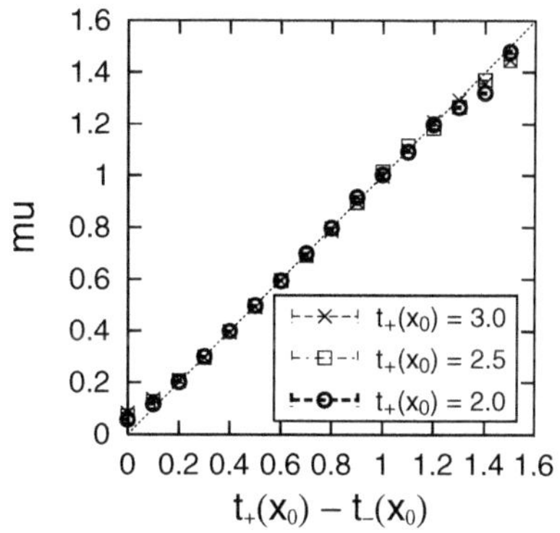

Fig. 4 The comparison with μ and $t_+(x_0) - t_-(x_0)$. Here, μ is estimated by the least square method of the simulation result, in which we take various values of $t_\pm(x_0)$

We also verify the reflection law $Q(R) = R^{-\mu-1}Q(R^{-1})$ derived in Ref. [2]. In the simulation, from Eqs.(5) and (6) with $\alpha = 0$, the reflection law is represented as

$$\mu = t_+(x_0) - t_-(x_0), \tag{11}$$

which is equivalent to the condition (9). Using this notation, the conditions (8) are expressed as $0 < \mu < 2$. Note that this equation (11) is consistent with the We observe this reflection law in various $t_\pm(x_0)$ (Fig. 4). profits data of Japanese firms[8].

From the above simulation results, we recognize that TST model holds the detailed balance under the Gibrat's law in the large scale region.

3 Simulation in the Middle Scale Region

In the middle scale region, the log-normal distribution must be deduced[8] if the detailed balance and the Non-Gibrat's law are satisfied. However, this scheme is not manifest analytically in the stochastic simulation because a multiplicative noise $b(t)$ in Eq. (7) must be modified as $b(x(t),t)$ under the Non-Gibrat's law. This extension is beyond the analytical framework of TST model. Hence, we examine this scheme numerically.

Now, we introduce $x_{\rm int}$ satisfying $t_+(x_{\rm int}) = t_-(x_{\rm int})$. As for the distribution of b, we employ Eqs. (5)–(6) with $\alpha = 0$ for $x > x_0$, $\alpha \neq 0$ for $x_{\rm int} < x < x_0$ and $\alpha = 0$ for $x < x_{\rm int}$ keeping $t_\pm(x)$ continuously (Fig. 5). The last parameterization is imposed to exclude immoderate slopes of growth rate distributions. As the middle scale region, we recognize the region $\max(x_{\rm min}, x_{\rm int}) < x < x_0$.

Figure 6 shows the typical scatter plot of the simulation in which we choose the following parameters: $t_+(x_0) = 2.5$, $t_-(x_0) = 1.5$, $x_0 = 10^6$, $k = 0.5$, $\lambda = 30$ and $\alpha = 0.1$. In order to judge the validity of the detailed balance, we apply the same K-S test in the previous section. Each p value is displayed in Fig. 7. The detailed balance is verified not only in the Gibrat's large scale region ($x > x_0 = 10^6$) but also in the Non-Gibrat's middle scale one ($\max(x_{\rm min}, x_{\rm int}) = 10^4 < x < x_0 = 10^6$).

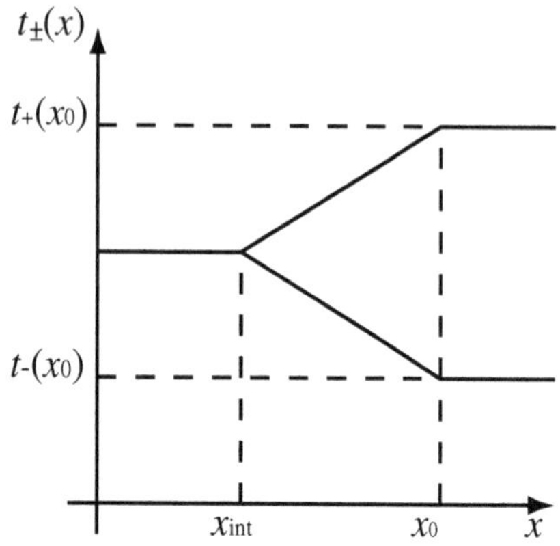

Fig. 5 The shape of continuous functions $t_\pm(x)$ defined as $t_\pm(x) = t_\pm(x_0) \pm \alpha \ln \frac{x}{x_0}$ with $\alpha = 0$ for $x > x_0$, $\alpha > 0$ for $x_{\text{int}} < x < x_0$ and $\alpha = 0$ for $x < x_{\text{int}}$. The horizontal axis is logarithmic scale

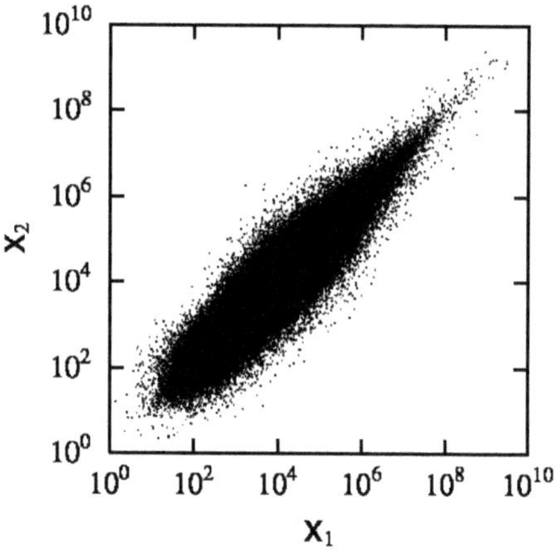

Fig. 6 The scatter plot of data points under Non-Gibrat's law, the number of which is "500,000". We employ the following parameters: $t_+(x_0) = 2.5$, $t_-(x_0) = 1.5$, $x_0 = 10^6$, $k = 0.5$, $\lambda = 30$ and $\alpha = 0.1$

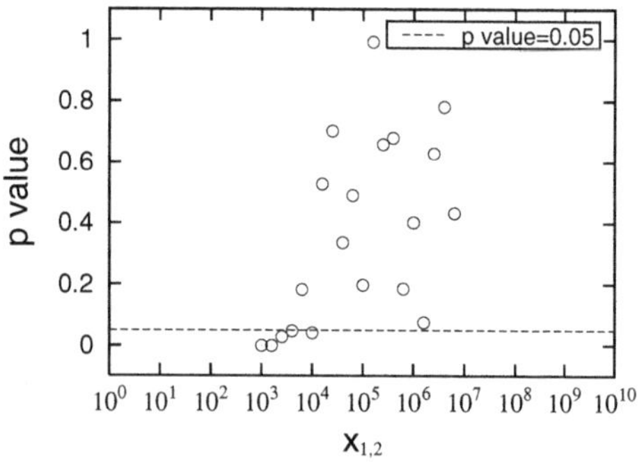

Fig. 7 Each p value of one-dimensional Kolmogorov-Smirnov test for Fig. 6

The resultant pdf of x appears in Fig. 8. This figure tells us that the log-normal distribution in the middle scale region is observed besides Pareto's law in the large scale one. In order to ensure the adequacy of the log-normal distribution, we compare the parameter α inputted in the simulation and that estimated by fitting the pdf

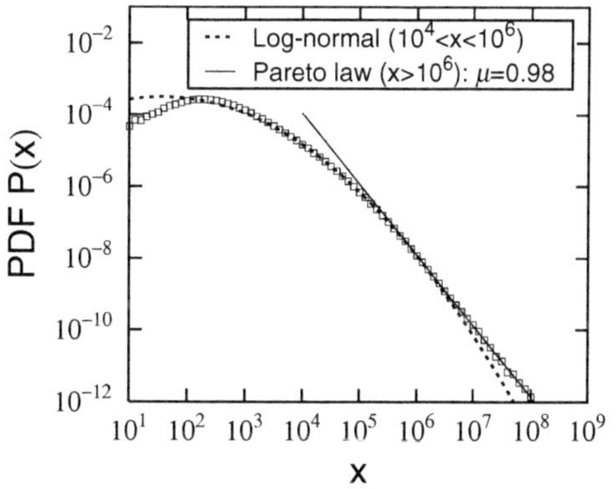

Fig. 8 The resultant pdf $P(x)$ with $t_+(x_0) = 2.5$, $t_-(x_0) = 1.5$, $x_0 = 10^6$, $k = 0.5$, $\lambda = 30$ and $\alpha = 0.1$. In the large scale region, Pareto's law is observed. In the middle scale one, the log-normal distribution is observed

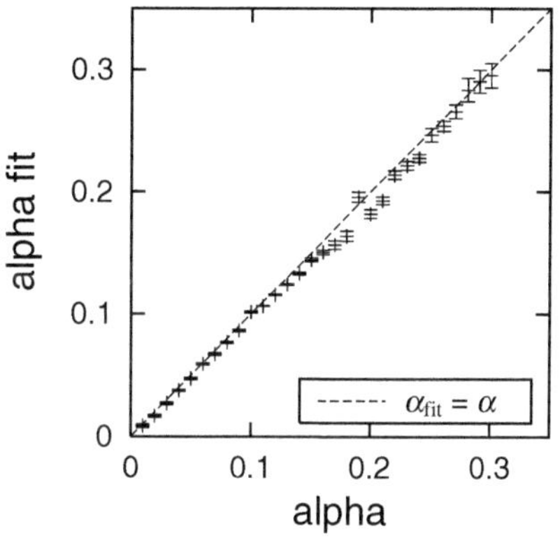

Fig. 9 The relation between α and α_{fit} for the case $t_+(x_0) = 2.5$, $t_-(x_0) = 1.5$, $x_0 = 10^6$, $k = 0.5$ and $\lambda = 30$

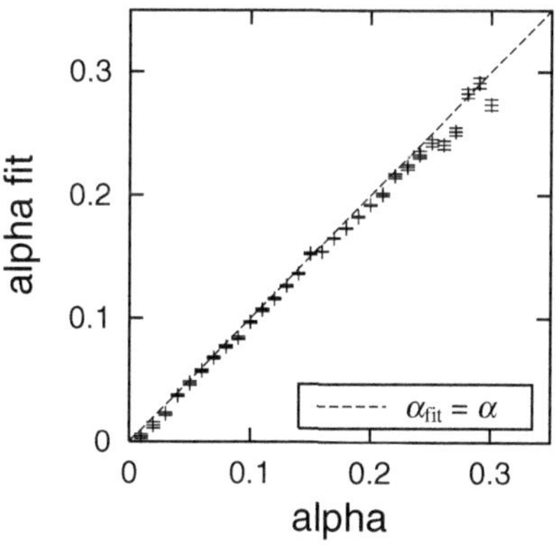

Fig. 10 The relation between α and α_{fit} for the case $t_+(x_0) = 3.5$, $t_-(x_0) = 2.0$, $x_0 = 10^6$, $k = 0.5$ and $\lambda = 30$

in the Non-Gibrat's region (denoted by α_{fit}). The result is displayed in Fig. 9 for the case $t_+(x_0) = 2.5$, $t_-(x_0) = 1.5$, $x_0 = 10^6$, $k = 0.5$ and $\lambda = 30$ and in Fig. 10 for the case $t_+(x_0) = 3.5$, $t_-(x_0) = 2.0$, $x_0 = 10^6$, $k = 0.5$ and $\lambda = 30$ for example. The significant correlation between α and α_{fit} exists. We can recognize form the above results that the simulation model also satisfies the detailed balance even under the Non-Gibrat's law in the middle scale region.

4 Issue of the Functional Forms of $b(t)$ and $f(t)$

In the above simulation, we have fixed the functional forms of the multiplicative noise $b(t)$ and the additive noise $f(t)$. A tent-shaped distribution and Weibull one are employed as the distribution of $b(t)$ and $f(t)$ respectively. There are several candidates for the distributions of $b(t)$ and $f(t)$. Moreover, b and f are non-negative. So, x is also non-negative. However, profits can be negative.

In this section, we consider another two cases. Normal distribution $\mathrm{N}(m, \sigma^2)$ is commonly utilized as the additive noise $f(t)$[11], and a tent-shaped distribution, an exponential one are employed as the multiplicative noise $b(t)$. In these two cases, the additive noise can become either positive or negative, and these might be asymmetric. Also, x can be positive and negative value. Hence, there are two indices of power-law (1) for positive and negative sides. Here, we confirm the dependence of two indices μ_+ and μ_- on m.

Firstly, a tent-shaped distribution is employed as the distribution of $b(t)$. The distribution of $f(t)$ is Normal distribution $\mathrm{N}(m, \sigma^2)$ with the standard deviation $\sigma = 20$. For the case $t_+(x_0) = 2.6, t_-(x_0) = 1.4, x_0 = 10^6$, the index μ is estimated 1.2 from Eq. (11). The result is shown in Fig. 11. The index μ_+ takes 1.20 and the index μ_- does 1.21 at $m = 0$. The index μ_+ hardly changes as m increases and takes 1.19 at $m = 30$. Namely, μ_+ does not depend on m. Also, μ_- hardly changes as m increases until around $m = 10$, and takes 1.26 at $m = 10$. However, μ_- increases for m over 10. This dependence is thought to be due to statistical effects because the number of negative x is much less than the positive one for m over 10. The similar results are obtained by using the following parameters sets; $t_+ = 3.0, t_- = 1.5$ and the index $\mu = 1.5$, and $t_+ = 2.9, t_- = 1.1$ and the index $\mu = 1.8$.

Next, an exponential distribution: $\gamma \exp(-\gamma b)$ is employed as $b(t)$. The distribution of $f(t)$ is same as the above. We simulate the cases that the parameter $\gamma = 1.1$, 1.2 and 1.3. For $\gamma = 1.1$, the result is shown in Fig. 12. The index μ_+ takes 1.24

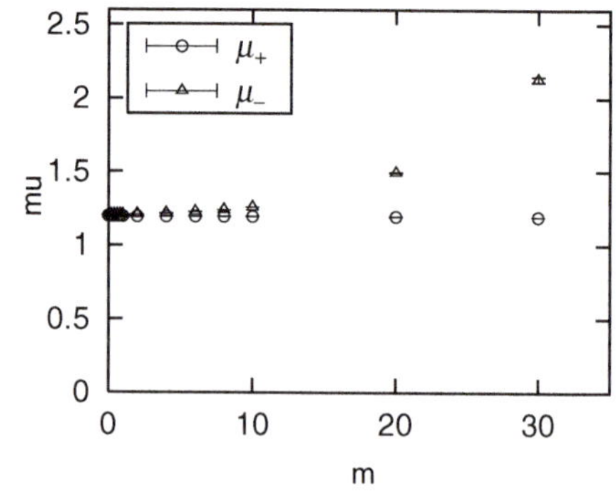

Fig. 11 The dependence of the positive side index μ_+ and the negative side index μ_- on m. A tent-shaped distribution and normal one $N(m, \sigma^2)$ with the standard deviation $\sigma = 20$ respectively

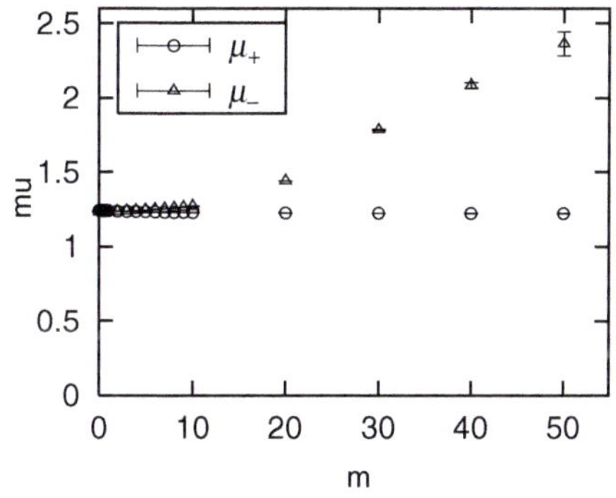

Fig. 12 The dependence of the positive side index μ_+ and the negative side indexμ_- on m. An exponential distribution and normal one $N(m, \sigma^2)$ with the standard deviation $\sigma = 20$ respectively

and the index μ_- does 1.23 at $m = 0$. The index μ_+ hardly changes as m increases and takes 1.22 at $m = 30$. Also, μ_- hardly changes as m increases until around $m = 10$, and takes 1.27 at $m = 10$, and increases for m over 10. This dependence is also thought to be due to statistical effects. Similar results are obtained for γ =1.2 and 1.3.

In the above two cases, the value of μ_+ and μ_- are consistent with results in section 2. Also, μ_+ does not depend on m. These facts suggest that results obtained in section 2 are independent on the functional form of additive noise.

5 Summary and Future Problems

We confirm that TST model satisfies not only Pareto's law but also the detailed balance under Gibrat's law, by using numerical simulation. We utilize a tent-shaped function the multiplicative noise $b(t)$. Equivalence of the reflection law and the equation which gives the Pareto index μ in TST model is also verified numerically. Furthermore, we extend the model modifying the stochastic coefficient under a Non-Gibrat's law. The detailed balance is also observed in the extended model. The obtained pdf of x is power-law in the large scale region, and is the log-normal distribution in the middle scale one.

We also simulate the case that the distribution of additive noise $f(t)$ is normal one. In this case, x become either positive and negative. We calculate the positive side index μ_+ and the negative side one μ_-. These values are consistent with the former results, and μ_+ does not depend on m. Apparently, the index μ_- seems to depend on m. This dependence is, however, considered to be artificial due to statistical effects. In this case, we merely calculate the indices μ_+ and μ_-. Thus, we must examine whether the detailed balance and Non-Gibrat's law hold or not in the middle scale region. To verify the independence of the above results on the functional form of b, other simulation is required by using another functional forms, for instance normal distribution. On the other hand, if the indices μ_+ or μ_- do depend on m using another functional form b, the simulation will has helpfulness for actual business situations. Because it is conceivable that the indices μ_+ and μ_- take different values in empirical data [12].

Acknowledgements. This work was supported in part by a Grant–in–Aid for Scientific Research (C) (No. 20510147) from the Ministry of Education, Culture, Sports, Science and Technology, Japan.

References

1. Pareto, V.: Cours d'Economique Politique. Macmillan, London (1897)
2. Fujiwara, Y., Souma, W., Aoyama, H., Kaizoji, T., Aoki, M.: Physica A 321, 59 (2003); Fujiwara, Y., Guilmi, C.D., Aoyama, H., Gallegati, M., Souma, W.: Physica A335, 197 (2004)
3. Gibrat, R.: Les inegalites economiques. Sirey, Paris (1932)
4. Badger, W.W.: In: West, B.J. (ed.) Mathematical Models as a Tool for the Social Science, p. 87. Gordon and Breach, New York (1980); Montrll, E.W., Shlesinger, M.F.: J. Stat. Phys. 32, 209 (1983)
5. Stanley, M.H.R., Amaral, L.A.N., Buldyrev, S.V., Havlin, S., Leschhorn, H., Maass, P., Salinger, M.A., Stanley, H.E.: Nature 379, 804 (1996)
6. Takayasu, H., Takayasu, M., Okazaki, M.P., Marumo, K., Shimizu, T.: cond-mat/000805. In: Novak, M.M. (ed.) Paradigms of Complexity, p. 243. World Scientific, Singapore (2000)
7. Aoyama, H.: Ninth Annual Workshop on Economic Heterogeneous Interacting Agents (WEHIA 2004) (2004); Aoyama, H., Fujiwara, Y., Souma, W.: The Physical Society of Japan 2004 Autumn Meeting (2004)
8. Ishikawa, A.: Physica A367 Physica A383/1, 79 (2007)
9. Tokyo Shoko Research, Ltd., `http://www.tsr-net.co.jp/`
10. Takayasu, H., Sato, A., Takayasu, M.: Phys. Rev. Lett. 79, 966 (1997)
11. Sato, A., Takayasu, H., Sawada, Y.: Phys. Rev. E61, 1081 (2000)
12. Fujiwara, Y.: Physica A337, 219 (2004)

Effect of Reputation on the Formation of Cooperative Network of Prisoners

Mieko Tanaka-Yamawaki and Taku Murakami

Abstract. We consider in this paper the effect of the player's reputation implemented in a multi-agent model of iterated prisoner's dilemma to develop cooperative networks. Our model assumes two separate strategies per agent to apply upon a cooperative partner and a defective partner. Starting from a randomly selected pair of strategies, (S_D , S_C) where S_D and S_C being the strategy on the defective partner and the cooperative partner, the agent autonomously learns a set of better strategies by imitating the better performing agent. Reputation is defined to be the rate of cooperative choice that the agent has chosen in the course of iteration. Each agent is given a fixed criterion on the minimum reputation to require upon the partner, so that this agent refuses to play the game with agents of bad reputation. We show by simulations that the model successfully develop cooperative networks of players by means of selecting the reputable partners as well as updating the strategies.

1 Introduction

It is of great interest to observe under the iterated prisoner's dilemma (IPD) that selfish players autonomously establish a cooperative relationship with their partners by means of co-evolution to learn global perspectives. Evolutional computation offers a self-learning system for agents to update their own knowledge under a certain criterion (maximizing, or minimizing the evaluation function, for example) in order to improve their own chromosomes. Co-evolution implies a competition among agents representing a certain population of strategies that drives the agents

---

Mieko Tanaka-Yamawaki
Department of Information and Electronics, Graduate School of Engineering,
Tottori University, Tottori 680-8552, Japan
e-mail: mieko@ike.tottori-u.ac.jp

Taku Murakami
Department of Information and Knowledge Engineering, Faculty of Engineering,
Tottori University, Tottori 680-8552, Japan

K. Nakamatsu et al. (Eds.): New Advan. in Intel. Decision Techno., SCI 199, pp. 615–623.
springerlink.com

toward a cooperative society, by means of properly chosen criterion, which is implemented by maximization or minimization of an evaluation function.

Lindgren obtained an interesting result by evolving the chromosomes of strategy of IPD by setting the length of chromosomes to grow by doubling and point mutations. [3]

Nowak and Sigmund [4] introduced the effect of players' reputation into the game in order to form a cooperative society. Robert and Sherratt [5] expanded the choice of actions to include intermediate actions between the perfect cooperation and the perfect defection. By combining both, Yao and Darwen [6] used a 3 layered neural network by adding the "reputation" as one of the input elements, in order to compute the best action out of four choices: {-1, -1/3, 1/3, 1}, corresponding to the defection, mild defection, mild cooperation, cooperation, and showed that the "reputation" indeed helps the agents to cooperate.

In this paper, we put forward the idea of including the effect of "reputation" by means of evolving network, in which agents of bad reputation have difficulty finding partners in the network of IPD. Namely, each player is given a right to refuse matching with the partner whose reputation is lower than his criterion. This selection process determines a network of IPD players as a subset of the fully connected network of N agents, which grows time to time. Can we achieve a cooperative society by doing this? Do such networks show any notable characteristics?

The rest of the paper is constructed as follows. We review the problem of prisoner's dilemma (PD) and the iterated prisoner's dilemma (IPD) in Section 2 and 3, respectively. Section 4 is devoted to present our model and Section 5 to show our result. We conclude our paper in Section 6 with some remarks.

2 Prisoner's Dilemma (PD)

Neumann and Morgenstein [7] presented a model called "prisoner's dilemma" for the model of social relationships where the cooperation is a difficult task when both sides cannot trust each other [8]. Namely one gets the mutual reward (R for both sides) by denying the crime under a severe fear of the partner's betrayal where the partner gets a traitor's reward (T, larger than anything else) and he suffers a sacrifice (S, smaller than anything else), while he betrays his partner by getting a traitor's reward (T) and giving the partner a sacrifice (S), and naturally suffers a penalty (P) if both sides betray. This condition $S<P<R<T$ is shown in Table 1. An additional condition $S+T<2R$ is applied in order to avoid alternating a C (Cooperation) and D (Defection). And help forming a mutual cooperation.

From the global point of view, the mutual cooperation is the best profitable solution for both sides objectively because a traitor eventually pays the debt by loosing the reputation as well as deteriorating the whole environment surrounding him by giving damage on the partners. Yet a player cannot choose a cooperative action easily, even though he realizes the cooperation is the best solution for both sides. Unless he perfectly trusts the partner, his final decision is to betray the partner. As shown in Table 1, his rational thinking concludes that the better choice is D if the partner chooses D to protect himself from the sacrificing situation when he receives S while the partner receives. In this case he is contend with a modest reward of P because this is still larger than S as shown in the right column of Table 1. The first column of the same table show that even if the partner chooses C the better choice of himself is again D since the reward as a traitor (T) is better than the reward as a cooperator (R). This is the reasoning for both players to choose the defective action, D, which is often called as the Nash equilibrium. In other words, the player chooses the larger reward out of the partner's possible choices, which is always the smaller reward to the player (min-max action).

Table 1 Payoff table of PD: Self (Partner)

Self/Partner	Cooperate	Defect
Cooperate	R(R)	S(T)
Defect	T(S)	P(P)

※ $S<P<R<T$, $S+T<2R$

3 Iterated Prisoner's Dilemma (IPD)

Repeated games can change the situation. By playing the game repeatedly, the past defective action would cause retaliation by the partner, or at least the fear on the potential retaliation makes the player to refrain from the defection. However, there is another problem. At the final matching when they have no more fear of retaliation, both sides would betray. This situation influences the second last matching, where both sides choose D because there is no point choosing C when the partner would definitely choose the defective action in the next matching. The same logic applies for the entire series of the iterated game and the entire population is trapped by the Nash equilibrium. This problem can be avoided by terminating the iteration by throwing dice, for example, so that the players cannot prepare for the final matching.

The real question is whether one can choose the cooperative action under a certain situation. One would do this under a tremendous mental suffering of trusting the unreliable partner, or disregarding his own profit completely.

If the prisoners are given a chance to communicate, they might reach mutual cooperation. This seems the current world situation encouraging various international conferences in various places such as Himeji, etc. Still, it is quite difficult for both sides to cooperate when they suffer from the fear of being betrayed by the partner.

Perhaps the most effective way of escaping from the Nash equilibrium is not to obey the voice of almighty but to prepare a certain environment under which the players autonomously choose the mutual cooperation.

We consider a multi-agent model over the evolving network of players in which the reputation of each player is implemented as the most important component to set up the network.

4 Reputable Players Network Model

We implement the players' reputation to our model in two ways. One is by means of updating the players' network where the players of bad reputation, measured by the rate of D that he/she had chosen in the IPD at the last generation, loose the chance to participate the game. Another element is the evaluation function, measured as the accumulation of payoffs to be used to update the players' strategies.

The model assumes the payoff parameters as shown in Table 2. The reputation (r) of the player i is defined as follows.

$$r_i = \frac{f_C(i) - f_D(i)}{f_C(i) + f_D(i)} \quad (-1 \le r_i \le 1) \tag{1}$$

Here $f_C(i)$ and $f_D(i)$ are the frequencies of action C and D, respectively, out of the number ($m = f_C(i) + f_D(i)$, $m = 1,..,M$) of the PD played so far.

4.1 Evolvable Network

Each PD game is played by the two players chosen from the entire population of N players and repeated M times. The first game is played by the entire population. The Second and the later PD are played only by the pairs who have agreed to match. Every player is given a fixed criterion of minimum reputation (given randomly) to require of the partner and is allowed to refuse playing the game if the partner does not satisfy the required criterion. This determines a network of the players of IPD.

This mechanism prompts the players toward cooperation. Each agent autonomously tries to keep its own reputation, to maintain the chance of joining the game for the sake of accumulating the payoffs. As a result, cooperative action is generated. However, the total setting is PD whose structure implies a potential sacrifice on cooperative agents.

Table 2 Payoff table of this model

Self/Partner	Cooperate	Defect
Cooperate	3(3)	0(5)
Defect	5(0)	1(1)

4.2 Evolvable Strategies

Each player holds two different strategies, one for cooperative partners and the other for defective partners. Those strategies are strings of length K, composed from the two actions C and D. We take K=4 in this article as a first step, thus we have 16 different types of strategies for each component. Initially those strategies are randomly given to the players and the players repeatedly use those strategies until the end of one generation. For example, some players are always cooperative by acting CCCC, tricky players would act CCCD, and wicked players would act CDDD for a cooperative partner, and persistent retaliators would choose DDDD for a traitor, or tactic player would choose CDCD to hint cooperation for a traitor. By doing so, we attempt to reflect various characters of the players chosen from the total 256 choices available in this model for a cooperative partner as well as defective partner.

We denote those strategies by ordered pairs whose first (second) element represents the action toward a defective (cooperative) partner.

For example, All_C strategy in the original IPD settings is written as <CCCC, CCCC> in our model. TFT (tit for tat) is written as <DDDD, CCCC> and All_D is written as <DDDD, DDDD>, etc.

Those pairs of strategies are updated by means of evolutional algorithm as follows.

1) The first generation

A pair of players is chosen from the entire population of N players and they play the game M times. Repeat the above process for all the possible pairs. The accumulated score and the number of D (defective action) are recorded and the reputation is counted for each player. Each player repeatedly uses the given strategy, in such a way that he uses the first (second) component of the ordered pair one by one if the partner's previous action was D(C). For example, a player having the strategy <CDDD, CCCD> would act CDD… when the partner acts DDD…, and the same player would act CCC… when the partner acts CDC…, until the chosen pair finish M iterated games.

2) Generation change

By sorting the N players in the descending order of the accumulated score, we keep the players beyond the average score unchanged and make the players below the average score imitating the strategies of upper scores.

3) Noise effect in imitating the strategies

We assume the effect of noise with a fixed rate. Using the Gaussian noise, we convert two D in the strategy string of length 4 to C for the output +2, and convert one C in the string to D for the output -1.

4) Reputable player's network

From the second generation to the end of simulation, the chosen pair of players plays the game only when the partner's reputation is higher than his/her criterion for both sides. Namely, the game is actually played only if

$$q_i < r_j \quad (-1 \leq q_i, r_j \leq 1) \tag{2}$$

Here q_i is a criterion of reputation for player i requires for his partners, which is given as a random number (between -1 and 1) as the initial condition. The pair does not play the game if one side refuses the match.

A question arises if the reputation of the players must be made public. In our model, all the players hold information on every other's reputation after the first generation when all the pairs have played M times of iterated games, thus the scores do not need to be publicized. From the second generation and after, not all the pairs match, and the players use the previous information on the reputation of partners if no mating was made with that partner in the previous generation. This situation corresponds to defining different reputations of the same player to different partners, so that the player i plays the game only if the following condition is satisfied.

$$q_i < r_{j,i} \quad (-1 \leq q_i, r_{j,i} \leq 1) \tag{3}$$

Here $r_{j,i}$ is the reputation of player i to the player j, defined similarly as Eq. (1) for a particular partner.

5 Simulation Result

We have obtained the formation of cooperative networks within a few hundred generations in all the cases we have tried, although there are some variances in terms of the onset of cooperation and the strategy distributions, as shown in Fig.1(a)-(d), where the vertical axis on the right hand side represent the average payoff, and the vertical axis on the left hand side represents the population of strategies represented by the numbers of C in the two different action of partners. For example, <DDDD, CCCD> is represented by (0,3) in those figures. The horizontal axis shows the generation.

All the 4 figures show that the average payoff curve rises up to 3, implying the cooperation, and stays around this value throughout the simulation. Also, it can be observed in all the figures that the regime where the strategy (3,3) dominates, the average payoff curve stays around the value 2.7, and as (3,3) decline, the curve rapidly reaches the final cooperation level at around 3.

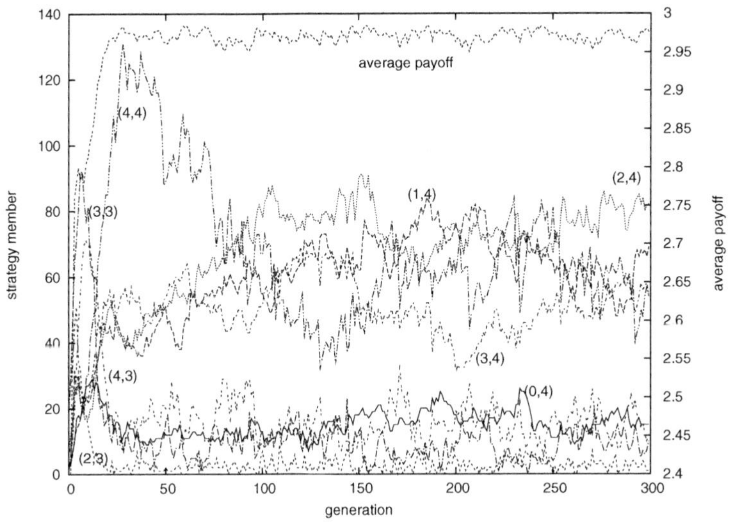

(a)

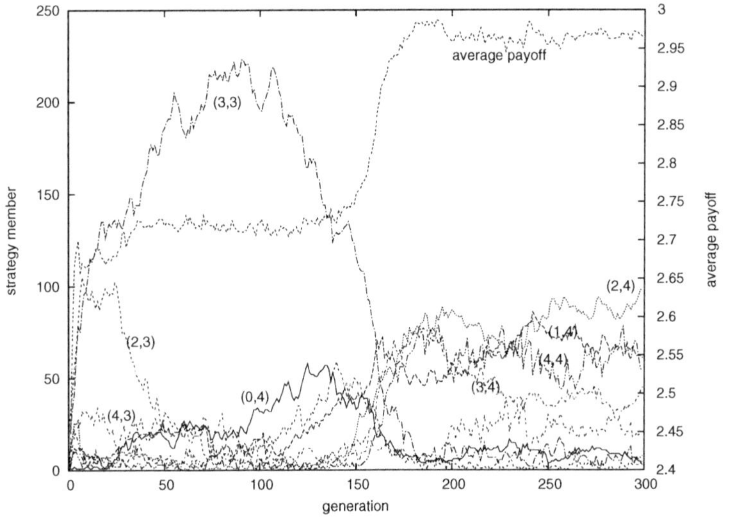

(b)

Fig. 1 (a)-(d) Average payoff and the population of strategies vs. the generation

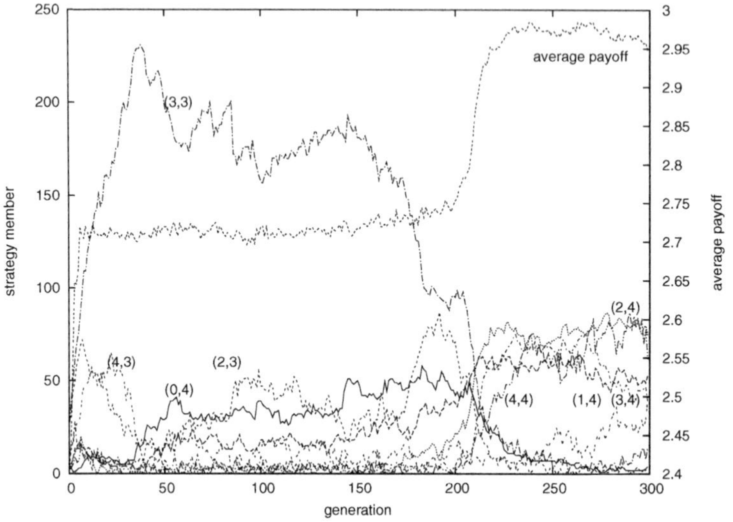

(c)

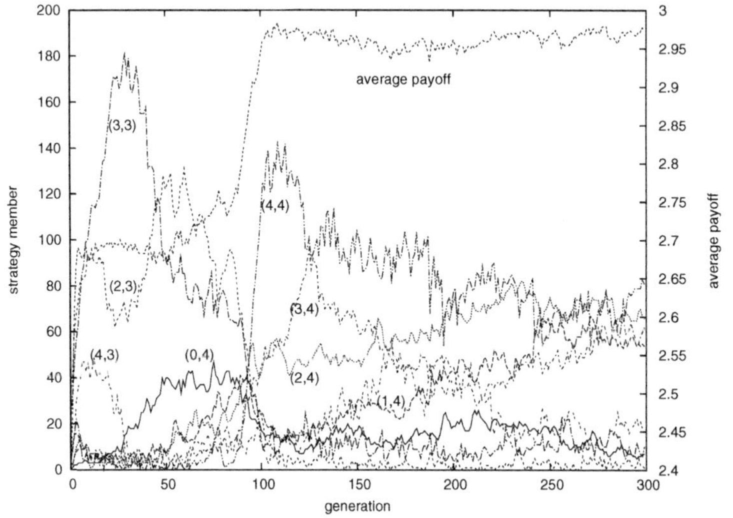

(d)

Fig. 1 (*continued*)

6 Conclusion and Future Perspectives

We have presented a multi-agent model of IPD including the effect of reputation, by considering two different strategies against two different actions, C and D of the partner. We also place the model on an evolvable network of reputable players. By co-evolving the strategies and the players' network, we have successfully realized cooperation among selfish prisoners.

We will explore this model in more detail by extending the length of the strategies from 4 to the larger to include more complicated strategies.

References

1. Axelrod, R.M.: The Evolution of Cooperation. Basic Books, New York (1984)
2. Axelrod, R.M.: The evolution of Strategies in the iterated prisoner's dilemma. In: Genetic Algorithms and Simulated Annealing. Morgan Kaufmann, San Francisco (1987)
3. Lindgren, K.: Evolutionary phenomena in simple dynamics. Artificial Life 2, 295 (1991)
4. Novak, M.A., Sigmund, K.: Evolution of indirect reciprocity by image scoring. Nature 393, 573 (1998)
5. Roberts, G., Sherratt, T.N.: Development of cooperative relationship through increasing investment. Nature 394, 175 (1998)
6. Yao, X., Darwen, P.J.: How important is your reputation in a multi-agent environment. In: Proceedings of 1999 International Conference on Systems, Man and Cybernetics (IEEE 1999), p. II-575 (1999)
7. von Neumann, J., Morgenstern, O.: Theory of Games and Economic Behavior. Princeton University Press, Princeton (1944)
8. Poundstone, W.: Prisoner's Dilemma: John von Neumann. Doubleday, New York (1992)

A Multi-agent Wideband Signal Detector

John Hefferan

Abstract. A group of intelligent agents exploring a time-frequency communications environment control a group of narrowband radio receivers. Multi-agent reinforcement learning is used in order to potentially increase the rate of processing wideband data and also to effectively co-ordinate the perceived narrowband information. The Multi-Agent System (MAS) attempts to detect the presence of a single tone frequency agile radio transmission contained within the wider received band. Differing learning policies are investigated and the performance of the detector is considered.

Keywords: Multi-Agent System (MAS), Frequency Hopping (FH).

1 Introduction

Modulated carrier electromagnetic transmission was originally accomplished by the transmission of a single modulated carrier frequency signal. This relatively simple approach made detecting the presence of information bearing signals also relatively easy. The carrier frequency, or one its signal components, were a constant feature of the power spectrum and could be detected by integrating the spectral region around the carrier frequency of the transmission. More recently, frequency hopping (FH) carrier transmissions have been developed. In these signals the carrier frequency is changed in a pseudo-random sequence after a fixed dwell time at each frequency. These FH signals have the advantage of a greater immunity to narrowband interfering signals than that of a single carrier frequency transmission. However, the difficulty in detecting and tracking a FH signal requires a much more complex receiver. Conventionally, FH signal reception has been carried out by the analysis of the output of a single wideband receiver [1] - using what could be thought of as a single agent approach.

John Hefferan
Command, Control, Communications and Intelligence Division
Defence Science and Technology Organisation,
Labs 203, PO Box 1500, Edinburgh, SA 5111, Australia
e-mail: john.hefferan@dsto.defence.gov.au

K. Nakamatsu et al. (Eds.): New Advan. in Intel. Decision Techno., SCI 199, pp. 625–633.
springerlink.com

In this work a multi-agent approach is introduced to the problem of detecting (and characterising) FH signals. The advantage of processing many regions of the time-frequency (t-f) environment concurrently, together with the opportunity to trial several learning policies have prompted this approach. Using many, digitally controlled, narrowband receivers is also now a feasible alternative to a single wideband tuner.

In this paper section 2 reviews the characteristics of the wideband communications band. Section 3 then introduces and defines the demodulator agent. This is followed, in sections 4 and 5, by a description of each of the two levels of learning that a group of demodulator agents undertake. Section 6 provides the investigation of the performance of the multi-agent detector in the task of detecting a FH signal in a simulated communications environment. Finally, some areas of future research and the conclusion are discussed.

2 Detection in a Wideband Environment

Figure 1 shows a representation of a wideband communications environment comprised of several communications carriers in a background of Additive White Gaussian Noise (AWGN). A FH signal will appear as very short bursts appearing at a predetermined sequence of frequencies and will, in general, only be visible in a power spectral plot such as figure 1 when the spectral plot is integrated over time and the FH signal revisits a set of fixed frequencies. Figure 2 shows a representation of a wide bandwidth communications environment as it varies in time.

3 The Demodulator Agent

In this work the concept of a demodulator agent is introduced. The agent, depicted in figure 3, is, in this work, a software frequency agile element capable of demodulating short bursts of modulated carrier wave transmissions at a given centre frequency

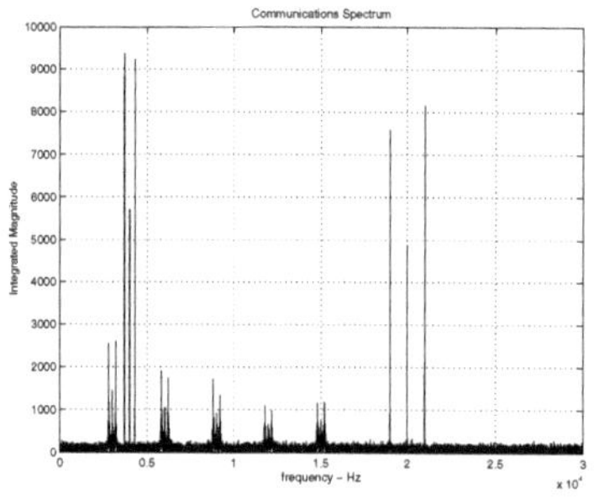

Fig. 1 A communications spectrum with various transmissions present

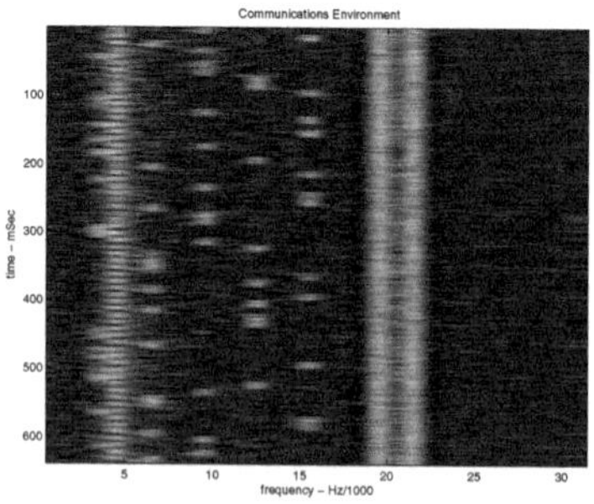

Fig. 2 A time-frequency view of the spectral environment

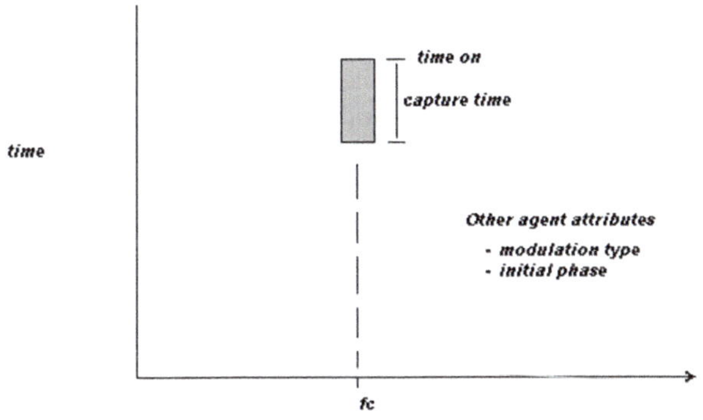

Fig. 3 A Demodulator Agent using a time-frequency (t-f) plot

(the carrier frequency f_c). Analog amplitude modulation (AM) may be used to convey information on the carrier. This modulation may be defined as in equation 1.

$$s(t) = m(t).cos(2\pi f_c t) \quad (1)$$

Here, the low frequency signal $m(t)$ modulates the higher frequency carrier. A key concept in this work is that of the time-frequency space as an environment for demodulator agents. The interaction of the agent with its environment is determined by the rule-set that each of the agents possess.

4 The Learning Algorithm - Layer 1 Reinforcement Learning

When using re-inforcement learning [2] in a multi-agent environment as described, the effects of agent interactions during learning is a complex process.[1] Independent learning by the multi-agent system is the simplest learning method to employ. Agents do not record the actions or effects of other agents in the environment thereby reducing the computational requirement. In cases where the multi-agent interactions effect the environment of other agents, this simplification cannot in general, be justified - the actions of other agents will have an effect on whether a particular agent can calculate the correct probability of entering a new state given its own action. In the first learning phase of this work however, agents are able to be treated as being independent and the results of their actions do not effect other agents directly. It is only in the subsequent cluster analysis phase that agent outcomes from the first learning phase are correlated.[2]

Each demodulator agent is initially placed randomly at a point on the time-frequency plane of interest. In this work the boundaries of the environment are a frequency boundary of $1000Hz \leq f \leq 15000Hz$ and a time boundary of $0 \leq t \leq 640msec$.

[1] See, for example [3] for an overview of some of the important issues.

[2] See [4] for an introduction to multi-agent intelligent clustering.

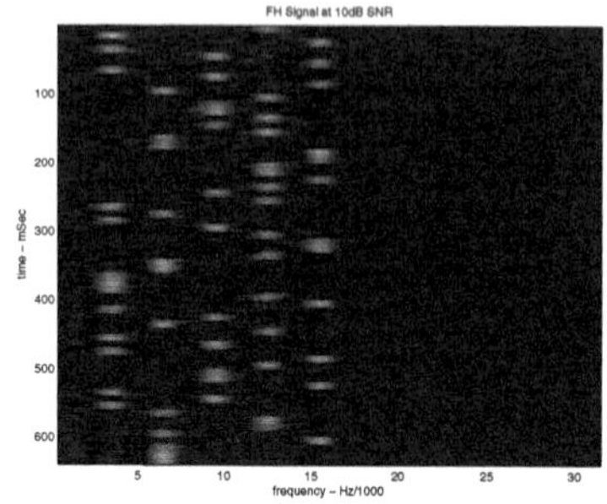

Fig. 4 Simulated Frequency Hopping signal at 10dB SNR

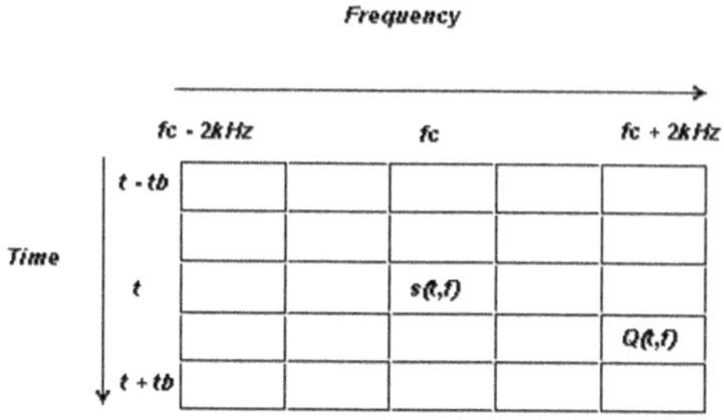

Fig. 5 Local Region

A FH signal consisting of a 10msec burst time AM modulated carrier is present in the environment together with a variable background level of AWGN. A t-f plot of a FH carrier modulated with a 200Hz sine wave in a background AWGN level of 10dB SNR is shown in figure 4

The environment that each agent finds itself in consists of a group of t-f cells surrounding its present location in t-f space. Figure 5 depicts the range of cells 'visible' to each agent during each iteration of the first learning phase.

4.1 Local Value

Each grid space (t, f) in the local environment is, in effect, a short duration narrowband segment of the wideband environment. Hence, agents are able to demodulate each of the segments in their range and thereby estimate a value for each. Here, the reward gained by agent (i) in each state $r_i(t, f)$ is defined in equation 2.

$$r_i(t,f) = S(200) / \sum_{f=f_{lo}}^{f=f_{hi}} S(f) \tag{2}$$

Here, the $S(f)$ is an estimate of the spectral energy at frequency f.

Hence, the particular state reward to an agent will be higher if the spectral component at 200Hz is relatively high. Other transmitted characteristics could be used for the value term such as demodulated speech characteristics, baud parameters or unique transmitter characteristics in the time-frequency space, for example the transmitter's RF signature.

In this work a reinforcement learning rule is used where different policies determine which action the agent will take depending upon its own current state and the state values that surround it. This is, effectively, a form of Q-Learning [2] where we can say that agent i, currently in state $s_k(t, f)$ at iteration k will determine its next state $s_{(k+1)}(t, f)$ using [5] in equation 3.

$$\hat{Q}(s,a) \leftarrow r(s) + \gamma \max_{a'} \hat{Q}(s', a') \tag{3}$$

Where $\hat{Q}$ is the current approximation to the Q function value for the state-action pair (s, a). γ is the *discount* factor devaluing the effect of future states, in this case $\gamma = 0.9$ to allow significant look-ahead influence.

4.1.1 Local Learning Policies (π_x)

Several different learning policies may be used here. The *greedy* policy is denoted by π_1. Here each agent updates its Q table for each state in range and will choose the path with the calculated value as its destination during that iteration.

Policy π_2 contains an element of exploration for each agent. Here during each iteration, the agent calculates the Q value for surrounding states as with π_1. However, if the highest value state is the current state and it has a local value $Q_k(t, f)$ less than some minimum value δ, then the agent will choose to move to the highest state in range other than its current state. This policy attempts to prevent agents becoming trapped at low value points in the environment by forcing a move to a different state.

Policy π_3 is an extension of π_2. It is similar except that instead of making a small exploratory move to a lower state when there are no higher states nearby, the agent is randomly reassigned to another point in the environment.

4.2 Global Value

The co-ordinated performance of the population of agents in detecting a FH wideband signal is determined by calculating the *global* value of the population at iteration k, Vg_k. In this work the Vg_k value is determined by scanning down the time axis of the given environment and testing the output of the agent population at time segments equal to half the expected burst time of the FH signal. Hence, the value of the current state of all agents can be quantified by calculating the global value. Here, this is defined in equation 4.

$$Vg_k = \sum_t V_i(200) \tag{4}$$

Where $V_i(200)$ is the local value of agent (i) if present at contiguous time intervals through the entire environmental timeframe. The global value is a measure of how well the agent population has been able to detect a FH transmission during the entire environment time window and may be used to update the local estimates of the Q function at each agent. The longer the contiguous sequence of high value states found by the population then the higher is the global value obtained.

5 The Learning Algorithm - Layer 2 Multi-agent Cluster Network Analysis

As a result of the initial layer of learning, the agent population converges to states of high local value whilst concurrently maximizing the global value of the population

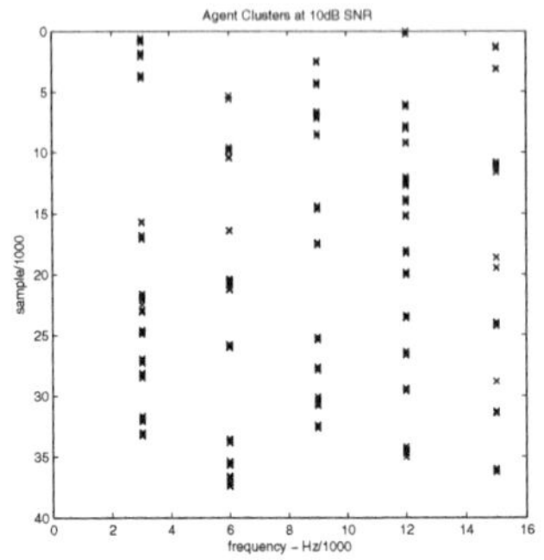

Fig. 6 Multi Agent Clusters at 10 dB SNR

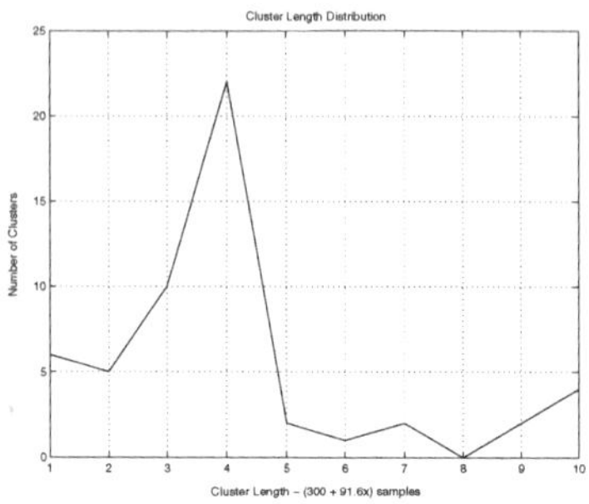

Fig. 7 Distribution of Cluster length

as a whole. As a result of this, the agent population effectively form clusters around each likely burst of modulated energy detected in the t-f environment. Figure 6 shows the clustering of the agent population following the initial learning layer when applied to the environment of figure 4.

The multi-agent clusters that have been formed are now analysed statistically and measures considered here are the cluster start time, cluster duration, cluster centre frequency and number of agent members.

We can further analyse the cluster data as depicted in figure 6 in an attempt to filter out clusters which are least likely to be part of a FH transmission.

5.1 *Burst Time Multiples*

Figure 7 is a histogram plot of the cluster duration of those depicted in figure 6. Here we can see a peak cluster duration of approximately four units which, when appropriately scaled equates to a mean cluster duration of 660 samples or 11mSec. The estimation of the likely burst time allows us to filter out clusters that do not conform to a burst time multiple derived from this value and are, hence, unlikely to be one the FH bursts.

5.2 *Burst Networks*

Each burst in the transmitted FH sequence links to a specific subsequent burst at the end of the burst time. These *linked* bursts are generally, but not necessarily, at different carrier frequencies and each burst follows on immediately from its predecessor. At the receiver however, detected energy bursts will potentially link to several clusters that have been detected. We can depict this as in figure 8 which assigns a probability to the links between successive clusters in the ambiguous burst network. The receiver is presented with the task of estimating the most probable path through the network of clusters and links. Burst length, inter-burst time and burst modulation

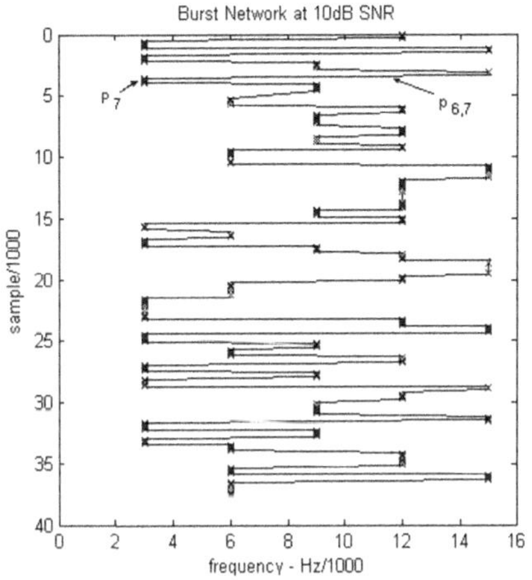

Fig. 8 Reconstructed Burst Network

all influence the path probability through clusters. As we now have an estimate of the burst structure and linkages we can estimate the path through the finite number of states by using Bellman's dynamic programming [6].

$$V^*(s_j) = \max_{a_j}\left(E[r_{j+1}] + \gamma \sum_{s_{j+1}} P(s_{j+1}|s_j, a_j)V^*(s_{j+1})\right) \quad (5)$$

Here the $P(s_j)$ term is derived from the estimated probability of a cluster at time offset j being from a burst distribution together with an estimate of the path probability ($p_{i,j}$) from cluster i to cluster j as depicted in figure 8. $V^*(s)$ indicates the path through the most likely clusters when the optimal policy is followed. It is evaluated effectively, from the likelihood of the particular cluster belonging to a distribution of FH bursts obtained from the prior cluster analysis (giving rise to estimates of a reward term $r(s_j)$). Specific knowledge of likely burst durations and frequency spacings will improve our confidence in the $P(s_j)$ estimates for each energy burst. We retain the most likely path in order to reconstruct the FH sequence as in figure 8.

6 Noise Performance

The performance of this approach to wideband signal detection in varying levels of background AWGN is studied in this section. Here, an AM FH signal with a constant 200Hz modulating tone was generated and a simulated transmission using a five carrier tone hopping transmission distributed within a frequency boundary of $3kHz - 15kHz$. The sampling frequency f_s was set at $60k$ samples per second. The t-f environments of the signal at the differing AWGN levels is shown in figure 9.

The resulting cluster distributions of 512 demodulator agents is shown for each of the noise environments also in figure 9. As the SNR is decreased it can be seen

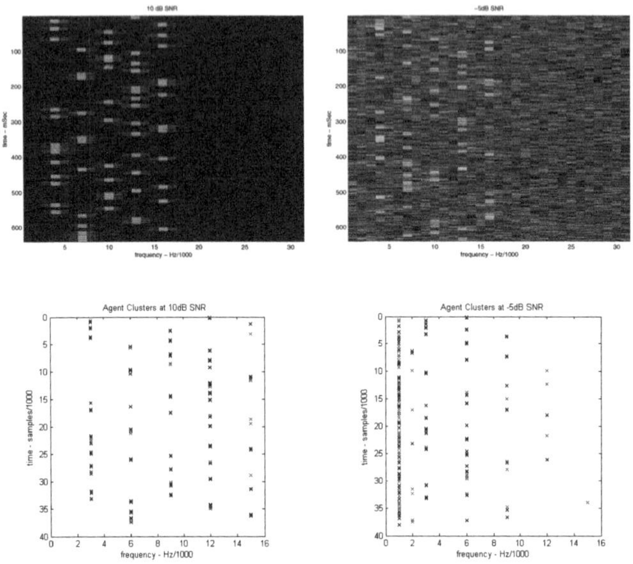

Fig. 9 Varying AWGN levels

that more agents were grouping at the lower frequency range (hence leaving fewer for the higher frequency bursts). This lower frequency signal is the result of noise floor components increased magnitude. Filtering of this noise component may occur in the cluster analysis phase of this algorithm. Increasing the number of demodulator agents would also leave a higher number available to associate with the higher frequency burst energy.

The presence of the 10mSec (600 sample) burst was detected in all cases with the longer duration interferer and noise clusters becoming more prominent as SNR decreased. Improved cluster filtering, local agent learning discriminants and distributions will facilitate better resolution of the burst network.

7 Future Work

Several areas of further investigation lend themselves here to improving this method in the detection of received signals. Firstly, the scalability of this approach to higher frequency transmissions needs to be tested. In an implementation of this method, the maximum sampling rate of the system's analog to digital (A/D) converter would be a limiting factor. A conventional 100M sample per second A/D would permit the capture of a 40MHz frequency band which is adequate to cover many satellite transponders in use currently.

Improved feature identification and analysis is possible. For example, the identification of specific transmitter characteristics (i.e.: the transmitter signature) would be potentially identifiable with this approach. Other features in the modulation space (for example speech spectra or envelope distribution) would also allow a greater application in the detection of active FH voice transmissions.

Finally, a directional antenna, providing a line of bearing to incident signal energy would also provide another important feature to aid the cluster filter.

8 Conclusion

This work has presented a multi-agent approach to the detection and recognition of wideband, frequency hopping communications signals. The two stage learning and analysis method employs a number of demodulator agents to each explore a region of the local time-frequency space and search for energy with features of interest. A collective measure of global value has been introduced which is able to provide quantitative reinforcement during the learning process. When used with an AM modulated FH carrier this method was shown to be able to detect the main burst duration. A method for determining the most likely burst sequence was also discussed. With further development this approach will provide a practical method of facilitating the analysis of frequency agile signals in the wideband time-frequency environment.

References

1. Horing, H.-C.: Probability of intercept for frequency hop signals using search receivers (1998), http://www.rohde-schwarz.com
2. Sutton, R.S., Barto, A.G.: Reinforcement Learning: An Introduction. The MIT Press, Cambridge (1998)
3. Vlassis, N.: A Concise Introduction to Multiagent Systems and Distributed Artificial Intelligence. Synthesis Lectures in Artificial Intelligence and Machine Learning, ch. 8.3. Morgan & Claypool Publishers (2007)
4. Park, J.-E., Oh, K.-W.: Multi-agent systems for intelligent clustering. Proc. World Academy Sc., Eng. and Tech. 11, 97–102 (2006)
5. Mitchell, T.: Machine Learning, ch. 13. McGraw-Hill, New York (1997)
6. Bellman, R.E.: Dynamic Programming. Princeton University Press, Princeton (1957)

Bayesian Estimation of GARCH Model with an Adaptive Proposal Density

Tetsuya Takaishi

Abstract. A Bayesian estimation of a GARCH model is performed for US Dollar/Japanese Yen exchange rate by the Metropolis-Hastings algorithm with a proposal density given by the adaptive construction scheme. In the adaptive construction scheme the proposal density is assumed to take a form of a multivariate Student's t-distribution and its parameters are evaluated by using the sampled data and updated adaptively during Markov Chain Monte Carlo simulations. We find that the autocorrelation times between the data sampled by the adaptive construction scheme are considerably reduced. We conclude that the adaptive construction scheme works efficiently for the Bayesian inference of the GARCH model.

Keywords: Markov Chain Monte Carlo, Bayesian inference, GARCH model, Metropolis-Hastings algorithm.

1 Introduction

In finance volatility of asset returns plays an important role to manage financial risk. To forecast volatility, various empirical models which mimic the properties of the volatility have been proposed. Engle[1] proposed Autoregressive Conditional Heteroskedasticity (ARCH) model where the present volatility is assumed to depend on the squares of past observations. Later Bollerslev[2] advocated Generalized ARCH (GARCH) model which is an extension of the ARCH model and includes additional past volatility terms to the present volatility estimate. It is known that the volatility of the financial assets exhibits clustering in the financial time series. The GARCH model can captures this property. Furthermore the return distribution generated from the GARCH process shows a fat-tailed distribution which is also seen in the real financial markets. There also exists extension of the GARCH model which incorporates the asymmetric property of the volatility[3, 4, 5, 6].

Tetsuya Takaishi
Hiroshima University of Economics
e-mail: takaishi@hiroshima-u.ac.jp

K. Nakamatsu et al. (Eds.): New Advan. in Intel. Decision Techno., SCI 199, pp. 635–643.
springerlink.com

A preferred algorithm to infer GARCH model parameters is the Maximum Likelihood (ML) method which estimates the parameters by maximaizing the corresponding likelihood function of the GARCH model. In this algorithm there is a practical difficulty in the maximization procedure when the output results are sensitive to starting values.

By the recent computer development the Bayesian inference by Markov chain Monte Carlo (MCMC) methods, which is an alternative approach to estimate GARCH parameters, has become popular. There exist a variety of methods proposed to implement the MCMC scheme[7]-[12]. In a recent survey[11] it is shown that Acceptance-Rejection/Metropolis-Hastings (AR/MH) algorithm works better than other algorithms. In the AR/MH algorithm the proposal density is assumed to be a multivariate Student's t-distribution and the parameters to specify the distribution are estimated by the ML technique. Recently a new method to estimate those parameters without relying on the ML technique was proposed[13]. In the method the parameters are determined by an MCMC simulation. During the MCMC simulation, the parameters are updated adaptively using the data sampled by the MCMC method itself. We call this method "adaptive construction scheme". The adaptive construction scheme was tested for artificial GARCH data and it is shown that the adaptive construction scheme can significantly reduce the correlation between sampled data[13]. In this study we apply the adaptive construction scheme to real financial data, US Dollar/Japanese Yen exchange rate and examine the efficiency of the adaptive construction scheme.

2 GARCH Model

The GARCH(p,q) model by Bollerslev[2] is given by

$$y_t = \sigma_t \varepsilon_t, \tag{1}$$

$$\sigma_t^2 = \omega + \sum_{i=1}^{q} \alpha_i y_{t-i}^2 + \sum_{i=1}^{p} \beta_i \sigma_{t-i}^2, \tag{2}$$

where the GARCH parameters are restricted to $\omega > 0$, $\alpha_i > 0$ and $\beta_i > 0$ to ensure a positive volatility, and the stationary condition $\sum_{i=1}^{q} \alpha_i + \sum_{i=1}^{p} \beta_i < 1$ is also required. ε_t is an independent normal error $\sim N(0,1)$.

In this study we focus on GARCH(1,1) model where the volatility σ_t^2 is given by

$$\sigma_t^2 = \omega + \alpha y_{t-1}^2 + \beta \sigma_{t-1}^2. \tag{3}$$

The likelihood function of the GARCH model is given by

$$L(y|\theta) = \Pi_{i=1}^{n} \frac{1}{\sqrt{2\pi\sigma_t^2}} \exp\,(-\frac{y_t^2}{\sigma_t^2}). \tag{4}$$

3 Bayesian Inference

Using Bayes' rule the posterior density $\pi(\theta|y)$ with n observations denoted by $y = (y_1, y_2, \ldots, y_n)$ is given by

$$\pi(\theta|y) \propto L(y|\theta)\pi(\theta), \tag{5}$$

where $L(y|\theta)$ is the likelihood function. $\pi(\theta)$ is the prior density which we have to specify depending on θ. In this study we assume that the prior density $\pi(\theta)$ is constant.

With $\pi(\theta|y)$ we infer θ as expectation values of θ. The expectation values are given by

$$\langle\theta\rangle = \frac{1}{Z}\int \theta\pi(\theta|y)d\theta, \tag{6}$$

where $Z = \int \pi(\theta|y)d\theta$ is the normalization constant. Hereafter we omit Z since this factor is irrelevant to MCMC estimations.

The MCMC technique gives a method to estimate eq.(6) numerically. The basic procedure of the MCMC method is as follows. First we sample θ drawn from a probability distribution $\pi(\theta|y)$. Sampling is done by a technique which produces a Markov chain. After sampling some data, we evaluate the expectation value as an average value over the sampled data $\theta^{(i)}$,

$$\langle\theta\rangle = \lim_{k\to\infty} \frac{1}{k}\sum_{i=1}^{k} \theta^{(i)}, \tag{7}$$

where k is the number of the sampled data. The statistical error for k independent data is proportional to $\frac{1}{\sqrt{k}}$. When the sampled data are correlated the statistical error will be proportional to $\sqrt{\frac{2\tau}{k}}$ where τ is the autocorrelation time between the sampled data. The autocorrelation time depends on the MCMC method we employ. Thus it is desirable to take an MCMC method which can generate data with a small τ.

4 Metropolis-Hastings Algorithm

The Metropolis-Hastings (MH) algorithm[14] is an MCMC simulation method which generates draws from any probability density. The MH algorithm is an extension of the original Metropolis algorithm[15]. Let us consider a probability distribution $P(x)$ from which we would like to sample data x. The MH algorithm consists of the following steps.

(1) First we set an initial value x_0 and $i = 1$.

(2) Then we generate a new value x_i from a certain probability distribution $g(x_i|x_{i-1})$ which we call proposal density.

(3) We accept the candidate x_i with a probability of $P_{MH}(x_{i-1}, x_i)$ where

$$P_{MH}(x_{i-1},x_i) = \min\left[1, \frac{P(x_i)}{P(x_{i-1})}\frac{g(x_i|x_{i-1})}{g(x_{i-1}|x_i)}\right]. \quad (8)$$

When x_i is rejected we keep x_{i-1}, i.e. $x_i = x_{i-1}$.

(4) Go back to (2) with an increment of $i = i+1$.

For a symmetric proposal density $g(x_i|x_{i-1}) = g(x_{i-1}|x_i)$, eq.(8) reduces to the Metropolis accept probability:

$$P_{Metro}(x_{i-1},x_i) = \min\left[1, \frac{P(x_i)}{P(x_{i-1})}\right]. \quad (9)$$

5 Adaptive Construction Scheme

Since the proposal density $g(x_i|x_{i-1})$ is dependent of the previous value x_{i-1}, usually the sampled data are correlated. One may use an independent proposal density $g(x_i)$ which does not depend on the previous value. Although in this case we can generate independent candidates x_i, it is important to choose the one close enough to the posterior density, in order to make the acceptance high enough.

The posterior density of GARCH parameters often resembles to a Gaussian-like shape. Thus one may choose a density similar to a Gaussian distribution as the proposal density. Following [10, 11], in order to cover the tails of the posterior density we use a (p-dimensional) multivariate Student's t-distribution given by

$$g(\theta) = \frac{\Gamma((\nu+p)/2)/\Gamma(\nu/2)}{\det\Sigma^{1/2}(\nu\pi)^{p/2}}\left[1+\frac{(\theta-M)^t\Sigma^{-1}(\theta-M)}{\nu}\right]^{-(\nu+p)/2}, \quad (10)$$

where θ and M are column vectors,

$$\theta = \begin{bmatrix}\theta_1\\ \theta_2\\ \vdots\\ \theta_p\end{bmatrix}, M = \begin{bmatrix}M_1\\ M_2\\ \vdots\\ M_p\end{bmatrix}, \quad (11)$$

and $M_i = E(\theta_i)$. Σ is the covariance matrix defined as

$$\frac{\nu\Sigma}{\nu-2} = E[(\theta-M)(\theta-M)^t]. \quad (12)$$

For later use we also define a matrix V as

$$V = E[(\theta-M)(\theta-M)^t]. \quad (13)$$

ν is a parameter to tune the shape of Student's t-distribution. When $\nu \to \infty$ the Student's t-distribution goes to a Gaussian distribution. In this study we take $\nu = 10$.

There are three parameters to be inferred for the GARCH(1,1) model. Therefore in this case $p = 3$ and $\theta = (\theta_1,\theta_2,\theta_3) = (\alpha,\beta,\omega)$, and Σ is a 3×3 matrix. The values of Σ and M are not known a priori. We determine these unknown parameters M and Σ through MCMC simulations. First we make a short run by the Metropolis algorithm and accumulate some data. Then we estimate M and Σ. Note that there is no need to estimate M and Σ accurately. Second we perform an MH simulation with the proposal density of eq.(10) with the estimated M and Σ. After accumulating more data, we recalculate M and Σ, and update M and Σ of eq.(10). By doing this, we adaptively change the shape of eq.(10) to fit the posterior density more accurately. We call eq.(10) with the estimated M and Σ "adaptive proposal density".

The random number generation for the multivariate Student's t-distribution can be done easily as follows. First we decompose the symmetric covariance matrix Σ by the Cholesky decomposition as $\Sigma = LL^t$. Then substituting this result to eq.(10) we obtain

$$g(X) \sim \left[1 + \frac{X^t X}{\nu}\right]^{-(\nu+p)/2}, \tag{14}$$

where $X = L^{-1}(\theta - M)$. The random numbers X are given by $X = Y\sqrt{\frac{\nu}{w}}$, where Y follows $N(0,I)$ and w is taken from the chi-square distribution ν degrees of freedom χ^2_ν. Finally we obtain the random number θ by $\theta = LX + M$.

6 Empirical Analysis

We make an empirical analysis based on daily data of the exchange rates for US Dollar and Japanese Yen. The sampling period of the exchange rates is 1 March 2000 to 29 February 2008, which gives 2007 observations. The exchange rates p_i are transformed to $r_i = 100[\ln(p_i/p_{i-1}) - \bar{s}]$ where $\bar{s}$ stands for the average value of $\ln(p_i/p_{i-1})$.

Our implementation of the adaptive construction scheme is as follows. First we make a short run by the Metropolis algorithm. We discard the first 3000 data as burn-in process. Then we accumulate 1000 data to estimate M and Σ. The estimated M and Σ are substituted to $g(\theta)$ of eq.(10). The shape parameter ν is set to 10. We re-start a run by the MH algorithm with the proposal density $g(\theta)$. Every 1000 update we re-calculate M and Σ using all accumulated data and update $g(\theta)$ for the next run. We accumulate 100000 data for analysis.

We also make a Metropolis simulation and accumulate 100000 data for analysis. The Metropolis algorithm in this study is implemented as follows. We draw a new θ' by adding a small random value $\delta\theta$ to the present value $\theta = (\theta_1, \theta_2, \theta_3) = (\alpha, \beta, \omega)$:

$$\theta'_j = \theta_j + \delta\theta_j, \tag{15}$$

where $\delta\theta_j = d(r - 0.5)$. r is a uniform random number in $[0,1]$ and d is a constant to tune the Metropolis acceptance. We choose d so that the acceptance becomes greater than 50%.

Fig. 1 compares the Monte Carlo time history of α sampled by the adaptive construction scheme with that by the Metropolis algorithm. It is clearly seen that the

Table 1 Results of parameters

	α	β	ω
Adaptive construction	0.03151	0.9403	0.01104
standard deviation	0.0078	0.017	0.0047
statistical error	0.00004	0.0001	0.00003
$2\tau_{int}$	2.8 ± 0.3	3.8 ± 0.4	4.1 ± 0.5
Metropolis	0.0318	0.9391	0.0114
standard deviation	0.0079	0.018	0.005
statistical error	0.0005	0.0014	0.0004
$2\tau_{int}$	400 ± 60	650 ± 100	620 ± 80

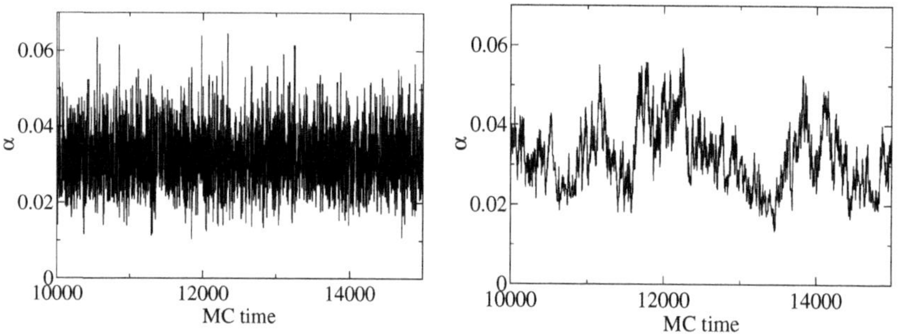

Fig. 1 Monte Carlo time histories of α sampled by the adaptive construction scheme (left) and the Metropolis algorithm(right)

data α produced by the Metropolis algorithm are very correlated. On the other hand the sampled data by the adaptive construction scheme seem to be well de-correlated. For other parameters β and ω we also see the similar behavior.

In order to see correlations between sampled data, we measure the autocorrelation function (ACF) defined as

$$ACF(t) = \frac{\frac{1}{N}\sum_{j=1}^{N}(x(j) - \langle x \rangle)(x(j+t) - \langle x \rangle)}{\sigma_x^2}, \tag{16}$$

where $\langle x \rangle$ and σ_x^2 are the average value and the variance of certain successive data x respectively.

Fig. 2 shows the ACF for the adaptive construction scheme and the Metropolis algorithm. The ACF of the the adaptive construction scheme decreases quickly as Monte Carlo time t increases. On the other hand the ACF of the Metropolis algorithm decreases very slowly which indicates that the correlation between the sampled data is very large.

We estimate the autocorrelation time by the integrated autocorrelation time τ_{int}. To calculate τ_{int} we define $\tau_{int}(T)$ as

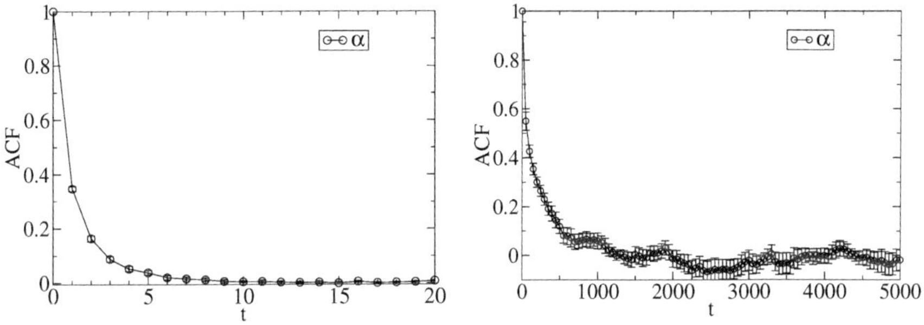

Fig. 2 Autocorrelation functions of α for the adaptive construction scheme (left) and the Metropolis algorithm (right)

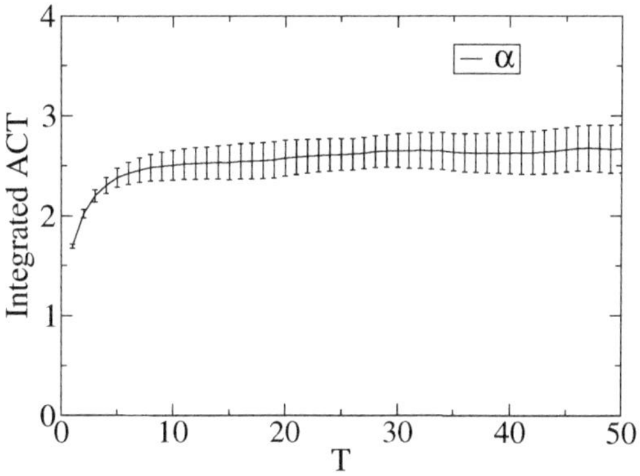

Fig. 3 The integrated autocorrelation time $\tau_{int}(T)$ of α sampled by the adaptive construction scheme

$$\tau_{int}(T) = \frac{1}{2} + \sum_{i=1}^{T} ACF(i). \tag{17}$$

τ_{int} is given by $\tau_{int}(T=\infty)$. In practice, however, it is impossible to sum up $ACF(t)$ to $T=\infty$. Since typically $\tau_{int}(T)$ increases with T and reaches a plateau we estimate τ_{int} at this plateau. Fig. 3 illustrates $\tau_{int}(T)$ of α sampled by the adaptive construction scheme. $\tau_{int}(T)$ increases with T and reaches a plateau around $T \geq 20$.

Results of τ_{int} are summarized in Table 1. The values of τ_{int} from the Metropolis simulations are very large, typically several hundreds. On the other hand τ_{int} from the adaptive construction scheme are very small, $2\tau_{int} \sim 2-3$[1]. This results in a factor of 10 reduction in terms of the statistical error. This reduction property is confirmed by the statistical errors of the sampled data (See Table 1). Thus it is concluded that the adaptive construction scheme is effectively working for reducing the correlations between the sampled data.

Fig. 4 shows the convergence property of the matrix V. The matrix elements V_{ij} are defined by $V = E[(\theta - M)(\theta - M)^t]$ with $\theta = (\theta_1, \theta_2, \theta_3) = (\alpha, \beta, \omega)$. For

[1] $2\tau_{int}$ is called an inefficiency factor.

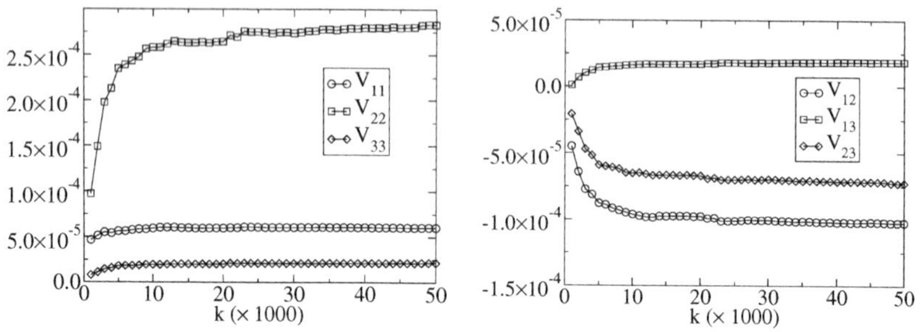

Fig. 4 The matrix elements of the symmetric covariance matrix V. Diagonal elements (left) and off-diagonal elements (right)

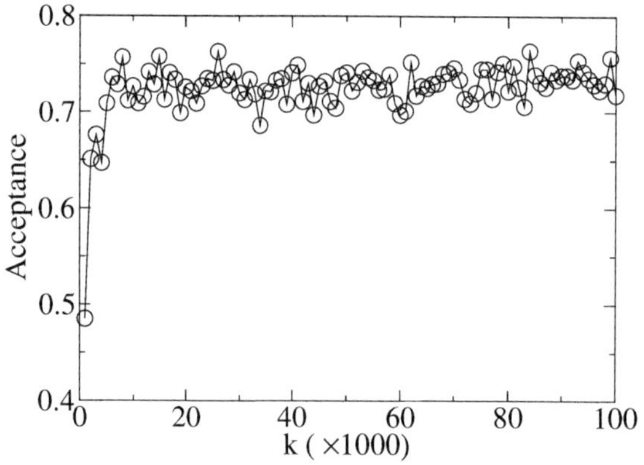

Fig. 5 Acceptance at MH step with the adaptive proposal density

instance $V_{12} = V_{\alpha\beta}$. All elements of V converge quickly to certain values as the simulations are proceeded.

Fig. 5 shows the acceptance at the MH algorithm with the adaptive proposal density of eq.(10). Each acceptance is calculated every 1000 updates and the calculation of the acceptance is based on the latest 1000 data. At the first stage of the simulation the acceptance is low. This is because at this stage M and Σ are not calculated accurately yet. However the acceptances increase quickly as the simulations are proceeded and reaches a plateau where the acceptance is more than 70%.

7 Summary

We proposed the adaptive construction scheme to construct a proposal density for the MH algorithm of the GARCH(1,1) model. The construction of the proposal density is performed using the data generated by MCMC methods. During the MCMC simulations the proposal density is updated adaptively. In this study we applied the adaptive construction scheme for the Bayesian inference of the GARCH(1,1) model by using US Dollar/Japanese Yen exchange rate. The numerical results show that the adaptive construction scheme significantly reduces the correlations between the sampled data. The autocorrelation time of the adaptive construction method is calculated to be $2\tau_{int} \sim 2-3$, which is comparable to that of the AR/MH method[11]. It is concluded that the adaptive construction scheme is an efficient method for the Bayesian inference of the GARCH(1,1) model. The adaptive construction scheme is not limited to the GARCH(1,1) model and can be applied for other GARCH-type models.

Acknowledgements. The numerical calculations were carried out on Altix at the Institute of Statistical Mathematics and on SX8 at the Yukawa Institute for Theoretical Physics in Kyoto University.

References

1. Engle, R.F.: Autoregressive Conditional Heteroskedasticity with Estimates of the Variance of the United Kingdom inflation. Econometrica 50, 987–1007 (1982)
2. Bollerslev, T.: Generalized Autoregressive Conditional Heteroskedasticity. Journal of Econometrics 31, 307–327 (1986)
3. Nelson, D.B.: Conditional Heteroskedasticity in Asset Returns: A New Approach. Econometrica 59, 347–370 (1991)
4. Glston, L.R., Jaganathan, R., Runkle, D.E.: On the Relation Between the Expected Value and the Volatility of the Nominal Excess on Stocks. Journal of Finance 48, 1779–1801 (1993)
5. Ding, Z., Granger, C.W.J., Engle, R.F.: A long memory property of stock market returns and a new model. Journal of Empirical Finance 1, 83–106 (1993)
6. Sentana, E.: Quadratic ARCH Models. Review of Economic Studies 62, 639–661 (1995)
7. Bauwens, L., Lubrano, M.: Bayesian inference on GARCH models using the Gibbs sampler. Econometrics Journal 1, c23–c46 (1998)
8. Kim, S., Shephard, N., Chib, S.: Stochastic volatility: Likelihood inference and comparison with ARCH models. Review of Economic Studies 65, 361–393 (1998)
9. Nakatsuma, T.: Bayesian analysis of ARMA-GARCH models: Markov chain sampling approach. Journal of Econometrics 95, 57–69 (2000)
10. Mitsui, H., Watanabe, T.: Bayesian analysis of GARCH option pricing models. J. Japan Statist. Soc. (Japanese Issue) 33, 307–324 (2003)
11. Asai, M.: Comparison of MCMC Methods for Estimating GARCH Models. J. Japan Statist. Soc. 36, 199–212 (2006)
12. Takaishi, T.: Bayesian Estimation of GARCH model by Hybrid Monte Carlo. In: Proceedings of the 9th Joint Conference on Information Sciences 2006, CIEF-214 arXiv:physics/0702240 (2007) doi:10.2991/jcis.2006.159
13. Takaishi, T.: An adaptive Markov Chain Monte Carlo method for GARCH model. arXiv:0901.0992v1 (2009)
14. Hastings, W.K.: Monte Carlo Sampling Methods Using Markov Chains and Their Applications. Biometrika 57, 97–109 (1970)
15. Metropolis, N., Rosenbluth, A.W., Rosenbluth, M.N., Teller, A.H., Teller, E.: Equations of State Calculations by Fast Computing Machines. J. of Chem. Phys. 21, 1087–1091 (1953)

Elucidation of Industrial Structure of the Japanese Economy through Visualization and Community Analysis

H. Iyetomi, K. Kamehama, T. Iino, Y. Ikeda, T. Ohnishi, H. Takayasu, and M. Takayasu

Abstract. We shed light on industrial structure of the economic system in Japan by combining visualization technique and community analysis. The production network consisting of submillion nodes (firms) and three million links (transactions) is visualized taking advantage of MD simulation technique. Also communities inherent in such a large-scale network is extracted through maximization of the modularity using both greedy (bottom-up) and bisection (top-down) algorithms; the bisection method works better. It is shown that nodes belonging to the same community are located close to each other in a visualization (three-dimensional) space.

1 Introduction

Various systems as encountered in informational, social, economical, and biological sciences are recognized as complex networks consisting of an enormous number of

H. Iyetomi
Department of Physics, Niigata University, Niigata 950-2181, Japan
e-mail: hiyetomi@sc.niigata-u.ac.jp

K. Kamehama
Department of Physics, Niigata University, Niigata 950-2181, Japan

T. Iino
Department of Physics, Niigata University, Niigata 950-2181, Japan

Y. Ikeda
Hitachi Research Institute, Tokyo 101-8010, Japan

T. Ohnishi
Graduate School of Law and Politics, the University of Tokyo, Tokyo 101-8010, Japan

H. Takayasu
Sony Computer Science Laboratory, Tokyo 141-0022, Japan

M. Takayasu
Department of Computational Intelligence and Systems Science, Tokyo Institute of Technology, Yokohama 226-8502, Japan

K. Nakamatsu et al. (Eds.): New Advan. in Intel. Decision Techno., SCI 199, pp. 645–652.
springerlink.com

linked elements [1, 2, 3, 4]. As an example, firms form a complex network through their business transactions for their production activities. A firm obtains materials from suppliers (upstream firms) and sell its products to customers (downstream firms). The business relationship so established between firms may also reflect how close are they geographically, financially, politically, historically, mentally and so on. Fortunately we can have access to empirical database on production network in which one million firms with four million links in Japan are listed. It almost exhausts the whole production activity over the nation.

The objective of the present work is to develop a visualization technique adapted for drawing such a large-scale network and to apply the method so obtained to the empirical data. Complex networks are highly difficult to analyze all connections because of their multitude and complexity. Visualization is one of useful techniques to illuminate structural properties of networks [5]. Of course visualization is not everything, but it helps us a lot to delve into complicated structures by stimulating our intuition. To visualize the network structure, we assume a spring-electrical model in which linked firms are connected with a spring and all of firms repels each other through a repulsive Coulomb interaction. We take full advantage of a molecular dynamics (MD) method [6] for an optimized structure in the model. The MD works well to reproduce ordered structures such as crystals of matter through slow cooling starting from any initial configurations. We thus expect it is also successful in visualizing network structures.

Here we also devote ourselves to development of a method for extracting community structures inherent in such a nationwide economic network. Finding communities in a network is to separate densely connected groups of nodes from the remainder. Then the network can be regarded as consisting of those groups which are sparsely connected to each other. The structural heterogeneity would play an important role in determining dynamics of macroeconomy such as business fluctuations, one of long-standing problems in the economic theory [7]. The quality function for communities is called modularity that was proposed by Newman [8]. The optimization of the modularity is one of promising approaches to identifying community structures involved in complex networks.

We thus demonstrate that combination of visualization and community analysis provides a useful tool for elucidating industrial structure and its evolution in the Japanese economic system.

2 Data on Transactions between Firms

We use a set of data on transactions between firms in Japan, compiled by Tokyo Shoko Research, Inc. It is almost the same dataset as analyzed in Ref. [9], where details of the data are described including how the research agency collected information. Here we just mention the dataset contains information in 2004 on individual firms exceeding one million in Japan and more than four million transactions between firms. The total number of firms in Japan is now estimated to be 2.5 million by the census due to the Ministry of Internal Affairs and Communications. By removing data irrelevant to the production network (isolated firms, incomplete data, etc.), we reduced the original dataset to the one consisting of $N = 780,544$ nodes (firms) and $M = 3,196,282$ links (transactions).

It has already been established that the connections between firms have characteristics of a scale-free network [9, 10]. Namely, the number of links (degree) originating from a node has a power-law distribution; some firms thus have a number of connections acting as hubs. The scale-free networks are observed in various places, including the topology of web-page links, the collaborative network of Hollywood actors, the power grid of U.S., and so on [1, 2, 3, 4]. Preferential attachment is a possible mechanism for formation of the scale-free networks; probability of linking to a node is proportional to the number of existing links that node has.

3 Network Visualization

The interaction potential between nodes in the MD simulation for an optimized network structure is given by

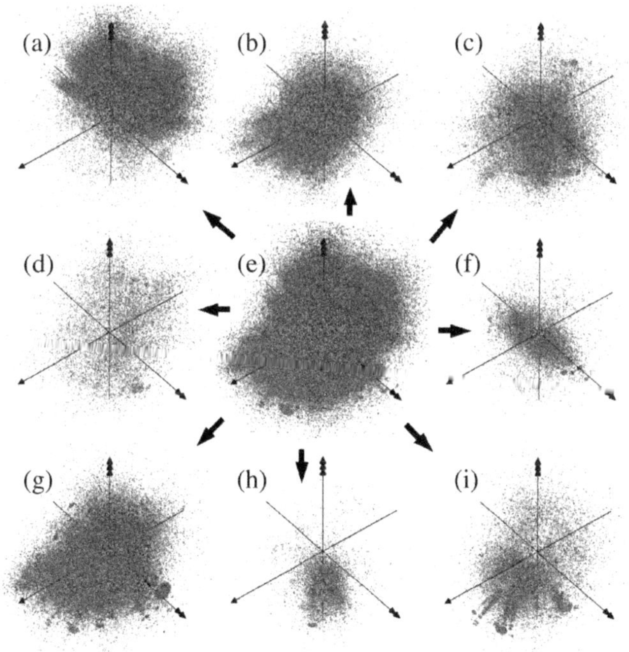

Fig. 1 An optimized network structure of whole economy in Japan (e); only nodes are drawn. And it is decomposed into various sectors: (a) construction, (b) manufacturing, (c) services, (d) real estate, (f) transport, (g) wholesales/retail trade, (h) information/communications, (i) others

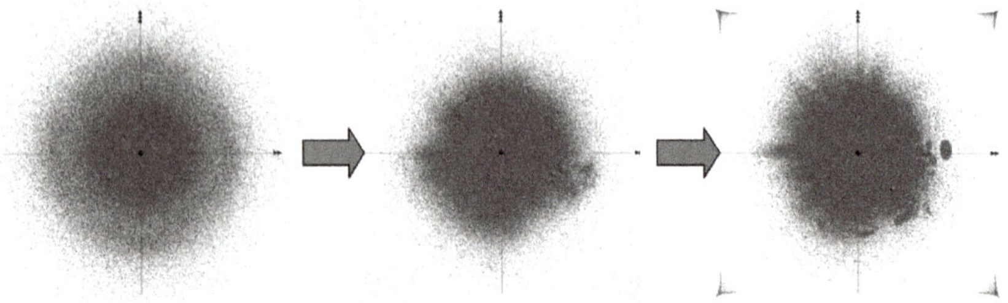

Fig. 2 Relaxation process for the optimized network structure, the rightmost drawing, identical to that in Fig. 1(e)

$$\Phi(r_{ij}) = \frac{Q_i Q_j}{r_{ij}} - \frac{1}{2} k_{ij} r_{ij}^2, \tag{1}$$

where Q_i is the Coulomb charge for node i and k_{ij} is the spring constant between nodes i and j. Here we neglect the direction of links (flow of goods or money), and assume that Q_i and k_{ij} take identical values for all of nodes and all of pairs, respectively.

The MD simulation on a system consisting of charged particles of million order is not a simple task because of the long-range nature of the Coulomb force; the computational task is order of N^2 without any tricks. To speed up the computation, we thereby implemented the hierarchical tree algorithm due to Barnes and Hut [11, 12] into our MD code. The whole space is divided by cells at hierarchical levels. The interactions of a given particle with other remote particles are replaced by the interactions with the cells to which those particles belong. As the distance between particles is larger, cells at higher level are used. On the other hand, the force on the particle is calculated directly for its neighbors residing in the same cell or the adjacent cells at the lowest level. This trick reduces the computational task to $O(N \log N)$.

We exemplify an optimized structure of the transaction network in Japan obtained by the MD simulation in Fig. 1 and its relaxation process in Fig. 2. We confirmed the final structure was essentially independent of initial configurations and cooling processes adopted. The distribution of firms is also decomposed into various sectors of industry in Fig. 1. We see that each component has its own characteristics. This figure thus constitutes a kind of industrial map in Japan. A partial network of firms belonging only to the manufacturing sector was visualized in Ref. [9]. Figure 3 shows how firms with different sizes are distributed on the optimized network; it is depicted that larger firms tend to concentrate at the center. We also prepared a picture of the optimized network coarse-grained at prefecture level as shown in Fig. 4; Tokyo occupies a very peculiar position linking over the whole nation. Figure 5 depicts how a shock due to a scandal over sale of contaminated rice by a food distribution company propagate through the network. The scandal was brought to light in fall, 2008. The shock is directed to the center of the network, indicating

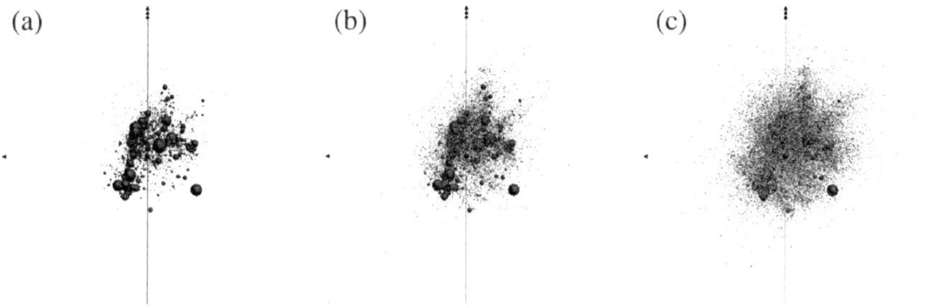

Fig. 3 Partial drawings of the optimized network structure in Fig. 2 where firms are selected according to their size. The size of firms is measured by the total number N_e of employees: (a) for $N_e \geq 500$, (b) for $N_e \geq 100$, and (c) for $N_e \geq 25$. The radius of each sphere representing a firm scales with the number of links it has

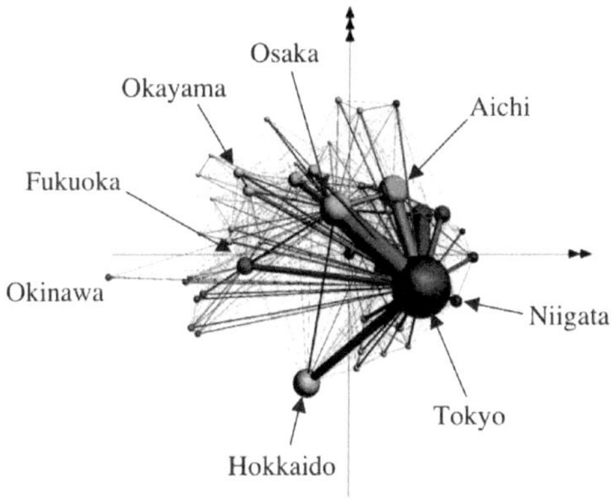

Fig. 4 Coarse-grained picture of the optimized network structure as shown in Fig 2 at prefecture level

amplified impact on the economic system; actually a number of downstream firms and consumers were affected.

4 Community Analysis

The modularity Q for a given network is defined by difference between the fraction of within-module links in the network and the expected fraction of within-module links in a randomly connected network with the same assignment of nodes into modules. It explicitly written as

$$Q = \sum_i \left(e_{ii} - \left(\sum_j e_{ij} \right)^2 \right) \tag{2}$$

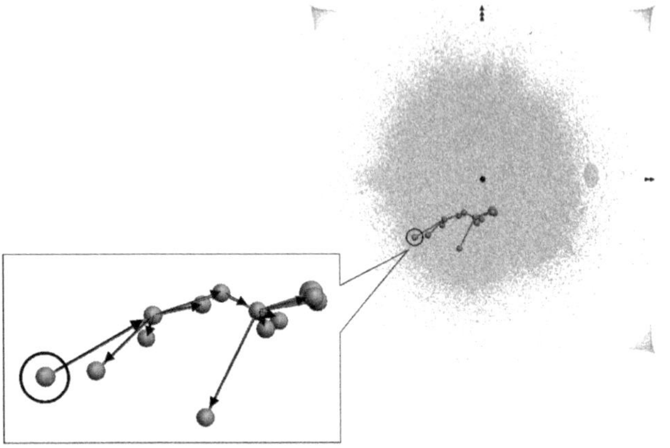

Fig. 5 Propagation of a shock due to the scandal over sale of contaminated rice by a food distribution company through the production network. The circle depicts the trigger company

where e_{ij} is the fraction of all edges connecting module V_i and module V_j given by

$$e_{ij} = \frac{1}{2M} \sum_{l \in V_i} \sum_{m \in V_j} a_{lm}, \tag{3}$$

with the adjacent matrix a_{lm} for the network. Thus $Q \simeq 0$ indicates that the network has no statistically significant communities as compared with the randomly connected network and $Q \simeq 1$ corresponds to a network which is partitioned into modules almost perfectly.

Table 1 Results of the community analysis on three networks

Network	Power Grid[a]	Transaction (all)[b]	Transaction ($N_e \geq 500$)[c]
# of nodes	4,941	780,544	4,851
# of links	6,594	3,196,282	25,832
Greedy method			
n_c	43	9,540	38
Q	0.934	0.539	0.504
Bisection method			
n_c	46	1,352	18
Q	0.936	0.653	0.548

[a] Power grid network in a western part of US.
[b] Transaction network of firms in Japan constructed using all of available data.
[c] Transaction network of large firms in Japan which have employees equal to or more than 500.

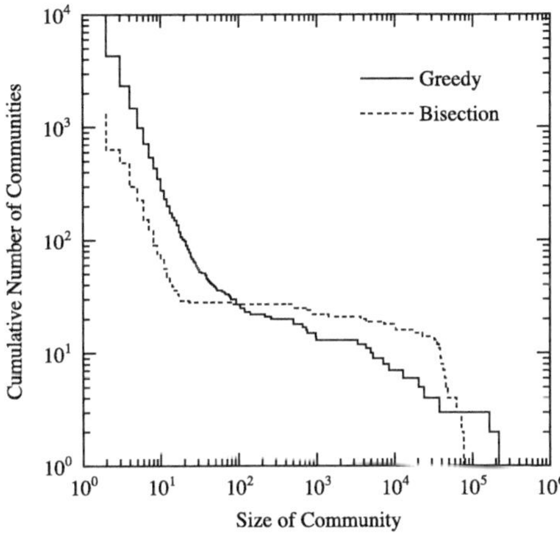

Fig. 6 Size distribution of communities in the transaction network of Japanese firms. Comparison between the greedy and the bisection algorithms

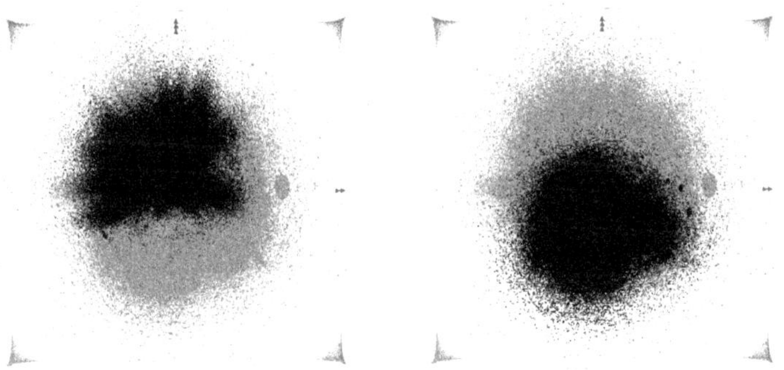

Fig. 7 The first (left) and second (right) largest communities in the optimized network as shown in Fig 2; those are illuminated by black color

We first use a bottom-up algorithm of Clauset *et al.* [13], which makes the optimization fast by using a greedy algorithm. However, this algorithm may not sufficiently optimize the modularity because the optimization tends to be trapped by a local maximum especially on large-scale networks. Alternatively we have worked out a top-down algorithm that implements the annealing method; bisection of the network is repeated to reach a maximum for the modularity. Table 1 compares effectiveness of the two algorithms; the bisection algorithm works better. As shown in Fig. 6, the bisection method decomposes three huge communities appeared in the greedy method into pieces. Also Newman has recently proposed a bisection method by utilizing the modularity matrix [14].

We compare the results of the community analysis with those of the visualization by the MD simulation. In Fig. 7, we highlight the first and second largest communities by discriminating from the remainder through change of the color. The nodes belonging to the same community are located close to each other as appreciated in those figures.

5 Conclusion

We obtained the optimized network structure for three million transactions between submillion firms by carrying out MD simulations, and then visualized the network in a three-dimensional space. Also we extracted communities inherent in such a large-scale network through maximization of the modularity using both greedy (bottom-up) and bisection (top-down) algorithms. The bisection method turned out to work better. Comparing the results for the community analysis with the visualized network, we demonstrated that firms belonging to the same community were located close to each other in a visualization space. Thus combination of visualization technique and community analysis gives a useful tool to elucidate industrial structure and its evolution in the economic system in Japan.

Acknowledgements. We thank the Research Institute of Economy, Trade and Industry (RIETI) for having provided us with the data on transaction relationship between firms in Japan. It was critcal for us to carry out the present work.

References

1. Watts, D.J.: Small Worlds: The Dynamics of Networks Between Order and Randomness. Princeton University Press, Princeton (1999)
2. Barabási, A.-L.: Linked: The New Science of Network. Perseus Publishing, Cambridge (2002)
3. Buchanan, M.: NEXUS: Small Wolds and the Groundbreaking Science of Networks. W.W. Norton & Co., New York (2002)
4. Newman, M.E.J.: SIAM Review 45, 167 (2003)
5. Hu, Y.: The Mathematica Journal 10, 37 (2006) (reference therein)
6. Allen, M.P., Tildesley, D.J.: Computer Simulation of Liquids. Oxford University Press, Oxford (1987)
7. Aoki, M., Yoshikawa, H.: Reconstructing Macroeconomics: A Perspective from Statistical Physics and Combinatorial Stochastic Processes. Cambridge Univ. Press, New York (2007)
8. Newman, M.E.J.: Phys. Rev. E 69, 066113 (2004)
9. Fujiwara, Y., Aoyama, H.: 0806.4280[e-print arXiv]
10. Fujiwara, Y., Aoyama, H., Souma, W.: In: Chakrabarti, B.K., Chakraborti, A., Chatterjee, A. (eds.) Econophysics and Sociophysics, p. 99. Wiley-VCH Verlag GmbH & Co. (2006)
11. Barnes, J., Hut, P.: Nature 324, 446 (1986)
12. Pfalzner, S.: Many-Body Tree Methods in Physics. Cambridge University Press, New York (1996)
13. Clauset, A., Newman, M.E.J., Moore, C.: Phys. Rev. E 70, 066111 (2004)
14. Newman, M.E.J.: PNAS 103, 8577 (2006)

Author Index

GPSR Compliance

The European Union's (EU) General Product Safety Regulation (GPSR) is a set of rules that requires consumer products to be safe and our obligations to ensure this.

If you have any concerns about our products, you can contact us on ProductSafety@springernature.com

In case Publisher is established outside the EU, the EU authorized representative is:

Springer Nature Customer Service Center GmbH
Europaplatz 3
69115 Heidelberg, Germany

Batch number: 10755036

Printed by Printforce, the Netherlands